SEMICONDUCTORS AND SEMIMETALS

VOLUME **82**

SPINTRONICS

SEMICONDUCTORS AND SEMIMETALS

A Treatise

Edited by

E. R. WEBER
*Fraunhofer Institute for
Solar Energy Systems ISE,
Freiburg, Germany*

SEMICONDUCTORS AND SEMIMETALS

VOLUME **82**

SPINTRONICS

Edited by

TOMASZ DIETL
Polish Academy of Sciences
Institute of Physics
Institute of Theoretical Physics
University of Warsaw, Warsaw, Poland

DAVID D. AWSCHALOM
Center for Spintronics and Quantum Computation
University of California
Santa Barbara, California, USA

MARIA KAMINSKA
Institute of Experimental Physics
University of Warsaw, Warsaw, Poland

HIDEO OHNO
Laboratory for Nanoelectronics and Spintronics
Research Institute of Electrical Communication
Tohoku University, Sendai, Japan

AMSTERDAM • BOSTON • HEIDELBERG • LONDON
NEW YORK • OXFORD • PARIS • SAN DIEGO
SAN FRANCISCO • SINGAPORE • SYDNEY • TOKYO
Academic Press is an imprint of Elsevier

Academic Press is an imprint of Elsevier
Radarweg 29, PO Box 211, 1000 AE Amsterdam, The Netherlands
32 Jamestown Road, London NWI 7BY, UK
30 Corporate Drive, Suite 400, Burlington, MA 01803, USA
525 B Street, Suite 1900, San Diego, CA 92101-4495, USA
360 Park Avenue South, New York, NY 10010-1710

First edition 2008

Copyright 2008 Elsevier Inc. All rights reserved.

No part of this publication may be reproduced, stored in a retrieval system or transmitted in any form or by any means electronic, mechanical, photocopying, recording or otherwise without the prior written permission of the publisher.

Permissions may be sought directly from Elsevier's Science & Technology Rights Department in Oxford, UK: phone (+44) (0) 1865 843830; fax (+44) (0) 1865 853333; email: permissions@elsevier.com. Alternatively you can submit your request online by visiting the Elsevier web site at http: //elsevier.com/locate/permissions, and selecting *Obtaining permission* to use Elsevier material

Notice

No responsibility is assumed by the publisher for any injury and/or damage to persons or property as a matter of products liability, negligence or otherwise, or from any use or operation of any methods, products, instructions or ideas contained in the material herein. Because of rapid advances in the medical sciences, in particular, independent verification of diagnoses and drug dosages should be made

ISBN: 978-0-08-044956-2
ISSN: 0080-8784

For information on all Academic Press publications
visit our Web site at elsevierdirect.com

Printed and bound in USA
08 09 10 11 10 9 8 7 6 5 4 3 2 1

Working together to grow
libraries in developing countries

www.elsevier.com | www.bookaid.org | www.sabre.org

ELSEVIER   BOOK AID International   Sabre Foundation

CONTENTS

Preface	ix

1. Single Spin Coherence in Semiconductors — 1

Maiken H. Mikkelsen, Roberto C. Myers, Gregory D. Fuchs, and David D. Awschalom

1. Introduction	2
2. Single Electron Spins in Quantum Dots	3
3. Few Magnetic Spins in Quantum Wells	18
4. Single Spins in Diamond	27
References	41

2. Theory of Spin–Orbit Effects in Semiconductors — 45

Jairo Sinova and A. H. MacDonald

1. Introduction	46
2. The Relativistic Origins of Spin–Orbit Coupling	48
3. Band Structure of Semiconductors: Effective $k \cdot p$ Hamiltonians	51
4. SHE and AHE	61
5. Topological Berry's Phases in Spin–Orbit Coupled Systems: ACE	79
References	85

3. Fermi Level Effects on Mn Incorporation in III-Mn-V Ferromagnetic Semiconductors — 89

K. M. Yu, T. Wojtowicz, W. Walukiewicz, X. Liu, and J. K. Furdyna

1. Introduction	90
2. Sample Preparation and Characterization	93
3. Effects of Mn Location on the Electronic and Magnetic Properties	98
4. Concluding Remarks	125
Acknowledgments	130
References	130

4. Transport Properties of Ferromagnetic Semiconductors   135

T. Jungwirth, B. L. Gallagher, and J. Wunderlich

1. Introduction 135
2. Basic Transport Characteristics 138
3. Extraordinary Magnetotransport 167
4. Summary 199
Acknowledgments 200
References 200

5. Spintronic Properties of Ferromagnetic Semiconductors   207

F. Matsukura, D. Chiba, and H. Ohno

1. Introduction 207
2. Spin-Injection and Detection of Spin-Polarization 209
3. Magnetic Tunnel Junction 212
4. Magnetic DW 218
5. Electric-Field Control of Ferromagnetism 228
6. Optical Control of Ferromagnetism 231
7. Summary 234
Acknowledgments 235
References 235

6. Spintronic Nanodevices   241

C. Gould, G. Schmidt, and L. W. Molenkamp

1. Introduction 241
2. Tunneling Anisotropic Magnetoresistance 244
3. Multi-TAMR Structures 248
4. Volatile and Nonvolatile Operation 250
5. Correlated Effects 252
6. Portability 259
7. Nanodevices 260
8. Lateral Nanoconstrictions 260
9. Current-Assisted Manipulation 264
10. Local Lithographic Anisotropy Control 268
11. Magnetic Characterization of Nanobars 269
12. Transport Characterization of Nanobars 271
13. Anisotropic Strain Relaxation 274
14. Memory Device Using Local Anisotropy Control 277
15. Device Operation 278
16. Magnetic States 279

17. Origin of the Resistance Signal	280
18. Conclusion and Outlook	283
Acknowledgments	284
References	284

7. Quantum Structures of II–VI Diluted Magnetic Semiconductors — 287

J. Cibert, L. Besombes, D. Ferrand, and H. Mariette

1. Magnetic and Electric Impurities in II–VI Nanostructures	287
2. Carrier-Induced Ferromagnetism in 2D DMSs	291
3. 0D Systems	298
4. Transport in Quantum II–VI DMS Structures	316
5. Summary	320
References	320

8. Magnetic Impurities in Wide Band-gap III–V Semiconductors — 325

Agnieszka Wolos and Maria Kaminska

1. Introduction	326
2. Diluted Magnetic Semiconductors	330
3. Nature of Mn Impurity in III–V Semiconductors	334
4. Magnetic Interactions in III–V DMSs with Mn	352
5. GaN-Based DMSs	357
6. Internal Reference Rule for Transition Metal Ions—Case of GaN	361
7. Summary and Conclusions	362
Acknowledgments	364
References	364

9. Exchange Interactions and Nanoscale Phase Separations in Magnetically Doped Semiconductors — 371

Tomasz Dietl

1. Introduction	372
2. Substitutional Transition Metal Impurities in Semiconductors	375
3. Origin of Exchange Interactions between Carriers and Localized Spins	381
4. Effects of sp–d(f) Exchange Interactions	382
5. Exchange Interactions between Effective Mass Carriers	392
6. Models of Ferromagnetic Spin–Spin Interactions in Semiconductors	397
7. p–d Zener Model of Carrier-Mediated Ferromagnetism	399
8. Effects of Disorder and Localization on Carrier-Mediated Ferromagnetism	407
9. Effects of Nonrandom Distribution of Magnetic ions	416

10. Is Ferromagnetism Possible in Semiconductors with
no Magnetic Elements? 422
11. Summary 423
Acknowledgments 425
References 425

10. Computational Nano-Materials Design for the Wide Band-Gap and High-T_C Semiconductor Spintronics 433

Hiroshi Katayama-Yoshida, Kazunori Sato, Tetsuya Fukushima, Masayuki Toyoda, Hidetoshi Kizaki, and An van Dinh

1. Introduction 433
2. Magnetic Mechanism, T_C, and Unified Physical Picture 436
3. Spinodal Nano-Decomposition and Nano-Spintronics Applications 444
4. New Class of Oxide Spintronics without a 3d TM 450
5. Conclusion 452
Acknowledgments 452
References 452

11. Properties and Functionalities of MnAs/III–V Hybrid and Composite Structures 455

Masaaki Tanaka, Masafumi Yokoyama, Pham Nam Hai, and Shinobu Ohya

1. Introduction 455
2. Fabrication and Structure of GaAs:MnAs Nano-Particles 456
3. Large Magnetoresistance at Room Temperature 458
4. Spin Dependent Tunneling Transport Properties in III–V Based Heterostructures Containing GaAs:MnAs 463
5. Properties of Zinc-Blende Type and NiAs-Type MnAs Nano-particles 471
6. Magneto-Optical Device Applications 478
Acknowledgments 483
References 484

Index 487
Contents of Volumes in This Series 499
See Color Plate Section in the back of this book

PREFACE

This volume presents recent progress and prospects of semiconductor spin electronics. It is aimed for postgraduate students and mature research workers in condensed matter physics, materials science, quantum informatics, and device engineering, who can find in a single volume a comprehensive and updated survey of this rapidly developing and controversial field. State-of-the-art materials growth and characterization as well as experimental, theoretical, and computational aspects of the electron spin physics in nonmagnetic and magnetically doped semiconductor films, heterostructures, and quantum dots as well as in composite ferromagnetic metal/semiconductor systems are described considering, from the one extreme, a behavior of a single spin confined in a quantum dot or localized on an impurity or defect and, on the other, striking effects of collective magnetic ordering in ferromagnetic and superparamagnetic semiconductor systems. Particular attention is paid to those findings which appear relevant in the context of search for discontinuous classical and quantum information and communication technologies.

Retrospectively, there is a variety of accomplishments, breakthrough, and challenges of semiconductor spintronics reviewed in this volume—a few examples follows.

- Spins in semiconductors offer a pathway toward integration of information storage and processing in a single material. These applications of semiconductor spintronics require techniques for the injection, detection, manipulation, transport, and storage of spins. Encoding and reading out spin information in single spins can be considered the ultimate limit for scaling magnetic information. Single spins in semiconductors also provide a solid state analog of atomic physics which may provide a pathway to quantum information systems in the solid state. As documented in Chapters 1 and 7, owing to the progress in achievable spatial and temporal resolution, single spins in semiconductors can now be optically probed and manipulated in a variety of systems over the time scale much shorter than the relevant coherence time. These demonstrations of coherent optical and optoelectronic spin control parallel elegant transport studies of semiconductor quantum dots, in which spin encoding, read out, and manipulation as well as the control over spin–spin coupling have recently been demonstrated.

- It has been long appreciated that the spin–orbit interaction, by controlling spin lifetimes and selection rules, has a profound influence on optical and transport properties of nonmagnetic semiconductors. Interestingly, there are a growing number of theoretical proposals and experimental demonstrations that this interaction can be exploited for generation of spin polarizations and spin currents by light as well as by electric fields and currents, even in the absence of an external magnetic field. Chapter 2, devoted to the theory of spin–orbit effects in semiconductors, presents chiefly recent rapid advance in the understanding of the intricate physics of the anomalous Hall effect, spin Hall effect, and Aharonov–Casher effect in semiconductors, the insight that may open doors for developing viable spin transistors and devices generating spin currents that will process and transmit information in the dissipation less fashion.
- Transition-metal doped III–V and II–VI compounds offer an unprecedented opportunity for considering physical phenomena and device concepts for previously unavailable combination of quantum structures and ferromagnetism in semiconductors. Since, at least in antimonides, arsenides, phosphides, and tellurides, the strength of ferromagnetic spin–spin interactions and the orientation of spontaneous magnetization vary with the hole density, the powerful methods developed to change the Fermi level in semiconductor structures can be employed to affect the magnetic ordering. The effect of co-doping by nonmagnetic acceptors in Mn-based III–V and II–VI diluted magnetic semiconductors (DMS) is discussed in Chapters 3 and 7, respectively, while Chapters 5 and 7 present ground-breaking demonstrations of magnetization manipulation by an electric field and light in appropriately engineered structures of these compounds. Furthermore, as discussed in Chapter 5, due to highly imbalanced hole spin concentrations, these materials can serve as efficient hole or electron spin injectors, the latter via interband tunneling in Esaki–Zener diodes, the advance accomplished in parallel to the observation of a sizable spin injection from ferromagnetic metals to semiconductors across a tunnel barrier.
- In addition to revealing unique functionalities mentioned above, ferromagnetic DMS, as described in Chapters 4–6, show the anomalous and planar Hall effects as well as allow engineering all-semiconductor single-crystal devices exhibiting anisotropic, giant, and tunneling magnetoresistance (AMR, GMR, TMR), well known from earlier studies and industrial applications of metal counterparts. Interestingly, however, specific properties of the hole-controlled ferromagnetic semiconductors, notably (Ga,Mn)As, such as the crucial role of the spin–orbit interaction, the possibility of anisotropy engineering by epitaxial strain, the importance of strain relaxation, and carrier depletion in nanostructures have lead to discoveries of novel spintronic effects described in Chapters 4 and 6.

Some of them, particularly tunneling and Coulomb blockade AMR, are now looked for in metallic nanostructures with a strong built-in spin–orbit interaction. Similarly, domain-wall displacement and magnetization switching by an electric current, the phenomena critical for the spread of magnetic random access memories and magnetic logic, have been first investigated theoretically and then experimentally in the context of metallic systems but, according to results presented in Chapter 5, the dissident character of semiconductor properties has allowed demonstrating that the spin-transfer is in fact a dominant mechanism accounting for the current-induced domain motion.

- In view of the massive progress outlined above, the major challenge, reviewed in Chapter 3 and Chapters 8–11, is to develop a functional semiconductor system with ferromagnetic capabilities persisting up to above room temperature. This issue is, from the one hand, related to the fundamental query why the Curie temperature of the hole-mediated ferromagnetism does not exceed 150–180 K (as found for (Ga,Mn)As in dozen of laboratories world-wide). On the other hand, it concerns the understanding and mastering high-temperature ferromagnetism detected in numerous DMS and diluted magnetic oxides without holes or even in some materials nominally undoped with magnetic elements. It appears that three phenomena are relevant here: (i) self-compensation which leads to the appearance of donor defects (like interstitial Mn) once the Fermi level reaches an appropriately deep position in the valence band, the issue elaborated in Chapter 3; (ii) strong coupling of holes and magnetic ions (increasing when cation–anion distance decreases), which according to results presented in Chapters 8 and 9 enhances hole localization and, thus, impedes carrier-mediated ferromagnetic ordering; (iii) spinodal decomposition, that is self-organized aggregation of magnetic ions in nanoscale regions (in the form of nanodots or nanocolumns), which—as argued in Chapters 8–10—account for the robust ferromagnetic features, persisting well above the room temperature and appearing when the solubility limit of the magnetic component is surpassed.

- It is more and more clear that spinodal decomposition, regarded initially as an unwanted outcome of quest for high T_C ferromagnetic DMS, constitutes an important and yet unexplored field of materials research. Actually, the use of embedded metallic and semiconducting nanocrystals is expected to revolutionize the performance of various commercial devices, such as flash memories and low current semiconductor lasers. As argued in Chapters 9–11, a number of spintronic, nanoelectronic, photonic, plasmonic, and thermoelectric applications can be envisaged for composite structures consisting of semiconductors and embedded magnetic, typically metallic, nanostructures. Importantly, these systems show magnetooptical and magnetotransport characteristics specific to ferromagnetic materials and, according to results presented in Chapter 11, particularly advanced in the development of optical isolators containing MnAs nanocrystal embedded in a semiconductor waveguide. As discussed in Chapters 9 and 10, it has been proposed and recently observed that the aggregation and solubility of magnetic ions can be controlled by growth conditions and co-doping. This is possible as states derived from the d shells reside typically in the band-gap, so that a shift of the Fermi level introduced by donor or acceptor states changes the valence and, thus, the chemistry of the magnetic constituent. This novel and controllable self-organized nanoassembly process may allow fabricating embedded nanostructures of a size, shape, and developed motif on demand.

- Despite the decade-long investigations, the theoretical understanding of ferromagnetism in DMSs has emerged as one of the most controversial and challenging fields of today's materials science and condensed matter physics. Among the unsettled issues is the question whether the holes mediating the spin–spin coupling reside in the d band (double-exchange scenario), in an acceptor band split of the valence band (impurity band scenario), or in the valence band (p-d Zener model). This unsatisfactory situation stems from the fact that in these systems conceptual difficulties of charge transfer magnetic insulators and strongly correlated disordered metals are combined with intricate properties of heavily doped semiconductors and semiconductor alloys, such as the Anderson–Mott quantum localization and the break down of the virtual-crystal approximation. Accordingly, DMSs have posed an enormous challenge to first principles methods, particularly those involving the local density approximation (LDA) and the coherent-potential approximation (CPA), the issue discussed in Chapters 9 and 10. It becomes increasingly obvious that inaccuracies of these approximations, such as an improper treatment of strong correlation at transition metal atoms, errors in band gaps and d-level positions, as well as an inadequate description of quantum localization, may result in incorrect predictions of the electronic and magnetic ground states, generally overestimating the tendency toward the metallic and ferromagnetic phases.

- In view of the above-mentioned complexity, a large variety of divergent model-Hamiltonian approaches have been proposed. Many of them assume that the holes accounting for the ferromagnetism reside in a Mn-acceptor impurity band. While future experiments, such as high-resolution tunneling or photoemission, will be decisive, a chain of arguments has been collected in Chapters 4 and 9 indicating that the relevant carriers actually occupy the valence band. Accordingly, the p-d Zener model can be applied, usually implemented in a spirit of second-principles approaches which employ experimentally available information whenever possible. In particular, the host band structure is parameterized within the multiband kp scheme (Chapters 4–7 and 9) or within the empirical multiorbital tight binding method (Chapter 9), the latter—together with the Landauer–Büttiker formalism—found to be particularly useful for describing properties of layered structures, like TMR junctions or Esaki–Zener diodes. As emphasized in Chapters 4–7 and 9, the p-d Zener model has made it possible to describe pertinent properties and functionalities, particularly in the case of (Ga,Mn)As, p-(Cd,Mn)Te, and related structures, including characteristics usually regarded as the domain of experimental studies, such as the magnitude of magnetic anisotropy. It is to be seen to what extend other model-Hamiltonian theories will provide a similarly successful description of experimental findings.

Tomasz Dietl
David D. Awschalom
Maria Kamińska
Hideo Ohno

CHAPTER 1

Single Spin Coherence in Semiconductors

Maiken H. Mikkelsen, Roberto C. Myers, Gregory D. Fuchs, and **David D. Awschalom**

Contents
1. Introduction — 2
2. Single Electron Spins in Quantum Dots — 3
   2.1. Optical selection rules and the Faraday effect — 4
   2.2. Measuring a single electron spin by Kerr rotation — 6
   2.3. Spin dynamics of a single electron spin — 9
   2.4. Ultrafast manipulation using the optical Stark effect — 13
   2.5. Conclusions — 17
3. Few Magnetic Spins in Quantum Wells — 18
   3.1. Mn-ions in GaAs as optical spin centers — 19
   3.2. Zero-field optical control of magnetic ions — 21
   3.3. Mechanism of dynamic polarization and exchange splitting — 23
   3.4. Spin dynamics of Mn-ions in GaAs — 25
   3.5. Conclusions — 27
4. Single Spins in Diamond — 27
   4.1. Introduction — 27
   4.2. NV basics — 28
   4.3. Anisotropic interactions of a single spin — 30
   4.4. Single NV spin manipulation and coherence — 35
   4.5. Coupled spins in diamond — 38
   4.6. Conclusions and outlook — 40
   References — 41

Center for Spintronics and Quantum Computation, University of California, Santa Barbara, California 93106

1. INTRODUCTION

Spins in semiconductors offer a pathway toward integration of information storage and processing in a single material (Wolf et al., 2001). These applications of semiconductor spintronics require techniques for the injection, detection, manipulation, transport, and storage of spins. Over the last decade, many of these criteria have been demonstrated on ensembles of spins. Encoding and reading out spin information in single spins, however, might be considered the ultimate limit for scaling magnetic information. Single spins in semiconductors also provide a solid state analog of atomic physics which may provide a pathway to quantum information systems in the solid state.

As we will see, single spins in semiconductors can be observed in a variety of systems. In general, the spin needs to be localized to a particular region in the host material in order to make it available for study. The confinement can be provided either by a quantum dot (Berezovsky et al., 2006; Besombes et al., 2004; Bracker et al., 2005; Hanson et al., 2007) or by an individual impurity (Epstein et al., 2005; Jelezko et al., 2004; Myers et al., 2008; Rugar et al., 2004; Stegner et al., 2006). Moreover, the confinement also determines the degree and nature of the coupling with the surrounding environment which is critical to the coherence time. In this chapter we focus on three varieties of single spin systems in semiconductors that can be optically probed and manipulated:

- Single electron spins in quantum dots (Section 2)
- Manganese acceptors in GaAs quantum wells (Section 3)
- Nitrogen-vacancy color centers in diamond (Section 4)

All three types of spins are detected using optical signals resolved in time and position. In the first two cases, the conservation of angular momentum allows for both spin-sensitive detection and injection. In the last case, spin-selective transition rates intrinsic to the color center conveniently provide the same abilities. By coupling these techniques with ultrafast optical pulses, it is possible to stroboscopically measure spin dynamics at bandwidths far exceeding state-of-the-art high speed electronics. The optical experiments discussed here are also spatially resolved using high numerical aperture microscope objectives. This reduces the focal spot size, and therefore measurement region, to ~ 1 μm in diameter or below. Such spatial resolution allows single quantum dots as well as individual impurities to be studied as in Section 2 and 4, respectively.

Section 2 discusses single electron spins in GaAs quantum dots. The single electron spin state can be sequentially initialized, manipulated, and readout using all-optical techniques. The spins are probed using time-resolved Kerr rotation which allows for the coherent evolution of a single electron spin to be observed, revealing a coherence time of ~ 10 ns at 10 K.

By applying off-resonant, picosecond-scale optical pulses, the spins can be manipulated via the optical Stark effect. Section 3 introduces magnetic ions as spin recombination centers in semiconductor quantum wells. The magnetization of few magnetic ions can be controlled at zero-field for these normally paramagnetic spins and can in principle be extended to the single-ion limit. These spins might be considered the most strongly coupled system in which strong exchange interactions couple the spin state of the magnetic ions to the band carriers. Despite this coupling, we will see that in the ultra dilute limit, magnetic ions exhibit coherence times near 10 ns at liquid He temperatures. Finally, Section 4 describes single nitrogen-vacancy centers in the wide bandgap semiconductor diamond. These spins represent the opposite end of the spectrum with a spin that is relatively uncoupled to the host band structure resulting in record room temperature spin coherence times of >350 µs (Gaebel et al., 2006). This enables traditional magnetic resonance techniques to be used for nitrogen-vacancy (NV) spin manipulation.

2. SINGLE ELECTRON SPINS IN QUANTUM DOTS

Semiconductor quantum dots (QDs) may be realized in a variety of ways, for example, by locally depleting a 2D electron gas with gate electrodes, through chemical synthesis and colloidal chemistry, and through self-assembly using molecular beam epitaxy (MBE). Each type of these quantum dot structures opens up a wide range of possibilities and allows for integration into different systems. The focus here is on optical measurements of MBE-grown GaAs QDs at a temperature of 10 K. A single electron spin state in an individual QD is probed using the magneto-optical Kerr effect. This technique allows one to probe the spin nonresonantly and thus minimally disturbing the system. Next, these measurements are extended to directly probe the time dynamics of the electron spin using pulsed pump and probe lasers. Finally, the ultrafast manipulation of a single electron spin state via the optical Stark effect is discussed.

In these structures the photoluminescence (PL) from individual QDs has been measured for more than a decade (Gammon et al., 1996). For an ensemble of dots, the inevitable size variations result in a spread of emission energies. Probing fewer dots by using small apertures, for example, allows this broad spectrum to be resolved into very narrow (~100 µeV) PL lines at different energies. An appropriately small aperture or low density of dots such that the PL lines from different dots can be spectrally resolved allows for individual dots to be studied (Gammon et al., 1996). The spin state may then be readout using polarized PL (Bracker et al., 2005; Ebbens et al., 2005) or polarization-dependent absorption (Högele et al., 2005; Li and Wang, 2004; Stievater et al., 2002). Recently, polarized PL measurements of single electron spins in an applied magnetic field have also been demonstrated revealing information about the spin-lifetime (Bracker et al., 2005). In the optical study of spin coherence, selection rules play a critical role and will briefly be reviewed below.

2.1. Optical selection rules and the Faraday effect

Electron spins in semiconductors may be initialized and readout using polarized light by exploiting the optical selection rules that exist in zinc-blende semiconductors with a sufficiently large spin–orbit coupling, such as GaAs. The band structure of a zinc-blende crystal is shown schematically in Fig. 1.1A, where E_g is the bandgap and Δ is the spin–orbit splitting. The conduction band is twofold degenerate ($S = 1/2$, $S_z = \pm 1/2$) and the valence band fourfold degenerate with the heavy holes ($J = 3/2$, $J_z = \pm 3/2$) and light holes ($J = 3/2$, $J_z = \pm 1/2$). It is critical that the split-off holes are at a lower energy, Δ, such that a pump energy may be chosen where only transitions from the heavy/light holes to the conduction band are allowed. As illustrated schematically in Fig. 1.1B, the selection rules are due to the conservation of angular momentum. The absorption of a circularly polarized photon transfers its angular momentum $L_z = \pm \hbar$ to the spin, and thus can only drive transition with $\Delta L_z = \pm 1$ (see Fig. 1.1B). For example, the absorption of a photon with $l = 1$ only allows transitions from the $J_z = -3/2$ heavy hole and $J_z = -1/2$ light hole. Conveniently, the

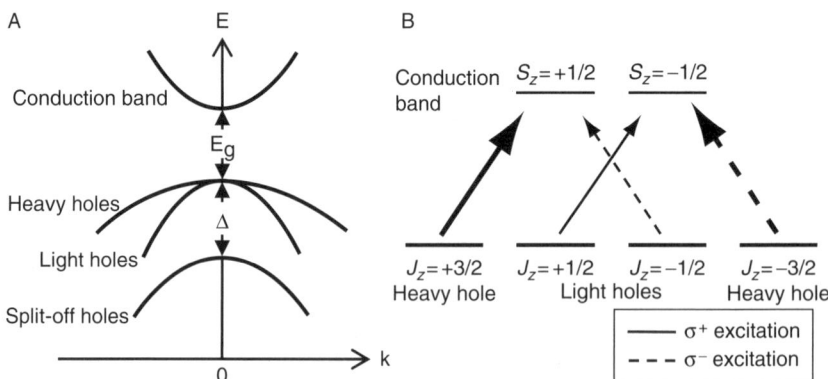

FIGURE 1.1 Band structure and optical selection rules. (A) Schematic of the band structure of a zinc-blende semiconductor. The bandgap, E_g, and the spin–orbit splitting, Δ, are indicated. (B) Optically allowed transitions from the heavy and light hole valance bands ($J = 3/2$) to the conduction band ($S = 1/2$). The thickness of the arrows indicates the strength of the transitions.

heavy hole transition is three times as likely as the light hole transition, which can be shown by calculating the dipole transition matrix elements. This means that three times as many electron spins with $S_z = +1/2$ rather than with $S_z = -1/2$ are injected giving a net spin polarization in the conduction band of 50%. The situation can be improved further if the semiconductor is strained, or if the electrons and holes are confined in one or more dimensions. In this case, the degeneracy of the heavy and light hole bands is lifted, and transitions can be pumped from the heavy hole band only, resulting in ideally 100% conduction band spin polarization.

The inverse process of this optical spin injection provides a means for detecting the spin polarization of carriers in a semiconductor. When an electron and hole recombine, light is emitted with circular polarization that reflects the spin state of the electron and hole. By measuring the degree of circular polarization of this luminescence, one can measure the spin polarization at the time of recombination. Additionally, if the g-factor is known, the spin lifetime can be obtained in a Hanle measurement. Here, the degree of circular polarization of the PL is measured as a function of an applied magnetic field perpendicular to the optically injected spins. The resulting curve has a Lorentzian lineshape, assuming a simple exponential decay process, with a half-width inversely proportional to the product of the g-factor and spin lifetime (Meier and Zakharchenya, 1984). The Hanle effect can even be used to measure the spin lifetime of a single electron spin as shown by (Bracker et al., 2005), where a spin lifetime of 16 ns was observed assuming a g-factor of ~0.2.

A more direct measurement of spin polarization can be obtained through the Faraday effect. Here, a net spin polarization in a material results in a different index of refraction for right and left circularly polarized light. When linearly polarized light is transmitted through the material, the two circularly polarized components acquire a relative phase shift, yielding a rotation of the polarization of the transmitted light. This rotation is proportional to the spin polarization along the direction of light. The Kerr effect is directly analogous, but refers to a measurement made in reflection rather than in transmission. A typical scheme is to use a circularly polarized pump laser to optically inject spins into the conduction band, and a linearly polarized probe laser to measure Faraday (or Kerr) rotation. In this type of pump-probe spectroscopy, the two lasers may have the same or different energies. If the pump and probe lasers are continuous wave (cw) then this provides information about the steady-state spin polarization, similar to the Hanle measurement described above. However, pulsed lasers allow this technique to be extended into the time domain to provide a more direct look at the spin dynamics.

2.2. Measuring a single electron spin by Kerr rotation

First, we are interested in the measurement of a single electron spin in a QD using Kerr rotation (KR) with cw pump and probe lasers (Berezovsky et al., 2006). For a single conduction-band energy level in a QD containing a spin-up electron in a state $|\psi_\uparrow\rangle$, optical transitions to the spin-up state are forbidden by the Pauli exclusion principle. Considering only transitions from a single twofold degenerate valence band level $|\psi_{v_0}\rangle$, the KR angle, θ_K, is given by:

$$\theta_K(E) = EC(|P_{\downarrow,v_0}^{\sigma+}|^2 - |P_{\downarrow,v_0}^{\sigma-}|^2)\frac{E - E_{0,v_0}}{(E - E_{0,v_0})^2 + \Gamma_{0,v_0}^2} \quad (1.1)$$

where E is the energy of the probe laser, C is a material-dependent constant, $P_{c,v}^{\sigma\pm} = \langle\psi_c|\hat{p}_x + i\hat{p}_y|\psi_v\rangle$ are the interband momentum matrix elements, E_{0,v_0} is the energy of the transition, and Γ_{0,v_0} is the linewidth of the transition.

For $\Gamma \ll |\Delta| \ll E$, where $\Delta = E - E_0$, we note that the KR angle θ_K, decays slower ($\sim\Delta^{-1}$) than the absorption line ($\sim\Delta^{-2}$) (Guest et al., 2002; Meier and Awschalom, 2005). Therefore, for a suitable detuning, Δ, KR can be detected while photon absorption is strongly suppressed. In a QD containing a single conduction band electron, the lowest-energy interband transition is to the negatively charged exciton state, X^-, with an energy E_{X^-}. Thus, a single electron spin is expected to produce a feature in the KR spectrum with the odd-Lorentzian lineshape given by Eq. (1.1), centered at the energy E_{X^-}.

The QDs studied here are MBE-grown simple interface fluctuation QDs, also called natural QDs. They consist of a 4.2 nm GaAs QW where a 2-min growth interruption at each QW interface allows large ($\sim$100 nm diameter; Guest et al., 2002) monolayer thickness fluctuations to develop that act as QDs (Gammon et al., 1996; Zrenner et al., 1994). In addition, the QDs are embedded within a diode structure enabling controllable charging of the dots with a bias voltage (Warburton et al., 2000) (see Fig. 1.2A). The front gate also acts as a shadow mask with 1 μm apertures which are used to isolate single dots as well as to identify the specific position of the dots. The spectrum of the PL as a function of applied bias voltage (Fig. 1.2B) is well established and can therefore be used to identify the different charging states (Bracker et al., 2005). Above 0.5 V, a single line is observed at 1.6297 eV which is caused by recombination from the negatively charged exciton (trion or X^-) state. Below 0.5 V, a bright line appears 3.6 meV higher in energy due to the neutral exciton (X^0) transition and in addition, a faint line at 1.6292 eV is visible from radiative decay of the biexciton (XX). To make the measurement of Kerr rotation from a single electron spin easier, the QD layer is centered within an optical microcavity with a resonance chosen to enhance the interaction of the

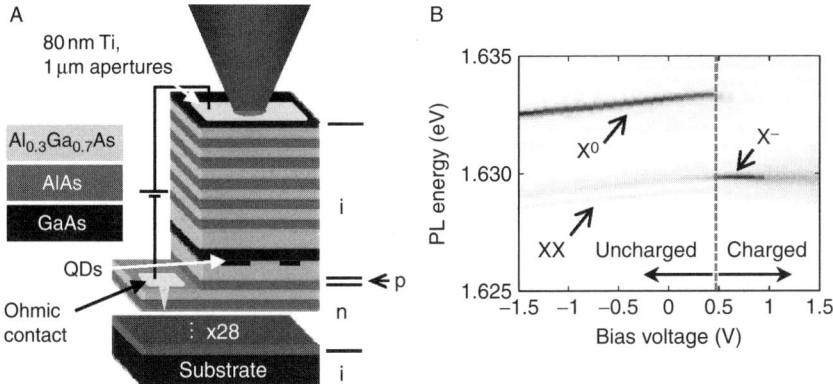

FIGURE 1.2 Sample structure and single dot PL. (A) Schematic of the sample structure. i, n, and p indicate intrinsic (undoped), n-doped, and p-doped regions of the sample, respectively. Twenty-eight repetitions of the AlAs/AlGaAs layers, indicated by × 28, are not shown. (B) PL of a single QD as a function of bias voltage; grayscale indicates PL intensity. A jump in the PL energy indicates the onset of QD charging. Adapted from Berezovsky et al. (2006).

optical field with the QD at energies well below the lowest interband transition. The front and back cavity mirrors are distributed Bragg reflectors (DBRs) composed of 5 and 28 pairs of AlAs/Al$_{0.3}$Ga$_{0.7}$As $\lambda/4$ layers, respectively (see Fig. 1.2A). This asymmetrical design allows light to be injected into and emitted from the cavity on the same side. The cavity has a quality factor of 120 and an expected enhancement of the KR by a factor of ~15 at the peak of the resonance (Li et al., 2006; Salis and Moser, 2005). As described above, a circularly polarized pump laser (1.654–1.662 eV) initializes the spins according to the optical selection rules and the spin polarization is verified through polarized PL measurements. To probe spins in the dot through KR, a second, linearly polarized, cw Ti:Sapphire laser is focused onto the sample, spatially overlapping the pump laser. The pump and probe beams are modulated using mechanical choppers, enabling lock-in detection of only spins injected by the pump. Furthermore, at each probe energy the pump excitation is switched between right and left circularly polarized light and the spin-dependent signal is obtained from the difference in the KR angle at the two helicities. The data in Fig. 1.3A show the KR signal and the PL as a function of probe energy at a bias $V_b = 0.2$ V when the QD is nominally uncharged. In this regime, the QD may contain a single spin-polarized electron through the capture of an optically injected electron, or spin-dependent X$^-$ decay. The X$^-$ energy coincides spectrally with a sharp feature observed in the KR data. We can fit these data to Eq. (1.1) including only a single transition in the sum, on top of a broad background (see Fig. 1.3B). The transition

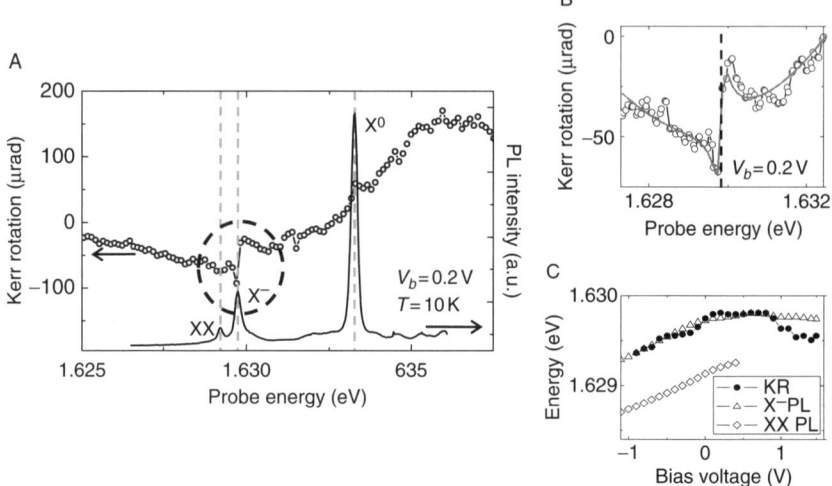

FIGURE 1.3 Single dot KR spectra. (A) KR (open circles) and PL (solid line) spectra at a bias voltage $V_b = 0.2$ V. The circled region is shown in more detail in (B). The solid line indicates a fit to the data and the energy of the X^- PL is indicated by the dashed line. (C) The center of the KR feature, E_0 (solid circles), and the energy of the X^- PL (open triangles) as a function of bias voltage; the two energies show a good agreement. The biexciton (XX) PL energy (open squares) is also shown for comparison. Adapted from Berezovsky et al. (2006).

energy E_0, as determined from the fit, is compared to the energy of the X^- PL line as a function of the applied bias in Fig. 1.3C. The two energies agree well and show the same quantum-confined Stark shift. For a single electron spin in the QD ground state, the lowest energy optical transition contributing in Eq. (1.1) is the X^- transition. From these observations we can conclude that the KR feature centered at the X^- energy is indeed from the measurement of a single electron spin in the QD. The measurement has also been repeated on other QDs and the same KR feature at the X^- PL energy has been observed. Additionally, the detection of a single electron spin in InAs QDs has recently been demonstrated using Faraday rotation (Atatüre et al., 2007). The large, broad KR background may be due to transitions involving excited electron and hole states, which are typically a few meV above the lowest transition (Gammon et al., 1996). If present, a KR feature due to the neutral exciton (X^0) spin should appear centered at the biexciton (XX) transition energy. The signal-to-noise in these measurements is not high enough to conclusively identify such a feature. Despite the large amplitude of the X^0 PL compared to the negatively charged exciton (X^-) PL in the uncharged bias regime (~10: 1), the short radiative lifetime of the X^0 state results in a low steady-state X^0 population, and therefore a low KR signal.

2.3. Spin dynamics of a single electron spin

In the experiments described above, only the steady-state spin polarization is measured, concealing information about the evolution of the spin state in time. However, using time delayed pump and probe pulses allows for the spin to be initialized and readout at different times, hence the coherent dynamics of the spin in the QD can be mapped out (Mikkelsen et al., 2007). For this a mode-locked Ti:Sapphire laser provides pump pulses with energy 1.653 eV, and duration ~150 fs at a repetition period $T_r = 13.1$ ns. The bandwidth of the spectrally broad pump pulses is narrowed to ~1 meV by passing the pump beam through a monochromator. The probe pulses are derived from a wavelength tunable cw Ti:Sapphire laser passing through an electro-optic modulator (EOM), allowing for electrical control of the pulse duration from cw down to 1.5 ns. This technique yields short pulses while maintaining the narrow linewidth and wavelength tunability of the probe laser. Also, it allows one to adjust the pulse duration so as to maintain enough average power to achieve good signal-to-noise, while keeping the instantaneous power low enough to avoid unwanted nonlinear effects. The EOM is driven by an electrical pulse generator triggered by the pump laser, allowing for electrical control of the time delay between the pump and the probe pulses. Additionally, in measurements with pump-probe delay of $\Delta t > 13$ ns, the pump beam has also been passed though an electro-optic pulse picker to increase the repetition period of the pulse train to $T_r = 26.2$ ns. For a fixed delay between the pump and the probe, the KR angle, θ_K, is measured as a function of probe energy. At each point, the pump excitation is switched between right and left circularly polarized light, as before, and the spin-dependent signal is obtained from the difference in θ_K at the two helicities. The resulting KR spectrum is fit to Eq. (1.1) plus a constant vertical offset, y_0. The amplitude, $\theta_0 = CE(|P^{\sigma+}_{\downarrow,v_0}|^2 - |P^{\sigma-}_{\downarrow,v_0}|^2)\Gamma^{-1}_{0,v_0}$, of the odd-Lorentzian is proportional to the projection of the spin in the QD along the measurement axis. By repeating this measurement at various pump-probe delays, the evolution of the spin state can be mapped out.

When a magnetic field is applied along the z-axis, transverse to the injected spin, the spin is quantized into eigenstates $|\uparrow\rangle$ and $|\downarrow\rangle$, with eigenvalues $S_z = \pm\hbar/2$. The pump pulse initializes the spin at time $t = 0$ into the superposition $|\psi(t=0)\rangle = (|\uparrow\rangle \pm |\downarrow\rangle)/\sqrt{2}$, for $\sigma\pm$ polarized excitation. If isolated from its environment, the spin state then coherently evolves according to $|\psi(t)\rangle = (\exp(-i\Omega t/2)|\uparrow\rangle \pm \exp(i\Omega t/2)|\downarrow\rangle)/\sqrt{2}$, where $\hbar\Omega = g\mu_B B_z$ is the Zeeman splitting. When the probe arrives at time $t = \Delta t$, the spin state is projected onto the x-axis, resulting in an average measured spin polarization of $\langle S_x(\Delta t)\rangle = \pm(\hbar/2)\cos(\Omega \cdot \Delta t)$. This picture has not included the various environmental effects that cause spin decoherence and dephasing, inevitably leading to a reduction of the measured spin polarization with time. The single spin KR amplitude as a function of delay, measured with a 3-ns duration probe pulse and a magnetic field $B_z = 491$ G, is shown in Fig. 1.4A, exhibiting the expected oscillations due to the coherent evolution described above. Figure 1.4B–F shows a sequence of KR spectra at several delays, and the fits from which the solid black data points in Fig. 1.4A are obtained. In the simplest case, the evolution of the measured KR amplitude can be described by an exponentially decaying cosine:

$$\theta(\Delta t) = A \cdot \Theta(\Delta t) \cdot \exp\left(-\Delta t/T_2^*\right) \cos(\Omega \cdot \Delta t) \quad (1.2)$$

where A is the overall amplitude, Θ is the Heaviside step function, and T_2^* is the effective transverse spin lifetime (though this measurement eliminates ensemble averaging, the observed spin lifetime may be reduced from the transverse spin lifetime, T_2, by inhomogeneities that vary in time). To model the data, the contributions from each pump pulse

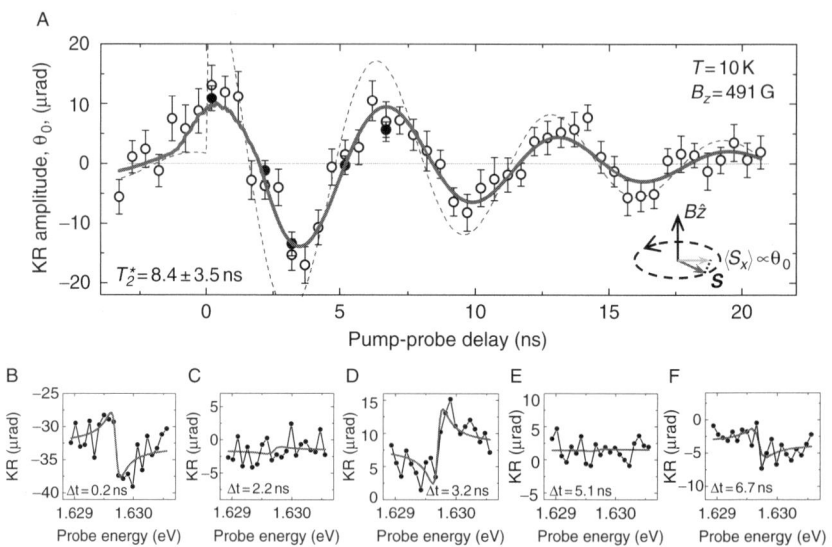

FIGURE 1.4 Coherent evolution of a single electron spin. (A) Single-spin KR amplitude, θ_0, as a function of time with 3-ns duration probe pulses and $B_z = 491$ G. The solid line is a fit to Eq. (1.3) and the dashed line shows Eq. (1.3) without the probe pulse convolution, plotted with the same parameters for comparison. The error bars indicate the standard error as obtained from the least-squares fit to the KR spectra. The solid circles indicate the values of θ_0 obtained from the fits shown in B–F. (B–F) KR angle as a function of probe energy at five different delays; solid lines are fits to the data. Adapted from Mikkelsen et al. (2007).

separated by the repetition period T_r are summed and convolved (denoted "*") with the measured probe pulse shape, $p(t)$:

$$\theta_0(\Delta t) = p^* \left[\sum_n \theta(\Delta t - nT_r) \right] \quad (1.3)$$

The solid line in Fig. 1.4A is a fit to Eq. (1.3), yielding a precession frequency $\Omega = 0.98 \pm 0.02$ GHz and $T_2^* = 8.4 \pm 3.5$ ns. The dashed line shows Eq. (1.3) without the probe pulse convolution, plotted with the same parameters for comparison. In Fig. 1.5A, the precession of the spin is shown at three different magnetic fields and as expected, the precession frequency increases with increasing field. The solid lines in Fig. 1.5A are fits to Eq. (1.3), and the frequency obtained from such fits is shown in Fig. 1.5B as a function of magnetic field. A linear fit to these data yields an electron g-factor of $|g| = 0.17 \pm 0.02$, consistent with the range of g-factors for these QDs found in previous ensemble or time-averaged measurements (Bracker et al., 2005; Gurudev-Dutt et al., 2005). At zero external magnetic field, as shown in Fig. 1.5C, the data fits well to a single exponential decay and the spin lifetime is found to be

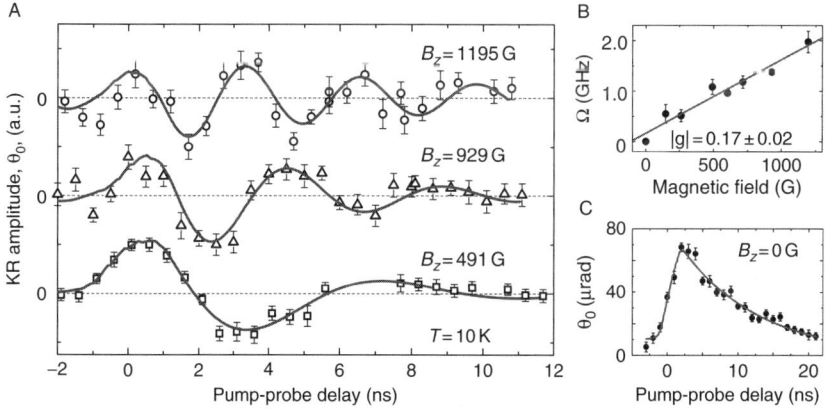

FIGURE 1.5 Magnetic field dependence. (A) Single-spin KR amplitude, θ_0, as a function of delay for $B_z = 1195$ G (circles), $B_z = 929$ G (triangles), and $B_z = 491$ G (squares). The probe-pulse duration is 1.5, 1.5, and 2 ns from top to bottom. Solid lines are fits to Eq. (1.3). The data are vertically offset for clarity. (B) Precession frequency, Ω, of the electron spin as a function of the applied magnetic field, as obtained from fits to the data. Each data point is the average of several delay scans of the same QD and the error bars indicate the root-mean-squared deviation of the measured frequencies. (C) KR amplitude, θ_0, as a function of delay at zero applied magnetic field with a 3 ns probe pulse duration. The solid line shows a fit to Eq. (1.3) yielding $T_2^* = 10.9 \pm 0.5$ ns. Adapted from Mikkelsen et al. (2007).

$T_2^* = 10.9 \pm 0.5$ ns. This value agrees with previous time-averaged (Bracker et al., 2005) and ensemble (Dzhioev et al., 2002; Gurudev-Dutt et al., 2005) measurements where the relevant decay mechanism is often suggested to be dephasing due to slow fluctuations in the nuclear spin polarization. However, these polarization fluctuations are not expected to result in a single exponential decay of the electron spin (Khaetskii et al., 2002; Merkulov et al., 2002). This suggests that other decay mechanisms than nuclear spin fluctuations might also be relevant in this case. In these QDs, the electronic level spacing of $\sim$1 meV (Merkulov et al., 2002) is of the same order as $k_B T$ for this temperature range. Therefore, thermally activated or phonon-mediated processes (Erlingsson et al., 2001; Golovach et al., 2004; Khaetskii and Nazarov, 2001; Semenov and Kim, 2004), which yield an exponential decay, might be significant in this regime.

This measurement technique is also sensitive to small nuclear spin polarizations. Ideally there should be no induced steady-state nuclear polarization in this experimental geometry. Since the magnetic field is applied perpendicular to the direction of the optically injected spins, nuclear spins that are polarized by the electron spins precess around the applied field, resulting in zero steady-state polarization. For any misalignment of the pump laser from the perpendicular, however, there is a projection of the spin along the magnetic field, and right (left) circularly polarized light induces a small dynamic nuclear polarization parallel (antiparallel) to the applied magnetic field (Meier and Zakharchenya, 1984; Salis et al., 2001). Because of the hyperfine interaction this acts on the electron spin as an effective magnetic field, increasing (decreasing) the total effective magnetic field, resulting in a slightly different precession frequency for right and left circularly polarized pump excitation. Since each data point is the difference of the KR signal with right and left circularly polarized excitation, a small deviation from perpendicular between the magnetic field and the electron spin yields a measured KR signal:

$$\theta(\Delta t) = A \cdot \Theta(\Delta t) \cdot \exp(-\Delta t / T_2^*)[\cos((\Omega + \delta)\Delta t) + \cos((\Omega - \delta)\Delta t)] \quad (1.4)$$

where $\delta = g\mu_B \overline{B_{nuc}}/\hbar$ is the frequency shift due to the steady-state effective nuclear field, $\overline{B_{nuc}}$. Insight into the build-up time for the dynamic nuclear polarization can be gained by varying the rate at which the pump helicity is switched. This reveals a nuclear polarization saturation time of $\sim$2 s (Mikkelsen et al., 2007), which agrees with what has previously been found in similar QDs (Gammon et al., 2001). Additionally, by measuring the effective magnetic field exerted by the nuclear spins, one can estimate the nuclear spin polarization.

2.4. Ultrafast manipulation using the optical Stark effect

In Section 2.3 it was discussed how the coherent evolution of a single electron spin can be readout. Here we turn our attention to the manipulation of spins. One way to control one or more spins is through the well-known phenomenon of magnetic resonance. By applying an oscillating magnetic field whose frequency matches the spin precession frequency, the spin state can be coherently controlled. This spin control through electron spin resonance (ESR) has recently been demonstrated on a single electron spin, performing complete rotations of the spin state on a timescale of tens of nanoseconds (Koppens et al., 2006; Nowack et al., 2007). Additionally, a variety of other optical manipulation schemes have been explored on ensembles of spin (Carter et al., 2007; Dutt et al., 2006; Greilich et al., 2007; Wu et al., 2007).

Using ultrafast optical pulses to coherently manipulate the spin state of a single electron is a key ingredient in many proposals for solid-state quantum information processing (Chen et al., 2004; Clark et al., 2007; Combescot and Betbeder-Matibet, 2004; Economou et al., 2006; Imamoglu et al., 1999; Pryor and Flatté, 2006). In the manipulation scheme discussed here (Berezovsky et al., 2008), the optical, or AC, Stark effect is exploited. Through this effect, an intense, nonresonant optical pulse creates a large effective magnetic field along the direction of the light for the duration of the pulse. Since the generation of short optical pulses is easy using mode-locked lasers, the optical Stark effect can be a very useful tool for ultrafast manipulation of spins. The optical Stark effect was first studied in atomic systems in the 1970s (Cohen-Tannoudji and Dupont-Roc, 1972; Cohen-Tannoudji and Reynaud, 1977; Suter et al., 1991) and subsequently explored in bulk semiconductors and in quantum wells (Combescot and Combescot, 1988; Joffre et al., 1989; Papageorgiou et al., 2004). In recent years, the optical Stark effect has been used to observe ensemble spin manipulation in a quantum well (Gupta et al., 2001), and to control orbital coherence in a QD (Unold et al., 2004).

Using perturbation theory, it is found that an optical field with intensity I_{tip}, detuned from an electronic transition by an energy δ, induces a shift in the transition energy:

$$\Delta E \approx \frac{D^2 I_{tip}}{\delta \sqrt{\varepsilon/\mu}} \quad (1.5)$$

where D is the dipole moment of the transition, and ε and μ are the permittivity and permeability of the material, respectively (Joffre et al., 1989). Because of the optical selection rules, for circularly polarized light, the optical Stark effect shifts only one of the spin sublevels and produces a spin splitting in the ground state which can be represented as an effective magnetic field, B_{Stark}, along the light propagation direction. By using ultrafast laser pulses with high instantaneous intensity to provide the Stark shift, large splittings can be obtained to perform coherent manipulation of the spin within the duration of the optical pulse (here, $B_{Stark} \sim 10$ T).

To perform this manipulation, a third beam is needed which is tuned below the exciton transition energy and synchronized with the pump and probe lasers. The third beam, which we will refer to as the tipping pulse, is derived from the same mode-locked Ti:Sapphire laser which provides the pump pulses. The relative time delay between the pump pulse and the tipping pulse is controlled by a mechanical delay line in the pump path. A schematic of the experimental setup is shown in Fig. 1.6A. Thus, we have three synchronized, independently tunable optical pulse trains that are focused onto the sample: the pump, the probe, and the tipping pulse. The circularly polarized tipping pulse (duration $\sim$30 ps, FWHM = 0.2 meV) is tuned to an energy below the lowest QD transition (see Fig. 1.6B) and is used to induce the Stark shift. Exactly as in the time-resolved measurements, the spin is initialized at time $t = 0$ along the growth direction (y-axis) and then in the case of a transverse magnetic field coherently precesses at the Larmor frequency. At time $t = t_{tip}$, the tipping pulse arrives and generates an additional spin splitting along the y-axis for the duration of the pulse. During this time, the spin precesses about the total effective field (which is typically dominated by B_{Stark}), and then continues to precess about the static applied field. The probe pulse then measures the resulting projection of the spin in the QD, $\langle S_y \rangle$ at time $t = t_{probe}$. This sequence is repeated at the repetition frequency of the laser (76 MHz), and the signal is averaged for several seconds for noise reduction. The spin dynamics can be described using a simple model including the effect of nuclear polarization (see Berezovsky et al., 2008). To map out the coherent dynamics of the spin in the QD, KR spectra are again measured as a function of pump-probe delay. Figure 1.7A shows the time evolution of a single spin in a transverse magnetic field with no tipping pulse applied. Each data point is determined from the fit to a KR spectrum at a given pump-probe delay, as in Fig. 1.4A. Convolving the expected spin dynamics with the measured profile of the probe pulse, a least-squares fit to the data can be performed and various parameters determined: Ω, T_2^*, and the effective field from the nuclear polarization, B_{nuc}. The solid curve in Fig. 1.7A shows the result of this fit, and the dashed line is the corresponding plot without the probe pulse convolution. The values obtained from the fit are later used to model the data including the effect of the tipping pulse. The data in Fig. 1.7B show the same coherent spin dynamics as in Fig. 1.7A, but with the tipping pulse applied at $t_{tip} = 1.3$ ns when the projection of the spin is mainly along the x-axis. The intensity of the tipping pulse is chosen to induce a $\sim\pi$ rotation

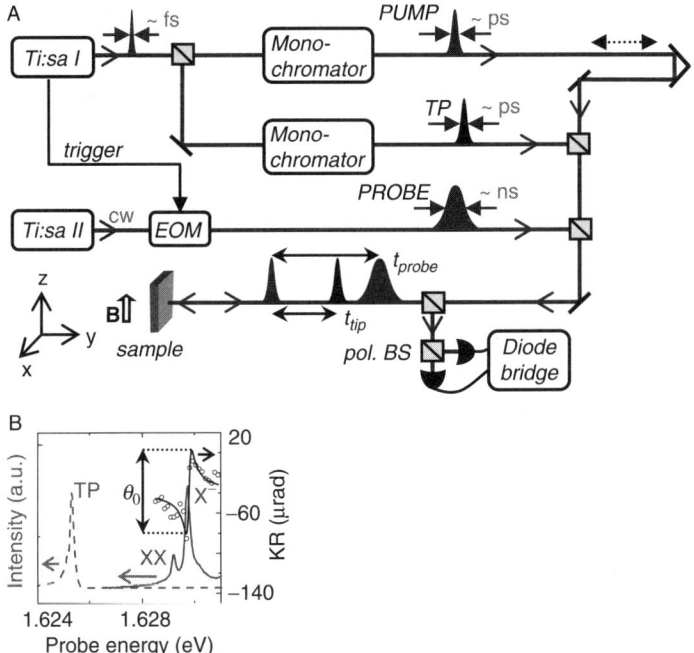

FIGURE 1.6 Setup for spin manipulation using the optical Stark effect. (A) Schematic of the experimental setup; pol. BS = polarizing beam-splitter. (B) The solid gray line shows negatively charged exciton (X$^-$) and biexciton (XX) PL from a single dot. The black data points show the corresponding single spin KR, and the black line is an odd–Lorentzian fit to this data from which the KR amplitude, θ_0, is obtained. The dashed line shows the tipping pulse (TP) spectrum at a detuning of 4.4 meV. Adapted from Berezovsky et al. (2008).

about the y-axis, which is determined as discussed below. After the tipping pulse, the spin has been flipped and the resulting coherent dynamics show a reversal in sign. This can be clearly seen by comparing the sign of the measured signal at the position indicated by the dashed, vertical line in Fig. 1.7. The predicted spin dynamics are shown in the dashed line, and the same curve convolved with the probe pulse is given by the solid line. Note that this curve is not a fit; all of the parameters are determined either in the fit to Fig. 1.7A, or as discussed below. Only the overall amplitude of the curve has been normalized.

Further details of this spin manipulation can be investigated by varying the tipping pulse intensity, I_{tip}, and the detuning, δ, of the tipping pulse from the QD transition energy for a fixed delay of the tipping and probe pulse as illustrated in Fig. 1.8B. In Fig. 1.8A, the KR amplitude, θ_0, as a function of the tipping pulse intensity is shown at a probe delay of 2.5 ns with the tipping pulse arriving at a delay of 1.3 ns, for three

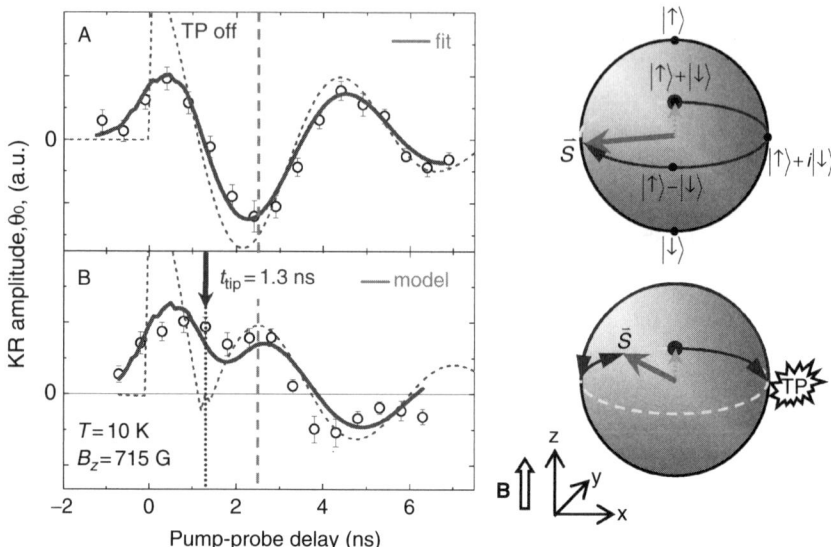

FIGURE 1.7 Ultrafast π rotations of a single spin. (A) Coherent single spin precession in a transverse magnetic field $B_z = 715$ G. The solid line shows a fit to the data, and the dashed line shows the corresponding plot without the probe pulse convolution for the same fit parameters. The diagrams on the right schematically show the evolution of the spin on the Bloch sphere. (B) Spin dynamics under the same conditions as in (A) but with the tipping pulse (TP) applied at $t_{tip} = 1.3$ ns so as to induce a $\sim$π rotation about the y-axis. Tipping pulse detuning: $\delta = 2.65$ meV and intensity: $I_{tip} = 4.7 \times 10^5$ W/cm². Adapted from Berezovsky et al. (2008).

different detunings of the tipping pulse from the transition. When the tipping pulse intensity is zero, the spin precesses undisturbed and yields a negative signal at $t_{probe} = 2.5$ ns as in Fig. 1.7A. As the intensity is increased, the spin is coherently rotated through an increasingly large angle and the observed signal at $t_{probe} = 2.5$ ns changes sign and becomes positive, as in Fig. 1.7B. Furthermore, the strength of the optical Stark effect is expected to decrease linearly as a function of the detuning, as seen in Eq. (1.5). The gray lines in Fig. 1.8A are plots of the predicted spin dynamics with parameters taken from the fit in Fig. 1.7A, and $\phi_{tip} = \beta I_{tip}$. The only parameter that is changed between the three curves in Fig. 1.8A is the strength of the optical Stark effect, β. From this, we can estimate the fidelity of a π-rotation to be $\sim$80%. The tipping pulse intensity required for a π-rotation, $I_\pi = \pi/\beta$, is shown in Fig. 1.8C as a function of detuning, displaying the expected linear dependence.

As explained in the previous section, a small misalignment from perpendicular between the pump beam and the static magnetic field gives rise to a small dynamic nuclear polarization. The data in Fig. 1.8A

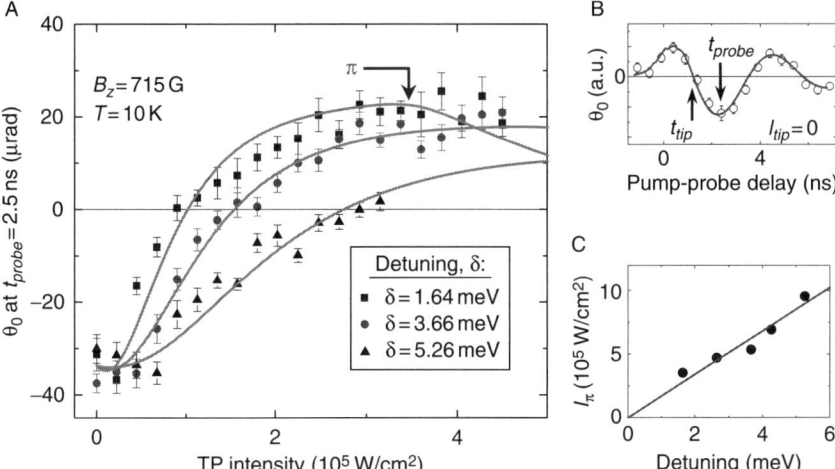

FIGURE 1.8 Intensity and detuning dependence. (A) Single spin KR amplitude, θ_0, as a function of tipping pulse (TP) intensity at three detunings from the X$^-$ transition. The tipping pulse arrives at $t_{tip} = 1.3$ ns, and the probe is fixed at $t_{probe} = 2.5$ ns, as illustrated in (B). The solid gray lines are fits to the data varying only a single parameter, the strength of the optical Stark effect, β. The tipping pulse intensity, I_π, required for a π rotation at $\delta = 1.64$ meV is indicated by the arrow. (C) I_π as a function of detuning, δ, as obtained from the fits. The solid line shows a linear fit to the data. Adapted from Berezovsky et al. (2008).

most clearly show the effects of this nuclear polarization on the observed spin dynamics. In the absence of dynamic nuclear polarization, one would expect the curves in Fig. 1.8A to be cosinusoidal, crossing zero at an intensity half that required for a π rotation. Dynamic nuclear polarization however, which is maximal when $\phi_{tip} \approx \pi/2$, distorts this ideal cosine form, as is well described by the model.

2.5. Conclusions

We have described a demonstration of sequential initialization, manipulation, and readout of the state of a single electron spin in a QD using all-optical techniques. First, a single electron spin in a QD is detected using a time-averaged magneto-optical Kerr rotation measurement at $T = 10$ K. This technique provides a means to directly probe the spin off-resonance, thus minimally disturbing the system. Next, this continuous single dot KR technique was extended into the time domain using pulsed pump and probe lasers, allowing observation of the coherent evolution of an electron spin state with nanosecond temporal resolution. The coherent single spin precession in an applied magnetic field directly revealed the electron g-factor and a transverse spin lifetime of $\sim$10 ns. Furthermore, the observed spin dynamics provided a sensitive probe of the local nuclear spin environment. Finally, a scheme to perform ultrafast coherent optical manipulation of a single electron spin state was described. By applying off-resonant, picosecond-scale optical pulses, the coherent rotation of a single electron spin in a QD through arbitrary angles up to π radians was shown. Measurements of the spin rotation as a function of laser detuning and intensity confirmed that the optical Stark effect is the operative mechanism and the results are well described by a model including the electron–nuclear spin interaction.

In principle, at most a few hundred single qubit flips could be performed within the measured spin coherence time. However, by using shorter tipping pulses and QDs with longer spin coherence times, this technique could be extended to perform many more operations before the spin coherence is lost. A mode-locked laser producing $\sim$100-fs-duration tipping pulses could potentially exceed the threshold ($\sim 10^4$ operations) needed for proposed quantum error correction schemes (Awschalom et al., 2002). Additionally, the spin manipulation demonstrated here may be used to obtain a spin echo (Rosatzin et al., 1990), possibly extending the observed spin coherence time. These results represent progress toward the implementation of scalable quantum information processing in the solid state.

3. FEW MAGNETIC SPINS IN QUANTUM WELLS

Detection and control of single magnetic atoms represents the fundamental scaling limit for magnetic information storage. In semiconductors, magnetic ion spins couple to the host semiconductor's electronic structure allowing for bandgap engineering of the exchange interactions (Awschalom and Samarth, 1999). The strong and electronically controllable spin–spin interactions existing in magnetic semiconductors offer an ideal laboratory for exploring single magnetic spin readout and control. The bandgap engineering and electronic control possible in semiconductor heterostructures allows for the localization and control of free carriers over nanometer (nm)-length scales, in particular for quantum wells and quantum dots (see Kaminska and Cibert). Magnetic ions doped within a semiconductor lattice can exhibit relatively long spin lifetimes.

Exploration of single magnetic spins in semiconductors was pioneered in the (II,Mn)-VI system (Besombes et al., 2004). Individual self-assembled ZnSe quantum dots are characterized using microscopically resolved photoluminescence measurements. These quantum dots are doped with Mn-ions at a density such that single magnetic ions occasionally occupy the center of a quantum dot. In these II–VI magnetic semiconductors, the magnetic ions are isoelectronic Mn^{2+} ions with spin-5/2. As will be discussed in later chapters (see Chapters 5 and 9), magnetic spins couple to the host semiconductor through the s–d and p–d exchange with electron and hole spins, respectively (Furdyna, 1988). The spin state of the magnetic ion is reflected through the exchange splitting of the exciton (electron–hole) spin states. Single magnetic ions in quantum dots are measured via the exchange splitting of the exciton states that results when a single Mn^{2+} ion is centered in the quantum dot. The coupling between exciton and magnetic ions can be controlled electrically by charging the dots with electrons or holes (Leger et al., 2006).

3.1. Mn-ions in GaAs as optical spin centers

A different situation arises in (III,Mn)-V magnetic semiconductors, where the Mn^{2+} ions contribute acceptor states within the bandgap causing the magnetic ions to behave as optical spin centers. As illustrated in Fig. 1.9A, photoexcited electrons in the bottom of the conduction band recombine with holes bound to Mn ions (e, A_{Mn}^0) and emit photoluminescence (PL) (Chapman and Hutchinson, 1967; Schairer and Schmidt, 1974). In GaAs, at low doping levels, the paramagnetic Mn^{2+} ions form a neutral acceptor configuration in which a spin-3/2 heavy hole state is loosely bound and antiferromagnetically coupled to the spin-5/2 Mn^{2+} ion. This neutral acceptor complex (A_{Mn}^0) has a total angular momentum state $J = 1$ and a g-factor $g_{A^0_{Mn}} = 2.77$ measured by electron paramagnetic resonance (EPR) and SQUID magnetometry (Frey et al., 1988; Schneider et al., 1987). This mixing of the Mn ion and valence band states opens the possibility for electrical manipulation of a single magnetic ion even in a bulk crystal (Tang et al., 2006). Individual Mn acceptors can be imaged at surfaces of GaMnAs via scanning tunneling microscopy (STM) as shown in Fig. 1.9B (Yakunin et al., 2004). Alternatively, micro-PL of Mn ions within single quantum wells allows for the spatial isolation of small numbers of Mn ions deep within bandgap engineered heterostructures. An important criterion to observe optical transitions near the bandgap edge in Mn doped GaAs is a proper choice of sample synthesis techniques. The growth technique of choice, MBE allows for atomic layer precision of heterostructures (Gossard, 1986). Most Mn-doped GaAs is grown at low substrate temperatures where defects such as arsenic antisites act as compensating defects and nonradiative traps (Erwin and Petukhov, 2002; Liu et al., 1995). Conversely, at high growth temperatures that are necessary for high quality GaAs/AlGaAs quantum structures, Mn tends to form interstitial defects and MnAs clusters (Ohno, 1998). At intermediate temperatures, a large range of Mn concentrations can be incorporated without forming the unwanted defects. This enables optical measurement of coherent electron spin precession in GaMnAs allowing for a precise measurement of the exchange coupling between electrons and Mn ions (Myers et al., 2005; Poggio et al., 2005).

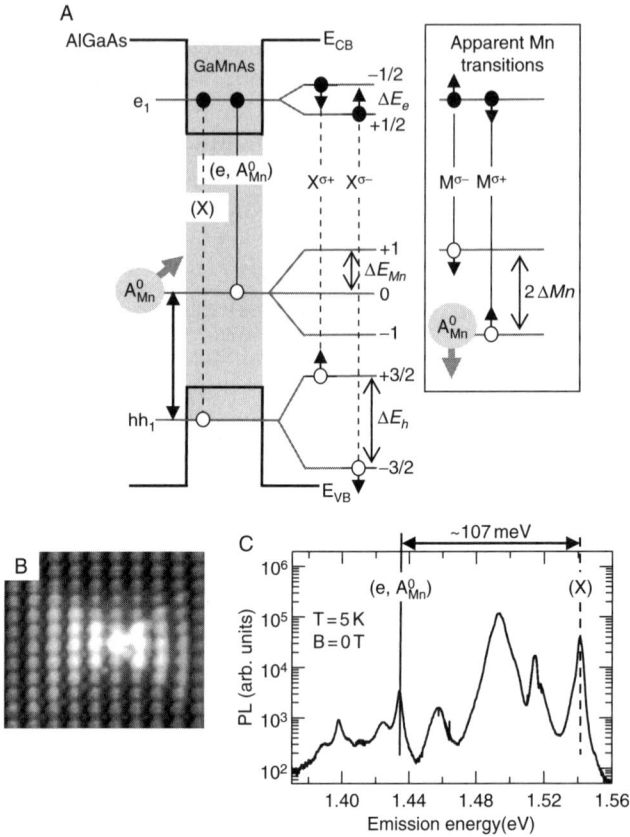

FIGURE 1.9 Optical transitions in (III,Mn-V) quantum wells. (A) Bandedge along the growth direction showing optical transitions of excitons and Mn acceptor emission. In addition to the exciton exchange splitting (ΔE_e and ΔE_h), an apparent spectroscopic splitting of Mn acceptor PL (ΔMn) occurs if the Mn spin states are split by ΔE_{Mn}. (B) Mn acceptors imaged by STM, adapted from Yakunin et al. (2004). (C) Full photoluminescence (PL) emission energy spectrum of a Mn:GaAs single QW with Mn doping density of 7.3×10^{18} cm^{-3}. Adapted from Myers et al. (2008).

Figure 1.9A shows a schematic of the optical transitions occurring within such Mn-doped quantum wells. As indicated in the PL spectrum (Fig. 1.9C), the Mn acceptor emission (e, A_{Mn}^0) is red-shifted by $\sim$107 meV from the quantum well exciton (X) PL. In bulk, the A_{Mn}^0 state lies $\sim$110 meV above the valence band edge.

Polarization-resolved PL yields the angular momentum of the emitted photons, and through conservation of angular momentum, gives information about the spin state of electrons, holes, and the Mn acceptors. In quantum wells of zinc-blende semiconductors, the optical selection rules yield optical transitions with specific circular polarity depending on the spin state of the recombining carriers (Meier and Zakharchenya, 1984). Figure 1.9A shows the spin-selective heavy hole exciton (X) transitions in the presence of a spin splitting generated by a positive magnetic field (either real or effective) along the measurement axis and near $k \sim 0$. The boxed region shows the circularly polarized Mn-related optical transitions, which occur due to overlap of the localized Mn acceptor state with the band tail. The circular polarization of the (e, A_{Mn}^0) emission is sensitive to both the spin polarization of the Mn ions and the electrons (Averkiev et al., 1988). The PL from (D^0, A_{Mn}^0) in bulk GaAs has been used to investigate the Mn ion polarization (Karlik et al., 1982; Kim et al., 2005). PL from recombination of hot electrons, away from $k \sim 0$, into the Mn acceptor state was also observed (Sapega et al., 2007). In these measurements, the polarization of the Mn acceptor state was tracked in multiple quantum wells showing a decrease in the polarization with increasing quantum confinement.

3.2. Zero-field optical control of magnetic ions

An example of the (e, A_{Mn}^0) PL spectra from an Mn-doped GaAs quantum well is shown in Fig. 1.10. The data are taken using a circularly polarized laser tuned to the heavy hole X transition in the quantum well, which injects spin-polarized electrons and holes into the quantum well as shown in the schematic. At zero magnetic field, the polarization resolved spectra show higher intensity for one detection helicity than the other implying a zero-field spin polarization, discussed below. By fitting the spectra to two Gaussians (solid lines), the peak position ($M^{\sigma\pm}$) and intensity ($I_M^{\sigma\pm}$) are extracted from which either a polarization $\left(P_{Mn} = (I_M^{\sigma+} - I_M^{\sigma-})/(I_M^{\sigma+} + I_M^{\sigma-})\right)$ or spectral splitting ($\Delta Mn = M^{\sigma+} - M^{\sigma-}$) are measured. When a linearly polarized (π) laser is used for excitation, the polarization (P_{Mn}) is zero at zero magnetic field and follows the paramagnetic alignment of Mn ions with longitudinal field (B_z) (Fig. 1.10B). If a $\sigma\pm$ polarized laser is used, the odd symmetry of (P_{Mn}) with magnetic field is broken. For $\sigma+$, the polarization saturates more quickly for $B_z < 0$, and shows slower saturation for positive field. The opposite trend is seen for the opposite helicity excitation. This implies that optical spin injection results in an effective magnetic field on the spin states of the Mn ions, which is positive for $\sigma-$ and negative for $\sigma+$. Such an effective magnetic field should generate a splitting of the Mn acceptor m_J states at zero field, as drawn schematically in Fig. 1.10A, resulting in a spectral splitting of the (e, A_{Mn}^0) emission lines. Such a spectral splitting (ΔMn) is observed at zero field, which traces the polarization (P_{Mn}) (Fig. 1.11).

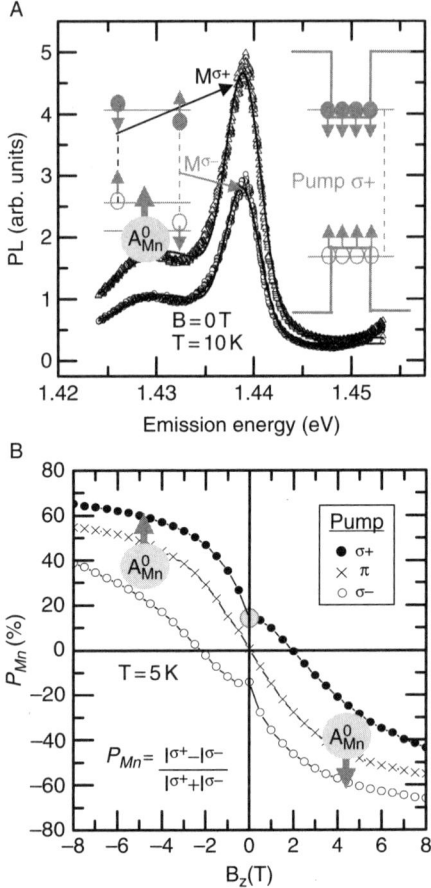

FIGURE 1.10 Optical readout of magnetic ion spin orientation. (A) Polarization-resolved Mn acceptor PL (P_{Mn}). A $\sigma+$ polarized laser injects spin polarized electrons and holes into the quantum well and generates an Mn-ion spin polarization. Solid lines are fits to the data (black points). (B) Longitudinal field (B_z) dependence of the Mn-ion spins. The field symmetry of the Mn paramagnetism under π polarized excitation is broken in the case of optical spin injection ($\sigma\pm$ polarized excitation). Adapted from Myers et al. (2008).

The angular momentum of the photons in the excitation beam is tuned by changing the polarization state smoothly from $\sigma-$ to π to $\sigma+$ using an electronically tuned waveplate. The orientation of the Mn-ions tracks the helicity in both polarization and spectral splitting corresponding to the Mn ion spin splitting and orientation shown schematically in Fig. 1.11. The optically induced magnetization, described above, occurs due to optical excitation of spin polarized carriers into the quantum well. By changing the energy of the excitation laser, it is possible to inject different configurations of electron and hole spins into the quantum well and

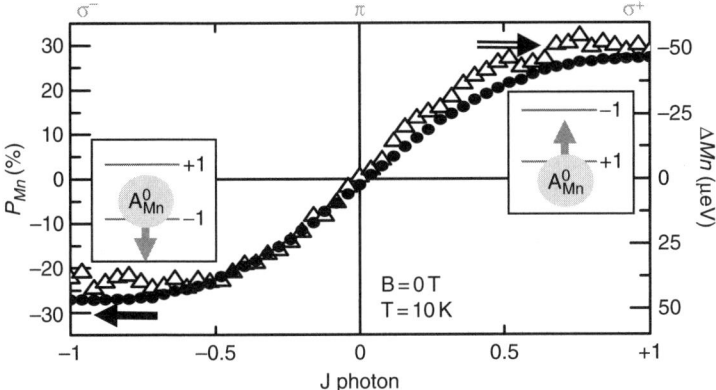

FIGURE 1.11 Zero-field optical control of magnetic ion spins. The circular polarization (left axis) and the associated spectral splitting (right axis) of the Mn acceptor emission are altered in sign and magnitude as the photon helicity and angular momentum are varied using a variable wave retarder. Adapted from Myers et al. (2008).

differentiate their effect on the magnetic ions. Figure 1.12 plots the intensity, polarization, and spectral splitting of the Mn acceptor emission as the laser energy is tuned near the absorption edge of the quantum well for a fixed helicity. At 1.546 eV, the σ+ polarized laser is resonant with the heavy hole exciton absorption generating spin down electrons and spin up heavy holes, as schematically shown. This excitation generates a maximum in the PL intensity of the Mn emission. A second PL maxima occurs at 1.557 eV where the laser coincides with the light hole exciton absorption which generates spin up electrons and light holes in the quantum well. For this resonance, the Mn polarization and spin splitting change sign compared to the heavy hole resonance. Thus, the orientation of the Mn spins at zero field is changed by tuning the laser excitation energy.

For a given photon helicity, the sign of the electron spin changes between the heavy and light hole exciton transitions, while the hole spins are parallel. This is drawn schematically in Fig. 1.12. Thus, the change in sign of the Mn ion polarization between the heavy and light hole transitions cannot occur due to interaction with holes alone, but rather the sign change implies that electron–Mn spin interaction generates the nonequilibrium magnetization.

3.3. Mechanism of dynamic polarization and exchange splitting

We now discuss how such a zero-field magnetization is possible in GaAs. For magnetic semiconductors, the mean-field interaction between charge carrier and magnetic ion spins is the so-called (s−d) and (p−d)

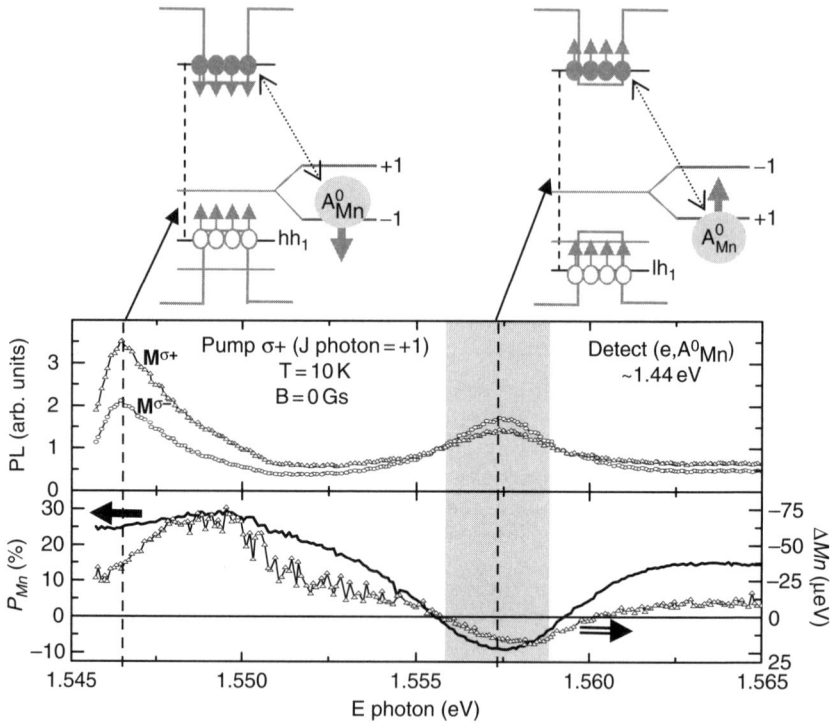

FIGURE 1.12 Photon energy dependence of the dynamic magnetic polarization. The photon helicity is fixed and its energy is varied. The detection energy is fixed at the Mn acceptor emission line. Optically injected electron spins change sign between the heavy hole (hh) and light hole (lh) exciton resonances and generate a change in sign of the Mn spin polarization and exchange splitting. Adapted from Myers et al. (2008).

exchange interactions between conduction and valence band spins and the magnetic ions (Dietl, 1994). The Mn spin splitting due to the s, p–d exchange coupling is the sum of both terms:

$$\Delta E_{Mn}^{s,p-d} = n\alpha\langle S_e\rangle + p\beta\langle S_h\rangle \tag{1.6}$$

where n and p are the optically excited electron and hole densities, and $\langle S_e\rangle$ and $\langle S_h\rangle$ are the electron and hole spin average values, respectively. When spin polarized electrons and holes are optically injected, $n\langle S_e\rangle$ and $p\langle S_h\rangle$ take on finite values. The data shown in Fig. 1.12 are taken using a pump power generating at most $n = p = 4 \times 10^{17}$ cm^{-3} electrons and holes spin polarized with $\langle S_e\rangle = 1/2$ and $\langle S_h\rangle = 3/2$. The average exchange field they exert on the magnetic ions decays with the spin lifetime of the carriers, where electron spins in these structures have measured spin lifetimes of ~100 ps (described below) and hole spins have spin lifetimes of ≤1 ps (Damen *et al.*, 1991). These parameters yield $\Delta E_{Mn}^{s,p-d} \leq 2\mu eV$. Thus, the s, p–d exchange field is too weak by more than an order of magnitude to explain the observed polarization and spin splitting (ΔMn ~ 50 μeV). The observed zero-field splitting of the Mn ion spin states originates from interactions with neighboring Mn ions. This is surprising considering the low doping level at which the effect is observed, corresponding to many atomic distances between Mn ions. Qualitatively, the loosely bound hole of the neutral Mn acceptor allows for longer range interactions between Mn spins than is otherwise possible. The interaction between neutral Mn acceptors has been treated using multi-band tight binding model of the Mn acceptor wave function (Tang and Flatté, 2004). This model has successfully predicted the shape of the acceptor wave function as imaged via scanning tunneling microscopy (Fig. 1.9B) and the interactions between pairs of Mn acceptors (Kitchen *et al.*, 2006). Using this model, the mean field effect of all the Mn ions in the crystal can be calculated which contributes an additional spin splitting to the Mn spin states:

$$\Delta E_{Mn} = \Delta E_{Mn}^{s,p-d} + \lambda \langle J_{Mn} \rangle \quad (1.7)$$

where λ is the Weiss molecular field due to neighboring Mn ions and $\langle J_{Mn} \rangle$ is the projection of angular momentum of the Mn ion along the magnetic field. λ is calculated by summing the interaction with neighboring Mn spins, yielding energy splittings that vary from 0.01 μeV for 6×10^{17} Mn/cm^3 to 400 μeV for 7×10^{18} Mn/cm^3 (Myers *et al.*, 2008). We now have a clear picture of how paramagnetic Mn ions become magnetized by spin injection. Electron–Mn spin interaction initiates a polarization of the Mn spins generating a finite $\langle J_{Mn} \rangle$. Once they are partially aligned, the Mn–Mn mean field interaction favors a parallel orientation of Mn spins and generates an exchange splitting as in Eq. (1.7).

3.4. Spin dynamics of Mn-ions in GaAs

Because Mn spins can be optically oriented at zero field, we can apply a transverse field (B_x) to precess the spins about the applied field. The time-averaged spin precession and decay is observed as a decrease in the PL polarization as the field is increased (Fig. 1.13A) called the Hanle effect:

$$P(B_x) = \frac{P(0)}{\left(g\mu_B B_x T_2^*/\hbar\right)^2 + 1} \quad (1.8)$$

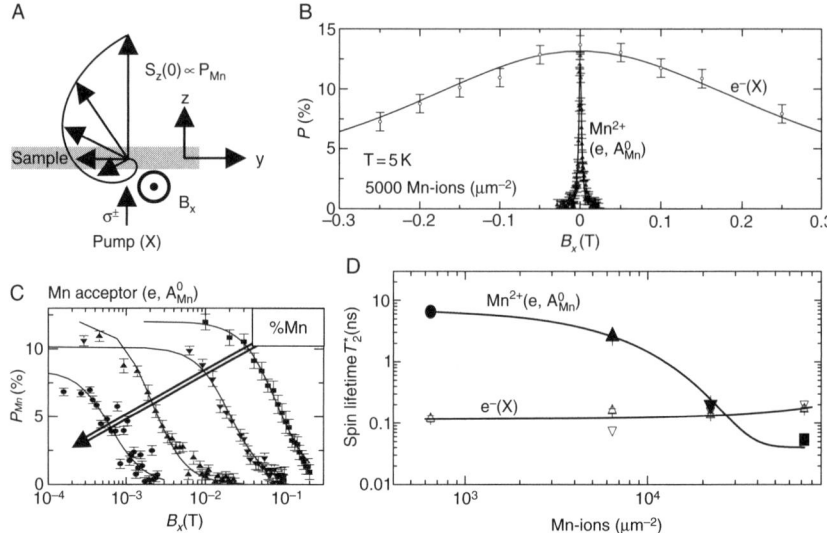

FIGURE 1.13 Spin lifetime measurement of Mn ions and electron spins. (A) Hanle effect measurement of time-averaged spin precession in a transverse field (B_x). (B) Hanle curves from the exciton PL (open circles) and the Mn acceptor emission (solid triangles), sensitive to the electron and Mn spin precession, respectively. (C) Hanle measurement of the Mn spin precession for samples with different Mn density. Lines in (B) and (C) are fits to Eq. (1.8). (D) Spin lifetime (T_2^*) of the Mn ions (closed) and the electrons (open) extracted from Hanle data are plotted as a function of Mn concentration. Lines guide the eye. Adapted from Myers et al. (2008).

and $1/T_2^* = 1/\tau_r + 1/\tau_s$, where T_2^* is the effective transverse spin lifetime of the spin ensemble, τ_r is the recombination lifetime and τ_s is the spin relaxation time (Meier and Zakharchenya, 1984). If the g-factor of the precessing spin is known, then the lifetime can be extracted. Electron spin precession in the QW is observed by measuring polarization of the (X) emission versus transverse field showing the Hanle effect. In a similar fashion, Mn spin precession is observed from Hanle measurements of the Mn acceptor emission. As seen in Fig. 1.13B, the Mn-related Hanle curve has a width at least one order of magnitude narrower than the electron (X)-related Hanle curve. From Eq. (1.8), this implies that gT_2^* for Mn acceptors is 10 times larger than for electrons in this particular QW. As the Mn density is reduced, a dramatic narrowing of the Hanle curves is observed, shown in a semi log plot in Fig. 1.13C. The g-factor of the electrons is known from optical measurements of these structures (Stern et al., 2007), and the g-factor of the neutral Mn acceptor is known to be $g_{A^0_{Mn}} = 2.77$ (Schneider et al., 1987) allowing the spin lifetimes to be extracted. These values are plotted in Fig. 1.13D as a function of the Mn ion density within single 10-nm thick GaAs quantum wells.

Using micro-PL technique, the collection spot is narrowed to a 1 µm diameter spot allowing for small numbers of Mn ions to be probed. Below 20,000 Mn ions within the detection spot, the spin lifetime dramatically increases and saturates near 10 ns at the lowest doping levels corresponding to the detection of several hundred magnetic ions within a square micron. This reflects the effect of inhomogeneous Mn–Mn spin interactions that dephase the spins. In contrast, electron spins maintain a roughly 100 ps lifetime over all Mn densities.

3.5. Conclusions

Micro-PL within single quantum wells provides a method to isolate small numbers of Mn ion emittors, which is scalable to the single ion limit. When spins are injected into the quantum well, a unique spin scattering mechanism generates zero-field, nonequilibrium polarization and magnetization of Mn ions even though they are paramagnetic at equilibrium. Optically induced magnetism in II–VI materials usually arises due to the mean field interaction of the ions with the electron and hole spins. In the case of Mn ions doped in GaAs, the neutral acceptor state results in a unique system where both electron–Mn and Mn–Mn interactions come into play and generate a dynamic exchange splitting of the magnetic ions. Mn ion spin coherence lifetimes can be at least as long as those of electron spins in quantum dots, allowing us to consider single Mn ion spins in III–V's as potentially useful for storage and readout of spin information. Because bandgap engineering is very well developed, a whole host of interesting structures for engineering the charge and spins of single Mn ions in GaAs-based heterostructures is possible.

4. SINGLE SPINS IN DIAMOND

4.1. Introduction

Diamond as a wide bandgap semiconductor has a number of useful and interesting properties including high mobilities (Isberg *et al.*, 2002) and excellent thermal conductivity (Wei *et al.*, 1993). Typical semiconducting diamond is p-type, with boron being the principle dopant. The most common impurity in diamond, however, is substitutional nitrogen that forms an impurity level ~1.7 eV below the conduction band (Farrier, 1969). Additionally, these nitrogen spins can form a different defect when they occur on a lattice site next to a vacancy (Fig. 1.14A). These defect centers, known as NV centers, are optically active single spins with robust quantum properties that extend up to room temperature. In this

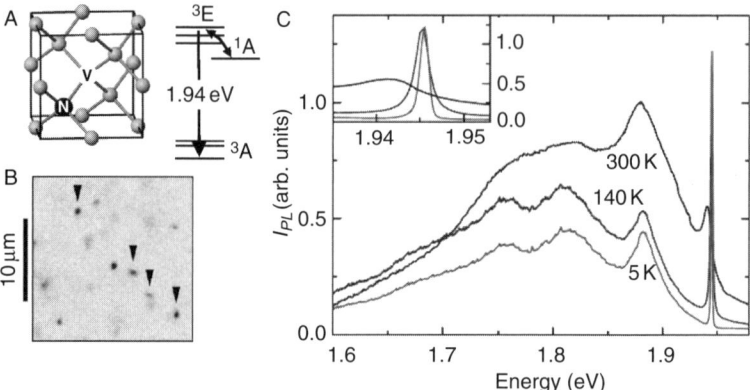

FIGURE 1.14 (A) A unit cell of the diamond lattice containing an NV center. The solid line connecting the nitrogen substitutional defect to the neighboring vacancy shows the symmetry axis of the NV center. Also shown is a summary diagram of the electronic levels of the NV center. (B) A spatial image of the photoluminescence intensity plotted using a linear gray scale. Several individual NV centers are marked. (C) An energy resolved photoluminescence spectra of an ensemble of NV centers at three temperatures. (Inset) Close-up of the zero phonon line. (A) and (B) are taken from Epstein et al. (2005) and (C) is taken from Epstein (2005).

section we will discuss diamond NV centers as individually addressable single spins that may form the basis for room temperature quantum information systems.

4.2. NV basics

An NV center can be thought of as a "solid-state" molecule with discrete energy levels inside the 5.5 eV gap of diamond. The electronic wave function is mainly centered on the dangling bonds of the carbon atoms adjacent to the vacancy, with very little overlap with the N nucleus (He et al., 1993). This results in a very small hyperfine splitting of the electronic spin levels due to the contact hyperfine interaction. Figure 1.14A shows a schematic of the electronic energy levels, including the ground state (^{3}A), the excited state (^{3}E), and a metastable spin singlet state (^{1}A) with intermediate energy (Nizovtsev et al., 2003). The electronic ground state (^{3}A) is a spin triplet (e.g., $S = 1$) with a zero-field splitting $D = 2.87$ GHz between the $m_s = 0$ and $m_s = \pm 1$ levels (Loubser and van Wyk, 1978; Nizovtsev et al., 2003). At zero applied field, the $m_s = \pm 1$ levels are degenerate in unstrained diamond, but they become Zeeman split in an externally applied magnetic field. The spin is quantized along the symmetry direction which is set by the orientation of the NV symmetry axis ($\langle 111 \rangle$) within the crystal lattice (Loubser and van Wyk, 1978). The excited state (^{3}E) is also a spin triplet but determination of its structure is an ongoing investigation (Manson and McMurtrie, 2007). The ^{3}A→^{3}E transition is phonon broadened at room temperature, with only a small portion of the photons emitted due to the direct transition (1.94 eV) known as the zero phonon line (ZPL). The ZPL, however, becomes much more prominent at low temperature (Fig. 1.1C) (Clark and Norris, 1971; Epstein, 2005).

Although the dipole allowed optical transitions between ^{3}A and ^{3}E are spin conserving (i.e., $\Delta m_s = 0$) in unstrained diamond, optical pumping polarizes the NV spin into the $m_s = 0$ state (Harrison et al., 2004). This is due to the presence of a spin-selective intersystem crossing. Transitions between ^{3}E and ^{1}A are dipole forbidden but still occur out of the $m_s = \pm 1$ states due to the spin–orbit interaction (Manson et al., 2006). Once in the ^{1}A state, the system eventually relaxes into the ground state. After several cycles, the probability for the spin to be in the $m_s = 0$ spin state will be high. In addition, since the ^{3}E→^{1}A transition is nonradiative, the NV center's PL rate under optical excitation is smaller when the spin is in the $m_s = \pm 1$ than when it is in the $m_s = 0$ state. This means the spin state of a single NV center can be determined by measuring the relative PL intensity. It should be noted, however, that measurements require illumination which, given enough time and intensity, will repolarize the spin and lose the spin information.

One major feature of diamond NV centers is that they can be imaged with confocal fluorescence microscopy techniques on an individual basis. Figure 1.14B shows an image of a region in a synthetic type 1b diamond that contains several individual NV centers. The image was taken with a scanning confocal microscope, using a 100 × objective with a numerical aperture of 0.73 and linearly polarized 532 nm light that is focused ~1 µm below the diamond surface. Since the focal spot is nearly diffraction-limited, individual NV centers can be resolved if they are spaced by more than ~1 µm. Although there is some thermal drift in the sample position over time on this length scale, using feedback loop tracking enables extended measurements of a single NV center for days or longer.

To insure that an individual NV center is being addressed rather than several at once, we use the fact that they are single-photon emitters. Using time correlated single photon counting we can measure the time-dependent auto-correlation function of the emitted photons. A schematic of the setup is shown in Fig. 1.15A. The PL intensity, I_{PL}, from the NV center(s) are passed through a nonpolarizing 50:50 beam-splitter in the Hanbury–Brown–Twiss arrangement and collected in two separate avalanche photo diode photon detectors. If there is only one NV center present, then at $t = 0$, the normalized autocorrelation function, $g^{(2)}(\tau) = \langle I_{PL}(\tau)I_{PL}(t+\tau)/I_{PL}(t)^2\rangle$ will approach zero (Becker, 2005). This effect is

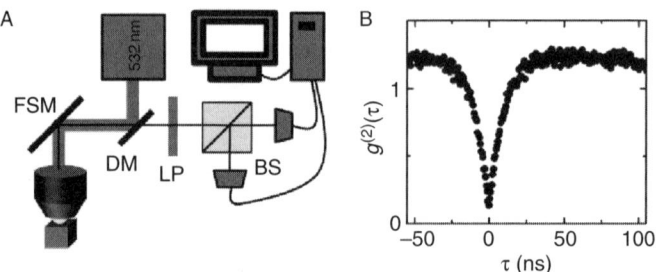

FIGURE 1.15 (A) Schematic of experimental apparatus used to study single NV centers. The abbreviated labels are: FSM = fast steering mirror, DM = dichroic mirror, LP = long pass filter, and BS = 50:50 nonpolarizing beam-splitter. In addition, the detectors are silicon avalanche photo-diodes that are connected to computer-based data acquisition. (B) A plot of the photoluminescence intensity correlation function $g^{(2)}(\tau)$ versus *tau* for NV1. The dip below $g^{(2)}(\tau) = 0.5$ indicates that the photons are emitted from a single source. (B) is taken from Epstein et al. (2005).

known as anti-bunching and is the signature of a single-photon emitter. Anti-bunching data for a single NV center is shown in Fig. 1.15B (Epstein et al., 2005).

4.3. Anisotropic interactions of a single spin

As mentioned previously, the symmetry axis of an NV center also determines its quantization axis. The Hamiltonian of the NV spin in the electronic ground state is:

$$H_{NV} = g_e \mu_B \vec{B} \cdot \vec{S}^{NV} + D[(S_z^{NV})^2 - S^{NV}(S^{NV}+1)/3] + \vec{S}^{NV} \bar{A}^{NV} \vec{I}^{NV} \quad (1.9)$$

where $g_e = 2.00$ is the electron g-factor, μ_B is the Bohr magneton, $\vec{B}$ is the magnetic field and $\vec{S}$ is the NV center spin, and D is the 2.87 GHz zero-field splitting (Charnock and Kennedy, 2001; Loubser and van Wyk, 1978). S_z for a single NV center is along one of the [111] crystal orientations. In addition, the last term in Eq. (1.9) describes the hyperfine interaction between the NV center electronic spin and the nuclear spin of the nitrogen. The components of $\bar{A}$ are a few MHz (He et al., 1993). Using this Hamiltonian, we can calculate the eigen-energies for different magnetic field orientations. This is plotted in Fig. 1.3B for the angle $\theta = 6°$ and $54.7°$, where θ is the angle between $\vec{B}$ and the [111] direction (neglecting the hyperfine term). When θ is nonzero, there are spin mixing terms in H_{NV}. Near $B = 1000$ G this causes a level avoided crossing (LAC) between $m_s = -1$ and $m_s = 0$ spin states. When θ is small, the spin mixing is only significant at fields near the LAC, while large values of θ create spin mixing even at fields far from the LAC (Epstein et al., 2005). This spin mixing can be seen in Fig. 1.16B where $|\alpha|^2$, the overlap of the each eigenstate with $|0\rangle$, which we define as the $m_s = 0$ spin state in the [111] basis, is plotted against B for the same two field angles. The other coefficients are given by $|m_s\rangle = \alpha|0\rangle + \beta|-1\rangle + \gamma|+1\rangle$ in that basis. Measurements of I_{PL} as a function of B can be used to investigate the spin mixing effects. Figure 1.16B plots these data for three different NV centers with B oriented 1° from [111]. NV1 shows two dips, one near $\sim$1000 G and another near $\sim$500 G. The $\sim$1000 G dip can be attributed to the LAC previously mentioned, while the $\sim$500 G dip has a different origin that we will address later. The PL intensities for NV2 and NV3 are significantly reduced as the field increases and show no other features above $B = 200$ G. The intensity is reduced due to enhanced spin mixing which indicates that these NV centers have one of the other $\langle 111 \rangle$ orientations (Epstein et al., 2005).

For NV centers that are aligned with [111], the width of the PL dip at $\sim$1000 G depends sensitively on the angle between B and [111] (Fig. 1.16D and E) (Epstein et al., 2005). As $\theta \to 0°$, the LAC dip becomes narrower, which indicates decreased spin mixing, and is consistent with the Hamiltonian in Eq. (1.9). While the LAC dip disappears altogether for some NV centers (e.g., NV4; Fig. 1.16E), this is not always the case. This residual spin mixing even at nearly perfect alignment can be attributed to strain and nuclear interactions (He et al., 1993), which can vary from center to center.

The LAC can be modeled as a pseudo spin-1/2 system for small values of θ. With B near the LAC, the $|+1\rangle$ state can be ignored since it has almost no overlap with $|0\rangle$ (i.e., see Fig. 1.16B). In this case, we can rewrite Eq. (1.9) as $H = g\mu_B(\vec{B} - \vec{B}_o)\vec{s}$ where $\vec{B}_o$ cancels the zero-field splitting and $\vec{s}$ is a spinor. Then, the Bloch equations for $\vec{B}$ in the $(1\bar{1}0)$ plane are taken to be:

$$\frac{ds_x}{dt} = -\Omega_z s_y - \frac{s_x}{T_2}$$

$$\frac{ds_y}{dt} = \Omega_z s_x - \Omega_x s_z - \frac{s_y}{T_2} \quad (1.10)$$

$$\frac{ds_z}{dt} = \Omega_x s_y - \frac{s_z}{T_1} + \Gamma$$

where $\hbar\vec{\Omega} = g\mu_B(\vec{B} - \vec{B}_o)$, $\hbar$ is the reduced Plank's constant and $\vec{\Omega}$ is the Larmor vector, T_1 is the longitudinal (spin-flip) relaxation time while T_2 is the transverse (dephasing) relaxation time (Epstein et al., 2005). Γ is the rate of optical polarization of the NV center, which is included since the

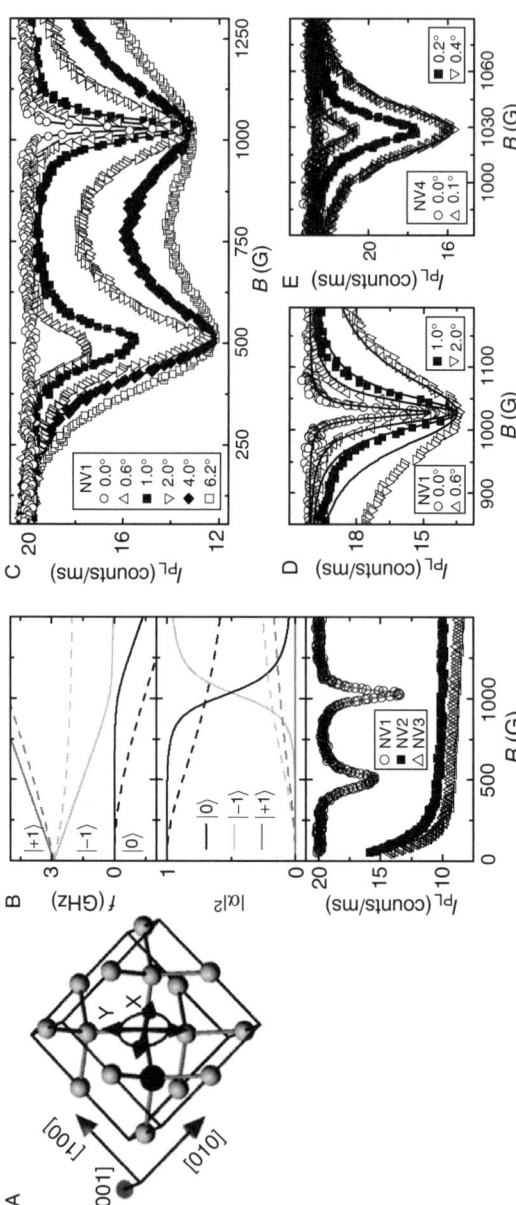

FIGURE 1.16 (A) Schematic of the measurement dipole transition X and Y. (B) Upper panel: Eigen-energies in frequency units (GHz) as a function of magnetic field for $\theta = 6°$ (solid lines) and $\theta = 54.7°$ (dashed lines). Middle panel: Overlap $|\alpha|^2$ for each spin level with $|0\rangle$ at the same two angles. (C) I_{PL} versus magnetic field at several angles for NV1. (D) Close-up of the data for B near 1000 G (points) for several angles along with fits (lines) to the model described in the text. (E) I_{PL} versus magnetic field at several angles (points) for NV4, along with fits (lines). This figure is modified from Epstein et al. (2005).

measurements are made under continuous illumination. Therefore, T_1 and T_2 depend on Γ as $(T_1)^{-1} = (T_z)^{-1} + 2\Gamma$ where T_z is the intrinsic longitudinal relaxation time, and there is a similar expression for T_2 (Epstein et al., 2005). Figure 1.16 illustrates the coordinate system, which is fixed relative to the lattice. Here we use $x \parallel [\bar{1}\bar{1}2], y \parallel [1\bar{1}0]$, and $B_0 \parallel z \parallel [111]$. Experimentally, the photoluminescence intensity I_{PL} can be related to the spin projection s_z. The steady-state solution for s_z is (Epstein et al., 2005):

$$s_z = \frac{T_1 \Gamma \left(1 + \Omega_z^2 T_2^2\right)}{1 + \Omega_x^2 T_1 T_2 + \Omega_z^2 T_2^2} \tag{1.11}$$

Furthermore, this model can be used to directly model the experimental data by taking $I_{PL} = An_0 + Bn_{-1} = A(1/2 + s_z) + B(1/2 + s_z)$. Here A and B are the photon emission rates from the $|0\rangle$ and $|-1\rangle$ spin states, respectively. Equation (1.11) combined with this simple model of the photoluminescence describes the $B \sim 1000$ G LAC data over a large range of angles. In addition, fits to the model can be used to extract lifetime for NV4 $T_1 = 64$ ns and $T_2 = 11$ ns using a laser power of 2.9 mW (Epstein et al., 2005). The large PL dip at ~ 500 G has origins that are not described by the ground-state spin Hamiltonian. This feature is observed in all the NV centers that were studied when suitably aligned with the [111] direction. This dip is a very broad feature, which suggests that it is related to a fast relaxation process. One possibility is that there is a LAC in the excited state of the NV center at this field and orientation. If that is the case, then using a similar analysis to the one described above for the ~ 1000 G LAC yields a $T_1 = 36$ ns and a $T_2 = 1.8$ ns for this state (Epstein et al., 2005).

Near the broad 500 G dip there are also signs of the NV center interacting with its environment. For every NV center in the diamond, there are 10^6–10^8 more nitrogen spins in these nitrogen rich type-Ib diamonds. The nitrogen impurities have electronic spin 1/2 and nuclear spin 1 (for ^{14}N) but are not optically active and hence "dark" to a direct optical probe. Nevertheless, these spins are present and interact with the NV center spin through dipolar coupling. The Hamiltonian for the nitrogen spins is:

$$H_N = g_N \mu_B B S_z^N + S^N \bar{A}^N I^N \tag{1.12}$$

where S^N is the nitrogen electronic spin, I^N is the nitrogen nuclear spin, $g_N = 2.00$ is the g-factor of the nitrogen electronic spin, and $\bar{A}$ is the nitrogen hyperfine tensor. The NV to N electronic dipolar coupling Hamiltonian is (Hanson et al., 2006b; Slicter, 1990):

$$H_{\text{dip}} = \frac{\mu_0 g_{\text{NV}} g_{\text{N}} \mu_B^2}{4\pi r^3} \left[\vec{S}^{\text{NV}} \cdot \vec{S}^{\text{N}} - 3\left(\vec{S}^{\text{NV}} \cdot \hat{r}\right)\left(\vec{S}^{\text{N}} \cdot \hat{r}\right) \right]. \qquad (1.13)$$

Figure 1.17A compares the spin-splitting as a function of magnetic field for both the NV centers and for an N electronic spin for B parallel to [111]. At 514 G, the spin splitting is the same for both of these spin species, and hence resonant coupling between the N spin bath and an NV center is possible for sufficiently high N concentrations (e.g., small enough value of $\langle r \rangle$). This dipolar interaction, therefore, creates another LAC at 514 G, which accounts for the PL dip observed at that field. Including the hyperfine interaction of the N electronic spin with its nuclear spin, the nitrogen levels are split into three (Fig. 1.17B), for $m_I = -1$, 0, and $+1$. Hence, there are actually three LACs, which can be observed separately for small values of θ (Fig. 1.17C) (Hanson et al., 2006b).

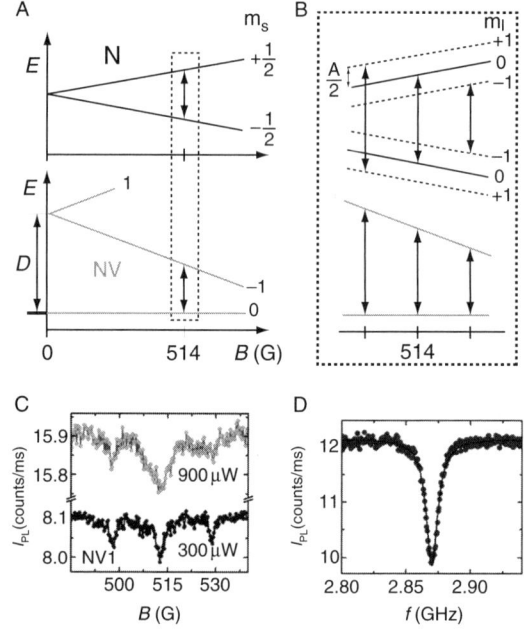

FIGURE 1.17 (A) Energy diagram versus magnetic field for the spin levels in nitrogen and NV centers. At 514 G, the spin splitting is the same for both centers. (B) Close-up of the energy diagram including the hyperfine splitting for ^{14}N. (C) I_{PL} versus magnetic field for NV1 at two laser powers where additional LACs due to the hyperfine splitting of N are evident as satellite dips. (D) I_{PL} versus frequency of applied microwave radiation. The PL dip is the result of a resonant spin transition. (A) and (B) are taken from Hanson et al. (2006b) and (C) and (D) are taken from Epstein et al. (2005).

4.4. Single NV spin manipulation and coherence

The techniques of ESR and nuclear magnetic resonance (NMR) that were developed for spin ensembles (Slicter, 1990) are also effective for single spin measurements. As with traditional ESR, the spin is manipulated with a microwave frequency magnetic field B_{rf}, but here the spin state is detected optically by measuring the PL intensity. When the microwave field is resonant with the spin splitting, the spin will undergo Rabi nutations. In order to understand these affects, it is helpful to think about the spin dynamics in a rotating reference frame (Slicter, 1990). In the lab frame without B_{rf}, the spin precesses about the static field at the Larmor frequency given by the energy splitting between the spin levels ΔE, or $f_L = \Delta E/h$, where h is the Plank's constant. By transforming into the reference frame that is rotating at f_L, the spin is static. When B_{rf} is applied, it appears as a static field in the rotating frame, and the spin's dynamic response is to precess[1] about B_{rf}. In this way, the spin can be manipulated between the various spin states, including coherent superpositions of the states.

The probability $P_{|0\rangle}$ of finding the spin in the $|0\rangle$ state will then be given by Rabi's formula (Sakurai, 1994):

$$P_{|0\rangle} = 1 - \frac{f_1^2}{f_1^2 + \Delta f^2} \sin^2\left(\pi\sqrt{f_1^2 + \Delta f^2}\right)t \quad (1.14)$$

where f_1 is the Rabi frequency, Δf is the detuning from the resonant frequency, and t is the time. The Rabi frequency depends on the amplitude of the B_{rf} as $f_1 = g\mu_B B_{rf}/(2h)$, where h is the Plank's constant. In addition, the overall oscillation decays as $e^{-t/T_2'}$ where T_2' is the inhomogeneous[2] dephasing time which is typically shorter than T_2, the transverse relaxation time.

ESR is easiest to observe in the frequency domain by continuously illuminating the NV center while sweeping the frequency f of B_{rf}. This measurement produces a dip in the PL signal as f approaches f_L since $\langle P_{|0\rangle}\rangle$ falls as the NV spin undergoes continuous oscillations between $|0\rangle$ and the other spin state on the time scale of the measurement (recall: the photon emission rate is highest for $|0\rangle$). Away from the resonance, the continuous illumination polarizes the spin into $|0\rangle$. Figure 1.17D shows typical ESR data taken of a single NV center for the $|0\rangle \rightarrow |\pm 1\rangle$ transition.

In the time domain, spin resonance is also known as Rabi oscillations, which is a direct measurement of the coherent evolution of the NV spin

---

[1] This is why we previously used the term nutation. In the lab frame, the spin dynamics include both the Larmor precession about the static field, and also a "wobble" due to the dynamical response to B_{rf}.
[2] Since we are measuring a single spin rather than an ensemble of spins, it might seem like we should use T_2 rather than T_2'; however, the need for signal averaging in the measurement means the signal is still sensitive to inhomogeneous broadening in time rather than in space as for an ensemble measurement.

under a resonant microwave drive. The measurement requires a pump-probe cycle that is outlined in Figure 1.18A. First, the laser illuminates the NV center for several microseconds in order to initialize it into the $|0\rangle$ state, after which the illumination is turned off using an acousto-optic modulator. Then, after letting the NV center relax into the electronic ground state (^3A) for 2–3 μs, a pulse of microwave radiation of varying lengths is applied. Finally, the spin state is read out with a short (2–3 μs) laser pulse during which I_{PL} is measured. This cycle is repeated many ($\sim 10^5$) times for each microwave pulse duration in order to build up statistics. Figure 1.18B shows a plot of I_{PL} as a function of microwave pulse duration for a single NV spin with three different values of microwave power (Hanson et al., 2006a). Each trace fits well to Eq. (1.14) with a single f_1 that is proportional to the square root of the microwave power. The oscillations in Fig. 1.18B decay with values of T_2' on a microsecond time scale.

In order to study the factors that contribute to T_2', Rabi measurements were performed to determine T_2' over a broad range of magnetic fields. The results are summarized in Fig. 1.18C and D (Hanson et al., 2006a). Most striking is the sharp peak in $1/T_2'$ located at 514 G that is evident in Fig. 1.18C. Figure 1.18D shows additional $1/T_2'$ data taken at fields near to 514 G. The fields where increases in $1/T_2'$ are observed closely track the PL dips that occur due to LACs where the NV and N levels are resonant.

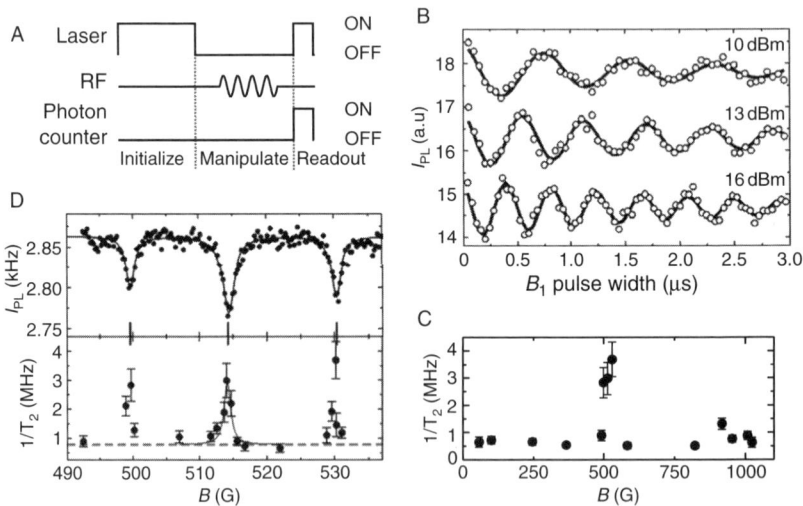

FIGURE 1.18 (A) Timing diagram that illustrates the pump-probe sequence used to measure Rabi oscillations. (B) Rabi oscillations plotted for NV14 for three values of the applied microwave power. (C) $1/T_2'$ plotted as a function of magnetic field. These data show increases in $1/T_2'$ at the fields where there is a LAC. (D) Close-up of $1/T_2'$ data near the 514 G LAC. Figures modified from Hanson et al. (2006a).

This strongly suggests that fluctuations in the N spins as they interact with the NV spin is an important source of decoherence (Hanson et al., 2006a). Another interesting feature of the Rabi data in Fig. 1.18B is that the T_2' also depends on the microwave power. This suggests that there is a refocusing effect, where a long microwave pulse can be viewed as a series of π-pulses (Hanson et al., 2006a; Vandersypen and Chuang, 2005). If true, then the transverse relaxation time T_2 is longer than T_2' due to dephasing caused by the environment of the NV spin. This was further investigated by using a Hahn spin-echo pulse sequence as outlined in Fig. 1.19A rather than a single microwave pulse. In this scheme, first a π/2-pulse brings the NV spin coherently into the $1/\sqrt{2}(|0\rangle + |-1\rangle)$ state. Then there is a period τ_1 of free evolution of the spin during which dephasing can occur. This is followed by a π-pulse which introduces a 180° phase shift so that the state is nominally $1/\sqrt{2}(|0\rangle - |-1\rangle)$ which is followed by another free evolution period, τ_2. Finally, another π/2-pulse is applied to bring the spin back to $|0\rangle$ for measurement of the state. If there is low-frequency dephasing that occurs during τ_1, it will occur with the opposite sign during τ_2, leading to a refocusing of the spin coherence (Slicter, 1990). First, this was checked by measuring I_{PL} as τ_2 was varied. It was found that there was a peak in I_{PL} at $\tau_1 = \tau_2 = \tau$ (Fig. 1.19B), which indicates that there are low-frequency interactions with the environment that cause dephasing of spin during the Rabi measurement. These effects reduce the nominal coherence time of the NV spin. Then, the Hahn echo at $\tau_1 = \tau_2 = \tau$ was measured as a function of τ, which is shown in Fig. 1.19C along with a fit to an exponential decay. The decay of the signal gives the value of T_2, which for this NV center is 6 μs. As expected, this is significantly longer than typical values of T_2' (see Fig. 1.18C) (Hanson et al., 2006a).

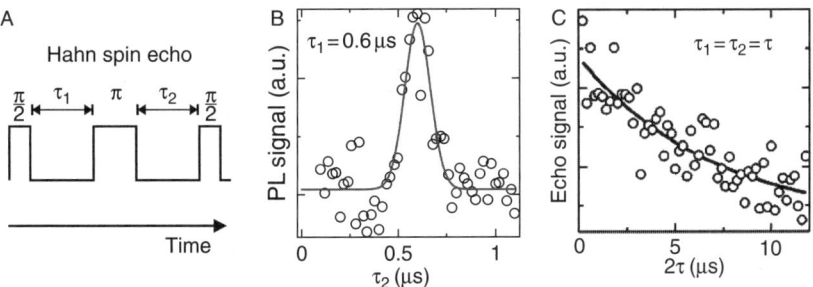

FIGURE 1.19 (A) Timing diagram of the spin echo pulses used in the Hahn measurement. (B) Plot of the PL signal versus τ_2 for a Hahn measurement where $\tau_1 = 0.6$ ms. The solid line is a guide to the eye. (C) Plot of the PL signal taken during a Hahn measurement where $\tau_1 = \tau_2 = \tau$. The data is fit to an exponential decay (solid line) giving a value of $T_2 = 6$ ms. (A) and (C) are modified from Hanson et al. (2006a).

4.5. Coupled spins in diamond

Previously, we have been addressing measurements and manipulations of a single NV electronic spin. These measurements have suggested that the spin environment created by a number of weakly coupled nitrogen spins has a major impact on the coherence properties of the NV center. Alternatively, the NV center may be strongly coupled to a single nitrogen spin, with dynamic properties that are dominated by the coupling (Gaebel et al., 2006; Hanson et al., 2006b).

In that situation, the ESR peak in the frequency domain is split by an amount proportional to the strength of the coupling (Fig. 1.20C and D). Since they are coupled by the dipolar interaction, the splitting depends strongly on the distance due to the factor of $1/r^3$ in Eq. (1.13). We can understand the splitting in terms of a local magnetic field exerted by the N spin. Away from the 514 G LAC, the spin projections (m_s^{NV} and m_s^{N}) remain good quantum numbers. Each of the two ESR peaks are due to the $|0\rangle \rightarrow |-1\rangle$ transition of the NV spin, but they occur at different frequencies depending on whether m_s^{N} is $+1/2$ or $-1/2$. That is because the nitrogen spin exerts a dipolar field $\vec{B}_{dip}$ on the NV center that depends on the relative position of the two spins with respect to their quantization axis.

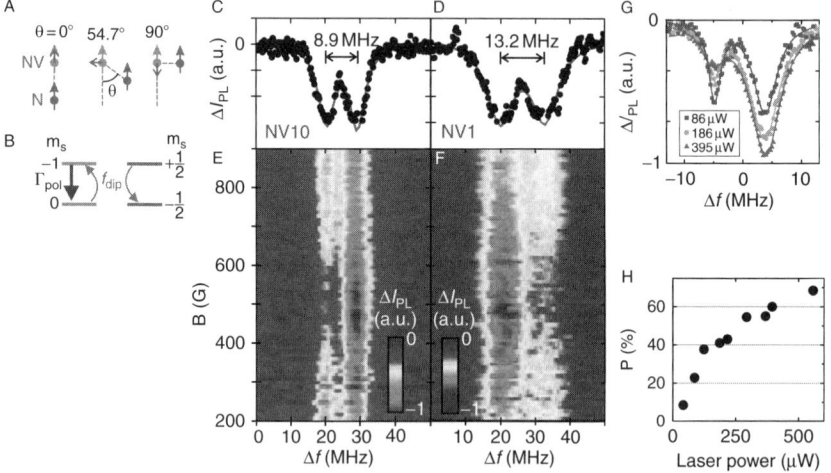

FIGURE 1.20 (A) Schematic diagram showing the direction of B_{dip} for an NV center coupled with a single N defect for the limiting cases of $\theta = 0°$, 54.7°, and 90°. (B) Energy diagram of the spin dynamics that results from dipolar coupling between an NV center and an N spin. (C) and (D) show ESR spectra taken as a function of B for NV10 and NV1, respectively. The spectra for each value of B are concatenated and plotted with a 2D color scale (E) and (F). (G) Plot of ESR spectra for three different values of laser illumination. (H) Plot of polarization versus laser power. Figure is modified from Hanson et al. (2006b) (See Color Plate.)

In this case, the NV center has a fixed quantization axis along [111] and the nitrogen's axis is set by the external field which is also aligned to the [111] direction. Figure 1.20A outlines the possibilities for the coupling. If the angle θ between the [111] direction and $\vec{r}$ is small (e.g., the spins are lined up front-to-back), then when the spins align with the external field, $\vec{B}_{dip}$ from the nitrogen is also aligned with the external field, making the total field on the NV center larger. If, on the other hand, θ is similar to 90° (e.g., the spins are lined up side-by-side) then $\vec{B}_{dip}$ is anti-aligned with the external field. As θ varies between these two angles, there is an increasing component perpendicular to the quantization axis, with θ = 54.7° being the "magic angle" where $\vec{B}_{dip}$ is exactly at 90° with respect to the NV spin orientation (Hanson et al., 2006b; Slicter, 1990). If θ is exactly at 0° or 90°, $\vec{B}_{dip}$ is parallel to the applied field, and there is no LAC between the NV and N eigenstates, with the only effect being to adjust the total field magnitude on the NV center. At angles in between, however, the spin levels of the NV center and N become mixed at the resonance condition ($B = 514$ G) to form a LAC (Hanson et al., 2006b). The NV center spin and the N spin can then undergo a spin flip-flop (Fig. 1.20B). Under continuous illumination, the NV center is polarized in $|0\rangle$ with the rate Γ_{pol}. Therefore, since the NV center is being preferentially driven into $|0\rangle$, the majority of the flip-flops will be to take the NV from $|0\rangle \rightarrow |-1\rangle$ and the N from $|+1/2\rangle \rightarrow |-1/2\rangle$. This results in the PL decrease associated with the LAC noted previously (e.g., see Fig. 1.17C and 1.18D).

Figure 1.20E and F shows ESR scans of two different NV centers (NV1 and 10) taken as a function of magnetic field and plotted together to form a 2D contour plot (Hanson et al., 2006b). As the applied field approaches the resonance point, one of the two ESR peaks disappears. This is the result of the flip-flop process described above, indicating that the peak that disappears is the $|+1/2\rangle$ state of N (Gaebel et al., 2006; Hanson et al., 2006b). This data also allows us to learn something of the angle θ between each of these NV centers and their respective N. The ESR peak that disappears for NV10 is at a lower frequency (Fig. 1.7E) indicating that $\vec{B}_{dip}$ tends to decrease the field magnitude for the $|+1/2\rangle$ state and therefore θ must be close to 90°. The peak that disappears for NV1, however, is at higher frequency so the opposite must be true; θ must be near to 0°. Given the sign of the coupling for each of these NV centers, it is possible to estimate the upper limit of the NV–N separation using Eq. (1.13) to be 2.3 nm for NV10 and 2.6 nm for NV1 (Hanson et al., 2006b).

Since the N spins are being polarized due to flip-flops with the NV center, the polarization of the N spins should also be proportional to the polarization rate Γ_{pol} of the NV center under illumination. This was investigated by measuring the ESR spectra for different values of the laser power. The data is plotted in Fig. 1.20G. As the laser power increases, the difference in the amplitude change, ΔI_{PL}, of the two peaks became greater. By fitting to the amplitudes, one can explicitly calculate the polarization of the N using:

$$P = \frac{\Delta I_{\text{PL}}(-1/2) - \Delta I_{\text{PL}}(+1/2)}{\Delta I_{\text{PL}}(-1/2) + \Delta I_{\text{PL}}(+1/2)} \quad (1.15)$$

where P is the N polarization and $\Delta I_{\text{PL}}(\pm 1/2)$ is the PL change in the optically detected ESR feature for $m_s^N = \pm 1/2$ (Hanson et al., 2006b). The results, plotted in Fig. 1.20H, show that the polarization becomes saturated near 70% as the rate of optical polarization is balanced against spin-relaxation processes (Hanson et al., 2006b). These measurements show a pathway toward initialization of coupled spins, which is an important measure of control for quantum information purposes.

4.6. Conclusions and outlook

The long spin lifetimes at room temperature (Gaebel et al., 2006; Kennedy et al., 2003), coupled with the ability to optically initialize and readout the spin state of diamond NV centers, make this system a strong candidate for applications in quantum information. Furthermore, they are quickly becoming a standard choice of room temperature, single-photon sources (Kurtsiefer et al., 2000) for applications in quantum optics. There have also been a number of theoretical proposals to use NV centers for decoherence free quantum computing (Brooke, 2007), quantum repeaters (Childress et al., 2006b), and quantum teleportation (Gottesman and Chuang, 1999) to name a few.

Recent observations of the coherent dynamics of NV centers coupled with individual nearby spins suggest a pathway toward useful room temperature quantum computers with multiple coupled qubits. These include N electronic spins as discussed here (Gaebel et al., 2006; Hanson et al., 2006b) and ^{13}C nuclear spins (Childress et al., 2006a; Dutt et al., 2007; Jelezko et al., 2004). Moreover, in the case of the coupled NV-^{13}C spin system, simple quantum computing gates have been successfully implemented, including a "CROT" gate (Jelezko et al., 2004) and a "swap gate" (Dutt et al., 2007).

One of the major challenges for the development of such quantum computers is the scaling of NV center-based qubits. Although nearby spins with sufficient coupling occur naturally, the probability of their formation is low. Not only does this make the likelihood of finding an NV center spin system composed of 3, 4, or 5 spin-based qubits prohibitively small, but it also essentially rules out the possibility of scaling to large arrays of coupled spins. Progress, however, has also been made on this front as well. Meijer and colleagues (Meijer et al., 2005; Rabeau et al., 2006) have successfully created regular arrays of single NV centers with the accuracy of 0.3 μm using ion implantation techniques. Although this level of control is not yet sufficient to customize the interaction between NV centers and surrounding spins, it is an important first step in scaling and engineering NV centers in diamond for quantum information applications.

We gratefully acknowledge Ronald Hanson and Jesse Berezovsky for helpful suggestions. This work was supported by AFOSR, ONR and NSF.

REFERENCES

Atatüre, M., Dreiser, J., Badolato, A., and Imamoglu, A. (2007). *Nat. Phys.* **3**, 101–105.
Averkiev, N. S., Gutkin, A. A., Osipov, E. B., and Reshchikov, M. A. (1988). *Sov. Phys. Solid State* **30**, 438.
Awschalom, D. D., and Samarth, N. (1999). *J. Magn. Magn. Mater.* **200**, 130–147.
Awschalom, D. D., Loss, D., and Samarth, N. (2002). "Semiconductor Spintronics and Quantum Computation." Springer-Verlag, Berlin.
Becker, W. (2005). "The bh TCSPC Handbook." Beckher & Hickl GmbH.
Berezovsky, J., Mikkelsen, M. H., Gywat, O., Stoltz, N. G., Coldren, L. A., and Awschalom, D. D. (2006). *Science* **314**, 1916–1920.
Berezovsky, J., Mikkelsen, M. H., Stoltz, N. G., Coldren, L. A., and Awschalom, D. D. (2008). *Science* **320**, 349–352.
Besombes, L., Leger, Y., Maingault, L., Ferrand, D., Mariette, H., and Cibert, J. (2004). *Phys. Rev. Lett.* **93**(20), 207403.
Bracker, A. S., Stinaff, E. A., Gammon, D., Ware, M. E., Tischler, J. G., Shabaev, A., Efros, A. l. L., Park, D., Gershoni, D., Korenev, V. L., and Merkulov, I. A. (2005). *Phys. Rev. Lett.* **94**(4), 047402.
Brooke, P. G. (2007). *Phys. Rev. A* **75**(2), 022320.
Carter, S. G., Chen, Z., and Cundiff, S. T. (2007). *Phys. Rev. B* **76**(20), 201308.
Chapman, R. A., and Hutchinson, W. G. (1967). *Phys. Rev. Lett.* **18**(12), 443–445.
Charnock, F. T., and Kennedy, T. A. (2001). *Phys. Rev. B* **64**(4), 041201.
Chen, P., Piermarocchi, C., Sham, L. J., Gammon, D., and Steel, D. G. (2004). *Phys. Rev. B* **69**, 075320.
Childress, L., Gurudev Dutt, M. V., Taylor, J. M., Zibrov, A. S., Jelezko, F., Wrachtrup, J., Hemmer, P. R., and Lukin, M. D. (2006a). *Science* **314**(5797), 281–285.
Childress, L., Taylor, J. M., Sørensen, A. S., and Lukin, M. D. (2006b). *Phys. Rev. Lett.* **96**(7), 070504.
Clark, C. D., and Norris, C. A. (1971). *J. Phys. C: Sol. Stat. Phys.* **4**(14), 2223–2229.
Clark, S. M., Fu, K. M. C., Ladd, T. D., and Yamamoto, Y. (2007). *Phys. Rev. Lett.* **99**, 040501.
Cohen-Tannoudji, C., and Dupont-Roc, J. (1972). *Phys. Rev. A* **5**, 968–984.
Cohen-Tannoudji, C., and Reynaud, S. (1977). *J. of Phys. B* **10**, 345–363.
Combescot, M., and Betbeder-Matibet, O. (2004). *Solid State Commun.* **132**, 129–134.
Combescot, M., and Combescot, R. (1988). *Phys. Rev. Lett.* **61**, 117–120.
Damen, T. C., Via, L., Cunningham, J. E., Shah, J., and Sham, L. J. (1991). *Phys. Rev. Lett.* **67**(24), 3432–3435.
Dietl, T. (1994). "Handbook on Semiconductors," Vol. 3B. North-Holland, Amsterdam.
Dutt, M. V., Gurudev, J., Cheng, J., Wu, Y., Xu, X., Steel, D. G., Bracker, A. S., Gammon, D., Economou, S. E., Liu, R. B., and Sham, L. J. (2006). *Phys. Rev. B* **74**, 125306.
Dutt, M. V., Gurudev, L., Childress, L., Jiang, L., Togan, E., Maze, J., Jelezko, F., Zibrov, A. S., Hemmer, P. R., and Lukin, M. D. (2007). *Science* **316**(5829), 1312–1316.

Dzhioev, R. I., Kavokin, K. V., Korenev, V. L., Lazarev, M. V., Meltser, B. Y., Stepanova, M. N., Zakharchenya, B. P., Gammon, D., and Katzer, D. S. (2002). *Phys. Rev. B* **66**(24), 245204.
Ebbens, A., Krizhanovskii, D. N., Tartakovskii, A. I., Pulizzi, F., Wright, T., Savelyev, A. V., Skolnick, M. S., and Hopkinson, M. (2005). *Phys. Rev. B* **72**(7), 073307.
Economou, S. E., Sham, L. J., Wu, Y., and Steel, D. G. (2006). *Phys. Rev. B* **74**, 205415.
Epstein, R. J. (2005). Controlled Interactions of Single Spins and Ensembles in Semiconductors. PhD thesis, University of California, Santa Barbara.
Epstein, R. J., Mendoza, F. M., Kato, Y. K., and Awschalom, D. D. (2005). *Nat. Phys.* **1**, 94.
Erlingsson, S. I., Nazarov, Y. V., and Fal'ko, V. I. (2001). *Phys. Rev. B* **64**, 195306.
Erwin, S. C., and Petukhov, A. G. (2002). *Phys. Rev. Lett.* **89**(22), 227201.
Farrier, R. G. (1969). *Solid State Commun.* **7**, 685.
Frey, T. h., Maier, M., Schneider, J., and Gehrke, M. (1988). *J. Phys. C: Sol. Stat. Phys.* **21**(32), 5539–5545.
Furdyna, J. K. (1988). *J. Appl. Phys.* **64**(4), R29–R64.
Gaebel, T., Domhan, M., Popa, I., Wittman, C., Neumann, P., Jelezko, F., Rabeau, J. R., Stavrias, N., Greentree, A. D., Prawer, S., Meijer, J., Twamley, J., et al. (2006). *Nat. Phys.* **2**, 408.
Gammon, D., Snow, E. S., Shanabrook, B. V., Katzer, D. S., and Park, D. (1996). *Phys. Rev. Lett.* **76**, 3005–3008.
Gammon, D., Efros, A. L., Kennedy, T. A., Rosen, M., Katzer, D. S., Park, D., Brown, S. W., Korenev, V. L., and Merkulov, I. A. (2001). *Phys. Rev. Lett.* **86**(22), 5176–5179.
Golovach, V. N., Khaetskii, A., and Loss, D. (2004). *Phys. Rev. Lett.* **93**, 016601–016604.
Gossard, A. C. (1986). *IEEE J. Quant. Electron.* **22**(9), 1649–1655.
Gottesman, D., and Chuang, I. L. (1999). *Nature* **402**(6760), 390–393.
Greilich, A., Wiemann, M., Hernandez, F. G. G., Yakovlev, D. R., Yugova, I. A., Bayer, M., Shabaev, A., Efros, A. l., Reuter, D., and Wieck, A. D. (2007). *Phys. Rev. B* **75**(23), 233301.
Guest, J. R., Stievater, T. H., Li, X., Cheng, J., Steel, D. G., Gammon, D., Katzer, D. S., Park, D., Ell, C., Thränhardt, A., Khitrova, G., and Gibbs, H. M. (2002). *Phys. Rev. B* **65**, 241310.
Gupta, J. A., Knobel, R., Samarth, N., and Awschalom, D. D. (2001). *Science* **292**(5526), 2458–2461.
Gurudev-Dutt, M. V., Cheng, J., Li, B., Xu, X., Li, X., Berman, P. R., Steel, D. G., Bracker, A. S., Gammon, D., Economou, S. E., Liu, R. B., and Sham, L. J. (2005). *Phys. Rev. Lett.* **94**, 227403–227404.
Hanson, R., Gywat, O., and Awschalom, D. D. (2006a). *Phys. Rev. B* **74**(16), 161203.
Hanson, R., Mendoza, F. M., Epstein, R. J., and Awschalom, D. D. (2006b). *Phys. Rev. Lett.* **97** (8), 087601.
Hanson, R., P Kouwenhoven, L., Petta, J. R., Tarucha, S., and Vandersypen, L. M. K. (2007). *Rev. Mod. Phys.* **79**, 1217.
Harrison, J., Sellars, M. J., and Manson Manson, N. B. (2004). *J. Lumin.* **107**, 245.
He, X. F., Manson, N. B., and Fisk, P. T. H. (1993). *Phys. Rev. B* **47**(14), 8816–8822.
Högele, A., Kroner, M., Seidl, S., Karrai, K., Atatüre, M., Dreiser, J., Imamoglu, A., Warburton, R. J., Badolato, A., Gerardot, B. D., and Petroff, P. M. (2005). *Appl. Phys. Lett.* **86**, 221905.
Imamoglu, A., Awschalom, D. D., Burkard, G., DiVincenzo, D. P., Loss, D., Sherwin, M., and Small, A. (1999). *Appl. Phys. Lett.* **83**, 4204–4207.
Isberg, J., Hammersberg, J., Johansson, E., Wikstrom, T., Twitchen, D. J., Whitehead, A. J., Coe, S. E., and Scarsbrook, G. A. (2002). *Science* **297**(5587), 1670–1672.
Jelezko, F., Gaebel, T., Popa, I., Domhan, M., Gruber, A., and Wrachtrup, J. (2004). *Phys. Rev. Lett.* **93**(13), 130501.
Joffre, M., Hulin, D., Migus, A., and Combescot, M. (1989). *Phys. Rev. Lett.* **62**, 74–77.

Karlik, I. Ya, Merkulov, I. A., Mirlin, D. N., Nikitin, L. P., Perel', V. I., and Sapega, V. F. (1982). *Sov. Phys. Solid State* **24,** 2022.
Kennedy, T. A., Colton, J. S., Butler, J. E., Linares, R. C., and Doering, P. J. (2003). *Appl. Phys. Lett.* **83**(20), 4190–4192.
Khaetskii, A. V., and Nazarov, Y. V. (2001). *Phys. Rev. B* **64,** 125316.
Khaetskii, A. V., Loss, D., and Glazman, L. (2002). *Phys. Rev. Lett.* **88,** 186802.
Kim, Y., Shon, Y., Takamasu, T., and Yokoi, H. (2005). *Phys. Rev. B* **71**(7), 073308.
Kitchen, D., Richardella, A., Tang, J. M., Flatté, M. E., and Yazdani, A. (2006). *Nature* **442** (7101), 436–439.
Koppens, F. H. L., Buizert, C., Tielrooij, K. J., Vink, I. T., Nowack, K. C., Meunier, T., Kouwenhoven, L. P., and Vandersypen, L. M. K. (2006). *Nature* **442**(7104), 766–771.
Kurtsiefer, C., Mayer, S., Zarda, P., and Weinfurter, H. (2000). *Phys. Rev. Lett.* **85**(2), 290–293.
Leger, Y., Besombes, L., Fernandez-Rossier, J., Maingault, L., and Mariette, H. (2006). *Phys. Rev. Lett.* **97**(10), 107401.
Li, J. B., and Wang, L. W. (2004). *Appl. Phys. Lett.* **84**(18), 3648–3650.
Li, Y. Q., Steuerman, D. W., Berezovsky, J., Seferos, D. S., Bazan, G. C., and Awschalom, D. D. (2006). *Appl. Phys. Lett.* **88,** 193126.
Liu, X., Prasad, A., Nishio, J., Weber, E. R., Liliental-Weber, Z., and Walukiewicz, W. (1995). *Appl. Phys. Lett.* **67**(2), 279–281.
Loubser, J., and van Wyk, J. A. (1978). *Rep. Prog. Phys.* **41,** 1201.
Manson, N. B., and McMurtrie, R. L. (2007). *J. Lumin.* **127,** 98.
Manson, N. B., Harrison, J. P., and Sellars, M. J. (2006). *Phys. Rev. B* **74**(10), 104303.
Meier, F., and Awschalom, D. D. (2005). *Phys. Rev. B* **71**(20), 205315–205319.
Meier, F., and Zakharchenya, B. P. (1984). Optical Orientation. Elsevier, Amsterdam.
Meijer, J., Burchard, B., Domhan, M., Wittmann, C., Gaebel, T., Popa, I., Jelezko, F., and Wrachtrup, J. (2005). *Appl. Phys. Lett.* **87**(26), 261909.
Merkulov, I. A., Efros, A. L., and Rosen, M. (2002). *Phys. Rev. B* **65.**
Mikkelsen, M. H., Berezovsky, J., Stoltz, N. G., Coldren, L. A., and Awschalom, D. D. (2007). *Nat. Phys.* **3,** 770–773.
Myers, R. C., Poggio, M., Stern, N. P., Gossard, A. C., and Awschalom, D. D. (2005). *Phys. Rev. Lett.* **95**(1), 017204.
Myers, R. C., Mikkelsen, M. H., Tang, J. M., Gossard, A. C., Flatte, M. E., and Awschalom, D. D. (2008). *Nat. Mater.* **7,** 203.
Nizovtsev, A. P., Kilin, S. Ya, Jelezko, F., Popa, I., Gruber, A., and Wrachtrup, J. (2003). *Phys. B* **340,** 106.
Nowack, K. C., Koppens, F. H. L., Nazarov, Yu. V., and Vandersypen, L. M. K. (2007). *Science* **318,** 1430–1433.
Ohno, H. (1998). *Science* **281**(5379), 951–956.
Papageorgiou, G., Chari, R., Brown, G., Kar, A. K., Bradford, C., Prior, K. A., Kalt, H., and Galbraith, I. (2004). *Phys. Rev. B* **69,** 085311.
Poggio, M., Myers, R. C., Stern, N. P., Gossard, A. C., and Awschalom, D. D. (2005). *Phys. Rev. B* **72**(23), 235313.
Pryor, C. E., and Flatté, M. E. (2006). *Appl. Phys. Lett.* **88,** 233108.
Rabeau, J. R., Reichart, P., Tamanyan, G., Jamieson, D. N., Prawer, S., Jelezko, F., Gaebel, T., Popa, I., Domhan, M., and Wrachtrup, J. (2006). *Appl. Phys. Lett.* **88**(2), 023113.
Rosatzin, M., Suter, D., and Mlynek, J. (1990). *Phys. Rev. A* **42,** 1839(R)–1841.
Rugar, D., Budakian, R., Mamin, J., and Chui, B. W. (2004). *Nature* **430,** 329.
Sakurai, J. J. (1994). "Modern Quantum Mechanics," rev. edn. Addison-Wesley.
Salis, G., and Moser, M. (2005). *Phys. Rev. B* **72,** 115540.
Salis, G., Awschalom, D. D., Ohno, Y., and Ohno, H. (2001). *Phys. Rev. B* **64**(19), 195304.
Sapega, V. F., Brandt, O., Ramsteiner, M., Ploog, K. H., Panaiotti, I. E., and Averkiev, N. S. (2007). *Phys. Rev. B* **75**(11), 113310.

Schairer, W., and Schmidt, M. (1974). *Phys. Rev. B* **10**(6), 2501–2506.
Schneider, J., Kaufmann, U., Wilkening, W., Baeumler, M., and Köhl, F. (1987). *Phys. Rev. Lett.* **59**(2), 240–243.
Semenov, Y. G., and Kim, K. W. (2004). *Phys. Rev. Lett.* **92**, 026601–026604.
Slicter, C. P. (1990). "Principles of Magnetic Resonance," 3rd edn. Springer.
Stegner, A. R., Boehme, C., Huebl, H., Stutzman, M., Lips, K., and Brandt, M. S. (2006). *Nat. Phys.* **2**, 835.
Stern, N. P., Myers, R. C., Poggio, M., Gossard, A. C., and Awschalom, D. D. (2007). *Phys. Rev. B* **75**(4), 045329.
Stievater, T. H., Li, X., Cubel, T., Steel, D. G., Gammon, D., Katzer, D. S., and Park, D. (2002). *Appl. Phys. Lett.* **81**(22), 4251–4253.
Suter, D., Klepel, H., and Mlynek, J. (1991). *Phys. Rev. Lett.* **67**, 2001–2004.
Tang, J. M., and Flatté, M. E. (2004). *Phys. Rev. Lett.* **92**(4), 047201.
Tang, J. M., Levy, J., and Flatté, M. E. (2006). *Phys. Rev. Lett.* **97**(10), 106803.
Unold, T., Mueller, K., Lienau, C., Elaesser, T., and Wieck, A. D. (2004). *Phys. Rev. Lett.* **92**, 157401.
Vandersypen, L. M. K., and Chuang, I. L. (2005). *Rev. Mod. Phys.* **76**(4), 1037–1069.
Warburton, R. J., Schäflein, C., Haft, D., Bickel, F., Lorke, A., Karrai, K., Garcia, J. M., Schoenfeld, W., and Petroff, P. M. (2000). *Nature* **405**, 926–929.
Wei, L., Kuo, P. K., Thomas, R. L., Anthony, T. R., and Banholzer, W. F. (1993). *Phys. Rev. Lett.* **70**(24), 3764–3767.
Wolf, S. A., Awschalom, D. D., Buhrman, R. A., Daughton, J. M., von Molnar, S., Roukes, M. L., Chtchelkanova, A. Y., and Treger, D. M. (2001). *Science* **294**(5546), 1488–1495.
Wu, Y., Kim, E. D., Xu, X., Cheng, J., Steel, D. G., Bracker, A. S., Gammon, D., Economou, S. E., and Sham, L. J. (2007). *Phys. Rev. Lett.* **99**, 097402.
Yakunin, A. M., Silov, A. Y. u., Koenraad, P. M., Wolter, J. H., Van Roy, W., De Boeck, J., Tang, J. M., and Flatté, M. E. (2004). *Phys. Rev. Lett.* **92**(21), 216806.
Zrenner, A., Butov, L. V., Hagn, M., Abstreiter, G., Böhm, G., and Weimann, G. (1994). *Phys. Rev. Lett.* **72**, 3382–3385.

CHAPTER 2

Theory of Spin–Orbit Effects in Semiconductors

Jairo Sinova[*,‡] and A. H. Macdonald[†]

Contents			
	1.	Introduction	46
	2.	The Relativistic Origins of Spin–Orbit Coupling	48
		2.1. Dirac equation	48
		2.2. From Dirac equation to spin–orbit coupling	49
	3.	Band Structure of Semiconductors: Effective k · p Hamiltonians	51
		3.1. 8-Band Kane model Hamiltonian	52
		3.2. 8-Band Kohn–Luttinger Hamiltonian	53
		3.3. Effective mass or envelope function approximation	56
	4.	SHE and AHE	61
		4.1. The semiclassical Boltzmann equation approach to anomalous Hall transport	64
		4.2. AHE in two dimensions	71
		4.3. Estimates of the role of various spin–orbit couplings contributing to skew scattering	73
		4.4. Intrinsic SHE: A heuristic picture	73
		4.5. Numerical studies of the spin Hall accumulation and SHE	77
	5.	Topological Berry's Phases in Spin–Orbit Coupled Systems: ACE	79
		5.1. Landauer–Büttiker calculations of ACE: Comparison to experiments	82
		References	85

---

[*] Department of Physics, Texas A & M University, College Station, Texas 77843-4242
[†] Department of Physics, University of Texas at Austin, Austin, Texas 78712-1081
[‡] Institute of Physics, ASCR v.v.i., Cukrovarnická 10, 162 53 Praha 6, Czech Republic

1. INTRODUCTION

Spin–orbit coupling is ubiquitous in the physics of semiconductor spintronics. It is one of the few echos from relativistic physics emerging in the field of condensed matter physics, governed for the most part by nonrelativistic physics and low energies.

Many studies in metallic spintronics have centered around the injection and generation of spin via optical and transport injection and external magnetic field manipulation. Within these studies, spin–orbit coupling becomes an important source of spin dephasing and in many instances it is important to minimize it. On the other hand, in semiconductor spintronics, spin–orbit coupling comes to the forefront as a source of spin manipulation and spin generation, becoming a necessary element in physical phenomena such as the spin Hall effect (SHE), anomalous Hall effect (AHE), and the study of Berry's phase effects in mesoscopic systems such as the Aharonov–Casher effect (ACE). One of the key aspects to the origin of this anomalous transport is the interband coherence present in the off-diagonal elements of the nonequilibrium density matrix of the system induced by the spin–orbit coupling. Dealing theoretically with such interband coherence within the usual linear transport theories has been at the source of its difficulty and fascinating aspects. In this chapter, we review the basic physics of spin–orbit coupling in the band structure of semiconductors starting from its origins from the Dirac equation to its emergence in the $\vec{k} \cdot \vec{p}$ description of simple semiconductors, and how it enters into the parameters of the different effective Hamiltonians. Some of this introductory aspects are treated more extensively in other reviews, for example, Winkler (2003), and we incorporate a small sketch of the derivations here for completeness and when relevant to important aspects of the physics of these effects.

We will focus our attention on these three transport effects, two of which have been recently observed for the first time in semiconductor spintronics systems. This choice is because it is spin–orbit coupling that is the reason for their existence at a fundamental level. spin–orbit coupling, as seen in the other chapters of this book, plays a major role in understanding quantitatively the properties of many semiconductor spintronic systems of interests, for example, diluted magnetic semiconductors and magnetization dynamics, but at many levels the basic physics of these systems can be understood without evoking spin–orbit coupling as their origin. This is not the case for the AHE, SHE, and ACE.

We will rely throughout the chapter in pedagogical descriptions present in many reviews, in particular the origins of spin–orbit couping and how it affects the band structure of semiconductors in the introductory sections (Winkler, 2003; Zutic et al., 2004). We hope that the topics covered here, although attempted to be self-contained, give a first general impression of the physics described and its origins and motivates the interested reader to follow up with a more in-depth reading of the recent extensive literature. As is the case with any summary of very recent developments, the topics covered are biased by the work of the authors and their approach to the subject. Hence, this chapter should not be viewed as an encompassing review of these topics but as a restricted view of some of the physical aspects of the effects that may be helpful to a researcher or graduate student interested in entering these topics, but keeping in mind that it covers topics that the authors have studied more in depth while leaving some other important topics uncovered. The SHE, AHE, and ACE have seen a major theoretical and experimental effort over the past few years and a well-balanced overreaching review will have to wait until the dust settles on some of these fascinating scientific endeavors. However, some recent reviews by Schliemann et al. (2006), focused on some aspect of the intrinsic SHE, and Engel et al. (2006), focusing on the extrinsic SHE and quantum SHE, give a good sketch of the developments up to that point. The physics of the AHE has been reviewed briefly by Sinova et al. (2004b) and Chien and Westgate (1980). Other review by Ganichev et al. (2008) on galvanich effects highlights the physics of semiconducting spin-orbit coupled systems. Although we mention some of the experimental results of these effects, we focus on this chapter on the theoretical aspects of the effects and on heuristic understanding of their origins.

The rest of the chapter is organized as follows: In Section 2, we derive the Pauli Hamiltonian by expanding the Dirac equation in powers of v/c in order to clarify the origins of the spin–orbit coupling in the electronics structure of semiconductors, which we treat in Section 3. In this section, we review the basics of the $\mathbf{k} \cdot \mathbf{p}$ approximation, the effective mass or envelope function approximation, the effective reduced 2-band Hamiltonian of the conduction electrons obtained from the 8-band Kane model, and the effective 2-d Hamiltonians that follow. In Section 4, we discuss some of the physics of the AHE and the SHE. We review within this section the basic mechanisms for anomalous and spin Hall transport derived from a semiclassical point of view. We apply these concepts to the 2d-graphene and 2d-Rashba system as an illustration and make the connection to the microscopic approach of the Kubo diagrammatic formalism where the equivalence between the two approaches is shown most clearly. Within the SHE, which follows from the basic phenomenology of the AHE, we review the main aspects of the intrinsic SHE within a clear heuristic approach to its understanding and also expand on the numerical results of bulk numerical calculations relating the effect to the spin-accumulation induced by the effect and some comparison to experiments in the strong spin–orbit coupling regime of 2d-hole systems. Many aspects of the SHE, such as mesoscopic SHE, quantum SHE, and the regime of weak spin–orbit coupling are not covered, in order to keep the chapter to a reasonable length since these topics deserve individual coverage by themselves. In Section 5, we review the recent results of the ACE, which illustrates a spin-dependent transport effect in the mesoscopic regime.

2. THE RELATIVISTIC ORIGINS OF SPIN–ORBIT COUPLING

2.1. Dirac equation

We begin our exploration of the spin–orbit effects at its very origin, the relativistic Dirac equation. To arrive at this fundamental equation, we reexamine the Schrödinger equation as a quantum mechanical equation of motion which is postulated in correspondence to classical mechanics. For instance for a free particle:

$$E = \frac{p^2}{2m} \text{ replacing } E \to i\hbar\frac{\partial}{\partial t}, p \to -i\hbar\nabla \tag{2.1}$$

$$i\hbar\frac{\partial}{\partial t}\Psi(\mathbf{x},t) = -\frac{\hbar^2}{2m}\Delta\Psi(\mathbf{x},t) \tag{2.2}$$

Following a similar procedure with the relativistic version of this equation, $E^2 = c^2p^2 + m^2c^4$, yields the Klein–Gordon equation, which is a wave equation with second-order derivatives in time and has the drawback that it does not yield a density operator with a connection to a nonnegative probabilistic function. In seeking an equation which obeys the relativistic relation and is linear in time derivatives of the form $i\hbar\partial\Psi/\partial t = H\Psi$, we need to choose H to have the form

$$H = c\boldsymbol{\alpha}\cdot\mathbf{p} + \beta mc^2$$

Coefficients α and β must be determined from the restriction that $H^2 = c^2(-\hbar^2\Delta) + m^2c^4 = E^2$. The condition above together with hermiticity of the Hamiltonian leads to conditions: $\{\alpha_j, \alpha_k\} = 2\delta_{jk}$, $\{\alpha_j, \beta\} = 0$ and $\beta^2 = 1$, which form the so-called Dirac algebra. Any matrices α and β fulfilling these conditions are connected by a unitary transformation. The most commonly used representation in condensed matter physics is the Dirac realization: $\boldsymbol{\alpha} = \sigma_1 \otimes \boldsymbol{\sigma}$, $\beta = \sigma_3 \otimes 1$, which allows a simple spin interpretation (σ_j are Pauli matrices).

The Dirac equation for a free particle reads:

$$i\hbar\frac{\partial}{\partial t}\Psi(\mathbf{x},t) = (c\boldsymbol{\alpha}\cdot\mathbf{p} + \beta mc^2)\Psi(\mathbf{x},t)$$

It is a vector equation with $\Psi(\mathbf{x},t)$ having four components ("bispinor"). The electro-magnetic field described by potentials A and $V = e\varphi$ enter the equation through the usual substitution: $p \to p - eA/c \equiv \pi$, $\varepsilon \to \varepsilon - V$, where $i\hbar(\partial/\partial t)\Psi(x,t) = \varepsilon\Psi(x,t)$. Here the charge of the electron is $e < 0$.

2.2. From Dirac equation to spin–orbit coupling

To obtain the weak relativistic limit of the Dirac equation, we rewrite the bispinor as

$$\Psi(x,t) = \begin{pmatrix} \varphi(x,t) \\ \chi(x,t) \end{pmatrix},$$

where it can be shown that χ is smaller than φ in the nonrelativistic limit by a factor of v/c. We then obtain two coupled equations for the spinors:

$$\left(i\hbar\frac{\partial}{\partial t} - V - mc^2\right)\varphi = c\boldsymbol{\sigma}\cdot\left(p - \frac{e}{c}A\right)\chi \qquad (2.3)$$

$$\left(i\hbar\frac{\partial}{\partial t} - V + mc^2\right)\chi = c\boldsymbol{\sigma}\cdot\left(p - \frac{e}{c}A\right)\varphi \qquad (2.4)$$

Here mc^2 is typically a very large energy scale and V is usually much smaller than this scale, so we look for energy solutions where the non-rest mass energy $E \equiv \varepsilon - mc^2$ is small: $|E| \ll mc^2$, that is, the total energy is very close to the rest energy in this scale. We next proceed in two steps. We first are going to obtain the Dirac equation to first order in v/c as a warm up and afterward we will do it to second order in v/c. Throughout this procedure, a useful identity is

$$(\boldsymbol{\sigma}\cdot X)(\boldsymbol{\sigma}\cdot Y) = X\cdot Y + i\boldsymbol{\sigma}\cdot[X\times Y] \qquad (2.5)$$

from which it can be shown that

$$\left(\boldsymbol{\sigma}\cdot\left(p - \frac{e}{c}A\right)\right)^2 = \left(p - \frac{e}{c}A\right)^2 - \frac{e\hbar}{c}\boldsymbol{\sigma}\cdot B. \qquad (2.6)$$

To obtain the Dirac equation to first order in v/c, we look for solutions of ε around mc^2. We do so first by rewriting Eq. (2.4) in the form

$$\chi = \frac{1}{2mc^2}\left[c\boldsymbol{\sigma}\cdot\boldsymbol{\pi}\varphi - \left(i\hbar\frac{\partial}{\partial t} - V - mc^2\right)\chi\right] \qquad (2.7)$$

This provides a recursive way to expand in powers of (v/c) since from the above equation and Eq. (2.3) yield χ to second order in v/c:

$$\chi = \frac{1}{2mc^2}\left[c\boldsymbol{\sigma}\cdot\boldsymbol{\pi}\varphi - \left(i\hbar\frac{\partial}{\partial t} - V - mc^2\right)\frac{1}{2mc}\boldsymbol{\sigma}\cdot\boldsymbol{\pi}\varphi\right] \qquad (2.8)$$

We can then substitute this into Eq. (2.3) for the upper spinor and obtain:

$$\left(i\hbar\frac{\partial}{\partial t} - V - mc^2\right)\varphi = \frac{(\boldsymbol{\sigma}\cdot\boldsymbol{\pi})^2}{2m}\varphi - \frac{1}{4m^3c^2}\boldsymbol{\sigma}\cdot\boldsymbol{\pi}\left(i\hbar\frac{\partial}{\partial t} - V - mc^2\right)\boldsymbol{\sigma}\cdot\boldsymbol{\pi}\varphi. \tag{2.9}$$

Manipulation of the above equation can simplify to:

$$\left(i\hbar\frac{\partial}{\partial t} - V - mc^2\right)\varphi = \left[\frac{(\boldsymbol{\sigma}\cdot\boldsymbol{\pi})^2}{2m} - \frac{1}{8m^3c^2}(\boldsymbol{\sigma}\cdot\boldsymbol{\pi})^4 - \frac{i\hbar e}{4m^2c^2}(\boldsymbol{\sigma}\cdot\boldsymbol{\pi})\boldsymbol{\sigma}\cdot\mathbf{E}\right]\varphi, \tag{2.10}$$

where we have used $\mathbf{E} = -\nabla V(\mathbf{r})/e$. One thing that may be noticed immediately with this equation is that the effective Hamiltonian for φ is not Hermitian which should make one worry. This has occurred because the particle effective Hamiltonian that we seek is a 2×2 matrix rather than the 4×4 which we really have since our particle state is being described by Ψ and not φ alone, that is, $\int \varphi^\dagger \varphi \neq 1$. The normalization condition of Ψ gives:

$$\int \Psi^\dagger \Psi = \int \varphi^\dagger \varphi + \chi^\dagger \chi \approx \int \varphi^\dagger \left(1 + \frac{(\boldsymbol{\sigma}\cdot\boldsymbol{\pi})^2}{4m^2c^2}\right)\varphi = \int \tilde{\varphi}^\dagger \tilde{\varphi} + O\left(\frac{v^3}{c^3}\right), \tag{2.11}$$

where we have defined $\tilde{\varphi} \equiv \left(1 + (\boldsymbol{\sigma}\cdot\boldsymbol{\pi})^2/4m^2c^2\right)\varphi \equiv \hat{G}\varphi$. Hence, what we would really want is an effective Hamiltonian for $\tilde{\varphi}$ rather than φ since it is $\tilde{\varphi}$ the bi-spinor which is normalized. This can be done by multiplying Eq. (2.9) by $\hat{G}$ on the left. The right-hand side (RHS) of Eq. (2.9) all commutes with $\hat{G}$ except for $\mathbf{E}$ but this term is already of order v^2/c^2, so we do not need to concern with it and $\hat{G}$ can be passed through to operate on φ by using the commutator of $\hat{G}$ and $[i\hbar(\partial/\partial t) - V - mc^2]$, which can be shown to be $-i\hbar(e/c)[(\boldsymbol{\sigma}\cdot\boldsymbol{\pi})(\boldsymbol{\sigma}\cdot\mathbf{E}) + (\boldsymbol{\sigma}\cdot\mathbf{E})(\boldsymbol{\sigma}\cdot\boldsymbol{\pi})]$, so the effective Hamiltonian defined by $i\hbar\partial\tilde{\varphi}/\partial t = H_{\text{eff}}\tilde{\varphi}$ is given by

$$H_{\text{eff}} = \frac{(\boldsymbol{\sigma}\cdot\boldsymbol{\pi})^2}{2m} - \frac{1}{8m^3c^2}(\boldsymbol{\sigma}\cdot\boldsymbol{\pi})^4 + \frac{i\hbar e}{8m^2c^2}[-(\boldsymbol{\sigma}\cdot\boldsymbol{\pi})(\boldsymbol{\sigma}\cdot\mathbf{E}) + (\boldsymbol{\sigma}\cdot\mathbf{E})(\boldsymbol{\sigma}\cdot\boldsymbol{\pi})]. \tag{2.12}$$

This expression can be simplified after further manipulation using the relation given in Eq. (2.5) to yield a more physically transparent equation:

$$H_{\text{eff}} = mc^2 + \frac{1}{2m}\pi^2 + V - \frac{e\hbar}{2mc}\boldsymbol{\sigma}\cdot\mathbf{B} - \frac{\hbar e}{8m^2c^2}\boldsymbol{\sigma}\cdot\left(2\mathbf{E}\times\boldsymbol{\pi} - i\frac{\hbar}{c}\frac{\partial \mathbf{B}}{\partial t}\right)$$

$$+ \frac{\hbar e}{8m^2c^2}\nabla^2 V - \frac{1}{8m^2c^2}\left(\pi^4 + \frac{e^2}{c^2}\hbar^2\mathbf{B}^2\right) \tag{2.13}$$

$$+ \frac{e\hbar}{8m^2c^3}[-\hbar^2\nabla^2(\boldsymbol{\sigma}\cdot\mathbf{B}) - i2\hbar\nabla(\boldsymbol{\sigma}\cdot\mathbf{B})\cdot\boldsymbol{\pi} + 2(\boldsymbol{\sigma}\cdot\mathbf{B})\pi^2]$$

Such effective Hamiltonian is sometimes called the Pauli Hamiltonian. The physical meaning of the constituents is clear. The expansion of the Dirac equation to second order around mc^2 yields naturally the Zeeman term with the correct g-factor of 2 as can be seen in the fourth term of H_{eff}, $[g_0\mu_B(1/2)\boldsymbol{\sigma} \cdot \boldsymbol{B}/2mc]$, where $g_0 = 2$, $\mu_B \equiv |e|\hbar/2mc > 0$. We have also obtained the spin–orbit coupling correction to the Schrödinger equation, given by the fifth term in the RHS:

$$\frac{\hbar}{4m^2c^2}\boldsymbol{\sigma} \cdot \nabla V \times \boldsymbol{\pi} \tag{2.14}$$

The sixth term in H_{eff} is a sort of contact interaction which can be seen by inserting the Coulomb potential in the expression $\hbar^2 \Delta V/8m^2c^2 \propto \delta(\mathbf{r})$; such term is felt only by the s-states. The last term quartic in momentum (Darwin term) can be interpreted as a fast shaking of the particle at a length scale corresponding to its Compton wavelength. $V(\mathbf{r})$ in Eq. (2.13) is the scalar potential present in the system which includes the periodic and scattering potential present in imperfect crystals, hence providing a starting point to discuss the relevant role of intrinsic and extrinsic spin–orbit coupling within the electronic structure of semiconductors.

As we will see below, the Pauli Hamiltonian will emerge again within the description of the effective Hamiltonian for a 2-band electron system influenced by other nearby bands in a semiconductor. However, the parameters appearing as coefficients of each term will change, for example, $m \to m_{\text{eff}}$ in the kinetic term, and in most instances this renormalization will dominate by several orders of magnitude the bare electron value present in Eq. (2.13).

3. BAND STRUCTURE OF SEMICONDUCTORS: EFFECTIVE k · p HAMILTONIANS

The band structure of an electron in a periodic potential $V(\mathbf{r}) = V(\mathbf{r} + \mathbf{R})$, where $\mathbf{R}$ is any lattice vector, can be obtained by evoking the Bloch theorem, which states that the general form of an eigenstate in the presence of such potential has the form $\psi_{n,\mathbf{k}}(\mathbf{r}) = \exp[i\mathbf{k} \cdot \mathbf{r}]\, u_{n,\mathbf{k}}(\mathbf{r})$, with $u_{n,\mathbf{k}}(\mathbf{r}) = u_{n,\mathbf{k}}(\mathbf{r} + \mathbf{R})$. Here n reflects the band index, $\mathbf{k}$ the wavevector of the electron, and $u_{n,\mathbf{k}}(\mathbf{r})$ obeys the Schrödinger equation:

$$\left\{ \frac{\hbar^2 k^2}{2m} + \hbar \mathbf{k} \cdot \boldsymbol{\pi} + \frac{1}{2m}\pi^2 + V - \frac{e\hbar}{2mc}\boldsymbol{\sigma} \cdot \mathbf{B} + \frac{\hbar}{4m^2c^2}\boldsymbol{\sigma} \cdot \nabla V(\mathbf{r}) \times \boldsymbol{\pi} \right.$$
$$\left. + \frac{\hbar^2}{4m^2c^2}\boldsymbol{\sigma} \cdot \nabla V(\mathbf{r}) \times \mathbf{k} \right\} u_{n,\mathbf{k}}(\mathbf{r}) = E_{n,\mathbf{k}} u_{n,\mathbf{k}}(\mathbf{r}), \tag{2.15}$$

where we have ignored (for the moment) the other terms in the Pauli Hamiltonian. Also, the last term in the LHS of Eq. (2.15) involving the crystal momentum is much smaller than the one involving the atomic momentum π and it can therefore be dropped.

The most widely used method to solve for $u_{n,\mathbf{k}}(\mathbf{r})$ is the $\mathbf{k} \cdot \mathbf{p}$ method. In this method, the Hamiltonian is expanded in wave-vector around a high symmetry point (typically the band edge) in the basis at such point. For direct band gap semiconductors, such high symmetry point is $\mathbf{k} = 0$ and for a finite wavevector is written as

$$|n\mathbf{k}\rangle = \sum_{v'\sigma} c_{v'\sigma}^n(\mathbf{k})|v'\sigma \mathbf{k} = 0\rangle, \quad (2.16)$$

where we have chosen the basis set at the band age $|v\rangle$ that contain no spin–orbit coupling. The algebraic equation for the expansion coefficients is given by Winkler (2003):

$$\sum_{v',\sigma'}\left\{\left[E_{v',0} - \frac{\hbar^2 k^2}{2m}\right]\delta_{v,v'}\delta_{\sigma,\sigma'} + \frac{\hbar}{m}\mathbf{k}\cdot\mathbf{P}_{v',\sigma';v,\sigma} + \Delta_{v',\sigma';v,\sigma}\right\}c_{v'\sigma'}^n = E_n(\mathbf{k})c_{v\sigma}^n,$$

(2.17)

where

$$\mathbf{P}_{v',\sigma';v,\sigma} \equiv \langle v'\sigma'|\pi|v\sigma\rangle, \quad (2.18)$$

$$\Delta_{v',\sigma';v,\sigma} \equiv \langle v'\sigma'|\frac{\hbar}{4m^2c^2}\boldsymbol{\sigma}\cdot\nabla V(\mathbf{r})\times\pi|v\sigma\rangle. \quad (2.19)$$

The operator π only couples states with opposite parity and, for example, in the case of a zinc blend material such as GaAs, it couples the valence to the conduction bands. The spin–orbit coupling term will split states with finite angular momentum into multiplets of the total angular momentum such as the sixfold degenerate valence band in GaAs which gets split into a fourfold degenerate (at $\mathbf{k} = 0$) total angular momentum $j = 3/2$ subspace and a twofold degenerate $j = 1/2$ subspace.

3.1. 8-Band Kane model Hamiltonian

Ignoring the other bands far away from the valence and conduction bands, we obtain the popular Kane model which yields a good approximation for describing the conduction band while missing, for example, the effective mass of the heavy hole which yields positive rather than a negative dispersion. In spite of this shortcoming, such a model has the advantage of simplicity by being described with only a few parameters: E_g, Δ, and P. For the case of the 8-band Kane model, which has spherical symmetry, we choose the basis sets $|1\rangle = |iS\uparrow\rangle, |2\rangle = |iS\downarrow\rangle, |3\rangle = |[(X-iY)/\sqrt{2}]\uparrow\rangle, |4\rangle = |Z\downarrow\rangle, |5\rangle = |[-(X+iY)/\sqrt{2}]\uparrow\rangle, |6\rangle = |[-(X+iY)/\sqrt{2}]\downarrow\rangle, |7\rangle = |Z\uparrow\rangle, |8\rangle|[(X-iY)/\sqrt{2}]\downarrow\rangle$ which gives the matrix Hamiltonian:

$$H_{8-K} = \begin{pmatrix} H_c & h \\ h^\dagger & H_v \end{pmatrix} = \frac{\hbar^2 k^2}{2m}I + \frac{1}{2}g_0\mu_B\sigma_{v,v'}\cdot\mathbf{B} + \begin{pmatrix} 0 & & & 0 \\ & 0 & & 0 \\ & & 0 & \\ 0 & 0 & & -E_g I \end{pmatrix}$$

$$+ \begin{pmatrix} 0 & 0 & Pk_- & 0 & -Pk_+ & 0 & Pk_z & 0 \\ 0 & 0 & 0 & Pk_z & 0 & -Pk_+ & 0 & Pk_- \\ Pk_+ & 0 & -\frac{2}{3}\Delta & \frac{\sqrt{2}\Delta}{3} & 0 & 0 & 0 & 0 \\ 0 & Pk_z & \frac{\sqrt{2}\Delta}{3} & -\frac{\Delta}{3} & 0 & 0 & 0 & 0 \\ -Pk_+ & 0 & 0 & 0 & 0 & 0 & 0 & 0 \\ 0 & -Pk_- & 0 & 0 & 0 & -\frac{2}{3}\Delta & \frac{\sqrt{2}\Delta}{3} & 0 \\ Pk_z & 0 & 0 & 0 & 0 & \frac{\sqrt{2}\Delta}{3} & -\frac{\Delta}{3} & 0 \\ 0 & Pk_+ & 0 & 0 & 0 & 0 & 0 & 0 \end{pmatrix}$$

(2.20)

where $k_\pm = (k_x \pm ik_y)/\sqrt{2}$, and as in the previous section $\mu_B = \hbar|e|/2mc$ is the Bohr magneton, $g_0 = 2$, $\sigma_{v,v'} = \langle v | \boldsymbol{\sigma} | v' \rangle$, and we use the fact that $e < 0$.

3.2. 8-Band Kohn–Luttinger Hamiltonian

If other bands are to be taken into account, then the method of invariants serves best (Winkler, 2003). This method consists of writing, within the bands of interest, all the terms allowed by the symmetry constraints of the crystal which are invariant under the appropriate symmetry operations, with their corresponding factors being phenomenologically adjusted as material parameters to best fit, for example, optical data. The 8-band Kane model is then transformed to a Kohn–Luttinger Hamiltonian where the basis functions corresponding to heavy, light, split-off holes, conduction-band electrons, and defect bands can be written, in this order, as $|1\rangle = (-1/\sqrt{2})(X_\uparrow + iY_\uparrow), |2\rangle = (1/\sqrt{6})(X_\uparrow - iY_\uparrow) + \sqrt{2/3}Z_\downarrow, |3\rangle = (-1/\sqrt{6})(X_\downarrow + iY_\downarrow) + \sqrt{2/3}Z_\uparrow, |4\rangle = (1/\sqrt{2})(X_\downarrow - iY_\downarrow), |5\rangle = (-1/\sqrt{3})(X_\downarrow + iY_\downarrow) - \sqrt{1/3}Z_\uparrow, |6\rangle = (-1/\sqrt{3})(X_\uparrow - iY_\uparrow) + \sqrt{1/3}Z_\downarrow, |7\rangle = S_\uparrow, |8\rangle = S_\downarrow$, where X, Y, Z are p-like orbitals, S and are the s orbitals associated with conduction bands.

In this basis, the 8-band Kohn–Luttinger Hamiltonian obtained through the method of invariants has the form

$$H_{KL} = \begin{pmatrix} H_{hh} & -c & -b & 0 & \frac{b}{\sqrt{2}} & c\sqrt{2} & e & 0 \\ -c^* & H_{lh} & 0 & b & -\frac{b^*\sqrt{3}}{\sqrt{2}} & -d & \frac{1}{\sqrt{3}}e^* & f^* \\ -b^* & 0 & H_{lh} & -c & d & -\frac{b\sqrt{3}}{\sqrt{2}} & f^* & \frac{e}{\sqrt{3}} \\ 0 & b^* & -c^* & H_{hh} & -c^*\sqrt{2} & \frac{b^*}{\sqrt{2}} & & \\ \frac{b^*}{\sqrt{2}} & -\frac{b\sqrt{3}}{\sqrt{2}} & d^* & -c\sqrt{2} & H_{so} & 0 & \frac{f}{\sqrt{2}} & \sqrt{\frac{2}{3}}e \\ c^*\sqrt{2} & -d^* & -\frac{b^*\sqrt{3}}{\sqrt{2}} & \frac{b}{\sqrt{2}} & 0 & H_{so} & -\sqrt{\frac{2}{3}}e^* & \frac{f^*}{\sqrt{2}} \\ e^* & \frac{1}{\sqrt{3}}e & f & 0 & \frac{f^*}{\sqrt{2}} & -\sqrt{\frac{2}{3}}e & H_{cb} & 0 \\ 0 & f & \frac{e^*}{\sqrt{3}} & e & \sqrt{\frac{2}{3}}e^* & \frac{f}{\sqrt{2}} & 0 & H_{cb} \end{pmatrix} \quad (2.21)$$

The quantities that appear in H_{KL} are:

$$H_{cb} = E_g + \frac{\hbar^2}{2m_c^*}(k_x^2 + k_y^2 + k_z^2); \quad (2.22)$$

$$H_{hh} = -\frac{\hbar^2}{2m_0}\left[\left(\gamma_1 + \gamma_2 - \frac{E_P}{2(E_g + \Delta_{so}/3)}\right)(k_x^2 + k_y^2) + (\gamma_1 - 2\gamma_2)k_z^2\right] \quad (2.23)$$

$$H_{lh} = -\frac{\hbar^2}{2m_0}\left[\left(\gamma_1 - \gamma_2 - \frac{E_P}{6(E_g + \Delta_{so}/3)}\right)(k_x^2 + k_y^2) \right.$$
$$\left. + \left(\gamma_1 + 2\gamma_2 - \frac{E_P}{3(E_g + \Delta_{so}/3)}\right)k_z^2\right]; \quad (2.24)$$

$$H_{so} = -\Delta_{so} - \frac{\hbar^2}{2m_0}\left(\gamma_1 - \frac{E_P}{3(E_g + \Delta_{so}/3)}\right)(k_x^2 + k_y^2 + k_z^2); \quad (2.25)$$

$$b = -\frac{\hbar^2}{2m_0}\sqrt{12}\left(\gamma_3 - \frac{E_P}{6(E_g + \Delta_{so}/3)}\right)k_z(k_x - ik_y) \quad (2.26)$$

$$c = -\frac{\hbar^2}{2m_0}\sqrt{3}\left[\left(\gamma_2 - \frac{E_P}{6(E_g + \Delta_{so}/3)}\right)(k_x^2 - k_y^2) - 2i\left(\gamma_3 - \frac{E_P}{6(E_g + \Delta_{so}/3)}\right)k_x k_y\right]; \quad (2.27)$$

$$d = -\frac{\hbar^2}{2m_0}\sqrt{2}\left[-\left(2\gamma_2 - \frac{E_P}{3(E_g + \Delta_{so}/3)}\right)k_z^2 + \left(\gamma_2 - \frac{E_P}{6(E_g + \Delta_{so}/3)}\right)(k_x^2 + k_y^2)\right], \quad (2.28)$$

$$p = \frac{iP}{\sqrt{2}}(k_x - ik_y), \quad f = \sqrt{\frac{2}{3}}iPk_z, \quad (2.29)$$

where m_0 is the bare electron mass, m_c^* is the effective mass of conduction electrons, for GaAs $m_c^* = 0.067 m_0$, and Δ_{so} is the energy of spin–orbit splitting. P is the defined as in the 8-band Kane model.

If one is interested only in the valence-band sector, for example, low-frequency conductivity in a hole-doped system, the 6-band KL model of the valence-band structure can be obtained by setting $E_P = 0$ and it is fully parametrized by the Luttinger parameters γ_1, γ_2, γ_3, and Δ_{so}. Note that in doing so one is not ignoring the effects of the conduction band on the valence band, as these are taken into account by the parameters of each term changing according to the equations above. For example, in GaAs the values are $\gamma_1 = 6.98$, $\gamma_2 = 2.06$, $\gamma_3 = 2.93$, and $\Delta_{so} = 341$ meV. For low hole-doped systems where only the heavy and light-hole bands are occupied ($E_F \ll \Delta_{so}$), the top valence band is well described by the 4-band Kohn–Luttinger Hamiltonian given by

$$H_{KL} = \frac{\hbar^2}{2m}\left[(2\gamma_1 + 5\gamma_2)k^2 - 2\gamma_3(\mathbf{k}\cdot\mathbf{J})^2 + 2(\gamma_3 - \gamma_2)\sum_i k_i^2 J_i^2\right], \quad (2.30)$$

where $\mathbf{J}$ is the angular momentum matrix of spin 3/2 which can be obtained from Eq. (2.21) by setting $\Delta_{so} \to \infty$. A further approximation is obtained by setting $\gamma_2 = \gamma_3$, obtaining in this way the spherical 4-band model which gives the simplest description of the heavy- and light-hole bands while retaining their dominant spin–orbit character. In the other direction, the 8-band Kohn–Luttinger model can be further augmented to the so-called 14-band extended Kane model which takes into account the antibonding p-bands in a zinc-blend semiconductor as well.

Also, an additional spin–orbit coupling term appears in a system of broken bulk inversion symmetry (BIA) such as II–VI and III–V semiconductors. These terms, called the Dresselhaus terms, have the form in the zinc blende systems:

$$H_{D-3d} = \mathcal{B}k_x(k_y^2 - k_z^2)\sigma_x + \text{cyclic permutations}. \quad (2.31)$$

These terms appear in the 14-band extended Kane model with values of $\mathcal{B} \approx 10\text{–}30\text{e V\AA}$ for GaAs and InAs although some other values are also used in the literature (Winkler, 2003).

3.3. Effective mass or envelope function approximation

One of the many successes of the $\mathbf{k} \cdot \mathbf{p}$ method is its ability to take into account the effects of electric and magnetic fields that vary slowly on the scale of the lattice constant of the crystal. These electric fields can be produced by impurities or by external means such as metallic gates or confining potentials in hetero-structures. In this case, one can write the expansion of the eigenstate as

$$\psi(\mathbf{r}) = \sum_{v'\sigma} c(\mathbf{r})_{v'\sigma} u(\mathbf{r})_{v'k=0}|\sigma'\rangle, \quad (2.32)$$

where now the expansion coefficients are space dependent with the restriction that $c(\mathbf{r})_{v'\sigma}$ vary slowly in the scale of the lattice constant, and will obey the coupled Schrödinger equation

$$\sum_{v',\sigma'}\left\{\left[E_{v',0} - \frac{\hbar^2\left(-i\hbar\nabla - \frac{e}{c}\mathbf{A}\right)^2}{2m}\right]\delta_{v,v'}\delta_{\sigma,\sigma'} + \frac{\hbar}{m}\left(-i\hbar\nabla - \frac{e}{c}\mathbf{A}\right)\right.$$

$$\left. \cdot \mathbf{P}_{v',\sigma';v,\sigma} + \Delta_{v',\sigma';v,\sigma} + \frac{g_0}{2}\mu_B\sigma_{\sigma,\sigma'}\cdot\mathbf{B}\delta_{v,v'}\right\}c(\mathbf{r})_{v'\sigma'} = E_n(\mathbf{k})c(\mathbf{r})_{v\sigma}. \quad (2.33)$$

The parameters $\Delta_{v\ell, \sigma';v,\sigma}$ and $\mathbf{P}_{v', \sigma';v, \sigma}$ of Eqs. 2.33 and 2.17 are the same and in fact the multiband envelope function Hamiltonian, Eq. (2.33), can be obtained by simply replacing $\hbar\mathbf{k} \to -i\hbar\nabla - (e/c)\mathbf{A}$ in Eq. (2.17) and adding the impurity or external potential $V(\mathbf{r})$ to the diagonal. With such consideration, it is important to note (see below) that in the presence of a magnetic field the kinetic crystal momentum $\mathbf{k}$ will no longer commute and that

$$\mathbf{k} \times \mathbf{k} = \frac{ie}{\hbar c}\mathbf{B} \text{ or } [k_\alpha, k_\beta] = \frac{ie}{\hbar c}\varepsilon_{\alpha\beta\gamma}B_\gamma, \quad (2.34)$$

and

$$[\mathbf{k}, V(\mathbf{r})] = -i\nabla V(\mathbf{r}) = ie\mathbf{E}. \quad (2.35)$$

Here, as above, we use the convention that the charge of an electron $e < 0$. The terms in the $\mathbf{k} \cdot \mathbf{p}$ Hamiltonian symmetric and antisymmetric in $\mathbf{k}$ and $V(\mathbf{r})$ will give rise to an effective mass (symmetric terms in $\mathbf{k}$ to second order), an effective g-factor (antisymmetric terms in $\mathbf{k}$ to second order), and an effective spin–orbit coupling strength α (antisymmetric terms in $\mathbf{k}$ and $V(\mathbf{r})$), when transforming the finite Hamiltonian to a block diagonal form up to a certain order in $1/E_g$ as we detail below in Fig. 2.1.

3.3.1. Effective mass 2-band model Hamiltonians: 3-d Conduction electrons

The general form of the finite $\mathbf{k} \cdot \mathbf{p}$ Hamiltonian couples the different (finite) bands. However, when studying the transport properties for the case where the Fermi energy resides in one or few bands separated by a large gap from the other bands, it is convenient to transform the Hamiltonian to a block diagonal form through a unitary transformation. This is

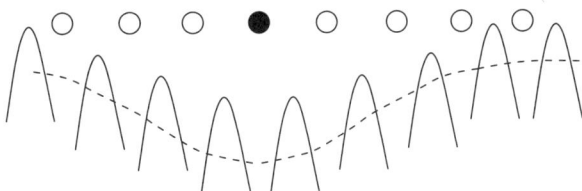

FIGURE 2.1 Illustration of the periodic and impurity potential. The impurity potential is a combination of the isolated impurity potential and the absence of the period part of the unperturbed periodic potential.

the so-called Löwdin partitioning and it amounts to taking into account the effects of the other bands up to a certain order in $1/E_g$.

The Lödwin procedure is equivalent to the one used by Nozieres and Lewiner (1973) in the study of the AHE in which, starting from the 8-band Kane model in Eq. (2.20), the effects of the valence band on the conduction bands through virtual transitions are taken into account to second order in $1/E_g$. The transformation is illustrated in Fig. 2.2. Since the 2-band effective Hamiltonian considered for the n-doped materials is widely used, we illustrate below some of the details of its derivation. In general, the Hamiltonian will be partitioned into a subspace A and B with h and $h^\dagger$ connecting the two subspaces. The blocks of the subspace A and B will contain a constant part and a $\mathbf{k}$ and $\mathbf{B}$ dependent part with the eigenvalues of the constant section of the block Hamiltonian, H_g in the illustration, A and B separated by an energy gap. For the 8-band Kane Hamiltonian, Eq. (2.20), H'_v and H'_c in Fig. 2.2 are given by the diagonal matrix:

$$H'_{c/v} = \frac{\hbar^2 k^2}{2m} + \frac{1}{2} g_0 \mu_B \sigma \cdot \mathbf{B}. \qquad (2.36)$$

The procedure to obtain the anti-Hermitian matrix S to yield the desired unitary matrix exp[S] that will transform H into the block diagonal form up to a given order in $1/E_g$ is illustrated in Appendix B of Winkler (2003) and, without the language of unitary transformations, by Nozieres and

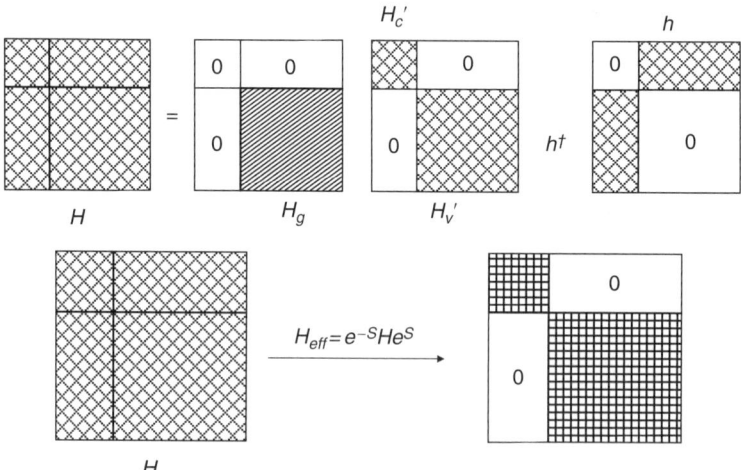

FIGURE 2.2 Illustration of the Löwdin transformation to obtain an effective block diagonal Hamiltonian. The Hamiltonian is partitioned in a diagonal, block diagonal, and off-block diagonal form. The two block diagonal parts are assumed to be separated by an energy gap E_g.

Lewiner (1973). The effective Hamiltonian for the conduction electrons that is obtained through this procedure is given by

$$H_{\text{eff}} = e^{-S} H e^{S} = H'_c - h H_g^{-1} h^\dagger - \frac{\Lambda H'_c + H'_c \Lambda}{2}$$
$$+ h H_g^{-1} H'_v h^\dagger + \frac{i}{2} [\dot{h} H_g^{-2} h^\dagger - h H_g^{-2} \dot{h}^\dagger], \quad (2.37)$$

with $\Lambda = h H_g^{-2} h^\dagger$. A similar transformation takes place for other operators to be calculated within this subspace, for example, velocity (Nozieres and Lewiner, 1973). Using the relation

$$h(k_\alpha) H_g^{-n} h^\dagger(k_\beta) = \frac{P^2}{3} k^2 \delta_{\alpha,\beta} \left(\frac{2}{(-E_g)^n} + \frac{1}{(-E_g - \Delta)^n} \right)$$
$$+ i \varepsilon_{\alpha,\beta\lambda} k_\alpha k_\beta \sigma_\gamma \frac{P^2}{3} \left(-\frac{1}{(-E_g)^n} + \frac{1}{(-E_g - \Delta)^n} \right), \quad (2.38)$$

yields the effective Hamiltonian

$$H_{\text{eff-3d}} = \frac{\hbar^2 k^2}{2m^*} + \frac{1}{2} g^* \mu_B \boldsymbol{\sigma} \cdot \mathbf{B} + V_c(\mathbf{r}) - \frac{\hbar^2 k^2}{4 m^* E_0} g^* \mu_B \boldsymbol{\sigma} \cdot \mathbf{B}$$
$$+ \lambda^* \boldsymbol{\sigma} \cdot \left(\mathbf{k} \times \nabla V(\mathbf{r}) \right) + \lambda^* g_0 \mu_B (\mathbf{k} \times \boldsymbol{\sigma}) \cdot (\mathbf{k} \times \mathbf{B}), \quad (2.39)$$

with

$$\frac{m}{m^*} = 1 + E_P \frac{(E_g + \frac{2}{3}\Delta)}{E_g (E_g + \Delta)} \propto \frac{1}{E_g}, \quad (2.40)$$

$$g^* = g_0 - \frac{2}{3} \frac{E_P \Delta}{E_g (E_g + \Delta)} \propto \frac{\Delta}{E_g^2}, \quad (2.41)$$

$$\lambda^* = \frac{P^2}{3} \left(\frac{1}{E_g^2} - \frac{1}{(E_g + \Delta)^2} \right) \propto \frac{\Delta}{E_g^3}, \quad (2.42)$$

$$E_0 = \frac{E_g (E_g + \Delta)(3 E_g + 2\Delta)}{9 E_g^2 + 11 E_g \Delta + 4 \Delta^2}, \quad (2.43)$$

where the optical matrix parameter $E_P \equiv 2 m P^2 / \hbar^2$ is of the order of 21–25 eV for most zinc-blend semiconductors. Above we have ignored Darwin type terms, that is, of order k^4 and independent of spin since they do not play a part on the physics discussed below. We note that the Hamiltonian Eq. (2.39) has the same symmetry as the Pauli Hamiltonian Eq. (2.13) but with renormalized parameters. Also note that, as sometimes is assumed, the simple replacement of m by m^* in Eq. (2.13) *does not* yield the correct effective Hamiltonian Eq. (2.39) and that each parameter renormalization is different.

The effective Hamiltonian Eq. (2.39) is the basic starting point of many of the discussions that follow with the external magnetic field set to zero, that is, $\mathbf{B} = 0$. For careful comparison with experiments at finite field, some of these effects have to be taken into account; but for the qualitative understanding of the AHE, SHE, and ACE, these can be ignored.

We should remark a few additional things about Eq. (2.39). The effects of the bands far away from the conduction band, for example the valence band, are quite strong and increase with decreasing gap. The spin-dependent effects in the conduction band are proportional to the spin-splitting present *in the valence bands*. In fact, although we have passed over some subtle point regarding differences between the scalar potentials felt by the different bands (Winkler, 2003), without the presence of a strong spin–orbit coupling within the Bloch valence bands, none of the effects discussed in what follows (both extrinsic and intrinsic) would exist.

3.3.2. Effective mass 2-d electron and hole model Hamiltonians

Confining electron or hole systems to two dimensions through a quantum well heterostructure will restrict the momentum to the plane. In two dimensions, the general form of the effective Hamiltonian is given by:

$$H_{\text{eff-2D}} = E_0(\mathbf{k}) + V_{\text{imp}}(\mathbf{r}) + \mathbf{b}(\mathbf{k}) \cdot \boldsymbol{\sigma} + \lambda^* \boldsymbol{\sigma} \cdot (\mathbf{k} \times \nabla V_{\text{imp}}(\mathbf{r})) \quad (2.44)$$

where $\mathbf{k} = (k_x, k_y, 0)$, λ^* is given by Eq. (2.42), E_0 is the normal effective mass 2-d dispersion in the plane, and $b(\mathbf{k})$ takes different forms, depending on the band structure and the axis of confinement.

Taking, for example, the confining direction to be [001] restricts the values of k_z to quantized values which in the low-doping regime or large confinement we can take to be only one with $\langle k_z \rangle = 0$ and $\langle k_z^2 \rangle \approx (\pi/d)^2$. The Dresselhaus term for conduction electrons confined to 2-d is then reduced to the form

$$H_{\text{D-2d}} = \beta(k_x \sigma_x - k_y \sigma_y), \quad (2.45)$$

with $\beta \approx -\mathcal{B}(\pi/d)^2$. Other terms, cubic in in-plane momentum, are also present in the effective 2-d spin–orbit coupling term but tend to be smaller than the linear one in the limit of small confinement and low density (Winkler, 2003). This of course corresponds to a $\mathbf{b}_D(\mathbf{k}) = \beta(k_x, -k_y, 0)$ in Eq. (2.44).

An additional important spin–orbit coupling term in the effective 2-d Hamiltonian is the Rasbha term which arises in systems with broken structure inversion asymmetry (SIA), for example, triangular confining potential, where $\langle \nabla_z V(\mathbf{r}) \rangle \neq 0$. This corresponds to a $\mathbf{b}_{R-e}(\mathbf{k}) = \lambda_{R-e}(k_y, -k_x, 0)$ (Winkler, 2003):

$$H_{R-e-2d} = \lambda_{R-e}(k_y \sigma_x - k_x \sigma_y) = \lambda_{R-e}(\mathbf{k} \times \hat{\mathbf{z}}) \cdot \boldsymbol{\sigma}. \quad (2.46)$$

Higher order contributions to this intrinsic spin–orbit coupling are smaller in the limit of strong confinement and need to be taken into account for wider quantum well structures. In describing the AHE, an additional exchange field $h\sigma_z$ is added which brakes the time-reversal symmetry, that is, $\mathbf{b}(\mathbf{k}) \neq \mathbf{b}(-\mathbf{k})$.

For a 2d-hole system, one typically replaces, in the expressions above, σ by J_i of the 3/2 angular momentum vector. Although the calculation of their dispersion in the presence of spin–orbit coupling couples all these bands, in the limit of strong confinement and low density, the heavy hole band is sufficiently decoupled to be unaffected by the light-hole band. In this limit, the allowed Rashba term, by symmetry, is cubic in momentum rather than linear:

$$H_{R-\text{hole}-2d} = i\lambda_{R-h}(k_-^3 \sigma_+ - k_+^3 \sigma_-). \quad (2.47)$$

Having now derived the necessary effective Hamiltonians which incorporate the spin–orbit coupling from the other nearby bands, we move on to discuss the three transport phenomena which feature spin–orbit coupling at their very origin. It is important, however, to always keep in mind the origin of the effective Hamiltonians to be used and their level of validity, for example, impurity potentials should be slow varying in the atomic length scale.

4. SHE AND AHE

Anomalous Hall Effect: In 1879, Edwin Hall ran a current through a gold foil and discovered that a transverse voltage was induced when the film was exposed to a perpendicular magnetic field. The ratio of this Hall voltage to the current is the Hall resistivity. For paramagnetic materials, the Hall resistivity is proportional to the applied magnetic field and Hall measurements give information about the concentration of free carriers and determine whether they are holes or electrons. Magnetic films exhibit both this ordinary Hall response and an extraordinary or AHE response that does not disappear at zero magnetic fields and which is proportional to the internal magnetization: $R_{\text{Hall}} = R_o H + R_s M$, where R_{Hall} is the Hall resistance, R_o and R_s are the ordinary and anomalous Hall coefficients, M is the magnetization perpendicular to the film, and H is the applied magnetic field. The AHE is the consequence of spin–orbit coupling in the system and allows an indirect measurement of the internal magnetization; the role of the broken time-reversal symmetry is to create an asymmetric spin-population which leads to a charge Hall voltage due to the asymmetric deflection of the majority and minority spins in the direction perpendicular to the current.

Spin Hall Effect: In systems with time-reversal symmetry and in the presence of spin–orbit coupling, in principle, the mechanisms that give rise to the AHE are also present but are not directly measurable through a charge voltage perpendicular to the flow of the current in the sample since there is no exchange field that induces the population asymmetry between spin up and down states. However, consistent with the global time-reversal symmetry, a spin-current will be generated perpendicular to an applied charge current which in turn will lead to spin accumulations with opposite magnetization at the edges. This is the SHE first predicted over three decades ago by invoking the phenomenology of the AHE in ferromagnets (Dyakonov and Perel, 1971; Hirsch, 1999).

Based on studies of intrinsic AHE (see below) in strongly spin–orbit coupled ferromagnetic materials, where scattering contributions do not seem to dominate, the possibility of an intrinsic SHE was put forward (Murakami *et al.*, 2003; Sinova *et al.*, 2004a), with scattering playing a minor role. This proposal has generated an extensive theoretical debate motivated by its potential as a spin injection tool in low dissipative devices. The interest in the SHE has also been dramatically enhanced by experiments by two groups reporting the first observations of the SHE in n-doped semiconductors (Kato *et al.*, 2004; Sih *et al.*, 2005) and in 2D hole gases (2DHG) (Wunderlich *et al.*, 2005). These observations based on optical experiments have been followed by metal-based experiments in which both the SHE and the inverse SHE have been observed (Kimura *et al.*, 2007; Saitoh *et al.*, 2006; Valenzuela and Tinkham, 2006) and by other experiments in semiconductors at other regimes, even at room temperature (Liu *et al.*, 2006; Stern *et al.*, 2006, 2007; Weber *et al.*, 2007).

Although the basic microscopic origins of the effects are still very actively being studied, there is a general agreement qualitatively on the three different mechanisms identified from the semiclassical treatment of transport theory that contribute to the AHE and SHE:

(a) *Intrinsic deflection*: The intrinsic mechanism is based solely on the topological properties of the Bloch states originating from the spin–orbit coupled electronic structure as first suggested by Karplus and Luttinger (1954). The spin–orbit coupling gives rise to an anomalous group velocity perpendicular to the applied electric field. This mechanism gives an anomalous Hall coefficient, R_s, proportional to the square of the ordinary resistivity, since the intrinsic AHE itself is insensitive to impurities.

(b) *Skew scattering*: The skew-scattering mechanism, as first proposed by Smit, relies on the asymmetric scattering of the electrons by impurities present in the material (Smit, 1955). Within this contribution, we can distinguished between (i) the skew scattering produced by the spin–orbit coupled disorder potential proportional to the strength of the effective spin–orbit coupling λ^* and (ii) the one produced by the scattering of spin–orbit coupled Bloch states from the pure disorder potential. As we will see below the later typically dominates. Not surprisingly, this skew-scattering contribution to R_s is sensitive to the type and range of the scattering potential and, in contrast to the intrinsic mechanism, it scales linearly with the diagonal resistivity.

(c) *Side-jump scattering*: The presence of impurities also leads to a side-step type of scattering that contributes to a net current perpendicular to the initial momentum. This is the so-called side-jump contribution, whose semiclassical interpretation was pointed out by Berger (1970). The incoming and outgoing trajectories of the quasiparticle upon scattering from an impurity are displaced in a side-step but without associated directional deflection.

These mechanisms are illustrated in Fig. 2.3. The correct accounting of such contributions are however nontrivial in the semiclassical procedure, which can lead to easy mistakes of several factors, hence the need (see below) to connect the microscopic and semiclassical approaches to the problem.

Much of the debate generated in the SHE has been inherited by long-standing debates in the theory of the AHE. The early theories of the AHE involved long and complex calculations, with results that were not easy to interpret, and often contradicting each other (Nozieres and Lewiner, 1973). The adversity facing these theories stems from the origin of the AHE: It appears due to the interband coherence and not just due to simple changes in the occupation of Bloch states, as was recognized in the early works of Kohn and Luttinger (1957) and Luttinger (1958). Besides this inherited discussions, the SHE has some issues of debate of its own not present in the AHE, such as the connection between the spin-current generated and the spin-accumulation induced in a system where the spin is not conserved and therefore the spin-current in a strongly spin–orbit coupled system is not determined uniquely by a continuity equation. Nowadays, most treatments of the AHE and the SHE either use the semiclassical Boltzmann transport theory, the diagrammatic approach based on the Kubo–Streda linear response formalism, or the nonequilibrium Green's function approach. The equivalence of the Kubo and semiclassical methods can be shown analytically for the two-dimensional Dirac-band graphene (Sinitsyn *et al.*, 2007), by explicitly identifying various diagrams of the more systematic Kubo–Streda treatment with the more physically

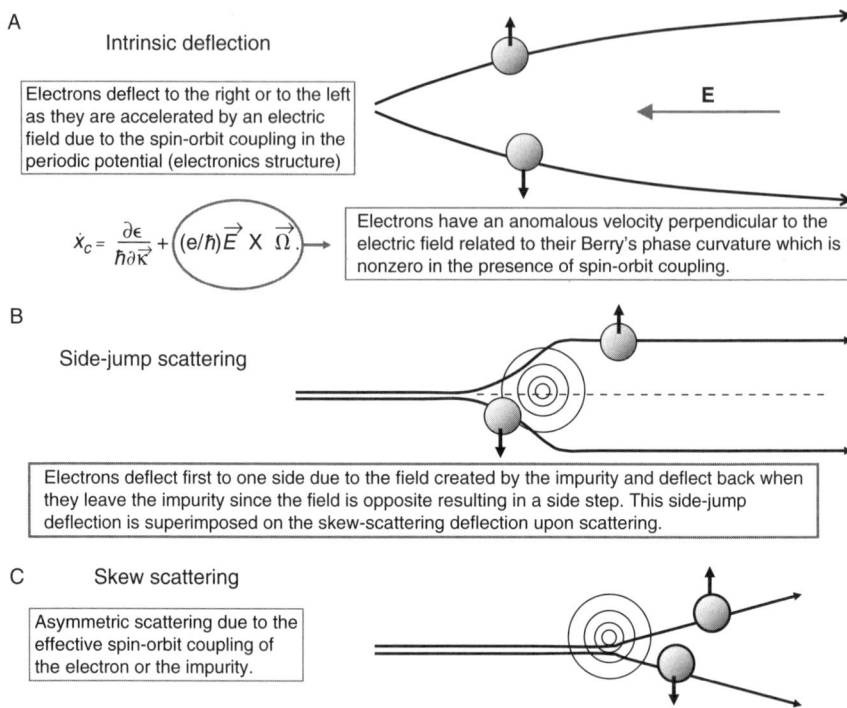

FIGURE 2.3 Illustration of the three mechanisms that give origin to the AHE and SHE.

transparent terms of the semiclassical Boltzmann approach. This identification of diagrams is believed to be correct in general although it has not been shown analytically for higher dimensional systems.

Since the semiclassical formalism is the more physically clear approach to linear transport, we follow it to illustrate the detailed origin of the mechanism mentioned above which give origin to the AHE and SHE. To do so, we focus our attention to the 2D effective Hamiltonian in Eq. (2.44) for simplicity.

4.1. The semiclassical Boltzmann equation approach to anomalous Hall transport

There is a substantial literature (Jungwirth *et al.*, 2002; Sinitsyn *et al.*, 2005, 2006b; Smit, 1955) on the application of the Boltzmann equation concepts to anomalous transport theory. However, stress was usually placed only on one of the many possible mechanisms for a Hall current. In this section we briefly review this approach, incorporating some new insights from recent work (Sinitsyn *et al.*, 2006b). The validity and justification of this approach has been shown by careful comparison with systematic microscopic approaches.

The semiclassical Boltzmann equation (SBE) describes the evolution of the electron distribution function as though electrons were classical particles labeled by a band index and living in the crystal's momentum space. A rigorous treatment of the electronic state evolution in the presence of disorder should generally consider the entire density matrix, including off-diagonal elements. Formal justifications of the Boltzmann description are usually made in terms of the properties of wave packets that have well-defined average momentum, center of mass coordinate, spin, intrinsic angular momentum, etc. The distribution functions then acquire the meaning of the probability distribution of wave packets that behave in many respects like classical particles (Marder, 2000).

The semiclassical distribution function evolves both due to hydrodynamical particle fluxes and due to collisions with impurities. Quantum mechanically, the elementary scattering process is described by the scattering matrix from one state to another in the presence of a perturbing potential. In the semiclassical approach, scattering can be accompanied by changes of coordinate. As we discuss below, these have to be taken into account when constructing a kinetic equation for the semiclassical distribution function that captures the anomalous transport. These effectively arise from the transformation of the position operator r to the r_{eff} within the Löwdin transformation in the effective 2-band model.

Below we present the construction of the Boltzmann theory from the golden rule used within the collision term which gives rise later to the skew scattering, to the equation of motion of the occupation function, and to the effective velocity, which includes the side-jump step mechanism. All of this is incorporated at the end, focusing for simplicity in the AHE although it translates to the SHE treatment, and the different terms are identified clearly from this semiclassical considerations. The identification with the microscopic diagrammatic approaches that was done in several specific model Hamiltonians in 2-d is also shown.

4.1.1. The golden rule

The golden rule connects the classical and quantum descriptions of a scattering event. It shows how the scattering rate $\omega_{l'l}$ between states with different quantum numbers l and l' due to an impurity potential is related to the so-called T-matrix elements:

$$\omega_{l'l} = 2\pi |T_{l'l}|^2 \, \delta(\varepsilon_{l'} - \varepsilon_l). \quad (2.48)$$

The scattering T-matrix is defined as $T_{l'l} = \langle l' | \hat{V} | \psi_l \rangle$, where $\hat{V}$ is the impurity potential operator and $| \psi_l \rangle$ is the eigenstate of the full Hamiltonian $\hat{H} = \hat{H}_0 + \hat{V}$ that satisfies the equation

$$|\psi_l\rangle = |l\rangle + \frac{\hat{V}}{\varepsilon_l - \hat{H}_0 + i\eta}|\psi_l\rangle. \quad (2.49)$$

For weak disorder, one can approximate the scattering state $|\psi_l\rangle$ by a truncated series in powers of $V_{ll'} = \langle l|\hat{V}|l'\rangle$ and take $l = (\mathbf{k}, \mu)$ as the combined (momentum, band) index of the eigenstate $|l\rangle = |\mathbf{k}, \mu\rangle$ of the unperturbed Hamiltonian $\hat{H}_0$:

$$|\psi_l\rangle \approx |l\rangle + \sum_{l''} \frac{V_{l''l}}{\varepsilon_l - \varepsilon_{l''} + i\eta}|l''\rangle + \ldots \quad (2.50)$$

From this, one can find the scattering rate up to the third order in the disorder strength $\omega_{ll'} = \omega_{ll'}^{(2)} + \omega_{ll'}^{(3)} + \ldots$, where $\omega_{ll'}^{(2)} = 2\pi\langle|V_{ll'}|^2\rangle_{\text{dis}}$ $\delta(\varepsilon_l - \varepsilon_{l'})$ and

$$\omega_{ll'}^{(3)} = 2\pi\left(\sum_{l''} \frac{\langle V_{ll'}V_{l'l''}V_{l''l}\rangle_{\text{dis}}}{\varepsilon_l - \varepsilon_{l''} - i\eta} + c.c.\right)\delta(\varepsilon_l - \varepsilon_{l'}). \quad (2.51)$$

The skew-scattering contribution to the Hall effect follows from the asymmetric part of the scattering rate (Smit, 1955): $\omega_{l'l}^{(a)} \equiv (\omega_{ll'} - \omega_{l'l})/2$. Since $\omega_{l'l}^{(2)}$ is symmetric, the leading contribution to $\omega_{l'l}^{(a)}$ appears at order V^3. The $\omega_{ll'}^{(3)}$ contribution to the scattering rate $\omega_{ll'}$ contains symmetric and antisymmetric parts. The symmetric part is not essential since it renormalizes only the second-order result of $\omega_{ll'}^{(2)}$. The sum in parentheses in Eq. (2.51) can be rewritten in the form

$$P\left(\sum_{l''} \frac{2\text{Re} - \langle V_{ll'}V_{l'l''}V_{l''l}\rangle_{\text{dis}}}{\varepsilon_l - \varepsilon_{l''}}\right) - 2\pi\sum_{l''}\delta(\varepsilon_l - \varepsilon_{l''})\text{Im}\langle V_{ll'}V_{l'l''}V_{l''l}\rangle_{\text{dis}}$$

$$(2.52)$$

The first term is symmetric under the exchange of indexes $l \leftrightarrow l'$ [note that this needs the fact that $\varepsilon_l = \varepsilon_{l'}$ due to the delta-function in Eq. (2.51)]; hence this sum does not contribute to the asymmetric part of the scattering amplitude and one can concentrate on the second one:

$$\omega_{ll'}^{(3a)} = -(2\pi)^2 \sum_{l''} \delta(\varepsilon_l - \varepsilon_{l''})\text{Im}\langle V_{ll'}V_{l'l''}V_{l''l}\rangle_{\text{dis}}\delta(\varepsilon_l - \varepsilon_{l'}), \quad (2.53)$$

with the superscript $3a$ meaning that this is the antisymmetric part of the scattering rate calculated at order V^3, a similar $\omega_{ll'}^{(4a)}$ term can be obtained at order V^4.

4.1.2. The coordinate shift: Origin of the side-jump mechanism

In a semiclassical description of wavepacket motion in a crystal, scattering produces both a change in the direction of motion and a separate coordinate shift (Adams and Blount, 1959; Berger, 1970; Nozieres and Lewiner, 1973; Sinitsyn et al., 2005, 2006b). In the lowest order Born approximation for a scalar disorder potential, one can derive an expression for the coordinate shift which accompanies scattering between 2-band states. The shift does not depend explicitly on the type of impurity and can be expressed in terms of initial and final states only (Sinitsyn et al., 2005, 2006b): for spin-independent scatterers:

$$\delta \mathbf{r}_{l'l} = \langle u_{l'} | i \frac{\partial}{\partial \mathbf{k}'} u_{l'} \rangle - \langle u_l | i \frac{\partial}{\partial \mathbf{k}} u_l \rangle - \hat{\mathbf{D}}_{\mathbf{k}',\mathbf{k}} \arg[\langle u_{l'} | u_l \rangle]. \quad (2.54)$$

where $\arg[a]$ is the phase of the complex number a and

$$\hat{\mathbf{D}}_{\mathbf{k}',\mathbf{k}} = \frac{\partial}{\partial \mathbf{k}'} + \frac{\partial}{\partial \mathbf{k}}$$

The topological interpretation of Eq. (2.54) has been explained in Sinitsyn et al. (2006b). The first two terms in Eq. (2.54) have been known for a long time. The last term was first derived only recently in Sinitsyn et al. (2006b) but is an essential contribution which makes the expression for the coordinate shift gauge invariant.

4.1.3. Kinetic equation of the semiclassical Boltzmann distribution

Equations. (2.48) and (2.54) contain the quantum mechanical information necessary to write down a SBE that takes into account both the change of momentum and the coordinate shift during scattering in a homogeneous crystal in the presence of a driving electric field $\mathbf{E}$. Keeping only terms up to the linear order in the electric field, the Boltzmann equation reads (Sinitsyn et al., 2006b):

$$\frac{\partial f_l}{\partial t} + e\mathbf{E} \cdot \mathbf{v}_{0l} \frac{\partial f_0(\varepsilon_l)}{\partial \varepsilon_l} = -\sum_{l'} \omega_{ll'} \left[f_l - f_{l'} - \frac{\partial f_0(\varepsilon_l)}{\partial \varepsilon_l} e\mathbf{E} \cdot \delta \mathbf{r}_{l'l} \right], \quad (2.55)$$

where $\mathbf{v}_{0l}$ is the usual velocity $\mathbf{v}_{0l} = \partial \varepsilon_l / \partial \mathbf{k}$. Note that if only elastic scatterings with static impurities are responsible for the collision term, the RHS. of Eq. (2.55) is linear in f_l and not in $f_l(1 - f_{l'})$. This property follows from the fact that Heisenberg operator time evolution equations for creation or annihilation operators in a noninteracting electron system are linear and can be mapped to the Schrödinger equation for amplitudes in a single particle system (Sinitsyn et al., 2006a). (For further discussion of this point, see Appendix B in Luttinger and Kohn, 1955.) This Boltzmann equation has the standard form except for the coordinate shift effect, which is taken into account in the last term in the collision integral on the RHS of Eq. (2.55). This term appears because the kinetic energy of a particle in the state l' before scattering into the state l is smaller than ε_l by the amount $e\mathbf{E}\cdot\delta\mathbf{r}_{ll'} = -e\mathbf{E}\cdot\delta\mathbf{r}_{l'l}$. The collision term does not vanish in the presence of an electric field when the occupation probabilities f_l are replaced by their thermal equilibrium values because $f_0(\varepsilon_l) - f_0(\varepsilon_l - e\mathbf{E}\cdot\delta\mathbf{r}_{ll'}) \approx -[\partial f_0(\varepsilon_l)/\partial\varepsilon_l]$ $e\mathbf{E}\cdot\delta\mathbf{r}_{l'l} \neq 0$. The last term in the collision integral in Eq. (2.55) (Sinitsyn et al., 2005) accounts for this interplay between coordinate shifts and spatial variation of local chemical potential in the presence of an electric field.

The total distribution function f_l in the steady state ($\partial f_l/\partial t = 0$) can be written as the sum of the equilibrium distribution $f_0(\varepsilon_l)$ and nonequilibrium contributions,

$$f_l = f_0(\varepsilon_l) + g_l + g_l^{\mathrm{adist}}, \tag{2.56}$$

where we have split the nonequilibrium contribution into two terms g_l and g_l^{adist} which solve independent self-consistent time-independent equations (Sinitsyn et al., 2006b):

$$e\mathbf{E}\cdot\mathbf{v}_{0l}\frac{\partial f_0(\varepsilon_l)}{\partial\varepsilon_l} = -\sum_{l'}\omega_{ll'}(g_l - g_{l'}) \tag{2.57}$$

and

$$\sum_{l'}\omega_{ll'}\left(g_l^{\mathrm{adist}} - g_{l'}^{\mathrm{adist}} + \frac{-\partial f_0(\varepsilon_l)}{\partial\varepsilon_l}e\mathbf{E}\cdot\delta\mathbf{r}_{l'l}\right) = 0. \tag{2.58}$$

(The label $adist$ stands for anomalous distribution.) A standard approach to solving these equations in 2D is to look for the solution of Eq. (2.57) in the form:

$$g_l = \left(-\frac{df_0(\varepsilon_l)}{d\varepsilon_l}\right)e\mathbf{E}\cdot(A_\mu\mathbf{v}_{0l} + B_\mu\mathbf{v}_{0l}\times\hat{\mathbf{z}}), \tag{2.59}$$

where $\hat{\mathbf{z}}$ is the unit vector in the out-of-plane direction. We will assume that the transverse conductivity is much smaller than the longitudinal one so that $A_\mu \gg B_\mu$. One then finds by the direct substitution of Eq. (2.59) into Eq. (2.55) that

$$A_\mu = \tau_\mu^{tr}, \quad B_\mu = (\tau_\mu^{tr})^2/\tau_\mu^\perp \tag{2.60}$$

where

$$1/\tau_\mu^{tr} = \sum_{\mu'}\int\frac{d^2\mathbf{k}'}{(2\pi)^2}\omega_{ll'}\left(1 - \frac{|v_{l'}|}{|v_l|}\cos(\phi-\phi')\right)$$

$$1/\tau_\mu^\perp = \sum_{\mu'}\int\frac{d^2\mathbf{k}'}{(2\pi)^2}\omega_{ll'}\frac{|v_{l'}|}{|v_l|}\sin(\phi-\phi') \tag{2.61}$$

and ϕ and ϕ' are the angles between $\mathbf{v}_{0l}$ or $\mathbf{v}_{0l'}$ and the x-axes. This completes the solution of Eq. (2.57) for g_l. We will show, how a similar *ansatz* can also solve Eq. (2.58) for the anomalous distribution g_l^{adist}.

4.1.4. Anomalous velocities and the AHE

To find the current induced (both charge and spin) by an electric field and hence the conductivity, we need to derive an appropriate expression for the velocity of semiclassical particles, in addition to solving the SBE for the state occupation probabilities. In considering the AHE, in addition to the band state group velocity $\mathbf{v}_{0l} = \partial \varepsilon_l / \partial \mathbf{k}$, one should also take into account velocity renormalizations due to the accumulations of coordinate shifts after many scattering events (the side-jump effect) and due to band mixing by the electric field (the anomalous velocity effect) (Nozieres and Lewiner, 1973; Sinitsyn et al., 2006b):

$$\mathbf{v}_l = \frac{\partial \varepsilon_l}{\partial \mathbf{k}} + \mathbf{\Omega}_l \times e\mathbf{E} + \sum_{l'} \omega_{l'l} \delta \mathbf{r}_{l'l}. \tag{2.62}$$

The second term in Eq. (2.62) captures changes in the speed at which a wave packet moves between scattering events under the influence of the external electric field and $\mathbf{\Omega}_l$ is the Berry curvature of the band (Sundaram and Niu, 1999)

$$\Omega_l(\mathbf{k}) = \varepsilon_{ijk} \text{Im} \left[\left\langle \frac{\partial u_l}{\partial k_j} \middle| \frac{\partial u_l}{\partial k_i} \right\rangle \right]. \tag{2.63}$$

The last term in Eq. (2.62) is due to the accumulation of coordinate shifts after many scatterings. This addition is justified by comparison with microscopic calculations since, by itself, the usual Boltzmann theory of transport ignores all events that occur in a short length from the scattering centers. Combining Eqs. (2.56) and (2.62), we obtain the total current $\mathbf{j} = e \sum_l f_l \mathbf{v}_l$.

4.1.5. Mechanisms of the AHE and SHE

Now we are ready to put all the above together. From the relation $\mathbf{j} = e \sum_l f_l \mathbf{v}_l$, we can extract the off-diagonal Hall conductivity which is naturally written as the sum of four contributions:

$$\sigma_{xy}^{\text{total}} = \sigma_{xy}^{\text{int}} + \sigma_{xy}^{\text{adist}} + \sigma_{xy}^{sj} + \sigma_{xy}^{sk}. \tag{2.64}$$

The first term is the so-called intrinsic contribution

$$\sigma_{xy}^{\text{int}} = -e^2 \sum_l f_0(\varepsilon_l) F_l, \tag{2.65}$$

which is due to the anomalous velocity of free electrons under the action of the electric field $\mathbf{v}_l^{(a)} = \mathbf{\Omega}_l(\mathbf{k}) \times e\mathbf{E}$. This term is called the intrinsic contribution because it is not related to impurity scattering, that is, its evaluation does not require knowledge of the disorder present in the system. The intrinsic AHE is completely determined by the topology of the Bloch band.

The next two contributions follow from coordinate shifts during scattering events:

$$\sigma_{xy}^{\text{adist}} = e \sum_l (g_l^{\text{adist}}/E_y)(v_{0l})_x \tag{2.66}$$

is the current due to the anomalous correction to the distribution function, while the direct side-jump contribution

$$\sigma_{xy}^{\text{sj}} = e \sum_l (g_l/E_y) \sum_{l'} \omega_{l'l}(\delta \mathbf{r}_{l'l})_x \tag{2.67}$$

is the current due to the side-jump velocity, that is, due to the accumulation of coordinate shifts after many scattering events. Since coordinate shifts are responsible both for $\sigma_{xy}^{\text{adist}}$ and for σ_{xy}^{sj} there is, unsurprisingly, an intimate relationship between those two contributions. Their equality can be demonstrated evaluating $\sigma_{xy}^{\text{adist}}$ for specific models. Because of this relationship, in most of the literature, $\sigma_{xy}^{\text{adist}}$ is usually considered to be part of the side-jump contribution, that is, $\sigma_{xy}^{\text{adist}} + \sigma_{xy}^{\text{sj}} \to \sigma_{xy}^{\text{sj-total}}$. However, they can be distinguished when connecting the SBE approach to the Kubo–Streda formalism, which is what ultimately justifies their appearance in the semiclassical theory.

Finally, σ_{xy}^{sk} is the skew-scattering contribution

$$\sigma_{xy}^{\text{sk}} = e \sum_l (g_l/E_y)(v_{0l})_x. \tag{2.68}$$

The skew-scattering contribution is independent of the coordinate shift and of the anomalous velocity. It is nonzero for an isotropic energy dispersion ε_l only when the scattering rate $\omega_{ll'}$ is asymmetric. Writing the transverse velocity in the form $(v_{0l})_x = |v_{0l}|\cos(\chi)$, then using Eq. (2.61) and substituting Eq. (2.59) for g_l into Eq. (2.68), one finds

$$\begin{aligned}\sigma_{xy}^{\text{sk}} &= e \sum_\mu \int \frac{d^2\mathbf{k}}{(2\pi)^2} (gl/E_y)|v_{0l}|\cos(\phi) \\ &= -e^2 \sum_\mu \frac{(\tau_\mu^{tr})^2}{\tau_\mu^\perp} \int \frac{d^2\mathbf{k}}{(2\pi)^2} \left(-\frac{df_0(\varepsilon_l)}{d\varepsilon_l}\right)|v_l|^2\cos^2(\phi).\end{aligned} \tag{2.69}$$

At zero temperature, this expression simplifies to

$$\sigma_{xy}^{sk} = -\sum_{\mu} \frac{e^2(\tau_\mu^{tr})^2 v_{F_\mu} k_{F_\mu}}{4\pi \tau_\mu^{\perp}} \quad (2.70)$$

There is potentially one more contribution which has not been included in Eq. (2.64) and which involves the product of g^{adist} and the side-jump velocity $\sum_{l'} \omega_{l'l} \delta \mathbf{r}_{l'l}$. Such a term does not contribute to the Hall conductivity in the present model calculation of isotropic scattering and band structure. It may however, in principle, give rise to a nonzero contribution given an appropriate anisotropic scattering and anisotropic Fermi distribution.

4.2. AHE in two dimensions

One of the simplest models where to illustrate the above theory is in the 2D Dirac Hamiltonian given by

$$\hat{H}_{\text{Dirac}} = v(k_x \sigma_x + k_y \sigma_y) + \Delta \sigma_z + V_{\text{imp}}(\mathbf{r}). \quad (2.71)$$

where σ_x and σ_y are Pauli matrices. This spin–orbit coupled model Hamiltonian breaks time-reversal symmetry and therefore has a nonzero anomalous Hall conductivity. The result of the semiclassical calculations outlined above for this model can be carried analytically due to the simplicity of its dispersion and yield the same result as Kubo formula calculations performed in the self-consistent non-crossing approximation with the following result (Sinitsyn et al., 2006a):

$$\sigma_{xy} = -\frac{e^2 \Delta}{4\pi \sqrt{(vk_F)^2 + \Delta^2}} \left[1 + \frac{4(vk_F)^2}{4\Delta^2 + (vk_F)^2} + \frac{3(vk_F)^4}{\left(4\Delta^2 + (vk_F)^2\right)^2} \right]$$
$$- \frac{e^2 V_1^3}{2\pi n V_0^4} \frac{(vk_F)^4 \Delta}{\left(4\Delta^2 + (vk_F)^2\right)^2}, \quad (2.72)$$

where k_F is the Fermi momentum and the parameters V_0 and V_1 characterizing the disorder distribution.

The various different contributions that have been identified in the semiclassical approach, can, for the particular case of two-dimensional Dirac bands, all be directly linked to particular Feynman-diagram subsums in the Kubo–Streda formalism. The correspondence is summarized in Fig. 2.4, in which the diagrammatic expansion of the Kubo–Streda formalism in the chiral (band-eigenstate) basis is summarized. In Fig. 2.4, Feynman diagrams are grouped in the three major contributions

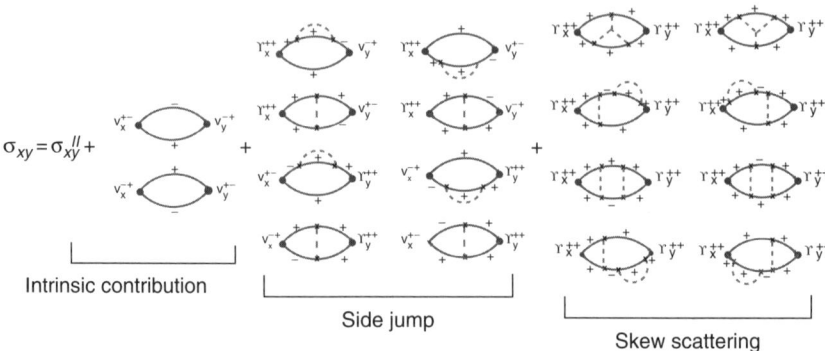

FIGURE 2.4 Graphical representation of the AHE conductivity within the clean band eigenstate representation. The subset of diagrams that correspond to specific terms in the semiclassical Boltzmann formalism is indicated.

that arise naturally in the semiclassical approach: intrinsic, side-jump, and skew scattering.

Another simple two-dimensional Hamiltonian is the 2-d electron gas with Rashba spin–orbit interaction as mentioned in Section 3.

$$H_{R-2d} = \frac{k^2}{2m}\sigma_0 + \alpha(\sigma_x k_y - \sigma_y, k_x) - h\sigma_z + V(\mathbf{r})\sigma_0 \qquad (2.73)$$

where m is the effective in-plane mass of the quasiparticles, α the spin–orbit coupling parameter, h the exchange field, and σ_i the 2 × 2 Pauli matrices. The eigenenergies of the clean system are

$$E_\pm = \frac{k^2}{2m} \pm \lambda_k \quad \text{with} \quad \lambda_k = \sqrt{h^2 + \alpha^2 k^2}. \qquad (2.74)$$

Although the combination of the parabolic dispersion and the linear spin–orbit coupling, as has been the case in the SHE, gives rise to the cancellation of the side-jump and intrinsic contribution to the AHE conductivity when both subbands are occupied and in the presence of simple delta-impurity scattering, when only one subband is occupied, the result is finite and in the small disorder limit and is the same where the exchange field is small $h \ll \alpha k_F$, but the spin–orbit interaction is now larger than the Fermi energy $\alpha k_F \gg \varepsilon_F$; the intrinsic and side-jump contributions are (Borunda et al., 2007; Nunner et al., 2007):

$$\sigma_{yx}^{AHE-R} = \frac{e^2}{2\pi}\frac{2h\varepsilon_F^3}{(\alpha k_F)^4} + \frac{e^2}{4\pi}\left(1 - \frac{2h\varepsilon_F}{(\alpha k_F)^2}\right) + \frac{e^2}{2\pi}\frac{V_1^3}{V_0^3}\frac{h}{n_i V_0}. \qquad (2.75)$$

The dependence on density, through ε_F, and Rashba spin–orbit coupling strength should provide an important test of the theory if the single occupation band regime can be reached in these materials.

4.3. Estimates of the role of various spin–orbit couplings contributing to skew scattering

In computing the above AHE response, we have ignored the extrinsic spin–orbit coupling term, which in principle contributes both to the skew and side-jump mechanisms, and have focused instead on the combination of the normal disorder and the intrinsic spin–orbit coupling. It can be shown that in the systems with intermediate gap strength, that is, GaAs, the former contribution dominates. The Rashba coupling in most experimental situations is of the order of $\alpha \sim 5 \times 10^{-11}$ eVm. The λ^* spin–orbit coupling parameter for GaAs is 5.3 Å^2 and for InAs is 120 Å^2, and we can therefore take as an upper estimate of $\lambda^* \sim 50$ Å^2, that is, an order of magnitude larger than in the mid-gap-based systems that we consider.

The next question is how to compare effects of the two contributions to the AHE? Since spin–orbit coupling cannot be compared directly, as shown by Sinitsyn et al. (2005), in the skew scattering induced by the topology of the Bloch band as we consider here, the Berry curvature plays the same role as parameter λ^* for extrinsic skew scattering. Note that it has also the same dimensionality.

Thus to compare the two effects, we should compare Berry curvature of the Rashba Hamiltonian near the Fermi surface with the parameter λ^*. The Berry curvature in this case is strongly momentum dependent:

$$\Omega - \frac{\alpha^2 \Delta}{2[(\alpha k_F)^2 + \Delta^2]^{3/2}} \quad (2.76)$$

This can be particularly large when all couplings (spin–orbit splitting, Zeeman and Fermi energy) are of the same order. In this parameter range, the Berry curvature is of the order of $1/k_{SO}^2$ where $k_{SO}=2m\alpha$.

This yields an estimate $k_{SO} = \approx 10^6\ m^{-1}$ which means that close to resonance, the Berry curvature would be about $\Omega_{max} \sim 10^7$ Å^2. This is many orders of magnitude larger than λ. This, however, would only be the peak value and would occur in a small interval of parameters. The more realistic situation is when $\varepsilon_F \sim \Delta \gg (\alpha k_F)$. In this case, the Berry curvature behaves as $F \sim \Omega_{max} (\alpha k_{SO}/\varepsilon_F)^2$.

So if we assume low end value of $(\alpha k_{SO}/\varepsilon_F) \sim 10^{-2}$ one finds that $\lambda^*/\Omega \sim 0.1$, so even then the Berry curvature effect dominates. For the case that we considered above, where λ^* is smaller (e.g., GaAs) by an order of magnitude, ignoring the extrinsic term is well justified.

4.4. Intrinsic SHE: A heuristic picture

The intrinsic contribution to the SHE can be understood clearly in a heuristic explanation presented by Sinova et al. (2004a). Although within the 2-d Rashba model, where this argument was first presented, the close connection between the spin-current generated and the spin-dynamics due to the linearity of the spin–orbit coupling in momentum forces the extrinsic contribution to cancel this intrinsic contribution (Inoue et al., 2004; Mishchenko et al., 2004), it still remains a valid picture for what is occurring physically and carries over to other systems with strong spin–orbit coupling where the side-jump contributions are known to vanish for the SHE.

The basic physics of the intrinsic SHE is illustrated schematically in Fig. 2.5. In a translationally invariant 2DES, electronic eigenstates have definite momentum and, because of spin–orbit coupling, a momentum-dependent effective magnetic field that causes the spins to align perpendicular to the momenta, as illustrated in Fig. 2.5A. In the presence of an electric field, taken to be in the $\hat{x}$ direction and indicated by blue arrows in

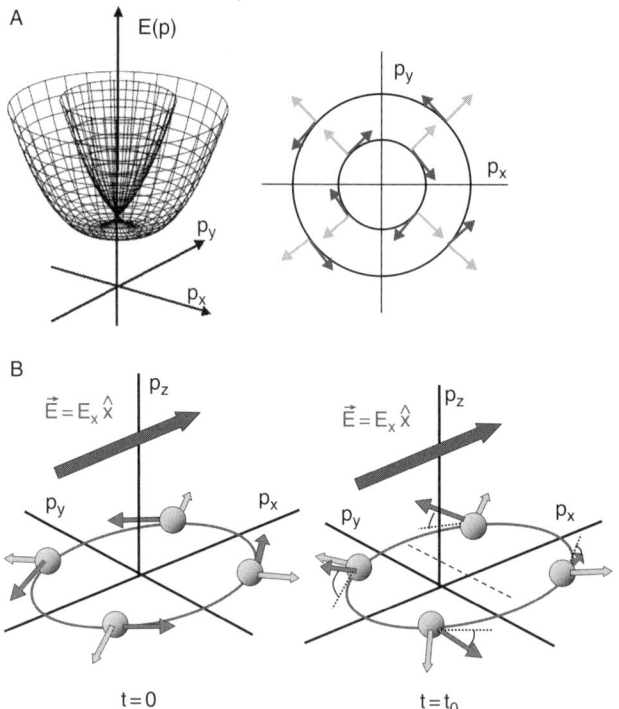

FIGURE 2.5 (A) The 2D electronic eigenstates in a Rashba spin–orbit coupled system are labeled by momentum (green arrows). For each momentum, the two eigenspinors point in the azimuthal direction. (B) In the presence of an electric field, the Fermi surface is displaced an amount $|eE_x t_0/\hbar|$ at time t_0 (shorter than typical scattering times). While moving in momentum space, electrons experience an effective torque which tilts the spins up for $p_y > 0$ and down for $p_y < 0$, creating a spin-current in the y-direction (Sinova et al., 2004a). (See Color Plate.)

Fig. 2.5B, electrons are accelerated and drift through momentum space at the rate $\dot{\vec{p}} = -eE\hat{x}$. The intrinsic SHE arises from the time dependence of the effective magnetic field experienced by the spin because of its motion in momentum space. For the Rashba Hamiltonian case of interest here, the effect can be understood most simply by considering the Bloch equation of a spin-1/2 particle. More generally, the effect arises from nonresonant interband contributions to the Kubo formula expression for the spin Hall conductivity that survive in the static limit.

The dynamics of an electron spin in the presence of time-dependent Zeeman coupling is described by the Bloch equation:

$$\frac{\hbar d\hat{n}}{dt} = \hat{n} \times \vec{\Delta}(t) + \alpha \frac{\hbar d\hat{n}}{dt} \times \hat{n}, \qquad (2.77)$$

where $\hat{n}$ is direction of the spin and α is a damping parameter that we assume is small. For the application, we have in mind the $\vec{p}$-dependent Zeeman coupling term in the spin-Hamiltonian is $-\vec{s} \cdot \vec{\Delta}/\hbar$, where $\vec{\Delta} = 2\lambda/\hbar(\hat{z} \times \vec{p})$. For a Rashba effective magnetic field with magnitude Δ_1 that initially points in the $\hat{x}_1$ direction then tilts (arbitrarily slowly) slightly toward $\hat{x}_2$, where $\hat{x}_1$ and $\hat{x}_2$ are orthogonal in-plane directions, it follows from the linear response limit of Eq.(2.77) that

$$\begin{aligned}\frac{\hbar d n_2}{dt} &= n_z \Delta_1 + \alpha d n_z / dt, \\ \frac{\hbar d n_z}{dt} &= -\Delta_1 n_2 - \alpha d n_2 / dt + \Delta_2,\end{aligned} \qquad (2.78)$$

where $\Delta_2 = \vec{\Delta} \cdot \hat{x}_2$. Solving these inhomogeneous coupled equations using a Greens function technique, it follows that to leading order in the slow-time dependences $n_2(t) = \Delta_2(t)/\Delta_1$, that is, the $\hat{x}_2$-component of the spin rotates to follow the direction of the field, and that

$$n_z(t) = \frac{1}{\Delta_1^2} \frac{\hbar d \Delta_2}{dt}. \qquad (2.79)$$

The intrinsic SHE follows from Eq. (2.79). When a Bloch electron moves through momentum space, its spin orientation changes to follow the momentum-dependent effective field and also acquires a momentum-dependent $\hat{z}$-component.

For a given momentum $\vec{p}$, the spinor originally points in the azimuthal direction. An electric field in the $\hat{x}$ direction ($\dot{p}_x = -eE_x$) changes the y-component of the $\vec{p}$-dependent effective field. Applying the adiabatic spin-dynamics expressions explained above, identifying the azimuthal direction in momentum space with $\hat{x}_1$ and the radial direction with $\hat{x}_2$, we find that the z-component of the spin direction for an electron in a state with momentum $\vec{p}$ is

$$n_{z,\vec{p}} = \frac{-e\hbar^2 p_y E_x}{2\lambda p^3}. \tag{2.80}$$

Summing over all occupied states, the linear response of the $\hat{z}$ spin-polarization component vanishes because of the odd dependence of n_z on p_y, as illustrated in Fig. 2.5, but the spin-current in the $\hat{y}$ direction is finite.

The Rashba Hamiltonian has two eigenstates for each momentum with eigenvalues $E_{\pm} = p^2/2m \mp \Delta_1/2$; the discussion above applies for the lower energy (labeled+for majority spin Rashba band) eigenstate, while the higher energy (labeled –) eigenstate has the opposite value of $n_{z,\vec{p}}$. Since Δ_1 is normally much smaller than the Fermi energy, only the annulus of momentum space that is occupied by just the lower energy band contributes to the spin-current.

$$j_{s,y}^{\text{int}} = \int_{\text{annulus}} \frac{d^2\vec{p}}{(2\pi\hbar)^2} \frac{\hbar n_{z,\vec{p}}}{2} \frac{p_y}{m} = \frac{-eE_x}{16\pi\lambda m}(p_{F+} - p_{F-}), \tag{2.81}$$

where p_{F+} and p_{F-} are the Fermi momenta of the majority and minority spin Rashba bands. When both bands are occupied, that is, when $n_{2D} > m^2\lambda^2/\pi\hbar^4 \equiv n_{2D}^*, p_{F+} - p_{F-} = 2m\lambda/\hbar$ and then the spin Hall conductivity is

$$\sigma_{\text{sH}} \equiv -\frac{j_{s,y}}{E_x} = \frac{e}{8\pi} \tag{2.82}$$

independent of both the Rashba coupling strength and of the 2DES density. For $n_{2D} < n_{2D}^*$, the upper Rashba band is depopulated. In this limit, p_{F-} and p_{F+} are the interior and exterior Fermi radii of the lowest Rashba split band, and σ_{sH} vanishes linearly with the 2DES density:

$$\sigma_{\text{sH}} = \frac{e}{8\pi} \frac{n_{2D}}{n_{2D}^*} \tag{2.83}$$

Similar arguments with the 2-d hole Rashba system give the correct intrinsic contribution in that case and the side-jump contribution in that case vanishes with only the skew-scattering contribution to the extrinsic SHE in that case.

The effects of disorder on the induced spin-current, within linear response, come in the form of self-energy lifetime corrections and vertex corrections. The lifetime corrections only reduce this induced current through a broadening of the bands without affecting its nature. On the other hand, vertex corrections and how affect the picture above have been the source of important debate since they make the intrinsic SHE vanish in the Rashba 2DEG system for any arbitrary amount of scattering (Chalaev and Loss, 2004; Inoue et al., 2004; Mishchenko et al., 2004). For p-type doping in both 3D and 2D hole gases, the vertex corrections vanish in the case of isotropic impurity scattering (Bernevig and Zhang, 2005; Khaetskii, 2006; Murakami, 2004; Shytov et al., 2006). As mentioned above, this result is now understood in the context of the specific relation of the spin-dynamics within this particular model as stated above (Chalaev and Loss, 2004; Dimitrova, 2005). This spin-dynamics is linked to the magneto-electric effect producing a homogeneous in-plane spin-polarization by an electric field in a Rashba 2DEG (Edelstein, 1990; Inoue et al., 2003) and is consistent with numerical treatments of the disorder through exact diagonalization finite size scaling calculations (Nomura et al., 2005a, b; Sheng et al., 2005).

4.5. Numerical studies of the spin Hall accumulation and SHE

The optical-based experiments measure directly the spin-accumulation induced at the edges of the examples through different optical techniques. On the other hand, most of the early theory has focused on the spin-current generated by an electric field which would drive such spin-accumulation. In most studies, this spin-current and its associated conductivity has been defined as $j_y^z \equiv \{v_y, s_z\}/2 = \sigma^{SHE} E_x$. This choice is a natural one but not a unique one in the presence of spin–orbit coupling since there is no continuity equation for spin density as is the case for charge density. The actual connection between the spin-accumulation and the induced spin-current is *not* straightforward in the situations where spin–orbit coupling is strong and this relation is the focus of current research and one of the key challenges ahead.

In order to resolve some of these issues, one can turn to numerical simulations of the systems of interest and directly calculate both spin-edge accumulation and spin-current and compare the results. This was done in the bulk numerical studies in a finite 2d hole system by Nomura et al. (2005b) where, unlike other numerical studies of spin-accumulation in mesoscopic SHE systems (Ma et al., 2004; Nikolić; et al., 2004; Onoda and Nagaosa, 2005; Wang et al., 2005), they considered a diffusive transport regime in which the system dimensions are large compared to scattering mean-free-paths, the regime that applies to current SHE devices. The experimental device which they compare their results to consisted of a 2DHG at a GaAs/AlGaAs hetero-junction bordered by two coplanar p–n junction light emitting diodes (LEDs) which are used as detectors of spin-polarization at the edges of the 2DHG channel (Wunderlich et al., 2005).

Within the exact diagonalization calculation disorder is treated exactly and the spin-accumulation and spin-current are computed independently. For example, the spin-accumulation in the geometry shown in Fig. 2.6 is obtained by

$$S_z(x) = -i\hbar E_y \int_0^{L_y} \frac{dy}{L_y} \sum_{\alpha,\alpha'} \frac{f(E_\alpha) - f(E_{\alpha'})}{E_\alpha - E_{\alpha'}} \frac{\langle \alpha|s^z(x,y)|\alpha'\rangle \langle \alpha'|j_y|\alpha\rangle}{E_\alpha - E_{\alpha'} + i\eta} \quad (2.84)$$

where $s^z(x, y)$ is the z-component of the local spin density operator, $j_y = (\pm|e|)\partial H/\partial \hbar k_y$ is the longitudinal electrical current operator for holes, and E_α and $|\alpha\rangle$ are the eigenenergies and eigenstates of the full Hamiltonian H which includes the spin–orbit coupling term, the disorder potential, and the hard-wall confining potential.

4.5.1. Comparing experimental and calculated magnitudes of the edge spin-polarization

Numerical data imply an approximate general form of the dimensionless edge polarization,

$$\frac{S^z_{\text{edge}}}{\hbar n} = g_s \frac{e}{4\pi v_F \hbar n} E, \quad (2.85)$$

where g_s is a numerical factor. In the measured system, the 2DHG density $n = 2 \times 10^{12}$ cm^{-2}, $v_F = 10^5$ ms^{-1}, $\Delta_{so}/E_F \approx 0.4$, and $(\hbar/\tau)/E_F \approx 0.1$. The system is therefore in the weak scattering, strong spin–orbit coupling regime and the corresponding numerical factor $g_s = 7.4$. From Eq. (2.85), one obtains that the expected polarization for the measured sample reaches ~8% and the width of the accumulation area is of order ~10 nm. The theoretical ~8% polarization is consistent with the measured value of ~1%, assuming an effective LED recombination width of the order ~100 nm. This number cannot be precisely determined experimentally because of the resolution limit set up by the wavelength of the emitted light (~700 nm). However, comparisons between experiments in the 1.5 μm and 10 μm 2DHG channels in Nomura et al. (2005b), analysis of the digital images of the active p–n junction area, and simulations of device I–V characteristics (Wunderlich et al., 2005) confirm a submicron width of the recombination region near the p–n junction.

Therefore, the edge spin-accumulation due to the SHE in a two-dimensional hole gas for the parameters used in the experimental study is expected to be about 10% spin-polarization. These predictions are consistent with the experimental finding of 2% optical polarization averaged over a distance ~$5L_{so}$ (Nomura et al., 2005b).

5. TOPOLOGICAL BERRY'S PHASES IN SPIN–ORBIT COUPLED SYSTEMS: ACE

In the early 1980s, it was shown that quantum mechanical particles propagating through a coherent nanoscale device acquire a quantum geometric which can have important physical consequences. This geometric phase, known as Berry phase (Berry, 1984), is acquired through the adiabatic motion of a quantum particle in the system's parameter space and can have strong effects on the transport properties due to self-interference effects of the quasiparticles when moving in cyclic motion. Its generalization to nonadiabatic motion is known as the Aharonov–Anandan phase (Aharonov and Anadan, 1987). A classical example of such geometric phases is the Aharonov–Bohm phase acquired by a particle going around a loop in the presence of a magnetic flux. Aside from the AB effect, the first experimental observation of the Berry phase was reported in 1986 for photons in a wound optical fiber (Tomita and Chiao, 1986). An important corollary to this phase is the Aharonov–Casher (AC) phase arising from the propagation of an electron in the presence of spin–orbit coupling (Aharonov and Casher, 1984; Nitta et al., 1999a).

This ACE can be calculated simply by considering two partial waves which move around the ring in counter propagating directions. They will acquire a phase difference which depends on the spin orientation with respect to the total magnetic field $\vec{B}_{tot} = \vec{B}_{ext} + \vec{B}_{eff}$ and the path of each partial wave. $\vec{B}_{eff}$ is the effective field induced by the spin–orbit interaction as illustrated in Fig. 2.7. The phase difference is approximately (Nitta et al., 1999a):

$$\Delta \varphi_{\psi_s^+ - \psi_s^-} = -2\pi \frac{\Phi}{\Phi_0} - b\pi(1 - \cos\theta) \qquad (2.86)$$

$$\Delta \varphi_{\psi_s^+ - \psi_s^-} = -2\pi \frac{\Phi}{\Phi_0} - b2\pi r \frac{m^*\alpha}{\hbar^2} \sin\theta \qquad (2.87)$$

where $s = \uparrow$ and $\downarrow$ denote parallel and antiparallel orientation to $\vec{B}_{tot}$, $b = +1$ for $s = \uparrow$ and $b = -1$ for $s = \downarrow$, and the superscript $-(+)$ denotes a clockwise (counterclockwise) evolution, respectively. In the above equations, α is the spin–orbit parameter, r the ring radius, m^* the effective electron mass, and θ the angle between the external ($\vec{B}_{ext}$) and the total magnetic field $\vec{B}_{tot}$. For both equations, the first term on the RHS can be identified as the AB phase. The second term of Eq. (2.86) is the geometric Berry or Aharonov–Anandan phase. The second term in Eq. (2.87) represents the dynamic part of the AC phase, that is, the phase of a particle with a magnetic moment that moves around an electric field. From the

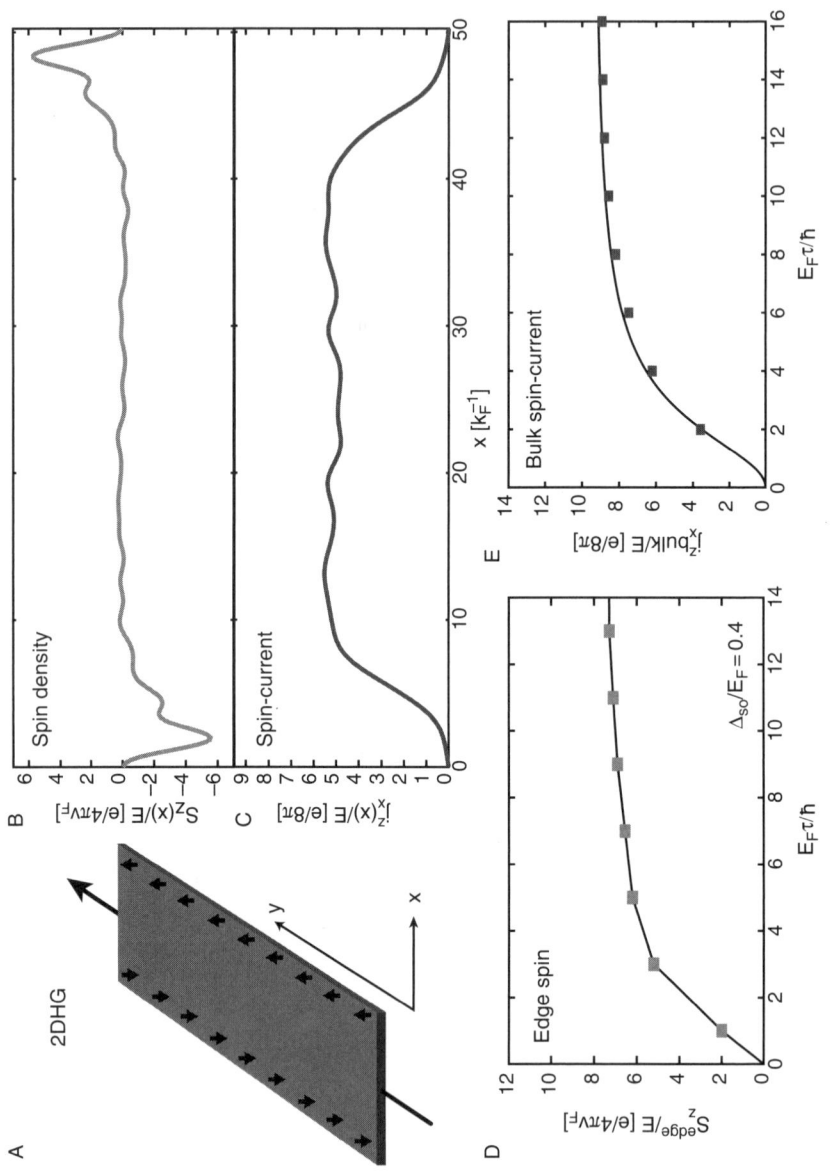

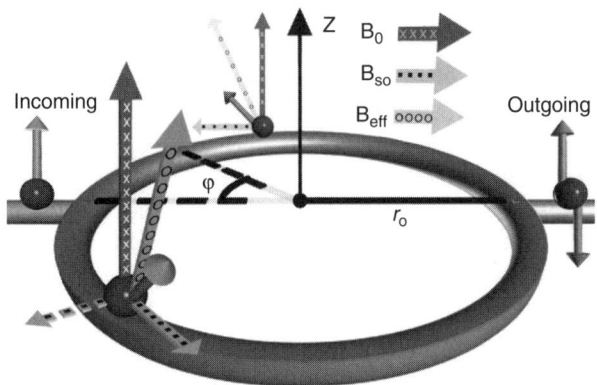

FIGURE 2.7 One channel ring of radius r_0 subject to Rashba coupling in the presence of an additional magnetic field B_o. Electron (hole) spin traveling around the ring acquires phase due to the applied out-of-plane magnetic field (gray arrow) and the Rashba in-plane magnetic field (momentum dependent, green arrows) caused by the spin–orbit interaction. (See Color Plate.)

expressions above, it can be seen that an increase in the AC phase will lead to a phase change that increases continuously with α, whereas the contribution due to the geometric phase results in a phase shift limited to $\Delta\varphi_{\text{geom}} \leq \pi$.

Both the AC phase (Mathur and Stone, 1992) and the geometric phase (Aronov and Lyanda-Geller, 1993; Qian and Su, 1994) depend on the spin–orbit interaction and therefore a complicated dependence on magnetic field and spin–orbit coupling strength is expected. The effects of the AC phase on transport through semiconducting ring structures can be tested in two-dimensional gas confined to an asymmetric potential well which enables an all electrical control of the spins via the type Rashba spin–orbit interaction by changing the gate voltage (Aronov and Lyanda Geller, 1993; Bychkov and Rashba, 1984; Choi et al., 1997; Mal'shukov et al., 1999; Nitta et al., 1999b). This spin-interference in a semi-conductor ring has been proposed as a way to control spin-polarized currents (Frustaglia and Richter, 2004b) and as a spin-filter (Molnar et al., 2004b).

Spin interference signals in square loop arrays have been recently reported by Koga et al. but their direct relationship to the ACE is not

---

FIGURE 2.6 Spin Hall effect in a two-dimensional hole gas. (A) Schematic of the geometry with a sample with edges. Spatial profile of (B) the z-component of spin and (C) the spin-current in the in the x-direction with a k-cubic Rashba model appropriate for a two-dimensional hole gas model at $\lambda k_F^3/E_F = 0.2$ and $E_F\tau/\hbar = 3$. Bottom: Disorder dependence of (D) edge spin and (E) bulk spin-current (Nomura et al., 2005b).

easily established. Direct signatures of the ACE have been experimentally detected (Bergsten et al., 2006; Konig et al., 2006). König et al. (2006) studied a single HgTe ring structure where the strength of the Rashba effect was controlled via a gate electrode by varying the asymmetry of the quantum well structure. Since HgTe is a narrow gap material, it has a large spin–orbit coupling and therefore is an ideal material to test the ACE. They observed systematic variations in the conductance of the device as a function of both external B-field and gate voltage. Bergsten et al. (2006) studied a multiple ring structure of InAlAs/InGaAs-based 2-d electron gas and used the gate-controlled Rashba strength to modulate the spin precession rate and hence follow the effects of the AC phase.

5.1. Landauer–Büttiker calculations of ACE: Comparison to experiments

The gate-voltage-dependent oscillations in the single ring experiments (Konig et al., 2006) exhibit a non-monotonic phase change, which is related to the dynamical part of the AC phase. To analyze ACE in a multichannel ring, numerical calculations within the Landauer–Büttiker (LB) formalism can be performed (Souma and Nikolic, 2004). The effective mass Hamiltonian for a two-dimensional ring under Rashba spin–orbit interactions and in a perpendicular magnetic field, B_z, is given by:

$$\hat{H}_r = \frac{\hat{\pi}^2}{2m_{\text{eff}}} + \frac{\alpha}{\hbar}(\hat{\sigma} \times \hat{\pi})_z + \hat{H}_Z + \hat{H}_{\text{conf}}(r) + \hat{H}_{\text{dis}} \qquad (2.88)$$

where $\hat{\pi} = \hat{\mathbf{p}} - e\hat{\mathbf{A}}$ and $\hat{H}_Z = (1/2)g\mu_B \sigma_z B_z$. The first term is the kinetic energy contribution, the second term corresponds to the spin–orbit Rashba interactions, the third is the Zeeman interaction, and $\hat{H}_{\text{conf}}$ and $\hat{H}_{\text{dis}}$ are the confinement and disorder components of the Hamiltonian, respectively. For an ideal 1D ring, the conductance in the magnetic and Rashba fields can be found analytically (Frustaglia and Richter, 2004a; Meijer et al., 2002; Molnar et al., 2004a; Nitta et al., 1999a). For a 2-d hole gas, the effective Hamiltonian is more complicated (Kovalev et al., 2007) and the frequency of oscillations with Rashba coupling strength changes in the strong coupling regime. To analyze the 2-d electron system in the wider rings with multi-channels, one uses the concentric tight-binding (TB) approximation to model the multichannel rings extending the calculations in Souma and Nikolic (2004) to include not only the Rashba interactions but also the effect of magnetic field.

Within this approximation, the Hamiltonian becomes:

$$\hat{H}_r = \sum_{n,m=1}^{N,M} \sum_{\sigma=\uparrow,\downarrow} \varepsilon_{nm} c^\dagger_{nm,\sigma} c_{n,m,\sigma} - \sum_{n,m=1}^{N,M} \sum_{\sigma=\uparrow,\downarrow} \left[t_\Theta^{n,n+1,m} e^{i\Phi r_m/\Phi_0} c^\dagger_{nm,\sigma} c_{n+1,m,\sigma'} + h.c. \right]$$
$$- \sum_{n=1}^{N} \sum_{m=1}^{M-1} \sum_{\sigma=\uparrow,\downarrow} \left[t_r^{m,m+1,n} c^\dagger_{nm,\sigma} c_{n,m+1,\sigma'} + h.c \right]$$

(2.89)

where n, m designate the sites in the azimuthal (Θ) and radial directions (r), respectively; $\varepsilon_{mn} = 4t\sigma_0 - \frac{1}{2}g\sigma_z\mu_B B_z$ is the on-site energy where $t = \hbar^2/(2ma^2)$ and a is the lattice constant along the radial direction. $t_\Theta^{n,n+1,m}$ and $t_r^{m,m+1,n}$ are the total nearest neighbor hopping parameters in azimuthal and radial directions, respectively, defined in Souma and Nikolic (2004). In the Landau gauge, the hopping parameter in azimuthal

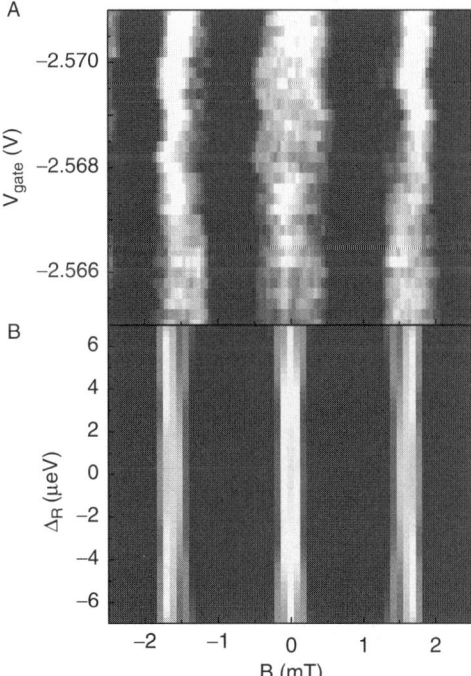

FIGURE 2.8 (A) Near the symmetry point where the Rashba coupling vanishes the interference pattern is unperturbed by the spin–orbit interaction. (B) The theoretical calculations for a 6-channel ring show consistent results for the corresponding range of Δ_{Rashba}. Yellow and blue correspond to conductance maxima and minima, respectively (Konig et al., 2006). (See Color Plate.)

direction is modified through the term $e^{i\Phi r_m/\Phi_0}$, where $\Phi = \pi B_z a$, $\Phi_0 = h/e$, and $r_m = r_1 + (m-1)a$ is the radius of a ring with m modes in radial direction. The innermost ring radius corresponds to $m = 1$ while $m = M$ stands for the outermost ring radius. The ring is attached to two semi-infinite, paramagnetic leads that constitute reservoirs of electrons at chemical potentials μ_1 and μ_2. The influence of semi-infinite leads in the mesoscopic regime is taken into account through the self-energy term and the total charge conductance is calculated as outlined in Souma and Nikolic (2004).

Using an effective electron mass and effective g-factor of n-doped HgTe parameters in the calculation, Fig. 2.8B shows the calculated conductance as a function of Rashba energy and magnetic field. For small Rashba energies, less than 5 µeV, the interference pattern is almost unperturbed by spin–orbit interactions and displays the multichannel

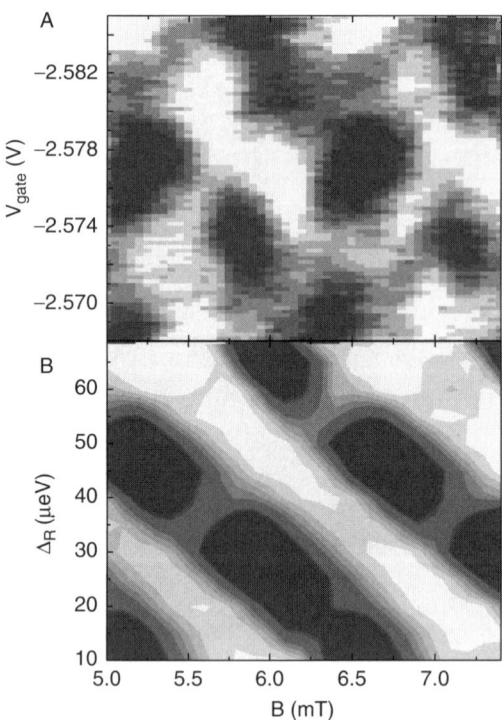

FIGURE 2.9 When the spin–orbit interaction is modified via the gate voltage, a shift of the conductance maxima (yellow) can be observed due to the Aharonov–Casher phase (A). In figure (B), the theoretical results for the conductance in a 6-channel ring as a function of the Rashba energy and B_{ext} are shown. The scaling of the y-axis allows a direct comparison of the experimental and theoretical data (Konig et al., 2006). (See Color Plate.)

Aharonov–Bohm oscillations, similar to the experimental data (Fig. 2.8A). Comparing the experimental and theoretical parameters leads to the conclusion that a change in the gate voltage of 10 mV leads to a change of 35 µeV in the Rashba energy, in good agreement with the results obtained from the Shubnikov–de Haas oscillations (Konig et al., 2006).

To verify the existence of AC phase, one has to perform the conductance calculations for much larger Rashba couplings. In the case of a strictly 1D ring, the appearance of the AC phase leads to periodic oscillations in conductance as a function of the Rashba energy (Frustaglia and Richter, 2004a; Molnar et al., 2004a). In contrast, the theoretically predicted interference pattern for conductance is more complex for a multichannel ring. In this case, the repetitive conductance minima and maxima move diagonally as a function of spin–orbit coupling, as can be seen in the theoretical simulation (Fig. 2.9B). The distinct interference pattern is a strong indication that there is only one conducting channel, presumably because of impurities and imperfections in the ring geometry. The theoretical model reproduces the main features of experimental data, that is, the diagonal position of maxima and minima of conductance (Fig. 2.9A). A more quantitative comparison of the experimental and theoretical data is difficult and should take into account incoherence effects as well as the change in width of the ring which is cumbersome to estimate.

REFERENCES

Adams, E., and Blount, E. (1959). *J. Phys. Chem. Solids* **10**, 286.
Aharonov, Y., and Anadan, J. (1987). *Phys. Rev. Lett.* **58**(16), 1593.
Aharonov, Y., and Casher, A. (1984). *Phys. Rev. Lett.* **53**(4), 319.
Aronov, A. G., and Lyanda-Geller, Y. B. (1993). *Phys. Rev. Lett.* **70**, 343.
Berger, L. (1970). *Phys. Rev. B* **2**, 4559.
Bergsten, T., Kobayashi, T., Sekine, Y., and Nitta, J. (2006). *Phys. Rev. Lett.* **97**(19), 196803.
Bernevig, B. A., and Zhang, S.-C. (2005). *Phys. Rev. Lett.* **95**, 016801.
Berry, M. V. (1984). *Proc. R. Soc. Lond.* **392**(1802), 45.
Borunda, M., Nunner, T. S., Luck, T., Sinitsyn, N. A., Timm, C., Wunderlich, J., Jungwirth, T., MacDonald, A. H., and Sinova, J. (2007). eprint arXiv:condmat/0702289.
Bychkov, Y. A., and Rashba, E. I. (1984). *J. Phys. C: Sol. State Phys.* **17**(33), 6039.
Chalaev, O., and Loss, D. (2004). *Phys. Rev. B* **71**, 245318.
Chien, L., and Westgate, C. R. (1980). "The Hall Effect and Its Applications." Plenum, New York.
Choi, T., Cho, S. Y., Ryu, C. M., and Kim, C. K. (1997). *Phys. Rev. B* **56**(8), 4825.
Dimitrova, O. V. (2005). *Phys. Rev. B* **71**, 245327.
Dyakonov, M. I., and Perel, V. I. (1971). *Sov. Phys. JETP* **33**, 467.
Edelstein, V. M. (1990). *Solid State Commun.* **73**, 233.
Engel, H.-A., Rashba, E. I., and Halperin, B. I. (2006). eprint arXiv:cond-mat/0603306.
Frustaglia, D., and Richter, K. (2004a). *Phys. Rev. B* **69**, 235310.
Frustaglia, D., and Richter, K. (2004b). *Phys. Rev. B* **69**(23), 235310.
Ganichev, S. O., and V. V. Belikov, preprint arXiv:0803, 0949.
Hirsch, J. E. (1999). *Phys. Rev. Lett.* **83**, 1834.

Inoue, J., Bauer, G. E. W., and Molenkamp, L. W. (2003). *Phys. Rev. B* **67**, 033104.
Inoue, J., Bauer, G. E. W., and Molenkamp, L. W. (2004). *Phys. Rev. B* **70**, 041303.
Jungwirth, T., Niu, Q., and MacDonald, A. H. (2002). *Phys. Rev. Lett.* **88**, 207208.
Karplus, R., and Luttinger, J. M. (1954). *Phys. Rev.* **95**, 1154.
Kato, Y. K., Myers, R. C., Gossard, A. C., and Awschalom, D. D. (2004). *Science* **306**, 1910.
Khaetskii, A. (2006). *Phys. Rev. B* **73**, 115323.
Kimura, T., Otani, Y., Sato, T., Takahashi, S., and Maekawa, S. (2007). *Phys. Rev. Lett.* **98**, 156601.
Kohn, W., and Luttinger, J. M. (1957). *Phys. Rev.* **108**, 590.
Konig, M., Tschetschetkin, A., Hankiewicz, E. M., Sinova, J., Hock, V., Daumer, V., Schafer, M., Becker, C. R., Buhmann, H., and Molenkamp, L. W. (2006). *Phys. Rev. Lett.* **96(7)**, 076804.
Kovalev, A. A., Borunda, M. F., Jungwirth, T., Molenkamp, L. W., and Sinova, J. (2007). eprint arXiv:cond-mat/0701534.
Liu, B., Shi, J., Wang, W., Zhao, H., Li, D., Zhang, S., Xue, Q., and Chen, D. (2006). eprint arXiv:cond-mat/0610150.
Luttinger, J. M. (1958). *Phys. Rev.* **112**, 739.
Luttinger, J. M., and Kohn, W. (1955). *Phys. Rev.* **97**, 869.
Ma, X., Hu, L., Tao, R., and Shen, S.-Q. (2004). *Phys. Rev. B* **70**, 195343.
Mal'shukov, A. G., Shlyapin, V. V., and Chao, K. A. (1999). *Phys. Rev. B* **60(4)**, R2161.
Marder, M. P. (2000). "Condensed Matter Physics." Wiley, New York, eprint Supplementary material by author.
Mathur, H., and Stone, A. D. (1992). *Phys. Rev. Lett.* **68**, 2964.
Meijer, F. E., Morpurgo, A. F., and Klapwijk, T. M. (2002). *Phys. Rev. B* **66**, 033107.
Mishchenko, E. G., Shytov, A. V., and Halperin, B. I. (2004). *Phys. Rev. Lett.* **93**, 226602.
Molnar, B., Peeters, F. M., and Vasilopoulos, P. (2004a). *Phys. Rev. B* **69**, 155335.
Molnar, B., Peeters, F. M., and Vasilopoulos, P. (2004b). *Phys. Rev. B* **69(15)**, 155335.
Murakami, S. (2004). *Phys. Rev. B* **69**, 241202.
Murakami, S., Nagaosa, N., and Zhang, S.-C. (2003). *Science* **301**, 1348.
Nikolić, B. K., Souma, S., Zârbo, L. P., and Sinova, J. (2004). *Phys. Rev. Lett.* **95**, 046601.
Nitta, J., Meijer, F. E., and Takayanagi, H. (1999a). *Appl. Phys. Lett.* **75**, 695.
Nitta, J., Meijer, F. E., and Takayanagi, H. (1999b). *Appl. Phys. Lett.* **75(5)**, 695.
Nomura, K., Sinova, J., Sinitsyn, N. A., and MacDonald, A. H. (2005a). *Phys. Rev. B* **72**, 165316.
Nomura, K., Wunderlich, J., Sinova, J., Kaestner, B., MacDonald, A. H., and Jungwirth, T. (2005b). *Phys. Rev. B* **72**, 245330.
Nozieres, P., and Lewiner, C. (1973). *Le Journal de Physique* **34**, 901.
Nunner, T. S., Sinitsyn, N. A., Borunda, M. F., Kovalev, A. A., Abanov, A., Timm, C., Jungwirth, T., Inoue, J.-I., MacDonald, A. H., and Sinova, J. 2007, Phys. Rev B76 235312 (2007).
Onoda, M., and Nagaosa, N. (2005). *Phys. Rev. B* **72**, 081301.
Qian, T. Z., and Su, Z. B. (1994). *Phys. Rev. Lett* **72**, 2311.
Saitoh, E., Ueda, M., Miyajima, H., and Tatara, G. (2006). *Appl. Phys. Lett.* **88**, 182509.
Schliemann, J., Loss, D., and Westervelt, R. M. (2006). *Phys. Rev. B* **73**, 085323.
Sheng, D. N., Sheng, L., Weng, Z. Y., and Haldane, F. D. M. (2005). *Phys. Rev. B* **72**, 153307.
Shytov, A. V., Mishchenko, E. G., Engel, H. A., and Halperin, B. I. (2006). *Phys. Rev. B* **73**, 075316.
Sih, V., Myers, R. C., Kato, Y. K., Lau, W. H., Gossard, A. C., and Awschalom, D. D. (2005). *Nat. Phys.* **1**, 31.
Sinitsyn, N. A., Niu, Q., Sinova, J., and Nomura, K. (2005). *Phys. Rev. B* **72**, 045346.
Sinitsyn, N. A., Hill, J. E., Min, H., Sinova, J., and MacDonald, A. H. (2006a). *Phys. Rev. Lett.* **97**, 106804.
Sinitsyn, N. A., Niu, Q., and MacDonald, A. H. (2006b). *Phys. Rev. B* **73**, 075318.

Sinitsyn, N. A., MacDonald, A. H., Jungwirth, T., Dugaev, V. K., and Sinova, J. (2007). *Phys. Rev. B* **75,** 045315.
Sinova, J., Culcer, D., Niu, Q., Sinitsyn, N. A., Jungwirth, T., and MacDonald, A. H. (2004a). *Phys. Rev. Lett.* **92,** 126603.
Sinova, J., Jungwirth, T., and Černe, J. (2004b). *Int. J. Mod. Phys. B* **18,** 1083.
Smit, J. (1955). *Physica* **21,** 877.
Souma, S., and Nikolic, B. (2004). *Phys. Rev. B* **70,** 195346.
Stern, N. P., Ghosh, S., Xiang, G., Zhu, M., Samarth, N., and Awschalom, D. D. (2006). *Phys. Rev. Lett.* **97,** 126603.
Stern, N. P., Steuerman, D. W., Mack, S., Gossard, A. C., and Awschalom, D. D. (2007). eprint arXiv:0706.4273.
Sundaram, G., and Niu, Q. (1999). *Phys. Rev. B* **59,** 14915.
Tomita, A., and Chiao, R. Y. (1986). *Phys. Rev. Lett.* **57,** 933.
Valenzuela, S. O., and Tinkham, M. (2006). *Nature* **442,** 176.
Wang, Q., Sheng, L., and Ting, C. S. (2005). *Intern. J. Mod. Phys. B* **19,** 4135.
Weber, C. P., Orenstein, J., Bernevig, B. A., Zhang, S.-C., Stephens, J., and Awschalom, D. D. (2007). *Phys. Rev. Lett.* **98,** 076604.
Winkler, R. (2003). Spin-orbit coupling in two dimensional electron systems. Spinger-Verlag, New York.
Wunderlich, J., Kaestner, B., Sinova, J., and Jungwirth, T. (2005). *Phys. Rev. Lett.* **94,** 047204.
Zutic, I., Fabian, J., and Das Sarma, S. (2004). *Rev. Mod. Phys.* **76,** 323.

CHAPTER 3

Fermi Level Effects on Mn Incorporation in III-Mn-V Ferromagnetic Semiconductors

K. M. Yu,* T. Wojtowicz,[†] W. Walukiewicz,* X. Liu,[‡] and J. K. Furdyna[‡]

Contents			
	1.	Introduction	90
		1.1. Diluted magnetic semiconductors	90
		1.2. Hole-mediated ferromagnetism in III$_{1-x}$Mn$_x$V alloys	91
		1.3. Saturation of hole concentration and Curie temperature	92
		1.4. Mn lattice site locations in GaMnAs	92
	2.	Sample Preparation and Characterization	93
		2.1. LT-MBE growth of GaMnAs	93
		2.2. Determination of the location of Mn in the GaMnAs lattice by ion channeling	95
		2.3. Charged dopant distribution by electrochemical capacitance-voltage profiling	96
		2.4. Magnetic characterization	97
	3.	Effects of Mn Location on the Electronic and Magnetic Properties	98
		3.1. Observation of Mn$_I$ in Ga$_{1-x}$Mn$_x$As	98
		3.2. Thermodynamic limits on hole concentration and on T_C in Ga$_{1-x}$Mn$_x$As	101
		3.3. T_C enhancement by elimination of Mn$_I$	103
		3.4. Fermi level manipulation	114
		3.5. Self-compensation in other III-Mn-V alloys	122

* Materials Sciences Division, Lawrence Berkeley National Laboratory, Berkeley, California 94720
[†] Institute of Physics, Polish Academy of Sciences, 02-668 Warsaw, Poland
[‡] Department of Physics, University of Notre Dame, Notre Dame, Indiana 46556

Semiconductors and Semimetals, Volume 82
ISSN 0080-8784, DOI: 10.1016/S0080-8784(08)00003-3

4. Concluding Remarks 125
  4.1. Alternative synthesis routes for III$_{1-x}$Mn$_x$V alloys 125
  4.2. Strategies for increasing T_C in III-Mn-V alloys 129
Acknowledgments 130
References 130

1. INTRODUCTION

1.1. Diluted magnetic semiconductors

Semiconductors and magnetic materials both play essential roles in modern electronics industry. Most electronic and optical semiconductor devices utilize the charge of electrons and holes to process information, and magnetic materials use the spin of magnetic ions for information storage. Thus, although the applications of semiconductors and magnetics have evolved independently, it appears logical to combine their properties for possible spin-electronic applications with increased functionalities (Prinz, 1990; Wolf et al., 2001). In this regard, the discovery of *long-range uniform* ferromagnetic order in (In,Mn)As (Ohno et al., 1992), and then in (Ga,Mn)As (Ohno et al., 1996), has been generally acknowledged as a landmark achievement since III–V semiconductor compounds are widely used in electronic, photonic, and microwave devices. Since then an enormous effort has been directed to the research and development of functional ferromagnetic semiconductors. Recent reports of ferromagnetic signatures (frequently above room temperature) in a number of other semiconductor thin layers (including oxides) doped with minute amounts of magnetic ions (Chambers et al., 2006; Liu et al., 2005; Theodoropoulou et al., 2002) have further increased the prospects for the practical realization of such spin-electronic devices.

Historically, coexistence of magnetism and semiconducting properties in diluted magnetic semiconductors (DMSs) has in fact already been realized by alloying the nonmagnetic semiconductor host with magnetic elements as early as the 1970s (Furdyna and Kossut, 1988). In those early years, the study of DMSs and their heterostructures was primarily focused on II–VI-based materials (such as those based on HgTe, CdTe, and ZnSe), where the valence of group-II cations is identical to that of many magnetic ions, such as Mn or Co (Furdyna and Kossut, 1988; Furdyna, 1988). The magnetic interaction in II–VI DMSs is dominated by antiferromagnetic exchange among the Mn spins, which results in paramagnetic, spin glass, and ultimately long-range antiferromagnetic behavior. Recent progress in doping technology of II–VI materials is, however, gradually extending these materials to broader forms of magnetism. For example, carrier-mediated ferromagnetism was recently discovered in *p*-type II–VI DMS heterostructures, although at the present time this occurs only at very low temperatures—the Curie temperatures T_C are typically below 2.0 K in these II–VI-based systems (Haury et al., 1997).

On the other hand, very significant strides have been made in developing ferromagnetic III–V-based semiconductors containing Mn, which remain ferromagnetic to much higher Curie temperatures (Furdyna et al., 2000, 2003; MacDonald et al., 2005; Ohno, 1998, 1999). Since their initial discovery much progress has been achieved especially in the fabrication of $Ga_{1-x}Mn_xAs$ and $In_{1-x}Mn_xAs$ alloys, with Curie temperatures T_C in $Ga_{1-x}Mn_xAs$ now reproducibly exceeding 150 K (Adell et al., 2005; Edmonds et al., 2004b; Ku et al., 2003; Wang et al., 2005). The progress made with $In_{1-x}Mn_xAs$ and $Ga_{1-x}Mn_xAs$ alloys has been subsequently followed by successful preparation of ferromagnetic III–V semiconductors $Ga_{1-x}Mn_xSb$ (Lim et al., 2004; Matsukura et al., 2000), $In_{1-x}Mn_xSb$ (Wojtowicz et al., 2003a), and most recently $Ga_{1-x}Mn_xP$ (Dubon et al., 2005; Scarpulla et al., 2005).

The above III–V-based ferromagnetics taken together have already opened a number of fundamental issues in magnetism and magnetotransport, as well as in the interrelationship between the two. Just as important, the development of these materials holds promise of integrating ferromagnetic and nonmagnetic semiconductors, with an eye on developing new devices that depend—as already noted—on electron charge as well as on its spin (Zutic et al., 2004). For example, $III_{1-x}Mn_xV$ alloys have already been integrated with III–V-based nonmagnetic semiconductor systems to form spin-injecting architectures (Ohno et al., 1999b), as well as structures that allow electrical (Ohno et al., 2000), optical (Koshihara et al., 1997), or other forms of external control (Goennenwein et al., 2004; Liu et al., 2004) of the ferromagnetism exhibited by the $III_{1-x}Mn_xV$ alloys.

1.2. Hole-mediated ferromagnetism in $III_{1-x}Mn_xV$ alloys

It is generally accepted that when Mn is incorporated at the cation site in both II–VI and III–V semiconductor materials, it enters the lattice as an Mn^{2+} ion that has a half-filled d-shell with an angular momentum $L=0$ and a spin of $S=5/2$. The exchange interaction between the localized Mn spins and band electrons is parameterized by exchange integrals $N_0\alpha$ and $N_0\beta$ for states in the conduction and in the valence band, respectively. This parameterization scheme has been spectacularly successful in describing various optical and magnetic phenomena observed in II–VI-based paramagnetic semiconductors. In III–V semiconductors, however, Mn^{2+} ions at the group-III cation site do not just serve as $S=5/2$ magnetic moments inserted into the III–V host, but also as acceptors, so that the resulting $III_{1-x}Mn_xV$ alloys automatically contain high concentrations of holes.

One should note at this point that—in spite of various theoretical attempts and approaches (Akai, 1998; Das Sarma et al., 2003; Dietl et al.,

2001; Inoue et al., 2000; König et al., 2000; Litvinov and Dugaev, 2001; Sanvito et al., 2001; Timm et al., 2002)—the understanding of the origin of ferromagnetism in III$_{1-x}$Mn$_x$V alloys is still rather incomplete. While the debate on this front continues, the theory of ferromagnetism based on the Zener model, the Ginzburg–Landau approach to the ferromagnetic phase transition in the III$_{1-x}$Mn$_x$V systems, and the Kohn-Luttinger $k \cdot p$ theory of the valence band in tetrahedrally coordinated semiconductors are now rather widely accepted, since they were often able to quantitatively describe thermodynamic, micromagnetic, transport, and optical properties of DMS with delocalized holes (Dietl, 2006; Dietl et al., 2000).

1.3. Saturation of hole concentration and Curie temperature

Although this chapter is intended to deal with III$_{1-x}$Mn$_x$V alloys generally, for specificity we will be using the results on Ga$_{1-x}$Mn$_x$As as illustrative examples, since at the present time, this material is by far the most studied and best understood in this family of materials. Calculations based on the Zener model (Dietl et al., 2000) predicted that T_C in Ga$_{1-x}$Mn$_x$ As could be improved by increasing the Mn content and/or the free hole concentration in the alloy. These predictions led to extensive experimental work aimed at achieving higher T_C for Ga$_{1-x}$Mn$_x$As. Experimentally, it has been established that T_C in Ga$_{1-x}$Mn$_x$As increases with increasing hole concentration (Ku et al., 2003). Although each Mn atom in the Ga sublattice is in principle expected to contribute a hole to the system, in practice it was found that the hole concentration in this material is significantly lower than the Mn concentration (by a factor of 2–3) (Ohno, 1998, 1999), especially as the Mn concentration increases. Potashnik et al. (2002) showed that in optimally annealed Ga$_{1-x}$Mn$_x$As alloys both the T_C and the conductivity saturate for $x > 0.05$, suggesting that, as x increases, more of the Mn acceptors are compensated by donors, and an increasing fraction of Mn spins ceases to participate in ferromagnetism. These self-compensation effects suggest that there exists a mechanism that limits the Curie temperature of Ga$_{1-x}$Mn$_x$As grown by the standard low-temperature molecular beam epitaxy (LT-MBE) methods. Thus, despite intense efforts, similar maximum values of T_C of ~150 K were found in thin Ga$_{1-x}$Mn$_x$As films prepared in different laboratories with rather different values of x (ranging from ~0.05 to 0.10) and annealed at temperatures in the range between 150 and 280 °C (MacDonald et al., 2005; Ohno, 1999).

1.4. Mn lattice site locations in GaMnAs

In the Zener model description of the ferromagnetism in Ga$_{1-x}$Mn$_x$As, only the substitutional Mn^{2+} ions are considered (Mn$_{Ga}$). These substitutional divalent Mn ions act as acceptors, so that the hole concentration

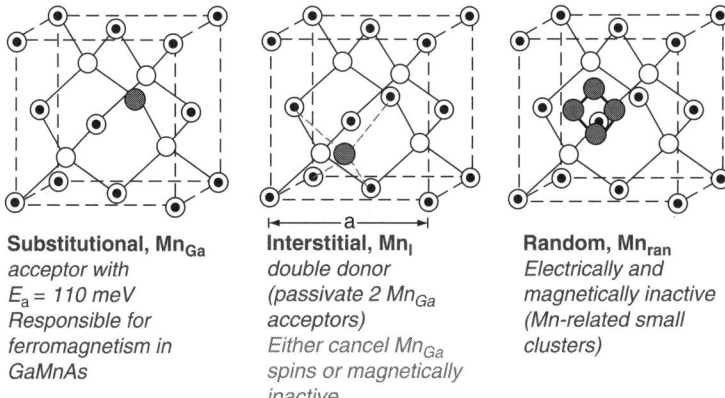

FIGURE 3.1 Schematics illustrating various lattice locations of Mn in a GaAs lattice.

p would ideally be expected to be given by $p = xN_o$, where $N_o = 2.2 \times 10^{22}$ cm^{-3} is the concentration of Ga sites in GaAs. However, in addition to the Mn$_{Ga}$ configuration already mentioned, Mn atoms in the Ga$_{1-x}$Mn$_x$As lattice can also occupy *interstitial* sites commensurate with the zinc-blende lattice structure, and some Mn atoms can precipitate out to form different phases (such as Mn clusters or MnAs inclusions). In the latter case, Mn resides at random sites that are incommensurate with the zinc blende lattice. The three types of Mn sites are illustrated in Fig. 3.1. Similar to all interstitials of metal atoms, Mn interstitials (Mn$_I$) act as donors, and thus tend to passivate the substitutional Mn acceptors. On the other hand, the Mn atoms that form precipitates (e.g., MnAs inclusions) do not contribute either holes or electrons, and are thus outside the Zener picture of magnetization. Therefore, to understand the saturation of holes and ferromagnetism in Ga$_{1-x}$Mn$_x$As, it is essential to address the issue of the lattice site location of Mn, and its relation to the concentration both of holes and of uncompensated Mn^{2+} spins.

2. SAMPLE PREPARATION AND CHARACTERIZATION

2.1. LT-MBE growth of GaMnAs

The equilibrium solubility of Mn in III–V semiconductors is known to be at most 10^{19} cm^{-3} (about 0.002 mole fraction). Thus, to incorporate the higher Mn concentrations into the III–V lattice that are required for achieving ferromagnetism, it is necessary to resort to LT-MBE, in which strong nonequilibrium growth conditions are realized. Typically such growth is carried out using MBE machines equipped with elemental sources Ga, Mn, In, Al, As, and Sb, along with Be and Si sources for doping. In our work, the fluxes of Ga, Be, Al, and Mn are supplied from standard effusion cells, and As_2 flux is produced by a cracker cell. Prior to the deposition of the Mn-containing film, we typically grow a 100–500 nm GaAs buffer layer at 590 °C (i.e., under normal GaAs growth conditions). The substrate is then cooled down for the growth of LT-GaAs, usually to a thickness of 3 nm, followed by a (Ga,Mn)As-based layer or heterostructure. The substrate temperatures used for the growth of these structures are typically 250–280 °C. The As_2:Ga beam equivalent pressure ratio of 20:1 is usually used during the LT-growth of $Ga_{1-x}Mn_xAs$, and the surface quality of the samples is monitored *in situ* by reflection high-energy electron diffraction (RHEED). During the LT-MBE, the RHEED pattern typically shows a (1 × 1) surface reconstruction for LT-GaAs and a (1 × 2) reconstruction for (Ga,Mn)As. During the growth of $Ga_{1-x}Mn_xAs$, one can monitor the oscillations of the intensity of the specular RHEED spot (so-called "RHEED oscillations"), which indicate that the growth proceeds layer by layer, with two-dimensional nucleation. These RHEED oscillations also provide information on the growth rate of the sample, and can also thus be used for estimating its Mn concentration (Kuryliszyn-Kudelska *et al.*, 2004).

Since the relaxed lattice constant a_0 of zinc blende, $Ga_{1-x}Mn_xAs$ is larger than that of GaAs by an amount that depends on the Mn concentration x; x-ray diffraction (XRD) measurements of a_0 can therefore also be used to determine x. For thin layers of $Ga_{1-x}Mn_xAs$, it is observed that $Ga_{1-x}Mn_xAs$ is coherently (fully) strained by the GaAs substrate (or by whatever buffer is used, e.g., $In_{1-x}Mn_xAs$). For example, XRD experiments reveal that a $Ga_{1-x}Mn_xAs$ film grown by LT-MBE is coherently strained throughout its thickness by the underlying layer (typically GaAs or $Ga_{1-y}In_yAs$), even for films as thick as ~7.0 μm (Welp *et al.*, 2004). This is probably due to the low growth temperature, which may prevent dislocations from nucleating. Given the measured vertical lattice parameter of the tetragonally distorted $Ga_{1-x}Mn_xAs$ layer, the relaxed lattice constant a_0 can then be closely estimated if one assumes that the elastic constants of GaMnAs are equal to those of GaAs (i.e., $C_{11} = 11.88$, $C_{12} = 5.32$, and $C_{44} = 5.94$ in 10^{11} dyn/cm^2; McSkimin *et al.*, 1967), and the Mn concentration x can be determined from the value of a_0 by using the equation proposed by Sadowski *et al.* (2001), $a_0(Ga_{1-x}Mn_xAs) = 5.65469 + 0.24661x$.

It is suggested that the nonequilibrium nature of LT-MBE growth and the solubility gap of Mn in GaAs lead to two phenomena that are crucial for elucidating the electronic and magnetic properties of ferromagnetic $Ga_{1-x}Mn_xAs$ films. The first of these is the nano-scale spinodal decomposition into regions with a low and a high concentration of Mn ions, resulting in spatial fluctuations of Mn concentration. Although such spinodal decomposition, which does not involve a precipitation of another crystallographic phase, often escapes detection, it has been observed by careful transmission electron microscopy in some $Ga_{1-x}Mn_x$ As films. Second, although in the LT-MBE process most of the incident Mn atoms are incorporated into the substitutional Mn_{Ga} positions, nevertheless a significant fraction of Mn also enters the system as Mn_I. Additionally, it is well known that low-temperature nonequilibrium MBE growth also favors the formation of arsenic antisites (As_{Ga}). Both of these defects—As_{Ga} and Mn_I—are double donors, and thus they play the undesirable role of compensating the holes produced in $Ga_{1-x}Mn_x As$ by Mn_{Ga}. In this chapter, we will primarily focus on the interconnection between Mn_I defects and the Fermi level in $Ga_{1-x}Mn_x As$ system and on the effect of Mn_I on both the electronic and the magnetic properties of this system.

2.2. Determination of the location of Mn in the GaMnAs lattice by ion channeling

Several studies have been carried out aimed at determining the local environment of Mn in $III_{1-x}Mn_x V$ alloys on atomic scale using extended x-ray absorption fine structure (EXAFS) (Shioda et al., 1998; Soo et al., 1996). While accurate determination of nearest and next-nearest neighbor distances (<±0.005Å), as well as a measure of local disorder, can be achieved using EXAFS, this technique is not well suited for measuring coordination numbers (~20%) and for chemically distinguishing neighbors with small atomic number differences (e.g., Ga and As). The location of impurities in the lattice can only be indirectly inferred from these studies by multiparameter fitting of the EXAFS spectra to calculated model structures.

Ion channeling techniques have been applied extensively for the determination of lattice site location of impurities in semiconductors and metals (Feldman et al., 1982; Tesmer and Nastasi, 1995). In the experiments, which we will describe, the location of Mn in the $Ga_{1-x}Mn_x As$ lattice was established by simultaneous channeled particle-induced x-ray emission (c-PIXE) and channeled Rutherford backscattering spectrometry (c-RBS) using a 2.0 MeV $^4He^+$ beam. The element-specific PIXE technique had to be used to pinpoint the Mn location since Mn atoms are lower in atomic number than the host Ga and As atoms, thus making RBS unsuitable for detecting small concentrations of Mn in GaAs. In this approach, the lattice location of Mn in the GaAs host can be determined by comparing the Mn (PIXE) and GaAs (RBS) angular scans, using the normalized yield as a function of tilt angle around the $\langle 100 \rangle$, $\langle 110 \rangle$, and $\langle 111 \rangle$ axial channels. Here the normalized yield for the RBS (χ_{GaAs}) or the PIXE Mn x-ray signals (χ_{Mn}) is defined as the ratio of the channeled yield to the corresponding unaligned "random" yield.

2.3. Charged dopant distribution by electrochemical capacitance-voltage profiling

One of the long-standing difficulties encountered in the study of the hole-mediated ferromagnetism in $III_{1-x}Mn_xV$ semiconductor alloys has been the difficulty to reliably determine the hole concentration. The Hall effect, which is commonly used to measure the concentration of free charge carriers in semiconductors, cannot be applied to materials like $Ga_{1-x}Mn_x$As at normally available magnetic fields because of the presence of AHE, a feature characteristic of conducting ferromagnets (Hurd, 1980; Potashnik et al., 2001). Ohno and Matsukura et al. (1999a) had performed Hall effect measurements at very high magnetic fields (ca. 27 T) and very low temperatures (ca. 50 mK), where the magnetization M saturates, and therefore the magnetic field dependence of the Hall resistivity ρ_{Hall} is dominated by the standard Hall term. However, even under such extreme conditions, a reliable determination of the free hole concentration is not straightforward since the presence of negative magnetoresistance in this regime indicates that magnetization M is still not completely saturated.

Capacitance-voltage (C-V) measurement is an alternative method for measuring the concentration of net charged dopants in semiconductors. This measurement can be carried out by forming a Schottky contact between the semiconductor of interest and a suitable electrolyte. In addition, this method permits one to obtain the distribution profile of charged dopants (up to several microns) by controlled electrochemical etching of the semiconductor (for a review of the ECV method, see, e.g., Blood, 1986). As an illustration, here we described such electrochemical capacitance-voltage (ECV) measurements carried out at room temperature using a BioRad PN4300 Semiconductor Profile Plotter. For all specimens discussed below, the electrolyte used to form the Schottky barrier was a 0.2M NaOH:EDTA solution.

The C-V technique provides information on the distribution of the net space charge in the depletion region of the semiconductor (Blood, 1986). In order to establish how deep states manifest themselves in ECV measurements in a heavily p-type doped GaAs-based films, we compared the space charge concentration measured by ECV with the free hole concentration obtained from Hall measurements on a series of *nonmagnetic* Be-doped LT-GaAs grown with different Be content (from 5 to ~15% Be). It is important to note that the growth conditions of the Be-doped films and undoped $Ga_{1-x}Mn_x$As films are very similar, and therefore the defects present in both types of films are expected to be similar both in nature and in quantity.

Figure 3.2 shows ECV profiles of the ionized acceptor concentrations in two $Ga_{1-y}Be_y$As thin films grown, respectively, with Be cell temperature of 1060 and 1100 °C. The Be concentrations in these films estimated from the change in the growth rate between GaAs and $Ga_{1-y}Be_y$As (as monitored by RHEED oscillations) were $y = 0.06$ and 0.17, respectively.

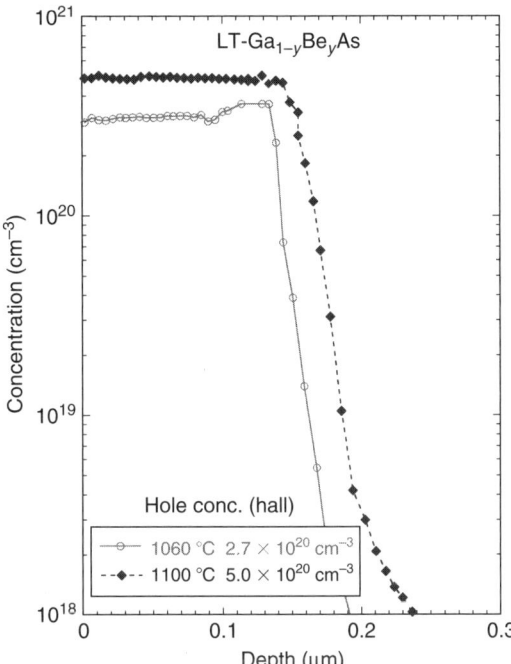

FIGURE 3.2 Net space-charge profiles measured at room temperature by electrochemical capacitance-voltage measurements for two nonmagnetic $Ga_{1-y}Be_yAs$ films grown by low-temperature molecular beam epitaxy (MBE). Values of hole concentration obtained by room temperature Hall measurements on the same $Ga_{1-y}Be_yAs$ films are also shown in the figure legend, together with Be cell temperatures used during the MBE growth.

The free hole concentrations measured by the Hall effect are also given in the figure. The results in Fig. 3.2 show that the net ionized acceptor concentration measured by ECV agrees to within 15% with the free hole concentration measured by the Hall effect, thus verifying that ECV provides a reliable tool for measuring carrier concentration.

2.4. Magnetic characterization

Magnetoresistance, Hall measurements, and superconducting quantum interference device (SQUID) magnetometry are most commonly used for electrical and magnetic characterization of $III_{1-x}Mn_xV$ ferromagnetics, including the determination of their Curie temperature T_C. It is now well established that electrical transport properties of these alloys are strongly linked to their magnetic properties, and can therefore be used to investigate them along with direct magnetic measurements. In this chapter, we illustrate this by magnetotransport measurements performed either in the Van der Pauw or in the six-probe Hall bar geometry, with an eye on the temperature and magnetic-field dependences of the resistivity and the anomalous Hall effect (AHE) of $Ga_{1-x}Mn_xAs$-type ferromagnetic systems. The magnetization measurements discussed here were performed using a Quantum Design MPMS XL system, with the aim of determining the directions of the easy and hard axes of magnetization, the Curie temperature, the coercive field, and other magnetic parameters.

3. EFFECTS OF Mn LOCATION ON THE ELECTRONIC AND MAGNETIC PROPERTIES

3.1. Observation of Mn_I in $Ga_{1-x}Mn_xAs$

As mentioned in Section 1.4, Mn atoms incorporated into ferromagnetic $Ga_{1-x}Mn_xAs$ can occupy three distinct types of lattice site: substitutional positions in the Ga sublattice (Mn_{Ga}), where Mn^{2+} ions act as acceptors and also contribute an uncompensated magnetic moment; interstitial positions (Mn_I), where they act as donors that passivate the Mn_{Ga} acceptors; and random locations (Mn_{rand}) that form clusters of Mn or MnAs. Figure 3.3 shows the c-PIXE and c-RBS angular scans about the $\langle 100 \rangle$, $\langle 110 \rangle$, and $\langle 111 \rangle$ axes for a 110-nm-thick $Ga_{1-x}Mn_xAs$ film with $x = 0.09$. The $\langle 100 \rangle$ and $\langle 111 \rangle$ angular scans of the Mn PIXE signals closely follow those of the GaAs RBS signals, suggesting that along these two directions, the Mn atoms are shadowed by the host Ga and/or As atoms. The slightly higher χ_{Mn} (normalized Mn PIXE yield) along these two axes indicates that a small fraction of Mn atoms are present in the form of random clusters. These clusters can probably act as precursors of larger precipitates when samples are annealed at high temperature (>400 °C).

In contrast to the $\langle 100 \rangle$ and $\langle 111 \rangle$ scans, along the $\langle 110 \rangle$ axis, the value of χ_{Mn} is much higher than χ_{GaAs}. This suggests that a fraction of the "non-random" Mn atoms in the sample illustrated in Fig. 3.3 is located at tetrahedral interstitial sites. Atoms in the these positions in a zinc blende crystal are shadowed by the host atoms when viewed along both the $\langle 100 \rangle$ and $\langle 111 \rangle$ directions, but are exposed in the $\langle 110 \rangle$ axial channel (Feldman et al., 1982; Tesmer and Nastasi, 1995). Schematics showing the various lattice locations as revealed by the projections for the axial channels are also shown in Fig. 3.3. The tetrahedral interstitials give rise to a double-peak feature in the $\langle 110 \rangle$ angular scan due to the flux peaking effect of the ion beam in the channel. A double-peak feature was indeed observable in the $\langle 110 \rangle$ scan data for the as-grown sample shown in Fig. 3.3 (indicated by the arrows). In the case of the

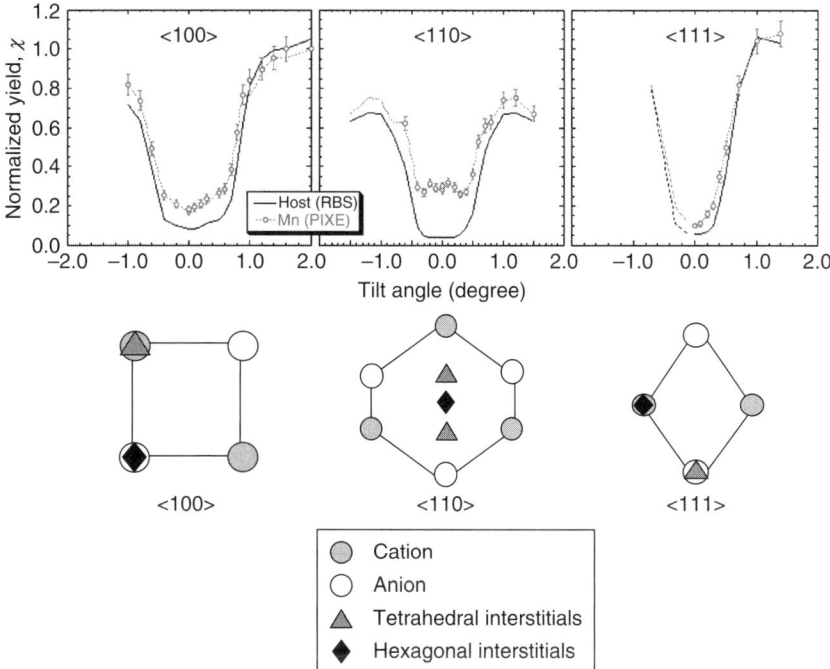

FIGURE 3.3 Mn and GaAs angular scans about the $\langle 100 \rangle$, $\langle 110 \rangle$, and $\langle 111 \rangle$ channels for a 110-nm-thick $Ga_{1-x}Mn_xAs$ film with $x = 0.09$ (top row). The lattice locations of Mn as revealed by the projections for various axial channels are shown in the lower row.

data in Fig. 3.3, the fraction of these intersititial Mn has been roughly estimated to be ~15% of the total Mn in the specimen (Feldman1 et al., 1982; Tesmer and Nastasi, 1995).

Ion channeling studies on a series of $Ga_{1-x}Mn_xAs$ samples with x ranging from 0.02 to 0.1 showed that the Mn_I concentration increases (from 5% to 15%) with the total Mn content. The upper panel of Fig. 3.4 shows the concentrations of Mn_{Ga} and Mn_I as measured by channeling experiments for increasing Mn concentrations x. A comparison of the net uncompensated Mn acceptors ($[Mn_{Ga}] - 2 \times [Mn_I]$) and the hole concentration obtained by ECV profiling, along with the total Mn concentration in the samples, is shown in the lower panel of Fig. 3.4. We have previously demonstrated the accuracy of determining free hole concentration in $Ga_{1-x}Mn_xAs$ alloys by the ECV technique (Yu et al., 2002b). The hole concentration agrees well with the net Mn acceptors for $Ga_{1-x}Mn_xAs$ samples with the exception of the $x = 0.09$ sample, where compensation by Mn_I alone cannot fully account for the low hole concentration. It is possible that in the latter case, As_{Ga} donors are also compensating some of the substitutional Mn acceptors (Sanvito and Hill, 2001).

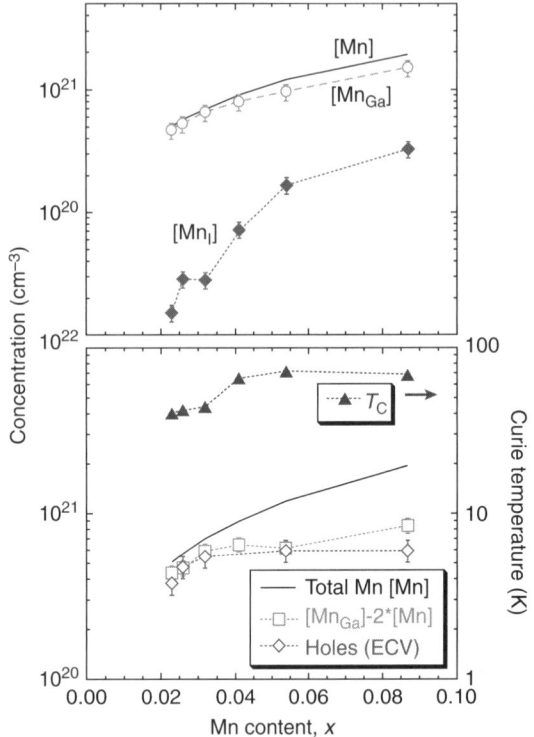

FIGURE 3.4 Concentrations of interstitial and substitutional Mn (top panel) and the net uncompensated Mn acceptor concentration ($[Mn_{Ga}]-2 \times [Mn_I]$) (lower panel) as measured by channeling experiments on as-grown $Ga_{1-x}Mn_xAs$ films. Hole concentrations obtained by electrochemical capacitance-voltage (ECV) profiling for x ranging from 0.02 to 0.09 and corresponding Curie temperatures T_C are shown in the lower panel for the same films.

The creation of Mn_I is extremely deleterious to the ferromagnetism of $III_{1-x}Mn_xV$ alloys for several reasons. First, compensation by the double Mn_I donors reduces the hole concentration (to an approximate value of $[Mn_{III}]-2 \times [Mn_I]$; second, it has been shown theoretically that interstitial Mn does not contribute to the Zener-type Mn–Mn exchange due to its negligible p–d coupling (Blinowski and Kacman, 2003). And finally, Mn_I—because they are both highly mobile and positively charged (Máca and Mašek, 2002; Blinowski and Kacman, 2003)—are expected to drift to interstitial sites *adjacent* to the negatively charged Mn_{III} acceptors to form antiferromagnetic Mn_I–Mn_{III} pairs, thus canceling the magnetic moment of Mn_{III} (Blinowski and Kacman, 2003; Yu et al., 2002a). The density of Mn ions that contribute to the ferromagnetism of the $III_{1-x}Mn_xV$ alloy is then reduced to the value of $[Mn_I]-[Mn_{III}]$. For these reasons, any increase in the Mn_I concentration will automatically lead to a lowering of the value of T_C.

3.2. Thermodynamic limits on hole concentration and on T_C in $Ga_{1-x}Mn_xAs$

3.2.1. The amphoteric defect model

The ECV measurements shown in Fig. 3.4 suggest that the hole concentration in $Ga_{1-x}Mn_xAs$ has a tendency to saturate at high Mn contents (~3–4%). To discuss this issue, it will be useful to invoke the amphoteric native defect model, which has been shown to provide good qualitative predictions of the maximum carrier concentration achievable by doping in a wide variety of semiconductors (including both III–V and II–VI materials) (Walukiewicz, 1989, 1993a, 2001). The model relates the type and concentration of native defects responsible for dopant compensation to the location of the Fermi level with respect to an internal energy reference, the so-called *Fermi level stabilization energy* E_{FS}. E_{FS} was found to be located at ~4.9 eV below the vacuum level, and is the same for all III–V and II–VI semiconductors. Semiconductor materials with the conduction band located close to E_{FS} can be easily doped n-type and, similarly, those whose valence band is close to E_{FS} are easily doped p-type. In III–V semiconductors, it is found that the maximum and minimum energy locations of the Fermi level typically do not deviate by more than 1.0 eV in either direction relative to E_{FS}.

In the specific case of GaAs, the conduction band is located at E_{FS} + 0.9 eV and the valence band at E_{FS} − 0.5 eV. The closer proximity of the valence band to E_{FS} suggests higher limits on the maximum hole concentration compared to that of conduction electrons. Experimentally, the maximum free electron concentration n_{max} in GaAs achievable under equilibrium growth conditions is limited to the mid 10^{18} cm^{-3} range, corresponding to the Fermi energy E_F located approximately at 0.1 eV above the conduction band edge (or at 1.0 eV above E_{FS}) (Walukiewicz, 1993b). We note that nonequilibrium techniques such as Se + Ga coimplantation (Inada et al., 1979a) and pulsed electron beam irradiation techniques (Inada et al., 1979b) have been used to achieve n_{max} values in GaAs up to 2×10^{19} cm^{-3}, corresponding to a maximum Fermi level of E_{FS} + 1.3 eV.

3.2.2. Maximum limit on the Curie temperature

For p-type GaAs, free hole concentrations as high as mid 10^{20}/cm^3 have been achieved via Zn diffusion in GaAs (Walukiewicz, 1993a), in good agreement with the prediction of the amphoteric defect model. This saturation of free holes was explained by an increase in the formation energy of substitutional Zn (Zn_{Ga}) and a decrease in the formation of interstitial Zn donors (Zn_I) as the Fermi energy approaches the valence band. When the Fermi energy reaches a critical value E_{Fmax} (i.e., the condition where the formation energy of Zn_I equals that of Zn_{Ga}), the compensation between these two types of defects results in a saturation of the hole concentration.

The observed dependence of [Mn_I] on the Mn content in $Ga_{1-x}Mn_xAs$ suggests that as the Mn content x increases, the concentration of Mn occupying the Ga sites (Mn_{Ga}) in $Ga_{1-x}Mn_xAs$ increases, pushing the Fermi energy to its maximum level E_{Fmax} (i.e., to the condition where the hole concentration p approaches p_{max}). This dependence of the defect formation energy on the Fermi level and the resulting free carrier saturation are illustrated in Fig. 3.5. Note that because the growth temperature of $Ga_{1-x}Mn_xAs$ is low (190–280 °C), the formation energy of Mn_{Ga} is lower than would be the case in equilibrium growth, which allows E_{Fmax} to occur at a lower energy with respect to the valence band edge than in GaAs grown under standard conditions. The relatively constant hole concentrations of about 6×10^{20} cm^{-3} (~1×10^{21} cm^{-3} for optimally annealed samples) observed for $Ga_{1-x}Mn_xAs$ with Mn concentrations higher than 3% (~6.6×10^{20} cm^{-3}) shown in Fig. 3.4 indicate that the hole concentration p in these $Ga_{1-x}Mn_xAs$ samples corresponds its saturation limit p_{max}. As this limit is reached, the formation energies of Mn_{Ga} acceptors and compensating Mn_I become comparable. Introducing additional Mn into the $Ga_{1-x}Mn_xAs$ beyond this point is expected to result in a downward shift of the Fermi energy that would in turn increase the formation energy of negatively charged Mn_{Ga} acceptors. As the formation of Mn_{Ga} acceptors becomes energetically unfavorable, an increasing

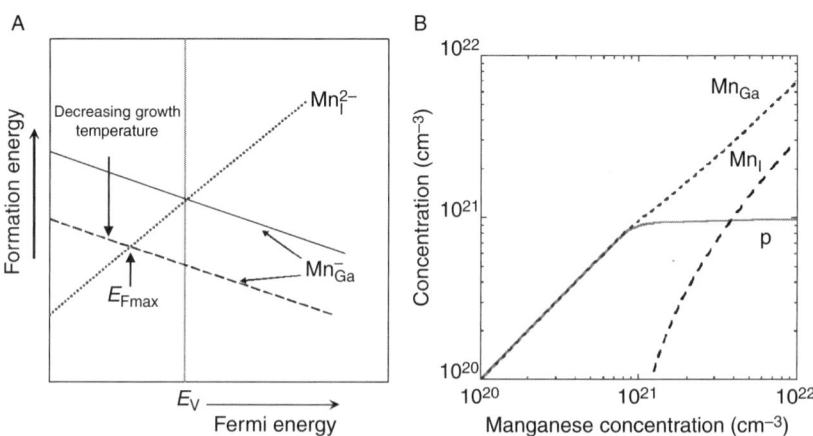

FIGURE 3.5 Schematics showing (A) the formation energy dependence of Mn_{Ga} and Mn_I on the position of the Fermi level and (B) the saturation of hole concentration resulting from compensation by Mn_I as a function of Mn concentration.

fraction of Mn will then be incorporated in the form of Mn_I donors and/or electrically inactive MnAs or Mn clusters. Given that the ferromagnetism in this system is related to the uncompensated Mn spins and is mediated by holes, such Fermi-level-induced hole saturation effect necessarily imposes a fundamental limit on the Curie temperature of the system.

The creation of Mn_I not only puts a limit on the maximum concentration of holes that is essential for mediating ferromagnetic order in $Ga_{1-x}Mn_xAs$ but also has a profound effect on the number of ferromagnetically active spins and—for a constant hole concentration—on the Zener coupling between these spins. Specifically, there are three mechanisms to note in this context. First, it has been shown theoretically that Mn_I on tetrahedral sites do not participate in the hole-mediated ferromagnetism because Mn_I d-orbitals do not hybridize with the p-states of the holes at the top of the valence band (Blinowski and Kacman, 2003). Second, the Mn_I donors may in fact form antiferromagnetically ordered Mn_I–Mn_{Ga} pairs (Blinowski and Kacman, 2003; Yu et al., 2002a), which automatically decrease the total number of uncompensated Mn spins participating in the ferromagnetism, thus further reducing T_C. Finally, when the number of active spins becomes approximately equal to the hole concentration, the average distance between the active Mn spins becomes larger than the first node in the oscillatory RKKY exchange coupling (at $\approx 1.17\ r_{hole}$, where r_{hole} is the average distance between holes) (Dietl et al., 2001). In this situation, some Mn_{Ga} ions may couple antiferromagnetically between themselves. This would at first lead to the drop in T_C, and eventually should drive the system into a spin-glass state (Dietl et al., 2001; Eggenkamp et al., 1995; Schliemann et al., 2001). We believe that some or all of the above factors manifest themselves in $Ga_{1-x-y}Mn_xBe_yAs$ with a high Be content that will be discussed in Section 3.4.2, contributing to the strong drop in T_C and to eventual disappearance of ferromagnetism as the Be content y increases.

3.3. T_C enhancement by elimination of Mn_I

3.3.1. Low-temperature annealing of $Ga_{1-x}Mn_xAs$

As discussed in previous sections (3.1 and 3.2), it is well established that point defects such as As_{Ga} and Mn_I play a crucial role in determining the magnetic properties of $Ga_{1-x}Mn_xAs$ (Yu et al., 2002a). Appropriate low-temperature annealing can alter (improve) the ferromagnetism of $Ga_{1-x}Mn_xAs$, most notably by increasing its Curie temperature, its total magnetic moment, and its hole concentration; by improving the homogeneity of the material; and by changing the temperature dependence of its magnetization to a more mean-field-like behavior (Hayashi et al., 2001; Potashnik et al., 2001). For example, Hayashi et al. (2001) reported an improvement in both the Curie temperature and the crystallinity of

$Ga_{1-x}Mn_xAs$ by heat treatment (annealing) after the MBE growth. Potashnik et al. (2001) have subsequently carried out a very systematic study of the effects of annealing of $Ga_{1-x}Mn_xAs$, showing in particular that the Curie temperature T_C and magnetization $M(T)$ both strongly depend on the annealing temperature and on its duration. To illustrate what physically happens in this process, in what follows we present our own study of postgrowth heat treatment on $Ga_{1-x}Mn_xAs$ epilayers, with special emphasis on the correlation between the location of Mn ions in the lattice and the magnetic properties of this material.

We have systematically investigated the influence of the annealing temperature on the electronic and magnetic properties of $Ga_{1-x}Mn_xAs$, using epilayers with a wide range of Mn concentrations (Kuryliszyn et al., 2003, Yu et al., 2002a). We have applied the following annealing temperatures (T_a) to each of the specimens: 260, 280, 300, 310, and 350 °C. The annealing was carried out in the atmosphere of N_2 gas. All the samples were annealed at the same fixed flow of N_2 gas of 1.5 SCFH (standard cubic feet per hour). The time of annealing was varied from 0.5 to 3 h. At the end of every annealing process, the samples were cooled by a rapid quench to room temperature under the flow of N_2 gas.

Figure 3.6 shows typical temperature dependences of the zero-field resistivity observed on as-grown samples and on samples annealed at different temperatures, for two different Mn contents ($x = 0.032$ and 0.083). We observe that resistivity ρ exhibits a distinctive peak at a certain temperature T_ρ. This peak in the resistivity is known to occur at a

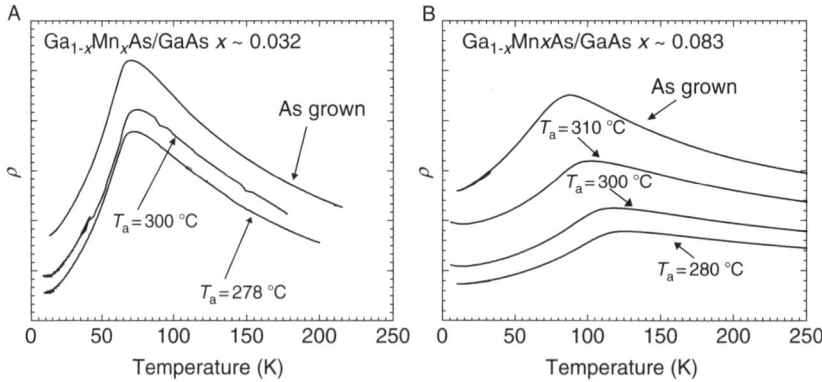

FIGURE 3.6 Temperature dependence of the zero-field resistivity for $Ga_{1-x}Mn_xAs$ samples, both as-grown and annealed at different temperatures (indicated in the figure), for (A) $x = 0.032$ and (B) $x = 0.083$. The peak in the resistivity ρ at temperature T_ρ occurs slightly above the Curie temperature T_C determined by magnetization measurements. The optimal annealing temperature is found to be around 280 °C, slightly higher than the temperature used in the low-temperature molecular beam epitaxy growth of the $Ga_{1-x}Mn_xAs$ epilayers.

temperature slightly above the Curie temperature T_C as determined from magnetization measurements. A theoretical magnetoimpurity model has been proposed to explain the resistance peak near the Curie temperature, as well as the negative magnetoresistance in magnetic semiconductors (Nagaev, 1996). According to this model, the Coulomb field of ionized Mn acceptors attracts free carriers, thereby enhancing indirect exchange interaction in the vicinity of these localized magnetic ions. Since the indirect exchange interaction tends to establish ferromagnetic order, local magnetic order is thereby enhanced in the vicinity of ionized Mn acceptors. Consequently, randomly distributed impurities create not only a fluctuating electrostatic potential but also static fluctuations of the magnetization, which in turn scatter charge carriers. Maximum fluctuations of magnetization are naturally expected to occur near the Curie temperature (i.e., just as the magnetic order begins to form), thus leading to a peak in resistance at or very near T_C.

Note that for samples with smaller content of Mn (e.g., $x = 0.032$), the influence of annealing on both the conductivity and on the Curie temperature is not particularly pronounced (see Fig. 3.6A). For $Ga_{1-x}Mn_xAs$ with a higher Mn content (e.g., $x = 0.083$), as-grown samples and those annealed at 260, 280, 300, and 310 °C show a clearly metallic behavior (see Fig. 3.6B), with the distinct resistivity peak that occurs around T_C. By modifying the process of annealing and subsequent transport characterization, we have established that best results were obtained by annealing at 280 °C for a time between 1.0 and 1.5 h (see Fig. 3.6B). After 1.5 h of annealing, a slight drop of the Curie temperature was observed, in agreement with the results of Potashnik et al. (2001). As the annealing temperatures exceeded 280 °C, the conductivity and the Curie temperature were both observed to decrease.

For the $Ga_{1-x}Mn_xAs$ epilayer with $x = 0.083$, the optimal annealing procedure ($T_a = 280$ °C, annealing time of 1.0 h) showed the resistivity peak at $T_\rho = 127$ K (see Fig. 3.6). SQUID magnetization measurements on this optimally annealed sample yielded a T_C of 115 K. The SQUID and the resistivity measurements have also revealed good agreement between magnetic and transport results for all the remaining samples, which we investigated. Figure 3.7 shows the temperature dependence of the magnetization M for a typical $Ga_{1-x}Mn_xAs$ sample with high content of Mn ($x = 0.081$), as-grown and after annealings at different temperatures for 1.0 h. The magnetization was measured by a SQUID magnetometer in weak magnetic field (i.e., 10 Gauss) after the sample has been magnetized at a higher field (ca. 1000 Gauss) applied parallel to the sample surface.

Figure 3.7 illustrates the three key points of the foregoing discussion. First, it shows the Curie temperature T_C of an as-grown $Ga_{1-x}Mn_xAs$ sample to be about 67 K. It then illustrates the dramatic increase in both the Curie temperature and the magnetization when annealing conditions

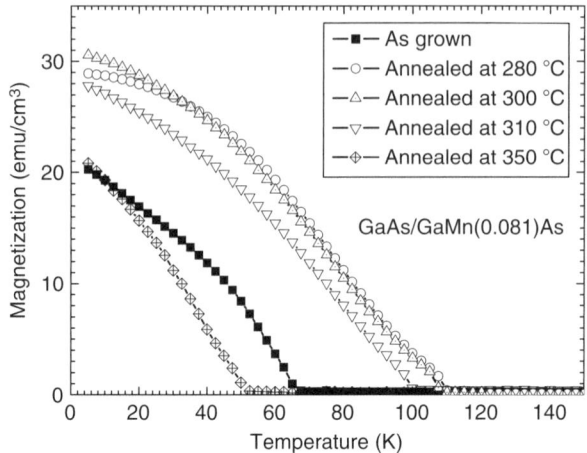

FIGURE 3.7 Magnetization of a $Ga_{1-x}Mn_xAs$ sample with a high Mn content ($x = 0.081$) as a function of temperature after various annealing procedures. The annealing temperatures are indicated in the figure. Note that the Curie temperature and the saturation magnetization are significantly enhanced by low-temperature annealing. When the annealing is carried out at temperatures higher than 300 °C, however, both the Curie temperature and the magnetization are seen to drop monotonically. The highest Curie temperature for this series is 111 K.

are optimally chosen (280 °C, 1.0 h). And, finally, Fig. 3.7 shows that annealing above the optimal temperature (in this case at 350 °C) produces an equally dramatic drop of T_C to a value below that of the as-grown sample.

Since we have demonstrated the presence of Mn_I in $Ga_{1-x}Mn_xAs$, and these Mn_I are expected to be rather mobile (there are many unoccupied interstitial positions), it is logical to ascribe the observed strong changes of magnetic and electrical properties upon low-temperature annealing to a redistribution of the interstitial Mn atoms within the alloy. We investigated the distribution of Mn atoms at various lattice sites for different annealing conditions by channeled RBS and PIXE techniques described earlier in the chapter. These results were correlated with the concentration of the holes and of the uncompensated magnetic moments of Mn^{2+}, giving us a better understanding of how the location of Mn within the $Ga_{1-x}Mn_xAs$ lattice affects the ferromagnetism of this material.

The PIXE and RBS angular scans about the ⟨110⟩ and ⟨111⟩ axes are plotted in the first and the second columns of Fig. 3.8 for three $Ga_{1-x}Mn_xAs$ samples: as-grown, annealed at 282 °C, and annealed at 350 °C. As described in detail in Section 3.1, the "double-peak" feature of the Mn scan in the ⟨110⟩ axis indicates the presence of Mn_I in the sample. Here the fraction of the Mn_I has been roughly estimated to be ~15% of the total Mn content of the sample. This double-peak feature is

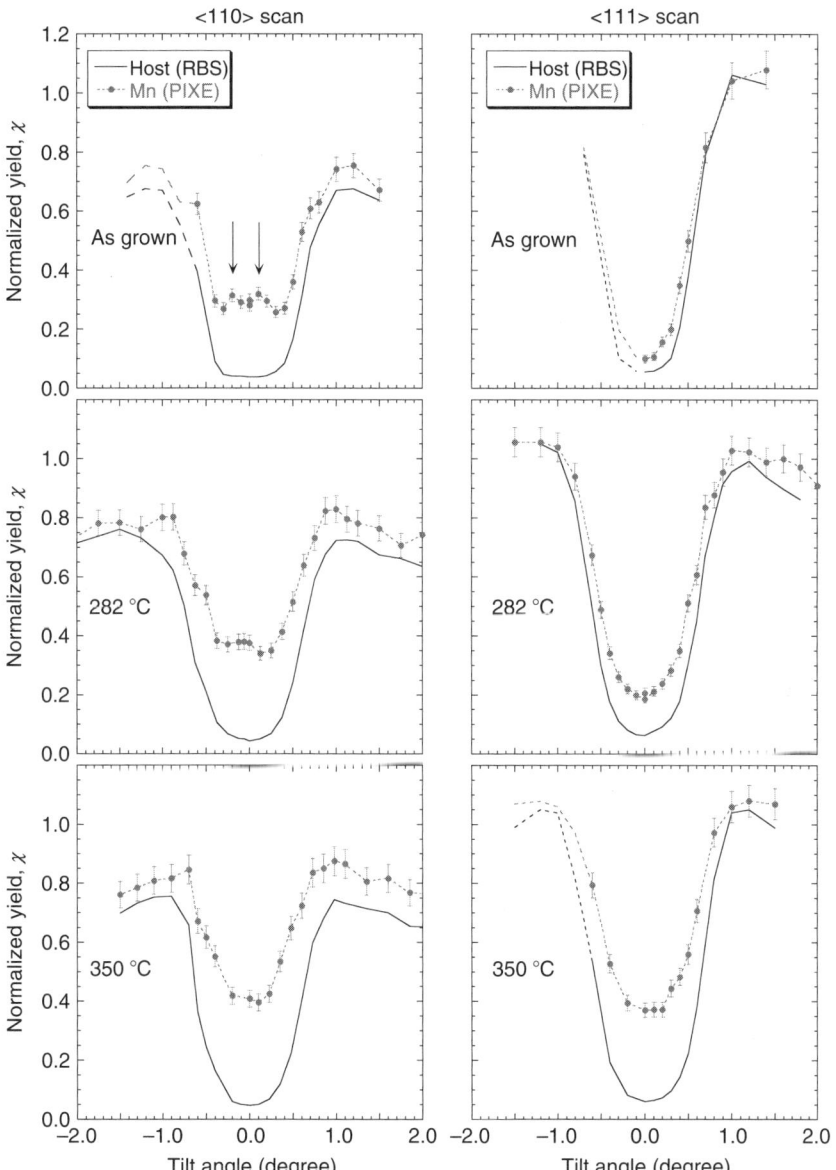

FIGURE 3.8 Angular scans about the $\langle 110 \rangle$ and $\langle 111 \rangle$ axes for three $Ga_{1-x}Mn_xAs$ samples originating from the same growth, but subjected to different annealing conditions.

considerably less prominent for the sample annealed at 282 °C, indicating that the concentration of Mn_I has been reduced by the annealing process. This suggests that the Mn_I are unstable, and precipitate out upon annealing to form Mn clusters and/or MnAs inclusions. The fact that channeled Mn signal in the sample annealed at 350 °C is high and nearly equal for all three channeling scan directions further suggests that, in addition to the intersititial Mn, also a significant fraction of the Mn atoms that were initially at substitutional Mn_{Ga} sites have left their original positions to form random precipitates.

We have discussed in Section 3.1 that the Mn_I act as compensating centers for the substitutional Mn_{Ga} acceptors, thus significantly reducing the concentration of holes. Furthermore, Mn_I are highly mobile and can drift preferentially to the interstitial sites immediately adjacent to the negatively charged substitutional Mn_{Ga} acceptors, thus forming antiferromagnetically aligned Mn_I–Mn_{Ga} pairs (Blinowski et al., 2002), thus canceling the magnetic contribution of Mn_{Ga} to the magnetization of the $Ga_{1-x}Mn_xAs$ system as a whole. The above channeling results clearly indicate that annealing the sample at 282 °C (only slightly above the growth temperature) can break these relatively weakly bound Mn_I–Mn_{Ga} pairs, releasing the highly mobile Mn_I to diffuse away, presumably to form precipitates and/or clusters. This then leaves the substitutional Mn_{Ga} ions to act as electrically active acceptors as well as uncompensated magnetic moments. The 282 °C annealing has therefore led both to a higher hole concentration and to a higher saturation magnetization, as confirmed by ECV and SQUID measurements, respectively.

A summary of the channeling, T_C, and hole concentration measurements on the set of samples which we have used is presented in Fig. 3.9. The upper panel of Fig. 3.9 shows fractions of Mn atoms at the various sites—substitutional (Mn_{Ga}), interstitial (Mn_I), and in random-cluster form (Mn_{rand})—obtained from the angular scans shown in Fig. 3.8. The lower panel shows the values of T_C measured by SQUID magnetization and of the hole concentrations measured by ECV for the same sample subjected to different annealing procedures. It is clear from these measurements that annealing-induced increases of hole concentration and of uncompensated spins both contribute to the increase in T_C. These results show, further, that the large changes in the electronic and magnetic properties induced by annealing can be attributed to rearrangement of the sites occupied by the highly unstable Mn_I within the $Ga_{1-x}Mn_xAs$ lattice. This will be discussed in more detail in Section 3.4.2.

3.3.2. Effects of reducing the film thickness

Recently, a number of reports of T_C over 150 K $Ga_{1-x}Mn_xAs$ have been made for very thin $Ga_{1-x}Mn_xAs$ films after the films were annealed at low temperature (Adell et al., 2005; Chiba et al., 2003; Edmonds et al., 2004b; Ku et al., 2003; Wang et al., 2005). However, there are as yet no reports of $T_C > 110$ K for samples thicker than 100 nm. These findings suggest that for very thin layers, surface and/or interfacial effects may play some role

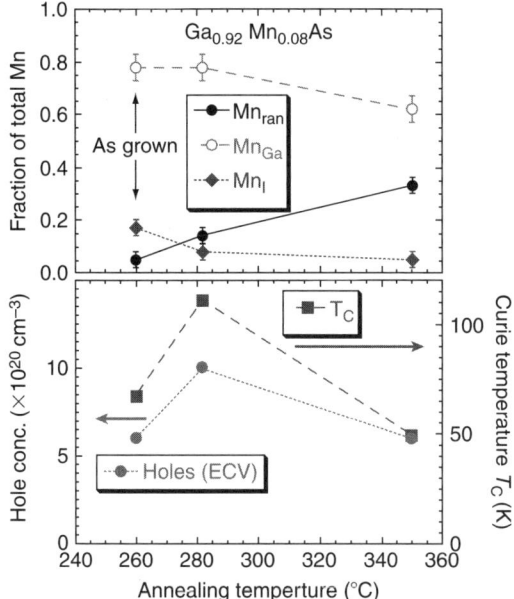

FIGURE 3.9 Upper panel shows fractions of Mn atoms at various sites—substitutional (Mn_{Ga}), interstitial (Mn_I), and in random-cluster form (Mn_{ran})—for the samples used in angular scans in Fig. 3.8. The lower panel shows the values of T_C measured by superconducting quantum interference device magnetization and of the hole concentration measured by electrochemical capacitance voltage (ECV) for the same samples.

in increasing T_C that has not yet been addressed in the preceding discussion. In this section, we explore this question by investigating the effects of Mn location in $Ga_{1-x}Mn_xAs$ as a function of film thickness.

A series of $Ga_{1-x}Mn_xAs$ films with thicknesses ranging from 14 to 200 nm and Mn concentration in the range from 7% to 10% were studied for this purpose. Figure 3.10 shows the angular scans of the Mn (PIXE) and GaAs (RBS) signals from 14 and 100 nm $Ga_{1-x}Mn_xAs$ films about the $\langle 110 \rangle$ and $\langle 111 \rangle$ axes. For the 14-nm-thick sample (upper panels), both the Mn and the host GaAs scans in the $\langle 110 \rangle$ and $\langle 111 \rangle$ directions are very similar, with the value of χ_{Mn} much higher than that of the host χ_{GaAs} for both orientations. This indicates that Mn atoms are incorporated in the 14-nm-thick $Ga_{1-x}Mn_xAs$ layer ($x = 0.1$) either at substitutional Ga sites (Mn_{Ga} fraction ~70%) or at random positions that are incommensurate with the lattice (Mn_{rand} fraction ~30%). The fraction of Mn_I in this sample is below the detection limit of the channeling technique (< 2% of the total Mn).

The channeling results for the 14 nm $Ga_{1-x}Mn_xAs$ layer contrast sharply with the results for $Ga_{1-x}Mn_xAs$ films thicker than 50 nm, where a substantial fraction of Mn is found in interstitial positions

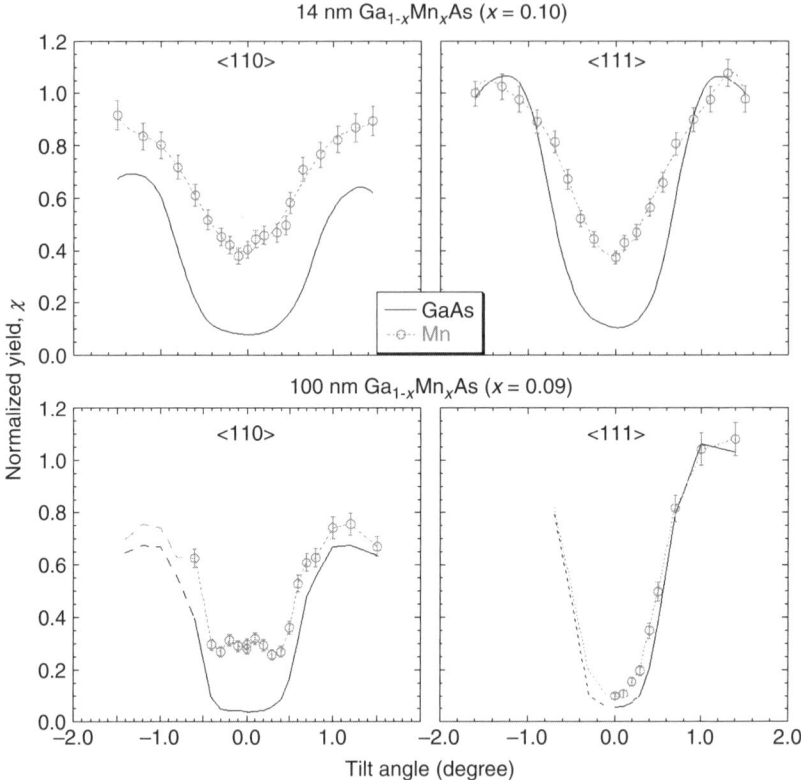

FIGURE 3.10 Angular scans of the Mn K_α x-rays and GaAs RBS signals observed about the $\langle 110 \rangle$ and $\langle 111 \rangle$ axes for two $Ga_{1-x}Mn_xAs$ samples: a 14-nm-thick film with $x = 0.10$ (upper panels) and a 100-nm-thick film with $x = 0.09$ (lower panels).

(Yu et al., 2002a). As an example, we show the angular scans for a 100-nm-thick $Ga_{1-x}Mn_xAs$ layer ($x = 0.09$) in the lower panels of Fig. 3.10. In this thick sample, the fraction of Mn_I is estimated to be ~15% (Yu et al., 2002a; Feldman et al., 2002). We note that SQUID magnetometry measurement shows T_C values of 110 and 65 K for the thin and the thick $Ga_{1-x}Mn_xAs$ samples shown in Fig. 3.10, respectively. We attribute the lower T_C in the thick sample to the presence of Mn_I (Yu et al., 2002a, 2003).

In a recent report, Edmonds et al. (2004b) found that Mn_I are relatively mobile and have a tendency to outdiffuse to the surface during post-growth low-temperature annealing. Such Mn_I outdiffusion was found to be governed by an energy barrier of ~0.7 eV. As a result of the Mn_I diffusion, an Mn-rich oxide layer was detected on the surface of $Ga_{1-x}Mn_xAs$ that can be etched off by HCl (Edmonds et al., 2004a). We believe that the large fraction of Mn_{rand} (~30%) that we observe in the

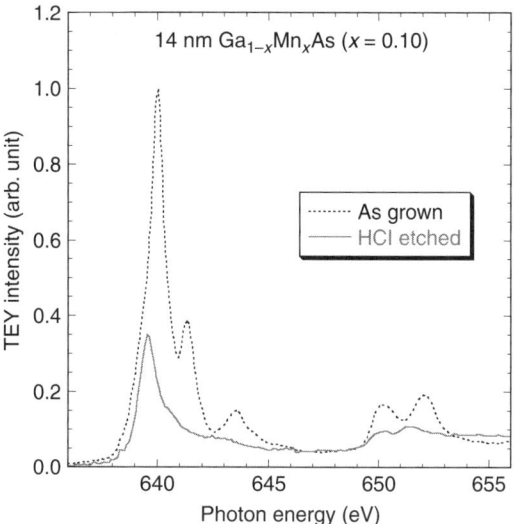

FIGURE 3.11 Mn 2p XAS spectra for a 14-nm-thick $Ga_{1-x}Mn_xAs$ sample, as-grown and after HCl etching.

14-nm $Ga_{1-x}Mn_xAs$ film, as described above, comes from the Mn in the surface oxide layer due to the outdiffusion of Mn_I. Because of the small thickness of the film, it is possible that this outdiffusion process occurred during growth and the Mn-rich layer subsequently oxidized when the film was exposed to air.

Channeling RBS and PIXE measurements on the 14 nm layer after etching in HCl shows that the removal of the ~2-nm-thick surface layer leads to a ~25% reduction in the total Mn concentration and a reduction of Mn_{rand} from ~30% to 15%. This suggests that the oxide surface layer is Mn-rich and contains $\sim 8 \times 10^{14}/cm^2$ of Mn_{rand}. The ~15% Mn_{rand} still present in the etched sample most probably exists in the form of small Mn-related clusters. Furthermore, magnetization measurements before and after the HCl etching show essentially identical results for both T_C and saturation magnetization, suggesting that the Mn_{rand} in the oxide layer do not participate in the carrier-mediated ferromagnetism of the sample.

This interpretation of the RBS/PIXE results is confirmed by x-ray absorption spectroscopy (XAS) measurements, which our team carried out in total electron yield mode at room temperature at the Advanced Light Source (Yu et al., 2005). Figure 3.11 shows the Mn $2p \rightarrow 3d$ XAS spectra for the 14-nm-thick sample before and after the HCl etching. As other researchers have demonstrated previously, the XAS spectrum for the as-grown sample shows a main peak for the L3 level at 640 eV, with a shoulder at ~0.5 eV lower in energy (Ishiwata et al., 2002; Yu et al., 2003). Ishiwata et al. (2002) interpreted this low-energy shoulder as being due to metastable paramagnetic defects due to coupling with excess As. Those authors argue that after low-temperature annealing these defects transform into the ferromagnetic component responsible for the higher energy peak. Figure 3.11 shows that only the low-energy peak remains after the Mn-rich oxide layer is removed by HCl etching. One should note here that XAS measurements on ZnMnOTe alloys also show the 640 eV peak, and have been found to indicate that this peak is produced by Mn–O bonds. This strongly suggests that the high-energy "major" peak in the XAS spectrum from the as-grown $Ga_{1-x}Mn_xAs$ film is due to Mn in the oxide layer, whereas the lower energy peak in the doublet arises from Mn_{Ga}. This is in agreement with the recent report by Edmonds et al. (2004a,b), who identified the presence of an Mn-rich oxide layer, and correlated this low-energy peak with strong x-ray magnetic circular dichroism, thus attesting to its ferromagnetic origin (Yu et al., 2003).

Figure 3.12 shows the distribution of Mn in various lattice sites in $Ga_{1-x}Mn_xAs$ films (x~0.07–0.10) with thicknesses ranging from 14 to 120 nm obtained by ion channeling studies. A monotonic increase in the Mn_I fraction and a corresponding decrease in the Mn_{rand} fraction are observed with increasing film thickness. Above a film thickness of 60 nm, however, the relative amounts of Mn_I and Mn_{rand} remain rather

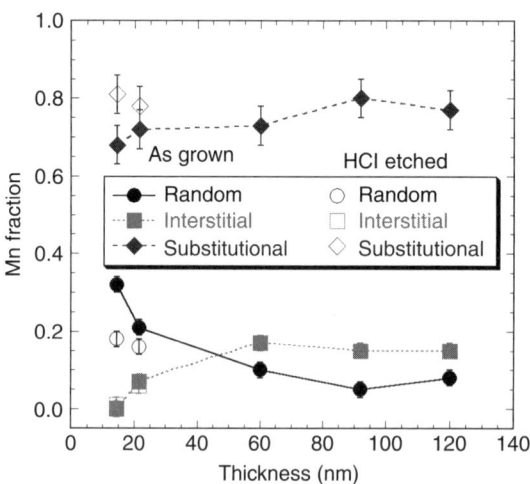

FIGURE 3.12 Fractions of Mn occupying various lattice sites—substitutional (Mn_{Ga}), interstitial (Mn_I), and in random-cluster form (Mn_{rand})—measured by channeling techniques for $Ga_{1-x}Mn_xAs$ samples with film thicknesses between 14 and 120 nm. Data for 14 and 21.7 nm $Ga_{1-x}Mn_xAs$ films etched by HCl are also shown as open symbols.

constant. To understand the origin of this thickness dependence, we note that for thin samples the density of Mn atoms per unit area in the surface oxide layer is relatively independent of film thickness, and is approximately equal to $\sim 8 \times 10^{14}$ cm^{-2}. This is close to the density of Ga sites on the (001) surface, suggesting that the outdiffusion of Mn$_I$ during growth, or after it is exposed to air, can be limited by the accumulation of approximately one monolayer of Mn on the surface. Note that for a 15% fraction of Mn$_I$ in a 15-nm-thick film with $x = 0.10$, the areal concentration of Mn$_I$ is $\sim 5 \times 10^{14}$ cm^{-2}. For films with thicknesses of ~ 10 nm, the diffusion length of Mn$_I$ in the film is comparable to the film thickness (Edmonds *et al.*, 2004b), and it is therefore conceivable that this Mn$_I$ outdiffusion can account for the elimination of Mn$_I$ in very thin films. This picture would suggest that at higher film thicknesses, Mn$_I$ outdiffuion affects only the outer thin layer of the film and is limited by the accumulation of approximately one monolayer of Mn on the surface, whereas the remaining Mn$_I$ stays incorporated in the bulk of the layer. The electronic and magnetic properties of these thick films are then determined by a balance between Mn$_{Ga}$, Mn$_I$, and Mn$_{rand}$.

The distribution of Mn in the 14 and 22 nm Ga$_{1-x}$Mn$_x$As films after HCl etching is also shown in Fig. 3.12. An increase in the fraction of Mn$_{Ga}$ is observed in both these thin samples due to the removal of the Mn-rich oxide layer. We also note that, except for the 14-nm-thick film, the net concentrations of Mn$_{Ga}$ in the Ga$_{1-x}$Mn$_x$As films are in the range of 1.0–1.3 $\times 10^{21}$ cm^{-3}. This is close to the maximum hole concentration of $\sim 1 \times 10^{21}$ cm^{-3} in Ga$_{1-x}$Mn$_x$As that corresponds to the maximum Fermi energy E_{Fmax} we reported previously (Yu *et al.*, 2002a). However, we estimate the concentration of Mn$_{Ga}$ to be $\sim 1.7 \times 10^{21}$ cm^{-3} for the 14-nm-thick film. This enhancement in the incorporation of Mn$_{Ga}$ could be partially explained by Fermi level pinning at the free surface and at the interface between the Ga$_{1-x}$Mn$_x$As layer and the underlying LT-GaAs. Such pinning would raise the Fermi energy, thus giving rise—by lowering the formation energy of Mn$_{Ga}$—to the higher T_C observed in thin layers without postgrowth annealing.

In summary, we found that for film thicknesses less than 60 nm, the surface on which Ga$_{1-x}$Mn$_x$As is grown acts as a sink that facilitates the outdiffusion of Mn$_I$, thus reducing their concentration in the film. As a case in point, for the Ga$_{1-x}$Mn$_x$As film thicknesses below 15 nm, no Mn$_I$ could be detected. One can thus conclude that, because of the absence of compensating Mn$_I$ defects, higher values of T_C can be achieved for such extremely thin Ga$_{1-x}$Mn$_x$As layers. Most of the Mn not incorporated as Mn$_{Ga}$ in this case accumulate as a surface oxide layer and do not participate in the ferromagnetism of the film. These results are fully consistent with our previously proposed model of Fermi-level-controlled upper limit of T_C in III$_{1-x}$Mn$_x$V ferromagnetic semiconductors.

3.4. Fermi level manipulation

It was established experimentally that T_C in $Ga_{1-x}Mn_xAs$ increases with increasing Mn_{Ga} concentration x and the hole concentration (Ohno, 1998, 1999). These observations are consistent with the Zener model of ferromagnetism proposed by Dietl et al. (2000), which predicts that

$$T_C = Cxp^{1/3}, \quad (3.1)$$

where x is the mole fraction of substitutional Mn^{2+} ions, p is the hole concentration, and C is a constant specific to the host material. Since the ferromagnetic ordering depends on the concentration of uncompensated holes, it would appear promising to use an additional acceptor impurity, such as Be, that has a shallower acceptor level and a higher solubility in GaAs (Specht et al., 1999) to increase the total hole concentration, thereby increasing T_C. However, as we have discussed in Section 3.2, the dependence of defect formation energy on the Fermi level results in saturation of the hole concentration. This in turn necessarily imposes a fundamental limit on the Curie temperature of the system. A number of systematic experiments have been undertaken to explore the effect of creating additional carriers on the properties of $Ga_{1-x}Mn_xAs$ by Be doping, which will be described below.

3.4.1. Extrinsic doping of GaMnAs with low Mn concentration

Since $Ga_{1-x}Mn_xAs$ is a strongly compensated material when it is heavily doped with Mn, it will be instructive to begin by discussing studies of extrinsic Be doping in $Ga_{1-x}Mn_xAs$ involving specimens with Mn concentration below the hole saturation level ($\sim 1 \times 10^{21}$ cm^{-3}), typically $x = 0.02$. A series of $Ga_{1-x}Mn_xAs$ samples was fabricated for this purpose with a fixed Mn concentration, with a systematically increasing Be doping level. Figure 3.13 shows the temperature dependence of zero-field resistivity for undoped and Be-doped $Ga_{1-x}Mn_xAs$ for $x = 0.02$ (Yuldashev et al., 2003). In the figure, A indicates an undoped $Ga_{1-x}Mn_xAs$ sample, and B and C indicate Be-doped $Ga_{1-x}Mn_xAs$ with increasing doping level. The sample without Be shows an insulating behavior, whereas the samples doped by Be show a metallic behavior. All the samples showed a peak in the zero-field resistivity around the Curie temperature T_C. It is seen from Fig. 3.13 that T_C of $Ga_{1-x}Mn_xAs$ without Be (sample A) is about 40 K, whereas samples doped by Be show higher Curie temperatures. As mentioned earlier, the Curie temperature of $Ga_{1-x}Mn_xAs$ in the Zener model is given by Eq. (3.1), where T_C is proportional to $p^{1/3}$. Thus, one can expect from Eq. (3.1) a monotonic increase of T_C by increasing Be doping, which is indeed observed in the resistivity measurements shown in Fig. 3.13.

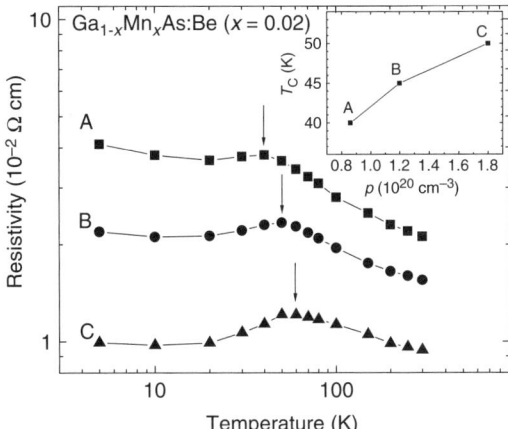

FIGURE 3.13 Temperature dependence of the resistivity for undoped (A) and Be-doped (B, C) samples of $Ga_{1-x}Mn_xAs$ with low Mn content ($x = 0.02$) at zero magnetic field. The dependence of T_C on carrier concentration of these samples is shown in the inset.

We now examine the effect of Be doping on the carrier concentration and on T_C in this series of $Ga_{1-x}Mn_xAs$ epilayers with low Mn content. The hole concentration for this series can be roughly estimated from room temperature Hall measurements. It is well known that determining the concentration of carriers from Hall measurements in ferromagnetic materials is complicated by the presence of AHE contribution, and we have shown in our ECV studies on $Ga_{1-x}Mn_xAs$ that the error produced by AHE can be very large (Yu et al., 2002b). In low Mn concentration samples, however, the increase of the hole concentration p measured by the Hall effect as the Be codoping level is increased does indeed indicate a true increase of p. The relationship between the Hall carrier concentration and T_C of this series of low Mn-concentration samples is presented in the inset of Fig. 3.13 (Yuldashev et al., 2003). Since the Mn concentration (and therefore also the holes produced by the Mn_{Ga} acceptors) in these samples is below the hole saturation level, the formation energies of compensating defects (Mn_I and As_{Ga}) remain higher than the formation energy of Mn_{Ga}. Hence, the introduction of Be acceptors during the growth of $Ga_{1-x}Mn_x$As at low values of x does increase the hole concentration, and consequently also the value of T_C.

3.4.2. Be doping of GaMnAs at high Mn concentration

Although the effect of Be codoping of $Ga_{1-x}Mn_xAs$ observed at low x—specifically the increase of hole concentration and the Curie temperature—is what one would logically expect, the effect observed for high Mn concentrations is, at first sight, quite counterintuitive. It is, however,

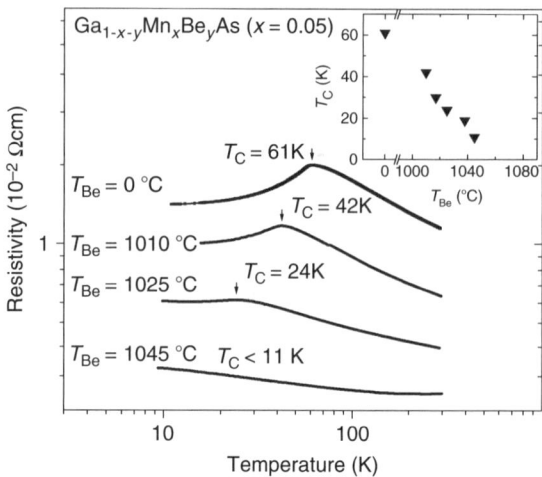

FIGURE 3.14 Temperature dependence of the resistivity for undoped and Be-doped samples of $Ga_{1-x}Mn_xAs$ with high Mn content ($x = 0.05$) at zero magnetic field. The dependence of T_C on Be doping of these samples is shown in the inset.

consistent with our Fermi-level-induced hole saturation model. Specifically, it was found that adding Be to $Ga_{1-x}Mn_xAs$ with $x > 0.05$ leads to a strong *decrease* in T_C (Furdyna et al., 2003; Lee et al., 2003; Wojtowicz et al., 2004b). This is clearly visible in Fig. 3.14 (Lee et al., 2003), where the maximum in the temperature dependence of the resistivity shifts to lower T with increasing Be level (where the level of Be is controlled by the Be cell temperature T_{Be} during MBE growth). This strong drop of T_C (see the inset in Fig. 3.14) was also confirmed by studies of AHE and by direct magnetization measurements using SQUID. Moreover, it was observed that, while T_C decreased with increased Be codoping, the hole concentration remained *nearly constant* (Furdyna et al., 2003; Wojtowicz et al., 2004b). This was determined by ECV for all samples, and was additionally corroborated by Hall measurements for samples with a high Be content (see Fig. 3.15B) that were not ferromagnetic, and thus not affected by AHE. Here the Be content shown in the figure was estimated from the lattice constant determined by XRD that was calibrated by RHEED intensity oscillations.

To obtain physical insight into the causes of this surprising behavior, the relationship between the location of Mn in the lattice and the ferromagnetism of the $Ga_{1-x-y}Mn_xBe_yAs$ layers was investigated using c-PIXE/RBS, similar to the studies described earlier in this chapter. Figure 3.16 shows the PIXE and RBS angular scans about the $\langle 110 \rangle$ and $\langle 111 \rangle$ axes for the $Ga_{1-x}Mn_xAs$ and $Ga_{1-x-y}Be_yMn_xAs$ films with increasing y, respectively (results from only three of six samples are shown for

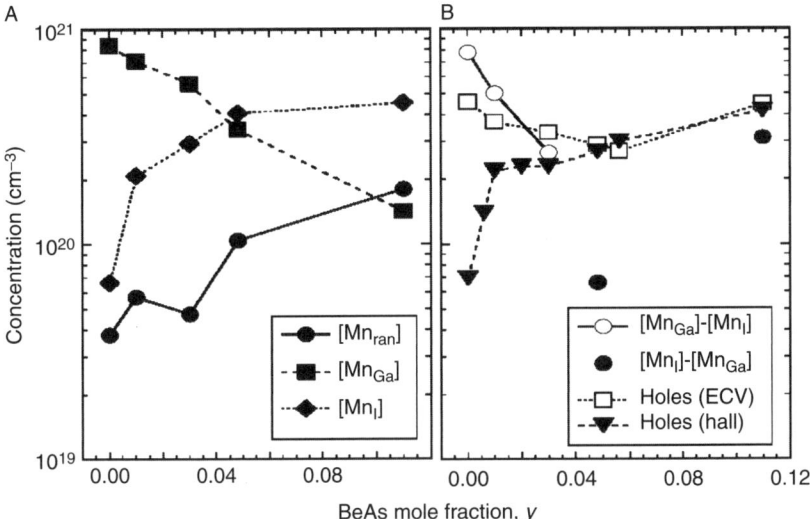

FIGURE 3.15 (A) Concentration of Mn atoms at various inequivalent sites in the GaMnAs epilayers obtained from the analysis of angular particle-induced x-ray emission scans. (B) Net concentration of uncompensated Mn_{Ga} ($Mn_{Ga}-Mn_I$), together with hole concentrations measured by electrochemical capacitance-voltage and by Hall resistivity for $Ga_{1-x-y}Mn_xBe_yAs$ films as a function of Be concentration y.

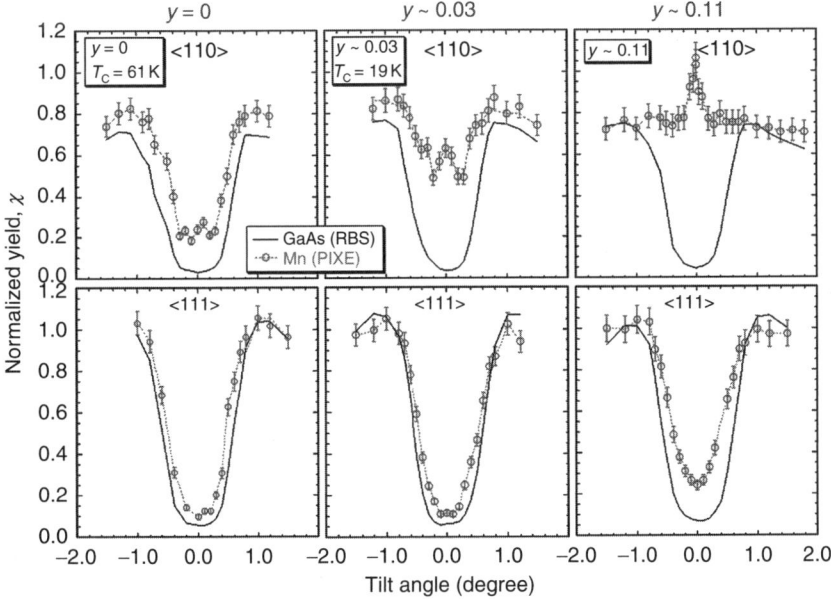

FIGURE 3.16 Angular scans about the $\langle 110 \rangle$ and $\langle 111 \rangle$ channeling axes for undoped and for Be-doped $Ga_{1-x}Mn_xAs$ samples.

simplicity; Yu et al., 2003, Yu et al., 2004b). As the Be content increases, the ⟨110⟩ Mn angular scans show a definite peak at the center of the [110] channel, that increases in intensity with increasing level of Be doping—a clear signature that the concentration of Mn_I increases as more Be is introduced into the alloy lattice (Yu et al., 2004b). At the same time, the normalized yields χ_{Mn} in the ⟨111⟩ scans also show a gradual increase, deviating from the corresponding host scans as the Be content increases, indicating that as the Be content increases, so does the formation of Mn-containing random clusters not commensurate with the GaAs lattice.

The fractions of Mn atoms at the various sites—substitutional (Mn_{Ga}), interstitial (Mn_I), and in random-cluster form (Mn_{ran})—as measured from the angular scans are shown in Fig. 3.15A. These results unambiguously reveal that the fraction of Mn_I as well as random Mn-based precipitates in the $Ga_{1-x-y}Mn_xBe_yAs$ increases monotonically with Be content. Notice that as the Be content y approaches the Mn content ($x = 0.05$), the concentration of Mn_I becomes larger than that of Mn_{Ga}. Consequently, the net concentration of uncompensated Mn ions that can participate in the ferromagnetism ($[Mn_{Ga}]$-$[Mn_I]$) decreases strongly with increasing y. This reduction of active spins will naturally lead to the disappearance of ferromagnetism.

The relatively constant hole concentration at $x \geq 0.05$ shown in Fig. 3.15B strongly suggests that the hole concentration in these $Ga_{1-x-y}Mn_xBe_yAs$ samples is at its saturation limit. We recall that at this limit, the formation energies of Mn_{Ga} acceptors and compensating Mn_I become comparable. Thus, the introduction of additional Be acceptors into $Ga_{1-x-y}Mn_xBe_yAs$ leads to a downward shift of the Fermi energy, that in turn increases the formation energy of negatively charged Mn_{Ga} acceptors and reduces the formation energy of Mn_I donors. As a result, an increasing fraction of Mn is incorporated into the system in the form of Mn_I donors and/or electrically inactive precipitates. Consequently, as argued earlier, there is a drastic drop in T_C with increasing y.

3.4.3. Modulation doping of $Ga_{1-x}Mn_xAs$ in heterostructures

Studies of low-temperature annealing and extrinsic Be codoping of $Ga_{1-x}Mn_xAs$ layers have clearly demonstrated that a strong increase of the Mn_I concentration occurs at the expense of Mn_{Ga} as E_F decreases, in a manner discussed earlier in this chapter. However, lowering the position of E_F can affect the creation of Mn_I only when the additional holes are already present during the $Ga_{1-x}Mn_xAs$ growth. Consider now modulation doping (MD) of a heterostructure in the form $Ga_{1-y}Al_yAs/Ga_{1-x}Mn_xAs/Ga_{1-y}Al_yAs$, where $Ga_{1-x}Mn_xAs$ acts as a quantum well (QW), as shown in Fig. 3.17. In this case, one expects an increase of T_C when the Be-doped barrier is grown *after* the $Ga_{1-x}Mn_xAs$ deposition because then the value of p increases *after* the ferromagnetically active region has already been formed. Similarly, one would expect a drop in T_C

when the Be-doped barrier is grown *before* the $Ga_{1-x}Mn_xAs$ QW is deposited, because in that case the holes will enter the well from the barrier as $Ga_{1-x}Mn_xAs$ well is being deposited, preventing the Mn from occupying the Mn_{Ga} acceptor sites, and inducing the creation of Mn_I instead.

We begin by discussing the experimental results that demonstrate that the Curie temperature of the $Ga_{1-x}Mn_xAs$ layer can be increased by MD when the acceptor-doped barrier is grown *after* the magnetic layer has already been deposited. The temperature dependence of zero-field resistivities ρ for MD heterostructures in which Be-doped $Ga_{1-y}Al_yAs$ was grown after depositing $Ga_{1-x}Mn_xAs$ is shown in Fig. 3.18. All samples

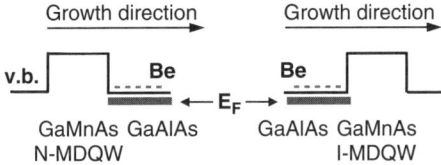

FIGURE 3.17 Schematic diagrams of the N-MDQW and I-MDQW geometries used in the modulation doping studies of $Ga_{1-x}Mn_xAs$ quantum wells adjacent to Be-doped $Ga_{1-y}Al_yAs$ barriers.

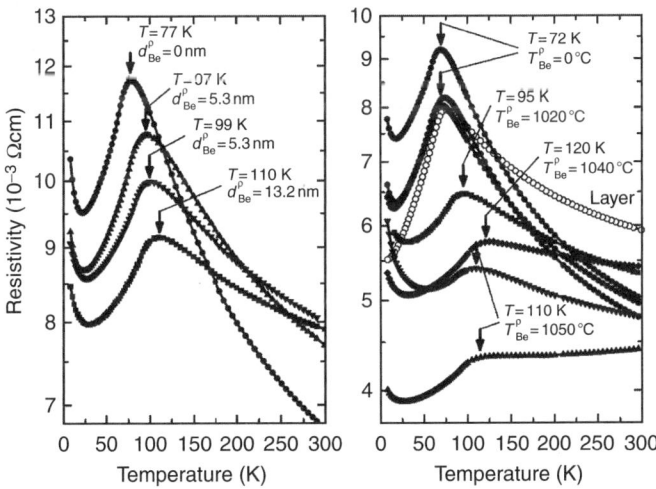

FIGURE 3.18 Temperature dependence of zero-field resistivities for $Ga_{1-x}Mn_xAs$/$Ga_{0.76}Al_{0.24}As$ heterostructures modulation doped by Be acceptors. Left-hand panel: sample series with $x = 0.062$, $T_{Be} = 1040\,°C$, and various d_{Be}. Right-hand panel: sample series with $x = 0.066$, $d_{Be} = 13.2$ nm, and various T_{Be}. In order to show reproducibility of the results, in the right-hand panel, we also present data for three undoped heterojunctions. Also shown as the open symbols are data for a $Ga_{0.94}Mn_{0.66}As$ epilayer with no GaAlAs barrier. Sample parameters and peak resistivity values ($T_\rho \approx T_C$) are indicated in the figure.

show a clear resistivity peak at a temperature T_ρ, providing a convenient marker of T_C (Lee et al., 2003; Wojtowicz et al., 2003b). The data for samples in which the thickness of the Be-doped region d_{Be} was varied while keeping the Be flux constant are shown in the left-hand panel. One can see that T_ρ increases from 77 K in the undoped structure ($d_{Be} = 0$) to ≈ 98 K for $d_{Be} = 5.3$ nm, and to 110 K for $d_{Be} = 13.2$ nm.

To further test the enhancement of T_C with increasing Be concentration in the barrier, we studied a series of specimens with a constant d_{Be} but grown with various T_{Be}. The results for this series are shown in the right-hand panel of Fig. 3.18. Additionally, data for a 67-nm-thick $Ga_{1-x}Mn_xAs$ epilayer with the same Mn concentration ($x = 0.066$) are also shown as the open symbols. While all of the undoped samples grown under the same conditions (three heterostructures and one layer) have very similar values of T_ρ (72 – 75 K), the doped structures all have higher T_ρ, ranging from 95 K (for $T_{Be} = 1020$ °C) to 110 K ($T_{Be} = 1050$ °C). The enhancement of T_C via MD is further corroborated by direct SQUID magnetization measurements (Wojtowicz et al., 2003b) and by the studies of AHE.

To demonstrate experimentally that the sequencing of layer growths plays a crucial role in the formation of Mn_I states—and hence in controlling T_C—we grew a series of low-temperature $Ga_{1-y}Al_yAs/Ga_{1-x}Mn_xAs/Ga_{1-y}Al_yAs$ QW structures in the following order: a structure with Be doping in the first barrier (so-called "inverted modulation doped QW," I-MDQW), an undoped structure, and a structure with Be doping in the second barrier ("normal MDQW"—N-MDQW) (Wojtowicz et al., 2003b). The left-hand panel in Fig. 3.19 shows the temperature dependence of the normalized magnetization obtained from the Arrot plot analysis of AHE data for this sample series. In the analysis, the side-jump scattering mechanism was assumed to be dominant in the AHE. This method of accessing the modulation-doping-induced variation of the magnetic properties of heterostructures is in good agreement with the results obtained from resistivity (Wojtowicz et al., 2003b) and SQUID magnetization measurements. The latter are presented in the right-hand panel of Fig. 3.19 in the form of temperature-dependent remanent magnetization. The values of T_C, taken to be the temperature at which the magnetization versus T drastically changes the slope before vanishing, are indicated with arrows in both panels. As compared to $T_C = 80$ K for the undoped sample, we find that—as in the previous series—T_C has increased (to 98 K) for the N-MDQW sample whose barrier was doped after the $Ga_{1-x}Mn_xAs$ QW was already grown. And, similarly as in the series discussed earlier, T_C has decreased (to 62 K) for the I-MDQW sample whose barrier was doped before growing the $Ga_{1-x}Mn_xAs$ QW, in agreement with our model of Fermi-energy-dependent creation of Mn_I.

We now focus on the Fermi-energy-dependent incorporation of Mn at various III–V lattice sites. The enhanced incorporation of Mn into

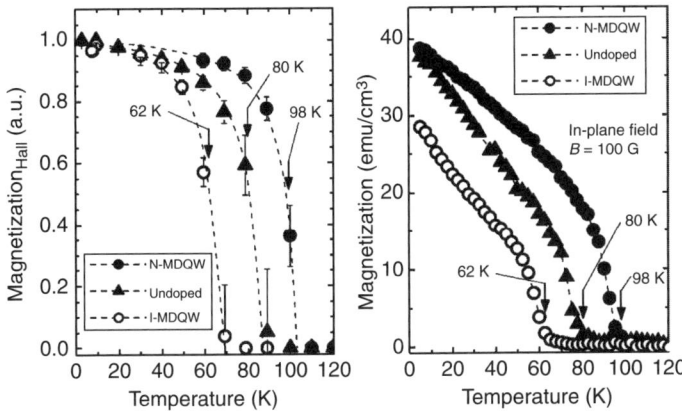

FIGURE 3.19 Left-hand panel: temperature dependence of normalized magnetization obtained from the Arrot plot analysis of the anomalous Hall effect data for three $Ga_{0.76}Al_{0.24}As/Ga_{0.938}Mn_{0.062}As/Ga_{0.76}Al_{0.24}As$ quantum well structures. Right-hand panel: temperature dependence of the remanent magnetization measured by superconducting quantum interference device (SQUID) with an in-plane magnetic field of 100 Gs for the same samples. Be acceptors were introduced either into the first barrier (that grown before the QW, I-MDQW) or into the second barrier (that grown after the QW, N-MDQW). The "undoped" sample represents a heterostructure with no doping in either of the barriers. The curves are guides for the eye. The values of T_C determined from SQUID data are marked in both panels.

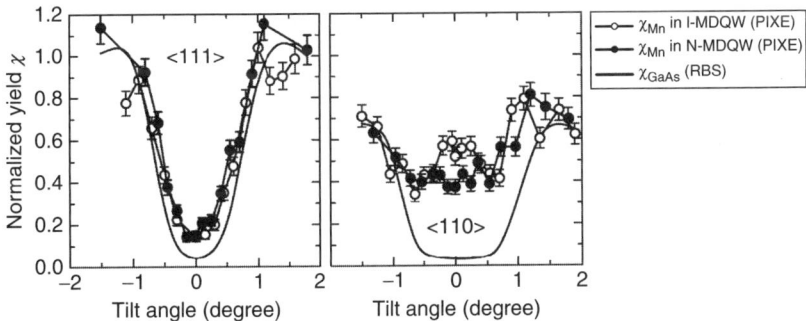

FIGURE 3.20 Normalized yields of the Mn (particle-induced x-ray emission, PIXE) and GaAs (Rutherford backscattering spectrometry, RBS) signals for a $Ga_{0.938}Mn_{0.062}As$ QW layer in both I-MDQW and N-MDQW geometries (the same as in Figures 3.17 and 3.19) as a function of the incident tilt angles for angular scans about the $\langle 110 \rangle$ and $\langle 111 \rangle$ channeling axes.

interstitial positions for structures where the Be-doped barrier was grown before the deposition of the $Ga_{1-x}Mn_xAs$ well was unambiguously confirmed by c-RBS/PIXE experiments (Yu et al., 2004a). Figure 3.20 shows the $\langle 110 \rangle$ and $\langle 111 \rangle$ angular scans for the $Ga_{1-x}Mn_xAs$ layers in both the I-MDQW and the N-MDQW geometries. The difference in the respective Mn locations for I-MDQW and N-MDQW is made clear by the Mn PIXE signal in angular scans about the $\langle 110 \rangle$ axial channel shown in the right-hand panel of Fig. 3.20. The much higher χ_{Mn} in the $\langle 110 \rangle$ scan for the I-MDQW indicates that the concentration of Mn_I is higher in the case when the Be-doped barrier layer is grown prior to depositing of the $Ga_{1-x}Mn_xAs$ QW. Assuming the flux peaking in the $\langle 110 \rangle$ channel to be between 1.5 and 2.0 (Yu et al., 2002a), we estimate the fractions of Mn_I to be ~20% for I-MDQW and ~11% for N-MDQW.

Results of RBS/PIXE studies presented in Fig. 3.20 provide direct experimental evidence that the manner of incorporation of Mn into $Ga_{1-x}Mn_xAs$ is directly controlled by the Fermi energy during the LT-MBE growth itself. These data also rule out the possibility that other effects—such as the competition between Mn and Be atoms to occupy the same substitutional sites—might be responsible for the increase in the number of Mn_I, since in modulation-doped structures, Mn and Be atoms are spatially separated. Our studies of inverted and normal MDQWs thus provide very strong support for the model of Fermi-level-induced limitation of T_C in $Ga_{1-x}Mn_xAs$ (Wojtowicz, 2003; Yu et al., 2002a).

3.5. Self-compensation in other III-Mn-V alloys

The findings concerning the lattice location of Mn in $Ga_{1-x}Mn_xAs$ described above have established that the Curie temperature of this material is limited by the formation of compensating Mn_I defects that are controlled electronically by the position of the Fermi level *while the material is being grown*. Further experiments described below will demonstrate that this *electronic* limitation on fabricating a ferromagnetic specimen also applies to other diluted ferromagnetic $III_{1-x}Mn_xV$ alloys. One of these is $In_{1-x}Mn_xSb$, representing an "extreme" among the $III_{1-x}Mn_xV$ alloys (Wojtowicz et al., 2003a, 2004a; Yanagi et al., 2004), since it has the *largest lattice constant*, the *smallest energy gap*, and the *smallest effective masses of carriers* in this family of ferromagnetic semiconductors. $In_{1-x}Mn_xSb$ is also important because it brings with itself new opportunities for applications in far-infrared spin-photonics and, due to its high carrier mobilities and to an especially strong spin-orbit interaction, also in devices that involve spin-dependent transport.

Figure 3.21 shows the c-PIXE and c-RBS angular scans about the $\langle 111 \rangle$ and $\langle 110 \rangle$ axes of two $In_{1-x}Mn_xSb$ samples, with $x = 0.02$ and 0.028, respectively. The higher χ_{Mn} observed in the $\langle 110 \rangle$ direction in both specimens clearly indicates the presence of Mn_I in this material, similar to the case of $Ga_{1-x}Mn_xAs$. Quantitative analysis of the data for $In_{1-x}Mn_xSb$ indicates that even at relatively small total Mn concentration of $x = 0.028$,

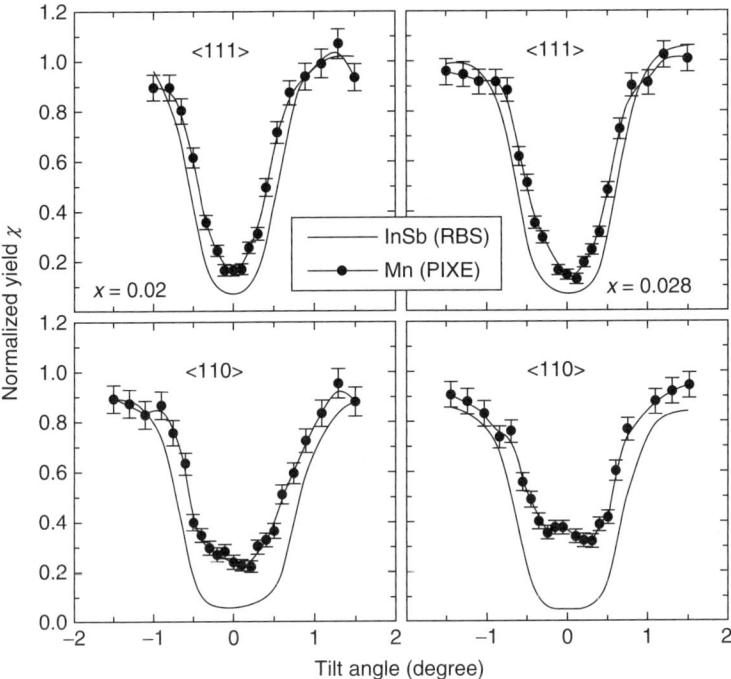

FIGURE 3.21 Mn (particle-induced x-ray emission, PIXE) and InSb (Rutherford backscattering spectrometry, RBS) angular scans about the $\langle 110 \rangle$ and $\langle 110 \rangle$ channels for $In_{1-x}Mn_xSb$ films with $x = 0.02$ (left panels) and $x = 0.028$ (right panels).

as much as 8% of the total Mn population already occupies interstitial sites, and additional 10% is in random sites. This means that, due to the creation of compensating donors, the maximum hole concentrations p_{max} that can be achieved in this material are of the order of 2×10^{20} cm^{-3}, significantly lower than the value of p_{max} in $Ga_{1-x}Mn_xAs$. Qualitatively, however, the electrical and magnetic compensation caused by the Mn_I ions in this material is believed to be similar to that observed in $Ga_{1-x}Mn_xAs$.

In Fig. 3.22, we show the calculated Curie temperatures for $In_{1-x}Mn_xSb$ obtained using the mean-field-theoretical (MFT) approach based on the eight-band effective band-orbital method (EBOM) (Vurgaftman and Meyer, 2001; Wojtowicz et al., 2003a, 2004a). Here we plot the theoretical Curie temperatures as a function of the concentration of ferromagnetically active Mn in $In_{1-x}Mn_xSb$ for three values of hole concentrations: $p = 1 \times 10^{20}$, 2.1×10^{20}, and 3×10^{20} cm^{-3}. Experimental values of T_C for the two $In_{1-x}Mn_xSb$ samples ($x = 0.02$ and 0.028) are also shown, marked with circles and rectangles, respectively. The open symbols are plotted using the *total* Mn concentrations x of the samples, while the full

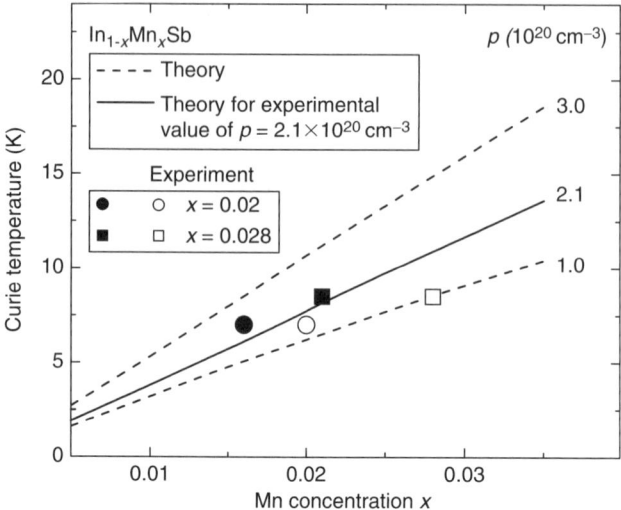

FIGURE 3.22 Curie temperature as a function of Mn concentration in ferromagnetic $In_{1-x}Mn_xSb$ layers. Lines represent theoretical results obtained using the effective band-orbital method–mean-field-theoretical formalism for three hole concentrations: $p = 1 \times 10^{20}$, 2.1×10^{20}, and 3×10^{20} cm^{-3}. Experimental values of T_C for two samples grown using Mn cell temperatures $T_{Mn} = 710\,°C$ and $720\,°C$ are marked with circles and rectangles, respectively. Full symbols are plotted using effective Mn concentrations $x_{eff} = 0.016$ and 0.021 on the x-axis (where x_{eff} is obtained by taking into account the Mn_{In}-Mn_I AF pairing). Open symbols are plotted using the total Mn concentration of the samples as x (0.02 and 0.028, respectively).

symbols correspond to the *effective* Mn concentrations for these two samples (0.016 and 0.021, respectively), obtained by taking into account the effect of Mn_{In}-Mn_I AF pairing, as determined from c-PIXE and c-RBS. As one can see, the agreement between experiment and theory is greatly improved when the effective Mn concentration is used in the calculations.

The effect of extrinsic Be doping in $In_{1-x}Mn_xSb$ alloys have also been investigated. Transport studies of Be codoped layers shown in Fig. 3.23 reveal that—when the Be content y is varied while keeping the Mn concentration x constant—the T_C progressively decreases for increasing Be doping level, as indicated by the shift of T_ρ toward lower temperatures. This is consistent with our previous experiments on Be codoping of $Ga_{1-x}Mn_xAs$ detailed in Section 3.4.2, and therefore can be understood in terms of the model of Fermi-level-induced formation of compensating defects (Yu et al., 2003). Specifically, the drop of T_C with increasing Be content observed in Fig. 3.23 is caused by the Fermi-level-induced reduction of active Mn spins due to the increase in the formation of Mn_I and random Mn clusters (Walukiewicz, 2001; Yu et al., 2003).

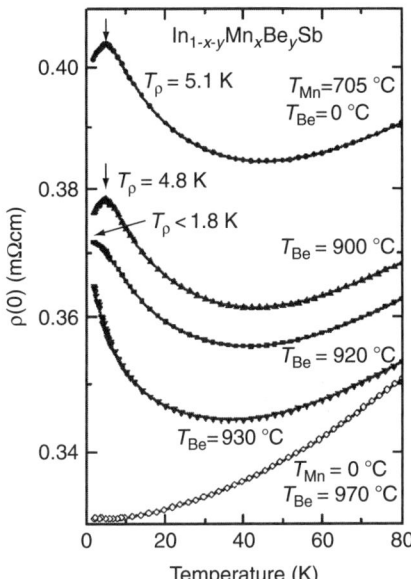

FIGURE 3.23 Resistivity as a function of temperature for $In_{1-x-y}Mn_xBe_ySb/InSb/CdTe$ layers grown with constant $T_{Mn} = 705\ °C$ (2.8% Mn) and with various levels of Be codoping determined by temperatures of the Be effusion cell (indicated in the figure as T_{Be}). The data for one nonmagnetic $In_{1-y}Be_ySb/InSb/CdTe$ layer with a similar hole concentration is also shown for comparison (open circles). The ρ of the nonmagnetic layer is shifted up by 0.08 mΩ cm.

4. CONCLUDING REMARKS

4.1. Alternative synthesis routes for $III_{1-x}Mn_xV$ alloys

To date, $Ga_{1-x}Mn_xAs$ is almost universally grown by LT-MBE. The low growth temperatures (typically < 300 °C) necessary to prevent the formation of thermodynamically more stable second phases such as MnAs result, however, in the formation of As_{Ga} and Mn_I defects, giving rise to self-compensation, and ultimately to a limitation of T_C. Although these materials are highly sensitive to preparation conditions, the small concentrations of magnetic impurities involved actually relax the constraints on materials synthesis, thus enabling the use of alternative synthesis routes. For example, several attempts have been reported on the synthesis of DMSs using ion implantation of magnetic ions and subsequent thermal annealing. However, solid phase regrowth of the amorphous layer caused by implantation damage requires thermal annealing up to 800 °C for a duration of several seconds. This process becomes thermodynamically similar to epitaxial growth at high temperature, thus making the formation of secondary phases (MnGa and MnAs) energetically favorable during this annealing schedule.

Dubon and coworkers have pioneered the synthesis of the III$_{1-x}$Mn$_x$V ferromagnetic semiconductors using the highly nonequilibrium method of combined ion implantation and pulsed laser melting (II-PLM) (Dubon et al., 2005, 2006; Farshchi et al., 2006; Scarpulla et al., 2003, 2005; Stone et al., 2006). The two processes—II and PLM—are inherently nonequilibrium: ion implantation allows the injection of elements into a semiconductor in concentrations above their equilibrium solubility limits, while in PLM, the transient heat flow following pulsed-laser irradiation induces melting and solidification at velocities exceeding 1 m/s. Using a pulsed excimer laser, the relatively heavy damage caused by ion implantation can be ameliorated by melting and recrystallization on timescales of the order of 10^{-7} s. Such rapid solidification prevents secondary phase formation even when the equilibrium solubility limit has been exceeded by orders of magnitude (White et al., 1980).

Ga$_{1-x}$Mn$_x$As films synthesized by the II-PLM process exhibit magnetic and transport properties consistent with material grown using MBE (Scarpulla et al., 2003; Scarpulla, 2006). Curie temperatures up to 137 K have been reached in Ga$_{1-x}$Mn$_x$As layers formed by II-PLM with a maximum x up to 0.08. Figure 3.24 shows results of ion channeling experiments carried out on an II-PLM Ga$_{1-x}$Mn$_x$As film implanted with 35 keV Mn$^+$ to 1×10^{16} cm^{-2}, irradiated at 0.4 J/cm^2 with the Lextra KrF laser. The $\langle 110 \rangle$ χ_{min} is 3.6%, which is indistinguishable from that obtained on an optimally channeled GaAs wafer. The insets of the plot show the RBS and PIXE angular scans about the $\langle 110 \rangle$ and $\langle 111 \rangle$ axes. For both the $\langle 110 \rangle$ and $\langle 111 \rangle$ directions, 80% of the total Mn content is observed to reside at substitutional sites. The fact that similar values of [Mn$_{Ga}$] are deduced from angular scans for the two directions indicates that this film is free of Mn$_I$, which is a general characteristic of II-PLM Ga$_{1-x}$Mn$_x$As. This is due to the high-temperature II-PLM processing (T near the melting temperature $T_M = 1238$ °C after solidification). Hence, further thermal annealing is not necessary for optimal T_C in these films. Typically, in II-PLM films, a higher fraction of Mn$_{rand}$ (>15%) is present. These Mn$_{rand}$ are expected to exist in the form of clusters incommensurate with the lattice that are too small to constitute a new phase.

Similar to the synthesis of Ga$_{1-x}$Mn$_x$As, implantation of Mn ions in semiinsulating GaP substrates followed by PLM results in ferromagnetic Ga$_{1-x}$Mn$_x$P films (Dubon et al., 2005; Scarpulla et al., 2005). An important difference between the Ga$_{1-x}$Mn$_x$As and Ga$_{1-x}$Mn$_x$P is the much larger hole binding energy in Ga$_{1-x}$Mn$_x$P, namely ~0.4 eV, compared to ~0.11 eV in Ga$_{1-x}$Mn$_x$As. Nevertheless, Ga$_{1-x}$Mn$_x$P exhibits ferromagnetism in spite of this large binding energy. The magnetic properties of a Ga$_{1-x}$Mn$_x$P film with an effective Mn concentration $x_{eff} \sim 0.032$

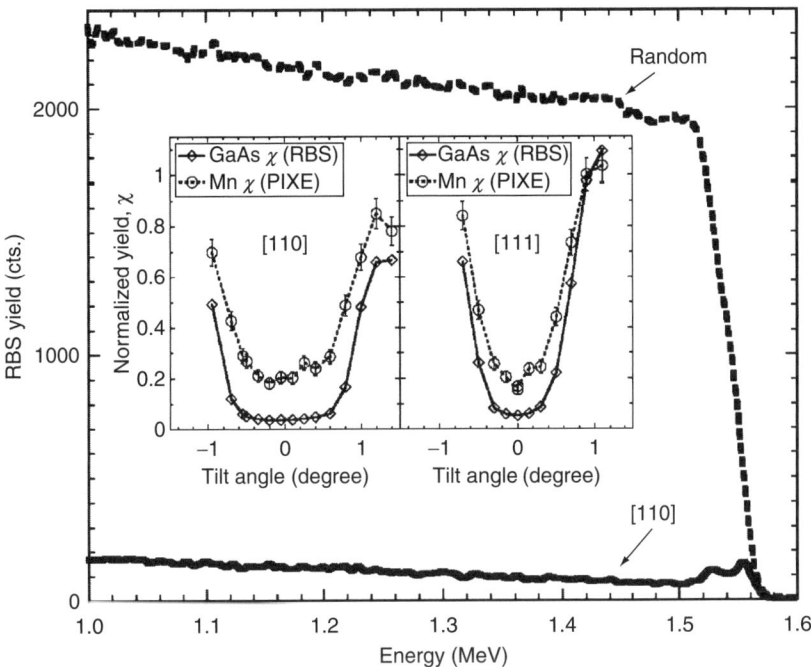

FIGURE 3.24 Main figure: Rutherford backscattering spectrometry (RBS) $\langle 110 \rangle$ channeling spectra for an ion implantation and pulsed laser melting (II-PLM) $Ga_{1-x}Mn_xAs$ sample. These RBS channeling data are indistinguishable from those obtained on an unimplanted GaAs wafer. Insets: RBS and particle-induced x-ray emission (PIXE) angular scans about the $\langle 110 \rangle$ and $\langle 111 \rangle$ channels for the II-PLM $Ga_{1-x}Mn_xAs$ film.

(Mn_{Ga} fraction as measured by channeling experiments) is shown in Fig. 3.25. Here the inset shows a hysteresis loop measured at 5.0 K for this material.

It should be pointed out that hole-mediated ferromagnetic phase in $Ga_{1-x}Mn_xP$ has so far not been demonstrated using LT-MBE. $Ga_{1-x}Mn_xP$ synthesized by II-PLM exhibits all of the major characteristics of a uniform hole-mediated ferromagnetic phase (Scarpulla, 2006). Figure 3.26 shows the T_C dependence on the substitutional Mn concentration for both LT-MBE $Ga_{1-x}Mn_xAs$ (Ohno, 1999; Jungwirth et al., 2005) and II-PLM $Ga_{1-x}Mn_xP$. Note that the values of T_C of both the $Ga_{1-x}Mn_xAs$ and the II-PLM $Ga_{1-x}Mn_xP$ materials increase linearly with magnetic ion concentration. Also, the saturation magnetization in $Ga_{1-x}Mn_xP$ films is in the range of 3.0–4.0 μ_B/Mn_{Ga} for $0.029 \leq x \leq 0.042$, which is close to the values of 4.0–4.5 μ_B/Mn_{Ga} reported for properly annealed $Ga_{1-x}Mn_xAs$. In addition to the linear dependence of T_C on the Mn_{Ga} concentration, extensive magnetotransport and x-ray magnetic dichroism (XMCD) measurements in these films (Farshchi et al., 2006; Scarpulla et al., 2005; Stone et al., 2006)

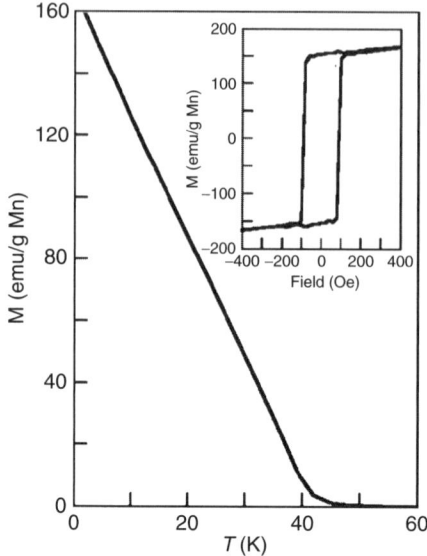

FIGURE 3.25 Magnetization as a function of temperature for a $Ga_{1-x}Mn_xP$ film after etching with HCl, using a measuring field of 50 Oe. The sample was produced using an Mn dose of 7.5×10^{15} cm^{-2}. The peak in the Mn distribution (as determined by secondary ion mass spectrometry) corresponds to an effective concentration $x_{eff}{\sim}0.03$. The inset shows a hysteresis loops measured at 5 K (Dubon et al., 2005).

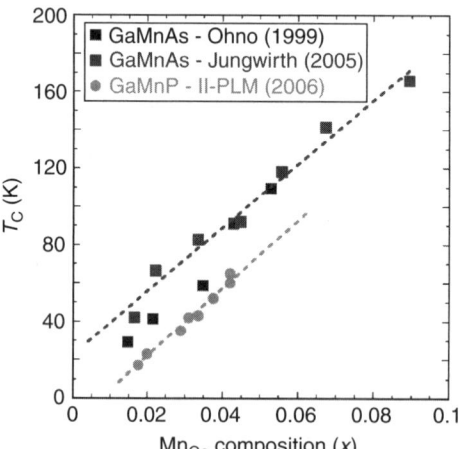

FIGURE 3.26 T_C for $Ga_{1-x}Mn_x$As films grown by low-temperature molecular beam epitaxy (Jungwirth et al., 2005; Ohno, 1999), and of $Ga_{1-x}Mn_xP$ fabricated using ion implantation and pulsed laser melting (II-PLM) plotted as functions of Mn_{Ga} concentration. The dashed lines indicating trends have slopes of ${\sim}1650$ K/x for the $Ga_{1-x}Mn_x$As data and ${\sim}1700$ K/x for the $Ga_{1-x}Mn_xP$ data (Scarpulla, 2006).

showed that they are indeed ferromagnetic semiconductors in which ferromagnetism is hole-mediated. We note finally that—based on Fourier-transform infrared (IR) absorption and far-IR photoconductivity measurements on II-PLM $Ga_{1-x}Mn_xP$ films—it was proposed that in this material the ferromagnetic exchange is mediated by holes localized in an Mn-derived band that is detached from the valence band (Scarpulla, 2006). This picture is consistent with the high binding energy of the holes in this wide bandgap $III_{1-x}Mn_xV$ alloy.

4.2. Strategies for increasing T_C in III-Mn-V alloys

The complexity of $III_{1-x}Mn_xV$ ferromagnetic semiconductors—and therefore much of their science—is determined primarily by the extensive role that defects play in determining their basic properties (MacDonald et al., 2005). The experimental results described in this chapter on low-temperature annealing, extrinsic and remote Be doping, and film thickness effects in $Ga_{1-x}Mn_xAs$ provide strong support for the model in which the incorporation of magnetically active substitutional Mn acceptors in $III_{1-x}Mn_xV$ alloys is governed by electronic processes taking place during the growth of the material. This suggests that one way to increase the Curie temperature of $III_{1-x}Mn_xV$ is to use MD in the barrier grown after depositing the ferromagnetic QW structures, in order to raise the hole concentration p above the value p_{max} that is allowed to occur while the $III_{1-x}Mn_xV$ layer is being deposited. The "proof of concept" for this strategy has already been provided by the experiments on MD described in Section 3.4.3.

Although it would appear that one might increase the Curie temperature by additional p-type doping, we have shown that (counterintuitively) simultaneous p-type doping by dopants other than Mn (e.g., Be) actually *lowers* T_C because adding holes prevents the incorporation of Mn_{Ga}. An alternative strategy that is being considered for increasing T_C is to employ *heavy n-type counter-doping* of $Ga_{1-x}Mn_xAs$ (using, e.g., Te or Si as dopants) in order to prevent the otherwise unavoidable creation of Mn_I at higher values of x. The reason for this is as follows. Once the maximum Fermi level is reached, we can add further substitutional Mn ions if we at the same time provide a donor, so that the total hole concentration does not exceed the allowed maximum value.

An alternative view of this process is that, when we use a Group-VI element as a dopant, we are actually providing a hexavalent "partner" ion for each additional Mn (which is divalent). We are thus effectively forming a $(III-V)_{1-x}(II-VI)_x$ alloy (see, e.g., Yim, 1976), taking advantage of the well-known miscibility of II-VI and III-V tetrahedrally bonded families, as long as the Grimm-Sommerfeld condition (an average of four valence electrons per atom) is obeyed (see, e.g., Miller et al. 1981).

In other words, we use the group-VI element as a kind of "Trojan horse" to introduce Mn where it, taken by itself, has little chance of admission. If, alternatively, we use Si as the n-type dopant in the growth of $Ga_{1-x}Mn_xAs$, we are in reality forming another alloy obeying the Grimm-Sommerfeld law, in this case in the form $(GaAs)_{1-2x}(MnSiAs_2)_x$, where $MnSiAs_2$ is essentially a chalcopyrite (i.e., a tetrahedrally bonded structure) (Miller et al., 1981). In such a mixed crystal, both Mn and Si will go into substitutional group-III sites—precisely where we want to place as much Mn as possible. Note that this mixed-crystal picture and the compensation of holes by electrons are essentially equivalent. While forming mixed-crystal such as those illustrated above one would, furthermore, need depart slightly from perfect stoichiometry by adjusting the constituent fluxes in such a way that the Mn concentration very slightly exceeds that of its "partner" element (i.e., either Te or Si in these examples), so as to provide the holes needed to achieve ferromagnetic coupling. Although the hole concentration will still remain "pinned" at p_{max} by the limit imposed on the Fermi level, the mixed-crystal strategy just presented would allow us to increase the number of active Mn proportion the incorporated "partner ion" (e.g., either Te or Si mentioned above), thus increasing T_C. If at present the highest Curie temperatures in the $Ga_{1-x}Mn_xAs$ alloy are around 170 K, in this picture increasing x by a factor of 2 should by itself (i.e., without increasing p) lead to T_C of above 300 K.

ACKNOWLEDGMENTS

We thank our numerous coworkers and coauthors (see References) for their stimulating and fruitful discussions, and their very significant contributions to many aspects of III-Mn-V reviewed in this chapter. The authors are grateful to the National Science Foundation and the Director, Office of Science, Office of Basic Energy Sciences, Division of Materials Sciences and Engineering, of the U.S. Department of Energy under Contract No. DE-AC02-05CH11231 for their support during the preparation of the manuscript. One of authors (T.W.) would like to acknowledge the support from Foundation for Polish Science through the Subsidy 12/2007.

REFERENCES

Adell, M., Ilver, L., Kanski, J., Stanciu, V., Svedlindh, P., Sadowski, J., Domagala, J. Z., Terki, F., Hernandez, C., and Charar, S. (2005). *Appl. Phys. Lett.* **86**, 112501.
Akai, H. (1998). *Phys. Rev. Lett.* **81**, 3002.
Blinowski, J., and Kacman, P. (2003). *Phys. Rev. B* **67**, 121204(R).
Blinowski, J., Kacman, P., Yu, K. M., Walukiewicz, W., Wojtowicz, T., and Furdyna, J. K. (2002). In *Proceedings of XXXI International School on the Physics of Semiconducting Compounds*, Jaszowiec, p. 33.
Blood, P. (1986). *Semicond. Sci. Technol.* **1**, 7.
Chambers, S. A., Droubay, T. C., Wang, C. M., Rosso, K. M., Heald, S. M., Schwartz, D. A., Kittilstved, K. R., and Gamelin, D. R. (2006). *Mater. Today* **9**, 28.

Chiba, D., Takamura, K., Matsukura, F., and Ohno, H. (2003). *Appl. Phys. Lett.* **82,** 3020.
Das Sarma, S., Hwang, E. H., and Kaminski, A. (2003). *Phys. Rev. B* **67,** 155201.
Dietl, T. (2006). *Physica E* **35,** 293.
Dietl, T., Ohno, H., Matsukura, F., Cibert, J., and Ferrand, D. (2000). *Science* **287,** 1019.
Dietl, T., Ohno, H., and Matsukura, F. (2001). *Phys. Rev. B* **63,** 195205.
Dubon, O. D., Scarpulla, M. A., Yu, K. M., and Walukiewicz, W. (2005). In *Proceedings of Compound Semiconductors 2004, IOP Conference Series* **184,** p. 399.
Dubon, O. D., Scarpulla, M. A., Farshchi, R., and Yu, K. M. (2006). *Physica B* **376–377,** 630.
Edmonds, K. W., Farley, N. R. S., Campion, R. P., Foxon, C. T., Gallagher, B. L., Johal, T. K., van der Laan, G., MacKenzie, M., Chapman, J. N., and Arenholz, E. (2004a). *Appl. Phys. Lett.* **84,** 4065.
Edmonds, K. W., Boguslawski, P., Wang, K. Y., Campion, R. P., Novikov, S. N., Farley, N. R. S., Gallagher, B. L., Foxon, C. T., Sawicki, M., Dietl, T., Nardelli, M. B., and Bernholc, J. (2004b). *Phys. Rev. Lett.* **92,** 037201.
Eggenkamp, P. J. T., Swagten, H. J. M., Story, T., Litvinov, V. I., Swüste, C. H. W., and de Jonge, W. J. M. (1995). *Phys. Rev. B* **51,** 15250.
Farshchi, R., Scarpulla, M. A., Stone, P. R., Yu, K. M., Sharp, I. D., Beeman, J. W., Sylvestri, H. H., Reichertz, L. A., Haller, E. E., and Dubon, O. D. (2006). *Solid State Commun.* **140,** 443.
Feldman, L. C., Mayer, J. W., and Picraux, S. T. (1982). "Materials Analysis by Ion Channeling." Academic Press, New York.
Furdyna, J. K. (1988). *J. Appl. Phys.* **64,** R29.
Furdyna, J. K., and Kossut, J. (1988). "Semiconductors and Semimetals." Vol. 25. Academic Press, New York.
Furdyna, J. K., Schiffer, P., Sasaki, Y., Potashnik, S. J., and Liu, X. Y. (2000). *In* "Optical Properties of Semiconductor Nanostructures, NATO Science Series" (M. L. Sadowski, M. Potemski, and M. Grynberg, eds), Vol. 81, p. 211. Kluwer, Dordrecht.
Furdyna, J. K., Liu, X., Lim, W. L., Sasaki, Y., Wojtowicz, T., Kuryliszyn, I., Lee, S., Yu, K. M., and Walukiewicz, W. (2003). *J. Korean Physical Soc.* **42,** S579.
Goennenwein, S. T. B., Wassner, T. A., Huebl, H., Brandt, M. S., Philipp, J. B., Opel, M., Gross, R., Koeder, A., Schoch, W., and Waag, A. (2004). *Phys. Rev. Lett.* **92,** 227202.
Haury, A., Wasiela, A., Arnoult, A., Cibert, J., Tatarenko, S., Dietl, T., and d'Aubigné, Y. M. (1997). *Phys. Rev. Lett.* **79,** 511.
Hayashi, T., Hashimoto, Y., Katsumoto, S., and Iye, Y. (2001). *Appl. Phys. Lett.* **78,** 1691.
Hurd, C. M. (1980). *In* "The Hall Effect and Its Applications" (C. L. Chienand and C. W. Westgate, eds), p. 43. Plenum Press, New York.
Inada, T., Kato, S., Hara, T., and Toyoda, N. (1979a). *J. Appl. Phys.* **50,** 4466.
Inada, T., Tokunaga, K., and Taka, S. (1979b). *Appl. Phys. Lett.* **35,** 546.
Inoue, J., Nonoyama, S., and Itoh, H. (2000). *Phys. Rev. Lett.* **85,** 4610.
Ishiwata, Y., Watanabe, M., Eguchi, R., Takeuchi, T., Harada, Y., Chainani, A., Shin, S., Hayashi, T., Hashimoto, Y., Katsumoto, S., and Iye, Y. (2002). *Phys. Rev. B* **65,** 233201.
Jungwirth, T., Wang, K. Y., Masek, J., Edmonds, K. W., König, J., Sinova, J., Polini, M., Goncharuk, N. A., MacDonald, A. H., Sawicki, M., Rushforth, A. W., Campion, R. P., *et al.* (2005). *Phys. Rev. B* **72,** 165204.
König, J., Lin, H.-H., and MacDonald, A. H. (2000). *Phys. Rev. Lett.* **84,** 5628.
Koshihara, S., Oiwa, A., Hirasawa, M., Katsumoto, S., Iye, Y., Urano, C., Takagi, H., and Munekata, H. (1997). *Phys. Rev. Lett.* **78,** 4617.
Ku, K. C., Potashnik, S. J., Wang, R. F., Chun, S. H., Schiffer, P., Samarth, N., Seong, M. J., Mascarenhas, A., Johnston-Halperin, E., Meyers, R. C., Gossard, A. C., and Awschalom, D. D. (2003). *Appl. Phys. Lett.* **82,** 2302.
Kuryliszyn, I., Wojtowicz, T., Liu, X., Furdyna, J. K., Dobrowolski, W., Broto, J.-M., Portugall, O., Rakoto, H., and Raquet, B. (2003). *J. Supercond.* **16,** 63.

Kuryliszyn-Kudelska, I., Domagala, J. Z., Wojtowicz, T., Liu, X., Lusakowska, E., Dobrowolski, W., and Furdyna, J. K. (2004). *J. Appl. Phys.* **95**, 603.
Lee, S., Chung, S. J., Choi, I. S., Yuldashev, Sh. U., Im, H., Kang, T. W., Lim, W. L., Sasaki, Y., Liu, X., Wojtowicz, T., and Furdyna, J. K. (2003). *J. Appl. Phys.* **93**, 8307.
Lim, W. L., Wojtowicz, T., Liu, X., Dobrowolska, M., and Furdyna, J. K. (2004). *Physica E* **20**, 346.
Litvinov, V. I., and Dugaev, V. K. (2001). *Phys. Rev. Lett.* **86**, 5593.
Liu, X., Lim, W. L., Titova, L. V., Wojtowicz, T., Kutrowski, M., Yee, K. J., Dobrowolska, M., Furdyna, J. K., Potashnik, S. J., Stone, M. B., Schiffer, P., Vurgaftman, I., et al. (2004). *Physica E* **20**, 370.
Liu, C., Yun, F., and Morkoc, H. J. (2005). *J. Mater. Sci. Mater. Electr.* **16**, 555.
Máca, F., and Mašek, J. (2002). *Phys. Rev. B* **65**, 235209.
MacDonald, A. H., Schiffer, P., and Samarth, N. (2005). *Nat. Mater.* **4**, 195.
Matsukura, F., Abe, E., and Ohno, H. (2000). *J. Appl. Phys.* **87**, 6442.
McSkimin, H. J., Jayaraman, A., and Andreatch, P., Jr. (1967). *J. Appl. Phys.* **38**, 2362.
Miller, A., MacKinnon, A., and Weaire, D. (1981). In "Solid State Physics," Vol. 36, p. 119. Academic Press, Boston.
Nagaev, E. L. (1996). *Phys. Rev. B* **54**, 16608.
Ohno, H. (1998). *Science* **281**, 951.
Ohno, H. (1999). *J. Magn. Magn. Mater.* **200**, 110.
Ohno, H., Munekata, H., Penney, T., von Molnár, S., and Chang, L. L. (1992). *Phys. Rev. Lett.* **68**, 2664.
Ohno, H., Shen, A., Matsukura, F., Oiwa, A., Endo, A., Katsumoto, S., and Iye, Y. (1996). *Appl. Phys. Lett.* **69**, 363.
Ohno, H., Matsukura, F., Omiya, T., and Akiba, N. (1999a). *J. Appl. Phys.* **85**, 4277.
Ohno, Y., Young, D. K., Beschoten, B., Matsukura, F., Ohno, H., and Awschalom, D. D. (1999b). *Nature* **402**, 790.
Ohno, H., Chiba, D., Matsukura, F., Omiya, T., Abe, E., Dietl, T., Ohno, Y., and Ohtani, K. (2000). *Nature* **408**, 944.
Potashnik, S. J., Ku, K. C., Chun, S. H., Berry, J. J., Samarth, N., and Schiffer, P. (2001). *Appl. Phys. Lett.* **79**, 1495.
Potashnik, S. J., Ku, C. K., Mahendiran, R., Chun, S. H., Wang, R. F., Samarth, N., and Schiffer, P. (2002). *Phys. Rev. B* **66**, 012408.
Prinz, G. A. (1990). *Science* **250**, 1092.
Sadowski, J., Mathieu, R., Svedlindh, P., Domagala, J. Z., Bak-Misiuk, J., Swiatek, K., Karlsteen, M., Kanski, J., Ilver, L., Åsklund, H., and Södervall, U. (2001). *Appl. Phys. Lett.* **78**, 3271.
Sanvito, S., and Hill, N. A. (2001). *Appl. Phys. Lett.* **78**, 3493.
Sanvito, S., Ordejón, P., and Hill, N. A. (2001). *Phys. Rev. B* **63**, 165206.
Scarpulla, M. A. (2006). Ph.D. Dissertation, University of California, Berkeley.
Scarpulla, M. A., Dubon, O. D., Yu, K. M., Monteiro, O., Pillai, M. R., Aziz, M. J., and Ridgway, M. C. (2003). *Appl. Phys. Lett.* **82**, 1251.
Scarpulla, M. A., Cardozo, B. L., Farshchi, R., Oo, W. M. H., McCluskey, M. D., Yu, K. M., and Dubon, O. D. (2005). *Phys. Rev. Lett.* **95**, 207204.
Schliemann, J., König, J., and MacDonald, A. H. (2001). *Phys. Rev. B* **64**, 165201.
Shioda, R., Ando, K., Hayashi, T., and Tanaka, M. (1998). *Phys. Rev. B* **58**, 1100.
Soo, Y. L., Huang, S. W., Ming, Z. H., Kao, Y. H., Munekata, H., and Chang, L. L. (1996). *Phys. Rev. B* **53**, 4905.
Specht, P., Lutz, R. C., Zhao, R., Weber, E. R., Liu, W. K., Bacher, K., Towner, F. J., Stewart, T. R., and Luysberg, M. (1999). *J. Vac. Sci. Technol. B* **17**, 1200.
Stone, P. R., Scarpulla, M. A., Farshchi, R., Sharp, I. D., Haller, E. E., Dubon, O. D., Yu, K. M., Beeman, J. W., Arenholz, E., Denlinger, J. D., and Ohldag, H. (2006). *Appl. Phys. Lett.* **89**, 012504.

Tesmer, J. R., and Nastasi, M. A. (1995). "Handbook of Modern Ion Beam Materials Analysis." Materials Research Society, Pittsburgh.
Theodoropoulou, N., Hebard, A. F., Overberg, M. E., Abernathy, C. R., Pearton, S. J., Chu, S. N. G., and Wilson, R. G. (2002). *Phys. Rev. Lett.* **89,** 107203.
Timm, C., Schäfer, F., and von Oppen, F. (2002). *Phys. Rev. Lett.* **89,** 137201.
Vurgaftman, I., and Meyer, J. R. (2001). *Phys. Rev. B* **64,** 245207.
Walukiewicz, W. (1989). *Appl. Phys. Lett.* **54,** 2094.
Walukiewicz, W. (1993a). *Mat. Res. Soc. Symp. Proc.* **300,** 421.
Walukiewicz, W. (1993b). In *Proceedings of the 17th International Conference on Defects in Semiconductors,* Material Science Forum, **143–147,** p.519.
Walukiewicz, W. (2001). *Physica B* **302–303,** 123.
Wang, K. Y., Campion, R. P., Edmonds, K. W., Sawicki, M., Dietl, T., Foxon, C. T., and Gallagher, B. L. (2005). *AIP Conference Proceedings* **772,** 333.
Welp, U., Vlasko-Vlasov, V. K., Menzel, A., You, H. D., Liu, X., Furdyna, J. K., and Wojtowicz, T. (2004). *Appl. Phys. Lett.* **85,** 260.
White, C. W., Wilson, S. R., Appleton, B. R., and Young, F. W., Jr. (1980). *J. Appl. Phys.* **51,** 738.
Wojtowicz, T. (2003). *Bull. Am. Phys. Soc.* **48,** 584.
Wojtowicz, T., Cywinski, G., Lim, W. L., Liu, X., Dobrowolska, M., Furdyna, J. K., Yu, K. M., Walukiewicz, W., Kim, G. B., Cheon, M., Chen, X., Wang, S. M., et al. (2003a). *Appl. Phys. Lett.* **82,** 4310.
Wojtowicz, T., Lim, W. L., Liu, X., Dobrowolska, M., Furdyna, J. K., Yu, K. M., Walukiewicz, W., Vurgaftman, I., and Meyer, J. R. (2003b). *Appl. Phys. Lett.* **83,** 4220.
Wojtowicz, T., Lim, W. L., Liu, X., Cywinski, G., Kutrowski, M., Titova, L. V., Yee, K., Dobrowolska, M., Furdyna, J. K., Yu, K. M., Walukiewicz, W., Kim, G. B., et al. (2004a). *Physica E* **20,** 325.
Wojtowicz, T., Furdyna, J. K., Liu, X., Yu, K. M., and Walukiewicz, W. (2004b). *Physica E* **25,** 171.
Wolf, S. A., Awschalom, D. D., Buhrman, R. A., Daughton, J. M., von Molnár, S., Roukes, M. L., Chtchelkanova, A. Y., and Treger, D. M. (2001). *Science* **294,** 1488.
Yanagi, S., Kuga, K., Slupinski, T., and Munekata, H. (2004). *Physica E* **20,** 333.
Yim, W. M. (1976). *J. Appl. Phys.* **10,** 948.
Yu, K. M., Walukiewicz, W., Wojtowicz, T., Kuryliszyn, I., Liu, X., Sasaki, Y., and Furdyna, J. K. (2002a). *Phys. Rev. B* **65,** 201303(R).
Yu, K. M., Walukiewicz, W., Wojtowicz, T., Lim, W. L., Liu, X., Sasaki, Y., Dobrowolska, M., and Furdyna, J. K. (2002b). *Appl. Phys. Lett.* **81,** 844.
Yu, K. M., Walukiewicz, W., Wojtowicz, T., Lim, W. L., Liu, X., Bindley, U., Dobrowolska, M., and Furdyna, J. K. (2003). *Phys. Rev. B* **68,** 041308(R).
Yu, K. M., Walukiewicz, W., Wojtowicz, T., Lim, W. L., Liu, X., Dobrowolska, M., and Furdyna, J. K. (2004a). *Appl. Phys. Lett.* **84,** 4325.
Yu, K. M., Walukiewicz, W., Wojtowicz, T., Lim, W. L., Liu, X., Dobrowolska, M., and Furdyna, J. K. (2004b). *Nucl. Instrum. Methods Phys. Res. B* **219–220,** 636.
Yu, K. M., Walukiewicz, W., Wojtowicz, T., Denlinger, J., Scarpulla, M. A., Liu, X., and Furdyna, J. K. (2005). *Appl. Phys. Lett.* **86,** 042102.
Yuldashev, S. U., Im, H., Yalishev, V. Sh., Park, C. S., Kang, T. W., Lee, S., Sasaki, Y., Liu, X., and Furduna, J. K. (2003). *Jpn. J. Appl. Phys.* **42,** 6256.
Zutic, I., Fabian, J., and Das Sarma, S. (2004). *Rev. Mod. Phys.* **76,** 323.

CHAPTER 4

Transport Properties of Ferromagnetic Semiconductors

T. Jungwirth,* B. L. Gallagher,[†] and J. Wunderlich[‡]

Contents			
	1.	Introduction	135
	2.	Basic Transport Characteristics	138
		2.1. Impurity-band to valence-band crossover	139
		2.2. Quantum interference and interaction effects	154
		2.3. Critical contribution to transport near the Curie point	162
	3.	Extraordinary Magnetotransport	167
		3.1. AMR in ohmic regime	167
		3.2. Tunneling AMR	177
		3.3. Coulomb blockade AMR and other MR transistors	184
		3.4. Current induced domain wall switching	192
	4.	Summary	199
		Acknowledgments	200
		References	200

1. INTRODUCTION

In this chapter we discuss transport properties of ferromagnetic semiconductors focusing on the prototype (Ga,Mn)As-based magnetotransport devices. The topic is reviewed from the perspective of research into new

---

* Institute of Physics ASCR, Cukrovarnická 10, 162 53 Praha 6, Czech Republic and School of Physics and Astronomy, University of Nottingham, Nottingham NG7 2RD, UK
[†] School of Physics and Astronomy, University of Nottingham, Nottingham NG7 2RD, UK
[‡] Hitachi Cambridge Laboratory, Cambridge CB3 0HE, UK and Institute of Physics ASCR, Cukrovarnická 10, 162 53 Praha 6, Czech Republic

spintronics technologies and basic physics of electronic transport in degenerate dilute-moment semiconductors and extraordinary magneto-resistance (MR) phenomena.

The fundamental principles of spintronics are best illustrated on the analogy with a classical resistor (see Fig. 4.1). In classical electronics the flow of electron charge is manipulated by external electric fields. A spintronic resistor, which in the early 1990s replaced classical magneto-inductive coil sensors in hard-drive read-heads and launched the era of spintronics, utilizes both the electron charge and spin for its functionality. While electric fields address the orbital degree of freedom of the electron charge, magnetic fields can be used to address the spin degree of freedom. These two independent ways of externally manipulating electrons must be complemented by an internal interaction between the orbital and spin degrees of freedom in the system. In solid state materials there are two distinct means of such an interaction. One is ferromagnetism in which the collective alignment of spins of electrons results, in general terms, from combined effects of the Pauli exclusion principle and Coulomb repulsion between the electron charges. Because magnetic order is associated with the strong Coulomb interactions between electrons, it can be very robust and persist to high temperatures, often to temperatures comparable to those at which crystalline order occurs. In the ferromagnetic state the collective spin coordinate can be manipulated by external fields comparable to the magnetic anisotropy fields in the material which are tunable and often weak.

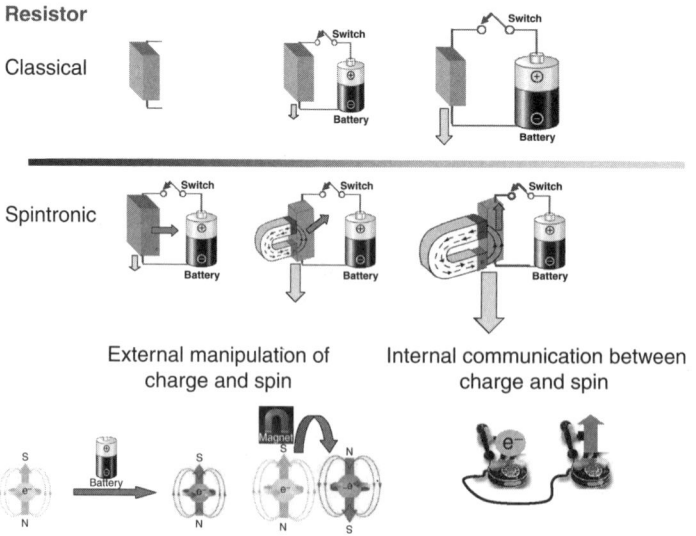

FIGURE 4.1 Schematic comparison of classical and spintronic resistors.

Spin–orbit (SO) coupling is the other internal interaction which is described by an effective single-particle potential, $H_{SO} = \mathbf{B}_{eff}\mathbf{s}$, where $\mathbf{B}_{eff} = 1/2m^2c^2 \, \nabla V \times \mathbf{p}$, m is the bare electron mass, c is the speed of light, $\mathbf{p}$ is the electron momentum, and ∇V is an electric field produced, primarily, by the atomic cores acting on the valence electrons. The SO interaction appears in the Hamiltonian when the relativistic Dirac equation is expanded around the nonrelativistic Schrödinger form to the second order in p/mc. (For a detailed discussion of the SO coupling see Chapter 2, this volume.) The spintronic resistor utilizes ferromagnetism to allow weak external magnetic fields change its internal magnetization orientation, and the SO-coupling induced response of the charge transport to the change in magnetization. The latter effect is called the anisotropic magnetoresistance (AMR).

In mid 1990s, the AMR-resistor sensors for hard-disk-drives were replaced with giant magnetoresistance (GMR) or tunneling magnetoresistance (TMR) spintronic elements comprising (at least) two magnetically decoupled ferromagnetic layers separated by a nonmagnetic conducting or insulating spacer. The resistance of these devices is controlled by relative orientations of magnetization in the magnetic layers. The TMR effect is as high as several 100's % which means that the device's MR has a diode-like characteristics. The high-conductance "on" and the low-conductance "off" states can represent logical "0" and "1" which is utilized in magnetoresistive random access memories (MRAMs). The first, 4 Mb commercial MRAM chip was introduced on market in 2006 with the prospect of evolving into a universal nonvolatile memory technology. (see Fig. 4.2 for comparison with other random access memories) (Slaughter et al., 2004).

Research of magnetotransport phenomena in ferromagnetic semiconductors, reviewed in this chapter, has several distinct motivation sources and general goals: (i) Among the fundamental physical constraints of current spintronic memory devices based on dense-moment metal ferromagnets are unintentional dipolar cross-links between densely integrated magnetic microelements and unintentional addressing of neighboring magnetic bits when writing is performed by external magnetic fields. (Ga,Mn)As and related ferromagnetic semiconductors doped with only ~1–10% of Mn magnetic moments have a ~100–10 weaker saturation magnetization than conventional ferromagnets. The dipolar interactions are comparatively suppressed and the low saturation moments also lead to relatively low critical currents for spin-transfer-torque magnetization switching which represents a fully local means of writing magnetic information. (ii) Spintronics research also looks beyond storage applications towards the integration with information processing technologies. The key device that will enable this integration is a spintronic transistor whose MR response is tuned by electrical gates.

	SRAM	DRAM	FLASH	FRAM	MRAM
Read	Fast	Moderate	Fast	Moderate	Moderate-fast
Write	Fast	Moderate	**Slow**	Moderate	Moderate-fast
Nonvolatile	**No**	**No**	Yes	**Partially**[b]	Yes
Endurance	Unlimited	Unlimited	**Limited**[c]	**Limited**[b]	Unlimited
Refresh	No	**Yes**	No	No	No
Cell size	**Large**	Small	Small	Medium	Small
Low voltage	Yes	Limited	**No**	Limited	Yes

[a]Bold letters indicate undesirable attributes.
[b]Destructive read and limited read/write endurance.
[c]Limited write endurance.

FIGURE 4.2 Comparisons of MRAM expected features with other memory technologies (from Slaughter et al., 2004).

Ferromagnetic semiconductors are natural materials of choice for realizing such a device. (iii) Many spintronic functionalities are underpinned by extraordinary MR effects, AMR being a prime example, in which charge carriers respond to changes in internal magnetization via SO interaction. Ferromagnetic semiconductors with tuneable, strongly SO-coupled yet relatively simple carrier bands are particularly favorable systems for exploring these subtle relativistic effects which still belong among the least understood ares of condensed matter physics.

In Section 2, we give an overview of the basic characteristics of conduction in (Ga,Mn)As ferromagnetic semiconductor including the discussion of the crossover from impurity-band to disordered valence-band transport (Section 2.1), of low-temperature quantum transport phenomena such as universal conductance fluctuations (UCFs) and weak localization (WL) (Section 2.2), and of the critical contribution to transport near the Curie point (Section 2.3). Section 3 starts with the discussion of the physics of AMR in (Ga,Mn)As standard ohmic devices (Section 3.1), then introduces new members of the family of anisotropic magnetotransport effects discovered in (Ga,Mn)As tunneling devices (Section 3.2) and Coulomb blockade (CB) devices (Section 3.3), and realizations of (Ga,Mn)As MR transistors. The section ends with a discussion of spin-transfer-torque phenomena in (Ga,Mn)As-based microdevices (Section 3.4). A brief summary of the chapter is given in Section 4.

2. BASIC TRANSPORT CHARACTERISTICS

The elements in the (Ga,Mn)As compound have nominal atomic structures $[Ar]3d^{10}4s^2p^1$ for Ga, $[Ar]3d^54s^2$ for Mn, and $[Ar]3d^{10}4s^2p^3$ for As. This circumstance correctly suggests that the most stable and, therefore, most common position of Mn in the GaAs host lattice is on the Ga site where its two 4s-electrons can participate in crystal bonding in much the same way as the two Ga 4s-electrons. Because of the missing valence 4p-electron, the Mn_{Ga} impurity acts as an acceptor. In the electrically neutral state the isolated Mn_{Ga} has the character of a local moment with zero angular momentum and spin $S = 5/2$ (Lánde g-factor $g = 2$) and a moderately bound hole.

The low-energy degrees of freedom in (Ga,Mn)As materials are the orientations of Mn local moments and the occupation numbers of acceptor levels near the top of the valence band. The number of local moments and the number of holes may differ from the number of Mn_{Ga} impurities in the GaAs host due to the presence of charge and moment compensating defects. Hybridization between Mn d-orbitals and valence As/Ga sp-orbitals, mainly the As p-orbitals on the neighboring sites, leads to an antiferromagnetic exchange interaction between the spins that they carry (Jungwirth et al., 2006).

At concentrations $\ll 1\%$ of substitutional Mn, the average distance between Mn impurities (or between holes bound to Mn ions) is much larger than the size of the bound hole characterized approximately by the impurity effective Bohr radius. These very dilute (Ga,Mn)As systems are insulating, with the holes occupying a narrow impurity band, and paramagnetic. Experimentally, ferromagnetism in (Ga,Mn)As is observed when Mn doping reaches approximately 1% (Campion et al., 2003; Ohno, 1999; Potashnik et al., 2002) and the system is near the insulator-to-metal transition. At these Mn concentrations, the localization length of the holes is extended to a degree that allows them to mediate, via the sp–d hybridization, ferromagnetic exchange interaction between Mn local moments, even though the moments are dilute. (For a detailed discussion of ferromagnetism in (Ga,Mn)As and other diluted magnetic semiconductor see Chapter 9, this volume.)

The understanding of magnetic and magnetotransport properties of ferromagnetic (Ga,Mn)As semiconductors which are close to the metal–insulator transition can only emerge from the understanding of the spectroscopic nature of the holes mediating the ferromagnetic coupling between Mn local moments. This key characteristics of Mn doped GaAs, as revealed by a survey of d.c. and a.c. transport studies over a wide range of Mn concentrations, is schematically illustrated in Fig. 4.3 and discussed in detail in the following section.

2.1. Impurity-band to valence-band crossover

GaAs is an intermediate band-gap III–V semiconductor, with $E_g = 1.5$ eV at low temperatures, in which an isolated Mn impurity substituting for Ga functions as an acceptor with an impurity binding energy of intermediate

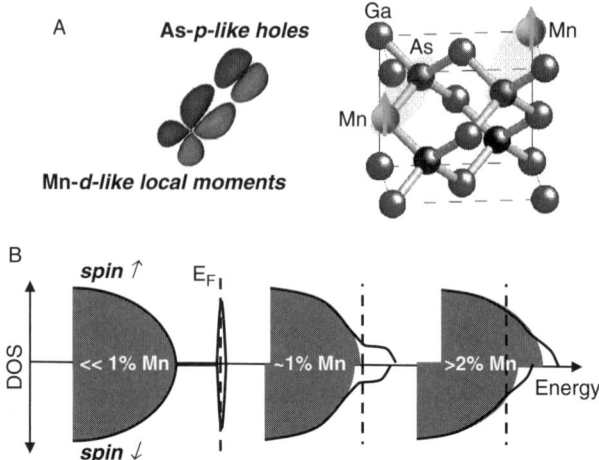

FIGURE 4.3 (A) Schematics of the (Ga,Mn)As crystal structure and of the Mn d-As p orbital hybridization which is the microscopic mechanism behind the hole mediated ferromagnetic coupling between Mn local moments. (B) Schematics of the onset of ferromagnetism at high Mn dopings near the metal–insulator transition.

strength, $E_a^0 = 0.11$ eV (Bhattacharjee and à la Guillaume, 2000; Blakemore et al., 1973; Chapman and Hutchinson, 1967; Madelung et al., 2003; Yakunin et al., 2004). Beyond a critical Mn concentration, Mn doped GaAs exhibits a phase transition to a state in which the Mn impurity levels overlap sufficiently strongly that the ground state is metallic, that is, that states at the Fermi level are not bound to a single or a group of Mn atoms but are delocalized across the system (Jungwirth et al., 2006; Matsukura et al., 2002). In the metallic regime Mn can, like a shallow acceptor (e.g., C, Be, Mg, Zn), provide delocalized holes with a low-temperature density comparable to Mn density, N_{Mn} (Jungwirth et al., 2005; MacDonald et al., 2005; Ruzmetov et al., 2004). ($x = 1$% Mn-doping corresponds to $N_{Mn} = 2.2 \times 10^{20}$ cm^{-3} in Ga$_{1-x}$Mn$_x$As.) The transition to the metallic state occurs at larger doping in GaAs when doped with Mn than when doped with shallow acceptors. Experimentally the transition appears to occur between $N_{Mn} \approx 1 \times 10^{20}$ cm^{-3} and 5×10^{20} cm^{-3}, as compared to the $\sim 10^{18}$ cm^{-3} critical density for the shallow acceptors in GaAs (Ferreira da Silva et al., 2004).

The hole binding potential of an isolated substitutional Mn impurity is composed of long-range Coulomb, short-ranged central-cell, and sp-d kinetic-exchange potentials (Bhattacharjee and à la Guillaume, 2000). Because of the kinetic-exchange and central cell interactions, Mn acceptors are more localized than shallow acceptors. A crude estimate of the critical metal–insulator transition density can be obtained with a short-range potential model, using the experimental binding energy and assuming an effective mass of valence-band holes, $m^* = 0.5m_e$. This model implies an isolated acceptor level with effective Bohr radius $a_0 = (\hbar^2/2m^* E_a^0)^{1/2} = 10\text{Å}$ The radius a_0 then equals the Mn impurity spacing scale $N_{Mn}^{-1/3}$ at $N_{Mn} \approx 10^{21}$ cm^{-3}. This explains qualitatively the higher metal–insulator transition critical density in Mn doped GaAs compared to the case of systems doped with shallow, more hydrogenic-like acceptors which have binding energies $E_a^0 \approx 30$ meV (Ferreira da Silva et al., 2004; Madelung et al., 2003).

Unlike the metal–insulator phase transition, which is sharply defined in terms of the temperature $T = 0$ limit of the conductivity, the crossover in the character of states near the Fermi level in semiconductors with increased doping is gradual (Dietl, 2007; Jungwirth et al., 2006; Lee and Ramakrishnan, 1985; Paalanen and Bhatt, 1991; Shklovskii and Efros, 1984). At very weak doping, the Fermi level resides inside a narrow impurity band (assuming some compensation) separated from the valence band by an energy gap of a magnitude close to the impurity binding energy. In this regime strong electronic correlations are an essential element of the physics and a single-particle picture has limited utility. Well into the metallic state, on the other hand, the impurities are sufficiently close together, and the long-range Coulomb potentials which contribute to the binding energy of an isolated impurity are sufficiently screened, that the system is best viewed as an imperfect crystal with disorder-broadened and shifted host Bloch bands. In this regime, electronic correlations are usually less strong and a single-particle picture often suffices. (Note that the short-range components of the Mn binding energy in GaAs, which are not screened by the carriers, move the crossover to higher dopings and contribute significantly to carrier scattering in the metallic state.)

Although neither picture is very helpful for describing the physics in the crossover regime which spans some finite range of dopings, the notion of the impurity band on the lower doping side from the crossover still have a clear qualitative meaning. The former implies that there is a deep minimum in the density-of-states between separate impurity and host band states. In the latter case the impurity band and the host band merge into one inseparable band whose tail may still contain localized states depending on the carrier concentration and disorder. Note that terms overlapping and merging impurity and valence bands describe the same basic physics in (Ga,Mn)As. This is because the Mn-acceptor states span several unit cells even in the very dilute limit and many unit cells as the impurity band broadens with increasing doping. The localized and the delocalized Bloch states then have a similarly mixed As-Ga-Mn spd-character. This applies to systems on either side of the metal–insulator transition. Also note that for randomly distributed Mn dopants in GaAs the impurity band is not properly modeled as a Bloch-like band with a weak dispersion (heavy effective mass).

Narrow impurity bands are expected and have been clearly observed in Mn doped GaAs samples with carrier densities much lower than the metal–insulator transition density, for example, in equilibrium grown bulk materials with $N_{Mn} = 10^{17}$–10^{19} cm^{-3} (Blakemore et al., 1973; Brown and Blakemore, 1972; Woodbury and Blakemore, 1973). The energy gap between the impurity band and the valence band, E_a, can be measured by studying the temperature dependence of longitudinal and Hall conductivities, which show activated behavior because of thermal excitation of holes from the impurity band to the much more conductive valence band (Blakemore et al., 1973; Marder, 2000; Woodbury and Blakemore, 1973). Examples of these measurements are shown in Figs. 4.4 and 4.6.

The activation energy decreases with increasing Mn density, following roughly the form (Blakemore et al., 1973)

$$E_a = E_a^0 \left[1 - \left(N_{Mn}/N_{Mn}^c \right)^{1/3} \right]. \quad (4.1)$$

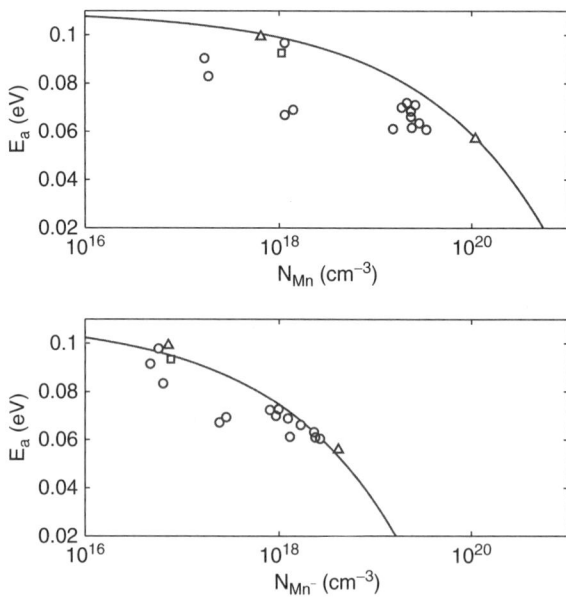

FIGURE 4.4 Mn doping dependent energy gap between the impurity band and valence band measured in the bulk equilibrium grown (Ga,Mn)As and plotted as a function of total Mn density (top panel) and ionized Mn denisty (bottom panel) (from Blakemore et al., 1973).

The lowering of impurity binding energies at larger N_{Mn}, which is expected to scale with the mean impurity separation as expressed in Eq. (4.1), is apparent already in the equilibrium grown bulk materials with $N_{Mn} = 10^{17}$–10^{19} cm^{-3}. The trend continues in epitaxially grown (at 400 °C) (Ga,Mn)As with $N_{Mn} \approx 1 \times 10^{19}$ and 6×10^{19} cm^{-3} (Poggio et al., 2005). When this trend is extrapolated using Eq. (4.1) it places the disappearance of the gap at Mn dopings of $\sim 1\%$. Note that this estimate has a large scatter depending also on whether all Mn or only ionized Mn impurities are considered when fitting the data by Eq. (4.1) (see Fig. 4.4) (Blakemore et al., 1973).

For $N_{Mn} \approx 1 \times 10^{19}$ cm^{-3} ($x \approx 0.06\%$), $E_a \approx 70$ meV and a sharp crossover from impurity band hopping conduction to activated valence-band conduction is observed at $T \approx 80$–100 K (see Figs. 4.5 and 4.6). Near the crossover, $N_{Mn}/p \sim 10^3$–10^5 where p is the valence-band hole density, consistent with the much higher mobility of valence band holes compared to impurity band holes in this regime for which the distinction is clearly

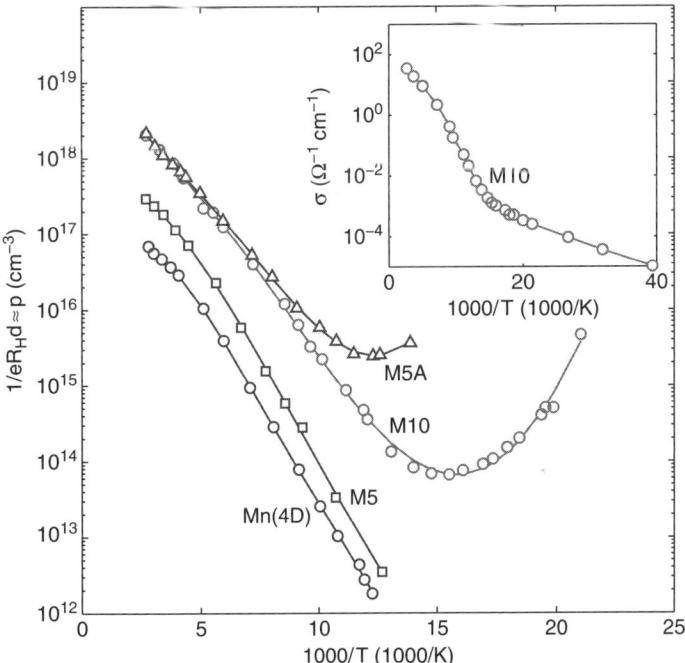

FIGURE 4.5 Hole densities determined from the Hall coefficient measurement as a function of temperature in bulk equilibrium grown (Ga,Mn)As. Inset: temperature-dependent longitudinal conductivity (from Blakemore et al., 1973; Brown and Blakemore, 1972). Sample Mn(4D) has Mn density 1.6×10^{17} cm^{-3}, M5: 1.1×10^{18} cm^{-3}, M10: 1.9×10^{19} cm^{-3}, Mn5A: 9.3×10^{19} cm^{-3}.

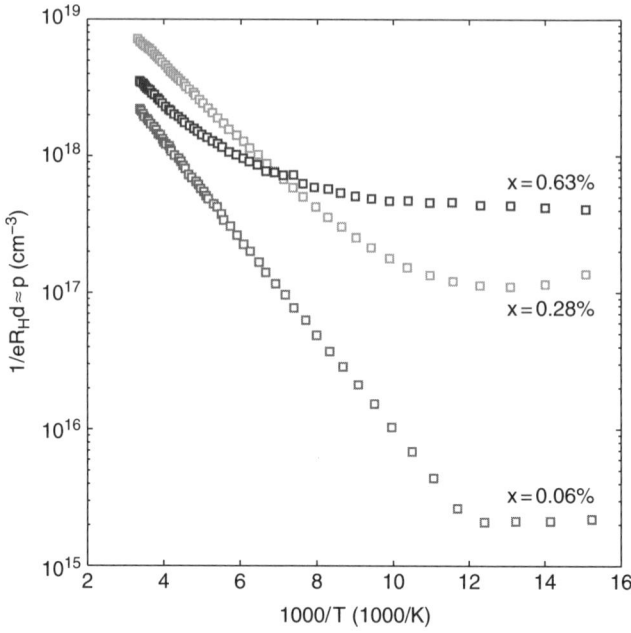

FIGURE 4.6 Hole densities determined from the Hall coefficient for MBE grown (at 400 °C) (Ga,Mn)As. From (Poggio et al., 2005).

still valid and useful. (Note that in the bulk equilibrium-grown samples the temperature at which the crossover occurs appears to be very sensitive to the detailed disorder configuration at this Mn doping level, with some samples being strongly insulating and others showing signatures of filamentary conducting channels (Brown and Blakemore, 1972; Poggio et al., 2005; Woodbury and Blakemore, 1973). For the $N_{Mn} \approx 6 \times 10^{19}$ cm^{-3} ($x \approx 0.3\%$) epitaxial sample (Poggio et al., 2005), the crossover from impurity band conduction to activated valence-band conduction can still be identified at $T \approx 140$ K (see Fig. 4.6) and $E_a \approx 50$ meV.

In the dilute limit where the system has a narrow insulating impurity band, the a.c. conductivity is expected to show a broad mid-infrared feature due to impurity band to valence band transitions, peaked at energy between E_a and $2E_a$ (Anderson, 1975; Brown and Blakemore, 1972; Fleurov and Kikoin, 1982). Lower energy peak positions correspond to shallow acceptors with weakly bound impurity states composed of a narrow range of valence-band wavevector Fourier components. The peak near $2E_a$ and a line shape $\sigma(\hbar\omega) \sim E_a^{1/2}(\hbar\omega - E_a)^{3/2}/(\hbar\omega)^3$ correspond to a short-range impurity potential for which a larger interval of valence-band states contributes to the transition amplitude. Consistent with these expectations a peak at $\hbar\omega \approx 200$ meV is observed in the weakly doped

(N_{Mn} ~ 10^{17} cm^{-3} (Ga,Mn)As samples (see Fig. 4.7) (Brown and Blakemore, 1972; Chapman and Hutchinson, 1967; Singley et al., 2002). At N_{Mn} ~ 10^{19} cm^{-3}, the peak is slightly red shifted ($\hbar\omega \approx 180$ meV) and significantly broadened (Brown and Blakemore, 1972). Both the red-shifting and the broadening are expected consequences of increasing overlap between acceptor levels, as illustarted in Fig. 4.8.

At higher Mn doping around ≈1%, the d.c. conductivity curves can no longer be separated into an impurity-band dominated contribution at low temperatures and an activated valence-band dominated contribution at higher temperatures, even though the samples are still insulators (see Fig. 4.9). In metallic materials with $x > 2\%$, no signatures of Mn-related hole activation is observed when temperature is swept to values as high

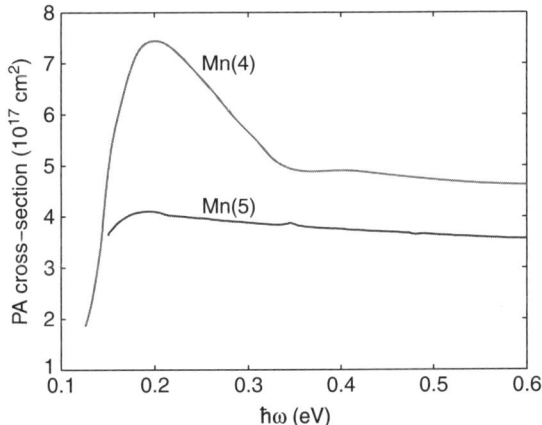

FIGURE 4.7 Infrared photoabsorption crossection measurements in bulk (Ga,Mn)As. From 20. Mn(4): 1.7×10^{17} cm^{-3}, Mn(5): 9.3×10^{18} cm^{-3}.

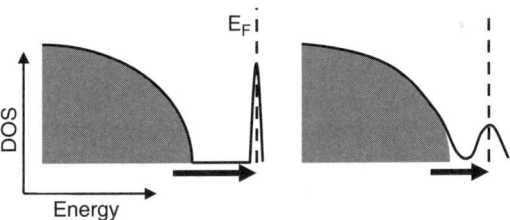

FIGURE 4.8 Schematic illustrations of the impurity band regime for low doped insulating (Ga,Mn)As. The cartoons are arranged from left to right according to increasing doping. Grey areas indicate delocalized states, white areas localized states. Arrows highlight the red shift in the impurity band ionization energy (from Jungwirth et al., 2007).

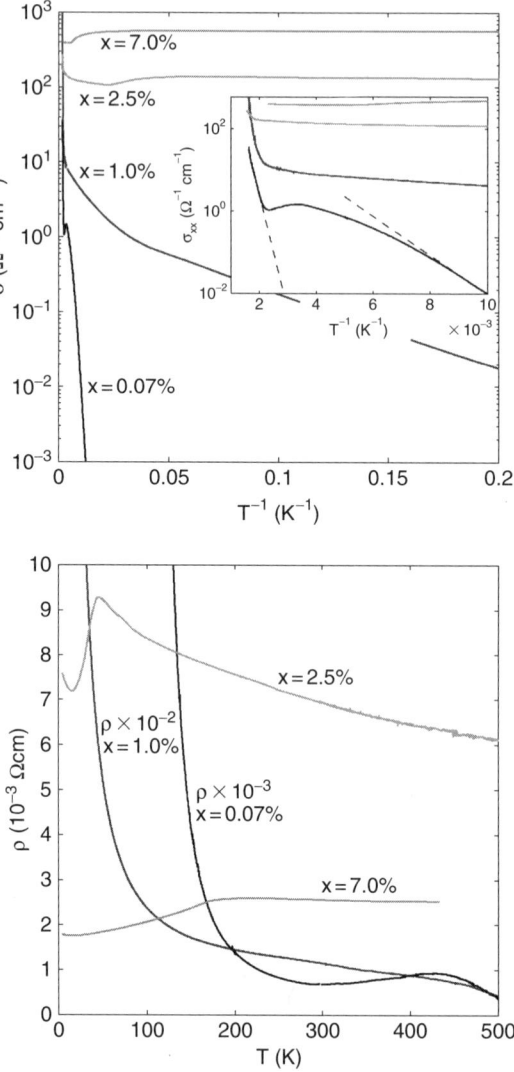

FIGURE 4.9 Comparison of longitudinal conductivities (upper panel) and resistivities (lower panel) of 0.07% and 1% doped insulating paramagnetic (Ga,Mn)As, and 2.5% and 7% Mn doped ferromagnetic (T_c approximately 40 and 170 K, resp.) metallic (Ga,Mn)As grown in the Prague MBE system with As_4 flux. The 0.07% doped material was grown at 530°, the other materials were grown by low-temperature MBE (240–200°). The 0.07% doped sample shows clear (Ga,Mn)As impurity band—valence band activation at low temperatures, impurity band exhaustion, and the onset of activated transport over GaAs band gap at high temperatures. The 2.5% and 7% doped samples show only weak conductance variations associated with the onset of ferromagnetism until the activated transport over GaAs band gap takes over (from Jungwirth et al., 2007).

as 500 K, that is, to $T \approx E_a^0/2$, and before the intrinsic conductance due to activation across the GaAs band-gap takes over (see Fig. 4.9).

Figure 4.10 shows the temperature-dependent resistivity curves for a set of metallic, annealed high conductivity ferromagnetic samples, which, within experimental uncertainty of 20%, show no compensation and

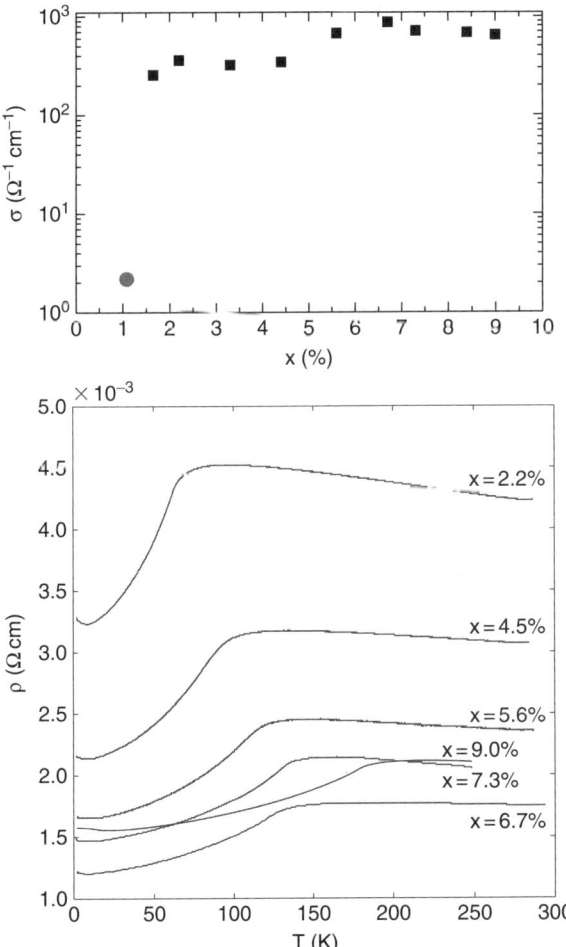

FIGURE 4.10 d.c. transport measurements in optimally annealed, metallic ferromagnetic (Ga,Mn)As (T_c is between approximately 50 and 170 K) grown by the Nottingham low-temperature MBE system with As_2 flux. Upper panel: conductivities at 10 K plotted as a function of Mn doping; insulating 1% doped material grown under the same conditions is included for comparison (grey dot). Lower panel: temperature dependence of longitudinal resistivities (from Jungwirth et al., 2007).

conductivities up to ≈ 900 Ω^{-1} cm^{-1}. Note that the high-doped metallic (Ga,Mn)As samples show only weak temperature-dependence of the conductivity associated with the onset of ferromagnetism and that the conductivity varies slowly with Mn composition in the metallic samples but changes dramatically in going from 1% to 1.5% Mn in these samples. Also note that no marked dependence of the hole density on temperature is observed for metallic samples (Ruzmetov et al., 2004).

The absence of the impurity band in high-doped (Ga,Mn)As is further evident by comparisons with other related materials (see Fig. 4.11) for which the high dopings have the disordered valence band character. The narrower-gap (In,Mn) As counterparts to (Ga,Mn)As, e.g., in which Mn acts as a shallower acceptor have similarly low magnitudes and similar temperature-dependence of the conductivity (Lee et al., 2007; Ohya et al., 2003; Schallenberg and Munekata, 2006). Comparably low conductivities (or mobilities) are also found in epitaxially grown GaAs doped with $\sim 10^{20}$ cm^{-3} of the shallow non-magnetic acceptor Mg (Kim et al., 2001). Note that the highest mobilities in p-type GaAs with the doping levels $\sim 10^{20}$–10^{21} cm^{-3} are achieved with Zn and C and these are only 5–10 times larger as compared to (Ga,Mn)As (see Fig. 4.11) (Glew, 1984; Yamada et al., 1989).

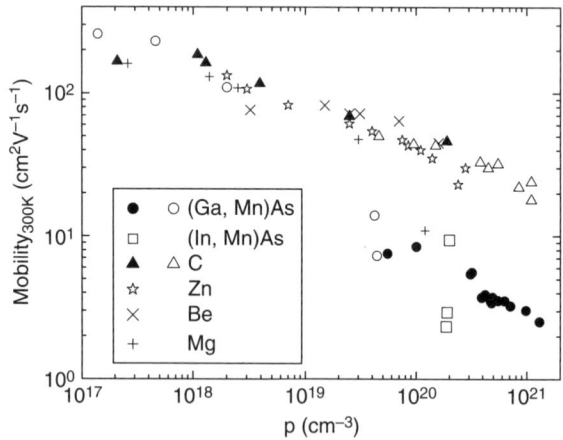

FIGURE 4.11 Room temperature Hall mobilities as a function hole concentration for GaAs doped with Mn (filled circles—(Jungwirth et al., 2007), open circles (Moriya and Munekata, 2003)), C (filled triangles (Jungwirth et al., 2007), open triangles (Yamada et al., 1989), Zn (stars—(Glew, 1984)), Be (X—(Parsons and Krajenbrink, 1983)), Mg (plus (Kim et al., 2001)), as well as InAs doped with Mn (square (Lee et al., 2007; Schallenberg and Munekata, 2006)). Hole concentrations were obtained from low temperature high field Hall measurements for the ferromagnetic metallic (Ga,Mn)As and InAs:Mn films, and from room temperature Hall measurements for the other films.

Another confirmation for the plausibility of the disordered valence band picture of metallic (Ga,Mn)As is provided by transport theory calculations. The valence band calculations treating disorder in the Born approximation overestimate the experimental d.c. conductivities of metallic (Ga,Mn)As by less than a factor of ten (Jungwirth et al., 2002a; Sinova et al., 2002) and any sizable discrepancy is removed by exact-diagonalization calculations (Yang et al., 2003) (see also Fig. 4.15) which account for strong disorder and localization effects.

The first infrared a.c. conductivity measurements in high-doped (Ga,Mn)As, reported in (Hirakawa et al., 2002; Nagai et al., 2001), were inconclusive. Only two samples were compared in both of these works, with approximately 3% and 5% nominal Mn-doping. The absence of a clear mid-infared peak in the 5% material in (Nagai et al., 2001) and the re-entrant insulating behavior of the higher doped sample in (Hirakawa et al., 2002) suggest that the growth at this early stage had not been optimized so as to minimize the formation of unintentional defects and to achieve appreciable growth reproducibility.

The more recent and more systematic infrared absorption measurements (Burch et al., 2006; Singley et al., 2002, 2003) for an insulating sample close to the metal–insulator transition and for a set of metallic highly doped (Ga,Mn)As materials are shown in Fig. 4.12. The a.c. conductivity of the 1.7% doped insulating sample is featureless in the mid-infrared region (Singley et al., 2002, 2003). This can be interpreted as a natural continuation of the trend, started in the low doped samples, in which the impurity band moves closer to the valence band and broadens with increasing doping, becoming eventually undetectable in the a.c. conductivity spectra. Note that the disappearance of the impurity band transitions can in principle result also from a strong compensation (Singley et al., 2003). A strong unintentional compensation in epitaxial (Ga,Mn)As at the ∼1% doping level is not typical, however. The 1.7% doped sample from (Singley et al., 2003) has a room temperature conductivity in the d.c. limit of about $5\,\Omega^{-1}\,cm^{-1}$, which presumably drops to a much lower value at 5 K (Singley et al., 2003). This is consistent with the $\sigma(T)$ dependence of the insulating 1% sample from the series of epitaxial materials in Fig. 4.9 which show no signatures of strong compensation over the entire range of doping from 0.07% to 7%.

A new mid-infrared feature emerges in the metallic >2% doped samples at frequencies, $\hbar\omega \approx 250$ meV, and is then red-shifted by ≈ 80 meV as the doping is further increased to approximately 7% without showing a marked broadening (see Fig. 4.12) (Burch et al., 2006; Singley et al., 2002). The association (Burch et al., 2006) of this peak to an impurity band, in analogy with the similar feature in the low doped insulating (Ga,Mn)As materials, is implausible because of (i) the absence of the activated d.c.-transport counterpart in the high-doped metallic samples, (ii) the blue-shift

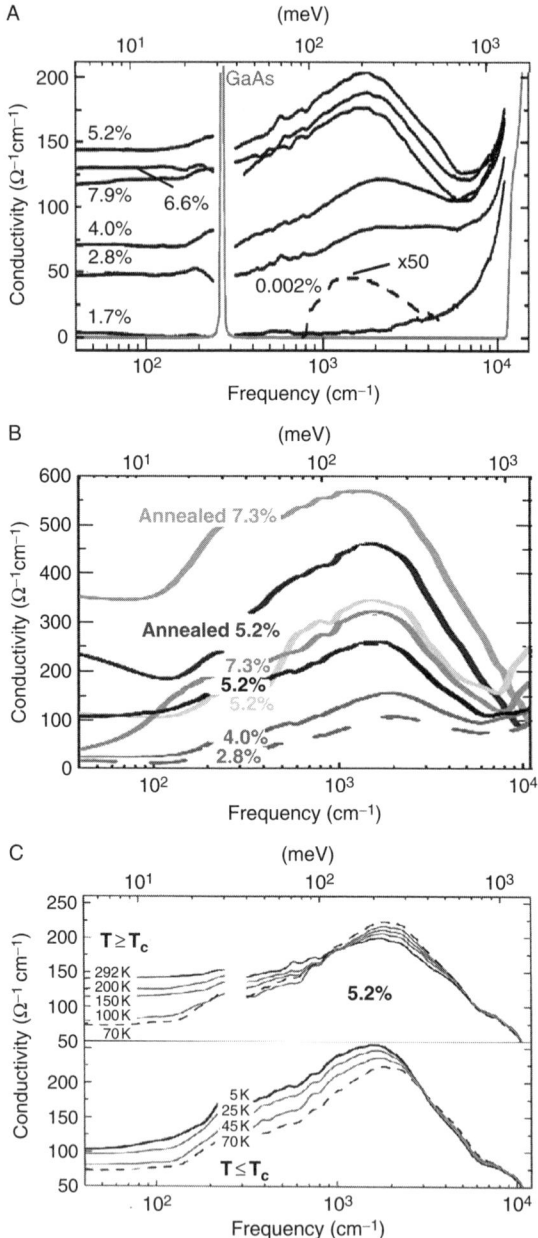

FIGURE 4.12 Infrared absorption measurements of (Ga,Mn)As. Top panel: comparison of GaAs, low-doped (Ga,Mn)As and high-doped as-grown (Ga,Mn)As materials at 292 K. Middle panel: Comparison of high-doped as-grown and annealed (Ga,Mn)As samples at 7 K. Bottom panel: Absorption measurements for the 5.2% as-grown (Ga,Mn)As below and above the ferromagnetic transition temperature (from Burch et al., 2006; Singley et al., 2002, 2003).

of this mid-infrared feature with respect to the impurity band transition peak in the $N_{Mn} \sim 10^{19}$ cm^{-3} sample, (iii) the appearance of the peak at frequency above $2E_a^0$, and (iv) the absence of a marked broadening of the peak with increased doping.

A mid-infrared peak has been observed in metallic GaAs heavily doped ($\sim 10^{19}$–10^{20} cm^{-3}) with shallow hydrogenic carbon acceptors and explained in terms of inter-valence-band transitions between heavy holes and light holes (Songprakob et al., 2002). This peak blue-shifts with increasing doping as the Fermi energy moves further into the valence and, consequently, the transitions move to higher energies (see Fig. 4.13) (Sinova et al., 2002; Songprakob et al., 2002). The mid-infrared peaks in the metallic (Ga,Mn)As materials are likely of a qualitatively similar origin. The red-shift seen in (Ga,Mn)As can be understood by recalling that while GaAs doped with $\sim 10^{19}$–10^{20} cm^{-3} C is already far on the metallic side, GaAs doped with $\sim 10^{20}$–10^{21} cm^{-3} of Mn is still close to the metal–insulator transition.

The merging of the impurity band and the valence band near the transition is illustrated schematically in Fig. 4.14. For metallic (Ga,Mn)As close to the metal–insulator transition, states in the band tail can still be expected to remain localized and spread out in energy. The red shift of the

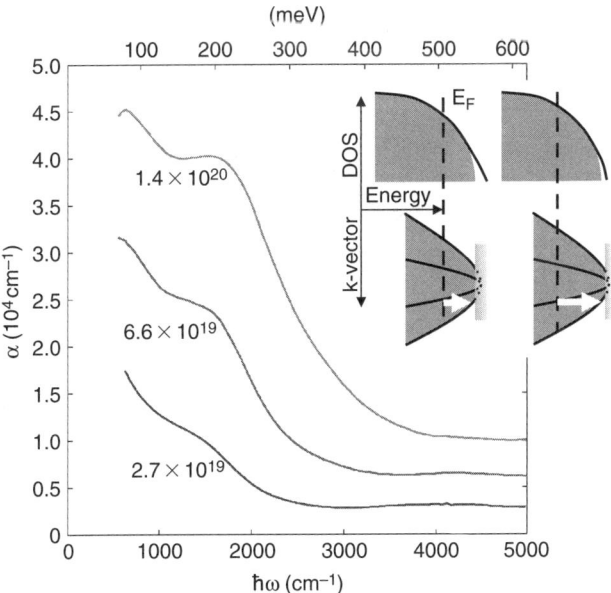

FIGURE 4.13 Infrared absorption measurements of metallic GaAs:C (from Songprakob et al., 2002). Inset illustrates the origin of the blue shift of the mid-infrared peak in these highly metallic systems.

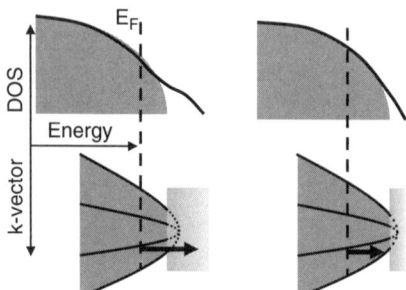

FIGURE 4.14 Schematic illustrations of the disordered valence band regime for the high doped metallic (Ga,Mn)As. Splitting of the bands in the ferromagnetic state is omitted for simplicity. The cartoons are arranged from left to right according to increasing doping. Grey areas indicate delocalized states, white areas localized states. Arrows highlight the red shift in the inter-valence-band transitions (from Jungwirth et al., 2007).

inter-valence-band transitions can result from an increased metallicity with increasing doping. At lower doping the abundance of localized states in the broad valence band tail enables transitions which take spectral weight from the low frequency region and provide a channel for higher-energy transitions. As doping increases, the valence band tail narrows because of increased screening. The inter-valence-band absorption peak then red-shifts as the low-frequency part of the spectrum adds spectral weight.

A detailed quantitative theoretical modeling of the infrared conductivity, or any other transport or micromagnetic characteristics, of high-doped GaAs:Mn is inherently difficult due to the strong disorder and due to the lack of accurate characterization of the doping parameters and unintentional disorder in the experimental materials. Nevertheless, a qualitative illustration of the complex behavior of the mid-infrared peak in strongly disordered valence bands can be found in finite-size exact-diagonalization calculations (Yang et al., 2003) which address the influence of strong disorder on the a.c. conductivity in metallic (Ga,Mn)As close to the metal–insulator transition. In these numerical data shown in Fig. 4.15, the mid-infrared feature is shifted to lower energies in the higher hole density more conductive system. In both experimental and numerical calculation cases, the red-shift of the peaks is accompanied by an increase of the d.c. conductivity. In experiment, this correlation is observed both by comparing different samples and, in a given sample, over the entire range of temperatures that was studied, below and above the ferromagnetic transition temperature (see Fig. 4.12; Burch et al., 2006; Singley et al., 2002, 2003).

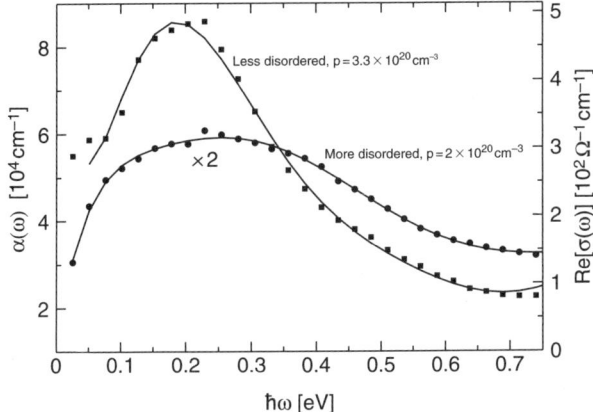

FIGURE 4.15 Exact diagonalization study showing the red shift of the valence band mid-infrared peak for the Kohn–Luttinger model assuming lower hole concentration and stronger disorder due to larger number of compensating defects (dots) and higher hole density and weaker disorder (rectangles). Lines are 10th-order polynomial fits to the numerical data to guide the eye (from Jungwirth et al., 2007; Yang et al., 2003).

As the metal–insulator transition is approached from the metal side in a multi-band system, it appears that the spectral weight that is lost from the disappearing Drude peak, centered on zero frequency, is shifted to much higher energies, perhaps due to transitions to localized valence band tail states as suggested by the cartoon in Fig. 4.14. There is a useful sum rule for the conductivity integrated over the energy range corresponding to the semiconductor band-gap, provided that the Fermi energy is not too deep in the valence band and that the valence band tails do not slide too deeply into the gap (Sinova et al., 2002; Yang et al., 2003). The sum-rule optical masses derived on the basis of the a.c. conductivity calculations for (Ga,Mn)As with the Fermi level in the valence band are consistent with experiment to within a factor of two to three (Burch et al., 2006; Sinova et al., 2002; Yang et al., 2003). (A more quantitative comparison is hindered by inaccuracies in the theoretical modeling of the (Ga,Mn)As valence bands and uncertainties associated with the high-frequency cut-off imposed on the experimental sum rules and with the determination of material parameters, hole densities in particular.)

Note that no general practical sum rule exists for the low-frequency part of the a.c. conductivity and that a Drude fit of the spectra near the d.c. limit would tend to strongly overestimate band effective masses as the metal–insulator transition is approached. In (Ga,Mn)As in particular, most of the experimental a.c. conductivity curves in Fig. 4.12 as well as the numerical simulations in Fig. 4.15 do not show a maximum at zero frequency (Burch et al., 2006; Singley et al., 2002, 2003; Yang et al., 2003) which makes the Drude fitting questionable.

The absolute conductance values of the mid-infrared conductivity peak found experimentally in metallic (Ga,Mn)As materials, which are expected to be relatively insensitive to disorder, agree with the values predicted theoretically by the inter-valence-band transition calculations (Sinova et al., 2002; Yang et al., 2003). This includes both the calculations treating disorder within the Born approximation (Sinova et al., 2002), which overestimate the d.c. conductivity, and the exact-diagonalization studies shown in Fig. 4.15 which account for strong disorder and localization effects (Yang et al., 2003).

In summary, a number of d.c. transport and optical measurements have been performed during the past four decades to elucidate the nature of states near the Fermi level in (Ga,Mn)As in the insulating and metallic regimes. A detail examination of these studies and comparisons with shallow acceptor counterparts to (Ga,Mn)As, show that impurity band markers are consistently seen in the insulating low-doped materials. Similar consistency is found when analyzing experiments in the high-doped metallic materials, in this case strongly favoring the disordered valence band picture. These conclusions are consistent with the established theoretical description of (Ga,Mn)As based on a variety of different *ab initio* and phenomenological approaches which explain many experimental magnetic and magnetotransport properties of these dilute moment ferromagnetic semiconductors.

2.2. Quantum interference and interaction effects

The range of quantum corrections to conductivity which arise in disordered conductors has been the subject of several reviews (Beenakker and van Houten, 1991; Bergmann, 1984; Kramer and MacKinnon, 1993; Lee and Ramakrishnan, 1985; Lee et al., 1987). Quantum interference effects become important when the phase coherence length, L_Φ, of electrons close to the Fermi energy is comparable with some relevant length scale.

In the Aharonov–Bohm (AB) effect one has division of amplitude of electron waves which then traverse the two arms of a ring before they recombine and interfere. Application of an external magnetic field produces phase shifts of opposite sign for the clockwise and anti-clockwise moving partial waves leading to the conductance of the ring varying periodically as a function of magnetic flux through the ring if L_Φ is comparable or larger than the ring diameter. The period corresponds to the flux through the ring changing by one flux quantum h/e.

Interference between partial waves from scattering centers within a conductor will also give rise to effects on the rsistivity. In the usual semi-classical theory of electron conduction this is neglected since it is assumed that such effects will be averaged away. However for conductors of size comparable with L_Φ the interference effects are intrinsically nonself-averaging. This leads to corrections to the conductivity of order e^2/h for a conductor of size comparable or smaller than L_Φ. Application of a magnetic field modifies the interference effects giving reproducible but aperiodic UCFs (Lee et al., 1987) of amplitude $\sim e^2/h$. One can think of a conductor with dimensions greater than L_Φ as made up of a number of independent phase coherent sub-units leading to averaging and UCFs of diminished amplitude. At finite temperature the UCF amplitude is reduced below $\sim e^2/h$ even for a conductor of size $\ll L_\Phi$ as the conductivity involves averaging over a range of energies $\sim k_B T$ about the Fermi energy. This effect, which is distinct from the decrease in L_Φ due to the increase in inelastic scattering with increasing temperature, introduces an additional length scale the "thermal length" $L_T = \hbar D/k_B T$ where D is the diffusion coefficient.

For conductors of size $\gg L_\Phi$ the UCFs average to zero. However, WL quantum corrections to the electrical transport coefficients are still present. These are due to constructive interference between partial waves undergoing multiple scattering from a state with wavevector k to a state $-k$ and partial waves traversing the time reversed trajectory. The effect is also referred to as coherent backscattering and it leads to a reduction of the conductivity. Convenient explicit expressions for the WL corrections can be obtained for $L_\Phi \gg \Lambda \gg \lambda_F$, where λ_F and Λ are the electron Fermi wavelength and mean free path. The second condition can be rewritten as $k_F \Lambda \gg 1$ where k_F is the Fermi wavevector. The corrections are of order $(k_F \Lambda)^{-1}$ and so become important for small $k_F \Lambda$. It has been argued that higher order corrections are small and that the condition $k_F \Lambda \gg 1$ can be relaxed to $k_F \Lambda > 1$. Application of a magnetic field can suppress the resistance enhancement due to WL as it removes time-reversal invariance leading to negative MR. In 3D the magnetic field begins to have a significant effect when $\ell_B \sim L_\Phi$, where $\ell_B = \left(\hbar/eB\right)^{1/2}$ is the magnetic length, and the magnetic field completely suppresses WL when $\ell_B \sim \Lambda$.

SO scattering from paramagnetic impurities in nonmagnetic metals can significantly modify the quantum corrections. It can reverse the sign of the localization correction, so it is said to produce a weak anti-localization (WAL) effect, which can result in a positive MR at weak magnetic fields (Dugaev et al., 2001). The situation is much less clear for the case of ferromagnetic metals. It has been argued (Lee and Ramakrishnan, 1985; Sil et al., 2005) that the processes leading to WAL in nonmagnetic systems are totally suppressed in ferromagnets, so that even in the presence of SO interaction one expect only a negative MR. However this is found to be only strictly true for the case of two-dimensional ferromagnets.

A distinct, electron–electron interaction quantum correction to the conductivity (Lee and Ramakrishnan, 1985) can arise in disordered conductors which often has a similar magnitude to the WL correction. This arises because electron–electron interactions cannot be treated independently of the disorder scattering for strong disorder. In the following paragraphs we review observations of quantum corrections to transport in (Ga,Mn)As.

Despite it's high crystalline quality, MBE grown high-doped (Ga,Mn)As is a strongly disordered conductor with approximately one negatively charged Mn ion per hole. Even in the highest quality (Ga,Mn)As materials with metallic conductivity the hole mean free path is comparable with the separation of the Mn impurities so the diffusivity is very low and typically $k_F \Lambda = \hbar \mu k_F^2 / e \approx 1 - 5$, as shown in Fig. 4.16. Note that these estimates, based on experimental mobilities μ and hole densities, are only qualitative as they are based on assuming Drude form of the conductivity and neglect

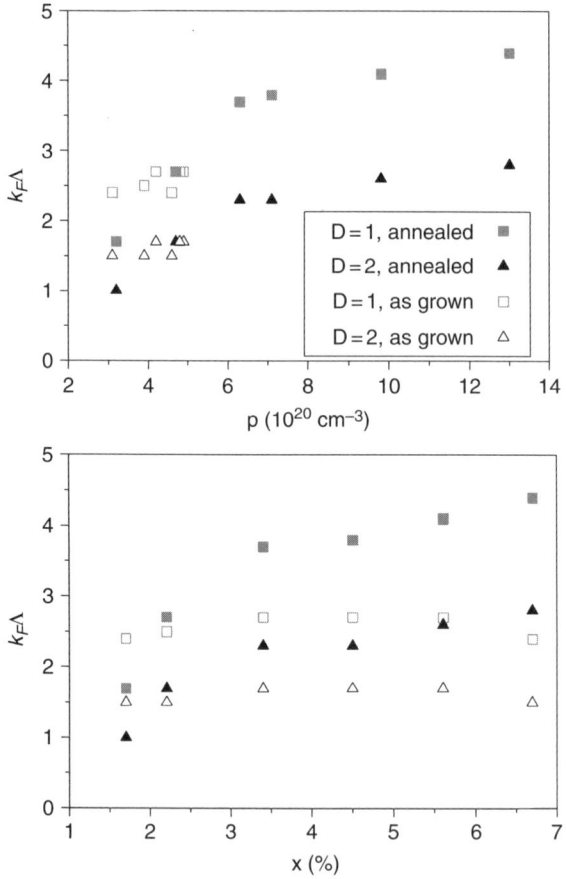

FIGURE 4.16 Estimates of the $k_F \Lambda$ parameter as a function of hole density (upper) panel and Mn doping (lower panel) for metallic (Ga,Mn)As epilayers obtained from the measured mobilities and hole densities (see Section 2.1) and assuming parabolic fully polarized ($D = 1$) and unpolarized ($D = 2$) heavy-hole valence bands.

non-parabolicity (Abolfath et al., 2001; Dietl et al., 2001) of the (Ga,Mn)As valence bands (k_F is approximated by $(6\pi^2 p/D)^{1/3}$). At temperatures which are a significant fraction of the Curie temperature one expects spin-disorder and SO scattering to lead to $L_\Phi \sim \Lambda$, strongly suppressing quantum corrections. However in high quality metallic (Ga,Mn)As it has been argued (Matsukura et al., 2004) that L_Φ need not be very small at low temperature because virtually all spins contribute to the ferromagnetic ordering and the large splitting of the valence band makes both spin-disorder and SO scattering relatively inefficient. The strong magneto-crystalline anisotropies also tend to suppress magnon scattering at low temperatures.

Recent observations (Vila et al., 2007; Wagner et al., 2006) of both large UCFs and evidence for AB oscillations in (Ga,Mn)As confirm that L_Φ can be large at low temperatures. Fig. 4.17 shows UCFs measured (Wagner et al., 2006) in (Ga,Mn)As wires of approximate width 20 nm and

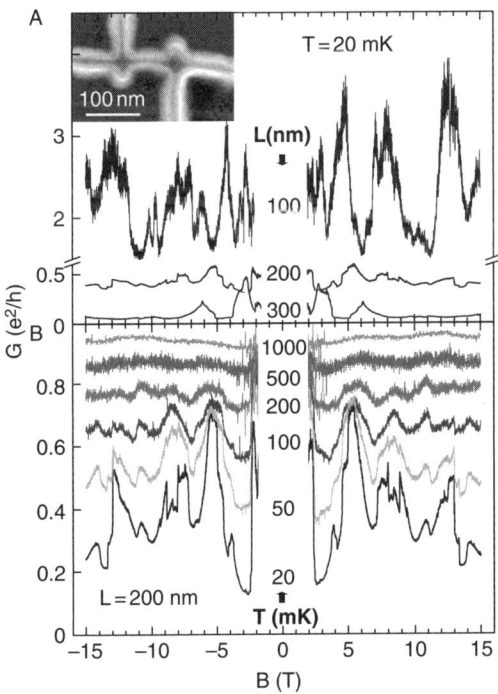

FIGURE 4.17 (A) Conductance fluctuations for three wires of different length L. For the shortest wire the amplitude of the conductance fluctuations is about e^2/h, expected for conductors with all spatial dimensions smaller or comparable to L_Φ. The inset shows an electron micrograph of a 20 nm wide wire with a potential probe separation of ~100 nm. (B) Conductance versus magnetic field of the 200 nm wire for different temperatures between 20 mK and 1 K (from Wagner et al., 2006).

thickness 50 nm. Panel (a) shows that the UCF amplitude is $\sim e^2/h$ in a 100 nm long wire at 20 mK. This directly demonstrates that $L_\Phi \sim 100$ nm. Similar measurements in higher conductivity (Ga,Mn)As give $L_\Phi \sim 100$ nm at 100 mK. These are large values corresponding to a phase relaxation time τ_Φ that is orders of magnitude larger than the elastic scattering time.

Figure 4.18 shows measurements (Wagner et al., 2006) of the magnetic field dependence of the conductivity of a lithographically defined 100 nm diameter (Ga,Mn)As ring compared to that of a 200 nm long (Ga,Mn)As wire. Additional small period oscillations are observed for the ring which the Fourier transform shows to be consistent with the expected AB period. This confirms the long L_Φ indicated by the large amplitude UCFs.

Figure 4.17B shows that the UCF amplitude decreases with increasing temperature and in Fig. 4.19 (Wagner et al., 2006) it is shown that the amplitude falls as T^{-n} with $n \sim 0.8$. The inset of Fig. 4.19 shows that the UCF amplitudes scale as $L^{3/2}$, where L is the wire length, as is expected for $L > L_\Phi$. A temperature dependence $L_\Phi \sim T^{-1/2}$ was obtained from the measured temperature dependence of the UCF amplitudes (Lee and Ramakrishnan, 1985). This was based on an analysis which assumes that the wires behave one dimensionally, that is, that L_Φ and L_T

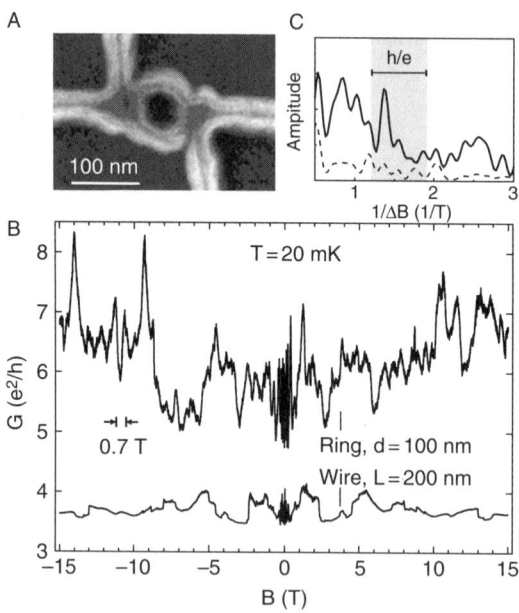

FIGURE 4.18 (A) Electron micrograph of a (Ga,Mn)As ring sample with a diameter of ~100 nm. (B) Comparison of the magnetoconductance trace of the ring sample with the conductance of a wire of comparable length and 20 nm width. (C) Corresponding FFT taken from the conductance of ring and wire. The region where AB oscillations are expected is highlighted (from Wagner et al., 2006).

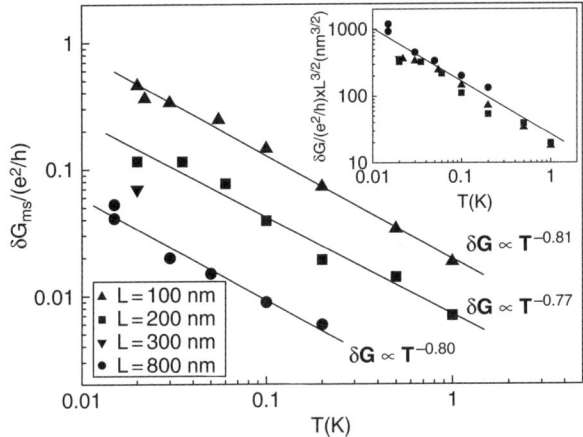

FIGURE 4.19 δG in units of e^2/h for wires of length 100, 200, and 800 nm. The measurements were carried out using currents of 10–500 pA. The maximum current at a given temperature was adjusted such that the fluctuation amplitude δG was not affected. Inset: $\delta G L^{3/2}/(e^2/h) = L_\Phi^{3/2}$ versus T (from Wagner et al., 2006).

are both large compared to the wire width and thickness. On the basis of the magnitude obtained in the analysis in (Wagner et al., 2006), this condition, however, is not necessarily fulfilled for the measured temperature range.

Figure 4.17 only shows data for large magnetic fields at which the magnetization of the (Ga,Mn)As is fully saturated. At lower fields non-reproducible fluctuations of the conductance are observed (Vila et al., 2007; Wagner et al., 2006). In this region the domains will most probably nucleate in the narrow wire structures modifying the UCFs and introducing additional varying AMR contributions. This is a complicated situation as one has no knowledge of the domain structure and the apparent "UCF correlation field" will depend upon the rate at which domains nucleate and domain walls move within the wires.

Since WL quantum corrections are suppressed by sufficiently large magnetic fields one expects a similar suppression by the internal magnetization. For dense moment ferromagnets like Iron, $\mu_0 M \sim 2$ T and the mean free path is usually quite large so WL is strongly suppressed. However, WL is observed for example in highly disordered Ni films (Aprili et al., 1997). For the dilute moment ferromagnet (Ga,Mn)As $\mu_0 M \sim 50$ mT, for high quality materials, while the field needed to suppresses WL, that is, when $\ell_B \sim \Lambda \sim 1$nm, is $\sim 1,000$ T. So since the UCF measurement firmly establish that $L_\Phi \gg \Lambda$ at the lowest temperatures one expects WL effects to be present, and since typically $k_F \Lambda \approx 2$–10, they may be large. However, one expects L_Φ to rapidly decrease at higher temperatures. Furthermore it is unclear if WAL effects should also be present in the ferromagnetic SO-coupled (Ga,Mn)As (Kramer and MacKinnon, 1993; Lee et al., 1987). UCFs have a very distinctive signature. Clear identification of WL effects is often much more problematic. The identification of WL contributions to the temperature dependence of resistance is difficult as they generally co-exist with other temperature dependent contributions and because the expected functional form can be very different for the different possible phase breaking mechanisms. The identification of the WL contributions to the measured MR may be particularly difficult for magnetic semiconductors as there are a number of effects, such as spin disorder scattering that can produce large MR in ferromagnets, particularly close to the localization boundary (Kramer and MacKinnon, 1993; Nagaev, 1998; Omiya et al., 2000).

For external magnetic field less than the coercive fields the MR response is usually dominated by AMR (see Section 3.1). At larger fields a negative isotropic MR is observed which can be very large for low conductivity material (Matsukura et al., 2004). This could be due to the suppression of spin disorder (Lee et al., 1987). However, as shown in Fig. 4.20 (Matsukura et al., 2004), the negative MR does not seem to saturate, even in extremely strong magnetic fields. It has been argued (Kramer and MacKinnon, 1993) that the negative MR arises from WL and, assuming a complete suppression of SO scattering, that the magnitude of the MR is consistent with the prediction of (Kawabata, 1980) which gives a correction of the form $-B^{1/2}$ (see Fig. 4.20).

Neumaier et al. (Neumaier et al., 2007) studied magnetotransport down to 20 mK in (Ga,Mn)As for arrays of nanowires as well as for films. These samples and materials are closely related to those in which UCFs and AB oscillations were observed (Wagner et al., 2006). The high field MR observed in these samples is considered to be due to suppression of spin disorder and is not analyzed. The main focus of this paper is the observation of a positive MR which emerges at the lowest temperatures. This is ascribed to WAL and is shown to be in reasonable agreement with theory. However, to try to isolate this contribution from the AMR response, which occurs in the same field range as the WAL, the authors have to subtract the AMR response from the measured MR assuming it to be temperature independent, which leads to additional uncertainty. In these samples the resistivity rises rapidly with decreasing temperatures at the lowest temperatures, as is usually observed in (Ga,Mn)As. Neumaier et al. (Neumaier et al., 2007) ascribe this to electron–electron interactions finding that the conductivity rises as $T^{1/2}$ and $\ln(T)$ in samples they describe as 1D and 2D. (As mentioned above it is not entirely clear if the samples can be regarded as 1D and 2D conductors.)

Rokhinson et al. (Rokhinson et al., 2007) have recently observed a $\sim 1\%$ negative MR at low temperature in a (Ga,Mn)As film which saturates at

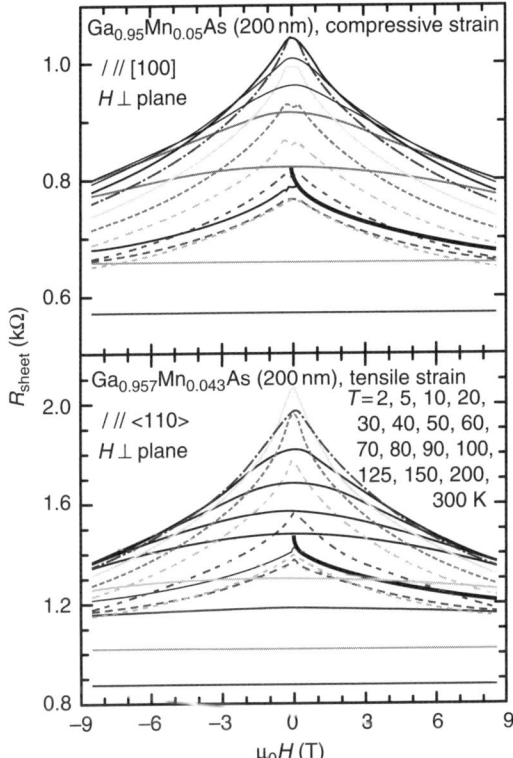

FIGURE 4.20 Field and temperature dependencies of resistance in $Ga_{0.95}Mn_{0.05}As/$ GaAs (compressive strain, upper panel) and in tensile strained $Ga_{0.957}Mn_{0.043}As/(In,Ga)As$ (lower panel) for magnetic field perpendicular to the film plane. Starting from up, subsequent curves at $B = 0$ correspond to temperatures in Kelvin: 70, 60, 80, 50, 90, 40, 100, 30, 125, 20, 2, 5, 10, 150, 200, 300 (upper panel) and to 50, 60, 40, 70, 30, 80, 90, 20, 100, 2, 10, 5, 125, 150, 200, 300 (lower panel) (from Matsukura et al., 2004).

$\sim$20 mT. As noted by the authors this feature does not have the behavior one would normally expect for WL. (Because of the small Λ one only expects saturation at very large magnetic fields.) However, the MR is still ascribed to WL and, furthermore, it is argued that the presence of WL is incompatible with the Fermi level being in an SO-coupled valence band. Given the state of the theory of quantum corrections in ferromagnets and the other studies in (Ga,Mn)As reviewed in this section, such an extrapolation is difficult to be justified.

In summary, the clear observation of UCFs in (Ga,Mn)As and the quite strong evidence for AB oscillations in (Ga,Mn)As ring structures provides clear evidence for phase coherence lengths of up to $\sim$100 nm at mK temperatures. This indicates that almost all spins are participating in the magnetic order with strong suppression of spin scattering. However, the dimensionality of the samples studied so far is marginal and so the temperature dependence of L_Φ obtained need to be viewed with some caution. On the basis of these relatively large L_Φ values, WL/WAL effects might well be significant in (Ga,Mn)As but their presence is yet to be firmly established. Given the number of open questions concerning quantum corrections in (Ga,Mn)As this will likely be an area of intense activity in the future research.

2.3. Critical contribution to transport near the Curie point

Since seminal works of de Gennes and Friedel (de Gennes and Friedel, 1958) and Fisher and Langer (Fisher and Langer, 1968), critical behavior of resistivity has been one of the central problems in the physics of itinerant ferromagnets. Theories of coherent scattering from long wavelength spin fluctuations, based on the original paper by de Gennes and Friedel, have been used to explain large peak in the resistivity $\rho(T)$ at the Curie temperature T_c observed in Eu-chalcogenide magnetic semiconductors (Haas, 1970). The emphasis on the long wavelength limit of the spin–spin correlation function, reflecting critical behavior of the magnetic susceptibility, is justified in these systems by the small density and corresponding small Fermi wavevectors of carriers.

As pointed out by Fisher and Langer (Fisher and Langer, 1968), the resistivity anomaly in high carrier density transition metal ferromagnets is qualitatively different and associated with the critical behavior of correlations between nearby moments. When approaching T_c from above, thermal fluctuations between nearby moments are partially suppressed by short-range magnetic order. Their singular behavior is like that of the internal energy and unlike that of the magnetic susceptibility. The singularity at T_c occurs in $d\rho/dT$ and is closely related to the critical behavior of the specific heat c_v. While Fisher and Langer expected this behavior for $T \to T_c^+$ and a dominant role of uncorrelated spin fluctuations at $T \to T_c^-$, later studies of elemental transition metals found a proportionality between $d\rho/dT$ and c_v on both sides of the Curie point (Joynt, 1984; Shacklette, 1974).

The nature of the critical contribution to resistivity of high-quality (Ga,Mn)As materials is illustrated in Fig. 4.21 (Novák et al., 2008). Fig. 4.21A shows magnetization $M(T)$ measured at 5 and 1,000 Oe along the $[1\bar{1}0]$-direction which is the easy-axis in this material over the whole studied temperature range. The 5 Oe field is applied to avoid domain reorientations when measuring the temperature dependent remanent magnetization. The remanence vanishes sharply at $T \to T_c^-$, with $T_c = 185$ K. Figure 4.21B shows results of corresponding zero-field and 1,000

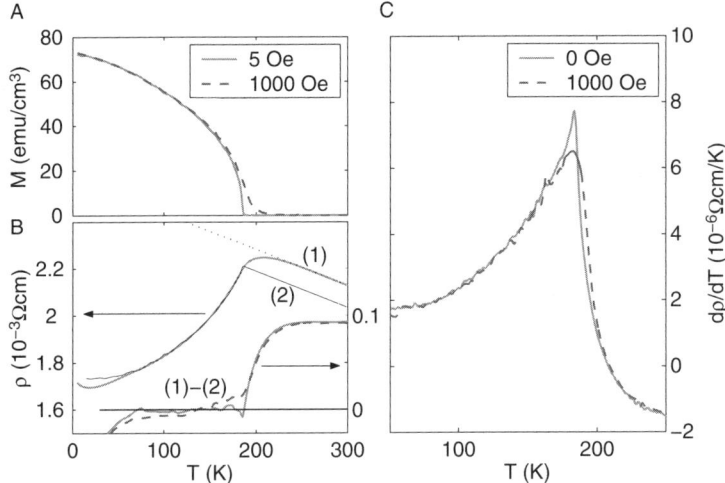

FIGURE 4.21 (A) Temperature dependent SQUID magnetization of an optimized 12.5% Mn-doped (Ga,Mn)As measured at 5 and 1,000 Oe. (B) Measured resistivity $\rho(T)$ at zero-field (1), the fit of $\rho(T)$ by $\rho_{fit}(T) = c_0 + c_{nm}T + c_{m2}M^2(T) + c_{m4}M^4(T)$ (2), and the difference $\rho(T) - \rho_{fit}(T)$ for the 0 and 1,000 Oe field measurements (right hand scale). For details on the fitting see the main text. (C) Temperature derivative of the measured resistivity at 0 and 1,000 Oe. The peak in zero-field $d\rho/dt$ coincides, within the experimental error, with the SQUID $T_c = 185$ K.

Oe-field measurements of $\rho(T)$. A useful insight into the physics of the shoulder in $\rho(T)$ near T_c is obtained by first removing the nonmagnetic part in the temperature dependence, $c_{nm}T$, which is approximated by linearly extrapolating from the high-temperature $\rho(T)$ data (see the straight dotted line in Fig. 4.1B). The measured $M(T)$ then allows to subtract the contribution from uncorrelated scattering, assuming a M^2 expansion dependence (Fisher and Langer, 1968; López-Sancho and Brey, 2003; Moca et al., 2007; Van Esch et al., 1997). On the ferromagnetic side, a very close fitting is achieved by $\rho_{fit}(T) = c_0 + c_{nm}T + c_{m2}M^2(T) + c_{m4}M^4(T)$. On the paramagnetic side, the nose dive of the remaining magnetic contribution to $\rho(T)$ suggests that the shoulder in the measured resistivity originates from a singular behavior at $T \to T_c^+$ rather than from disorder broadening effects.

The singularity is revealed by numerically differentiating the experimental zero-field $\rho(T)$ curve, as shown in Fig. 4.21C. The position of the sharp peak in $d\rho/dT$ coincides with the Curie temperature determined from $M(T)$. The critical nature of the transport anomaly is confirmed by measurements in the 1,000 Oe field: $\rho(T)$ at $T < T_c$ can be closely fitted by $\rho_{fit}(T)$ using the 1,000 Oe $M(T)$ and the same values of the fitting constants as in zero field. The field removes the singular behavior at T_c and makes the onset of the magnetic contribution to $d\rho/dT$ from the paramagnetic side more gradual than in zero field.

The $d\rho/dT$ singularity described in Fig. 4.21 is a generic characteristic of thin (Ga,Mn)As films spanning a wide range of Mn-dopings, prepared by reproducible growth and post-growth annealing procedures which had been optimized separately for each doping level. This is illustrated in Fig. 4.22 which shows $\rho(T)$ and $d\rho/dT$ data measured in optimized 4.5–12.5% Mn-doped materials. The samples show sharp $d\rho/dT$ singularities at Curie temperatures ranging from 81 to 185 K.

We now discuss the phenomenology of the singular $d\rho/dT$ behavior in (Ga,Mn)As in more detail. Ferromagnetism in (Ga,Mn)As originates from spin–spin coupling between local Mn-moments and valence band holes, $J\sum_i \delta(\mathbf{r} - \mathbf{R}_i)\mathbf{s}\mathbf{S}_i$ (Jungwirth et al., 2006; Matsukura et al., 2002). Here $\mathbf{S}_i$ represents the local spin and $\mathbf{s}$ the hole spin-density operator. This local-itinerant exchange interaction plays a central role in theories of the critical transport anomaly. When treated in the Born approximation, the interaction yields a carrier scattering rate from magnetic fluctuations which is proportional to the static spin–spin correlation function, $\Gamma(\mathbf{R}_i, T) \sim$

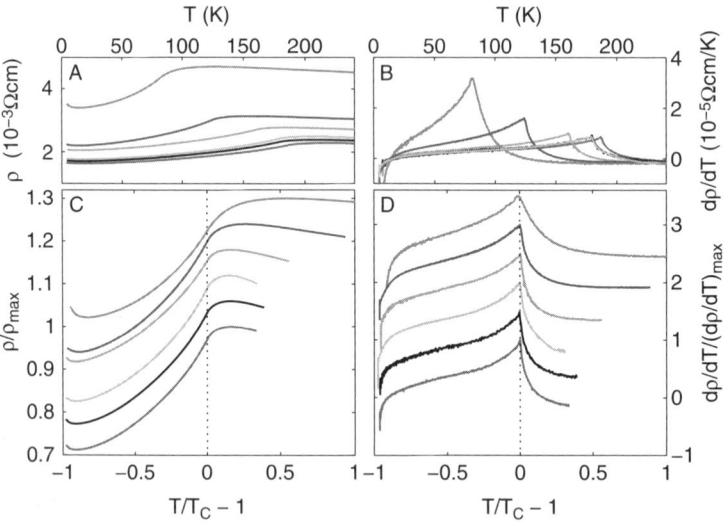

FIGURE 4.22 (A) Resistivities $\rho(T)$ and (B) temperature derivatives $d\rho/dT$ for optimized (Ga,Mn)As films (of thicknesses between 13 and 33 nm) prepared in Prague and Nottingham MBE-systems. Data normalized to maximum $\rho(T)$ and $d\rho/dT$ are plotted in (C) and (D), respectively. Curves in (A),(C), and (D) are ordered from top to bottom according to increasing Mn-doping and T_c: 4.5% Mn-doped sample with $T_c = 81$ K, 6% with 124 K, 10% with 161 K, 12% with 179 K (Prague and Nottingham samples), and 12.5% doped sample with 185 K Curie temperature. In (C) and (D) curves are offset for clarity.

$J^2[\langle S_i S_0 \rangle - \langle S_i \rangle \langle S_0 \rangle]$ (de Gennes and Friedel, 1958). Typical temperature dependences of the uncorrelated part, $\Gamma_{\text{uncor}}(\mathbf{R}_i, T) \sim \delta_{i,0} J^2 [S(S+1) - \langle S_i \rangle^2]$, and of the Fourier components, $\Gamma(\mathbf{k}, T) = \sum_{i \neq 0} \Gamma(\mathbf{R}_i, T) \exp(\mathbf{k} \mathbf{R}_i)$, are illustrated in the lower inset of Fig. 4.23 (Fisher and Langer, 1968). At small wavevectors, $\Gamma(\mathbf{k}, T)$ has a peak near T_c; at k similar to the inverse separation of the local moments ($kd_{\uparrow - \uparrow} \sim 1$) the peak broadens into a shoulder while the singular behavior at T_c is in the temperature derivative of the spin–spin correlator.

The M^2 expansion, providing a good fit to the magnetic contribution to the resistivity $\rho_m(T) = \rho(T) - c_{nm} T$ at $T < T_c$ in Fig. 4.21, corresponds to the dominant contribution from Γ_{uncor} on the ferromagnetic side of the transition. A more detailed assessment of the temperature dependence of magnetization and resistivity is obtained from $\log(M)$ versus $\log(t)$ and $\log(d\rho_m/dt)$ versus $\log(t)$ plots shown in Fig. 4.23. Examples of low-noise magnetization data recorded in the 10%, 12%, and 12.5% doped materials from Fig. 4.22 are shown in the left panel of Fig. 4.23. In high-quality films one finds a power-low dependence with the exponent $\beta \approx 0.4 \pm 20\%$. The $\log(d\rho_m/dt)$ versus $\log(t)$ plots are shown in the right panel of Fig. 4.23 for the 6–12.5% doped materials. All data sets collapse to a common temperature dependence at $T < T_c$ and at $T > T_c$. On the ferromagnetic side the

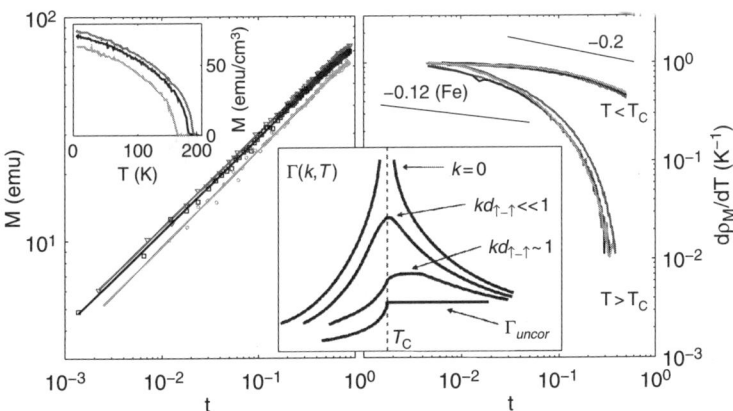

FIGURE 4.23 Left panel: $\log(M) - \log(t)$ plot of the 5 Oe magnetization data measured in 10, 12, and 12.5% doped samples of Fig. 4.2. Fitted straight lines show a power-law with critical exponent $\beta = 0.42 \pm 0.01$. Measured $M(T)$ curves are plotted in the top inset. Right panel: $\log(d\rho/dT) - \log(t)$ plots for the 6–12.5% doped samples of Fig. 4.2. Lines corresponding to exponent $-0.2(2\beta - 1)$ and -0.12 (c_v exponent in Fe) are included for comparison. Lower inset: Schematics of the uncorrelated part and Fourier components of the spin–spin correlation function for different ratios between wavevector and inverse of the spin separation.

dependence is, as discussed above, similar to $(2\beta-1)\log(t) \approx -0.2\log(t)$. The comparison is, however, only qualitative as no clear power-low behavior is observed in the $d\rho/dt$ data on either side of the phase transition.

The shoulder in $\rho_m(T)$ and the presence of the singularity in $d\rho_m/dT$ on the paramagnetic side suggests that large wavevector components of $\Gamma(\mathbf{k}, T)$ dominate the temperature dependence of the scattering in the $T \to T_c^+$ critical region. This is consistent with the ratio between hole and Mn local-moment densities approaching unity, that is, with large carrier Fermi wavevectors in annealed (Ga,Mn)As materials (Jungwirth et al., 2005). The property makes (Ga,Mn)As distinct from the dense-moment magnetic semiconductors (Haas, 1970) and makes its critical transport anomaly reminiscent of elemental transition metal ferromagnets (Joynt, 1984; Shacklette, 1974).

The strength of the temperature dependence of $d\rho_m/dT$ as T approaches T_c from above appears very large. In iron, for example, $d\rho_m/dT$ follows closely the specific heat, $c_v \sim t^\alpha$, with the critical exponent $\alpha \approx -0.12$. Since the estimated value of β in (Ga,Mn)As is rather normal (between 0.5 and 0.3) and measurements of the magnetic susceptibility (Matsukura et al., 2002; Wang et al., 2005a) do not indicate anomalous deviations from the Curie–Weiss exponent $\gamma = -1$, the Widom scaling $\alpha = 2\beta-\gamma-2$ suggests also a normal value of $|\alpha| \ll 1$ in (Ga,Mn)As. Higher precision studies of the critical region are necessary to establish whether the stronger temperature dependence of $d\rho_m/dT$ reflects a more intriguing relationship between magnetism and transport (Timm et al., 2005) or whether the proportionality between $d\rho/dT$ and c_v is recovered in the closer vicinity of T_c (Kim et al., 2003; Klein et al., 1996).

We point out that the common characteristics of transport in the critical region, illustrated in Figs. 4.2, and 4.4, bear important implications for studies of the Curie temperature limit in (Ga,Mn)As. The record $T_c = 185$ K, 12.5% Mn-doped material represents the current highest doping for which reproducible optimization of growth and annealing conditions has been achieved (which becomes exceedingly tedious for concentrations above 10%). Comparing the $d\rho/dT$ singularity in this epilayer with the lower-doped samples provides an experimentally simple yet physically appealing probe which suggests that despite the increase of T_c by more than 100 K the physical nature of the magnetic state has not significantly changed up to this largest Mn-doping studied so far.

Finally we remark that the possibility to study the critical transport anomaly in the class of ferromagnetic semiconductors is rather unique as they have a much simpler band structure and larger variability of key parameters compared to metal ferromagnets. It also provids a tool for direct transport measurement of T_c in bulk (Ga,Mn)As and in microdevices in which standard magnetometry is not feasible.

3. EXTRAORDINARY MAGNETOTRANSPORT

MR effects comprise the ordinary responses of carriers to the Lorentz force produced by the external magnetic field and the extraordinary responses to internal magnetization in ferromagnets via SO interaction. The interference effects discussed in the previous section are an example of the ordinary MR. In this section we focus on the extraordinary MR.

Advanced computational techniques and experiments in new unconventional ferromagnets have recently led to a significant progress in coping with the subtle, relativistic nature of the extraordinary effects. There are two distinct extraordinary MR coefficients, the anomalous Hall effect (AHE) and the AMR. The AHE is the antisymmetric transverse MR coefficient obeying $\rho_T(\mathbf{M}) = -\rho_T(-\mathbf{M})$, where the magnetization vector $\mathbf{M}$ is pointing perpendicular to the plane of the Hall bar sample. The AMR is the symmetric extraordinary MR coefficient with the longitudinal and transverse resistivities obeying, $\rho_L(\mathbf{M}) = \rho_L(-\mathbf{M})$ and $\rho_T(\mathbf{M}) = \rho_T(-\mathbf{M})$, where $\mathbf{M}$ has an arbitrary orientation.

The AHE has, so far, attracted more interest and (Ga,Mn)As has become one of the favorable test bed systems for its investigation. Here the unique position of (Ga,Mn)As ferromagnets stems from their tunability and the relatively simple, yet strongly SO coupled and exchange split carrier Fermi surfaces (Jungwirth et al., 2006; Matsukura et al., 2002). The principles of the microscopic description of the AHE in (Ga,Mn)As have been successfully applied to explain the effect in other itinerant ferromagnets (Dugaev et al., 2005; Fang et al., 2003; Haldane, 2004; Kötzler and Gil, 2005; Lee et al., 2004; Sinitsyn et al., 2005; Yao et al., 2004), including the conventional transition metals such as iron and cobalt, a pattern that has since then been repeated for other extraordinary MR coefficients. The advances in the understanding of the AHE are discussed in several reviews (Chien and Westgate, 1980; Dietl et al., 2003; Jungwirth et al., 2006; Sinova et al., 2004a; see also Chapter 2, this volume) and therefore we focus here on the AMR coefficients.

3.1. AMR in ohmic regime

The AMR was discovered in conventional ferromagnetic metals 150 years ago (Thomson, 1857) but despite its historical importance in condensed matter physics and magnetic recording technologies the effect remains relatively poorly understood. Phenomenologically, AMR has a noncrystalline component, arising from the lower symmetry for a specific current direction, and crystalline components arising from the crystal symmetries (Döring, 1938; van Gorkom et al., 2001). A physically appealing picture has been used to explain the positive sign of the noncrystalline AMR observed in most transition metal ferromagnets (McGuire and Potter, 1975; Smit, 1951). The interpretation is based on the Mott's model of the spin-up and spin-down two-channel conductance corrected for perturbative SO coupling effects. In the model most of the current is carried by the light-mass s-electrons which experience no SO coupling and a negligible exchange splitting but can scatter to the heavy-mass d-states. AMR is then explained by considering the SO potential which mixes the exchange-split spin-up and spin-down d-states in a way which leads to an anisotropic scattering rate of the current carrying s-states (McGuire and Potter, 1975; Smit, 1951). Controversial interpretations, however, have appeared in the literature based on this model (Potter, 1974; Smit, 1951) and no clear connection has been established between the intuitive picture of the AMR the model provides and the successful, yet fully numerical *ab initio* transport theories (Banhart and Ebert, 1995; Ebert et al., 2000; Khmelevskyi et al., 2003). Experimentally, the noncrystalline and, the typically much weaker, crystalline AMR components in metals have been indirectly extracted from fitting the total AMR angular dependencies (van Gorkom et al., 2001).

Among the remarkable AMR features of (Ga,Mn)As ferromagnetic semiconductors are the opposite sign of the noncrystalline component, compared to most metal ferromagnets, and the crystalline terms reflecting the rich magnetocrystalline anisotropies (Baxter et al., 2002; Goennenwein et al., 2005; Jungwirth et al., 2003; Limmer et al., 2006; Matsukura et al., 2004; Tang et al., 2003; Wang et al., 2005b). In Fig. 4.24 we show an example of AMR data replotted from (Limmer et al., 2006) which reports a systematic experimental and phenomenological study of the AMR coefficients in (Ga,Mn)As films, grown on (001)- and (113) A-oriented GaAs substrates, for general, non-saturating and saturating in-plane and out-of-plane magnetic fields. As further illustrated in Figs. 4.25–4.27, microscopic numerical simulations (Jungwirth et al., 2002a, 2003) consistently describe the sign and magnitudes of the noncrystalline AMR in the standard (Ga,Mn)As materials with metallic conductivities and capture the more subtle crystalline terms associated with, for example, growth-induced strain (Jungwirth et al., 2002a; Matsukura et al., 2004). In the following paragraphs we describe some of the rich AMR phenomenology in (Ga,Mn)As in more detail and explain the basic microscopic physics of the AMR in dilute moment ferromagnets. For simplicity we discuss only AMR in saturating magnetic fields, that is, for **M** fully aligned with the external field, and the pure AMR geometry with zero (antisymmetric) Hall signal, that is, for **M** oriented in the plane of the device. (Ga,Mn)As films grown on the (001)-GaAs substrate will be considered.

The phenomenological decomposition of the AMR of (Ga,Mn)As into various terms allowed by symmetry is obtained by extending the standard phenomenology (Döring, 1938), to systems with cubic (Jungwirth

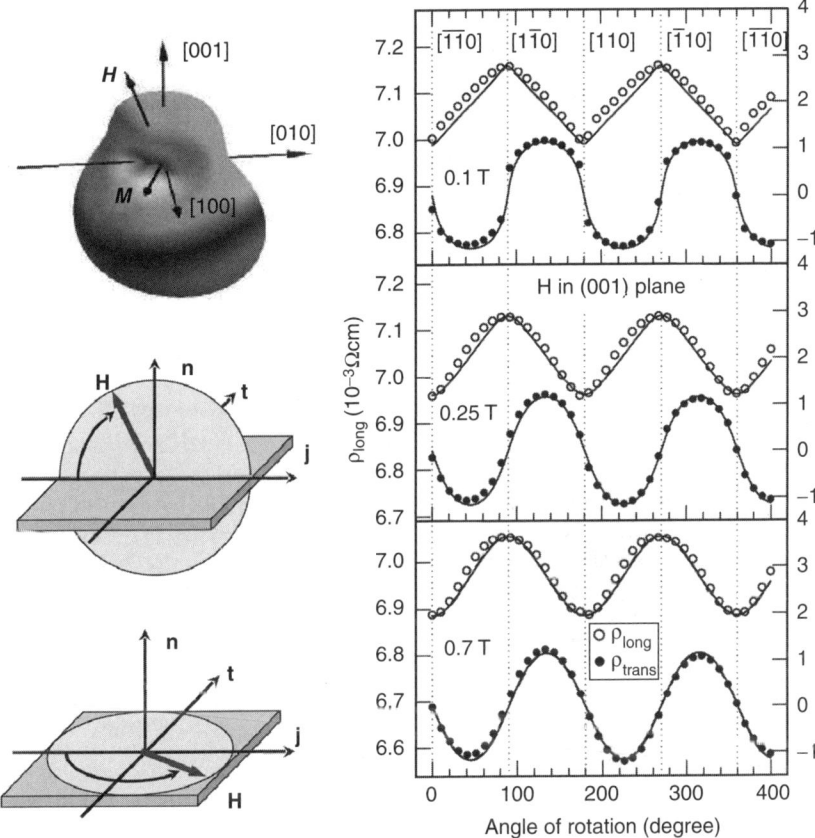

FIGURE 4.24 Left panels: Magnetization and external magnetic field vectors with different orientations below saturation fiels. The out-of-plane and in-plane AMR measurement geometries. Right panel: measured longitudinal and transverse in-plane AMR curves at external fields smaller than the saturation field (0.1 and 0.25 T) and larger than the saturation field (0.7 T). The solid lines represent fits to the experimental data (from Limmer et al., 2006).

et al., 2002b) plus uniaxial (Rushforth et al., 2007) anisotropy. With this the longitudinal AMR is written as (Rushforth et al., 2007),

$$\frac{\Delta\rho_{xx}}{\rho_{av}} = C_I \cos 2\phi + C_U \cos 2\psi + C_C \cos 4\psi + C_{I,C} \cos(4\psi - 2\phi). \quad (4.2)$$

where $\Delta\rho_{xx} = \rho_{xx} - \rho_{av}$, ρ_{av} is the ρ_{xx} averaged over 360° in the plane of the film, ϕ is the angle between the magnetization unit vector $\hat{M}$ and the current I, and ψ the angle between $\hat{M}$ and the [110] crystal direction.

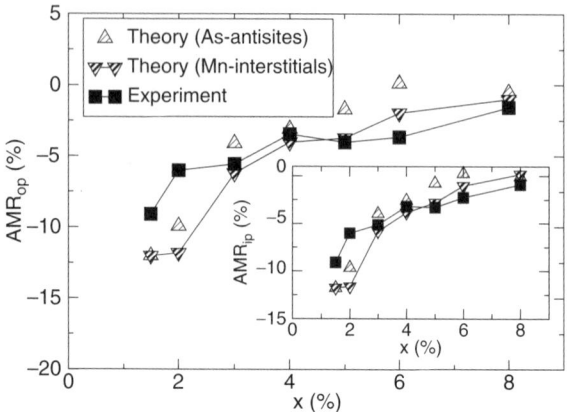

FIGURE 4.25 Experimental (filled symbols) AMR coefficients and theoretical data obtained assuming As-antisite compensation (open symbols) and Mn-interstitial compensation (semi-filled symbols). AMR$_{op}$ is the relative difference between longitudinal resistivities for **M** parallel to the current and **M** out-of-plane and perpendicular to the current. AMR$_{ip}$ is the relative difference between longitudinal resistivities for **M** parallel to the current and **M** in-plane and perpendicular to the current (from Jungwirth et al., 2003).

The four contributions are the noncrystalline term, the lowest order uniaxial and cubic crystalline terms, and a crossed noncrystalline/crystalline term. The purely crystalline terms are excluded by symmetry for the transverse AMR and one obtains,

$$\frac{\Delta \rho_{xy}}{\rho_{av}} = C_I \sin 2\phi - C_{I,C} \sin(4\psi - 2\phi). \tag{4.3}$$

Microscopically the emergence of the AMR components has been explained starting from the disordered valence-band description of (Ga,Mn)As whose applicability to materials with metallic conductivities has been thoroughly justified in Section 2.1 (Jungwirth et al., 2006, 2007). A practical implementation of this description is based on the canonical Schrieffer–Wolff transformation of the Anderson Hamiltonian which for (Ga,Mn)As replaces hybridization of Mn d-orbitals with As and Ga sp-orbitals by an effective spin–spin interaction of $L = 0$, $S = 5/2$ local-moments with host valence band states. These states, which carry all the SO-coupling, can be described by the **k·p** Kohn–Luttinger Hamiltonian (Dietl et al., 2001; Jungwirth et al., 2006).

In the dilute moment systems like (Ga,Mn)As ferromagnets, two distinct microscopic mechanisms lead to anisotropic carrier lifetimes, as illustrated in Fig. 4.28A (Rushforth et al., 2007): One combines the SO-coupling in the carrier band with polarization of randomly distributed magnetic scatterers and the other with polarization of the carrier band

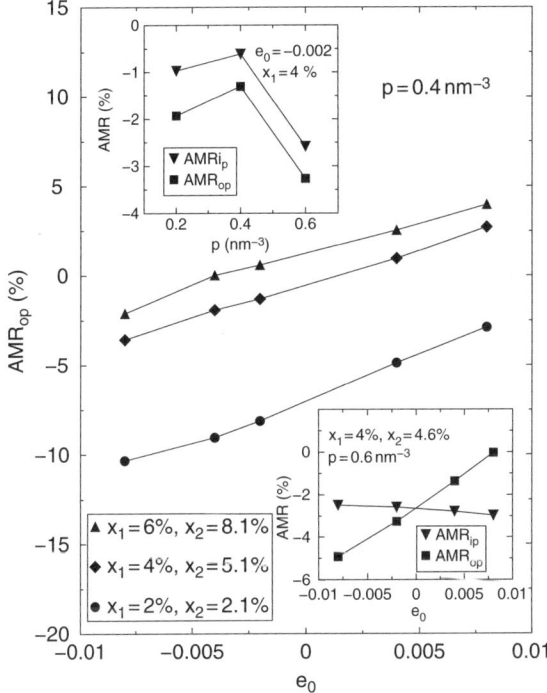

FIGURE 4.26 Theoretical AMR effects in (Ga,Mn)As obtained from the KL kinetic-exchange model and Boltzmann transport theory. Curves correspond to total Mn doping x1 and compensation due to As antisites or total Mn doping x2 and compensation due to Mn interstitials. Main panel: Out-of-plane AMR coefficient as a function of strain for several Mn dopings. Lower inset: Out-of-plane and in-plane AMR co-efficients as a function of strain. Upper inset: Out-of-plane AMR coefficient as a function of the hole density (from Jungwirth et al., 2002a).

itself resulting in an asymmetric band-spin-texture. Although acting simultaneously in real systems, theoretically both mechanisms can be turned on and off independently. Since the former mechanism clearly dominates in (Ga,Mn)As, the spin-splitting of the valence band is neglected in the following qualitative discussion. This is further simplified by focusing on the noncrystalline AMR in the heavy-hole Fermi surfaces in the spherical, $s \| k$, spin-texture approximation (Jungwirth et al., 2002b) (see Fig. 4.28A) and considering scattering off a δ-function potential $\infty (\alpha + \hat{M}s)$. Here s and k are the carrier spin-operator and wavevector, and α represents the ratio of nonmagnetic and magnetic parts of the impurity potential. Assuming a proportionality between conductivity and lifetimes of carriers with $k \| I$ gives the folowing qualitative analytical expression for the AMR in this class of materials (Rushforth et al., 2007).

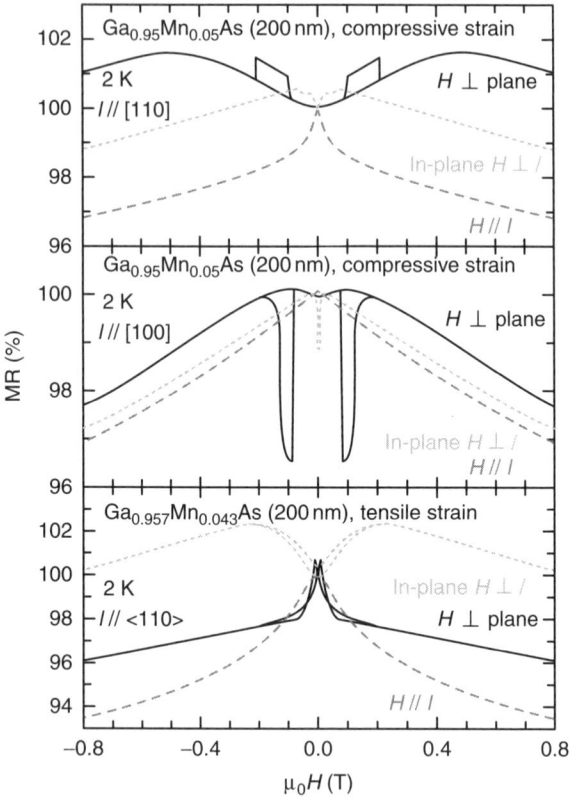

FIGURE 4.27 Upper and middle panels: Experimental field-induced changes in the resistance of $Ga_{0.95}Mn_{0.05}As$ grown on a GaAs substrate under compressive strain for current along the [110] crystal direction (upper panel) and [100] crystal direction (middle panel) for three different orientations of the magnetization. Lower panel: Experimental magnetoresistance (MR) curves in $Ga_{0.957}Mn_{0.043}As$ grown on (In,Ga)As substrate under tensile strain (current along the [110] crystal direction) (from Matsukura et al., 2004).

$$\frac{\sigma(\hat{\mathbf{M}}||I)}{\sigma(\hat{\mathbf{M}}\perp I)} = \left(\alpha^2 + \frac{1}{4}\right)\left(\alpha^2 + \frac{1}{12}\right)\left(\alpha^2 - \frac{1}{4}\right)^{-2}. \quad (4.4)$$

Therefore when $\alpha \ll 1$, one expects $\sigma(\hat{\mathbf{M}}||I) < \sigma(\hat{\mathbf{M}}\perp I)$ (as is usually observed in metallic ferromagnets). But the sign of the noncrystalline AMR reverses at a relatively weak nonmagnetic potential ($\alpha = 1/\sqrt{20}$ in the model), its magnitude is then maximized when the two terms are comparable ($\alpha = 1/2$), and, for this mechanism, it vanishes when the magnetic term is much weaker than the nonmagnetic term ($\alpha \to \infty$).

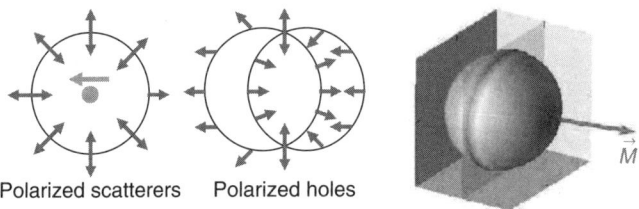

FIGURE 4.28 Noncrystalline AMR in spherical bands: 2D Cartoons of AMR mechanisms and calculated anisotropic scattering rate on the 3D Fermi surface of the minority heavy-hole band in $Ga_{0.95}Mn_{0.05}As$. From (Rushforth et al., 2007).

Physically, carriers moving along $\hat{\mathbf{M}}$, that is, with s parallel or antiparallel to $\hat{\mathbf{M}}$, experience the strongest scattering potential among all Fermi surface states when $\alpha = 0$, giving $\sigma(\hat{\mathbf{M}} \| I) < \sigma(\hat{\mathbf{M}} \perp I)$. When the nonmagnetic potential is present, however, it can more efficiently cancel the magnetic term for carriers moving along $\hat{\mathbf{M}}$, and for relatively small α the sign of AMR flips. Since $\alpha < 1/\sqrt{20}$ is unrealistic for the magnetic acceptor Mn in GaAs (Jungwirth et al., 2002a, 2006) one obtains $\sigma(\hat{\mathbf{M}} \| I) > \sigma(\hat{\mathbf{M}} \perp I)$, consistent with experiment. Note that the analysis based on Eq. (4.4) also predicts that when the SO-coupling in the host band is of the form $s \perp k$, as in the Rashba-type 2D systems, or when Mn forms an isovalent pure magnetic impurity, for example, in II–VI semiconductors, the sign of the noncrystalline AMR will be reversed.

Numerical simulations of hole scattering rates, illustrated in Fig. 4.28 on a color-coded minority heavy-hole Fermi surface, were obtained (Rushforth et al., 2007) within the spherical approximation but including the hole spin polarization, light-hole and split-off valence bands, and realistic nonmagnetic and magnetic Mn impurity potentials (Jungwirth et al., 2002a). The simulations confirm the qualitative validity of the analytical, noncrystalline AMR expressions of Eqs. (4.4).

Differences among experimental AMRs for current along the [100], [110], and $[1\bar{1}0]$ directions show that cubic and uniaxial crystalline terms are also sizable. This phenomenology is systematically observed in experimental AMRs of weakly or moderately compensated metallic (Ga,Mn)As films. Typical data for such systems, represented by the 25 nm $Ga_{0.95}Mn_{0.05}As$ film with 3.6% AMR, are shown in Fig. 4.29B for the Hall bars patterned along the [100], [010], [110], and $[1\bar{1}0]$ directions (Rushforth et al., 2007).

The high crystalline quality metallic (Ga,Mn)As samples allow to produce low contact resistant Hall bars accurately orientated along the principle crystallographic axes, from which it is possible to extract the independent contributions to the AMR. It is also possible to fabricate low contact resistance Corbino disk samples for which the averaging over the radial current lines eliminates all effects originating from a specific

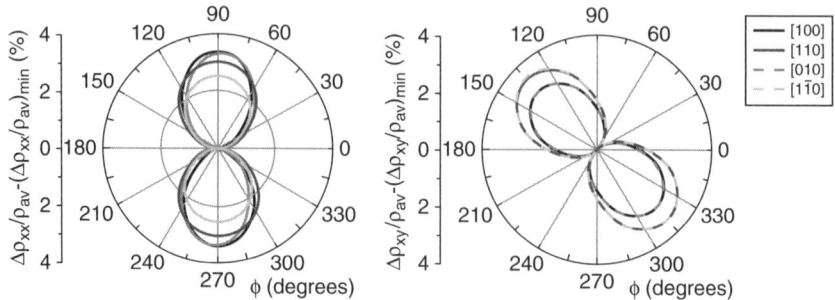

FIGURE 4.29 (Measured (at 4.2 K) longitudinal and transverse AMR for $Ga_{0.95}Mn_{0.05}As$ as a function of the angle between $\hat{M}$ and **I**. The legend shows the direction of the current. The y-axes show $\Delta\rho/\rho_{av}$ shifted such that the minimum is at zero (from Rushforth et al., 2007).

direction of the current (Rushforth et al., 2007). Corbino measurements are possible in these materials because they are near perfect single crystals but with low carrier density and mobility (compared with single crystal metals) and so can have source-drain resistances large compared with the contact resistances.

Measured results for a Corbino device fabricated from the same 25 nm $Ga_{0.95}Mn_{0.05}As$ film as used for the Hall bars are shown in Fig. 4.30 (Rushforth et al., 2007). The AMR signal is an order of magnitude weaker than in the Hall bars and is clearly composed of a uniaxial and a cubic contribution. Figure 4.30 also shows the crystalline components of the AMR extracted by fitting the Hall bar data to the phenomenological longitudinal and transverse AMR expressions. Figure 4.30D shows the consistency for the coefficients $C_{I,C}$, C_U, and C_C when extracted from the Hall bar and Corbino disk data over the whole range up to the Curie temperature (80 K). The uniaxial crystalline term, C_U, becomes the dominant term for $T \geq 30$ K. This correlates with the uniaxial component of the magnetic anisotropy which dominates for $T \geq 30$ K as observed by SQUID magnetometry measurements (Rushforth et al., 2007; Sawicki et al., 2005).

A unique AMR phenomenology has been observed on ultra thin (5 nm) $Ga_{0.95}Mn_{0.05}As$ films (Rushforth et al., 2007). Measurements on the Hall bars plotted in Fig. 4.31A show that the AMR is very different from that observed in the 25 nm film. Application of the phenomenological analysis to the Hall bar data shows that this behavior is a consequence of the crystalline terms dominating the AMR with the uniaxial crystalline term being the largest. SQUID magnetometry on 5 nm $Ga_{0.95}Mn_{0.05}As$ films consistently shows that the uniaxial component of the magnetic anisotropy dominates over the whole temperature range (Rushforth et al., 2006). The Corbino disk AMR data for a nominally identical 5 nm

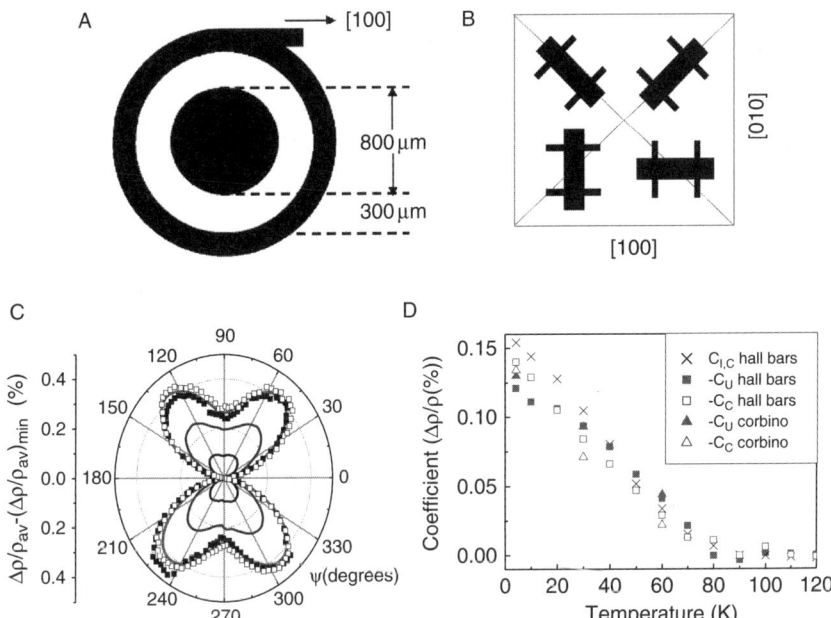

FIGURE 4.30 (A) and (B) Cartoons of the Corbino disk and Hall bar devices, respectively. (C) AMR of the 25 nm $Ga_{0.95}Mn_{0.05}As$ film in the Corbino geometry at 4.2 K (largest signal), 30 K, and 60 K and the crystalline component extracted from the Hall bars (AMR[110] + AMR[1$\bar{1}$0])/2 (closed points) and (AMR[100] + AMR[010])/2 (open points). (D) Temperature dependence of the crystalline terms extracted from the Hall bars and Corbino devices (from Rushforth et al., 2007).

film, shown in Fig. 4.31B, confirmed the observation of the highly unconventional 6% AMR totally dominated by the uniaxial crystalline term.

The 5 nm films have lower Curie temperatures ($T_c \approx 30$ K) than the 25 nm films and become highly resistive at low temperature indicating that they are close to the metal–insulator transition. The strength of the effect in the 5 nm films is remarkable and it is not captured by theory simulations assuming weakly disordered, fully delocalized (Ga,Mn)As valence bands. It might be related to the expectation that magnetic interactions become more anisotropic with increasing localization of the holes near their parent Mn ions as the metal–insulator transition is approached (Jungwirth et al., 2006).

The crystalline AMR terms in (Ga,Mn)As can be tuned by the use of lithographic patterning to induce an additional uniaxial anisotropy in very narrow Hall bars (Rushforth et al., 2007). In studies of magnetocrystalline anisotropy (Hümpfner et al., 2007; Wenisch et al., 2007; Wunderlich et al.,

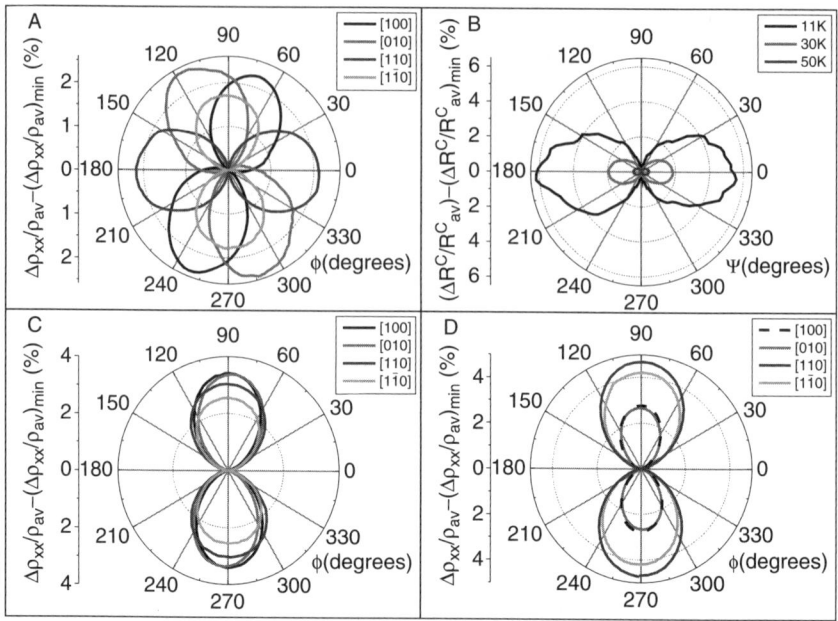

FIGURE 4.31 (A) Longitudinal AMR of the 5 nm $Ga_{0.95}Mn_{0.05}As$ Hall bars. $T = 20$ K. (B) AMR of a 5nm $Ga_{0.95}Mn_{0.05}As$ film in the Corbino geometry. (C) AMR for macroscopic Hall bars and (D) narrow (1 μm wide) Hall bars (from Rushforth et al., 2007).

2007a) it has been found that the patterning allows the inplane compressive strain in the (Ga,Mn)As film to relax in the direction along the width of the Hall bar and this can lead to an additional uniaxial component in the magnetocrystalline anisotropy for bars with widths on the order of 1 μm or smaller. (The effects of strain relaxation on magnetocrystalline anisotropy and current induced domain wall switching are discussed in Section 3.4 and in Chapter 6, this volume.) Fig. 4.31C and D show the AMR of 45 μm wide bars and 1μm wide bars fabricated from nominally identical 25 nm $Ga_{0.95}Mn_{0.05}As$ wafers. For the 45 μm bars, the cubic crystalline symmetry leads to the AMR along [100] and [010] being larger than along [110] and $[1\bar{1}0]$ (Rushforth et al., 2007). For the narrow bars the opposite relationship is observed. This is consistent with the addition of an extra uniaxial component, whose presence in the magnetocrystalline anisotropy is confirmed by SQUID magnetometry measurements, which adds 0.8% to the AMR when current is along [110] and $[1\bar{1}0]$ and subtracts 0.4% when the current is along [100] and [010]. These post-growth lithography induced modifications are significant fractions of the total AMR of the parent (Ga,Mn)As material.

3.2. Tunneling AMR

As mentioned in the Introduction, the current response to changes in the magnetic state is strongly enhanced in layered structures consisting of alternating ferromagnetic and nonmagnetic materials. The GMR and TMR effects which are widely exploited in current metal spintronics technologies reflect the large difference between resistivities in configurations with parallel and antiparallel polarizations of ferromagnetic layers in magnetic multilayers, or trilayers like spin-valves and magnetic tunnel junctions (Gregg et al., 2002). The effect relies on transporting spin information between layers and therefore is sensitive to spin-coherence times in the system. Despite strong SO coupling which reduces spin coherence in (Ga,Mn)As, functional spintronic trilayer devices can be built, as demonstrated by the measured large TMR effects in (Ga,Mn)As based tunneling structures (Brey et al., 2004; Chiba et al., 2004a,b; Elsen et al., 2007; Mattana et al., 2005; Ohya et al., 2007; Saffarzadeh and Shokri, 2006; Saito et al., 2005; Sankowski et al., 2007; Tanaka and Higo, 2001).

In this section we review the physics of a tunneling anisotropic magnetoresistance (TAMR) which is an extraordinary MR effect first observed in (Ga,Mn)As based tunnel devices (Brey et al., 2004; Ciorga et al., 2007; Elsen et al., 2007; Giraud et al., 2005; Gould et al., 2004; Rüster et al., 2005a; Saito et al., 2005; Sankowski et al., 2007). TAMR, like AMR, arises from SO coupling and reflects the dependence of the tunneling density of states of the ferromagnetic layer on the orientation of the magnetization with respect to the current direction or crystallographic axes. The effect does not rely on spin-coherence in the tunneling process and requires only one ferromagnetic contact.

In Fig. 4.32 we show a typical TAMR signal which was measured in a (Ga,Mn)As/AlO$_x$/Au vertical tunnel junction (Gould et al., 2004; Rüster et al., 2005b). For the in-plane magnetic field applied at an angle 50° off the [100]-axis the MR is reminiscent of the conventional spin-valve signal with hysteretic high resistance states at low fields and low resistance states at saturation. Unlike the TMR or GMR, however, the sign changes when the field is applied along the [100]-axis. Complementary SQUID magnetization measurements confirmed that the high resistance state corresponds to magnetization in the (Ga,Mn)As contact aligned along the [100]-direction and the low resistance state along the [010]-direction, and that this TAMR effect reflects the underlying magnetocrystalline anisotropy between the **M** || [100] and **M** || [010] magnetic states. Since the field is rotated in the plane perpendicular to the current, the Lorentz force effects on the tunnel transport can be ruled out. Also, in this geometry, a pure crystalline TAMR is detected. Microscopic calculations consistently showed that the SO-coupling induced density-of-states anisotropies with respect to the magnetization orientation can produce TAMR effects in (Ga,Mn)As of the order ~1–10% (Gould et al., 2004; Rüster et al., 2005b).

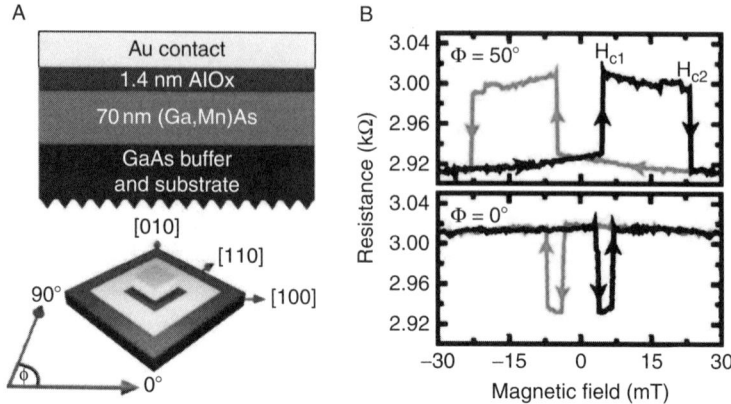

FIGURE 4.32 (A) Schematic showing layer structure, contact geometry, and crystallographic directions in the TAMR device. (B) MR hysteresis curves for $T = 4.2$ K and 1 mV bias with in-plane magnetic field H along 0°, and 50°. MR is spin-valve-like with two abrupt switching events at H_{c1} and H_{c2}. Depending on the angle, the width of the feature and, more important, its sign can change. The high resistance state corresponds to magnetization along [100]-axis and the low resistance state along [010]-axis (from Rüster et al., 2005b).

All-semiconductor TAMR devices with a single ferromagnetic electrode were realized in p-(Ga,Mn)As/n-GaAs Zener–Esaki diodes (Ciorga et al., 2007; Giraud et al., 2005). For magnetization rotations in the (Ga,Mn)As plane (Ciorga et al., 2007) a comparable TAMR ratios were detected as in the (Ga,Mn)As/AlO$_x$/Au tunnel junction. About an order of magnitude larger TAMR (40%) was observed when magnetization was rotated out of the (Ga,Mn)As plane towards the current direction (Giraud et al., 2005). These experiments suggest that the noncrystalline TAMR is stronger than the crystalline terms.

Unlike the AMR in the ohmic regime, a systematic microscopic analysis of the basic physics of the noncrystalline and crystalline TAMR terms has not been reported yet. However, several detailed numerical studies have been performed based on microscopic tight-binding or $\mathbf{k} \cdot \mathbf{p}$ kinetic-exchange models of the (Ga,Mn)As valence band and the Landauer–Büttiker quantum transport theory (Brey et al., 2004; Elsen et al., 2007; Giddings et al., 2005; Sankowski et al., 2007). Besides the Zener–Esaki diode geometry (Sankowski et al., 2007) the simulations consider magnetic tunnel junctions with two ferromagnetic (Ga,Mn)As contacts and focus on comparison between the TMR and TAMR signals in structures with different barrier materials and (Ga,Mn)As parameters (Brey et al., 2004; Elsen et al., 2007; Sankowski et al., 2007). Figure 4.33 shows the theoretical dependence of the TMR ratio for parallel and antiparallel configurations of the two (Ga,Mn)As contacts and **M** along the [100]-direction

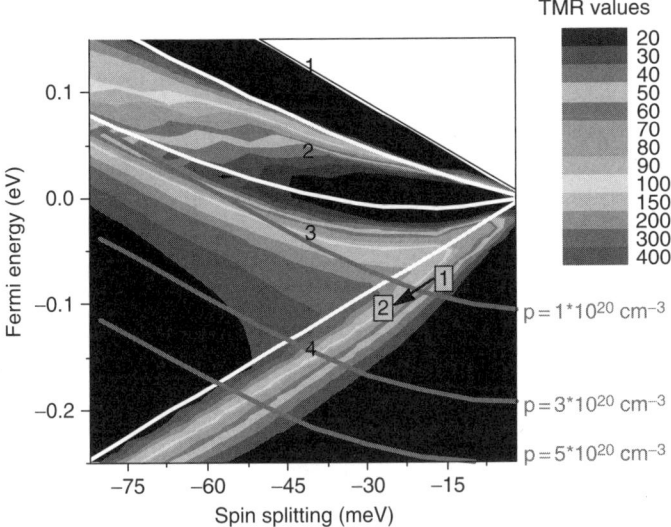
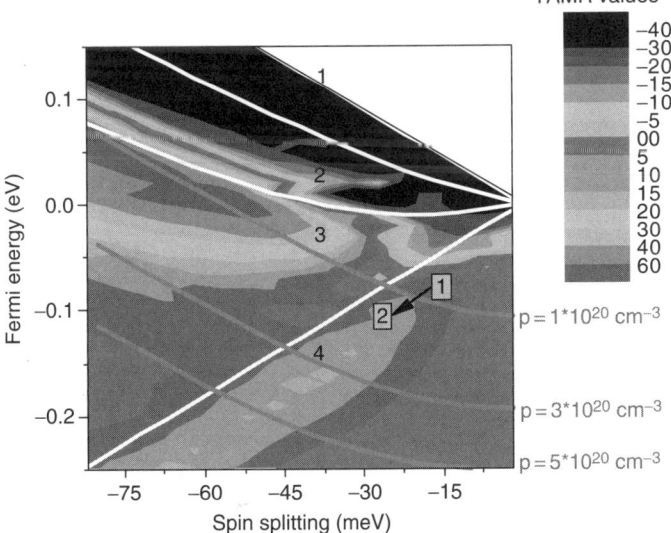

FIGURE 4.33 Tunnel MR values (A) and tunnel AMR values (B) represented as a function of the Fermi and spin splitting energy for a 6 nm (In,Ga)As barrier with a band offset of 450 meV. White lines represent the four bands at the center of the Brillouin zone. Gray lines indicate the Fermi energy for different hole concentrations (from Elsen et al., 2007). (See Color Plate.)

and the TAMR ratio for parallel magnetizations in the (Ga,Mn)As films and **M** along the [100]-direction and the [001]-direction (current direction) in a tunneling device with an InGaAs barrier (Elsen *et al.*, 2007). The corresponding experimental measurements are shown in Fig. 4.34. There is an overall agreement between the theory and experiment, seen also in tunnel junctions with other barrier materials, showing that the TMR is typically 10 times larger than the TAMR. Both the theory and experiment also find that the TMR signal is always positive, that is, the MR increases as the field is swept from saturation to the switching field, while the TAMR can have both signs depending on the field angle but also depending on the parameters of the (Ga,Mn)As film such as the hole concentration and polarization, on the barrier characteristics, or on the temperature (Elsen et al., 2007; Gould et al., 2004).

At very low temperatures and bias voltages huge TAMR signals were observed (Rüster et al., 2005b) in a (Ga,Mn)As/GaAs/(Ga,Mn)As tunnel junction, as shown in Fig. 4.35, which are not described by the one-body theories of anisotropic tunneling transmission coefficients. The observation has been interpreted as a consequence of electron–electron correlation effects near the metal–insulator transition (Pappert et al., 2006). Large AMR effects were also measured in lateral nano-constriction tunnel devices

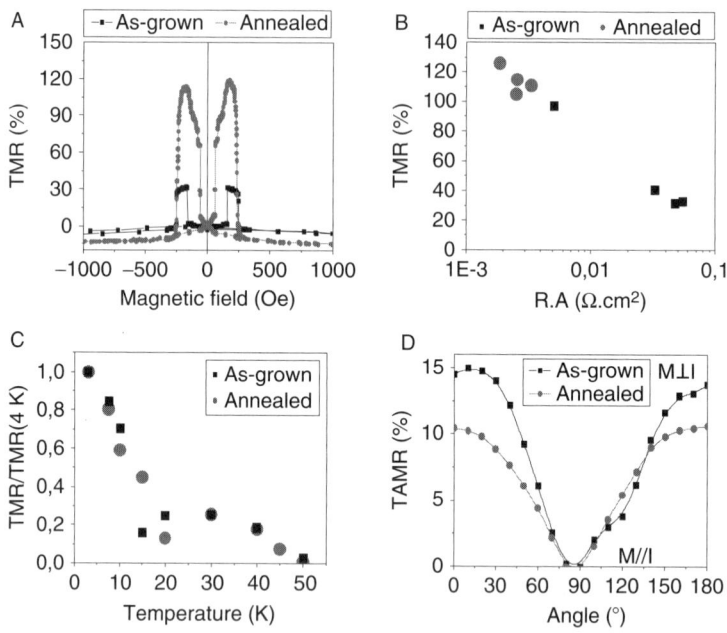

FIGURE 4.34 (A) Tunnel MR measurements as a function of the magnetic field at 1 mV and 3 K for a 128 μm² junction. (B) Tunnel MR measurements as a function of Resistance. Area product at 3 K for 4 (un)annealed junctions. (C) Tunnel MR at 1 mV as a function of the temperature before and after annealing. (D) Tunnel AMR measurements as a function of the magnetic field at 1 mV and 3 K (from Elsen et al., 2007).

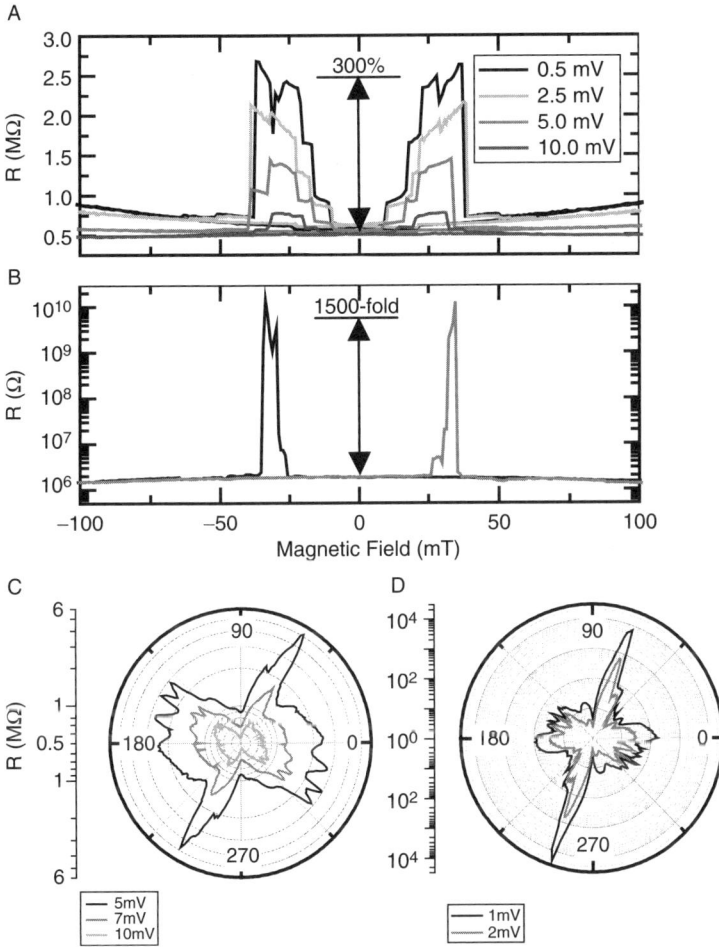

FIGURE 4.35 Amplification of the TAMR effect at low bias voltage and temperatures. (A) TAMR along in-plane angle 65° off the [100]-axis at 4.2 K for various bias voltages. (B) Very large TAMR along in-plane angle 95° at 1.7 K and 1 mV bias. (C), (D) TAMR at various bias voltages at 1.7 K (from Rüster *et al.*, 2005b).

fabricated on ultra-thin (Ga,Mn)As materials. (For other (Ga,Mn)As nano-constriction MR studies see (Rüster *et al.*, 2003; Schlapps *et al.*, 2006).) The AMR signals in the unstructured part of the device and in the nano-constriction are compared in Figs. 4.36 and 4.37, showing a significant enhancement of the signal in the tunneling regime. Subsequent studies of these nano-constrictions with an additional side-gate patterned along the constriction, discussed in detail in the following section, indicate that single-electron charging effects might be responsible for the observed large AMR signals.

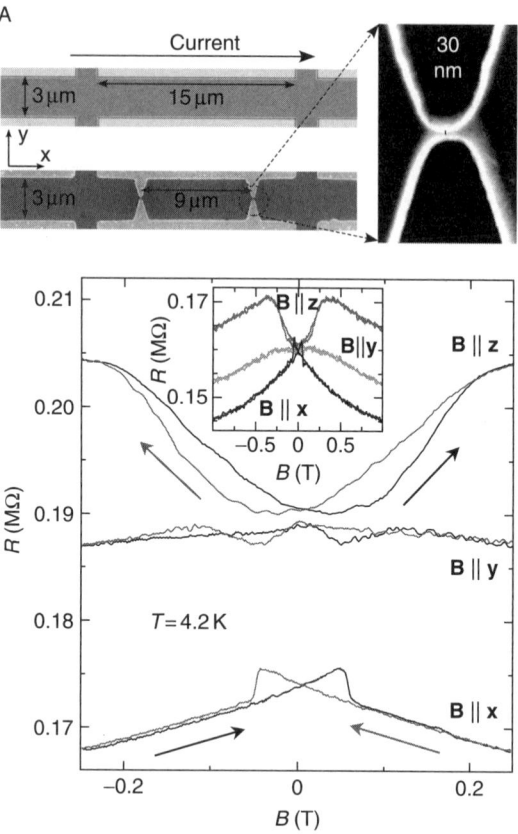

FIGURE 4.36 (A) Schematic of the unstructured bar and scanning electron microscopy image of a double constricted nanodevice. Lower panel: Low-field a.c. MR measurements for the unstructured bar with applied field in three orthogonal orientations at a temperature of 4.2 K. The inset shows d.c. MR measurements for a wider field range (from Giddings et al., 2005).

Before moving on to the (Ga,Mn)As-based field effect transistors we conclude this section with a remark on the impact of the TAMR discovery in (Ga,Mn)As on spintronics research in metal ferromagnets. Ab initio relativistic calculations of the anisotropies in the density of states predicted sizable TAMR effects in room-temperature metallic ferromagnets (Shick et al., 2006). This opens prospect for spintronic devices analogous to current MRAMs but with a simpler geometry as the TAMR tunnel junctions do not require antiferromagnetically coupled ferromagnetic contacts on either side of the tunnel junction and exchange biasing which produces different coercive fields in the two ferromagnets. Predicted density-of-states anisotropies for several model systems ranging

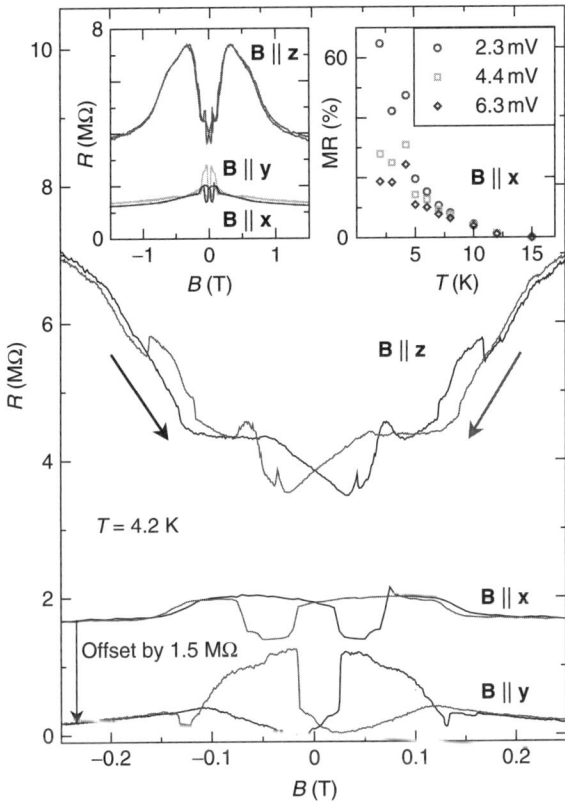

FIGURE 4.37 Low-field a.c. MR measurement of the 30 nm constriction with applied field in the three orthogonal directions. Left inset: Measured a.c. MRs for a wider field range. Right inset: the temperature dependence of the hysteretic low-field MR for three different voltages (from Giddings et al., 2005).

from simple hcp-Co to more complex ferromagnetic structures with enhanced SO coupling, namely bulk and thin film L10-CoPt ordered alloys and a monatomic-Co chain at a Pt surface step edge are shown in Fig. 4.38. Landauer–Büttiker transport theory calculations for a Fe/vacuum/Cu structure pointed out that apart from the density-of-states anisotropies in the ferromagnetic metal itself, the TAMR in the tunnel devices can arise from SO-coupling induced anisotropies of resonant surface or interface states (Chantis et al., 2007). Experimentally, several reports of metal TAMR devices have already appeared in the literature including Fe, Ni, and Co lateral break-junctions (Bolotin et al., 2006; Viret et al., 2006) which showed comparable (∼10%) low-temperature TMR and TAMR signals, Fe/GaAs/Au vertical tunnel junctions (Moser et al., 2007) with a 0.4% TAMR at low temperatures, a $Co/Al_2O_3/NiFe$ magnetic tunnel

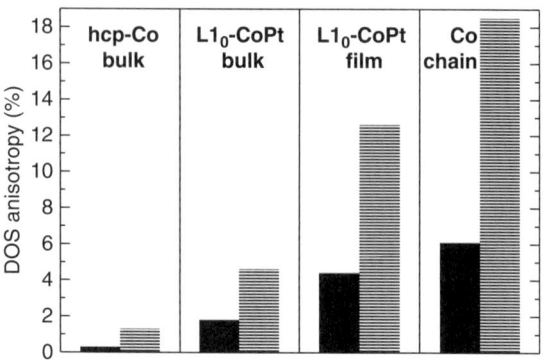

FIGURE 4.38 Density of states anisotropies in ferromagnetic metal systems. See (Shick et al., 2006) for the detailed definitions of the particular density of states plotted in the figure.

junction with a 15% TAMR at room temperature (Grigorenko et al., 2006), and reports of strongly bias dependent TAMRs in devices with CoFe (Gao et al., 2007) and CoPt electrodes (Park et al., 2008).

3.3. Coulomb blockade AMR and other MR transistors

(Ga,Mn)As and related dilute moment ferromagnets qualify as ferromagnetic semiconductors to the extent that their magnetic and other properties can be altered by usual semiconductor electronics engineering variables such as doping and external electric fields. For example, (In,Mn)As and (Ga,Mn)As based field effect transistors were built to study electric field control of ferromagnetism. It has been demonstrated that changes in the carrier density and distribution in thin ferromagnetic semiconductor film due to an applied gate voltage can reversibly induce the ferromagnetic/paramagnetic transition (Chiba et al., 2006a; Ohno et al., 2000). Another remarkable effect observed in this transistor is the electric field assisted magnetization reversal (Chiba et al., 2003, 2006a). This novel functionality is based on the dependence of the width of the hysteresis loop on bias voltage, again through the modified charge density profile in the ferromagnetic semiconductor thin film. For a functional spintronic transistor, however, it is the MR which is the key characteristics which needs to be controlled by the gate electric field. Such a functionality was demonstrated in a (Ga,Mn)As single-electron transistor (SET), shown in Fig. 4.39, through a Coulomb blockade anisotropic magnetoresistance (CBAMR) effect (Wunderlich et al., 2006, 2007a,b) which is discussed in detail in this section. (Other realizations of ferromagnetic semiconductor transistors with gate-controlled MRs include the hybrid structures with piezoelectric transducers attached to (Ga,Mn)As epilayers (Goennenwein et al., 2008, Rushforth et al., 2008, De Ranien et al., 2008, Overby et al., 2008) or all-semiconductor n-GaAs/AlGaAs/(Ga,Mn)As p-n junction transistors (Olejnik et al., 2008, Owen et al., 2008).)

In the conventional single-electron transistor (SET), the transfer of an electron from a source lead to a drain lead via a small, weakly-coupled island is blocked due to the charging energy of $e^2/2C_\Sigma$ where C_Σ is the total capacitance of the island. Applying a voltage V_G between the source lead and a gate electrode changes the electrostatic energy function of the charge Q on the island to $Q^2/2C_\Sigma + QC_GV_G/C_\Sigma$ which has a minimum at $Q_0 = -C_GV_G$. By tuning the continuous external variable Q_0 to $(n + 1/2)e$, the energy associated with increasing the charge Q on the island from ne to $(n + 1)e$ vanishes and electrical current can flow between the leads. Changing the gate voltage then leads to CB oscillations in the source-drain current (see Fig. 4.39B) where each period corresponds to increasing or decreasing the charge state of the island by one electron.

The schematic cartoons in Fig. 4.40A and B indicate the contributions to the Gibbs energy associated with the transfer of charge Q from the lead to the island. The energy can be written as a sum of the internal, electrostatic charging energy term and the term associated with, in general, different chemical potentials of the lead and of the island:

$$U = \int_0^Q dQ' \Delta V_D(Q') + Q\Delta\mu/e, \qquad (4.5)$$

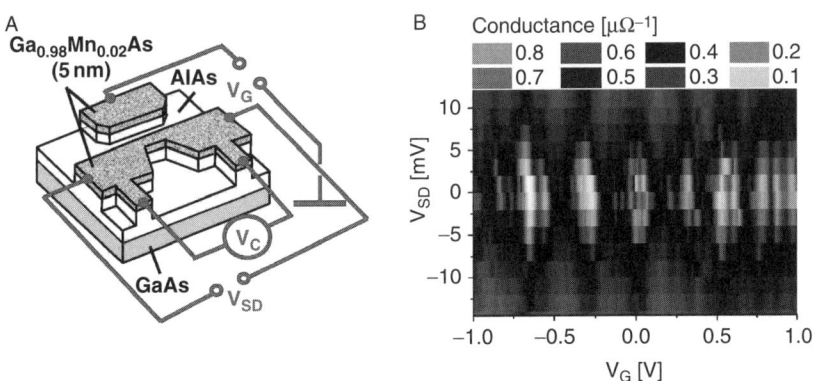

FIGURE 4.39 (A) Schematics of our ferromagnetic SET. Trench isolated side-gate and channel aligned along the [110] direction were patterned by e-beam-lithography and reactive ion etching in a 5 nm Ga$_{0.98}$Mn$_{0.02}$As epilayer. (B) CB conductance (I/V_C) oscillations with gate voltage for different source-drain bias. The diamond patterns in this 2D plot are clear finger-prints of single-electron transport (from Wunderlich et al., 2006).

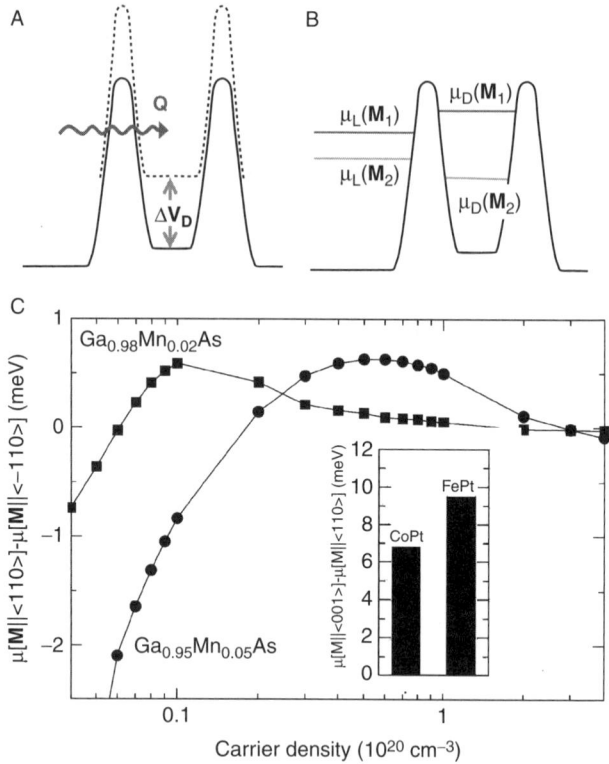

FIGURE 4.40 (A,B) Schematics of the charging energy contribution (A) to the Gibbs energy and the contribution proportional to the difference $\Delta\mu(\mathbf{M})$ in chemical potentials of the lead, $\mu_L(\mathbf{M})$, and of the island, $\mu_D(\mathbf{M})$ (B); (C) Calculated chemical potential anisotropy in (Ga,Mn)As as a function of carrier density for Mn local moment concentrations of 2% and 5%. The uniaxial anisotropy is modeled by introducing a weak shear strain $e_{xy} = 0.001$. Insets: calculated chemical potential anisotropies in L1$_0$ FePt and CoPt ordered alloys (from Wunderlich et al., 2006).

where $\Delta V_D(Q) = (Q + C_G V_G)/C_\Sigma$. The Gibbs energy U is minimized at $Q_0 = -C_G(V_G + V_M)$.

The ferromagnetic SET in Fig. 4.39 consists of a trench-isolated side-gated 40 nm wide channel patterned in an ultra-thin (5 nm) Ga$_{0.98}$Mn$_{0.02}$As epilayer. The narrow channel technique is a simple approach to realize a SET and has been used previously to produce nonmagnetic thin film Si and GaAs-based SETs in which disorder potential fluctuations create small islands in the channel without the need for a lithographically defined island (Kastner, 1992; Tsukagoshi et al., 1998). The nonuniform carrier concentration near the channel produces differences between chemical potentials Dm of the lead and of the island in the constriction.

There are two mechanisms through which Dm depends on the magnetic field. One is caused by the direct Zeeman coupling of the external magnetic field and leads to the ordinary CB MR previously observed in ferromagnetic metal SETs (Ono et al., 1997). Measurements of the ordinary CB MR in the (Ga,Mn)As SET are shown in Fig. 4.41.

The extraordinary CBAMR effect, observed in the (Ga,Mn)As SET, is attributed to the SO coupling induced anisotropy of the carrier chemical potential, that is, to magnetization orientation dependent differences between chemical potentials of the lead and of the island in the constriction (Wunderlich et al., 2006). For the CBAMR effect, the magnetization orientation dependent shift of the CB oscillations is given by $V_M = C_\Sigma/C_G$ $\Delta\mu(\mathbf{M})/e$. Since $|C_G V_M|$ has to be of order $|e|$ to cause a marked shift in the oscillation pattern, the corresponding $|\Delta\mu(\mathbf{M})|$ has to be similar to e^2/C_Σ, that is, of the order of the island single-electron charging energy. The fact that CBAMR occurs when the anisotropy in a band structure derived parameter is comparable to an independent scale (single-electron charging energy) makes the effect distinct and potentially much larger in

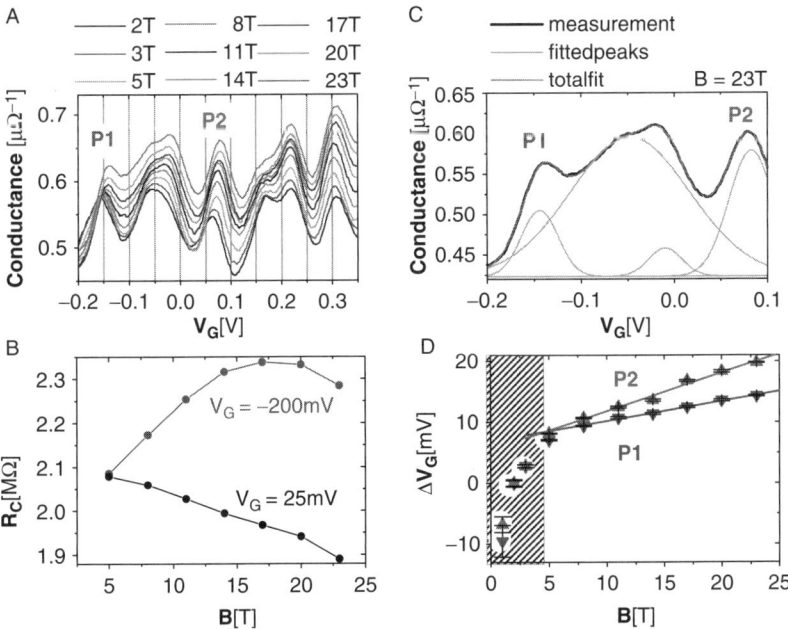

FIGURE 4.41 Coulomb blockade (CB) effect at high fields: (A) CB oscillations at $B =$ 2–23 T. Successive measurements are offset by 7.5 nΩ^{-1}. (B) Positive and negative MR at gate voltages $V_G = -200$ and 25 mV. (C) Gaussian multipeak fitting for individual peak position evaluation (D) Peak positions P1 and P2 versus magnetic field. The magnetization is saturated above 5 T (from Wunderlich et al., 2007a).

magnitude as compared to the previously observed AMR in the ohmic regime (normal AMR) and in the tunneling regime (TAMR). In the terminology introduced in Section 3.1, the CBAMR is a purely crystalline anisotropic magnetotransport coefficient.

Microscopic calculations of chemical potential anisotropies in ferromagnetic (Ga,Mn)As are shown in Fig. 4.40C. They are sensitive to the carrier and local moment densities and $|\Delta\mu|$ can be as large as several millielectron volts. Fig. 4.40C also shows relativistic *ab initio* calculations of the chemical potential anisotropies in model ferromagnetic metal systems, FePt and CoPt. The anisotropies are roughly an order of magnitude larger than in (Ga,Mn)As. This suggests that, building on the existing expertise in room-temperature nonmagnetic single-electronic devices, metal based ferromagnetic SETs may offer a route to room-temperature CBAMR applications. So far, however, the CBAMR has only been reported in the (Ga,Mn)As SET and the corresponding experimental data are shown in Figs. 4.42–4.44.

The extraordinary CB MR was measured in magnetic fields applied in the plane of the film at an angle θ to the current I direction (B_θ), or in the perpendicular-to-plane direction (B_{ptp}). Examples of large, hysteretic and gate-voltage dependent MR of the SET are shown in Fig. 4.42 for up and down sweeps of the magnetic field B_{90}. At about 20 mT the MR is ∼100% and negative for $V_G = 0.94$ V but is larger than 1,000% and positive for $V_G = 1.15$ V. Fig. 4.42C and D show resistance variations during magnetic field B_{ptp} sweeps, with a particularly large MR at $V_G = 0.935$ V, due to the rotation of magnetization out of the plane of the (Ga,Mn)As epilayer. Resistance variations by more than three orders of magnitude are observed with the high resistance state realized at saturation.

The sensitivity of the MR to the orientation of the applied magnetic field is an indication of the AMR origin of the effect. This is confirmed by the observation of comparably large and gate-controlled MR in the devices when saturation magnetization is rotated with respect to the crystallographic axes. The field-sweep and rotation measurements are presented in Fig. 4.43C and D and compared with analogous measurements of the normal AMR in the unstructured part of the (Ga,Mn)As bar, plotted in Fig. 4.43A and B. In the unstructured bar, higher or lower resistance states correspond to magnetization along or perpendicular to the current direction. Similar behavior is seen in the SET part of the device at, for example, $V_G = -0.4\ V$, but the AMR is now hugely increased and depends strongly on the gate voltage.

Figure 4.44 demonstrates that the dramatic AMR effects in the SET are due to shifts in the CB oscillation pattern caused by the changes in magnetization orientation. The shifts are seen in Fig. 4.44A, which shows the CB oscillations for several magnetization angles. Blue curves indicate shifts in the oscillation pattern due to magnetization rotations.

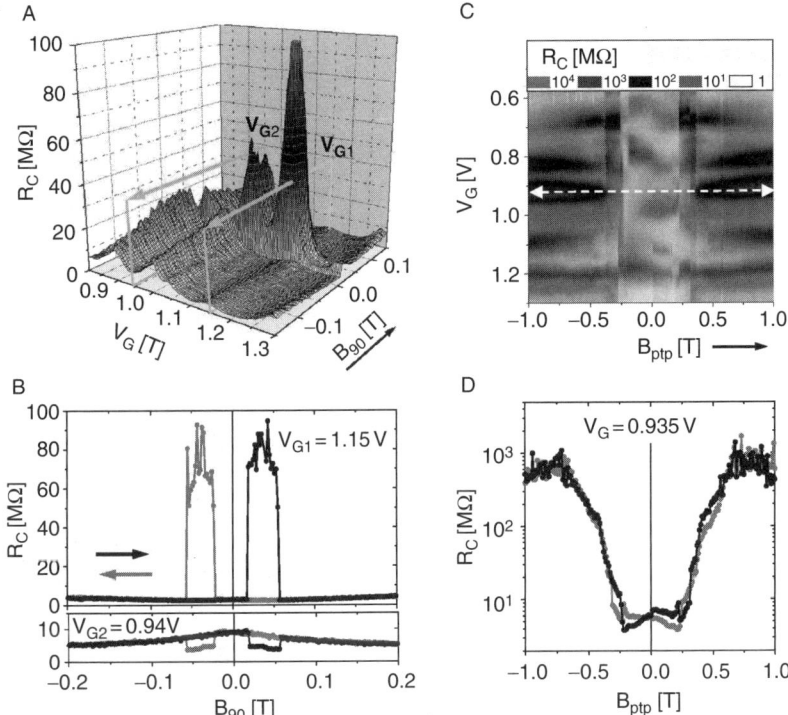

FIGURE 4.42 Coulomb-blockade conductance oscillations measured at $V_{SD} = 3$ mV bias versus gate voltage and in-plane, perpendicular to the channel direction (A), and perpendicular-to-plane magnetic field (C). (B) Spin-valve like MR signals of different magnitudes and opposite signs for gate voltages 1.15 and 0.94 V. (D) Huge MR signal for up and down sweeps of the perpendicular-to-plane magnetic field at $V_G = 0.935$ V (from Wunderlich et al., 2007b).

For example, at $V_G = -0.4$ the oscillations have a peak for $\theta = 0°$ that moves to higher V_G with increasing θ until, for $\theta = 40°$, a minimum in the oscillatory resistance occurs at $V_G = -0.4$ V. Figure 4.44A and B show that the magnitude of the resistance variations with θ at fixed V_G and with V_G at fixed θ is comparable. As pointed above in the theory discussion, the key parameters for the CBAMR effect are the magnetization rotation induced electrochemical shifts relative to the CB charging energy. Consistent with theory, the charging energy obtained from experimental data in Fig. 4.44C is ~4 meV.

The gate voltage dependent MR in the (Ga,Mn)As SET device can be combined also with the electric control of the magnetization switching. In Figure 4.45A the gate voltage dependence of the switching field is highlighted by the dashed line. Electrical field assisted magnetization reorientation is illustrated in Fig. 4.45B.

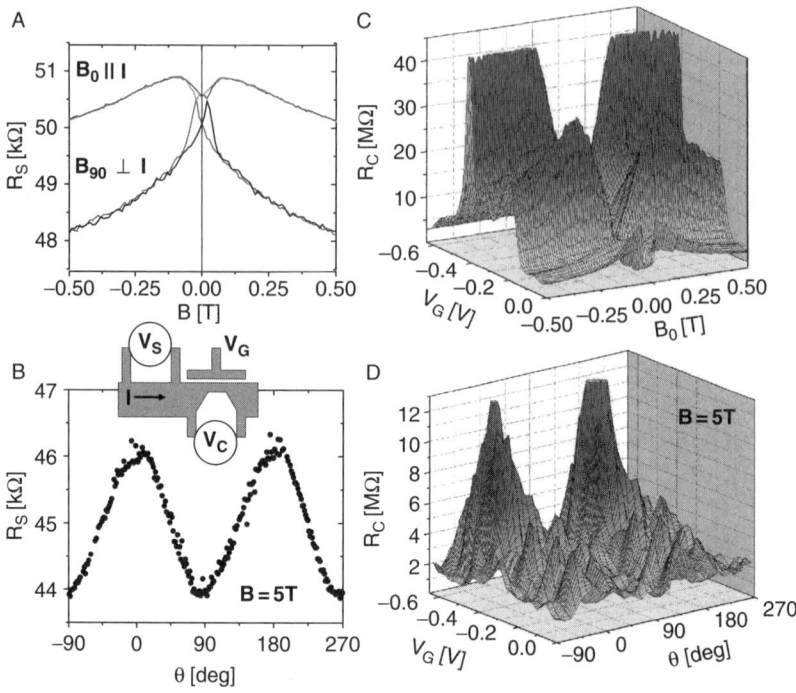

FIGURE 4.43 (A) Resistance $R_S = V_S/I$ of the unstructured bar (see schematic diagram) versus up and down sweeps of in-plane magnetic field parallel and perpendicular to the current direction. (B) R_S versus the angle between the current direction and an applied in-plane magnetic field of 5 T, at which $\mathbf{M}\|\mathbf{B}$. (C) Channel resistance R_C versus gate voltage and down sweep of the magnetic field parallel to current. (D) R_C versus gate voltage and the angle between the current direction and an applied in-plane magnetic field of 5 T (from Wunderlich et al., 2006).

We conclude this section with a discussion of the new functionality concept offered by the CBAMR effect. The huge and hysteretic MR signals show that CBAMR-SETs can act as nonvolatile memories. In nonmagnetic SETs, the CB "on" (low-resistance) and "off" (high-resistance) states are used to represent logical "1" and "0" and the switching between the two states can be realized by applying a gate voltage, in analogy with a standard field-effect transistor. The CBAMR-SET can be addressed also magnetically with comparable "on" to "off" resistance ratios in the electric and magnetic modes. The functionality is illustrated in Fig. 4.46. The inset of panel A shows two CB oscillation curves corresponding to magnetization states $\mathbf{M_0}$ and $\mathbf{M_1}$. As evident from panel B, $\mathbf{M_0}$ can be achieved by performing a small loop in the magnetic field, $B \rightarrow B_0 \rightarrow 0$ where B_0 is

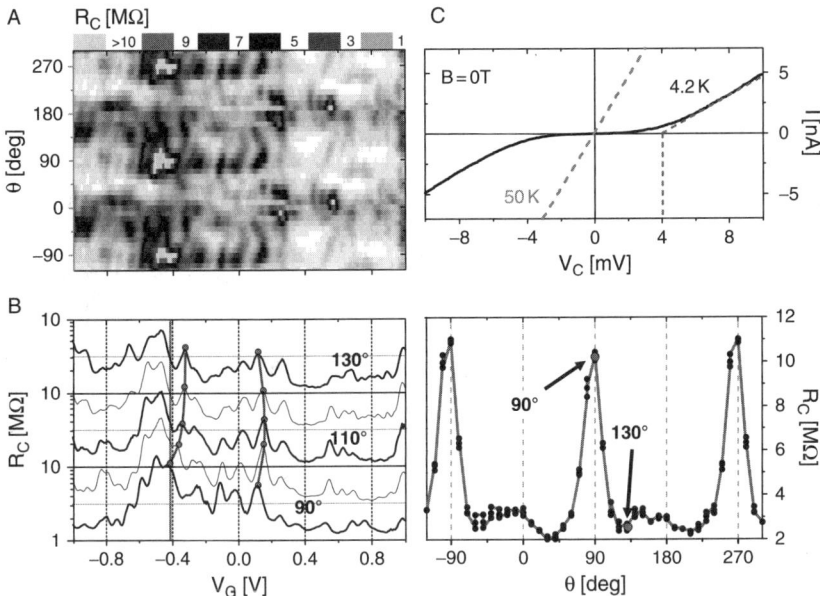

FIGURE 4.44 (A) Channel resistance versus gate voltage and the angle between the current direction and applied in-plane saturation field of 5 T. (B) θ sweep for fixed $V_G = -0.4$ V. (C) $I - V_C$ characteristic of the channel at 4.2 K (CB regime) and 50 K (Ohmic regime) used to determine the SET charging energy. This inferred value remains constant with decreasing temperature below 4.2 K (from Wunderlich et al., 2007b).

larger than the first switching field B_{c1} and smaller than the second switching field B_{c2}, and $\mathbf{M}_1$ is achieved by performing the large field-loop, $B \to B_1 \to 0$ where $B_1 < -B_{c2}$. The main plot of Fig. 4.46A shows that the high resistance 0 state can be set by either the combinations ($\mathbf{M}_1$, V_{G0}) or ($\mathbf{M}_0$, V_{G1}) and the low resistance 1 state by ($\mathbf{M}_1$, V_{G1}) or ($\mathbf{M}_0$, V_{G0}). One can therefore switch between states 0 and 1 either by changing V_G in a given magnetic state (the electric mode) or by changing the magnetic state at fixed V_G (the magnetic mode). Due to the hysteresis, the magnetic mode represents a nonvolatile memory effect. The diagram in Fig. 4.46C illustrates one of the new functionality concepts the device suggests in which low-power electrical manipulation and permanent storage of information are realized in one physical nanoscale element. Panel D then highlights the possibility to invert the transistor characteristic; for example, the system is in the low-resistance "1" state at V_{G1} and in the high-resistance "0" state at V_{G0} (reminiscent of an n-type field effect transistor) for the magnetization $\mathbf{M}_1$ while the characteristic is inverted (reminiscent of a p-type field effect transistor) by changing magnetization to $\mathbf{M}_0$.

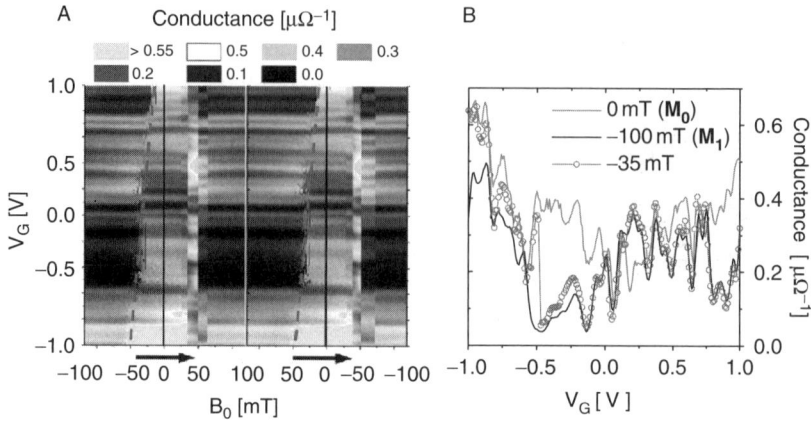

FIGURE 4.45 (A) Channel conductance versus gate voltage and in-plane, parallel to current magnetic field. The dashed line highlights the critical reorientation field which is gate bias dependent and decreases from about 40 mT at $V_G = -1$ V to less than 20 mT at $V_G = 1$ V. (B) CB oscillations at zero magnetic field where the system remains at saturation magnetization M_0 (grey) and at $B_0 = -100$ mT where the system remains at saturation magnetization M_1 (black) over the gate voltage range between $V_G = -1$ to 1 V. Conductance measurements at intermediate field strengths of -35 mT (circles) show a transition from M0 to M1 at critical gate voltages of about -0.5 V (from Wunderlich et al., 2007b).

3.4. Current induced domain wall switching

Spin-transfer-torque current induced switching discussed in this section can be regarded as an inverse extraordinary MR effect as in this case spin-polarized currents induce magnetization reorientations in a ferromagnet. As already mentioned in the Introduction, (Ga,Mn)As dilute moment ferromagnets have ~100–10 times weaker saturation magnetization, M_s, than conventional dense-moment metal ferromagnets. One particularly important implication of the low M_s is the orders of magnitude lower critical current in the spin-transfer-torque magnetization switching (Chiba et al., 2004b; Sinova et al., 2004b) than observed for conventional ferromagnets, which follows from the approximate scaling of $j_c \sim H_a M_s$. Here H_a is the typical magnetic anisotropy field which is similar in (Ga,Mn)As and metal ferromagnets. Critical currents for domain wall switching of the order 10^5 Acm^{-2} have been reported and the effect thoroughly explored in perpendicularly magnetized (Ga,Mn)As thin film devices (Chiba et al., 2006b; Yamanouchi et al., 2004, 2006). The perpendicular magnetization geometry is particularly useful for the direct magneto-optical Kerr-effect imaging of the domains, as illustrated in Fig. 4.47. (For a detailed discussion of these experiments see Chapter 5,

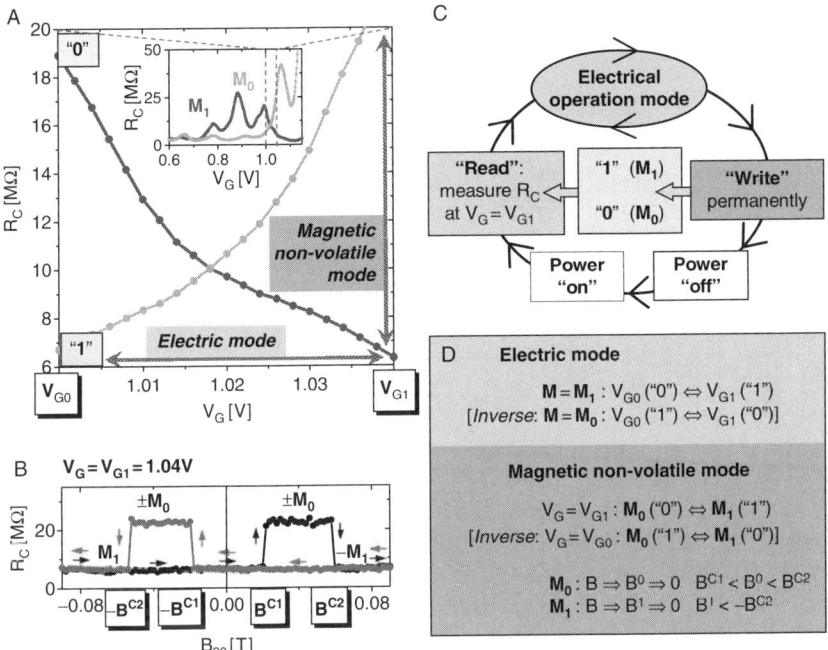

FIGURE 4.46 (A) Two opposite transistor characteristics (blue and green) in a gate-voltage range V (V_{G0}) to 1.04 V (V_{G1}) for two different magnetization orientations M_0 and M_1; corresponding CB oscillations in a larger range of $V_G = 0.6$ to 1.15 V are shown in the inset. Switching between low-resistance ("1") and high-resistance ("0") states can be performed electrically or magnetically. (B) Hysteretic MR at constant gate voltage V_{G1} illustrating the nonvolatile memory effect in the magnetic mode. (C) Illustration of integrated transistor (electric mode) and permanent storage (magnetic mode) functions in a single nanoscale element. (D) The transistor characteristic for $M = M_1$ is reminiscent of an *n*-type field effect transistor and is inverted (reminiscent of a p-type field effect transistor) for $M = M_0$; the inversion can also be realized in the nonvolatile magnetic mode (from Wunderlich *et al.*, 2007b).

this volume.) In an in-plane magnetized (Ga,Mn)As material, on the other, one can exploit the lithographically induced strain relaxation to control the micromagnetic characteristics of the conducting channels containing the domain walls. These experiments are shown in Figs. 4.48–4.51 (Wunderlich *et al.*, 2007c).

Figure 4.48A shows scanning electron micrographs of one of the devices studied. The structure consists of an L-shaped channel the arms of which are Hall-bars aligned along the $[1\bar{1}0]$ and [110] directions. Magnetization orientations in the individual microbars are monitored locally by measuring longitudinal and transverse components of the AMR at in-plane magnetic fields. Devices A and B, whose magnetic

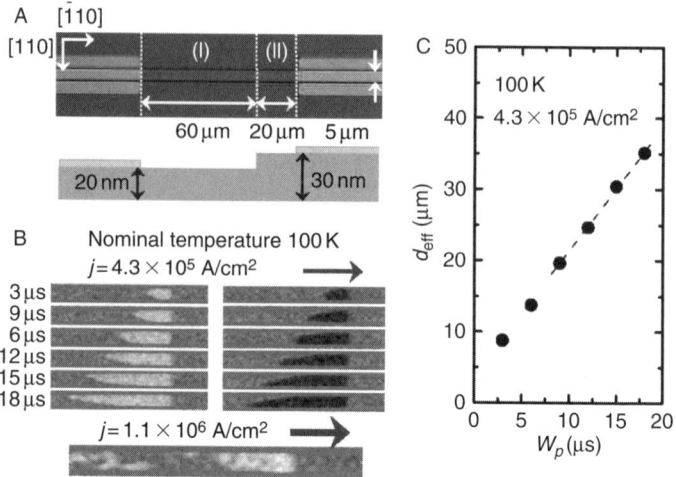

FIGURE 4.47 (A) Layout of the device showing the 5 μm mesa and step for DW pinning in perpendicular magnetic anisotropy (Ga,Mn)As film. (B) 7 μm wide magneto-optical images with a 5 μm mesa in the center show that DW moves in the opposite direction to current independent of the initial magnetization orientation, and that DW displacement is proportional to pulse duration (C). The lowest panel in (B) shows destruction of ferromagnetic phase by Joule heating (from Yamanouchi et al., 2006).

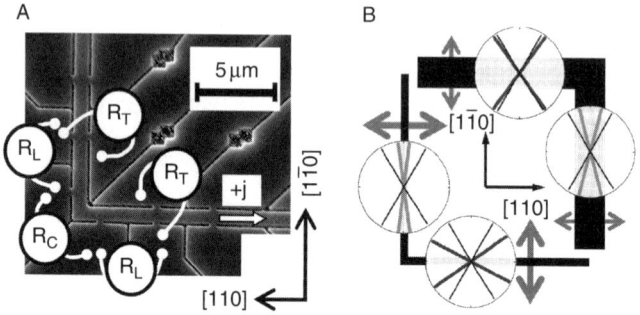

FIGURE 4.48 (A) Scanning electron micrograph of the L-shaped microdevice B with the longitudinal (L) and transverse (T) resistance contacts in the bars and the corner (C) resistance contacts (from Wunderlich et al., 2007c).

easy-axes are shown in Fig. 4.48B, have 4 and 1 μm wide, Hall bars, respectively. The easy-axes in the microdevices are rotated from their bulk positions towards the direction of the respective bar and the effect increases with decreasing bar width.

The L-shaped geometry of these devices is well suited for a systematic study of the link between the locally adjusted magnetic anisotropies in the individual micro-bars and their current induced switching characteristics.

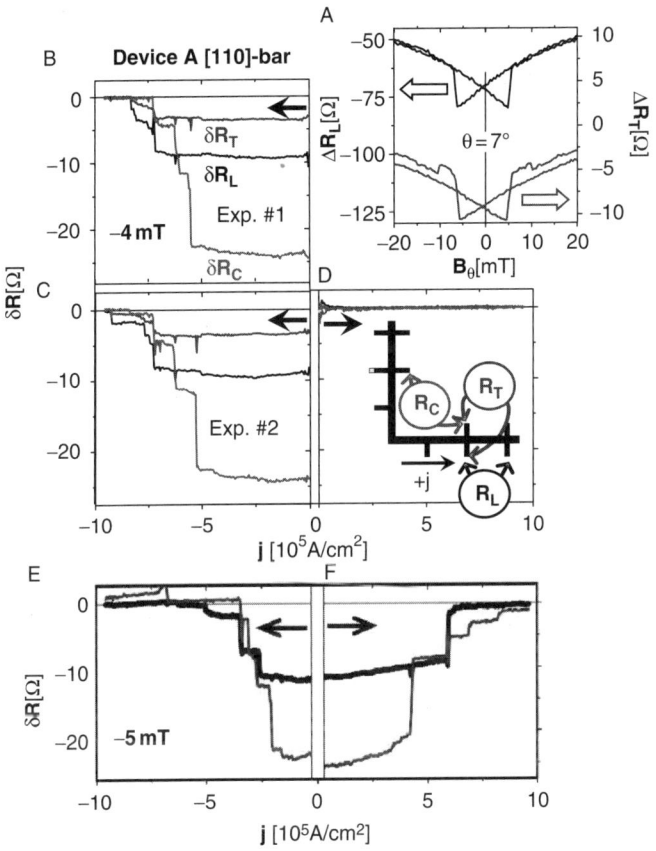

FIGURE 4.49 (A) Field-sweep measurements at $\theta = 7°$ in the [110]-bar of device A. (B) Differences between the first and second negative current ramps for the longitudinal and transverse resistance in the [110]-bar and in the corner of device A at -4 mT external field applied along $\theta = 7°$. Arrows indicate the current ramp direction. (C) Same as (B) for the second independent experiment. (D) Same as (B) and (C) for positive current ramps. The inset shows contacts used for measurements of R_L, R_T, and R_C in all panels. (E),(F) -5 mT field assisted current induced switching experiments (from Wunderlich et al., 2007c).

Apart from the distinct magnetocrystalline anisotropy fields, the two bars in each device have identical material parameters and lithographical dimensions. They also share a common domain-wall nucleation center at the corner of the L-shaped channel since in this region the lattice relaxation effects and the corresponding enhancement of the magnetocrystalline anisotropies are less pronounced. Apart from this effect, the domain wall nucleation at the corner can be expected to be supported by an enhanced current induced heating in this part of the device.

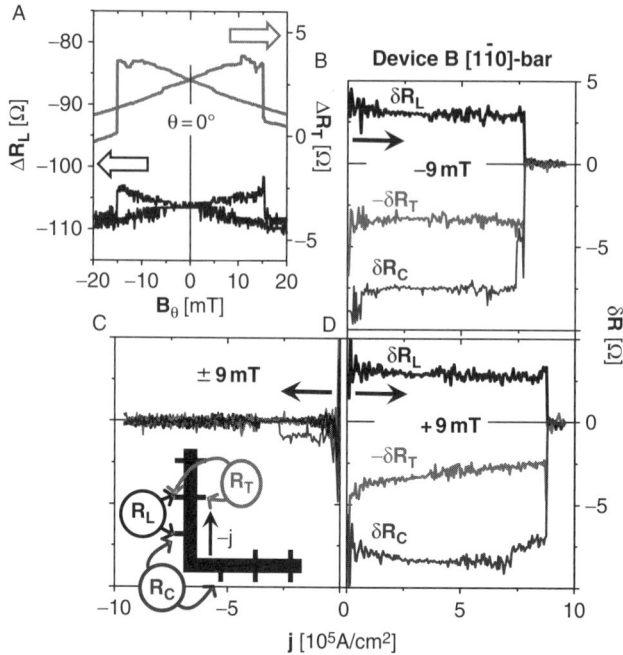

FIGURE 4.50 (A) Field-sweep measurements at $\theta = 0°$ in the $[1\bar{1}0]$-bar of device B. (B) Difference between the first and second positive current ramps in the $[1\bar{1}0]$-bar of device B at -9 mT field applied along $\theta = 0°$. Note that $-\delta R_T$ is plotted for clarity. (C) Same as (B) at negative current ramps at ± 9 mT. The inset shows contacts used for measurements of R_L and R_T in all panels. (D) Same as (B) at $+9$ mT field (from Wunderlich et al., 2007c).

The basic phenomenology of current induced switchings observed in the L-shaped microbars is illustrated in Fig. 4.49. The particular field-assisted switching data plotted in the figure were measured in the [110]-bar of device A at $\theta = 7°$. At this off-easy-axis angle the current induced switching can be easily induced and detected due to the hysteretic bistable character of the low field magnetization and the clear AMR signal upon reversal (see Fig. 4.49A). The measurements were performed by first applying a saturation field and then reversing the field and setting it to a value close to but below the switching field in the field-sweep experiment (see Fig. IIIDA). Then, the first current ramp was applied which triggered the reversal, followed by subsequent control current ramps of the same polarity which showed no further changes in the magnetization. Fig. 4.49B–F show the difference, δR, between resistances of the first and the subsequent current ramps.

First we discuss data in Fig. 4.49B and C taken at -4 mT external field and negative current ramps. The two independent experiments (panels B

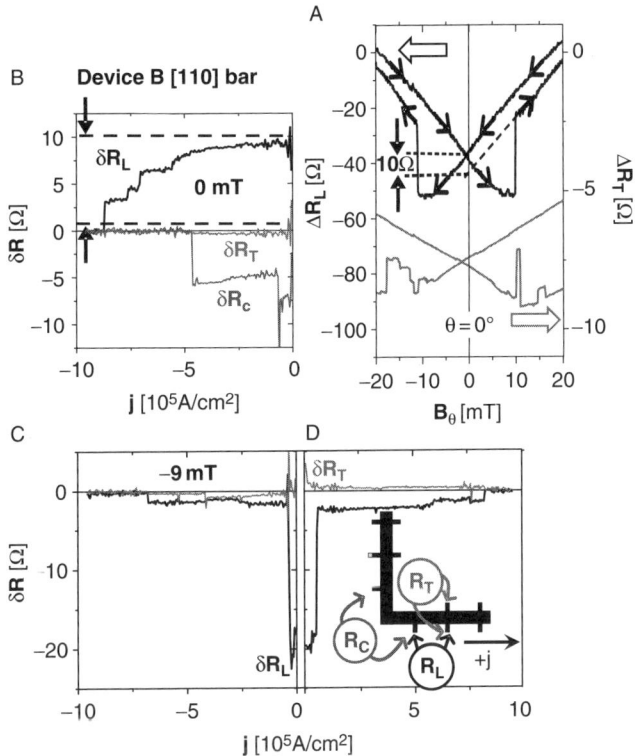

FIGURE 4.51 (A) Field-sweep measurements at $\theta = 0°$ in the [110]-bar of device B. (B) Difference between the first and second negative current ramps in the [110]-bar of device B at zero field. (C) Difference between the first and second negative current ramps at -9 mT field applied along $\theta = 0°$. (D) Same as (C) at for positive current ramps. The inset shows contacts used for measurements of R_L and R_T in all panels (from Wunderlich et al., 2007c).

and C, respectively) performed at nominally identical conditions demonstrate the high degree of reproducibility achieved in these devices. This includes the step-like features which is associated with domain wall depinning/pinning events preceding full reversal. To understand this process in more detail, the longitudinal and transverse resistance measurements in the [110]-bar are complemented with the resistance measurements at the corner of the L-shaped channel. The schematic plot of the respective voltage probes is shown in the inset. The first magnetization switching event at $j \approx -5 \times 10^5$ Acm^{-2} is detected by the step in the δR_C signal, that is, occurs in the corner region between the R_C contacts. For current densities in the range between $j \approx -5 \times 10^5$ Acm^{-2} and $j \approx -6 \times 10^5$ Acm^{-2} the domain wall remains pinned in the corner region. The next domain wall propagation and pinning event in δR_C is observed between

$j \approx -6 \times 10^5$ Acm^{-2} and $j \approx -7 \times 10^5$ Acm^{-2} and for $|j| > 7 \times 10^5$ Acm^{-2} the region between the R_C contacts is completely reversed. The depinning events at $j \approx -5 \times 10^5$ Acm^{-2} and $j \approx -6 \times 10^5$ Acm^{-2} are also registered by the R_L and R_T contacts through noise spikes in the respective δR_L and δR_T signals. However, beyond these spikes, δR_L and δR_T remain constant for $|j| < 7 \times 10^5$ Acm^{-2} indicating that the domain wall has not reached the section of the [110]-bar between the R_L contacts at these current densities. Constant δR_C and step-like changes in δR_L and δR_T at $|j| > 7 \times 10^5$ Acm^{-2} are signatures of the domain wall leaving the corner section and entering the part of the [110]-bar between the R_L contacts. The reversal of this part is completed at $j \approx -8 \times 10^5$ Acm^{-2}.

The -4 mT field assisted current induced switching is not observed at positive current ramps up to the highest experimental current density of $j = 1 \times 10^6$ Acm^{-2} which indicates that spin-transfer-torque effects can be contributing to the reversal. The domain wall propagates in the direction opposite to the applied hole current, in agreement with spin-transfer-torque studies of perpendicularly magnetized (Ga,Mn)As films (Chiba et al., 2006b). This direction of the domain wall propagation is assigned to the antiferromagnetic alignment of hole spins with respect to the total moment in (Ga,Mn)As (Chiba et al., 2006b; Yamanouchi et al., 2004, 2006).

A suppression of the role of the spin-transfer-torque relative to the thermally assisted switching mechanism is expected at fields closer to the coercive field. The data taken at -5 mT field shown in Fig. 4.49E and F are consistent with this expectation. Current induced switchings are observed here at lower critical currents and for both current polarities. Nevertheless, the asymmetry between the negative and positive critical currents is still apparent and consistent with a picture of cooperative effects of heating and spin-transfer-torque for negative currents and competing effects of the two mechanisms for positive currents.

The distinct current induced switching characteristics achieved by patterning one bar along the [110] direction and the other bar along the $[1\bar{1}0]$ direction are illustrated in Figs. 4.50 and 4.51 on a set of experiments in device B. The measurements shown in Fig. 4.50B–D were taken on the $[1\bar{1}0]$-bar in an external field of a magnitude of -9 mT applied along $\theta = 0°$ (see corresponding field sweep measurements in Fig. 4.50A). Up to the highest experimental current densities, the switching is observed only for the positive current polarity. Since for the opposite magnetic field sweep the current induced switching at $+9$ mT is also observed at positive currents (compare Fig. 4.50B and D), the Oersted fields are unlikely to be the dominant switching mechanism. (The Oersted fields generated by the experimental currents are estimated to be two orders of magnitude weaker than the anisotropy fields (Yamanouchi et al., 2006).)

The character of the current induced switching in device B at -9 mT is completely different in the [110]-bar compared to the $[1\bar{1}0]$-bar, as shown in Fig. 4.51C and D. The switching occurs at much lower current densities due to the lower coercive field of the [110]-bar at $\theta = 0°$ (compare Figs. 4.50A and 4.51A), and the asymmetry between the positive and negative switching currents is small, suggesting that heating plays an important role in this experiment. Although clear jumps are seen in δR_L, which are consistent with the field-sweep data in Fig. 4.51A, the switching signal of the δR_T in the [110]-bar is absent. This feature is ascribed to a fabrication induced strong pinning at the R_T contacts.

In Fig. 4.51B the pinning at the R_T contacts is exploited to study current induced switching at zero magnetic field. (If the switching of the whole bar was complete the zero-field 180° rotation from negative to positive easy-axis directions would be undetectable by the AMR measurement.) Again no switching signal is seen in δR_T but a clear step in δR_L. As for all field-assisted experiments, the sense and magnitude of the jump in δR_L for zero field correlates well with the field sweep measurements (see the dashed line in Fig. 4.51A). Also consistent with the trends in the field-assisted experiments, the switching occurs at larger current than in the -9 mT field assisted switching. Up to the highest experimental current density, the zero-field switching is observed only in the negative current ramp, as expected for the domain wall propagation from the corner (see the δR_c signal in Fig. 4.51B) to the [110]-bar due to spin-transfer-torque.

We conclude that (Ga,Mn)As microchannels with locally controlled magnetocrystalline anisotropies and inherently weak dipolar fields represent a new favorable class of systems for exploring extraodinary magnetore-sistance effects at micro- and nano-scale. Easy-axes rotations have been observed which depend on the width and crystal orientation of the microchannels and it has been demonstrated that domain-wall spin-transfer-torque effects are sensitive to these engineerable magnetocrystalline anisotropies and can occur at much smaller current densities than in the dense-moment metal ferromagnet channels.

4. SUMMARY

This chapter has covered several areas of the rich physics of transport effects in (Ga,Mn)As materials and microdevices. (Ga,Mn)As is a remarkable system which allows to study basic transport phenomena in extrinsic semiconductors over many decades of doping densities. Particularly intriguing in these systems is the interplay between charge doping by Mn-acceptors, strong SO-coupling in the hole bands, and the interaction between hole spins and Mn d-shell local moments. While the conduction in (Ga,Mn)As has a well understood impurity band character at low dopings and a disordered valence band character at high dopings, the intermediate dopings around 1% of Mn are intriguing and difficult to model. On the other hand it is exactly this doping regime in which ferromagnetism sets on and where future research will likely uncover new interesting physics which is undetectable in the more conventional systems with shallow, hydrogenic-like nonmagnetic dopands. Another potentially fruitful and only partially understood area of research is related to the role played by the internal exchange field and SO-coupling in low-temperature, quantum-coherent magnetotransport. Ferromagnetic semiconductor materials and microdevices in which these fields and other doping characteristics are largely tunable represent ideal systems to study these phenomena.

The most extensively explored transport characteristics of (Ga,Mn)As and related ferromagnetic semiconductors are the extraordinary MR effects. (Ga,Mn)As has provided an unprecedented physical insight into the anomalous Hall and AMR effects in standard ohmic devices and led to the discovery of AMR in the tunneling and CB regimes. Many results of the extraordinary MR studies in ferromagnetic semiconductors may be directly relevant to room-temperature metal ferromagnet systems with strong SO-coupling and may therefore lead to new technological applications even prior to the realization of high Curie temperature ferromagnetic semiconductors. Other favorable characteristics, such as the sensitivity of micromagnetic and magnetotransport coefficients to doping and electrical gating, and the higher integrability due to low saturation moments are specific to the class of dilute moment ferromagnetic semiconductors among which (Ga,Mn)As has become a prototypical material.

ACKNOWLEDGMENTS

This review is based on numerous helpful discussions with our colleagues. In particular we acknowledge contributions by Dimitri Basov, Kenneth Burch, Richard Campion, Elbio Dagotto, Tomasz Dietl, Laurence Eaves, Kevin Edmonds, Alexander Finkelstein, Tom Foxon, Devin Giddings, Charles Gould, Andrew Irvine, Konstantin Kikoin, Chris King, Jan Kučera, Yuri Kusraev, Allan MacDonald, Jan Mašek, Laurens Molenkamp, Vít Novák, Adriana Moreo, Hideo Ohno, Kamil Olejník, Elisa Ranieri, Andrev Rushforth, Victor Sapega, Alexander Shick, Jairo Sinova, Carsten Timm, Karel Výborný, Dieter Weiss, and Jan Zemen.

REFERENCES

Abolfath, M., Jungwirth, T., Brum, J., and MacDonald, A. H. (2001). *Phys. Rev. B* **63**, 054418, arXiv:cond-mat 0006093.
Anderson, W. W. (1975). *Solid-State Electron.* **18**, 235.
Aprili, M., Lesueur, J., Dumoulin, L., and Nédellec, P. (1997). *Solid State Commun.* **102**, 41.
Banhart, J., and Ebert, H. (1995). *Europhys. Lett.* **32**, 517.
Baxter, D. V., Ruzmetov, D., Scherschligt, J., Sasaki, Y., Liu, X., Furdyna, J. K., and Mielke, C. H. (2002). *Phys. Rev. B* **65**, 212407.
Beenakker, W. J., and van Houten, H. (1991). In "Solid State Physics" (H. Ehrenreich and D. Turnbull, eds), vol. 44. Academic Press, Boston.

Bergmann, G. (1984). *Phys. Rep.* **107**, 1.
Bhattacharjee, A. K., and à la Guillaume, C. B. (2000). *Solid State Commun.* **113**, 17.
Blakemore, J. S., Brown, W. J., Stass, M. L., and Woodbury, D. A. (1973). *J. Appl. Phys.* **44**, 3352.
Bolotin, K. I., Kuemmeth, F., and Ralph, D. C. (2006). *Phys. Rev. Lett.* **97**, 127202.
Brey, L., Tejedor, C., and Fernández-Rossier, J. (2004). *Appl. Phys. Lett.* **85**, 1996.
Brown, W. J., and Blakemore, J. S. (1972). *J. Appl. Phys.* **43**, 2242.
Burch, K. S., Shrekenhamer, D. B., Singley, E. J., Stephens, J., Sheu, B. L., Kawakami, R. K., Schiffer, P., Samarth, N., Awschalom, D. D., and Basov, D. N. (2006). *Phys. Rev. Lett.* **97**, 087208, arXiv:cond-mat/0603851.
Campion, R. P., Edmonds, K. W., Zhao, L. X., Wang, K. Y., Foxon, C. T., Gallagher, B. L., and Staddon, C. R. (2003). *J. Cryst. Growth* **251**, 311.
Chantis, A. N., Belashchenko, K. D., Tsymbal, E. Y., and van Schilfgaarde, M. (2007). *Phys. Rev. Lett.* **98**, 046601, arXiv:cond-mat/0610061.
Chapman, R. A., and Hutchinson, W. G. (1967). *Phys. Rev. Lett.* **18**, 443.
Chiba, D., Yamanouchi, M., Matsukura, F., and Ohno, H. (2003). *Science* **301**, 943.
Chiba, D., Matsukura, F., and Ohno, H. (2004a). *Physica E* **21**, 966.
Chiba, D., Sato, Y., Kita, T., Matsukura, F., and Ohno, H. (2004b). *Phys. Rev. Lett.* **93**, 216602, arXiv:cond-mat/0403500.
Chiba, D., Matsukura, F., and Ohno, H. (2006a). *Appl. Phys. Lett.* **89**, 162505.
Chiba, D., Yamanouchi, M., Matsukura, F., Dietl, T., and Ohno, H. (2006b). *Phys. Rev. Lett.* **96**, 096602, arXiv:cond-mat/0601464.
Chien, L., and Westgate, C. R. (1980). "The Hall Effect and Its Applications." Plenum, New York.
Ciorga, M., Einwanger, A., Sadowski, J., Wegscheider, W., and Weiss, D. (2007). *Phys. Stat. Sol. A* **204**, 186.
de Gennes, P. G., and Friedel, J. (1958). *J. Phys. Chem. Solids* **4**, 71.
De Ranieri, E., Rushforth, A. W., Vyborny, K., Rana, U., Ahmed, E., Campion, R. P., Foxon, C. T., Gallgher, B. L., Irvine, A. C., Wunderlich, J., and Jungwirth, T. (2008). *New J Phys.* **10**, 065003.
Dietl, T. (2007). *J. Phys.: Condens. Matter.* **19**, 165204, arXiv:0711.0340.
Dietl, T., Ohno, H., and Matsukura, F. (2001). *Phys. Rev. B* **63**, 195205, arXiv:cond-mat/0007190.
Döring, W. (1938). *Ann. Phys. (Leipzig)* **424**, 259.
Dietl, T., Matsukura, F., Ohno, H., Cibert, J., and Ferrand, D. (2003). In "Recent Trends in Theory of Physical Phenomena in High Magnetic Fields" (I. Vagner, ed.), p. 197. Kluwer, Dordrecht, arXiv:cond-mat/0306484.
Dugaev, V. K., Crépieux, A., and Bruno, P. (2001). *Phys. Rev. B* **64**, 104411.
Dugaev, V. K., Bruno, P., Taillefumier, M., Canals, B., and Lacroix, C. (2005). *Phys. Rev. B* **71**, 224423, arXiv:cond-mat/0502386.
Ebert, H., Vernes, A., and Banhart, J. (2000). *Solid State Commun.* **113**, 103.
Elsen, M., Jaffrès, H., Mattana, R., Thevenard, L., Lemaître, A., and George, J. M. (2007). *Phys. Rev. B* **76**, 144415, arXiv:0706.0109.
Fang, Z., Nagaosa, N., Takahashi, K. S., Asamitsu, A., Mathieu, R., Ogasawara, T., Yamada, H., Kawasaki, M., Tokura, Y., and Terakura, K. (2003). *Science* **302**, 5642.
Ferreira da Silva, A., Pepe, I., Sernelius, B. E., Person, C., Ahuja, R., de Souza, J. P., Suzuki, Y., and Yang, Y. (2004). *J. Appl. Phys.* **95**, 2532.
Fisher, M. E., and Langer, J. S. (1968). *Phys. Rev. Lett.* **20**, 665.
Fleurov, V. N., and Kikoin, K. A. (1982). *J. Phys. C* **15**, 3523.
Gao, L., Jiang, X., Yang, S.-H., Burton, J. D., Tsymbal, E. Y., and Parkin, S. S. P. (2007). *Phys. Rev. Lett.* **99**, 226602.
Giddings, A. D., Khalid, M. N., Jungwirth, T., Wunderlich, J., Yasin, S., Campion, R. P., Edmonds, K. W., Sinova, J., Ito, K., Wang, K. Y., Williams, D., Gallagher, B. L., and Foxon, C. T. (2005). *Phys. Rev. Lett.* **94**, 127202, arXiv:cond-mat/0409209.

Giraud, R., Gryglas, M., Thevenard, L., Lemaître, A., and Faini, G. (2005). *Appl. Phys. Lett.* **87**, 242505, arXiv:cond-mat/0509065.
Glew, R. W. (1984). *J. Cryst. Growth* **68**, 44.
Goennenwein, S. T. B., Russo, S., Morpurgo, A. F., Klapwijk, T. M., Van Roy, W., and De Boeck, J. (2005). *Phys. Rev. B* **71**, 193306, arXiv:cond-mat/0412290.
Goennenwein, S. T. B., Althammer, M., Bihler, C., Brandlmaier, A., Geprags, S., Opel, M., Schoch, W., Limmer, W., Gross, R., and Brandt, M. S. (2008). *Phys. Stat. Sol.* (RRL) **2**, 96.
Gould, C., Rüster, C., Jungwirth, T., Girgis, E., Schott, G. M., Giraud, R., Brunner, K., Schmidt, G., and Molenkamp, L. W. (2004). *Phys. Rev. Lett.* **93**, 117203, arXiv:cond-mat/0407735.
Gregg, J. F., Petej, I., Jouguelet, E., and Dennis, C. (2002). *J. Phys. D: Appl. Phys.* **35**, R121.
Grigorenko, A. N., Novoselov, K. S., and Mapps, D. J. (2006). arXiv:cond-mat/0611751.
Haas, C. (1970). *Crit. Rev. Solid State Sci.* **1**, 47.
Haldane, F. D. M. (2004). *Phys. Rev. Lett.* **93**, 206602, arXiv:cond-mat/0408417.
Hirakawa, K., Katsumoto, S., Hayashi, T., Hashimoto, Y., and Iye, Y. (2002). *Phys. Rev. B* **65**, 193312.
Hümpfner, S., Sawicki, M., Pappert, K., Wenisch, J., Brunner, K., Gould, C., Schmidt, G., Dietl, T., and Molenkamp, L. W. (2007). *Appl. Phys. Lett.* **90**, 102102, arXiv:cond-mat/0612439.
Joynt, R. (1984). *J. Phys. F: Met. Phys.* **14**, 2363.
Jungwirth, T., Abolfath, M., Sinova, J., Kučera, J., and MacDonald, A. H. (2002a). *Appl. Phys. Lett.* **81**, 4029, arXiv:cond-mat/0206416.
Jungwirth, T., Niu, Q., and MacDonald, A. H. (2002b). *Phys. Rev. Lett.* **88**, 207208, arXiv:cond-mat/0110484.
Jungwirth, T., Sinova, J., Wang, K. Y., Edmonds, K. W., Campion, R. P., Gallagher, B. L., Foxon, C. T., Niu, Q., and MacDonald, A. H. (2003). *Appl. Phys. Lett.* **83**, 320, arXiv:cond-mat/0302060.
Jungwirth, T., Wang, K. Y., Mašek, J., Edmonds, K. W., König, J., Sinova, J., Polini, M., Goncharuk, N. A., MacDonald, A. H., Sawicki, M., et al. (2005). *Phys. Rev. B* **72**, 165204, arXiv:cond-mat/0505215.
Jungwirth, T., Sinova, J., Mašek, J., Kučera, J., and MacDonald, A. H. (2006). *Rev. Mod. Phys.* **78**, 809, arXiv:cond-mat/0603380.
Jungwirth, T., Sinova, J., MacDonald, A. H., Gallagher, B. L., Novák, V., Edmonds, K. W., Rushforth, A. W., Campion, R. P., Foxon, C. T., Eaves, L., et al. (2007). *Phys. Rev. B* **76**, 125206, arXiv:0707.0665.
Kastner, M. A. (1992). *Rev. Mod. Phys.* **64**, 849.
Kawabata, A. (1980). *Solid State Commun.* **34**, 431.
Khmelevskyi, S., Palotás, K., Szunyogh, L., and Weinberger, P. (2003). *Phys. Rev. B* **68**, 012402.
Kim, D., Zink, B. L., Hellman, F., McCall, S., Cao, G., and Crow, J. E. (2003). *Phys. Rev. B* **67**, 100406.
Kim, J. S., Lee, D. Y., Bae, I. H., Lee, J. I., Noh, S. K., Kim, J. S., Kim, C. P., Ban, S., Kang, S.-K., Kim, S. M., et al. (2001). *J. Korean Phys. Soc.* **39**, S518.
Klein, L., Dodge, J. S., Ahn, C. H., Snyder, G. J., Geballe, T. H., Beasley, M. R., and Kapitulnik, A. (1996). *Phys. Rev. Lett.* **77**, 2774.
Kötzler, J., and Gil, W. (2005). *Phys. Rev. B* **72**, 060412(R).
Kramer, B., and MacKinnon, A. (1993). *Rep. Prog. Phys.* **56**, 1469.
Lee, P. A., and Ramakrishnan, T. V. (1985). *Rev. Mod. Phys.* **57**, 287.
Lee, P. A., Stone, A. D., and Fukuyama, H. (1987). *Phys. Rev. B* **35**, 1039.
Lee, S., Trionfi, A., Schallenberg, T., Munekata, H., and Natelson, D. (2007). *Appl. Phys. Lett.* **90**, 032105, arXiv:cond-mat/0608036.
Lee, W.-L., Watauchi, S., Miller, V. L., Cava, R. J., and Ong, N. P. (2004). *Science* **303**, 1647, arXiv:cond-mat/0405584.

Limmer, W., Glunk, M., Daeubler, J., Hummel, T., Schoch, W., Sauer, R., Bihler, C., Huebl, H., Brandt, M. S., and Goennenwein, S. T. B. (2006). *Phys. Rev. B* **74**, 205205, arXiv:cond-mat/0607679.
López-Sancho, M. P., and Brey, L. (2003). *Phys. Rev. B* **68**, 113201.
MacDonald, A. H., Schiffer, P., and Samarth, N. (2005). *Nat. Mater.* **4**, 195, arXiv:cond-mat/0503185.
Madelung, O., Rössler, U., and Schulz, M. (2003). "Impurities and Defects in Group IV Elements, IV-IV and III-V Compounds. Part b: Group IV-IV and III-V Compounds, vol. 41A2b of Landolt-Börnstein—Group III Condensed Matter." Springer-Verlag.
Marder, M. P. (2000). "Condensed Matter Physics." Wiley, New York, Supplementary material by author.
Matsukura, F., Ohno, H., and Dietl, T. (2002). *In* "Handbook of Magnetic Materials," (K. H. J. Buschow, ed.), Elsevier Amsterdam, vol. 14, p. 1, From Ohno Lab Homepage.
Matsukura, F., Sawicki, M., Dietl, T., Chiba, D., and Ohno, H. (2004). *Physica E* **21**, 1032.
Mattana, R., Elsen, M., George, J. M., Jaffrès, H., Dau, F. N. V., Fert, A., Wyczisk, M. F., Olivier, J., Galtier, P., Lépine, B., *et al.* (2005). *Phys. Rev. B* **71**, 075206.
McGuire, T., and Potter, R. (1975). *IEEE Trans. Magn.* **11**, 1018.
Moca, C. P., Sheu, B. L., Samarth, N., Schiffer, P., Jankó, B., and Zaránd, G. (2007). arXiv:0705.2016.
Moriya, R., and Munekata, H. (2003). *J. Appl. Phys.* **93**, 4603, arXiv:cond-mat/0301508.
Moser, J., Matos-Abiague, A., Schuh, D., Wegscheider, W., Fabian, J., and Weiss, D. (2007). *Phys. Rev. Lett.* **99**, 056601, arXiv:cond-mat/0611406.
Nagaev, E. L. (1998). *Phys. Rev. B* **58**, 816.
Nagai, Y., Junimoto, T., Nagasaka, K., Nojiri, H., Motokawa, M., Matsujura, F., Dietl, T., and Ohno, H. (2001). *Jpn. J. Appl. Phys.* **40**, 6231.
Neumaier, D., Wagner, K., Geissler, S., Wurstbauer, U., Sadowski, J., Wegscheider, W., and Weiss, D. (2007). *Phys. Rev. Lett.* **99**, 116803, arXiv:cond-mat/0703053.
Novák, V., Olejník, K., Wunderlich, J., Cukr, M., Výborný, K., Rushforth, A. W., Campion, R. P., Gallagher, B. L., Sinova, J., and Jungwirth, T. (2008). *Phys. Rev. Lett.* **101**, 077201, arXiv: 0804.1578.
Ohno, H. (1999). *J. Magn. Magn. Mater.* **200**, 110.
Ohno, H., Chiba, D., Matsukura, F., Omiya, T., Abe, E., Dietl, T., Ohno, Y., and Ohtani, K. (2000). *Nature* **408**, 944.
Ohya, S., Kobayashi, H., and Tanaka, M. (2003). *Appl. Phys. Lett.* **83**, 2175, arXiv:cond-mat/0303333.
Ohya, S., Hai, P. N., Mizuno, Y., and Tanaka, M. (2007). *Phys. Rev. B* **75**, 155328, arXiv:cond-mat/0608357.
Omiya, T., Matsukura, F., Dietl, T., Ohno, Y., Sakon, T., Motokawa, M., and Ohno, H. (2000). *Physica E* **7**, 976.
Ono, K., Shimada, H., and Ootuka, Y. (1997). *J. Phys. Soc. Jpn.* **66**, 1261.
Overby, M., Chernyshov, A., Rokhinson, L. P., Liu, X., and Furdyna, J. K. (2008). *Appl. Phys. Lett.* **92**, 192501.
Paalanen, M. A., and Bhatt, R. N. (1991). *Physica B* **169**, 223.
Pappert, K., Schmidt, M. J., Hümpfner, S., Rüster, C., Schott, G. M., Brunner, K., Gould, C., Schmidt, G., and Molenkamp, L. W. (2006). *Phys. Rev. Lett.* **97**, 186402, arXiv:cond-mat/0608683.
Park, B. G., Wunderlich, J., Willams, D. A., Joo, S. J., Jung, Shin, K. H., Olejnik, K., Shick, A. B., and Jungwirth, T. (2008). *Phys. Rev. Lett.* **100**, 087204.
Parsons, J. D., and Krajenbrink, F. G. (1983). *J. Electrochem. Soc.* **130**, 1782.
Poggio, M., Myers, R. C., Stern, N. P., Gossard, A. C., and Awschalom, D. D. (2005). *Phys. Rev. B* **72**, 235313.

Potashnik, S. J., Ku, K. C., Mahendiran, R., Chun, S. H., Wang, R. F., Samarth, N., and Schiffer, P. (2002). *Phys. Rev. B* **66**, 012408, arXiv:cond-mat/0204250.
Potter, R. I. (1974). *Phys. Rev. B* **10**, 4626.
Rokhinson, L. P., Lyanda-Geller, Y., Ge, Z., Shen, S., Liu, X., Dobrowolska, M., and Furdyna, J. K. (2007). *Phys. Rev. B* **76**, 161201, arXiv:0707.2416.
Rushforth, A. W., De Ranieri, E., Zemen, J., Wunderlich, J., Edmonds, K. W., King, C. S., Ahmad, E., Campion, R. P., Foxon, C. T., Gallagher, B. L., Výborný, K., and Kučera, J., et al. (2008). *Phys. Rev. B* **78**, 085314.
Rushforth, A. W., Giddings, A. D., Edmonds, K. W., Campion, R. P., Foxon, C. T., and Gallagher, B. L. (2006). *Phys. Stat. Sol. (c)* **3**, 4078, arXiv:cond-mat/0610692.
Rushforth, A. W., Výborný, K., King, C. S., Edmonds, K. W., Campion, R. P., Foxon, C. T., Wunderlich, J., Irvine, A. C., Vašek, P., Novák, V., et al. (2007). *Phys. Rev. Lett.* **99**, 147207, arXiv:cond-mat/0702357.
Rüster, C., Borzenko, T., Gould, C., Schmidt, G., Molenkamp, L. W., Liu, X., Wojtowicz, T. J., Furdyna, J. K., Yu, Z. G., and Flatté, M. E. (2003). *Phys. Rev. Lett.* **91**, 216602, arXiv:cond-mat/0308385.
Rüster, C., Gould, C., Jungwirth, T., Sinova, J., Schott, G. M., Giraud, R., Brunner, K., Schmidt, G., and Molenkamp, L. W. (2005a). *Phys. Rev. Lett.* **94**, 027203, arXiv:cond-mat/0408532.
Rüster, C., Gould, C., Jungwirth, T., Girgis, E., Schott, G. M., Giraud, R., Brunner, K., Schmidt, G., and Molenkamp, L. W. (2005b). *J. Appl. Phys.* **97**, 10C506.
Ruzmetov, D., Scherschligt, J., Baxter, D. V., Woj-towicz, T., Liu, X., Sasaki, Y., Furdyna, J. K., Yu, K. M., and Walukiewicz, W. (2004). *Phys. Rev. B* **69**, 155207.
Saffarzadeh, A., and Shokri, A. A. (2006). *J. Magn. Magn. Mater.* **305**, 141, arXiv:cond-mat/0608006.
Saito, H., Yuasa, S., and Ando, K. (2005). *Phys. Rev. Lett.* **95**, 086604.
Sankowski, P., Kacman, P., Majewski, J. A., and Dietl, T. (2007). *Phys. Rev. B* **75**, 045306, arXiv:cond-mat/0607206.
Sawicki, M., Wang, K.-Y., Edmonds, K. W., Campion, R. P., Staddon, C. R., Farley, N. R. S., Foxon, C. T., Papis, E., Kaminska, E., Piotrowska, A., et al. (2005). *Phys. Rev. B* **71**, 121302, arXiv:cond-mat/0410544.
Schallenberg, T., and Munekata, H. (2006). *Appl. Phys. Lett.* **89**, 042507.
Schlapps, M., Doeppe, M., Wagner, K., Reinwald, M., Wegscheider, W., and Weiss, D. (2006). *Phys. Stat. Sol. A* **203**, 3597.
Shacklette, L. W. (1974). *Phys. Rev. B* **9**, 3789.
Shick, A. B., Máca, F., Mašek, J., and Jungwirth, T. (2006). *Phys. Rev. B* **73**, 024418, arXiv:cond-mat/0601071.
Shklovskii, B. I., and Efros, A. L. (1984). "Electronic Propreties of Doped Semiconductors." Springer-Verlag, New York.
Sil, S., Entel, P., Dumpich, G., and Brands, M. (2005). *Phys. Rev. B* **72**, 174401.
Singley, E. J., Kawakami, R., Awschalom, D. D., and Basov, D. N. (2002). *Phys. Rev. Lett.* **89**, 097203.
Singley, E. J., Burch, K. S., Kawakami, R., Stephens, J., Awschalom, D. D., and Basov, D. N. (2003). *Phys. Rev. B* **68**, 165204.
Sinitsyn, N. A., Niu, Q., Sinova, J., and Nomura, K. (2005). *Phys. Rev. B* **72**, 045346, arXiv:cond-mat/0502426.
Sinova, J., Jungwirth, T., Yang, S. R. E., Kučera, J., and MacDonald, A. H. (2002). *Phys. Rev. B* **66**, 041202, arXiv:cond-mat/0204209.
Sinova, J., Jungwirth, T., and Černe, J. (2004a). *Int. J. Mod. Phys. B* **18**, 1083, arXiv:cond-mat/0402568.
Sinova, J., Jungwirth, T., Liu, X., Sasaki, Y., Furdyna, J. K., Atkinson, W. A., and MacDonald, A. H. (2004b). *Phys. Rev. B* **69**, 085209, arXiv:cond-mat/0308386.

Slaughter, J. M., Dave, R. W., DeHerrera, M., Durlam, M., Engel, B. N., Janesky, J., Rizzo, N. D., and Tehrani, S. (2004). *J. Supercond.* **15,** 19.
Smit, J. (1951). *Physica* **17,** 612.
Songprakob, W., Zallen, R., Tsu, D. V., and Liu, W. K. (2002). *J. Appl. Phys.* **91,** 171.
Tanaka, M., and Higo, Y. (2001). *Phys. Rev. Lett.* **87,** 026602.
Tang, H. X., Kawakami, R. K., Awschalom, D. D., and Roukes, M. L. (2003). *Phys. Rev. Lett.* **90,** 107201, arXiv:cond-mat/0210118.
Thomson, W. (1857). *Proc. R. Soc. Lond.* **8,** 546.
Timm, C., Raikh, M. E., and von Oppen, F. (2005). *Phys. Rev. Lett.* **94,** 036602, arXiv:cond-mat/0408602.
Tsukagoshi, K., Alphenaar, B. W., and Nakazato, K. (1998). *Appl. Phys. Lett.* **73,** 2515.
Van Esch, A., Van Bockstal, L., De Boeck, J., Verbanck, G., Van Steenbergen, A. S., Wellmann, P. J., Grietens, B., Herlach, R. B. F., and Borghs, G. (1997). *Phys. Rev. B* **56,** 13103.
van Gorkom, R. P., Caro, J., Klapwijk, T. M., and Radelaar, S. (2001). *Phys. Rev. B* **63,** 134432.
Vila, L., Giraud, R., Thevenard, L., Lemaître, A., Pierre, F., Dufouleur, J., Mailly, D., Barbara, B., and Faini, G. (2007). *Phys. Rev. Lett.* **98,** 027204, arXiv:cond-mat/0609410.
Viret, M., Gabureac, M., Ott, F., Fermon, C., Barreteau, C., and Guirado-Lopez, R. (2006). *Eur. Phys. J. B* **51,** 1, arXiv:cond-mat/0602298.
Wagner, K., Neumaier, D., Reinwald, M., Wegscheider, W., and Weiss, D. (2006). *Phys. Rev. Lett.* **97,** 056803, arXiv:cond-mat/0603418.
Wang, K. Y., Campion, R. P., Edmonds, K. W., Sawicki, M., Dietl, T., Foxon, C. T., and Gallagher, B. L. (2005a). In "Proceedings of the 27th International Conference on the Physics of Semiconductors" (J. M. C. G. V. de Walle, ed.), Vol. 772, p. 333. AIP. Springer, arXiv:cond-mat/0411475.
Wang, K. Y., Edmonds, K. W., Campion, R. P., Zhao, L. X., Foxon, C. T., and Gallagher, B. L. (2005b). *Phys. Rev. B* **72,** 085201, arXiv:cond-mat/0506250.
Wenisch, J., Gould, C., Ebel, L., Storz, J., Pappert, K., Schmidt, M. J., Kumpf, C., Schmidt, G., Brunner, K., and Molenkamp, L. W. (2007). *Phys. Rev. Lett.* **99,** 077201, arXiv:cond-mat/0701479.
Woodbury, D. A., and Blakemore, J. S. (1973). *Phys. Rev. B* **8,** 3803.
Wunderlich, J., Jungwirth, T., Kaestner, B., Irvine, A. C., Wang, K. Y., Stone, N., Rana, U., Giddings, A. D., Shick, A. B., Foxon, C. T., et al. (2006). *Phys. Rev. Lett.* **97,** 077201, arXiv:cond-mat/0602608.
Wunderlich, J., Jungwirth, T., Novák, V., Irvine, A. C., Kaestner, B., Shick, A. B., Foxon, C. T., Campion, R. P., Williams, D. A., and Gallagher, B. L. (2007a). *Solid State Commun.* **144,** 536.
Wunderlich, J., Jungwirth, T., Irvine, A. C., Kaestner, B., Shick, A. B., Campion, R. P., Williams, D. A., and Gallagher, B. L. (2007b). *J. Magn. Magn. Mater.* **310,** 1883.
Wunderlich, J., Irvine, A. C., Zemen, J., Holý, V., Rushforth, A. W., Ranieri, E. D., Rana, U., Výborný, K., Sinova, J., Foxon, C. T., et al. (2007c). *Phys. Rev. B* **76,** 054424, arXiv:0707.3329.
Yakunin, A. M., Silov, A. Y., Koenraad, P. M., Wolter, J. H., Van Roy, W., De Boeck, J., Tang, J. M., and Flatté, M. E. (2004). *Phys. Rev. Lett.* **92,** 216806, arXiv:cond-mat/0402019.
Yamada, T., Tokumitsu, E., Saito, K., Akatsuka, T., Miyauchi, M., Konagai, M., and Takahashi, K. (1989). *J. Cryst. Growth* **95,** 145.
Yamanouchi, M., Chiba, D., Matsukura, F., and Ohno, H. (2004). *Nature* **428,** 539.
Yamanouchi, M., Chiba, D., Matsukura, F., Dietl, T., and Ohno, H. (2006). *Phys. Rev. Lett.* **96,** 096601, arXiv:cond-mat/0601515.
Yang, S. R. E., Sinova, J., Jungwirth, T., Shim, Y. P., and MacDonald, A. H. (2003). *Phys. Rev. B* **67,** 045205, arXiv:cond-mat/0210149.
Yao, Y., Kleinman, L., MacDonald, A. H., Sinova, J., Jungwirth, T., sheng Wang, D., Wang, E., and Niu, Q. (2004). *Phys. Rev. Lett.* **92,** 037204.

CHAPTER 5

Spintronic Properties of Ferromagnetic Semiconductors

F. Matsukura,*,† D. Chiba,*,† and H. Ohno*,†

Contents		
	1. Introduction	207
	2. Spin-Injection and Detection of Spin-Polarization	209
	3. Magnetic Tunnel Junction	212
	3.1. Magnetic tunnel junction	212
	3.2. Current-induced magnetization switching	215
	4. Magnetic DW	218
	4.1. DW observation	218
	4.2. DW resistance	221
	4.3. DW motion	224
	5. Electric-Field Control of Ferromagnetism	228
	6. Optical Control of Ferromagnetism	231
	7. Summary	234
	Acknowledgments	235
	References	235

1. INTRODUCTION

Ferromagnetic semiconductors exhibit semiconducting and ferromagnetic properties simultaneously. The study of ferromagnetic semiconductor started in the 1960s, dealing with chalcogenides and spinels (Methfessel and Mattis, 1968). Spintronic device structures were also

---

* Laboratory for Nanoelectronics and Spintronics, Research Institute of Electrical Communication, Tohoku University, Katahira 2-1-1, Aoba-ku, Sendai 980-8577, Japan
† ERATO Semiconductor Spintronics Project, Japan Science and Technology Agency, Sanbancho 5, Chiyoda-ku, Tokyo 102-0075, Japan

Semiconductors and Semimetals, Volume 82 © 2008 Elsevier Inc.
ISSN 0080-8784, DOI: 10.1016/S0080-8784(08)00005-7 All rights reserved.

investigated (Esaki et al., 1967). Most of the studies being done in the 1980s and thereafter involve magnetic semiconductors made by introducing a sizable amount of magnetic elements into nonmagnetic III–V and II–VI host compound semiconductors (Dietl and Ohno, 2003; Furdyna and Kossut, 1988). They are referred to as diluted magnetic semiconductors. The successful epitaxial growth of (In,Mn)As (Munekata et al., 1989) by molecular-beam epitaxy and subsequent discovery of ferromagnetism in (In,Mn)As and in (Ga,Mn)As (Ohno et al., 1992, 1996) revived the research of spintronic structures using ferromagnetic semiconductors, as they can readily be integrated in well-established semiconductor devices (Matsukura et al., 2002). Owing to its spontaneous spin polarization of carriers, ferromagnetic semiconductors show a variety of spin-dependent phenomena that can form a basis of new functionalities. The device structures based on ferromagnetic semiconductors can thus be used as a proof-of-concept prototyping to demonstrate new schemes of device operation, as a vehicle to unveil physics and gain understanding involved in the functional operation of spintronic devices, or as a tool to expand our knowledge about the operation of existing devices for further improvement.

In this chapter, we review the properties of spintronic device structures based on ferromagnetic semiconductors. Because of spontaneous splitting of spin-states, ferromagnetic semiconductors can be used as a source of spin-polarized carrier from which a spin-polarized current can be injected into an adjacent layer. The spin-polarization of such a spin-polarized current can be detected by optical means using integrated light-emitting diodes, which emit circularly-polarized light reflecting spin-polarization of the current (Ohno et al., 1999). High tunnel magnetoresistance (TMR) ratio reported in fully epitaxial (Ga,Mn)As-based magnetic tunnel junctions (MTJs) (Chiba et al., 2004a; Tanaka and Higo, 2001) is another consequence of high spin-polarization, which can be used to switch magnetization direction of the ferromagnetic layer in an MTJ (Chiba et al., 2004b).

(Ga,Mn)As shows a clear magnetic domain structure, as observed by a scanning microscope (Shono et al., 2000) and by a magneto-optical microscope (Welp et al., 2003). Magnetic domains extending almost a millimeter despite the dilute concentration of magnetic ions (Mn concentration of approximately 5%) reveals the long-range nature of ferromagnetic interaction in these materials. Interface of two domains called a domain-wall (DW) has a finite width. Manipulation of the position of a DW by spin-polarized current is of recent focus not only because of the nontrivial physics involved in the process, but also because of its importance in nonvolatile memory application (Parkin, 2004) and for the operation of new logic scheme (Allwood et al., 2002). The position of a DW in (Ga,Mn)As can be controlled electrically at current density of $j \sim 10^5 \, \text{A/cm}^2$

(Yamanouchi et al., 2004), which is 2–3 orders of magnitude smaller than the current density required to observe similar motion in NiFe nanowires. The DW velocity as the function of j and temperature has been investigated for (Ga,Mn)As (Yamanouchi et al., 2006), which was the first systematic measurement of DW velocity versus j in any material. The analysis shows that the motion of the DW is governed by spin-transfer from carriers to local magnetic spins, in the entire current range experimentally investigated; that is, above as well as below the threshold current density, where DW creep motion has been observed.

Ferromagnetism of (Ga,Mn)As and (In,Mn)As is brought about by the existence of holes, where the energy gain by the repopulation of holes among the spin split states stabilizes ferromagnetism (Dietl et al., 2000). Control of the magnetic phase transition under a constant temperature has been shown possible by changing the hole concentrations by external means. The electrical control of magnetism has been demonstrated using a metal–insulator–semiconductor structure with either a (Ga,Mn)As or (In,Mn)As channel by applying electric-fields to the channel, where isothermal and reversible change of the Curie temperature as well as the coercivity was observed (Chiba et al., 2006a; Ohno et al., 2000). Similar effect but not a reversible one has been observed by the light irradiation (Koshihara et al., 1997), where photo-generated carriers play a critical role in changing the magnetic properties. The possibility of magnetization direction control by irradiation of circularly polarized lights has been investigated (Oiwa et al., 2002). Theoretical aspects of ferromagnetism has been dealt with in a separate chapter by Dietl in this volume.

2. SPIN-INJECTION AND DETECTION OF SPIN-POLARIZATION

For spintronic devices based on semiconductors, one of the technical issues to pursue is the injection of the spin-polarized carriers (spin injection) into and their transport in nonmagnetic semiconductors. Realization of efficient electrical spin injection, spin transport, and spin detection are the processes one needs to fully control for such devices as spin field-effect transistors (FET) (Datta and Das, 1990) and spin metal-oxide-semiconductor FET (Sugahara and Tanaka, 2004).

Optical spin injection by circularly polarized light taking advantage of the selection rules of optical transitions has been a standard way to inject spin-polarized carriers into semiconductors (Meier and Zakharchenya, 1984). Using time-resolved Faraday rotation measurements combined with drift transport under an applied electric field in GaAs, it has been shown that spin relaxation time can be over 100 ns and lateral spin transport can exceed 100 μm without loss of spin coherence (Kikkawa and Awshalom, 1999). Spin injection via vacuum has also been shown possible using GaAs, where spin-polarized carriers were injected from a ferromagnetic Ni tip through vacuum and detected by circularly polarized photoluminescence (PL) (Alvarado and Renaud, 1992).

Electrical spin-injection from a ferromagnet into an adjacent nonmagnetic semiconductor layer has been achieved in a device structure integrating an light-emitting diode (LED) for spin-detection and a ferromagnetic semiconductor layer as a spin-injector (Ohno et al., 1999; Young et al., 2002). Note that (Ga,Mn)As is p-type due to the acceptor nature of Mn in GaAs, and has a spontaneous imbalance in the spin population of the holes in the absence of magnetic fields. Spin-polarization of the injected carriers was detected by measuring the degree of circular polarization of the emitted light from the LED. This, so-called spin-LED structure for detection of spin polarization has also been used to demonstrate electrical spin-injection into nonmagnetic semiconductors from paramagnetic semiconductor under magnetic field (Filderling et al., 1999) and from ferromagnetic metals (Hanbicki et al., 2002; Zhu et al., 2001).

Injection of electron spins is usually preferable because electrons have longer spin life time than holes. In order to make it possible to inject spin-polarized electrons from a p-type ferromagnetic semiconductor, band-to-band tunneling has been employed (Johnston-Halperin et al., 2002; Kohda et al., 2001). The device consists of an Esaki diode (ED) (p-(Ga,Mn)As/ n-GaAs) and a spin-LED for detection of spin polarization of injected carriers. Initial experiments showed polarization change ΔP of 1–8%, several times larger than the case of the hole spin injection. Later, a much higher spin-polarization of injected electrons, 80%, was detected (Van Dorpe et al., 2004). Here, the spin relaxation time and the recombination time were first determined from the PL depolarization under an applied magnetic field (the Hanlé effect) and then used to obtain the spin-polarization of the injected current. A strong bias dependence of spin polarization was also found, which was explained by the calculation combining the p–d Zener model with tight-binding approximation and the Landauer–Büttiker scheme to take into account the interface and inversion symmetry effects for ballistic transport (Van Dorpe et al., 2005).

Two-terminal devices used in these earlier experiments for band-to-band injection and spin-LED detection do not allow separate biases for the injection part and the detection part. To control the biases independently, three-terminal devices were fabricated, where the emitter contact was made on the top of (Ga,Mn)As, the base on the n^+-GaAs, and the emitter on the p-GaAs substrate (Kohda et al., 2006). Figure 5.1 shows the electroluminescence polarization P_{EL} dependence on the bias V_{EB} between the emitter and base as a function of V_{CB} (bias between the collector and base) at 10 K and $\mu_0 H = 1$ T (μ_0: the permeability of vacuum),

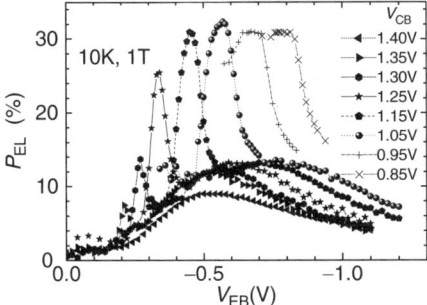

FIGURE 5.1 Bias dependence of electroluminescence polarization for three terminal device based on Esaki diode (ED) and light-emitting diode (Kohda et al., 2006). The ED consisted of 20 nm $Ga_{0.943}Mn_{0.057}As$ and 15 nm n^+-GaAs doped with [Si] = 10^{19} cm^{-3} and the LED consisted of 30 nm graded n-$Al_yGa_{1-y}As$ with $y = 0 - 0.1$ and [Si] = 10^{18} cm^{-3}, 60 nm n-$Al_{0.1}Ga_{0.9}As$ spacer with [Si] = 10^{17} cm^{-3}, 100 nm p-GaAs active layer with [Be] = 2×10^{18} cm^{-3}, and p-$Al_{0.3}Ga_{0.7}As/p$-GaAs on p-GaAs substrate.

where H is applied parallel to the growth direction (Faraday configuration). The strong V_{EB} dependence of P_{EL} indicates that the spin-polarization of the valence bands in the (Ga,Mn)As emitter is reflected on the spin-injection efficiency. The highest obtained value of $P_{EL} = 32.4\%$ corresponds to a spin-polarization over 85% of the injected electrons, after correcting spin relaxation in the LED part obtained by the Hanlé effect measurements.

We note that (Ga,Mn)As can be used as the electrical spin-detection layer in the electrical spin-injection measurements (Chen et al., 2006).

Electrical spin-injection from ferromagnetic metals to semiconductors is also possible. Conductivity mismatch between metal and semiconductor (neglecting interface effects) makes an efficient injection in the diffusive regime difficult (Schimidt et al., 2000). To overcome this difficulty associated with the diffusive transport, one needs to employ a ferromagnetic metal with spin polarization close to 100% (Schimidt et al., 2000) and/or use a tunnel junction contact (Fert and Jaffrès, 2001; Rashba, 2000). It has been shown that the natural Schottky barrier at semiconductor–metal interface can improve the spin injection efficiency from Fe to GaAs-based LED through the barrier (Hanbicki et al., 2002; Zhu et al., 2001). The contact material dependence of the experimental results, including metal injector with tunnel junction, is summarized elsewhere (Van Roy et al., 2007). The spin injection in lateral device configuration is also possible, which was demonstrated in n-GaAs channel with source and drain contacts of Fe, where the scanning Kerr microscope was utilized to image the spin polarization of carriers injected from the source and accumulated near the drain (Crooker et al., 2005). A review on spin injection from metal electrodes into semiconductors is available (Jonker and Flatte, 2006).

3. MAGNETIC TUNNEL JUNCTION

3.1. Magnetic tunnel junction

MTJs consist of two ferromagnetic layers separated by a thin insulator barrier, through which carriers tunnel in current-perpendicular-to-the-plane (CPP) configuration, and exhibit TMR, because the resistance of junctions depends on the relative magnetization orientation of the two ferromagnetic layers (Julliere, 1975; Miyazaki and Tezuka, 1995; Moodera et al., 1995). MTJs are now regarded as one of the most important building blocks for a new generation of spintronic devices, such as sensors, magnetic heads, and magnetic random access memory elements.

(Ga,Mn)As makes it possible to realize fully epitaxial single-crystal MTJ structures with high-quality interfaces (Chiba et al., 2000, 2004a; Hayashi et al., 1999; Mattana et al., 2003; Saito et al., 2005; Tanaka and Higo, 2001; Uemura et al., 2005; Watanabe et al., 2006), which provide useful information on the fundamental properties of MTJs. These (Ga,Mn)As-based MTJs also allow one to explore novel device structures such as resonant tunneling diodes (RTD) (Elsen et al., 2007; Ohno et al., 1998, Ohya et al. 2007) and hot-carrier transistors (Mizuno et al., 2007) with a TMR effect. The high structural integrity with abrupt interfaces of (Ga,Mn)As/GaAs and (Ga,Mn)As/AlAs superlattices has been confirmed by X-ray diffraction measurement (Hayashi et al., 1997; Shen et al., 1997a). In metallic systems such as Fe/Cr superlattices, the coupling between ferromagnetic layers (Fe) separated by nonmagnetic metal layers (Cr) was found to oscillate as a function of the thickness of the nonmagnetic layers, due to the Ruderman–Kittel–Kasuya–Yosida (RKKY) interaction (Parkin et al., 1990). A spin-valve structure (Dieny et al., 1991) as well as an antiferromagnetically coupled multilayer (Baibich et al., 1988) are important ingredients for a giant magnetoresistance (GMR) spin-valve devices. For ferromagnetic semiconductors, a spin-valve type GMR signal has been observed in a (Ga,Mn)As/(Al,Ga) As/(Ga,Mn)As trilayer structure (Chiba et al., 2000) and ferromagnetic interlayer coupling has been reported (Akiba et al., 1998; Chiba et al., 2000). No clear oscillatory interlayer coupling has been observed in ferromagnetic semiconductor structures other than the one inferred from transition temperature of superlattices (Mathieu et al., 2002). High carrier spin polarization P of (Ga,Mn)As has been predicted both by first principal calculations (Ogawa et al., 1999) and by the p–d Zener model (Dietl et al., 2001a). Experimentally, P over 85% has been reported by using an Andreev reflection spectroscopy in (Ga,Mn)As/superconducting junctions (Braden et al., 2003; Panguluri et al., 2005). With such a high P together with high quality epitaxial interfaces, a high TMR ratio is expected for (Ga,Mn)As structures.

The first TMR effect with (Ga,Mn)As electrodes was reported in an MTJ with a 3-nm AlAs barrier with top and the bottom (Ga,Mn)As layers (Hayashi et al., 1999). The AlAs barrier is high enough (0.55 eV for GaAs–AlAs in the valence band) to ensure the tunnel transport of holes. A TMR ratio defined as a ratio of the resistance difference to the parallel resistance (the low resistance) of 44% at 4.2 K was reported, although negative magnetoresistance, most likely coming from effects other than the spin-valve effect, was included in the calculation of the ratio. A TMR of 5.5% having clear high and low resistance states, that is, having clear parallel and anti-parallel magnetization configurations, has been reported in the (Ga,Mn)As/AlAs/(Ga,Mn)As MTJ grown on an (In,Ga)As buffer layer (Chiba et al., 2000), in which the magnetization direction was perpendicular to the plane due to tensile strain. The highest TMR ratio in (Ga,Mn)As/AlAs/(Ga,Mn)As MTJs, so far, is ~75% at 8 K in an MTJ with a 1.5 nm AlAs barrier, where the quality of the interface between the AlAs and (Ga,Mn)As layers was improved by inserting 1 nm GaAs layers into between the layers to avoid the diffusion of Mn into the barrier. The thickness of the barrier was varied by the use of a wedged AlAs layer (Tanaka and Higo, 2001). The TMR was found to decrease rapidly with increasing barrier thickness, which was explained by a dominant contribution of the zone center carriers as the barrier thickness increases because of faster decay for tunneling carriers with larger $k_\parallel$, wave vector parallel to film plane.

A TMR ratio of 290%, which corresponds to 77% of P according to the Jullière formula, TMR ratio $= 2P^2/(1-P^2)$ (Julliere, 1975), at 0.39 K was reported in a (Ga,Mn)As MTJ with a 6 nm GaAs as a barrier layer as shown in Fig. 5.2 (Chiba et al., 2004a). GaAs is known to act as a barrier for holes in (Ga,Mn)As; the barrier height is 0.1 eV measured from the hole Fermi energy in (Ga,Mn)As (Ohno et al., 2002). When the thickness of GaAs layer is less than 6 nm, the two (Ga,Mn)As layers started to couple

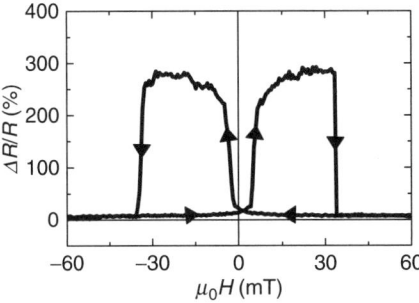

FIGURE 5.2 TMR curve of 20 nm (Ga,Mn)As/6 nm GaAs/20 nm (Ga,Mn)As at 0.39 K. Magnetic field was applied along [100] direction (Chiba et al., 2004a).

ferromagnetically making it difficult to observe a clear TMR effect (Sato et al., 2005). TMR ratio of GaAs-barrier MTJs was found to decrease monotonically as the bias voltage V increased and V_{half}, where the TMR ratio becomes 0.5 of the one at zero bias, is in the range of 40–100 mV, similar to the result of AlAs-barrier MTJs (Mattana et al., 2003, 2005). The magnitude of V_{half} is much smaller than that of metal-based MTJs (300–400 mV) (Moodera et al., 1995). Qualitative explanation of the observed small V_{half} is the following (Chiba et al., 2004a). Because the free carrier concentration of (Ga,Mn)As is in the range of 10^{19}–10^{21} cm^{-3}, the Fermi energy of (Ga,Mn)As is of the order of a few hundreds of meV (Dietl et al., 2001a). When the bias becomes greater than the Fermi energy of (Ga,Mn)As, the spin polarization of the empty states in the collector is reduced, resulting in a rapid reduction of TMR ratio. This bias dependence was reproduced theoretically by employing the Landauer–Büttiker formalism combined with a tight-binding transfer matrix method (Sankowski et al., 2006, 2007). The TMR ratio of up to 100% has been reported in (Ga,Mn)As MTJ with II–VI ZnSe barrier, which indicates that high quality interfaces can also be achieved in such a III–V/II–VI heterostructure (Saito et al., 2005) having a small lattice mismatch (ZnSe lattice constant ($a = 0.5669$ nm) and GaAs ($a = 0.5655$ nm)) and a similar growth temperature of ~250 °C. Using an MTJ structure, a high current gain in a (Ga,Mn)As MTJ based hot-carrier device was observed (Mizuno et al., 2007), indicating that active (Ga,Mn)As devices using the spin degree of freedom is possible.

The anisotropic TMR behavior has been reported by several groups (Chiba et al., 2004a; Higo et al., 2001; Tanaka and Higo, 2002; Uemura et al., 2005). For (Ga,Mn)As-based MTJ with a 1.5-nm AlAs barrier, when the magnetic field was applied along the [100] axis in the film plane, TMR ratio as high as 75% was obtained. The TMR ratio decreased when the applied magnetic field direction was along the [1$\bar{1}$0] and [110] (Tanaka and Higo, 2002; Uemura et al., 2005). This anisotropic behavior can be explained by the coherent magnetization rotation model with a cubic magnetic easy axis of $\langle 100 \rangle$ in (Ga,Mn)As (Hamaya et al., 2004; Liu et al., 2003; Sawicki et al., 2005; Welp et al., 2003). To explain the angular dependences of both the switching field and the TMR ratio, the DW displacement model is shown to be more appropriate rather than the coherent rotation model (Uemura et al., 2005).

Tunneling anisotropic magnetoresistance (TAMR) is a spin-valve like magnetoresistance which depends on the direction of the magnetization of (a) ferromagnetic electrode(s) and is observed in the MTJ devices, which was discovered in (Ga,Mn)As/AlO$_x$/Au layer structure (Gould et al., 2004). Anisotropic TMR described above is explained mainly by magnetocrystalline anisotropy, whereas TAMR seems to be related closely to the anisotropic density of states in electrodes. A very large

TAMR over 150,000% was then observed in a (Ga,Mn)As/(2 nm) GaAs/ (Ga,Mn)As structure (Rüster et al., 2005). A strong correlation between the TAMR effect, of the order of 10%, and the magnetic anisotropy energy of (Ga,Mn)As, was pointed out suggesting that the TAMR effect might originate from spin–orbit interaction in the valence band (Saito et al., 2005). A calculation using the Landauer–Büttiker formalism combined with the tight-binding transfer matrix method has predicted that the smaller the hole concentration greater the TAMR effect (Sankowski et al., 2007). Although the origin of the very large TAMR is still under debate, it may be related to the anisotropic valence band structure induced by spin–orbit interaction as well as the lattice strain in combination with the metal–insulator transition (Pappert et al., 2006).

The TMR effect in a QW combines both magnetism and semiconductor quantum physics. Spin accumulation in a structure with AlAs/GaAs/ AlAs QW sandwiched by (Ga,Mn)As layers was studied and hole spin relaxation time of 100 ps was inferred from the results (Mattana et al., 2003). A double-barrier AlAs/GaAs/AlAs RTD with a (Ga,Mn)As emitter showed resonant tunneling characteristics manifesting spontaneous spin splitting of the valence band (Ohno et al., 1998). An oscillatory TMR with bias was observed in a structure with AlAs/(In,Ga)As/AlAs QW with (Ga,Mn)As emitter and collector (Ohya et al., 2005). In the bias dependence, an enhancement of TMR ratio was observed at LH1 resonant state in a structure with AlAs/(Ga,Mn)As/AlAs QW and GaAs:Be bottom electrode instead of (Ga,Mn)As (Ohya et al., 2007).

It has been shown that the TMR effect can be observed in hybrid MTJ structures with a ferromagnetic metal electrode in combination with a (Ga,Mn)As electrode (Chun et al., 2002; Saito et al., 2006). Other ferromagnetic semiconductors have also been explored as an electrode material for MTJ, such as (Ga,Cr)N/AlN/(Ga,Cr)N (Kim et al., 2006) and (Ti,Co)O$_2$/ Al–O/Fe$_{0.1}$Co$_{0.9}$ (Toyosaki et al., 2005).

3.2. Current-induced magnetization switching

Current-induced magnetization switching (CIMS) by spin–torque exerted from the interaction between spin current and local magnetic moments (Berger, 1996; Slonczewski, 1996) is an important and interesting subject both from the fundamental as well as from the technological viewpoints. A number of experimental confirmations (Albert et al., 2000; Katine et al., 2000; Myers et al., 1999; Sun, 1999; Tsoi et al., 1998) have been reported. Since the observation of spin-wave excitation using a point contact to a Co/Cu multiplayer (Tsoi et al., 1998) and a hysteretic current-driven switching in Co/Cu/Co nanopillar at room temperature (Albert et al., 2000; Katine et al., 2000), many experiments have been carried out in lithographically patterned CPP giant-magnetoresistance (CPP-GMR) nanopillars.

There have recently been many reports on CIMS on MTJ with low junction resistance (Fuchs et al., 2004; Hayakawa et al., 2005; Huai et al., 2004). Typical critical current density J_C required for magnetization reversal in these systems is of the order of 10^6 A/cm^2 or higher.

(Ga,Mn)As has a small magnetization of 0.1 T or less (Matsukura et al., 2002) and high P (Dietl et al., 2001a; Ogawa et al., 1999), which are expected to result in reduction of J_C according to the Slonczewski's spin-transfer torque model (Sinova et al., 2004). On the other hand, spin–orbit interaction has to be taken into account to describe the valence band structure of GaAs (spin–orbit splitting at the top of valence band is 0.34 eV, whereas the Fermi energy is 0.2 eV or less), which may mix the spin states of carriers. It is thus interesting to see how the CIMS manifests itself in ferromagnetic III–V semiconductors. CIMS reported in (Ga,Mn)As MTJs (Chiba et al., 2004b, 2006b; Elsen et al., 2006) showed that J_C for switching is indeed much lower than the metallic MTJs.

The MTJ structure used by Chiba et al. (2004b) consists of an 80 nm (Ga,Mn)As/6 nm GaAs/15 nm (Ga,Mn)As epitaxial stack (from the surface side). Different thickness for the top and the bottom (Ga,Mn)As layers is employed to identify the role of the total magnetic moment of the layers. Rectangular devices having lateral dimensions of a($/\!/$ [$\bar{1}$10]) × b($/\!/$ [110]) = 1.5 × 0.3, 2.0 × 0.4, and 2.5 × 0.5 µm^2 are made. All devices show a square pseudo spin-valve resistance-magnetic field (R–H) curves at 30 K, indicating that parallel and antiparallel magnetization configurations are realized. Circular devices made as a reference showed rounded R–H curves, showing the importance of the shape of the devices. $\mu_0 H_C$ determined from the R–H curves increased as the device dimension was reduced, suggesting that in-plane crystalline anisotropy and/or shape anisotropy begun to play a role in magnetization reversal process. $\mu_0 H_C$ of the top layer of the 1.5 × 0.3 µm^2 device was about an order of magnitude greater than that determined from the magnetization measurement on a larger sample (25 mm^2).

The result of CIMS measurement of the 1.5 × 0.3 µm^2 device is shown in Fig. 5.3, Here ΔR is the difference between the resistance measured at $V_d = +10$ mV after application of I_{pulse} (horizontal axis) with the pulse duration $w_p = 1$ ms and the resistance with parallel magnetization M at the external field $H = 0$. Positive V_d (and thus current) is defined as biasing the top layer positive with respect to the bottom layer. Each data point was measured after setting the magnetization configuration shown as the legends to the initial ones [parallel (A) or antiparallel (B)] by H. A clear switching from the initial low resistance state to a high resistance state is observed in the positive current direction for configuration A at the current density $J_C^{\text{AP}} = 2.0 \times 10^5$ A/cm^2. Opposite switching is observed in the negative current direction for configuration B at $J_C^{\text{P}} = -1.5 \times 10^5$ A/cm^2. These low and high resistances after switching are in good

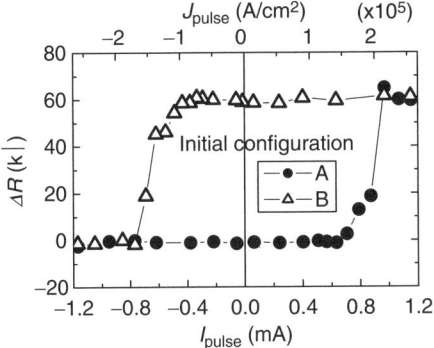

FIGURE 5.3 The results of current induced magnetization switching in (Ga,Mn)As/GaAs/(Ga,Mn)As MTJ at 30 K. ΔR is the resistance (measured at + 10 mV) difference between the resistance of MTJ after application of I_{pulse} (1 ms) and that at parallel magnetization configuration at $H = 0$. Closed circles show the I_{pulse} dependence of ΔR for initial configuration A (parallel M), and open triangles show the results for initial configuration B (antiparallel M) (Chiba et al., 2004b).

agreement with the values seen in R–H measurements. R–H measurements after current pulse are used to determine $\mu_0 H_C$ of the free layer, from which we can determine which layer is the free layer. The effect of Oersted fields generated by current pulse can be ruled out, because the bottom layer was always the free layer independent of the initial configurations. Effect of heating did not play a major role in the observed behavior, because one can control the magnetization direction of the free layer by the current pulse direction. The observed reversal was thus attributed to the spin-transfer torque. The current direction for switching is consistent with what we expect from the way the bands spin-split in (Ga,Mn)As due to the negative p–d exchange interaction (Ogawa et al., 1999). Parenthetically, the current direction of Chiba et al. (2004b) is the same as Elsen et al. (2006), whereas it is opposite in the report by Moriya et al. (2004). The fact that the bottom thin layer is the free layer is also consistent with the spin-transfer torque model as the total magnetic moment of the bottom layer is less than that of the thick top layer.

The observed J_C is compared with the one calculated by available phenomenological formula based on spin-transfer torque applicable to metallic CPP-GMR systems. It is not clear yet whether one should regard the present device as a CPP-GMR device or as a MTJ device, because of the combination of a small barrier (~0.1 eV) and a large bias (~1 V) required to see the switching. Under the assumption of coherent rotation of magnetization with a uniaxial anisotropy, the critical current densities for switching are calculated to be $J_P = -1.2 \times 10^6$ A/cm^2 and $J_{AP} = 3.9 \times 10^6$ A/cm^2 for our devices under $H = 0$ using the Slonczewski's formulae for CPP-GMR devices (Slonczewski, 1996). Here, 0.04 T was used for saturation magnetization (determined at 10 K), the spin polarization $P = 0.26$ was calculated from the TMR ratio of 15% at 30 K using the Julliere's formula (Julliere, 1975), and the damping constant $\alpha = 0.02$ from the reported ferromagnetic resonance data and theories (Liu et al., 2003; Matsuda et al., 2006; Sinova et al., 2004; Tserkovnyak et al., 2004). The anisotropy field was set to 0.1 T (Hamaya et al., 2004; Sawicki et al., 2005; Welp et al., 2003); a small shape anisotropy field (of the order of 0.01 T) is neglected (Hamaya et al., 2004). The calculated J_C is an order of magnitude smaller than those typically observed in metal systems despite the fact that the volume of the free layer is much greater, because of the small magnetization of (Ga,Mn)As. However, it is almost an order of magnitude greater than the observed value, that is, CIMS takes place in ferromagnetic III–V structures at lower current density than the metal counterparts even with significant spin–orbit interaction. To understand the data, one might have to consider the total angular momentum carried by the holes (Elsen et al., 2006; Levy and Fert, 2006). Incoherent processes during reversal might also be taking place, as the transition at the lower H_C is not as sharp as the higher one, suggesting the possibility of nucleation of magnetic domains during switching (Chiba et al., 2006c; Moriya et al., 2004; Ravelosona et al., 2006).

J_C is known to be a function of temperature as well as pulse duration described as $J_C = J_{C0}[1-(k_BT/KV)\ln(w_p/\tau_0)]$, where J_{C0} is J_C derived from Slonczewski's model, k_B is the Boltzmann constant, V the volume of the free layer, K the anisotropic energy constant, and τ_0 the inverse of the attempt frequency (Koch et al., 2004). For (Ga,Mn)As-based device, J_C shows almost no dependence on pulse duration probably due to sufficiently large KV comparing with the thermal energy k_BT (Chiba et al., 2006c). Because T_C of (Ga,Mn)As is much lower than that of metal ferromagnets, the Joule heating could affect the magnitude of J_C through the temperature dependence of M, K, and P, as was observed in (Ga,Mn)As/(In,Ga)As/(Ga,Mn)As MTJ (Elsen et al., 2006). Application of a bias higher than the Fermi energy strongly reduces the magnitude of the TMR ratio. This reduction suggests that P under high biases is much diminished. However, the spin transfer torque could increase as current increases, if the total spin momenta are conserved.

4. MAGNETIC DW

4.1. DW observation

Magnetic domains having uniform magnetization but with compensating directions are formed in macroscopic ferromagnets, in order to reduce magnetostatic energy. The state with zero magnetostatic energy is a demagnetized state, where net magnetization of the samples is compensated by the presence of domains. The boundary between two domains is called a magnetic DW, within which the direction of magnetic moments changes gradually over many lattice, which is a competition between exchange and anisotropy energies (Chikazumi, 1997). The overall domain structure is determined so as to reduce the total energy, including magnetostatic energy, exchange energy at DW, and other energies such as magnetocrystalline anisotropy energy and magnetoelastic energy.

In spite of the low Mn concentration of the order of 5%, (Ga,Mn)As exhibits a well defined magnetic domain structure, which is one of the evidences of the long-range nature of ferromagnetic interaction in the material. The observation has been done by using scanning microscopic techniques (Fukumura *et al.*, 2001; Shono *et al.*, 2000) and magneto-optical microscopes (Thevenard *et al.*, 2006; Welp *et al.*, 2003). Magnetic force microscope (Martin *et al.*, 1988), which has quite a high spatial resolution, is not suitable for ferromagnetic semiconductors due to the invasiveness of the magnetic tip, which usually has a much higher magnetization than the material itself. Scanning Hall probe microscope (SHPM) (Chang *et al.*, 1992) and scanning SQUID (superconducting quantum interference device) microscope (SSQM) (Vu *et al.*, 1993) have, therefore, utilized for the domain structure observation for (Ga,Mn)As. SHPM has a higher spatial resolution than SSQM but its sensitivity is lower (Fukumura *et al.*, 2001). Figure 5.4 shows the domain structure of $Ga_{0.957}Mn_{0.043}As$ with a perpendicular easy-axis taken by SHPM as a function of temperature (Shono *et al.*, 2000) and $Ga_{0.97}Mn_{0.03}As$ with an in-plane easy-axis (Welp *et al.*, 2003). Here, the perpendicular z direction of the stray field H_z from the sample was measured by a 1 μm × 1 μm Hall probe 0.5 μm above the sample surface. A stripe domain structure was clearly observed. The domain width of 1.5 μm at 9 K increased to 2.5 μm at 30 K. The stripes became less regular above 60 K. The maximum $\mu_0 H_z \sim 4$ mT decreased as the temperature was raised up to T_C of 80 K. The domain width of (Ga,Mn)As having perpendicular easy axis has been calculated by the *p–d* Zener model combined with micromagnetic theory (Dietl *et al.*, 2001b) and shown to result in reasonable agreement with the experimentally obtained width (Shono *et al.*, 2000). In contrast to the domain structure, observation of DW has so far not been reported, primarily because of its thin width. The DW width $\delta = \pi\sqrt{A_s/K_u}$ is calculated to be 10–20 nm, where A_s is the magnetic stiffness (9×10^{-14} J/m) and K_u is the magnetic anisotropy (3×10^3 J/m^3), again using the *p–d* Zener model (Dietl *et al.*, 2001b).

Snapshots of the domain structures of (Ga,Mn)As during magnetization reversal were recorded by the magneto-optical microscope (Chiba *et al.*, 2006a; Thevenard *et al.*, 2006; Welp *et al.*, 2003). The microscope images show that the reversal of (Ga,Mn)As films with in-plane easy axis

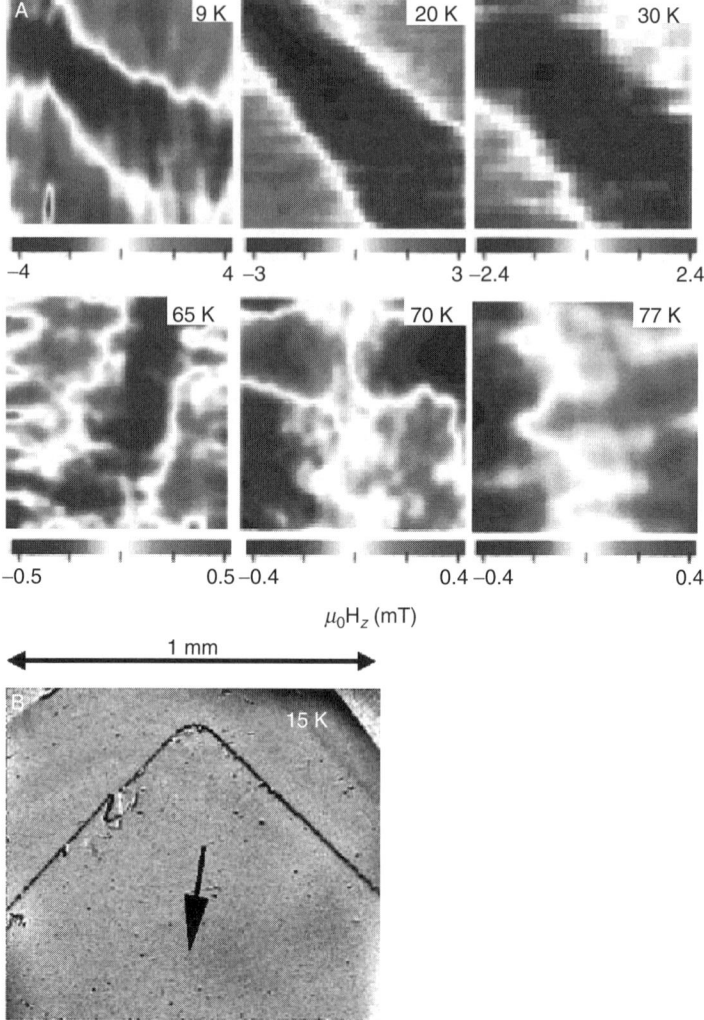

FIGURE 5.4 (A) Magnetic domain structure obtained by scanning Hall probe microscope for (Ga,Mn)As with perpendicular easy axis. The scanned area are 4.8×4.8 μm^2 for 9–30 K and 7.3×7.3 μm^2 for 63–77 K (Shono et al., 2000). (B) Magnetic domain for (Ga,Mn)As with in-plan easy axis at 15 K obtained by magneto-optical microscope (Welp et al., 2003).

proceeds with domain nucleation and subsequent expansion. They also show that the domains can be as extended as ~1 mm. These (Ga,Mn)As layers exhibit 90° or 180° domains, depending on the magnitude of the in-plane magnetic anisotropies (Welp et al., 2003).

In (Ga,Mn)As layers with perpendicular easy axis, the nucleation and expansion of domains are often perturbed by the cross-hatched surface morphology. To have a perpendicular easy axis in the relevant carrier concentration range, tensile strain needs to be introduced (Dietl et al., 2001a; Shen et al., 1997b). This is done by inserting a thick lattice-relaxed (In,Ga)As or (In,Al)As buffer layer that has a greater lattice constant than the (Ga,Mn)As layer grown on top of it. Thus, at the interface between the buffer layer and the substrate GaAs a high density of misfit dislocation is generated, which results in a cross-hatched surface morphology. When the morphology exceeds a certain roughness, DW propagation is perturbed in such a way that DWs propagate along the cross-hatched lines (Chiba et al., 2006a; Thevenard et al., 2006) and filamentary domain structure with 360° DW is formed (Chiba et al., 2006a; Thevenard et al., 2006). The in-plane magnetic anisotropies of compressive (Ga,Mn)As obtained from domain images compare well with those obtained from the magnetization measurements; they both indicate the coexistence of cubic and uniaxial magnetocrystalline anisotropies.

A combination of magneto-optical microscope and scanning technique, scanning laser magneto-optical microscope: SLMOM, has also been utilized to observe magnetic domain structure in (Ga,Mn)As (Kondo et al., 2006). By irradiating a local area by a focused circularly polarized laser light, change in the domain structure was observed, showing possible control of domain structure by an optical means.

4.2. DW resistance

DWs have effects on carrier transport. This has attracted much attention and a number of studies have been done on ferromagnetic metals (Marrows, 2005), because it reveals the interaction between carrier spins and spatially varying localized spins. In order to measure a small DW resistance (DWR), a large number of DWs was prepared in a samples, using dense domain patterns at demagnetization states (Danneau et al., 2002; Gregg et al., 1996; Marrows and Dalton, 2004; Ruediger et al., 1998; Viret et al., 1996) or geometrically confined patterns (Buntinx et al., 2005; Lepadatu and Xu, 2004; Otani et al., 1998; Taniyama et al., 1999). Either positive (Gregg et al., 1996; Taniyama et al., 1999; Viret et al., 1996) or negative (Otani et al., 1998; Ruediger et al., 1998; Tang et al., 2004) contribution of DWR has been found.

The DWR of 180° DW has been investigated systematically for (Ga,Mn)As with perpendicular easy axis. The DWR has been found to be positive (Chiba et al., 2006d). Magnetotransport and magneto-optical Kerr effect (MOKE) microscope measurements were used to determine the DW characteristics. Figures 5.5A and B show optical micrographs of a patterned device prepared for DWR measurements, and a schematic

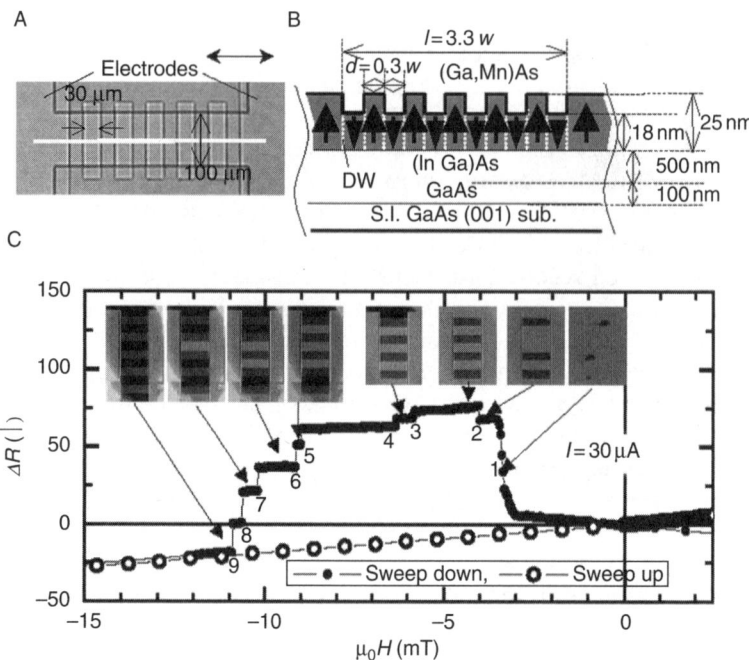

FIGURE 5.5 (A) Optical micrographs of patterned device for domain-wall (DW) resistance measurements. (B) Schematic cross-sectional view under the white line of (A). (C) Magnetoresistance of the device at 45 K. Insets show MOKE images in various magnetic fields disclosing relation between number of DWs and resistance (Chiba et al., 2006d).

cross-sectional view under the white line of (a), respectively. The stepped boundaries were prepared by surface etching, which introduce coercivity H_C difference of the etched and unetched regions (Yamanouchi et al., 2004). The DWs were prepared at each boundary by the application of external fields H by using the H_C difference. From the in-plane magnetic anisotropy energies determined by ferromagnetic-resonance spectra of similar films, the Bloch wall is calculated to be energetically stabler than the Néel wall. The width is evaluated to be 17 nm (Dietl et al., 2001b). Figure 5.5C summarizes the magnetoresistance MR and the corresponding MOKE images measured simultaneously at 45 K. The increase and the decrease of the number of DWs seen in MOKE resulted in the staircase-like resistance increase and decrease in the MR curve. In the case of perpendicular easy axis, the zig-zagging current caused by the abrupt polarity change of the Hall electric field at each DW leads to an extrinsic contribution to the apparent DWR (Partin et al., 1974; Tang and Roukes, 2004). Using the conductivity tensors obtained from separate Hall measurements, this extrinsic contribution, R_{ext}, is evaluated from the continuity equation and obtained $R_{\text{ext.}} = 6.3\,\Omega$ per DW, by employing a standard finite element method. This is in good agreement with the experimental value of the average DWR per DW of $+6.9\,\Omega$, indicating that the apparent DWR is dominated by this extrinsic effect.

In order to separate the intrinsic effects proportional to the DW area from the extrinsic contribution arising from the polarity change of the Hall electric field across DW, which depends on the length l to width w ratio of the device, a series of structures with varying the DW area but keeping the l-to-w ratio were designed and fabricated. Measured DWR at 55 K as a function of w^{-1} clearly shows that DWR can be decomposed into a contribution independent of w ($R_{\text{ext.}}$) and linear in w^{-1} (the intrinsic DWR, $R_{\text{int.}}A_d \sim 0.5\,\Omega\mu m^2$). The anisotropic magnetoresistance (AMR) contribution (Miyake et al., 2002) was evaluated and shown to be negligible under the experimental conditions. The intrinsic contribution is in good agreement with the computed value from the theory with a mistracking effect (Levy and Zhang, 1997). The finite-element calculations reproduce quantitatively the magnitude and temperature dependence of $R_{\text{ext.}}$ in this series of the structure. On the other hand, a negative DWR has been reported for (Ga,Mn)As with in-plane easy axis and 90° DW (Tang et al., 2004).

The first model of DW scattering was suggested by Cabrera and Falicov (1974) who considered the reflection of electrons by an effective potential created by spatially varying magnetization within the DW. Because the reflection probability depends on the ratio of the DW width to the Fermi wavelength and is exponentially small for large ratios, the DWR is small. A model based on a mixing of spin-channels by carrier spin-mistracking has been put forward by Gregg (1996), Viret (1996), and developed further by Levy and Zhang (1997). Here, the mixing is caused by a small nonadiabaticity of carrier spins and spin-dependent carrier scattering due to the disorder in DW, which increases higher-resistivity spin channel contribution and produces a positive DWR. This model appears to explain the experimental results on (Ga,Mn)As as well as DWR measurements on ferromagnetic metals. A negative DWR due to the destruction of quantum localization by DW was put forward by Tatara and Fukuyama (1997). It is possible that the two experiments done on (Ga,Mn)As differ in disorder and measurement temperature so that one is probing the positive DWR regime and the other probing the negative DWR regime. Temperature dependence of DWR spanning a wider range of temperature together with detailed measurements and analysis to separate the intrinsic DWR from DWR caused by AMR and other extrinsic mechanisms are required to probe the nature of DWR of (Ga,Mn)As and related ferromagnetic semiconductors.

One of the effects often neglected but could play a critical role, in particular in p-type ferromagnetic semiconductors, is the spin–orbit interaction.

According to the calculations under disorder-free condition, the spin–orbit interaction leads to nonzero value of intrinsic contribution on DWR even in the adiabatic limit (Nguyen et al., 2007a; Oszwałdowski et al., 2007). They are one order or more smaller than the value observed experimentally. It is possible that disorder plays a major role in the physics of intrinsic DWR in (Ga,Mn)As.

4.3. DW motion

Electrical manipulation of magnetization direction is one of the key technologies for realizing high-density magnetic memories. Interaction between DW and spin-polarized current results in current-induced DW motion. A nonvolatile serial memory based on current-induced DW motion has been proposed (Parkin, 2004). However, the physics current-induced DW motion has not been fully understood. Ferromagnetic semiconductor, (Ga,Mn)As, is different from ferromagnetic metals in a number of ways: for example, it has a small magnetization (1/100 of Fe) while maintaining high spin polarization of carriers; magnetic interaction is due to p-electrons rather than d-electrons making sd models more appropriate. With these distinct features, (Ga,Mn)As provides a unique set of experimental data revealing the physics of interaction between localized spins and conduction carrier spins that cannot easily be accessed using ferromagnetic metals.

The possibility of current-induced DW motion was first pointed out by Berger as early as the 1970s (Berger, 1978, 1984, 1992) and subsequently a number of experimental investigations were done (Freitas and Berger, 1985; Gan et al., 2000; Hung and Berger, 1988). In a macroscopic sample, however, a high current required to observe the effect generated a high Oersted field at the same time, preventing a straightforward interpretation of the experimental results. The recent advances of nanotechnology has made it possible to fabricate devices small enough so that the Oersted fields generated by current are no longer a concern. In recent years, a growing number of experiments mostly on NiFe nanowires showed that the DW moves by an application of current pulse with current density of $10^7 \sim 10^8$ A/cm^2 in the absence of magnetic fields (Kläui et al., 2005; Thomas et al., 2006; Vernier et al., 2004; Yamaguchi et al., 2004). At about the same time, the DW position in (Ga,Mn)As was electrically been manipulated in the absence of magnetic field by the application of a current pulse across the DW in a device that has an order of magnitude greater cross-section (Yamanouchi et al., 2004). This is because (Ga,Mn)As requires 10^5 A/cm^2 current density to observe the effect.

The (Ga,Mn)As device used in the experiment had a 20-µm wide electrical channel. The easy axis of the (Ga,Mn)As channel layer was made perpendicular to the sample surface by inserting an (In,Ga)As or

(In,Al)As buffer layer (Shen et al., 1997b). This easy axis direction was chosen to monitor the DW position through the anomalous Hall effect and by the MOKE microscope. The DW switching between two positions was observed for the channel with three regions with different thickness (25, 22, and ~18 nm), where the thinnest region was set to the center of the channel. This stepped structure allows patterning of coercivity H_C due to slight nonuniformity along growth direction in the film (Chiba et al., 2003a; Koeder et al., 2003). Each step acts as a confinement potential for DW as the DW energy increases with its area. The DW position was initialized to the one of the stepped boundaries of the thinnest region by H. After setting $H = 0$, a current pulse of 10^5 A/cm² for 100 ms at 80 K (T_C of this film was 90 K) was applied. Both the anomalous Hall signals and MOKE images indicated that the DW moved from one side of the step boundary to another side, in the direction opposite to the current direction. The application of a subsequent current pulse in the opposite direction switched back the DW to its initial position as shown in Fig. 5.6. In this experiment, the increase of the temperature was about 0.4 K by the application of the current of 10^5 A/cm².

The temperature dependence as well as the current density j dependence of the DW velocity in (Ga,Mn)As was measured systematically using a 5-μm wide channel with a single step (Yamanouchi et al., 2006). The thickness of the thick and thin region was 30 and 20 nm, respectively. A DW was prepared at the step boundary using the coercivity difference of the two regions by sweeping an external magnetic field. Then current pulses were applied. The position of the DW was monitored by MOKE. The DW speed, v_{eff}, was calculated from the average distance DW traveled for a given pulse width. The device temperature was monitored by measuring the device resistance during pulse and by comparing it with the temperature dependent device resistance measured separately.

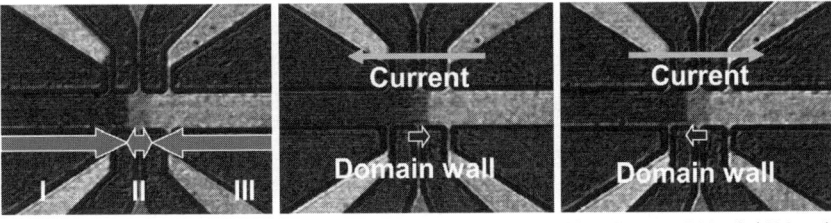

FIGURE 5.6 Current induced DW motion in (Ga,Mn)As in stepped structure with central thinnest region II observed by MOKE. A DW prepared at the boundary of regions I and II (left) can be moved to the opposite boundary regions of II and III by a current pulse of −300 μA (middle) and be moved back to its original position by the current with the opposite direction (left) (Yamanouchi et al., 2004).

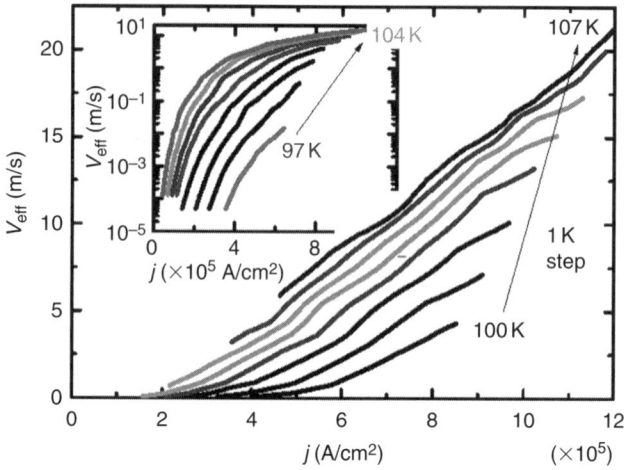

FIGURE 5.7 DW velocity as a function of current density at various device temperatures. The inset presents the same data but in logarithm plot to show the DW activity in sub-threshold regime (Yamanouchi et al., 2006a).

The maximum temperature rise was $\sim$16 K at the highest current density of 1.2×10^6 A/cm^2. The j dependence of v_{eff} at a fixed temperature showed that there are two distinct regimes separated by a critical current density j_C, which is a few 10^5 A/cm^2, as shown in Fig. 5.7. Above j_C, v_{eff} increases linearly with j, while below j_C the DW velocity is low and its functional form is more complex. The direction of the motion, linear dependence of v_{eff} on j above j_C, and the slope A were found to be in quantitative agreement with what is expected from spin-transfer. The slope A given by $A = Pg\mu_B/(2eM)$, where P is the current polarization, g the Lande g-factor, μ_B Bohr magneton, and e the elementary charge, is within a factor of 2 of the experimentally observed slope. A theoretical study has indicated that spin–orbit interaction in (Ga,Mn)As results in spin accumulation and nonadiabacity inside the DW, which can make the DW mobility high when transport is in the ballistic regime (Nguyen et al., 2007b). We cannot expect ballistic transport in (Ga,Mn)As, as the mean-free path is of the order of nm. The intrinsic threshold current density, $j_C^i = 2eK\delta_W/(\pi\hbar P)$, proposed by Tatara and Kohno (2004) explains the magnitude and the temperature dependence determined from the experiment very well. Here, K is the hard-axis anisotropy and is given by $M^2/(2\mu_0)$. The good agreement of the threshold current density alone, however, cannot settle the discussion on the origin of j_C. The extrinsic threshold current density due to pinning proposed by Barnes and Maekawa (2005) can also show similar temperature dependence as it is related to the anisotropy (Barnes, 2007, private communication). Parenthetically, no current-induced

DW has been observed in (Ga,Mn)As layers having in-plane easy axis (Tang et al., 2006; Yamanouchi, 2006, private communication). Although not conclusive, the anisotropy involved in the in-plane easy axis samples is crystalline anisotropy and is 10 times greater than the one relevant in the perpendicular easy-axis ones under ordinary strain and carrier concentration conditions. It is, therefore, tempting to explain the contrast between the two easy-axis directions by the difference in anisotropies.

For ferromagnetic metals, this intrinsic j_C is calculated to be too high to explain experiments, although the calculation is not simple as in most cases a number of DW structures are involved. In order to reconcile the discrepancy, various mechanisms for a field-like term (often called the β-term) have been proposed (Tatara and Kohno, 2004; Thiaville et al., 2005; Zhang and Li, 2004). The field like term is proportional to j but acts on the DW as an effective field, exerting torque in perpendicular direction to the one from spin-transfer. Tatara and Kohno predicted that an intrinsic DWR is a source of such a term (Tatara and Kohno, 2004), supplying linear momentum to the DW by carrier reflection. So far, the observed DWR in (Ga,Mn)As is too large to explain the observed velocity (Chiba et al., 2007) by this mechanism.

An important finding in the current-induced DW motion in (Ga,Mn)As is the DW motion below j_C (the inset of Fig. 5.7). The j dependence of v_{eff} below j_C obeys an empirical scaling law, showing the existence of current-induced DW creep (Lemerle et al., 1998). In this regime, competition between the driving force and disorder results in a slow activated motion of the DW, which could be used to address the presence or absence of the field-like term in this material. If the field-like term is present, it should be possible to observe a creep phenomenon similar to the one induced by an external magnetic field. Comparison between current-induced and field-induced cases has shown that they obey similar scaling laws but the scaling exponents of the two are incompatible, indicating that current-driven creep is fundamentally different from field-driven creep (Yamanouchi et al., 2007). A model based on spin-transfer torque has been put forward to explain the current-induced creep, which explains the observed critical exponent well. Thus, in the case of (Ga,Mn)As, current-induced DW motion appears to be able to be understood by the spin-transfer effect alone in the entire range of velocities so far accessed by experiment. It is in contrast with the study of the effect of current on the field-induced DW motion in permalloy, which suggested that the field-like term cannot be neglected (Hayashi et al., 2006). It is perhaps worth noting that the critical exponent observed in the field-driven creep is different from what one observed in metallic ferromagnets (Lemerle et al., 1998) but rather similar to the ones reported in ferroelectric creep (Tybell et al., 2002). This is tentatively attributed to the difference in the length-scales of DW width δ and the pinning potential l; in ferromagnets

$\delta \ll l$, whereas in (Ga,Mn)As $\delta \gg l$. Spin-transfer torque assisted thermal activation process below j_C with rigid wall approximation (Duine et al., 2007; Tatara et al., 2005) has also been proposed; the predicted temperature and/or current density dependence of velocity appears to be different from the experimental findings.

5. ELECTRIC-FIELD CONTROL OF FERROMAGNETISM

Because ferromagnetism of (Ga,Mn)As and (In,Mn)As is induced by the presence of holes (Dietl et al., 2000, 2001a; König et al., 2000), the magnetic properties of these materials show strong dependence on the hole concentration p. The control of Curie temperature T_C and coercivity H_C by applying external electric fields E to modulate p has been reported in several ferromagnetic semiconductors (Boukari et al., 2002; Chiba et al., 2003b, 2006a; Nazmul et al., 2004; Ohno et al., 2000; Park et al., 2002). This unique technique can add a new dimension to the future magnetism usage, because the properties of magnetic materials can be altered after they have been prepared and put into use. The first electrical control of T_C was reported for a 5-nm thick (In,Mn)As channel in a FET structure, using spin-coated SiO_2 (dielectric constant $\kappa = 3$–4) as an insulator, where application of $E = \pm 1.5$ MV/cm resulted in a change of T_C by 2 K (Ohno et al., 2000). Electrical control of ferromagnetism in GaAs-based structure has been shown in Mn δ-doped GaAs/AlGaAs heterostructures (Nazmul et al., 2004). It was, however, difficult to unambiguously observe the modulation of magnetism in (Ga,Mn)As thin films (Chiba et al., 2003c) for some time. The (Ga,Mn)As layers used in earlier studies were thick compared to the (In,Mn)As layer used for the (In,Mn)As experiment, because (Ga,Mn)As thin films show a tendency to become insulating at a thickness below 5 nm. The maximum amount of hole modulation Δp is proportional to κ, E, and the inverse of the channel thickness t as $\Delta p = \kappa \varepsilon_0 E / (et)$, a thick layer does not allow a large modulation ratio $\Delta p / p$. Employing Al_2O_3 with a high-κ and a high dielectric breakdown field E_{BD}, a clear T_C and H_C modulation in a (Ga,Mn)As film has been observed (Chiba et al., 2006a).

The properties of FETs with (In,Mn)As channels and polyimide insulators are first described in the following. The ferromagnetic channel layers were grown by molecular beam epitaxy at low temperature (~250 °C) on 5 nm InAs/200–400 nm $Al_{0.8}Ga_{0.2}Sb$/50–100 nm AlSb from the surface side onto a semi-insulating GaAs (001) substrate. The samples were processed into FET devices with a Hall-bar geometry. The channel layer was covered with a 0.8–1.0 µm thick spun-on polyimide gate insulator and then by a Cr/Au metal gate electrode. Hall resistance R_{Hall} is proportional to M at low T and low $\mu_0 H$, due to the anomalous Hall effect.

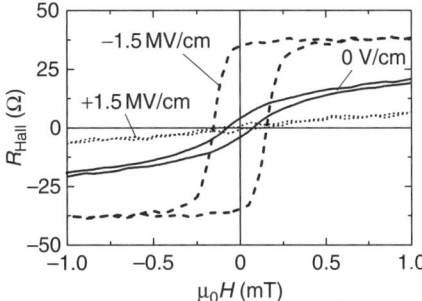

FIGURE 5.8 Hall resistance versus magnetic field under three different gate biases for a FET with a (In,Mn)As channel layer measured at 22.5 K (Ohno et al., 2000).

In the vicinity of T_C, the magnetization curves (in this case $R_{Hall}-\mu_0 H$ curve) show more square shape under negative E, which is the direction for the accumulation of holes indicating enhanced ferromagnetic order, while paramagnetic-like response under positive E, as shown in Fig. 5.8. In typical samples, the application of E of ± 1.5 MV/cm results in $+1.0 - 2.0$ K change of T_C, which was determined by the Arrott-plots. The applied E modulates ± 3–4% of sheet resistance R_{sheet}, where corresponding change in the sheet hole concentration Δp_t is calculated from the gate capacitance to be $\mp 2.7 \times 10^{12}$ cm^{-2}. Under the assumption of constant mobility, p can be determined from $R_{sheet}^{-1}-E$ curve. T_C for samples under different E as a function of the obtained p is in reasonable agreement with the calculated results by the p–d Zener model (Dietl et al., 2000, 2001a).

Employing a gate-insulator with high-κ and high E_{BD} is important to achieve a large Δp. As mentioned earlier, Al$_2$O$_3$ is one of such high-k insulators. Atomic layer deposition (ALD) is a powerful method to grow those insulators at low temperature (<200 °C) without damaging the low-temperature grown ferromagnetic semiconductors, where thermal treatment has significant effects on their magnetic and electrical properties. Using Al$_2$O$_3$ deposited by ALD, a clear T_C modulation in (Ga,Mn)As has been observed (Chiba et al., 2006a). A 50-nm-thick Al$_2$O$_3$ gate-insulator was deposited at 100 °C by using ALD on Hall-bar shaped 7-nm Ga$_{0.953}$Mn$_{0.047}$As channel layer on a 7 nm GaAs/30 nm Al$_{0.80}$Ga$_{0.20}$As/500 nm In$_{0.13}$Ga$_{0.87}$As/100 nm GaAs buffer layer structure, which was grown on a semi-insulating GaAs (001) substrate by molecular beam epitaxy. The (In,Ga)As buffer layer introduces tensile strain in the (Ga,Mn)As layer, making the magnetic easy-axis perpendicular to the plane. The dielectric constant κ and the breakdown field of the Al$_2$O$_3$ were determined to be 6.7 and above 8 MV/cm, respectively, from separate measurements on square-shaped 100 nm Au/50 nm Al$_2$O$_3$/100 nm

Au capacitors having various areas (2.5×10^{-3} mm^2–4.0×10^{-2} mm^2). Electric fields of $|E| = 5$ MV/cm resulted in a gate leakage current of less than 70 nA/cm^2. T_C at $E = +5$ (-5) MV/cm was 65 (70) K, which was 3.5 (1.5) K lower (higher) than that at $E = 0$. The application of $E = +5$ (-5) MV/cm modulates 43.5 (-12.3)% of R_{sheet} at 70 K (near T_C). From the gate capacitance, $|E|=5$ MV/cm is estimated to produce a sheet hole concentration change Δpt of 1.8×10^{13} cm^{-2}. The curve calculated by the p–d Zener model is again in reasonable agreement with the observed magnitude and modulation of T_C. One also notices a deviation from the p–d Zener model when $x > 0.08$, where the magnitudes of T_C and ΔT_C are smaller than those expected from the model. Although the origin is not understood, for most of the devices, a linear relationship was found between $\Delta T_C/T_C$ and $\Delta p/p$. The proportionality constant is about 0.2, different from 1/3 expected from a simple parabolic band.

The application of electric fields at low temperatures has a significant effect on $\mu_0 H_C$. For instance, $\mu_0 H_C$ can be modified by a factor of 5 at 40 K, from 1 mT at -1.5 MV/cm to 0.2 mT at $+1.5$ MV/cm in 5 nm (In,Mn)As, while virtually keeping the square shape of hysteresis loops unchanged (Chiba et al., 2003b, 2004c, 2006b). To approach the mechanisms governing the E dependence of $\mu_0 H_C$, we measured the T dependence of $\mu_0 H_C$ under different E. $\mu_0 H_C$ increases monotonically as temperature is reduced below T_C. When an effective transition temperature T_C^* is defined as the temperature at which $\mu_0 H_C$ becomes zero, all the measured $\mu_0 H_C$ are found to collapse onto a single curve by plotting them against the reduced temperature, T/T_C^*. The presence of such a scaling curve indicates that $\mu_0 H_C$ is determined by T/T_C^* alone and other factors such as carrier concentration do not play a critical role except through T_C^*. The scaling curve turned out to be proportional to $\{1-(T/T_C^*)\}^\eta$ with a scaling exponent η ($\eta = 2.1$ from $1-T/T_C^* = 0.05-0.2$, and $\eta = 1.7$ above $1-T/T_C^* = 0.4$). To our knowledge no theory is available that is capable of explaining the observed temperature dependence and the scaling shown here. Meanwhile, the MOKE images for both 5 nm (In,Mn)As and 7 nm (Ga,Mn)As films indicate that the nucleation field $\mu_0 H_n$ of domains determines $\mu_0 H_C$ of such films in measurement time scale (sweeping rate of $\mu_0 H$) (Chiba et al., 2006a,b). This reveals that the modulation of $\mu_0 H_C$ by E is a result of the modulation of $\mu_0 H_n$, although further studies are clearly necessary to address the mechanism of this modulation as well as scaling of $\mu_0 H_C$.

Since the magnitude of $\mu_0 H_C$ is a function of E, the magnetization reversal can be triggered by switching of E (Chiba et al., 2003b). Figure 5.9 shows that this is indeed possible, where the results for sample with 4 nm (In,Mn)As channel are presented. We first saturated the magnetization of the (In,Mn)As channel under $E = -1.5$ MV/cm in positive magnetic field, and then reduced the field through 0 mT to a small negative bias magnetic field of $\mu_0 H_0 = -0.2$ mT. This is the initial state

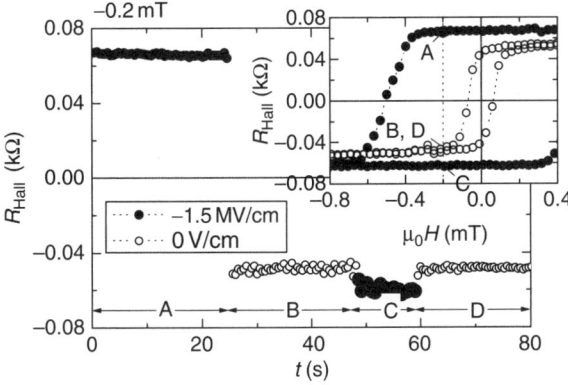

FIGURE 5.9 Time evolution of Hall resistance resulting from a sequence of applied electric fields in a FET with a (In,Mn)As channel layer, showing an electrically assisted magnetization reversal. Inset shows the hysteresis curves under two different electric fields, where the state labeled A–D correspond to the states in main panel (Chiba et al., 2003b).

indicated by point A in the inset of Fig. 5.9, where two $R_{Hall} - \mu_0 H$ curves under $E = 0$ and -1.5 MV/cm are shown. R_{Hall} as a function of time t is displayed in the main panel of Fig. 5.9. The state remains at point A until we switches E to 0 at $t = 25$ s because the magnitude of $\mu_0 H_0 = -0.2$ mT is smaller than the coercivity $\mu_0 H_C = -0.5$ mT under $E = -1.5$ MV/cm. In response to the switching of E, the sign of R_{Hall} changes from positive to negative, showing that the electrical switching brings the state to point B of the inset because $|\mu_0 H_C| < |\mu_0 H_0|$ at $E = 0$ V/cm. This demonstrates that electrical switching can trigger the magnetization reversal. Once magnetization is reversed, the sign of R_{Hall} and thus M remains negative and shows only a small variation when E is switched back and forth between 0 and -1.5 MV/cm, as shown by regions C and D of Fig. 5.9; corresponding to points C and D in the inset, respectively. This electrically assisted magnetization reversal without changing applied magnetic fields or temperature demonstrates the possibility of an electrical Curie point writing, where magnetization reversal is assisted by making the system closer to (Nazmul et al., 2005) or beyond its Curie temperature.

6. OPTICAL CONTROL OF FERROMAGNETISM

Because carriers mediate ferromagnetic interaction, photo-generated carriers change their magnetic properties and thus optical control of ferromagnetism in these materials is possible. Light illumination onto a 12-nm thick $In_{0.06}Mn_{0.94}As$ layer at 5 K was found to turn the magnetic state from

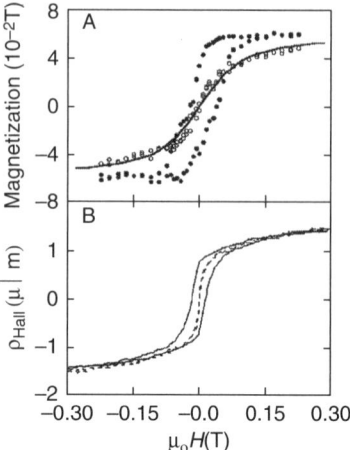

FIGURE 5.10 (A) Magnetization curves at 5 K of (In,Mn)As observed before (open circles) and after (closed circles) right irradiation. Solid line shows a curve obtained from the Brillouin function. (B) Hall resistivity at 5 K before (dashed line) and after (solid line) light irradiation (Koshihara et al., 1997).

no coercivity to a state with magnetic hysteresis, which was probed by both magnetization and magnetotransport measurements as shown in Fig. 5.10 (Koshihara et al., 1997). The photon energy required to see the change of the magnetic state was found to be greater than 0.8 eV, where persistent photoconductivity was also observed. The ferromagnetic state persisted even after switching the light off until the sample temperature was raised above 45 K (Munekata et al., 2002). These results show that the internal electric-field at the (In,Mn)As/GaSb interface separate the photo-generated electron-hole pairs spatially, and holes accumulate in the (In,Mn)As layer and change the magnetic interaction. The high-field Hall measurement (up to 15 T) showed that the hole concentration was 3.76×10^{19} cm^{-3} before and 3.90×10^{19} cm^{-3} after light illumination. Quantitative understanding of the change of hole concentration (1.4×10^{18} cm^{-3}) and the magnitude of the change in magnetic properties remains an open issue. Similar photo-induced effects, enhancement of magnetization in ferromagnetic state and its persistence, were observed in In$_{0.95}$Mn$_{0.05}$As$_{0.8}$Sb$_{0.2}$/InSb, where lower threshold photon energy ~0.6 eV is observed (Kanamura et al., 2002).

Photo-induced change of H_C was observed in a ferromagnetic 30 nm In$_{0.93}$Mn$_{0.07}$As/GaSb layer, whose T_C was 40 K (Oiwa et al., 2001). The laser diode light was irradiated for 30 min, and the magnetization curve was recorded in the dark using the anomalous Hall effect. A clear decrease of $\mu_0 H_C$ was observed after irradiating the sample. The change of H_C was about 45% at 35 K and decreased with temperature becoming

2% at 4.2 K. The original H_C before irradiation was recovered after raising the sample temperature above 180 K, at which persistent photoconductivity also vanished. The effect was observed only for the sample with square hysteresis and not for the sample with soft hysteresis, indicating the effect depends on the way the magnetization reversal takes place. Here, no significant change of magnetic anisotropy constant was observed, whereas other group observed a change of the magnetic easy axis of (In,Mn)As by incident light (Liu et al., 2004). The number of holes accumulated in the (In,Mn)As layer during the light irradiation was estimated to be about 3% of the background hole concentration (Munekata et al., 2002). Again, further study is necessary to clarify why a small change of hole concentration leads to a dramatic effects on H_C.

Magnetization enhancement of (Ga,Mn)As by circularly polarized light illumination has been observed for (Ga,Mn)As (Oiwa et al., 2002). A 200 nm $Ga_{0.89}Mn_{0.11}As$ layer with in-plane magnetic easy axis was illuminated by a 600 mW/cm^2 circularly polarized light beam normal to the sample plane to inject spin-polarized carriers. A decrease in sheet resistance was observed under illumination, which did not depend on the polarity of the light. On the other hand, the Hall resistance R_{Hall} at zero magnetic field, dominated by the anomalous Hall effect, showed a sign change depending on the polarity of the light; the change corresponds to about 15% of the saturation magnetization. After irradiation, a small residual component of R_{Hall} remained (less than 0.1% of the saturation magnetization) showing a change in domain structure in which a small perpendicular component was included. The domain structure of (Ga,Mn)As was later observed by using scanning laser magneto-optical microscope (Kondo et al., 2006). The light polarization dependence of R_{Hall} was observed also in Mn δ-doped GaAs, where the enhancement (reduction) of R_{Hall} under right (left) circularly polarization was observed in its magnetic field dependence and linearly polarized light had no effect on it (Nazmul et al., 2004).

The dynamics of photo-induced effect was probed by time-resolved Kerr-rotation (TRKR) (Mitsumori et al., 2004). The observed TRKR signals showed the sign change upon changing the polarization of light. Its decay and temporal change in reflectivity were different above and below T_C. Above T_C, a single exponential decay of ~25 ps was observed, which was attributed to the trapping process of photo-generated carriers. Below T_C, the two decay processes were superimposed; the fast decay was attributed to the same trapping observed above T_C and the slow decay to the one with magnetic origin reflecting the magnetization dynamics. The transients were interpreted as an instantaneous magnetization rotation (tilting away from its original direction) with the generation of photo-carriers followed by its decay, suggesting a possibility of ultrafast control of the magnetization of ferromagnetic semiconductors by photo-carrier injection.

A softening of the hysteresis loop (Wang et al., 2004), quenching of ferromagnetism (Wang et al., 2005) and enhancement of ferromagnetism (Wang et al., 2007) have also been reported. These effects were explained by the flow of spins between carriers and Mn, and rapid distribution of photogenerated holes in spin-split valence bands. In the time-resolved signals for (Ga,Mn)As below T_C, a slow demagnetization component was also observed (Kojima et al., 2003; Wang et al., 2007), which was interpreted as an effect due to thermal isolation between the charge and spin systems due to the half-metallic nature of (Ga,Mn)As (Kojima et al., 2003). The coherent Mn spin precession was observed in the time-resolved Kerr rotation signal when (Ga,Mn)As was pumped by linear polarized light at zero magnetic field; it was shown that the precession can be analyzed by the Landau–Lifshtiz–Gilbert equation (Qi et al., 2007).

Optical control of ferromagnetism by light irradiation was also observed in II–VI Mn-doped semiconductors, indicating that the effects are not limited to III–V ferromagnetic semiconductors. By designing the structure, it has was shown that one could either enhance or destroy ferromagnetism in (Cd,Mn)Te quantum well (QW) with (Cd,Zn,Mg)Te barriers (Boukari et al., 2002). Illumination of light enhances the ferromagnetism in QW in an undoped region of a *p-i-n* structure, whereas illumination destroys that in a *p-i-p* structure. In *p-i-n* structure, photogenerated holes, separated by internal electric-field from electrons, migrate toward the QW to enhance the ferromagnetism. By contrast in the *p-i-p* structure, the photo-generated electrons are collected by the QW due to band bending, where they diminish hole-induced ferromagnetism.

7. SUMMARY

Ferromagnetic semiconductors show a variety of spintronic properties, and most of which have not been accessible in conventional semiconductor devices and magnetic devices based on ferromagnetic metals. The devices based on ferromagnetic semiconductors can deal with semiconducting and magnetic functionalities simultaneously, and thus are being used to explore their potential application as well as physics involved in the device application. Since the operating temperature above room temperature is crucial for the application, a large effort is continuing to established ferromagnetic semiconductors with the Curie temperature well above room temperature. Our knowledge of the spintronic properties will be useful when appropriate ferromagnetic semiconductors in usage are once discovered. A number of new experimental findings have emerged and new ideas have embodied in the device operation, however, they are still in the elementary level from the technological point of view. In order to advance the area, we should clarify the architecture to realize the systems for special and even general purposes by making the most of the advantage of the devices based on ferromagnetic semiconductors.

ACKNOWLEDGMENTS

The authors thank T. Dietl, M. Sawicki, Y. Ohno, K. Ohtani, Y. Nishitani, M. Endo, M. Kohda, and T. Kita for their collaboration and useful discussions.

REFERENCES

Akiba, N., Matsukura, F., Shen, A., Ohno, Y., Ohno, H., Oiwa, A., Katsumoto, S., and Iye, Y. (1998). *Appl. Phys. Lett.* **73**, 2122.
Albert, F. J., Katine, J. A., Buhrman, R. A., and Ralph, D. C. (2000). *Appl. Phys. Lett.* **77**, 3809.
Allwood, D. A., Xiong, G., Cooke, M. D., Faulkner, C. C., Atkinson, D., Vernier, N., and Cowburn, R. P. (2002). *Science* **296**, 2003.
Alvarado, S. F., and Renaud, P. (1992). *Phys. Rev. Lett.* **68**, 1387.
Baibich, M. N., Broto, J. M., Fert, A., Nguyen Van Dau, F., Petroff, F., Etienne, P., Creuzet, G., Friederich, A., and Chazelas, J. (1988). *Phys. Rev. Lett.* **61**, 2472.
Barnes, S. E., and Maekawa, S. (2005). *Phys. Rev. Lett.* **95**, 107204.
Berger, L. (1978). *J. Appl. Phys.* **49**, 2156.
Berger, L. (1984). *J. Appl. Phys.* **55**, 1954.
Berger, L. (1992). *J. Appl. Phys.* **71**, 2721.
Berger, L. (1996). *Phys. Rev. B* **54**, 9353.
Boukari, H., Kossacki, P., Bertolini, M., Ferrand, D., Cibert, J., Tatarenko, S., Wasiela, A., Gaj, J. A., and Dietl, T. (2002). *Phys. Rev. Lett.* **88**, 207204.
Braden, J. G., Parker, J. S., Xiong, P., Chun, S. H., and Samarth, N. (2003). *Phys. Rev. Lett.* **91**, 056602.
Buntinx, D., Brems, S., Volodin, A., Temst, K., and Haesendonck, C. V. (2005). *Phys. Rev. Lett.* **94**, 017204.
Cabrera, G. G., and Falicov, L. M. (1974). *Phys. Stat. Sol. (b)* **61**, 59.
Chang, A. M., Hallen, H. D., Harriot, L., Hess, H. F., Loa, H. L., Kao, J., Miller, R. E., and Chang, T. Y. (1992). *Appl. Phys. Lett.* **61**, 1974.
Chen, P., Moser, J., Kotissek, P., Sadowski, J., Zenger, M., Weiss, D., and Wegschneider, W. (2006). *Phys. Rev. B* **74**, 241302(R).
Chiba, D., Akiba, N., Matsukura, F., and Ohno, H. (2000). *Appl. Phys. Lett.* **77**, 1873.
Chiba, D., Takamura, K., Matsukura, F., and Ohno, H. (2003a). *Appl. Phys. Lett.* **82**, 3020.
Chiba, D., Yamanouchi, M., Matsukura, F., and Ohno, H. (2003b). *Science* **301**, 943.
Chiba, D., Yamanouchi, M., Matsukura, F., Abe, E., Ohno, Y., Ohtani, K., and Ohno, H. (2003c). *J. Supercond.* **16**, 179.
Chiba, D., Matsukura, F., and Ohno, H. (2004a). *Phys. E* **21**, 966.
Chiba, D., Sato, Y., Kita, T., Matsukura, F., and Ohno, H. (2004b). *Phys. Rev. Lett.* **93**, 216602.
Chiba, D., Yamanouchi, M., Matsukura, F., and Ohno, H. (2004c). *J. Phys. Condens. Matter* **16**, S5693.
Chiba, D., Matsukura, F., and Ohno, H. (2006a). *Appl. Phys. Lett.* **89**, 162505.
Chiba, D., Matsukura, F., and Ohno, H. (2006b). *J. Phys. D* **39**, R215.
Chiba, D., Kita, T., Matsukura, F., and Ohno, H. (2006c). *J. Appl. Phys.* **99**, 08G514.
Chiba, D., Yamanouchi, M., Matsukura, F., Dietl, T., and Ohno, H. (2006d). *Phys. Rev. Lett.* **96**, 096602.

Chiba, D., Yamanouchi, M., Matsukura, F., Dietl, T., and Ohno, H. (2007). *J. Magn. Magn. Mater.* **310**, 2078.
Chikazumi, S. (1997). "Physics of Ferromagnetism," 2nd edition, Oxford University Press, Oxford.
Chun, S. H., Potashnik, S. J., Ku, K. C., Schiffer, P., and Samarth, N. (2002). *Phys. Rev. B* **66**, 100408.
Crooker, S. A., Furis, M., Lou, X., Adelmann, C., Smith, D. L., Palmstrøm, C. J., and Crowell, P. A. (2005). *Science* **309**, 2191.
Datta, S., and Das, B. (1990). *Appl. Phys. Lett.* **56**, 665.
Danneau, R., Warin, P., Attane, J. P., Petej, I., Beigné, C., Fermon, C., Klein, O., Marty, A., Ott, F., Samson, Y., and Viret, M. (2002). *Phys. Rev. Lett.* **88**, 157201.
Dieny, B., Speriosu, V. S., Parkin, S. S. P., Gurney, B. A., Wilhoit, D. R., and Mauri, D. (1991). *Phys. Rev. B* **43**, 1297.
Dietl, T., and Ohno, H. (2003). *MRS Bulltein* **October**, 714.
Dietl, T., Ohno, H., Matsukura, F., Cibert, J., and Ferrand, D. (2000). *Science* **287**, 1019.
Dietl, T., Matsukura, F., and Ohno, H. (2001a). *Phys. Rev. B* **63**, 195205.
Dietl, T., König, J., and MacDonald, A. H. (2001b). *Phys. Rev. B* **64**, 241201(R).
Duine, R. A., Núñez, A. S., and MacDonald, A. H. (2007). *Phys. Rev. Lett.* **98**, 056605.
Elsen, M., Boulle, O., George, J.-M., Jaffrès, H., Mattana, R., Cros, V., Fert, A., Lemaitre, A., Giraud, R., and Faini, G. (2006). *Phys. Rev. B* **73**, 035303.
Elsen, M., Jaffrès, H., Mattana, R., Tran, M., George, J.-M., Miard, A., and Lemaître, A. (2007). *Phys. Rev. Lett.* **99**, 127203.
Esaki, L., Stiles, P. J., and von Molnar, S. (1967). *Phys. Rev. Lett.* **19**, 852.
Fert, A., and Jaffrès, H. (2001). *Phys. Rev. B* **64**, 184420.
Filderling, R., Keim, M., Reuscher, G., Ossau, W., Schmidt, G., Waag, A., and Molenkamp, L. W. (1999). *Nature* **402**, 787.
Freitas, P. P., and Berger, L. (1985). *J. Appl. Phys.* **57**, 1266.
Fuchs, G. D., Emley, N. C., Krivorotov, I. N., Braganca, P. M., Ryan, E. M., Kiselev, S. I., Sankey, J. C., Ralph, D. C., Buhrman, R. A., and Katine, J. A. (2004). *Appl. Phys. Lett.* **85**, 1205.
Fukumura, T., Shono, T., Inaba, K., Hasegawa, T., Koinuma, H., Matsukura, F., and Ohno, H. (2001). *Phys. E* **10**, 135.
Furdyna, J. K., and Kossut, J., eds. (1988). Semiconductors and semimetals, In "Diluted Magnetic Semiconductors," Vol. 25, Academic Press, San Diego.
Gan, L., Chung, S. H., Aschenbach, K. H., Dreyer, M., and Gomez, R. D. (2000). *IEEE Trans. Magn.* **36**, 3047.
Gould, C., Rüster, C., Jungwirth, T., Girgis, E., Schott, G. M., Giraud, R., Brunner, K., Schmidt, G., and Molenkamp, L. W. (2004). *Phys. Rev. Lett.* **93**, 117203.
Gregg, J. F., Allen, W., Ounadjela, K., Viret, M., Hehn, M., Thompson, S. M., and Coey, J. M. D. (1996). *Phys. Rev. Lett.* **77**, 1580.
Hamaya, K., Moriya, R., Oiwa, A., Taniyama, T., Kitatomo, Y., Yamazaki, Y., and Munekata, H. (2004). *Jpn. J. Appl. Phys.* **43**, L306.
Hanbicki, A. T., Jonker, B. T., Itskos, G., Kioseoglou, G., and Petrou, A. (2002). *Appl. Phys. Lett.* **80**, 1240.
Hayakawa, J., Ikeda, S., Lee, Y. M., Sasaki, R., Meguro, T., Matsukura, F., Takahashi, H., and Ohno, H. (2005). *Jpn. J. Appl. Phys.* **44**, L1267.
Hayashi, M., Thomas, L., Bazaliy, Y. B., Rettner, C., Moriya, R., Jiang, X., and Parkin, S. S. (2006). *Phys. Rev. Lett.* **96**, 197207.
Hayashi, T., Tanaka, M., Seto, K., Nishinaga, T., and Ando, K. (1997). *Appl. Phys. Lett.* **71**, 1825.
Hayashi, T., Shimada, H., Shimizu, H., and Tanaka, M. (1999). *J. Cryst. Growth* **201/202**, 692.
Higo, Y., Shimizu, H., and Tanaka, M. (2001). *J. Appl. Phys.* **89**, 6745.
Huai, Y., Albert, F., Nguyen, P., Pakala, M., and Valet, T. (2004). *Appl. Phys. Lett.* **84**, 3118.

Hung, C.-Y., and Berger, L. (1988). *J. Appl. Phys.* **63,** 4276.
Johnston-Halperin, E., Lofgreen, D., Kawakami, R. K., Young, D. K., Coldren, L., Gossard, A. C., and Awshalom, D. D. (2002). *Phys. Rev. B* **65,** 041306.
Jonker, B. T., and Flatte, M. E. (2006). *In* "Nanomagnetism: Ultrathin films, Multilayer and Nanostructures" (Bland and Milles, eds.), pp. 227–272. Elsevier, New York.
Julliere, M. (1975). *Phys. Lett.* **54A,** 225.
Kanamura, M., Zhou, Y. K., Okumura, S., Asami, K., Nakajima, M., Harima, H., and Asahi, H. (2002). *Jpn. J. Appl. Phys.* **41,** 1019.
Katine, J. A., Albert, F. J., Buhrman, R. A., Myers, E. B., and Ralph, D. C. (2000). *Phys. Rev. Lett.* **84,** 3149.
Kikkawa, J. M., and Awshalom, D. D. (1999). *Nature* **397,** 139.
Kim, M.-S., Zhou, Y.-K., Funakoshi, M., Emura, S., Hasegawa, S., and Asahi, H. (2006). *Appl. Phys. Lett.* **89,** 232511.
Kläui, M., Vaz, C. A. F., Bland, J. A. C., Wernsdorfer, W., Faini, G., Cambrill, E., Heyderman, L. J., Notting, F., and Rüdiger, U. (2005). *Phys. Rev. Lett.* **94,** 106601.
Koch, R. H., Katine, J. A., and Sun, J. Z. (2004). *Phys. Rev. Lett.* **92,** 088302.
Koeder, A., Frank, S., Schoch, W., Avrutin, V., Limmer, W., Thonke, K., Sauer, R., Waag, A., Krieger, M., Zuem, K., Ziemann, P., Brotzmann, S., and Bracht, H. (2003). *Appl. Phys. Lett.* **82,** 3278.
Kohda, M., Ohno, Y., Takamura, K., Matsukura, F., and Ohno, H. (2001). *Jpn. J. Appl. Phys.* **40,** L1274.
Kohda, M., Kita, T., Ohno, Y., Matsukura, F., and Ohno, H. (2006). *Appl. Phys. Lett.* **89,** 012103.
Kojima, E., Shimano, R., Hashimoto, Y., Katsumoto, S., Iye, Y., and Kuwata-Gonokami, M. (2003). *Phys. Rev. B* **68,** 193203.
Kondo, T., Nomura, K., Koizumu, G., and Munekata, H. (2006). *Phys. Stat. Sol. (c)* **3,** 4263.
König, J., Lin, H.-H., and MacDonald, A. H. (2000). *Phys. Rev. Lett.* **84,** 5628.
Koshihara, S., Oiwa, A., Hirasawa, M., Katsumoto, S., Iye, Y., Urano, C., Takagi, H., and Munekata, H. (1997). *Phys. Rev. Lett.* **78,** 4617.
Lemerle, S., Ferre, J., Chappert, C., Mathet, V., Giamarchi, T., and Le Doussal, P. (1998). *Phys. Rev. Lett.* **80,** 849.
Lepadatu, S., and Xu, Y. B. (2004). *Phys. Rev. Lett.* **92,** 127201.
Levy, P., and Zhang, S. (1997). *Phys. Rev. Lett.* **79,** 5110.
Levy, P. M., and Fert, A. (2006). *Phys. Rev. B* **74,** 224446.
Liu, X., Sasaki, Y., and Furdyna, J. K. (2003). *Phys. Rev. B* **67,** 205204.
Liu, X., Lim, W. L., Titova, L. V., Wojtowicz, T., Kutrowski, M., Yee, K. J., Dobrowolska, M., Furdyna, J. K., Potashnik, S. J., Stone, M. B., Schiffer, P., Vurgaftman, I., *et al.* (2004). *Physica E* **20,** 370–373.
Marrows, C. H. (2005). *Adv. Phys.* **54,** 585.
Marrows, C. H., and Dalton, B. C. (2004). *Phys. Rev. B* **92,** 097206.
Martin, Y., Rugar, D., and Wickramasinghe, H. K. (1988). *Appl. Phys. Lett.* **52,** 244.
Mathieu, R., Svedlindh, P., Sadowski, J., Świątek, K., Karlsteen, M., Kanski, J., and Ilver, L. (2002). *Appl. Phys. Lett.* **81,** 3013.
Matsuda, Y. H., Oiwa, A., Tanaka, K., and Munekata, H. (2006). *Physica B* **376–377,** 668.
Matsukura, F., Ohno, H., and Dietl, T. (2002). *In* "Handbook of Magnetic Materials" (Buschow, ed.), Vol. 14, pp. 1–87. Elsevier, Amsterdam.
Mattana, R., George, J.-M., Jaffrès, H., Nguyen Van Dau, F., and Fert, A. (2003). *Phys. Rev. Lett.* **90,** 166601.
Mattana, R., Elsen, M., George, J.-M., Jaffrès, H., Nguyen Van Dau, F., Fert, A., Wyczisk, M. F., Olivier, J., Galtier, P., Lépine, B., Guivarc'h, A., and Jézéquel, G. (2005). *Phys. Rev. B* **71,** 075206.
Methfessel, S., and Mattis, D. C. (1968). Magnetic semiconductors. *In* "Encyclopedia of Physics, Vol. XVIII/1, Magnetism" (Wijn, ed.), pp. 389–562, Springer, Berlin.

Meier, F., and Zakharchenya, B. P. (1984). "Optical Orientation." Elsevier, Amsterdam.
Mitsumori, Y., Oiwa, A., Słupinski, T., Maruki, H., Kashimura, Y., Minami, F., and Munekata, H. (2004). *Phys. Rev. B* **69**, 033203.
Miyake, K., Shigeto, K., Mibu, K., Shinjo, T., and Ono, T. (2002). *J. Appl. Phys.* **91**, 3468.
Miyazaki, T., and Tezuka, N. (1995). *J. Magn. Magn. Mater.* **139**, L231.
Mizuno, Y., Ohya, S., Hai, P. N., and Tanaka, M. (2007). *Appl. Phys. Lett.* **90**, 162505.
Moodera, J. S., Kinder, L. R., Wong, T. M., and Meservey, R. (1995). *Phys. Rev. Lett.* **74**, 3273.
Moriya, R., Hamaya, K., Oiwa, A., and Munekata, H. (2004). *Jpn. J. Appl. Phys.* **43**, L825.
Munekata, H., Ohno, H., von Molnar, S., Segmüller, A., Chang, L. L., and Esaki, L. (1989). *Phys. Rev. Lett.* **63**, 1849.
Munekata, H., Oiwa, A., and Slupinski, T. (2002). *Phys. E* **13**, 2002.
Myers, E. B., Ralph, D. C., Katine, J. A., Louie, R. N., and Buhrman, R. A. (1999). *Science* **285**, 867.
Nazmul, A. M., Kobayashi, S., Sugahara, S., and Tanaka, M. (2004). *Jpn. J. Appl. Phys.* **43**, L223.
Nazmul, A. M., Amemiya, T., Shuto, Y., Sugahara, S., and Tanaka, M. (2005). *Phys. Rev. Lett.* **95**, 017201.
Nguyen, A. K., Shchelushkin, R. V., and Brataas, A. (2007a). *Phys. Rev. Lett.* **97**, 136603.
Nguyen, A. K., Skadsem, H. J., and Brataas, A. (2007b). *Phys. Rev. Lett.* **98**, 146602.
Ogawa, T., Shirai, M., Suzuki, N., and Kitagawa, I. (1999). *J. Magn. Magn. Mater.* **196–197**, 428.
Ohno, H., Munekata, H., Penny, T., von Molnár, S., and Chang, L. L. (1992). *Phy. Rev. Lett.* **68**, 2664.
Ohno, H., Shen, A., Matsukura, F., Oiwa, A., Endo, A., Katsumoto, S., and Iye, Y. (1996). *Appl. Phys. Lett.* **69**, 363–365.
Ohno, H., Akiba, N., Matsukura, F., Shen, A., Ohtani, K., and Ohno, Y. (1998). *Appl. Phys. Lett.* **73**, 363.
Ohno, H., Chiba, D., Matsukura, F., Omiya, T., Abe, E., Dietl, T., Ohno, Y., and Ohtani, K. (2000). *Nature* **408**, 944.
Ohno, Y., Young, D. K., Beshoten, B., Matsukura, F., Ohno, H., and Awshalom, D. D. (1999). *Nature* **402**, 790.
Ohno, Y., Arata, I., Matsukura, F., and Ohno, H. (2002). *Phys. E* **13**, 521.
Ohya, S., Hai, P. N., and Tanaka, M. (2005). *Appl. Phys. Lett.* **87**, 012105.
Ohya, S., Hai, P. N., Mizuno, Y., and Tanaka, M. (2007). *Phys. Rev. B* **75**, 155328.
Oiwa, A., Mitsumori, Y., Moriya, R., Słupinski, T., and Munekata, H. (2002). *Phys. Rev. Lett.* **88**, 137202.
Oiwa, A., Słupinski, T., and Munekata, H. (2001). *Appl. Phys. Lett.* **78**, 518.
Oszwałdowski, R., Majewski, J. A., and Dietl, T. (2007). *Phys. Rev. B* **74**, 153310.
Otani, Y., Kim, S. G., and Fukamichi, K. (1998). *IEEE Trans. Magn.* **34**, 1096.
Panguluri, R. P., Ku, K. C., Wojtowicz, T., Liu, X., Furdyna, J. K., Lyanda-Geller, Y. B., Samarth, N., and Nadgorny, B. (2005). *Phys. Rev. B* **72**, 054510.
Pappert, K., Schmidt, M. J., Hümpfner, S., Rüster, C., Schott, G. M., Brunner, K., Gould, C., Schimidt, G., and Molenkamp, L. W. (2006). *Phys. Rev. Lett.* **97**, 186402.
Park, Y. D., Hanbicki, A. T., Erwin, S. C., Hellberg, C. S., Sullivan, J. M., Mattson, J. E., Ambrose, T. F., Wilson, A., Spanos, G., and Jonker, B. T. (2002). *Science* **295**, 651.
Parkin, S. S. P. (2004). U.S. Patent No. 6834005.
Parkin, S. S. P., More, N., and Roche, K. P. (1990). *Phys. Rev. Lett.* **64**, 2304.
Partin, D. L., Karnezos, M., deMenezes, L. C., and Berger, L. (1974). *J. Appl. Phys.* **45**, 1852.
Qi, J., Xu, Y., Tolk, N. H., Liu, X., Furdyna, J. K., and Perakis, I. E. (2007). *Appl. Phys. Lett.* **91**, 112506.
Rashba, E. I. (2000). *Phys. Rev. B* **62**, R16267.
Ravelosona, D., Mangin, S., Lemaho, Y., Katine, J. A., Terris, B. D., and Fullerton, E. E. (2006). *Phys. Rev. Lett.* **96**, 186604.
Ruediger, U., Yu, J., Zhang, S., Kent, A. D., and Parkin, S. S. P. (1998). *Phys. Rev. Lett.* **80**, 5639.

Rüster, C., Gould, C., Jungwirth, T., Sinova, J., Schott, G. M., Giraud, R., Brunner, K., Schmidt, G., and Molenkamp, L. W. (2005). *Phys. Rev. Lett.* **94**, 027203.
Saito, H., Yuasa, S., and Ando, K. (2005). *Phys. Rev. Lett.* **95**, 086604.
Saito, H., Yuasa, S., Ando, K., Hamada, Y., and Suzuki, Y. (2006). *Appl. Phys. Lett.* **89**, 232502.
Sankowski, P., Kacman, P., Majewski, J., and Dietl, T. (2006). *Physica E* **32**, 375.
Sankowski, P., Kacman, P., Majewski, J., and Dietl, T. (2007). *Phys. Rev. B* **75**, 045306.
Sato, Y., Chiba, D., Matsukura, F., and Ohno, H. (2005). *J. Superconduct.* **18**, 345.
Sawicki, M., Wang, K.-Y., Edmonds, K. W., Campion, R. P., Staddon, C. R., Farley, N. R. S., Foxon, C. T., Papis, E., Kamińska, E., Piotrowska, A., Dietl, T., and Gallagher, B. L. (2005). *Phys. Rev. B* **71**, 121302.
Schimidt, G., Ferrand, D., Molenkamp, L. W., Flip, A. T., and van Wees, B. J. (2000). *Phys. Rev. B* **62**, R4790.
Shen, A., Ohno, H., Matsukura, F., Sugawara, Y., Ohno, Y., Akiba, N., and Kuroiwa, T. (1997a). *Jpn. J. Appl. Phys.* **36**, L73.
Shen, A., Ohno, H., Matsukura, F., Sugawara, Y., Akiba, N., Kuroiwa, T., Oiwa, A., Endo, A., Katsumoto, S., and Iye, Y. (1997b). *J. Cryst. Growth* **175/176**, 1069.
Shono, T., Hasegawa, T., Fukumura, T., Matsukura, F., and Ohno, H. (2000). *Appl. Phys. Lett.* **77**, 1363.
Sinova, J., Jungwirth, T., Liu, X., Sasaki, Y., Furdyna, J. K., Atkinson, W. A., and MacDonald, A. H. (2004). *Phys. Rev. B* **69**, 085209.
Slonczewski, J. C. (1996). *J. Magn. Magn. Mater.* **159**, L1.
Sugahara, S., and Tanaka, M. (2004). *Appl. Phys. Lett.* **84**, 2307.
Sun, J. Z. (1999). *J. Magn. Magn. Mater.* **202**, 157.
Tanaka, M., and Higo, Y. (2001). *Phys. Rev. Lett.* **86**, 026602.
Tanaka, M., and Higo, Y. (2002). *Phys. E* **13**, 495.
Tang, H. X., Masmanidis, S., Kawakami, R. K., Awschalom, D. D., and Roukes, M. L. (2004). *Nature* **431**, 52.
Tang, H. X., and Roukes, M. L. (2004). *Phys. Rev. B* **70**, 205213.
Tang, H. X., Kawakami, R. K., Awschalom, D. D., and Roukes, M. K. (2006). *Phys. Rev. B* **74**, 041310(R).
Taniyama, T., Nakatani, I., Namikawa, T., and Yamazaki, Y. (1999). *Phys. Rev. Lett.* **82**, 2780.
Tatara, G., and Fukuyama, H. (1997). *Phys. Rev. Lett.* **78**, 3773.
Tatara, G., and Kohno, H. (2004). *Phys. Rev. Lett.* **92**, 086601.
Tatara, G., Vernier, N., and Ferré, J. (2005). *Appl. Phys. Lett.* **88**, 252509.
Thevenard, L., Largeau, L., Mauguin, O., Partriarche, G., Lemaître, A., Vernier, N., and Ferré, J. (2006). *Phys. Rev. B* **73**, 195331.
Thiaville, A., Nakatani, Y., Miltat, J., and Suzuki, Y. (2005). *Europhys. Lett.* **69**, 990.
Thomas, L., Hayashi, M., Jiang, X., Moriya, R., Rettner, C., and Parkin, S. S. P. (2006). *Nature* **443**, 197.
Toyosaki, H., Fukumura, T., Ueno, K., Nakano, M., and Kawasaki, M. (2005). *Jpn. J. Appl. Phys.* **28**, L896.
Tserkovnyak, Y., Fiete, G. A., and Halperin, B. I. (2004). *Appl. Phys. Lett.* **84**, 5234.
Tsoi, M. M., Jansen, A. G., Bass, J., Chiang, W.-C., Seck, M., Tsoi, V., and Wyder, P. (1998). *Phys. Rev. Lett.* **80**, 4281.
Tybell, T., Paruch, P., Giamarchi, T., and Triscone, J. M. (2002). *Phys. Rev. Lett.* **89**, 097601.
Uemura, T., Sone, T., Matsuda, K., and Yamamoto, M. (2005). *Jpn. J. Appl. Phys.* **44**, L1352.
Van Dorpe, P., Liu, Z., Van Roy, W., Motsnyi, V. F., Sawicki, M., Borghs, G., and De Boeck, J. (2004). *Appl. Phys. Lett.* **84**, 3495.
Van Dorpe, P., Van Roy, W., De Boeck, J., Borghs, G., Sankowski, P., Kacman, P., Majewski, J. A., and Dietl, T. (2005). *Phys. Rev. B* **72**, 205322.
Van Roy, W., Van Dorpe, P., Vanheertum, R., Vandormael, P.-J., and Borghs, G. (2007). *IEEE Trans. Electron Dev.* **54**, 933.

Vernier, N., Allwood, D. A., Atkinson, D., Cooke, M. D., and Cowburn, R. P. (2004). *Europhys. Lett.* **65**, 526.
Viret, M., Vignoles, D., Cole, D., Coey, J. M. D., Allen, W., Daniel, D. S., and Gregg, J. F. (1996). *Phys. Rev. B* **53**, 8464.
Vu, L. N., Wistrom, M. S., and Van Haelingen, D. J. (1993). *Appl. Phys. Lett.* **63**, 1693.
Wang, J., Khodaparast, G. A., Kono, J., Slupinski, T., Oiwa, A., and Munekata, H. (2004). *Phys. E* **20**, 412; Corrigendum (2005). *ibid.* **25**, 681.
Wang, J., Sun, C., Kono, J., Oiwa, A., Munekata, H., Cywiński, Ł., and Sham, L. J. (2005). *Phys. Rev. Lett.* **95**, 167401.
Wang, J., Cotoros, C., Dani, K. M., Liu, X., Furdyna, J. K., and Chelma, D. S. (2007). *Phys. Rev. Lett.* **98**, 217401.
Watanabe, M., Toyao, H., Okabayashi, J., and Yoshino, J. (2006). *Phys. Stat. Sol. (c)* **3**, 4180.
Welp, U., Vlasko-Vlasov, V. K., Liu, X., Furdyna, J. K., and Wojtowicz, T. (2003). *Phys. Rev. Lett.* **90**, 167206.
Yamaguchi, A., Ono, T., Nasu, S., Miyake, K., Mibu, K., and Shinjo, T. (2004). *Phys. Rev. Lett.* **92**, 077205.
Yamanouch, M., Ieda, J., Matsukura, F., Barnes, S. E., Maekawa, S., and Ohno, H. (2007). *Science* **317**, 1726.
Yamanouchi, M., Chiba, D., Matsukura, F., and Ohno, H. (2004). *Nature* **428**, 539.
Yamanouchi, M., Chiba, D., Matsukura, F., Dietl, T., and Ohno, H. (2006). *Phys. Rev. Lett.* **96**, 096601.
Young, D. K., Johnston-Halperin, E., Awschalom, D. D., Ohno, Y., and Ohno, H. (2002). *Appl. Phys. Lett.* **80**, 1598.
Zhang, S., and Li, Z. (2004). *Phys. Rev. Lett.* **93**, 127204.
Zhu, H. J., Ramsteiner, M., Kostial, H., Wassermeier, M., Schönherr, H.-P., and Ploog, K. H. (2001). *Phys. Rev. Lett.* **87**, 016601.

CHAPTER 6

Spintronic Nanodevices

C. Gould, G. Schmidt, and L. W. Molenkamp

Contents		
	1. Introduction	241
	2. Tunneling Anisotropic Magnetoresistance	244
	3. Multi-TAMR Structures	248
	4. Volatile and Nonvolatile Operation	250
	5. Correlated Effects	252
	6. Portability	259
	7. Nanodevices	260
	8. Lateral Nanoconstrictions	260
	9. Current-Assisted Manipulation	264
	10. Local Lithographic Anisotropy Control	268
	11. Magnetic Characterization of Nanobars	269
	12. Transport Characterization of Nanobars	271
	13. Anisotropic Strain Relaxation	274
	14. Memory Device Using Local Anisotropy Control	277
	15. Device Operation	278
	16. Magnetic States	279
	17. Origin of the Resistance Signal	280
	18. Conclusion and Outlook	283
	Acknowledgments	284
	References	284

1. INTRODUCTION

Following the discovery of the Giant Magneto Resistance (GMR) effect in 1988 (Baibich *et al.*, 1988; Binasch *et al.*, 1989), magnetoelectronic device elements with operational functionalities relying on the magnetic properties

Physikalisches Institut (EP3), Universität Würzburg, Am Hubland, D-97074 Würzburg, Germany of the metal constituents of the structures entered mainstream and commercially relevant electronics in record time. Field sensors operating using GMR were already on the market as early as 1995, and IBM introduced a GMR-based hard drive read head in 1997 (Theis and Horn, 2003), less than a decade after the first laboratory demonstration of the effect.

An additional spin-electronics application, Magnetoresistive Random Access Memory (MRAM), will soon be available to the marketplace. Built out of magnetic tunnel junctions (MTJ) it provides nonvolatile data storage: information is not lost, when the instrument is powered down and does not need to be refreshed during operations (Åkerman, 2005). This is especially interesting for portable electronic devices promising instant on/off functionality and reduced battery consumption. MRAM combines nonvolatility and low power consumption with other advantages such as relatively fast read and write times and no wear out during write cycles. This makes it a promising candidate for the "universal" memory device, eliminating the need for multiple memory categories, for example, for fast access or cost effective data storage. However, MRAM storage cells, being made of metallic elements, are part of the metal interconnect layer, far away and largely independent from the semiconductor devices below. Magnetic material is used to store data and semiconductor devices to process the information.

Ferromagnetic semiconductors offer the promise of overcoming this restriction by combining magnetic and semiconducting properties within the same material. This was highlighted in a recent review (Awschalom and Flatté, 2007) describing the tremendous advantages of semiconductor spintronics offering significant gain in integration potential of memory, information transport, and logic operations into the same material system. This enhanced potential for integration suggests that semiconductor-based memory devices could become successors/alternatives to technologies such as MRAM. Moreover, fundamental issues in the interfacing of metallic and semiconducting elements, such as the conductance mismatch issue (Schmidt et al., 2000), which impose strict limits on the device architectures accessible to these classic metallic magnetoelectronic devices, would be naturally circumvented using all semiconductor architectures.

It is therefore obvious that the development of a functional technology based on magnetic semiconductors would immediately gain industrial relevance. Moreover, as we will soon show, in addition to having the capability of emulating the functionalities of their metallic counterparts, ferromagnetic semiconductor (FS)-based spintronic devices offer a plethora of novel mechanisms which can be harnessed into new device paradigms with the potential to drive progress in the post-CMOS era of information technology.

In previous chapters, the authors gave a theoretical description (B. Gallagher and T. Jungwirth), and experimental demonstration

(H. Ohno) of how the blending of magnetic and semiconducting properties within the same material leads to incredibly rich transport phenomena, the investigation of which has spawned multiple branches of spintronics research (Wolf et al., 2001; Zutic et al., 2004).

In the present chapter, we turn our attention to how these transport phenomena can be harnessed into spintronics nanodevices, the potential of which is nearly limitless. Ultimately, such devices could evolve towards the fully quantum mechanical extension of spintronics focused on the use of individual electron spins and wavefunctions to store and manipulate information as a way to potentially implement a scheme for theoretically predicted quantum computation, encryption, and information transfer (DiVincenzo et al., 2000; Loss and DiVincenzo, 1998). Devices whose operation is based on the manipulation of wavefunctions of individual particles are currently still the subject of long-term fundamental research, and any eventual implementation is still in its infancy.

Here, we will focus mostly on a more mature class of FS devices where the aim is either to expand on currently used magnetoelectronic functionalities, or more interestingly, to achieve new device functionalities based on the properties of FS, but where the device still functions with macroscopic current levels, and the physics can be described by studying an ensemble of current carrying particles without needing to attend to their individual character.

Before we proceed to a discussion of devices, we believe a short note about their eventual implementation is in order. All the structures to be described in this chapter have only been demonstrated at cryogenic temperatures. As will be clear from the description, we emphasize that this is not a fundamental limitation of the operation principles of the devices, but rather purely a materials issue. The underlying requirement for the materials to be used in these devices is that they exhibit strong spin–orbit coupling and appropriate magnetic anisotropies. This behavior is most prevalent in FS, none of which have to date been convincingly demonstrated to exhibit ferromagnetism at room temperature. It is important to note however that not only is significant materials development work steadily making progress in the hunt for a room temperature FS, but also, as we will discuss in more detail, many of the device schemes presented here may eventually be portable to novel metal based ferromagnets. Indeed, some of the effects described below, after having been discovered in FS, were proven to also exist in certain metallic structures despite having previously been overlooked. As such, while the current need for cryogenic temperatures for the study of these devices is a short-term obstacle to their commercialization, it is in no way a fundamental criticism of the devices. For this reason, both the scientific and industrial communities remain confident of a bright commercial future for FS type spintronics devices.

With the above caveat about materials in mind, the decision of which material system to use for spintronic nanodevice investigations is no longer a question of industrial relevance, but rather more of scientific expedience. For this reason, the material of choice is (Ga,Mn)As, which is widely viewed as the prototypical FS, having risen to this position following a string of successes in the late 1990s beginning with the demonstration of a 110 K Curie temperature by Ohno et al. (1996) and Matsukura et al. (1998). The material was described in detail in the chapter by Ohno. Briefly, it is produced by introducing typically 2–6% Mn into a GaAs lattice. Because the valence 2 Mn replaces valence 3 Ga in the lattice, the resulting material is strongly p-type, and exhibits a strong carrier-mediated ferromagnetism. This magnetism originates in the fact that the spins of the itinerant holes couple to each Mn ion they encounter, and act as a mediating agent to provide an indirect exchange which acts to align the Mn ions. Moreover, because the material has the underlying zincblende structure of GaAs, and because of strong spin–orbit coupling of the valence band, this hole-mediated ferromagnetism is strongly anisotropic with symmetry characteristics taken from the lattice. A detailed theoretical description of this sophisticated ferromagnetic state is given in the chapter by Gallagher and Jungwirth.

The present chapter will discuss in detail two categories of devices. In the first part of the chapter, we will discuss microdevices that operate based on tunneling anisotropic magnetoresistance (TAMR), as an example of the potential for fundamentally novel transport phenomena that is offered by FS. We will then turn to proper nanodevices, starting with a device which shows how very large effects can be obtained by confining the domain walls (DW) between various regions of FS on the nanometer scale. This theme will be expanded upon in a discussion of very recent devices demonstrating a newly developed technique to locally engineer the anisotropy of (Ga,Mn)As, as well as a description of functional devices produced using such a scheme.

2. TUNNELING ANISOTROPIC MAGNETORESISTANCE

TAMR was first discovered in 2004 (Gould et al., 2004) in an experiment investigating the injection of tunneling current into a FS. This was accomplished by measuring transport in a structure consisting of a single ferromagnetic (Ga,Mn)As layer fitted with a tunnel barrier and a nonmagnetic metal contact.

The FS layer in this sample is a 70 nm thick epitaxial (Ga,Mn)As film grown using low temperature (270 °C) molecular beam epitaxy onto a GaAs (001) substrate. The (Ga,Mn)As layer incorporates 6% Mn, and thus has a p-type doping concentration of $\sim 10^{21}$ cm^{-3}. It has a Curie temperature of 70 K. After growth of the semiconductor layer, a 1.4 nm Al layer is deposited on its surface, and oxidized in situ for 8 h using 100 mbar of pure oxygen, thus producing the closed AlOx layer which forms the tunnel barrier. An electrical contacting layer consisting of 5 nm of Ti as a sticking layer followed by 300 nm of Au is then evaporated onto the layer stack. The device is patterned as shown in Fig. 6.1 using standard optical lithography and chemically assisted ion beam etching (CAIBE). In a first step material is etched away, leaving only the 100 × 100 μm² square pillar comprised of a metal contact on a tunnel barrier. The surrounding backside contact is then made by depositing a metal sticking layer and an Au layer.

The typical device resistance of such structures is of the order of 3 kΩ or higher, which is more than two orders of magnitude larger than the resistance of ~10 Ω measured for identically patterned structures where the tunnel barrier is omitted. This confirms that the resistive properties of the device are fully dominated by the barrier, and that magnetoresistance effects within the FS are completely negligible.

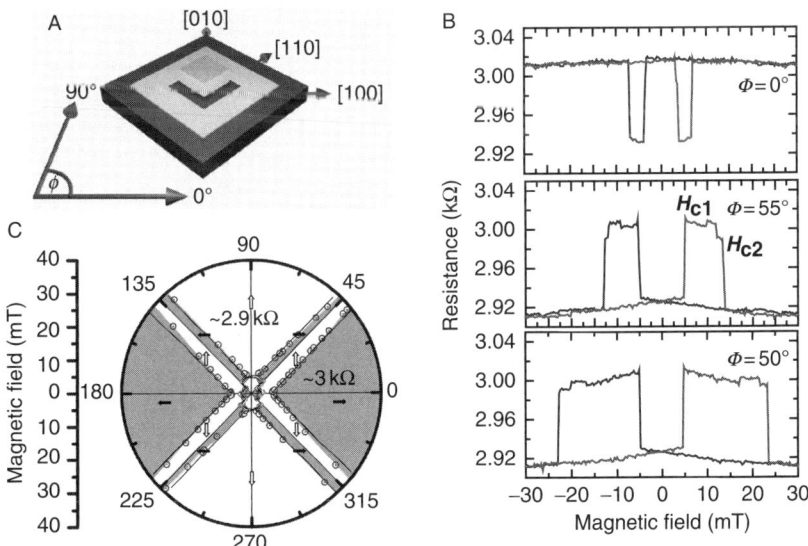

FIGURE 6.1 (A) Schematic of the first TAMR device. (B) Magnetoresistance curves at various in-plane field angles, showing the spin-valve–like behavior characteristic of TAMR. (C) Polar plot summarizing both the data and the model. The device exhibits high (low) resistance if the magnetization is along ±[100] (±[010]). The open triangles are experimental extracted switching fields H_{c1} and H_{c2} data and the lines fits to the model. The arrows indicate the magnetization direction, and the dark shading indicates regions of high resistance. (See Color Plate.)

Our TAMR samples are characterized in a variable temperature magnetocryostat equipped with a vector magnet allowing for the application of magnetic fields of up to 300 mT in any direction. Since the (Ga,Mn)As layers used in these studies have a hard axis perpendicular to plane, the results presented here are mostly focused on fields applied in the plane of the magnetic layer, in which case the field direction is characterized by a single angle ϕ which is defined with respect to the [100] direction, as indicated in Fig. 6.1A.

Some representative 4.2 K magnetoresistance curves at various angles are given in Fig. 6.1B. For each curve, the magnetic field is swept from negative saturation[1] fields to positive saturation and back, with the figure only showing the interesting region between -30 and $+30$ mT. All magnetoresistance curves exhibit spin-valve–like behavior with a signal amplitude of $\sim$3% delimited by two switching events (labeled H_{c1} and H_{c2}). The width and sign of the TAMR feature, however, strongly depends on the applied field angle ϕ. Despite this change of sign of the signal, note that the device has (at low magnetic fields) only two distinct resistance states, a low one of $\sim$2,920 Ω and a high one of $\sim$3,000 Ω.

Combining similar measurements for many values of ϕ, we compile the polar plot of Fig. 6.1C. The open triangles are the experimental values of the switching events H_{c1} and H_{c2} extracted from individual magnetoresistance curves. These symbols are plotted in the angular direction ϕ at which the particular magnetoresistance curve was obtained, and the magnetic field values of the switching event are given by the radial coordinate in the polar plot. The shaded areas correspond to conditions where the sample is in its high resistance state. The switching events in the figure form a highly organized pattern reminiscent of switching previously observed in magneto-optical studies of epitaxial Fe films. This is confirmed by the solid lines in the figure, which are fits to the data using the model developed by Cowburn et al. (1995) to describe magnetization studies of epitaxial Fe grown on GaAs. The model consists of calculating the magnetostatic energy for a single magnetic domain (or macrospin) subjected to both a ([100] + [010]) biaxial anisotropy and a [010] uniaxial anisotropy, in the presence of an external magnetic field, and allowing for magnetization reorientation through a DW nucleation and propagation process. Even though the [010] anisotropy term is rather small, it plays a primary role because it breaks the symmetry between the [100] and [010] directions which would otherwise be degenerate. It is important to note that while (Ga,Mn)As is known to also have a [110] uniaxial anisotropy which indeed is significantly larger than the [010] energy term, it plays no significant role in the present effect, and can to

---

[1] The term saturation is used here to indicate a magnetic field sufficiently large to overcome all anisotropies in the sample and force the magnetization parallel to the external field.

first order be neglected in the modeling. The reason for this is that the [110] anisotropy component only breaks the symmetry between the [110] and [1$\bar{1}$0] directions which are hard magnetization directions and not energetically favorable at zero external field. For the low-field nonvolatile TAMR effect discussed here, only magnetization configurations that are stable in the absence of an applied field are significant, and the symmetry between these configurations is not affected by the [110] anisotropy term.

Given an understanding of the switching mechanism, the curves of Fig. 6.1B can be easily interpreted. Take the 50° data as an example. At sufficiently large negative fields, the magnetization points in the 230° direction, parallel to the field. As the field is reduced to –30 mT, the magnetization relaxes to the nearest easy axis, which is the –[010] direction. This configuration remains until, at $H_{c1} = 5$ mT, the energy gained by a reorientation of the magnetization to a direction closer to that of the applied external field is greater than the energy cost of nucleating and propagating a DW. At this field, the magnetization switches to the [100] direction. As the field is swept still higher, the external field applies more force on the magnetization, and at $H_{c2} = 23$ mT, it switches a second time, to the [010] direction before gradually rotating towards its final position parallel to the magnetic field at high fields (of the order of 200 mT).

We thus clearly see that the apparent magnetoresistance of the sample at low fields is more accurately described as a dependence of the sample resistance on its magnetization state, and more specifically on whether the magnetization points along ±[100] or ±[010]. This is emphasized in the polar plot of Fig. 6.1C, where the arrows indicate the direction of the magnetization within each region according to the magnetization model. Comparing this magnetization direction to the shading of the various regions makes clear that the sample is in a high resistance state whenever M lies along the ±[100] crystallographic direction, and has a lower resistance when M is along ±[010].

To understand why the resistance of such structures is governed by the magnetization direction of the (Ga,Mn)As layer, we consider that in the simplest model, the transmittance of a tunnel barrier is proportional to a product of the tunneling matrix element describing the properties of the barrier and the density of states (DOS) at the Fermi energy on either side of the barrier. Given that the barrier properties, and thus the tunneling matrix element, are to a very good approximation magnetic field independent, as is the DOS of the metal contact, it is clear that the transmission through the tunnel barrier should be proportional to the (Ga,Mn)As DOS. The theoretical treatment of the (Ga,Mn)As DOS presented in the chapter by Gallagher and Jungwirth shows an interdependence between the DOS and the magnetization. This leads to the existence of an anisotropy in the (Ga,Mn)As DOS with respect to the magnetization orientation which, for the anisotropies observed in this device, is sufficiently large to explain the TAMR effect.

The electronic structure of (Ga,Mn)As is described within a k.p approximation of the GaAs host valence bands, with the Mn treated on a mean field level as effective exchange field (Abolfath et al., 2001; Dietl et al., 2000, 2001). The breaking of the [100]–[010] symmetry due to the presence of the [010] uniaxial term is accounted for theoretically by introducing a phenomenological uniaxial strain of order 0.1%. This strain, in conjunction with the very strong spin–orbit interaction in the valence band, leads to values of the [010] uniaxial anisotropy consistent with that used to fit the switching behavior in Fig. 6.1C.

The calculation indicates that the DOS at the Fermi level indeed changes slightly depending on the direction of the magnetization, and in particular is different for magnetization along [100] or [010]. For appropriate material parameters, the difference is however only a few per mil, and thus too small to account for the observed resistance effect. This discrepancy is lifted by taking the momentum distribution of the states within the DOS into consideration. The full DOS obviously contains states with momentum in all space directions. This implies that holes traveling at the Fermi velocity will have a large spread in their momentum normal to the barrier. Holes with momentum perpendicular to the barrier of course have a much greater tunneling probability than those which arrive at glancing incidence. Transport through the barrier is thus dominated by a subset of the DOS with high momentum perpendicular to the barrier. By including this in the model, we find that as we increase the strictness of this momentum conservation condition the anisotropy increases significantly, and that for reasonable parameters, it reaches a few percent, consistent with the experimentally observed resistance change.

A subtle point worth noting is that while the magnetic and DOS anisotropies both play important roles in the observation of TAMR, these two anisotropies are not inextricably linked, and can be independently influenced. This fact is experimentally manifested in the temperature dependence of the effect. As the temperature is increased above 20 K, a reversal of the DOS anisotropy is observed and the resistance values in Fig. 6.1C are all inverted, with the high resistance regions becoming low resistance, and vice versa. This is *not* accompanied with a reversal of the uniaxial magnetic anisotropy, which remains constant throughout the experiment. Such a result is consistent with the above theoretical modeling which correctly reproduces a change in the DOS anisotropy as the magnetization is reduced upon increasing temperature (Gould et al., 2004).

3. MULTI-TAMR STRUCTURES

The functionality of the above described single junction TAMR device can be expanded by using structures with multiple TAMR interfaces. We now turn to a device with a double TAMR geometry (Pappert et al., 2006;

Rüster et al., 2005a). The full layer structure consists of a $Ga_{0.94}Mn_{0.06}As$ (10 nm)/GaAs (2 nm)/$Ga_{0.94}Mn_{0.06}As$ (100 nm) trilayer. The two (Ga,Mn) As layers are highly p-type and nominally identical. They are separated by a low-temperature grown GaAs layer which is insulating and forms an epitaxial tunnel barrier between the two ferromagnetic layers. The layer stack is structured into devices similar to those of the single TAMR case and studied using similar transport techniques.

Figure 6.2B shows that the devices have, for the most part, similar TAMR behavior as their single-sided counterparts with the obvious exception that the amplitude of the effect has increased from 3% to 40%. This increase in amplitude is not unexpected. One would expect an approximate doubling of the amplitude from putting two interfaces in series, and a further increase is predicted from the improved momentum conservation afforded by using an epitaxial barrier instead of an amorphous one (Gould et al., 2004; Rüster et al., 2005a). The 4.2 K behavior of this device is thus fully consistent with the previously described model. The two layers are nominally identical, and thus have very similar magnetization behavior. In most measurements they act in unison, and the response of the single and double interface devices to the external magnetic field is, with the exception of the increased amplitude, similar.

In practice however, the two layers are not perfectly identical, and in this section we focus on one important result of this difference which serves as a demonstration of the possibilities offered by TAMR in future devices where layers are engineered to intentionally have different anisotropy properties. In Fig. 6.3, we show a measurement of the magnetoresistance of this device

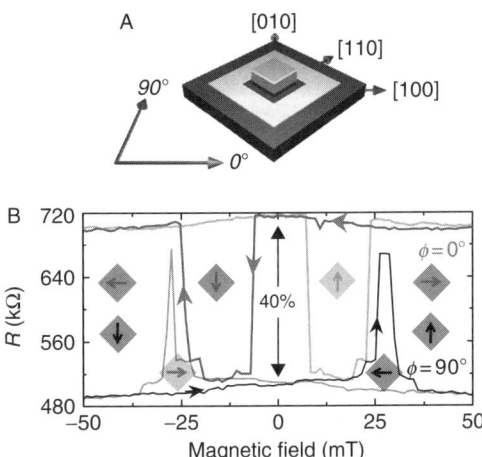

FIGURE 6.2 (B) Magnetoresistance curves along the 40° direction for the double TAMR structure depicted schematically in (A). The arrows in the diamonds indicate the direction of magnetization of *both* layers at each point in the spin-valve curve. (See Color Plate.)

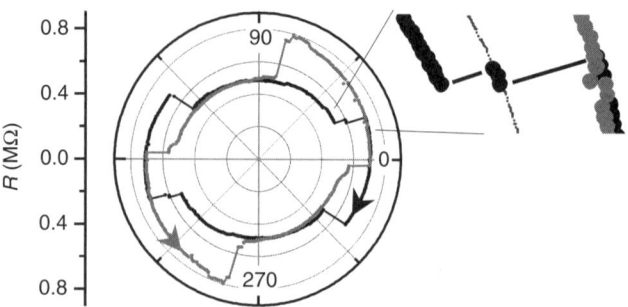

FIGURE 6.3 Resistance of the sample as a function of sweeping the field direction at a field strength of 25 mT, and a zoom of the interesting region.

as a function of sweeping the magnetic field direction while maintaining its amplitude at 25 mT. This field is sufficiently strong to cause DW-mediated reorientations of the magnetization between the various easy axes, but still so weak as to not allow the magnetization to deviate significantly from these easy axes. As the field is rotated, the magnetization of the layers thus switches by 90° four times per field revolution. The figure shows data for both clockwise and counterclockwise rotation, which are of course offset due to hysteretic effects. Because the two layers are slightly different, the reorientation in each occurs at a slightly different field direction such that for the small region emphasized in the zoomed in part of the figure, the magnetization vectors of the two layers are at an angle of 90° with respect to each other. The two TAMR interfaces are thus in different states, with one in its high resistance state, and the other in the low state. The total resistance of the device thus takes an intermediate value. In this device, the range of this intermediate state is very small because the two layers are nominally identical. This is however not an intrinsic requirement and by properly designing the properties of the ferromagnetic layers, the device performance could be engineered in order to enhance the angular range in which this state is active.

While so far, devices with a maximum of only two interfaces have been experimentally demonstrated, yielding a ternary device, it is easy to, at least in principle, expand upon this idea to multiple layers and interfaces, which would lead to a device with a large number of resistance states that may have implications to eventual device architectures which go well beyond the current binary standard.

4. VOLATILE AND NONVOLATILE OPERATION

So far, we have limited our discussion to the case of nonvolatile TAMR. The effect also has an implementation in a volatile mode. To make the distinction clear, in Fig. 6.4, we show the results of a "write–read"

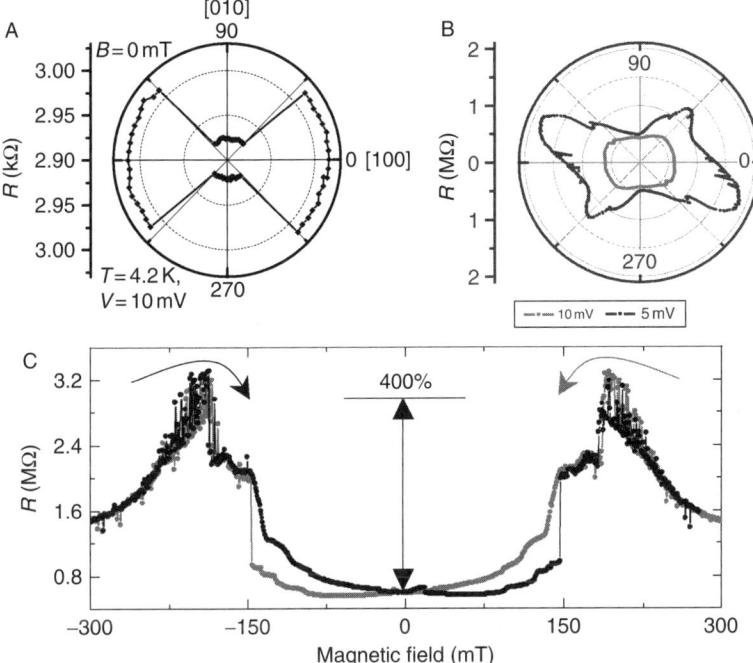

FIGURE 6.4 Volatile and nonvolatile operations modes: (A) Write–read memory demonstration. The state of the device is written by applying a magnetic field of 50 mT in various directions. It is then read, in the absence of magnetic field, by measuring the zero field resistance of the device. (B) Magnetoresistance as a function of field angle in a saturation magnetic field at two different bias voltages. (C) Magnetoresistance of the device in response to a perpendicular to plane magnetic field.

experiment (Rüster et al., 2005b) for operating the structure as a nonvolatile memory element, and compare it to a high field ϕ-scan demonstrating the effect as a field direction sensor.

In the write–read experiment, we first "write" the single layer TAMR device into a desired state by applying a sufficiently large magnetic field ($\sim$50 mT) in a given direction ϕ. As the field is swept back down to zero, the magnetization relaxes to the easy axis nearest to the direction of this writing field. As such, if the applied field was within 45° of the ±[100] direction, the magnetization relaxes to that axis, whereas it otherwise relaxes to the ±[010] state. Once the state has been written, the device is perfectly stable, with the information stored in the magnetic state.

The storage is nonvolatile, and the information can be maintained indefinitely without power consumption. It can be read by simply measuring its resistance, which, as seen in the figure, will be either high or low depending on the direction of writing. This reading process is nonperturbative and can be repeated as often as desired without affecting the integrity of the stored information. Write–read experiments on two interface devices yield similar results, including a third nonvolatile information storage state associated with the intermediate resistance state of the 90° configuration.

The origin of the volatile effect in the field sensor experiment is identical in nature to that of the nonvolatile mode (Brey et al., 2004; Rüster et al., 2005a). In each case, it is a manifestation of the fact that the tunneling resistance is dependent on the direction of magnetization. The only difference between the two effects is that in the volatile case, the direction of magnetization is not energetically stable, and thus requires an external force, such as an applied magnetic field, to maintain it.

Figure 6.4B and C show two demonstrations of operating a device in volatile TAMR mode. Panel C shows the magnetoresistance of the device as a function of applying an external magnetic field perpendicular to the device interfaces. As the field is made sufficiently large, the magnetization is forced out of the plane, and the magnetoresistance increases causing an increase of the device resistance of up to 400%. It should be noted that in this case, the resistance increase is not necessarily pure TAMR in its original form. In addition to the previously described significance of the direction of the magnetization with respect to the crystal axis, in this geometry, the changing angle between magnetization and current also will play a role (Giddings et al., 2005).

In Fig. 6.4B, a pure in-plane volatile TAMR demonstration is presented in an experiment where the resistance is measured for a 300 mT field applied in various directions ϕ in the plane of the layers. Because this field is large enough to overcome all anisotropies in the (Ga,Mn)As layer, it forces the magnetization direction parallel to the applied field. As is clear from the figure, this leads to a smooth, nonmonotonic, change in device resistance as a function of the magnetic field direction. The exact dependence of the resistance on direction results from a very complex interplay between the transport properties and the anisotropies, and depends critically on exact Mn and charge carrier concentrations and the strain conditions. As such, it differs from sample to sample, and cannot be determined from first principle. Nevertheless, it is stable and reproducible for a given sample.

5. CORRELATED EFFECTS

The results presented so far on multi-TAMR structures are primarily extensions of the physical phenomena already present in the original TAMR structure. We now look at how a surprising additional physical process not only begins to play a role but also in some senses becomes dominant, as the temperature and bias voltage across the device are decreased.

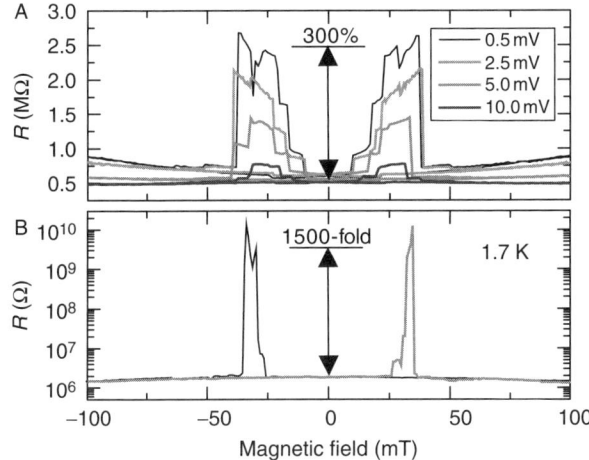

FIGURE 6.5 Amplification of the TAMR effect with decreasing bias voltage (A) and lowering of the temperature from 4.2 to 1.7 K (B).

Figure 6.5 shows that as bias voltage is decreased from the 10 mV used in Fig. 6.2B to 0.5 mV, the amplitude of the spin valve increases by roughly an order of magnitude, from 40% up to 300%. When the temperature is then lowered from 4.2 K down to 1.7 K as in part B of the figure, the amplitude of the effect completely diverges, reaching a measured value of 150,000%. Indeed, the measured value merely represents a lower limit for the resistance ratio. In this configuration, at 1.7 K and 1 mV of applied bias, no current can be detected, and the reported value of 10 GΩ corresponds to the 0.1 pA detection floor of the current amplifier.

While the model described previously correctly describes the 4.2 K, high bias behavior of this sample, it simply cannot, for reasonable parameters, account for this massive amplification. An additional physical process must be at work here, and in the following we provide evidence that it originates in correlation effects related to crossing the metal-to-insulator transition in FSs with strong spin–orbit coupling (Pappert et al., 2006).

The differential conductance data of Fig. 6.6 support the conclusion that charge carrier correlations play a primary role. The $G(V) = dI/dV$ curves of Fig. 6.6B are calculated from current–voltage-characteristics taken in the remanent magnetization state after preparation of the sample at an angle ϕ. The $G(V)$ curves at 1.7 K are segregated into two groups (see Fig. 6.6A) according to their resistance states identified in a write–read experiment similar to that of Fig. 6.4. That is, the magnetization of both (Ga,Mn)As layers relaxes into the low resistance state corresponding to

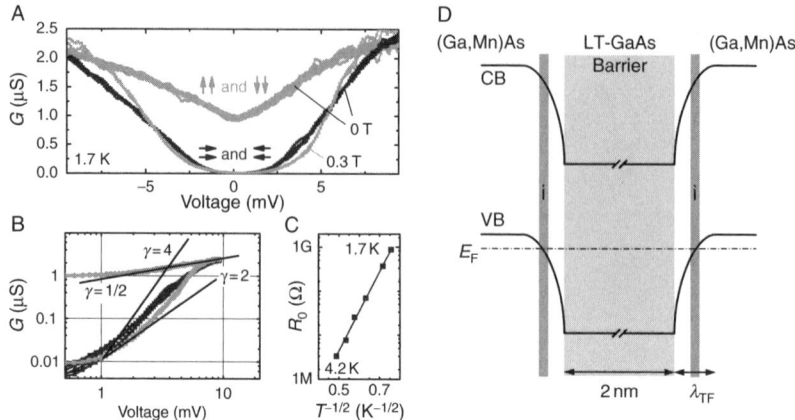

FIGURE 6.6 (A) Differential conductance–voltage curves at 1.7 K including sets of curves belonging to the two magnetization states at $B = 0$, and one set at $B = 300$ mT along $\phi = 3$ and $6°$. (B) Log–log plot of same data. (C) Temperature dependence of the zero bias resistance of the high resistance state. (D) Schematic of the band bending caused by the barrier. (See Color Plate.)

magnetization along [010] for all preparation angles between $45°$ and $135°$ (upper curves in Fig. 6.6A), and relaxes into the [100]-high-resistance state for preparation fields within $45°$ of the [100] direction (lower curves).

Each set of $G(V)$ curves has a distinct shape. The high conductance curves are square-root–like and show nonvanishing conductance at zero voltage. This is typical for the metallic behavior of highly doped semiconductors taking part in the tunnel process. The square root dependence stems from weak localization in metals (Altshuler and Aronov, 1979). The high resistance state, on the other hand, exhibits insulating behavior with the conductance vanishing at zero voltage. These curves follow a higher power law as seen in the log–log plot of Fig. 6.6B.

The shape of these $G(V)$ curves is reminiscent of those observed by (Lee et al., 1999) in their investigations of tunneling from a metal into disordered boron-doped silicon near the metal–insulator transition (MIT). Their theory describes the formation of a soft ES-gap (Efros and Shklovskii, 1975). Such a gap can be formed because of an increase in electron–electron Coulomb interaction as screening is reduced when the MIT is approached from the metallic side. This gap around the Fermi energy is observable in the low bias tunneling conductance. It manifests itself in a different power law behavior $G(V) \propto V_m$ with $m = 1/2$ for metallic (just above the MIT) and $m \geq 2$ for insulating (below the MIT) material. This metal-to-insulator transition is thermally activated as seen in Fig. 6.6C, consistent with the exponential activation following $R \propto \exp(1/T)^{1/2}$ expected for an ES gap material under the usual assumption of single hop tunneling (Efros and Shklovskii, 1975; Sandow et al., 2001). We note however that because the accessible temperature range spans less then one decade, an $R \propto \exp(1/T)^{1/4}$ dependence expected from a non correlated Mott transition cannot be confidently excluded.

Figure 6.7 offers a further hint into the origin of the effect. It presents a 300 mT ϕ-scan as in Fig. 6.4B showing strong and apparently random oscillations reminiscent of quantum interference effects. These presumably come from a statistically defined electronic state in the sample. Figure 6.7A–C shows that the amplitude of these oscillations grows significantly with reduced bias for angular ranges (gray regions in the figure) associated with ES behavior. Figure 6.7C–E shows a significant effect upon thermal cycling to temperatures of some tens of K (above the ES activation temperature). The three curves are nominally identical measurements after subsequent thermal cycling. While the main behavior of the device remains consistent, the details of the fluctuations change significantly, as one would expect from a randomizing of the impurity configuration. Note that such a change in the fluctuation pattern occurs due to the thermal cycling. On any given cooldown, the oscillations are

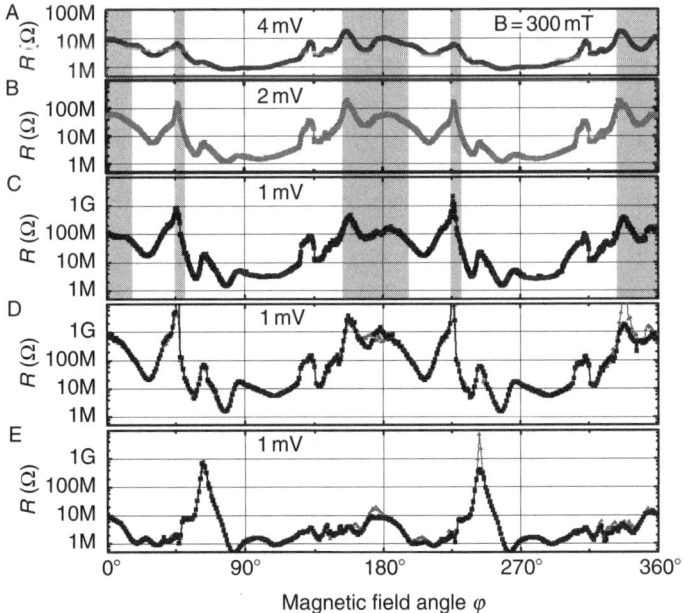

FIGURE 6.7 Resistance of the sample at 1.7 K and 300 mT as a function of magnetic field direction under various bias and on different cooldowns. (D) and (E) each show the results of two separate measurement, confirming reproducibility. (See Color Plate.)

fully reproducible as confirmed by the two, almost indistinguishable, sets of measurements included in each of Fig. 6.7D and E.

These fluctuations are attributed to quantum interference effects in variable range hopping, which is the main transport mechanism in the insulating state (Raikh and Ruzin, 1987). The behavior of the device is thus consistent with the (Ga,Mn)As injection layer undergoing a MIT triggered by a reorientation of the magnetization from [010] to [100], and the formation of an ES-gap. In order to understand how a change in magnetization direction can trigger a MIT, we need to consider the criterion for the passage from metallic to insulating properties.

Metallic transport properties require the charge carriers to be in long range itinerant states. This condition implies significant overlap of the wavefunctions between acceptors (Mn atoms). As such, the passage through the MIT depends not only on the carrier density but also on the volume occupied by the wavefunctions of the states of the individual dopants (Pollak, 1965).

We have already discussed that details of the (Ga,Mn)As DOS are influenced by the magnetization direction. We now focus on a numerical calculation of the wavefunctions of the bound hole states to show that this dependence also leads to a renormalization of the volume occupied by the wavefunction as a function of the orientation of the magnetization.

Luttinger and Kohn (1955) showed that a charged impurity in a semiconductor host can be described by treating the impurity wavefunction as an envelope wavefunction $B(k)$ of Bloch waves. Using the six-band $k \cdot p$ framework with the full Luttinger Hamiltonian $H(k)$ (Dietl et al., 2001), the time-independent Schrödinger equation is:

$$H(\mathbf{k})\mathbf{B}(\mathbf{k}) + \int d^3\mathbf{k}' U(\mathbf{k} - \mathbf{k}')\mathbf{B}(\mathbf{k}') = E_b \mathbf{B}(\mathbf{k}) \tag{6.1}$$

where E_b is the binding energy. U is the Fourier transform of the Coulomb limit of the Yukawa interaction potential between the hole and impurity-center. $H(\mathbf{k})$ is a 6×6 matrix acting on the six dimensional vector $\mathbf{B}(\mathbf{k})$ locally in k-space. It describes the band structure including pd-exchange, strain and SO-coupling. Equation (6.1) is solved numerically for the acceptor ground state in a magnetic zinc blende semiconductor. The obtained k-space wavefunctions are then fitted to hydrogen ground state wavefunctions in order to extract an effective Bohr radius for each k direction and for various magnetization directions (Schmidt et al., 2007).

As above, the experimentally observed [010] uniaxial anisotropy is treated phenomenologically by introducing a uniaxial strain term ε_u along [010]. An important result of the bulk band structure calculation is that the energy ordering of the bands depends on the pd-exchange, the relative strain and the magnetization direction. In Fig. 6.8A, the Γ point energy of the top four valence bands is plotted as a function of the amplitude of ε_u and the

Spintronic Nanodevices 257

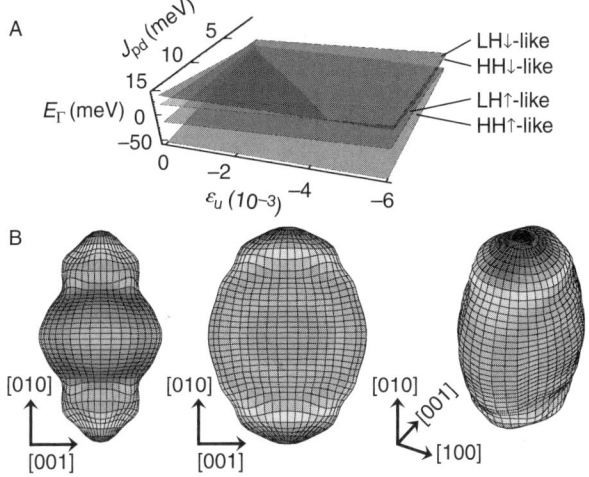

FIGURE 6.8 (A) Γ point energy of the four top bands as a function of strain ε_u and pd-exchange. (B) Extents of a hole bound to a Mn-impurity. [001] is the growth direction and ε_u is along [010]. Left: Magnetization (**M** || [100]). Middle: (**M** || [010]). Right: Perspective view for **M** || [010]. (See Color Plate.)

pd-exchange. For zero strain, we obtain the expected nearly equidistant spacing between the four bands, while increasing strain shifts the relative energy position of the light-hole–like bands with respect to the heavy-hole–like bands, leading to a change in ordering of the bands at the line defining the crossing point between the top two planes in the figure, and thus to a profound change in the character of the wavefunction at a given energy. This is directly reflected in the wavefunction of the impurity hole state.

Figure 6.8B shows the extent of a bound hole for magnetization along different in-plane easy axes. The amplitude of ε_u is set beyond the crossing point for the bulk energy at the Γ-point. There is a significant difference in the size of the wavefunction between the $\vec{M} \parallel \varepsilon_u$ and the $\vec{M} \perp \varepsilon_u$-state. The result is that the wave function overlap between neighboring dopant sites depends on the direction of the magnetization, and for reasonable parameters is consistent with the observation of the magnetization reorientation-induced MIT.

We emphasize that the above calculation does *not* predict a MIT for the bulk (Ga,Mn)As layer which, as typical for high quality (Ga,Mn)As, has nearly temperature independent resistivity in the relevant temperature range. Our prediction is limited to the thin (Ga,Mn)As layer near the interface with the tunnel barrier. The barrier is made from LT-GaAs, a material with many mid-gap traps. This leads to a gradual spatial depletion of the (Ga,Mn)As near the barrier on the length scale of the

Thomas-Fermi screening length of $\sim 2\text{Å}$, and thus to a much lower effective local carrier density as depicted in Fig. 6.6D.

When this thin depleted layer is sufficiently close to the MIT, a transition can be triggered by a change in wavefunction extent. A key point to note is that, while this thin depleted layer does not play a significant role in determining the magnetic properties of the (Ga,Mn)As, it has a dominant effect on the transport. This comes from the very short mean free path of holes in (Ga,Mn)As (of the order of a few Å). Given that by definition, the transition from diffusive to tunneling transport takes place at a carrier density where hole motion no longer takes place through classical diffusion, this extremely short mean free path leads to a very thin effective injector layer from which the tunneling originates. This thin injection layer has considerably reduced carrier density compared to the bulk and is consistent with our model.

The relationship between a MIT and the development of an ES-gap is the reduction in screening caused by the reduction of mobile carriers. With unscreened Coulomb interactions between localized states, the DOS at the Fermi-energy for carrier concentrations just below the MIT vanishes with a dimension-dependent power law (Efros and Shklovskii, 1975). Of course, we cannot rule out that the change in screening also causes a slight shift in position of the effective injector thus modifying the tunneling distance, and accounting for part of the observed resistance change.

Larkin and Shklovskii (2002) derived a power law behavior for conductance versus voltage (G–V) curves for tunneling between two three dimensional localized materials where the parabolic DOS $D(\varepsilon) \propto \varepsilon^2$ on each side multiply and lead to $G \propto V^6$. In our case, given the very thin nature of the injector, a 2D description of the DOS of the tunneling reservoirs is likely more appropriate. Thus, the DOS is linear in energy, and the expected power of the G–V curves is 4. Moreover, it is unclear in our experiment whether both sides of the barrier play an equivalent role, or whether the effects are dominated by one electrode. Depending on the relative role of the two barriers, one would thus expect a power law somewhere between $\propto V^2$ and $\propto V^6$. Neglecting very low voltages where thermal smearing is important, this prediction is in good agreement with the experimental observation of Fig. 6.6.

In summary, the amplification of the TAMR signal resides in a magnetization reorientation induced MIT stemming from a modification of the wave functions of individual dopants due to the coupling of the Mn dopants to the magnetization direction in the bulk. This transition is accompanied by the opening of an ES-gap at the Fermi energy which manifests itself as a change in the power law behavior in conductance–voltage characteristics in tunneling. This is the first observation of a MIT induced by a change in magnetization *direction* in any material and is not only of fundamental interest but may also have technological relevance in opening up new possibilities of controlling the transport properties of devices by magnetization reversal.

6. PORTABILITY

In the preceding sections, we have described in depth the novel transport phenomena of TAMR that was discovered by investigating FS-based devices. Earlier we mentioned that one route towards the eventual commercialization of such activities is to find effects which have previously been overlooked by the metallic-based magnetoelectronics industry, but which may, under the right circumstances, also be present in metal-based devices. In the case of TAMR, the necessary ingredients for an appropriate material system are that it exhibits appropriate magnetic anisotropies, as well as strong spin–orbit coupling which will link the DOS to the magnetic configuration. While these conditions are well met in FSs, they can also often be present in certain magnetic metals (Shick and Mryasov, 2003)* or in appropriate sample geometries.

Indeed TAMR has already begun its move towards a metal-based implementation with the first observation of TAMR related effects appearing in metallic devices only a couple of years after the first report of TAMR in 2004.

For example in early 2006, two groups independently reported demonstrations of a TAMR in ferromagnetic metal break junctions (Bolotin et al., 2006; Viret et al., 2006). In these experiments, a tunneling device is produced by fashioning a break junction from traditional ferromagnetic metal material, and the magnetoresistance of the device is investigated in magnetic fields large enough to ensure that the magnetization of all parts of the structure are forced parallel to the external field, and thus that the tunneling is occurring between two regions of parallel magnetization. The result is a magnetoresistance behavior comparable to the volatile version of TAMR described above. The effect is attributed to spin–orbit coupling which drives anisotropies locally in the DOS as a result of mesoscopic effects associated with the shape of the tunneling constriction.

More recently, TAMR phenomena were observed in a traditional tunneling geometry such as that of Fig. 6.1, but consisting of a Co/Fe/GaAs/Au stack (Moser et al., 2006). This particular sample showed a difference in tunneling rates for magnetization of the epitaxial Fe layer along [110] and [1$\bar{1}$0], which allowed operation of the device in volatile mode. The origin of the breaking of the symmetry in the DOS in this case originates from Rashba (Chantis et al., 2007) and possibly also Dresselhaus (Matos-Abiague and Fabian, 2007) splitting in the system.

* Indeed, an experimental demonstration of TAMR in such a system was reported while this book was in press: B. G. Park et al., Phys. Rev. Lett. 100, 087204 (2008).

While these demonstrations were, for a myriad of technical reasons, still performed at cryogenic temperatures, they clearly demonstrate the effects in materials where the Curie temperature is well above room temperature, and thus constitute an important first step, and a strong indication that novel mechanisms discovered in work on FSs may one day soon be portable to industrially relevant temperatures.

7. NANODEVICES

Many common themes for the design of novel spintronics devices are based on using DW (Allwood et al., 2005; Parkin, 2003) or interfaces between regions of different magnetization [such as GMR (Baibich et al., 1988; Binasch et al., 1989), or Tunneling magnetoresistance (TMR) (Julliere, 1975)], to store and manipulate information. Magnetoresistance effects of DW in metals are quite small (Kent et al., 2001), and large effects have only been observed in mechanically manipulated nanojunctions (García et al., 1999), where magnetostriction can play an important role (Wegrowe et al., 2003). Much larger effects were however predicted (Flatté and Vignale, 2001; Vignale and Flatté, 2002) from DW in magnetic semiconductors due to the large spin polarization in these materials.

8. LATERAL NANOCONSTRICTIONS

The first demonstration of making use of such DW in FSs was reported in 2003 (Rüster et al., 2003), using lateral nanofabricated constrictions in (Ga,Mn)As wires to strongly pin the DW and thus reduce their thickness (Bruno, 1999). This approach facilitates ballistic hole transport through the DW, while the lateral fabrication technology excludes any influence of magnetostriction.

Devices as in Fig. 6.9 consist of a central island of 100 nm width and 500 nm length connected to two 400 nm wide and 10 µm long leads by constrictions with widths of ~10 nm. These constrictions form pinning centers for DW. The 400-nm wide leads are contacted by voltage and current probes to allow four-probe transport measurements as indicated in the figure. An important and oft overlooked point is that special care is taken in the lithography to align the long axis of the island to the [100] (or equivalent) direction, and not parallel to a ([110]) cleaving edge. The current, magnetic field, and magnetic easy axis of the layer are thus all collinear, leading to a device where the various elements can be reliable described in a single domain model. After processing, the resist is left on the sample in order to allow further etching steps and thus subsequent narrowing of the constrictions.

Spintronic Nanodevices 261

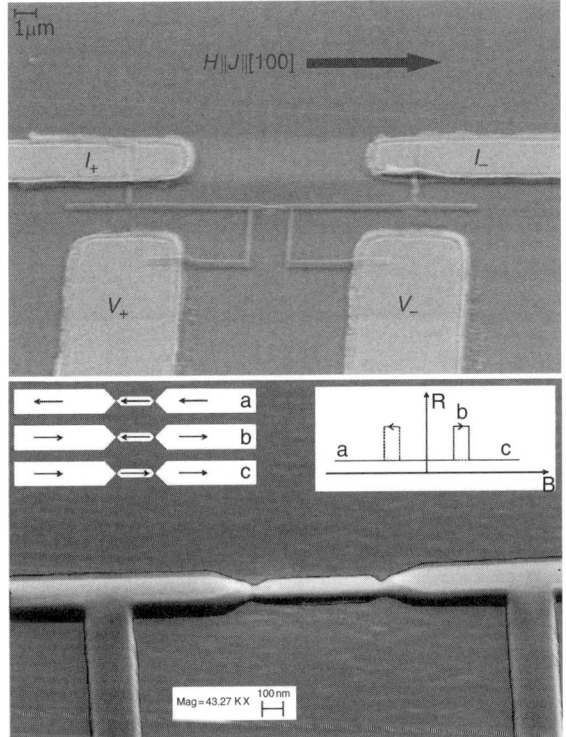

FIGURE 6.9 SEM micrographs of the nanoconstrictions device, with insets schematically depicting the expected behavior. (See Color Plate.)

While the effect we aim to investigate in principle would only require a single constriction, the choice of the double constriction geometry has two important practical advantages. First, the symmetric design avoids complications associated with thermoelectric voltages. Second, size effects due to the inner island being near the single domain limit cause the (magnetically isolated) island to switch at significantly higher fields than the outer leads. In fact, control experiments using Superconducting Quantum Interference Device (SQUID) magnetometry on arrays of islands and leads confirmed that while the wide leads reverse their magnetization at around 15–20 mT (depending on lithographic parameters), the island exhibits switching fields of order 60–90 mT. The coercive field of the unpatterned epilayer is ~8 mT.

The insets of Fig. 6.9, illustrates schematically the expected magnetoresistance for such a device. As the field is swept from large negative values where all three regions are magnetized in parallel (a) towards positive fields, the leads switch first, causing an antiparallel alignment of the island and the leads (b). In this state, DW are trapped in each constriction and the device resistance is increased. At larger positive fields the magnetization of the island also reverses, realigning the magnetization of all three regions (c) and the resistance returns to its original value. This magnetization reversal is hysteretic, leading to the depicted spin-valve–like Magnetoresistance (MR).

Typical experimental results are presented in Fig. 6.10A, for a sample with four terminal resistance of about 80 kΩ showing a spin-valve signal of 8%. The figure shows multiple nominally identical measurements, and while the basic spin-valve signal is reproducible, the fine structure varies from sweep to sweep. This is indicative of multiple pinning sites near the constrictions. The various pinning sites yield different geometrical confinement of the DW, thus altering their resistance. Given the extremely small dimensions of the constrictions, individual impurities and side wall damage from the etching are likely causes of these pinning centers.

Transport in the previous device was in the diffusive regime, with linear current–voltage characteristics of the device in the parallel state. By taking advantage of the resist having been left on the sample, the device can be re-etched for a few additional seconds. This further narrows the constrictions and produces devices with much higher constriction resistance. Such a device with a 4 MΩ resistance and showing a 2,000% spin-valve signal is shown in Fig. 6.10B. In this device, the current–voltage characteristics are no longer linear, and the constrictions have clearly been pushed into a tunneling transport regime.

To understand the behavior of both these samples in a uniform picture requires a realization that etching causes a gradual depletion of the carrier density at the constrictions. This is because dry etching of semiconductors damages the walls of the epilayer, and the resulting charged impurities induce sidewall depletion of the interior of the semiconductor. This mechanism will clearly be most effective at the narrowest parts of the structure, that is, at the constrictions. In the following calculation, we assume for simplicity that both constrictions have equal resistance.

As a first order approximation, we assume that the DW resistance is given by the expression of Valet and Fert (1993) for the spin-accumulation–induced resistance ΔR at an abrupt junction between two regions of opposite magnetization:

$$\Delta R \approx 2\beta^2 \rho^* l \qquad (6.2)$$

where ρ^* is the spin-symmetric bulk resistivity in the magnetic material, l is the length of the constriction, and β is the spin polarization in the constriction. For reasonable material parameters, this model is consistent with the data of Fig. 6.10A, but not the larger effect of Fig. 6.10B.

In the case of the larger effect, the constrictions clearly are in the tunneling regime, and one could be tempted to model the observed MR in terms of Julliere's TMR model (Julliere, 1975). In order to explain a

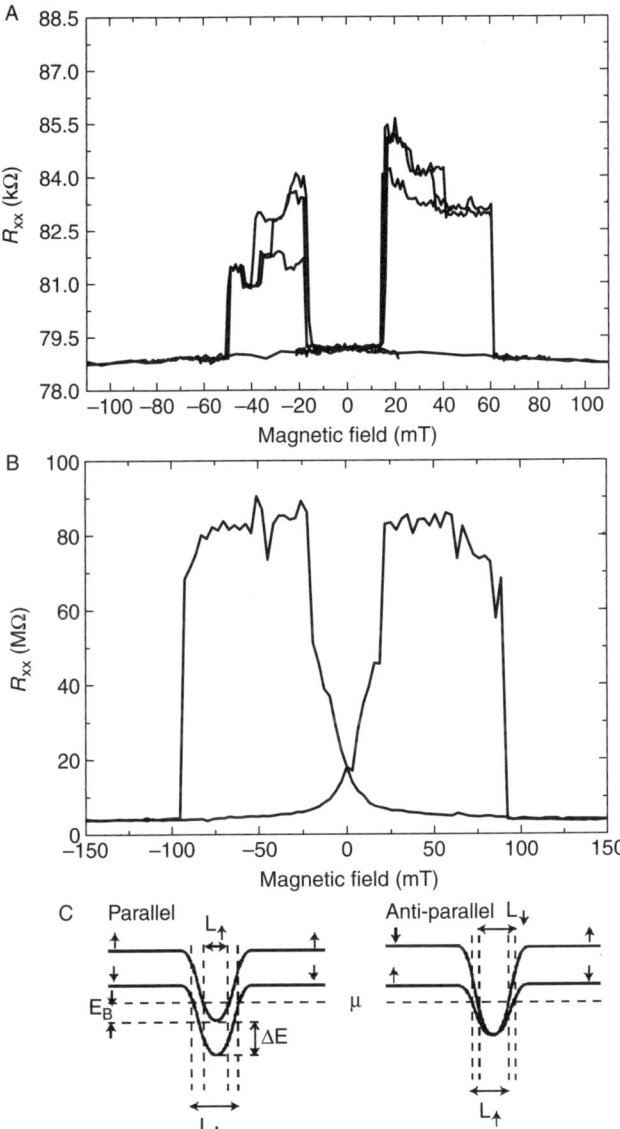

FIGURE 6.10 Magnetoresistance of nanoconstrictions devices in the linear (A) and nonlinear (B) regime. (C) Schematic of the model invoked to explain the effect in the nonlinear regime.

2,000% MR signal, the Julliere model however requires a spin polarization of the contacts of ca. 95%, much larger than the value of $\beta \approx 20\%$ suggested by the results of fitting Eq. (6.2) of the low resistance sample, and also inconsistent with the generally accepted polarization in (Ga,Mn)As.

This large discrepancy suggests that the model from Julliere is not applicable here and that a different description is needed. We also note in passing that a model of the MR effect due to any joint action of both constrictions can be ruled out because of the long distances between the barriers compared to the extremely short mean free path. We can therefore safely consider a model based on two independent barriers.

The model supposes that the etching process creates a shallow barrier of parabolic shape between the two regions of (Ga,Mn)As, as shown in Fig. 6.10C. The barrier height for the majority-spin holes above the chemical potential μ is E_B. If the barrier is very thin, such that the hole wave functions can penetrate into the barrier region and continue to couple to the Mn spins, then it is reasonable to assume that ΔE in the barrier region is the same as in the bulk. This results in a barrier for minority-spin holes that is higher ($E_B + \Delta E$) than for majority-spin holes (E_B). As a consequence of the non-abrupt barrier, the thickness of the barrier for minority-spin holes, $L_{P\downarrow}$, will be greater than that for majority spin holes, $L_{P\uparrow}$. For the antiparallel situation depicted on the right, however, the barriers for the two spin channels are the same and approximately $E_B + \Delta E/2$, and their thicknesses are also the same, $L \approx (L_{P\uparrow} + L_{P\downarrow})/2$. Calculating the transmission through such parabolic barriers is straightforward, and for the material and constrictions parameters of this device, yields a spin-valve signal in excellent agreement with the observed model.

This work therefore clearly shows that by properly confining DW between various regions of a (Ga,Mn)As sample, significant magnetoresistance effects are possible, and that if information were to be stored in the position (or in the presence or absence) of such DW, it could readily be retrieved by transport techniques.

9. CURRENT-ASSISTED MANIPULATION

A next important step towards realizing useful devices is of course the manipulation/control of such DW. As we have already seen, domains can be manipulated by the application of external magnetic field, which, for the purposes of laboratory demonstrations, are perfectly adequate. However, in any serious large-scale implementation scheme, the addressing of individual elements by external magnetic fields would be a severe limitation, and the ability to control the state of the device through electrical means is infinitely more desirable.

How such a control can be achieved has already been discussed in detail in the chapter by Ohno when covering the classic experiment of Yamanouchi et al. (2004). Here we show how the effects demonstrated there in macroscopic structures can equally be put to work in nanoconstriction devices.

The device (Gould et al., 2006) is conceptually similar to the one of Fig. 6.9, and again consists of a lithographically defined nanostructure patterned in a similar 20 nm thick (Ga,Mn)As layer. The left part of Fig. 6.11 shows an SEM image and schematic diagram of the device, which is comprised of a small central island, separated from large triangular leads by a pair of nanoconstrictions of 10 nm in width. Again, the orientation is such that the current path is along a (Ga,Mn)As magnetic easy axis. The typical magnetoresistance curve of the device presented in Fig. 6.11 shows similar spin-valve behavior as seen in Fig. 6.10, the origin of which is of course identical.

In Fig. 6.12, we demonstrate that if the device is set to the high resistance configuration where the central region is anti-parallel to the leads, the magnetization of the inner island can be reversed using a current-assisted switching mechanism. The figure shows four magnetoresistance measurements. The first, indicated by the solid line, is a normal magnetoresistance measurement taken with 5 mV of bias voltage as in Fig. 6.11. The second scan, with the data displayed as dots, is a similar measurement, except that the scan is stopped at 8 mT (indicated by the vertical dotted line), and several minutes are allowed to elapse before the scan is continued. This interruption has no effect on the measured resistance. For the final two curves, we again repeat similar magnetoresistance measurements, which are stopped at 8 mT. Now, however, we increase the bias voltage up to +100 mV (dark thick line) or −100 mV (light thick line), and ramp back down to 5 mV before continuing the sweep. The time taken for this process is shorter than the waiting time used for the second curve. Nevertheless, when the curves are continued

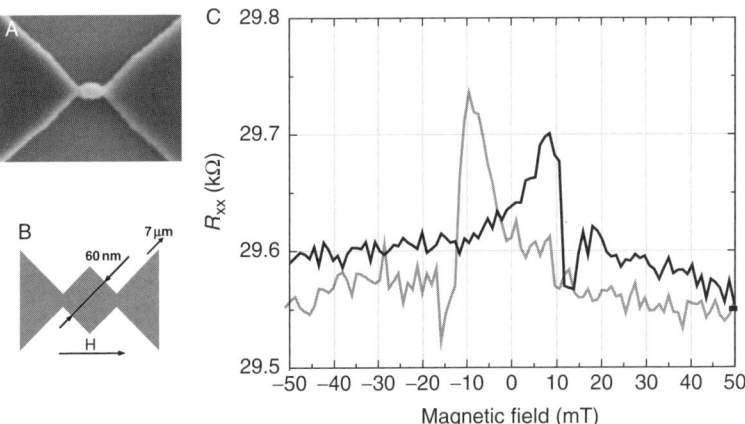

FIGURE 6.11 Left: SEM and schematic of the nanoconstriction device used for current-induced switching. Right: Typical spin-valve signal from this device.

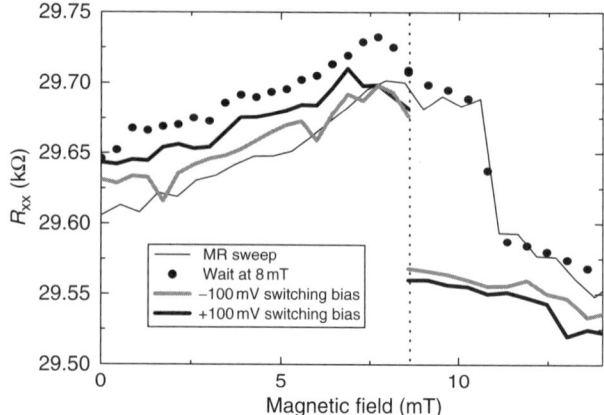

FIGURE 6.12 Current-assisted switching behavior of the device. The thin solid line is the unperturbed spin-valve signal with the other curves showing how the valve can be switched using a bias voltage of ±100 mV. See text for details.

the central region has already reversed its magnetization as the device is in the low magnetoresistance state. The sign of the magnetization reversing voltage is unimportant as the device and magnetic configuration are symmetric.

More information can be gained by monitoring the resistance during the time where the field sweep is halted and the bias voltage is being ramped up and down. We present such data, obtained on a second, nominally identical sample, in Fig. 6.13. For each pair of measurements, the sample is first prepared in the anti-parallel configuration. After interrupting the magnetoresistance scan the magnetic field is kept constant.

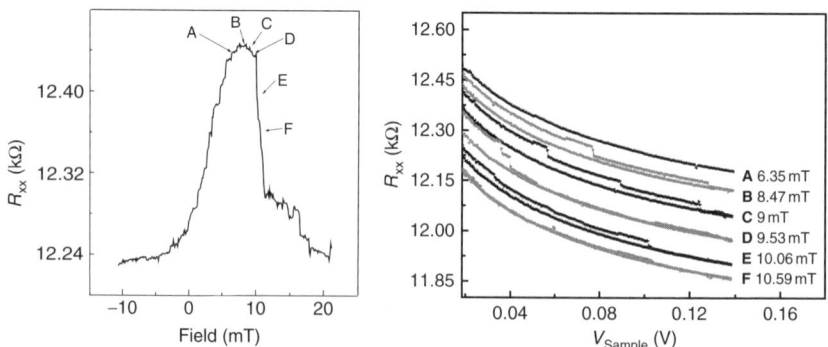

FIGURE 6.13 Right: Resistance vs. voltage curves showing the abrupt resistance change at the current-induced switching events for measurements taken at the various magnetic fields indicated in the left part of the figure. See text for details.

We then measure the resistance while increasing the bias up to 180 mV, and notice, for example in set C, that in addition to the smooth and monotonic decrease expected from the nonlinear nature of the nanoconstriction resistance, the curve contains some abrupt resistance decreases. We continue to measure the resistance as we sweep the bias back down, and find only a smooth increase, without any jumps. At low bias, the two curves differ in resistance by the amplitude of the spin-valve signal. If we now repeat similar bias scans, we obtain the lower (and smooth) current–voltage characteristic for both up- and down- sweep. This is because the sample is now in the parallel state, and the high bias voltage can have no further effect. While it is difficult to give exact numbers for the dimensions of the constrictions, using reasonable estimates, we get a typical switching current density of $\sim 10^6$ A cm^{-2}.

Similar measurements were performed at various magnetic fields as indicated in the left part of the figure. The curves at 8.47, 9.00, 9.53, 10.06, and 10.59 mT show clear switching behavior. No abrupt switching was observed at magnetic fields below 6.5 mT or above 11 mT (not shown). Interestingly, for the entire field range between 6.5 and 11 mT where switching does occur, there is no correlation between the position (or number) of switching events, and the field at which the experiment was performed. The total amplitude of the sum of all switching events on a given curve does of course depend on the magnetic field at which the measurement is performed since the total amplitude must be consistent with the amplitude of the magnetoresistance spin-valve signal.

The mechanism driving the current-assisted switching in these devices cannot be unambiguously determined. While the results of Fig. 6.13 preclude the idea of heating above T_c, since the discontinuous changes in resistance must take place while the magnetization is still finite, we cannot completely exclude any role of heating. One could speculate that local heating of the sample to a temperature below its Curie temperature, but sufficiently high to change its anisotropy or coercive field, causes a lowering of the switching field and thus permits reversal of the central region under the applied magnetic field. We find this explanation unsatisfactory however, both because we would not expect significant heating at the power level of 100 nW used in the studies, and because under this model, for measurements done at lower field, a correspondingly greater reduction of the coercive field would be needed. This should lead to a strong dependence of the switching bias on the magnetic field, which is clearly incompatible with the results of Fig. 6.13.

For this reason, we believe that the mechanism most likely at work here is the same as discussed elsewhere in this book by Ohno (Yamanouchi et al., 2004); namely a pd exchange-mediated spin angular momentum transfer between the current carrying itinerant holes, and the localized Mn spins in the central region (Tatara and Kohno, 2004).

10. LOCAL LITHOGRAPHIC ANISOTROPY CONTROL

While the preceding results certainly provide possibilities for basic device concepts, the elaboration of more sophisticated device architectures requires an additional degree of control. All the results presented so far consisted of devices where the magnetic anisotropy of the structures derived directly from their host layers. While this method is sufficient for basic demonstrations, more complex device designs clearly require the ability to locally control and engineer the magnetic properties of individual components in order to allow for the fabrication of structures where various constituent elements have varied anisotropies.

Local control of anisotropies in ferromagnetic metals is well established, making successful use of an approach based on shape anisotropy. This approach has also been tried in (Ga,Mn)As with somewhat lackluster results. Hamaya et al. (2003, 2006) reported the observation of shape induced anisotropy in (Ga,Mn)As wires of 100 nm thickness $\times$ 1.5 $\times$ 200 μm^2, but only over a limited temperature range. Our own experience in attempting to use wires of similar dimensions have however yielded inconsistent results with the wires having irreproducible anisotropy, with either biaxial or uniaxial easy axes in rather arbitrary directions.

Moreover, a simple calculation of the expected shape anisotropy term in such wires indicates that it should not play a significant role. While the infinite rod model used by Hamaya et al. (2006) does predict an appreciable shape anisotropy field given by $\mu_0 M_S/2$, where M_S is the saturation magnetization, it is not applicable to structures which are much thinner than their lateral dimensions. A more exact rectangular prism calculation (Aharoni, 1998) for typical nanopatterned structures with dimensions of 20 $\times$ 200 $\times$ 1,000 nm gives a five times weaker shape anisotropy with an anisotropy energy density of 80 J/m^3 which is much too small to compete with the typical crystalline anisotropy of 3,000 J/m^3 in this material.

An alternative to shape anisotropy is the use of strain imposed anisotropy. In (Ga,Mn)As, growth strain is known to reduce the cubic symmetry and create a uniaxial anisotropy with an easy (hard) magnetic axis in growth direction when the layer is tensile (compressively) strained. This growth strain influences the strength of the perpendicular component of the anisotropy of the *whole* layer. Here we discuss (Ga,Mn)As grown on a (001) oriented GaAs substrate with an out-of-plane hard magnetic axis which confines the magnetization in the plane.

As mentioned earlier, the net in-plane magnetic anisotropy results from a competition of two primary contributions: the crystal symmetry-induced biaxial anisotropy with [100] and [010] easy axes, and a uniaxial anisotropy with the direction of its easy axis assuming either of the in-plane ⟨110⟩ directions. The anisotropy of (Ga,Mn)As is further enriched by a second uniaxial along [010] discussed in the TAMR section (Gould et al., 2004). The interplay of these three, temperature-dependent (Wang et al., 2005) anisotropies, leads to a material with a very complex magnetic behavior which can be difficult to control or reproduce from one layer to the next.

In (Ga,Mn)As, it was recently shown (Hümpfner et al., 2007) that an additional agent, that is, lithographically induced strain relaxation, also plays a significant role in nanopatterned structures and is the only reasonable means by which to properly exercise *local* control of the anisotropy. We demonstrate that patterning-imposed relaxation effects can not only be observed in ferromagnetic (Ga,Mn)As but also these effects can be made to dominate the magnetic anisotropy in the entire temperature range, up to the Curie temperature T_c.

11. MAGNETIC CHARACTERIZATION OF NANOBARS

Standard 20 nm thick (Ga,Mn)As layers with a T_c of 70 K are nanolithographically patterned into arrays of either [100] or [010] oriented bars, termed nanobars, for magnetic investigation and also into equivalent individual nanobars contacted for transport investigations. Figure 6.14 shows SEM photographs of the structures. Each individual bar has lateral dimensions of 200 nm by 1 µm. The full array of them is defined using electron beam lithography with a negative resist. After developing, the defined pattern is transferred into the (Ga,Mn)As layer using CAIBE. As many as 8 million nanobars are laid out to provide sufficient total magnetization for the magnetic anisotropy studies carried out by SQUID magnetometery.

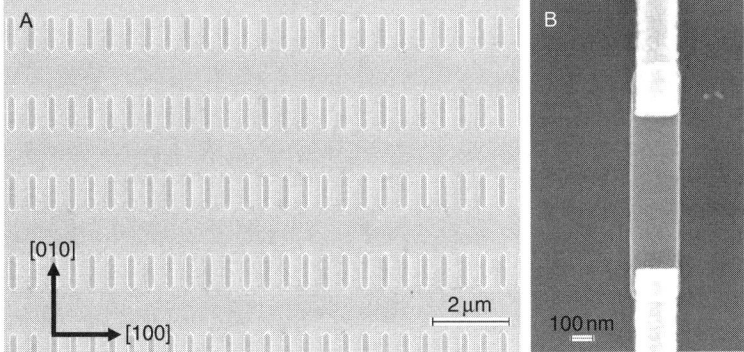

FIGURE 6.14 (A) SEM photograph of a small section of a typical 8 million nanobar array. The individual bars have lateral dimensions of 200 nm by 1 µm. (B) An individual nanobar contacted for transport characterization.

In SQUID investigations on these nanobar arrays we examine the magnetization M versus H dependencies of the sample in applied magnetic fields of up to ± 100 mT for the four major in-plane orientations. All background signals originating from the substrate and sample holders are subtracted from the presented data.

The salient features of the SQUID investigations are summarized in Fig. 6.15. We start with the unpatterned, "parent" layer (top panels) to show that, as is typical for (Ga,Mn)As, it exhibits equivalent behavior along [100] and [010], both at 5 K (Fig. 6.15A), and close to T_c (Fig. 6.15B). This is a simple manifestation of the fact that the presence of a $\langle 110 \rangle$ uniaxial anisotropy, which bisects the fourfold $\langle 100 \rangle$ easy directions and acts equivalently on [100] and [010] does not break the symmetry between these directions. The [010] uniaxial anisotropy is too weak to measurably break the symmetry.

This behavior is in stark contrast to that of the patterned array, as shown in the bottom panels of Fig. 6.15 showing magnetization studies of an array of nanobars oriented such that their long axis is along the [010] direction. This axis remains a magnetic easy axis, similar to that of the

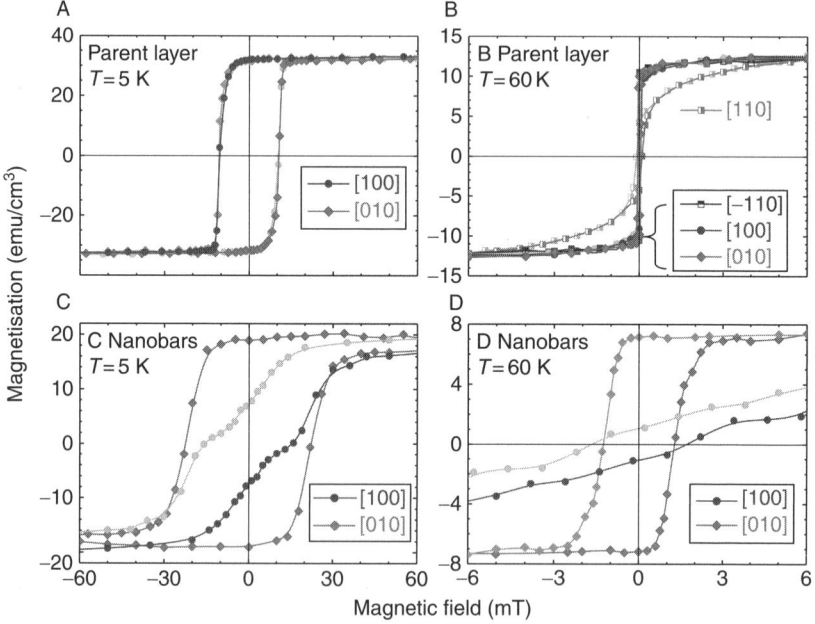

FIGURE 6.15 SQUID magnetization data for (A–B) the parent layer and (C–D) the array of nanobars having their long side aligned to [010]. Light shades are used to mark the high resolution data obtained by numerical reflection to mimic the full hysteresis after confirming hysteretic symmetry with coarse measurements.

host. The magnetic response along the [100] direction, which is along the short side of the nanobars has however been completely modified, and now exhibits pronounced hard axis behavior. From the hard axis measurements, and using for simplicity a purely uniaxial model which provides a lower-bound estimate, we determine the lithographically imposed anisotropy field $\mu_0 H_L$ produced by our nanopatterning to be 25 and 20 mT at 5 and 60 K, respectively. This field is comparable to the crystalline fourfold anisotropy field ($\sim$ 100 mT) which dominates the behavior of the parent layer at 5 K, and is much larger than the $\langle 110 \rangle$ uniaxial term ($\sim$2 mT) which dominates the behavior of the unpatterned layer at 60 K. The shape anisotropy field is too weak to play a significant role, contributing only 4 and 1.4 mT at these respective temperatures.

The magnetic anisotropy has thus been transformed from a strongly temperature-dependent mixture of fourfold and uniaxial contributions into a well defined, nearly temperature independent, uniaxial behavior imposed locally along the long axis of the nanobars. Arrays of nanobars patterned with their long axis along [100] give fully equivalent results.

The submicron dimensions of the nanobars have also allowed us to reach the single domain limit in (Ga,Mn)As at low temperatures. As seen in Fig. 6.15C, the magnetization reversal along the easy axis of the nanobars takes place roughly at the uniaxial anisotropy field of 25 mT, indicating a nearly fully coherent behavior of the magnetization inside the nanobar. This again indicated that the patterning has led to a fundamental change in the nature of the magnetic response of the material, and can thus be effectively used for device engineering purposes.

12. TRANSPORT CHARACTERIZATION OF NANOBARS

Having achieved the desired anisotropy control in the arrays, we now turn to electrical investigations, for which individual nanobars are prepared using similar lithography as in the patterning of the arrays. The main challenge in this case is finding a nonperturbative way of contacting the nanobar. This is nontrivial as it requires the formation of ohmic contacts with a $\sim (100 \text{ nm})^2$ length scale. Moreover, our experience has shown that improperly optimized contacts exert uncontrolled strain onto the layer, significantly altering its anisotropy (Gould et al., 2007). We succeeded by using a Ti layer patterned by lift-off as a mask. After etching, the Ti mask is removed, and Ti/Au contacts are applied by e-beam lithography and lift-off. This yields contacts with a resistance-area product of below 10^{-6} $\Omega \cdot \text{cm}^2$. Transport characterization of such nanobars patterned along either the [100] or [010] directions on the same chip are discussed.

Magnetoresistance scans along multiple ϕ directions are shown in Fig. 6.16. The observed behavior is due to Anisotropic magnetoresistance

(AMR), that is, the fact that the resistivity of a ferromagnetic material depends on the angle between the current and its magnetization (Jan 1957; McGuire and Potter, 1975). The angle θ between magnetization and current can then be inferred from the resistance R at any field value through:

$$\rho_{xx} = \rho_\perp - (\rho_\perp - \rho_\parallel)\cos^2(\theta) \qquad (6.3)$$

and from the magnetization behavior, the anisotropy of the (Ga,Mn)As stripes can be deduced. The left part of Fig. 6.16 presents MR scans on the nanobar along the [010] crystal direction at various temperatures, while the right column shows the same set of measurements on a [100]-oriented nanobar. All of the panels in Fig. 6.16 exhibit uniaxial magnetic anisotropy along the current direction.

We first focus on a description of the panels on the left side of the figure. The thick lines are the field sweeps along 0°, which is the long axis of the [010]-oriented nanobar, and which is obviously also the current direction. Magnetic field sweeps along 0° yield a low resistance curve at all temperatures, indicating through Eq. (6.3) that M remains collinear with the current direction throughout the whole field sweep, as expected for a measurement along the uniaxial magnetic easy axis.

When the field is swept perpendicular to the nanobar (top curve in each panel), the large values of the resistivity at high magnetic fields confirm that the magnetization is forced perpendicular to the bar. The resistance decreases monotonically as the field is swept down to zero, because the magnetization rotates towards the easy axis direction.

MR scans along other angles between 0° and 90° (in 10° steps) are also shown as black lines. In all these curves, we observe the lowest resistance state at zero external field, confirming that the nanobar axis is the magnetic easy axis in the whole temperature range. The observed switching fields are consistent with the SQUID measurements. The linear background is due to the isotropic magnetoresistance (Matsukura et al., 2004). It is more noticeable in the high temperature data because the change of scales in the figure and the reduction of the AMR signal has enhanced its relative contribution.

The right column of Fig. 6.16 presents results for the [100]-oriented nanobar. Since the coordinate system is fixed to the crystallographic axes, and not the axis of the nanobar, the fully opposite MR properties clearly indicate that the uniaxial behavior is related to the elongated shape of the nanobar. The thick line, where the field was swept orthogonal to the nanobar (again along 0°) is a typical hard axis magnetoresistance scan at all temperatures. The parent layer easy axis perpendicular to the wire has thus been overwritten by the patterning process and the lithographically imposed uniaxial anisotropy is the dominant anisotropy up to T_c, as was seen in the magnetization investigations.

Spintronic Nanodevices 273

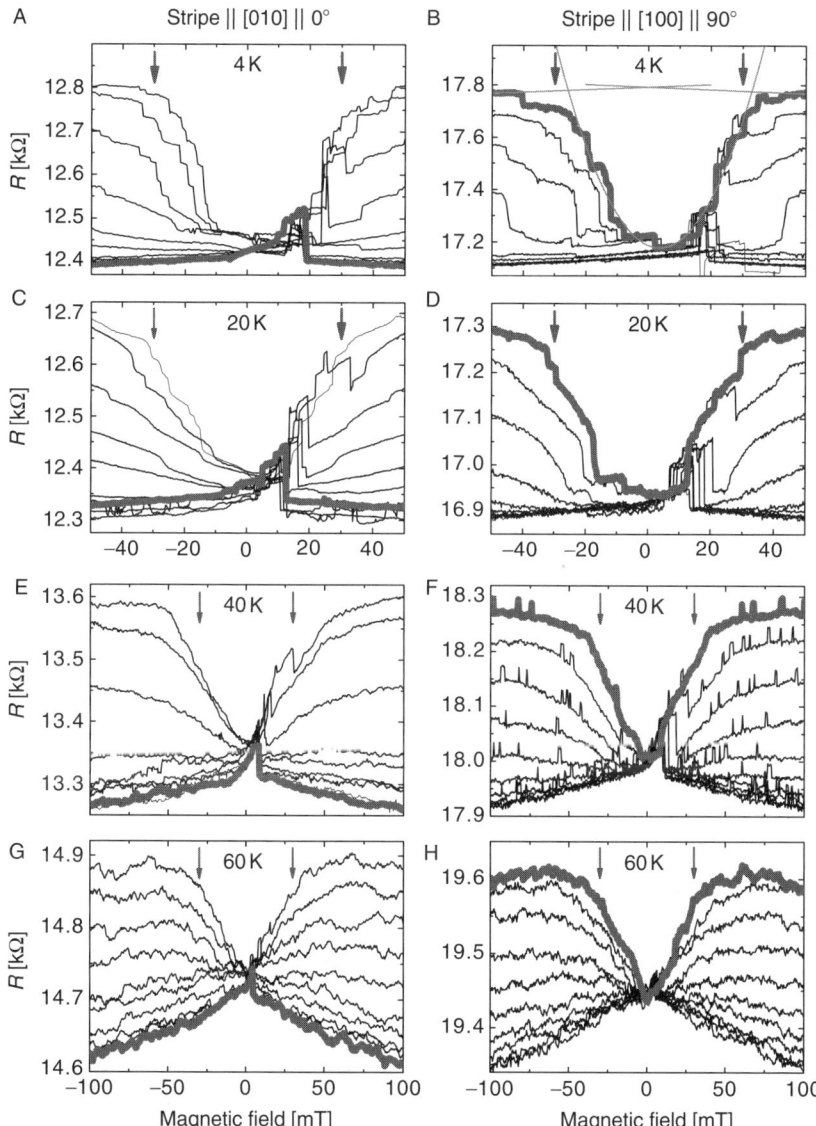

FIGURE 6.16 Magnetoresistance scans for angles between 0° and 90° in 10° steps on the bars along [010] (left column) and [100] (right) at various temperatures and 5 mV bias voltage. The thick line indicates the field sweep along the [010] crystal direction. The arrows indicate 30 mT an estimate for the uniaxial anisotropy field.

We can now estimate the strength of this lithographically imposed anisotropy. The hard axis MR-scan would be parabolic if only a pure uniaxial anisotropy was present and the field necessary to force the magnetization perpendicular to the easy axis a direct measure for the strength of the anisotropy. To estimate this anisotropy field, we fit a parabola (West, 1960) to the low field data of the perpendicular field scan in Fig. 6.16B and interpolate the isotropic magnetoresistance of this scan back to the origin (thin grey lines). The fitted parabola is slightly shifted towards positive fields, which indicates the presence of a small biaxial anisotropy contribution. The intersections between the grey lines and the parabola give $\mu_0 H_a \sim 30$ mT. The same number (marked with arrows) is a reasonable estimate for the anisotropy field at all temperatures and for both nanobar orientations. This indicates that indeed the lithography induced uniaxial anisotropy is almost unchanged between 4 and 60 K. The latter is a strong indication that the present effect is fundamentally different from classic shape anisotropy, which depends on the volume magnetization, and thus decreases with increasing temperature until it vanishes at T_c. Moreover, while size effects may play a role in the observed increase of the coercive field, they would play no role in modifying the anisotropy.

13. ANISOTROPIC STRAIN RELAXATION

The results presented above are not direct evidence of strain relaxation. Such direct confirmation requires X-ray diffraction measurements which are not possible on the small structures investigated here. We can nevertheless verify that strain relaxation is the important agent using X-ray diffraction measurements on long and narrow etched (Ga,Mn)As stripes (Wenisch et al., 2006a,b).

A SEM picture of such long stripes is shown in Fig. 6.17 along with a finite element simulation of the lattice displacement, that suggests the presence of anisotropic strain relaxation.

Transport measurements on an ensemble of ca. 250 long stripes contacted in parallel are shown in Fig. 6.18A. The MR measurements along in-plane angles between 0° and 90° in 10° steps clearly confirm that the dominating magnetic anisotropy component is uniaxial and has its easy axis along the bars.

The uniaxial anisotropy of the long stripes is also confirmed by the SQUID measurements shown in Fig. 6.18B. Here the scan along the bar is clearly an easy axis loop and the hysteresis loop perpendicular to the nanobar in the layer plane shows typical hard axis behavior. Again, the parent layer anisotropy, where both these directions are equivalent magnetic easy axes (as shown in the inset of Fig. 6.18B, the two curves are

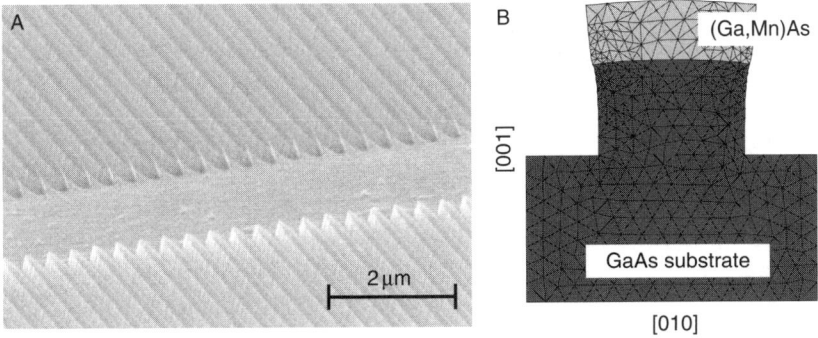

FIGURE 6.17 (A) SEM image of 1 μm long and 200 nm wide (Ga,Mn)As stripes for X-ray investigations. (B) Finite element simulation of the lattice displacement (100 times exaggerated) after strain relaxation in a cross-section of a 200 nm wide stripe.

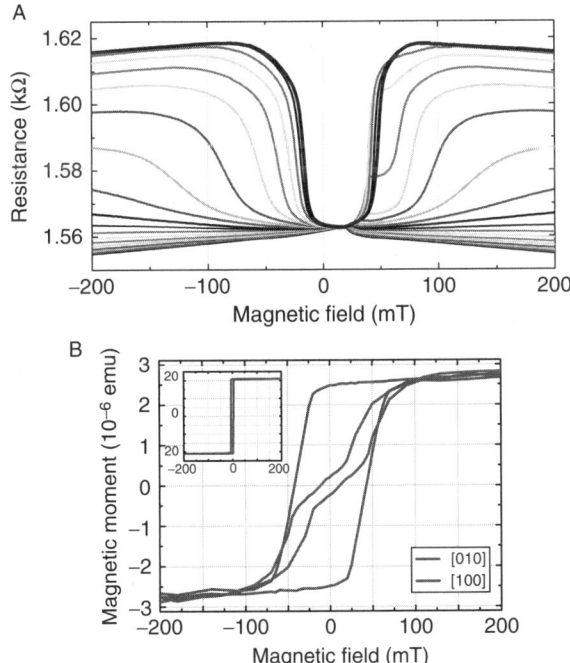

FIGURE 6.18 (A) Transport and (B) SQUID confirmation of the dominating uniaxial magnetic anisotropy term induced by lithographic patterning in the long stripes. The inset in (B) shows the same SQUID measurements on the unpatterned (Ga,Mn)As parent layer for comparison.

identical on this scale) has obviously been overwritten by the patterning process. Additionally, as in the nanobars in the previous section, the coercivity in the stripes is significantly increased compared with the unpatterned layer.

Grazing incidence X-ray diffraction (GIXRD) studies were performed on these stripes using the Hamburg Synchrotronstrahlungslabor (HASY-LAB) (Wenisch et al., 2006b). The strain relaxation in the (Ga,Mn)As stripes was then determined from GIXRD reciprocal space mapping in the vicinity of the (3 3 3) Bragg reflection. These measurements showed that the lattice constant parallel to the stripes is unchanged with respect to that of the fully strained parent layer. Perpendicular to the stripe direction, the lattice however shows a large degree of relaxation.

To further elucidate the relationship between the observed strain relaxation and the magnetic properties, we present the results of k.p calculations of the magnetization-direction dependence of the mean energy per hole. Assuming nanobars oriented along the [100] direction such that they permit only strain relaxation along [010], Fig. 6.19 shows the evolution of this energy as a function of magnetization angle for various degrees of uniaxial strain relaxation. The equivalent energy minima along [100] and [010] of the fully biaxially strained (Ga,Mn)As parent layer gradually morphs into a uniaxial behavior as the degree of strain relaxation in the [010] direction is increased. Assuming homogeneous strain (ε) distribution in the stripes and a reasonable carrier density of

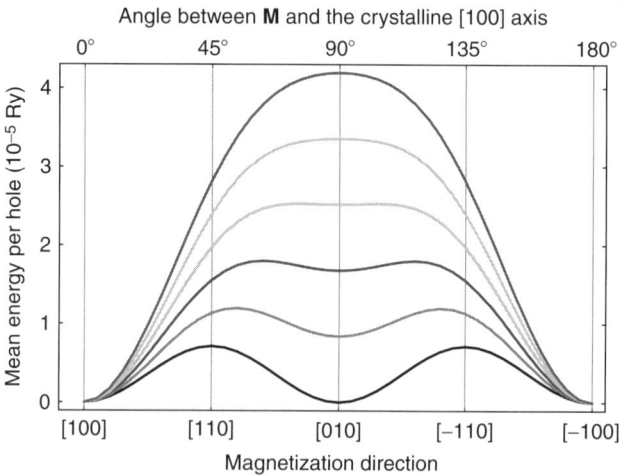

FIGURE 6.19 Energy per valence band hole for a layer with various degrees of strain in the [010] direction ranging from the pseudomorphic biaxial case (bottom line) to the fully relaxed case (top line). The [100] direction, along the long axis of the bars, is kept fully strained.

$4 \times 10^{20}\,\text{cm}^{-3}$, we can deduce from the calculations that the uniaxial contribution from strain dominates over the biaxial term to such a degree, that above $\varepsilon = -0.6 \times 10^{-3}$, only a single stable axis remains; along [100]. This is characteristic of primarily uniaxial behavior along the nanobars. Given that that X-ray data of our patterned samples exhibit strain levels that exceed this number, this is highly suggestive that the strain induced relaxation is responsible for the modification of the anisotropy.

The transport and magnetization results from both the small and large bars, when combined with the X-ray studies and theoretical modeling, form a compelling set of evidence that lithographically induced strain relaxation can be a powerful agent in controlling the anisotropy in FSs. We stress that this effect is not an extension of a previously known anisotropy term such as shape or size effects, but rather is a fundamentally novel contributor, which stems from the deformation of the lattice structure of the material during strain relaxation. Due to the unique material properties which couple the magnetization states in FSs to their lattice, these locally induced anisotropic lattice deformations are reflected in the magnetic properties. We believe this will prove a useful tool for studying novel spintronics effects related to transport between regions of different anisotropies or unique magnetization configurations within a layer.

14. MEMORY DEVICE USING LOCAL ANISOTROPY CONTROL

This lithographic anisotropy control greatly enhances the scope of possible device paradigms open to investigation as it allows for devices where the functional element involves transport between regions with different magnetic anisotropy properties. A first such device is shown in Fig. 6.20. It is comprised of two (Ga,Mn)As nanobars, oriented perpendicular to each other and with each nanobar exhibiting strong uniaxial magnetic anisotropy. These two nanobars are electrically connected through a constriction whose resistance is determined by the relative magnetization states of the nanobars. We show that the anisotropic magnetoresistance effect yields different constriction resistances depending on the relative orientation of the two nanobar-magnetization vectors.

This structure is a very basic FS memory device that operates in the nonvolatile regime. It offers a new method of storing information in a magnetic semiconductor by encoding it into the relative magnetization orientation of the two orthogonal bars. The information can be read-out through a measurement of the resistance of the constriction which, in the case of partially depleted constrictions, offers a tuneable On/Off-resistance ratio with values up to 280% having been achieved to date. In combination with the above mentioned progress in current-assisted DW motion, this

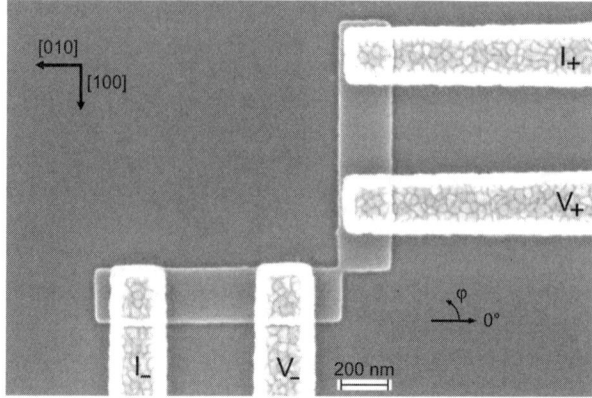

FIGURE 6.20 SEM photograph of the device identifying the orientation of the nanobars with respect to the crystal directions, and the definition of the current (I_+,I_-) and voltage (V_+,V_-) leads and the writing angle ϕ.

scheme constitutes a significant step towards an all electrical semiconductor memory architecture.

15. DEVICE OPERATION

Fabrication of the device is as described for the individual nanobars. The bars are 200 nm wide and 1μm long and oriented along the [100] and [010] crystal direction. They form a 90°-angle and touch each other in one corner, where a constriction with a width of some tens of nm is formed. We discuss characteristics of two typical structures that are representative for two limits of device behavior that we have observed as the constriction resistance varies from the diffusive transport regime to one where hopping transport dominates the conductance.

Transport measurements are carried out at 4.2 K. The sample state is first "written" by an in-plane magnetic field of 300 mT along a writing angle ϕ (as defined in Fig. 6.20). The field is then slowly swept back to zero while ensuring that the magnetic field vector never deviates from the ϕ-direction. The four-terminal resistance of the constriction in the resulting remanent state is determined by applying a voltage V_b to the current leads (I_+ and I_-), and recording both the voltage drop between contacts V_+ and V_- and the current flowing from I_+ to I_-. The polar plot of Fig. 6.21 shows the constriction resistance of the remanent magnetization state as a function of the writing angle ϕ. This resistance of course includes a small contribution from the ends of the nanobars, but is dominated by the constriction. It shows a higher value upon writing the

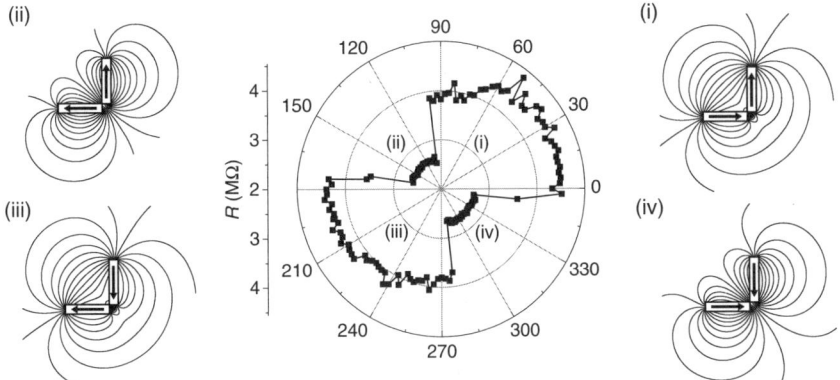

FIGURE 6.21 Polar plot showing the results of a "write–read" experiment. The state of the device is written by applying a magnetic field of 300 mT in the ϕ direction. This field is then swept back to zero, and the resistance of the device is measured. The insets sketch the magnetic configuration of the device in each quadrant and the corresponding field line patterns.

sample in the (extended) first quadrant ($-3° \leq \phi < 98°$) and a lower value upon writing in the (shrunken) second quadrant ($98° < \phi < 167°$). The third and forth quadrant obviously behave as their point symmetric counterparts.

16. MAGNETIC STATES

In order to explain these results, we first confirm that our method for lithographically inducing uniaxial anisotropy also works for coupled nanobars. To do so, we perform two-terminal resistance measurements on each of the nanobars, and determine that indeed they exhibit the same uniaxial character as shown for the individual bars in Fig. 6.16.

Knowing that each bar shows a uniaxial magnetic easy axis along its long axis, the structure clearly has four possible magnetic states at zero magnetic field as sketched in the insets of Fig. 6.21. In sectors (i) and (iii) the nanobars are magnetized "in series," that is, the magnetization vectors meet in a configuration which we will call *head-to-tail*. In (ii) and (iv) on the other hand, both magnetization vectors point away from (*tail-to-tail*) or towards (*head-to-head*) the constriction.

When, during the writing process, the sample is magnetized along a given direction at 300 mT, the magnetization of both bars is brought parallel to the magnetic field. As the field is subsequently lowered to zero, the magnetization of each nanobar relaxes to the respective nanobar easy axis, selecting the direction which is closest to the writing angle ϕ.

For a nanobar along $0°$ this means, assuming no interaction between the bars, that **M** relaxes to $0°$ upon writing the bar along any angle between $+90°$ and $-90°$; otherwise **M** relaxes to $180°$. If the bars in our device were completely noninteracting, one would thus expect the magnetization configuration in each quadrant to be as depicted in the sketches of Fig. 6.21, with each quadrant accounting for exactly one fourth of the total plot.

The deviation from this behavior is due to magnetostatic interactions between the two bars, which cause a preference for head-to-tail configurations. A simple magnetostatic calculation shows that the repulsive field felt by the tip of one bar due to being near the wrong pole of the other bar is of the order of 2 mT, which is $\sim 5\%$ of the uniaxial anisotropy field. The energy density of this field is thus strong enough to overcome a small fraction of the energy barrier against rotation towards the opposite magnetization direction, which corresponds to an angle of $\sim 3°$. The head-to-tail quadrants thus increase commensurably.

Magnetic field line patterns for the four magnetization configurations calculated using a simple bar magnet model are sketched in the insets. At the constriction, the field lines are close to parallel to the current in the head-to-tail configuration of quadrants i and iii, whereas in the tail-to-tail and the head-to-head configuration of quadrants ii and iv, the field lines are approximately perpendicular to the current.

17. ORIGIN OF THE RESISTANCE SIGNAL

Having understood the magnetic configurations of the device, we now discuss why these should lead to two very distinct resistance states. From the above magnetostatic arguments and associated field line sketches, in connection with the AMR for metallic (Ga,Mn)As described by Eq. (6.3), we can explain a few percent resistance difference between the head-to-tail and the head-to-head configuration, much smaller and of a different sign than the effect in Fig. 6.21. We have indeed observed such a small AMR related effect in similar structures (Pappert et al., 2007), which have a wider constriction (and correspondingly, 100 times lower constriction resistance).

The much larger effect shown in Fig. 6.21 can only be explained by taking into consideration effects beyond AMR, and as it turns out, may be related to a phenomenon which in some ways is connected to the large amplification of TAMR discussed previously. It results from depletion in the narrow constriction which drives the transport in the critical constriction region into the hopping regime. We suggest that in the hopping regime the AMR coefficient *changes sign*, leading to the observed changes in magnetoresistance. Important evidence for this claim comes from the angle-dependent magnetoresistance behavior of the samples at a field of 300 mT, strong enough to force the magnetization close to parallel to the external field. This data is given in Fig. 6.22 for the high-resistance sample.

Similar data on low-resistance devices exhibit typical AMR behavior as expected for metallic (Ga,Mn)As with the lowest resistance occurring when **M** is forced parallel to the current through the constriction ($\phi \sim 45°$) and ca. 3% higher for **M**⊥**J**. In contrast, the high-resistance constriction of the device in Fig. 6.22 shows a huge and inverted AMR signal. The resistance at $\phi \sim 45°$, where **M** ∥ **J**, is more than five times larger than for **M**⊥**J**.

This is not the first observation of an inverted AMR signal; the same effect has recently been reported in thin (Ga,Mn)As devices (Rushforth *et al.*, 2006; Wunderlich *et al.*, 2006) in which the transport is in the hopping regime. This situation is similar to our device, where from the resistance one already can infer that the constriction is partially depleted. Evidence for hopping transport comes also from the current–voltage characteristics of the high-resistance constriction, which are very nonlinear, and with a degree of nonlinearity which depends on the magnetization direction. Fields in directions far away from the current cause the strongest nonlinearity whereas fields nearly parallel to the current produce the least nonlinear IV-curve.

The strong dependence of the IV-characteristic and the resistance on the magnetization direction are characteristic of transport going through an MIT from the diffusive into the hopping regime depending on the angle of the magnetization as previously discussed in our TAMR device and occurs in partly depleted samples due to the wave-function geometry

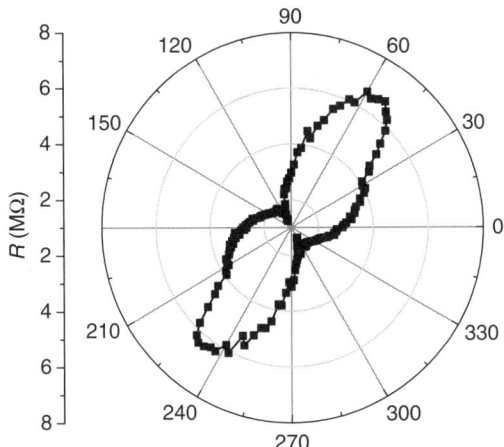

FIGURE 6.22 Constriction resistance in a rotating 300 mT external magnetic field.

change depending on the magnetization direction. The localized hole wave-function has an oblate shape with the smaller axis pointing in the magnetization direction as shown in Fig. 6.23A (Schmidt et al., 2007). Consider the overlap of such oblate shapes statistically distributed with respect to the direction of the current in connection with the Thouless localization criterion. As depicted schematically in Fig. 6.23B, the wave-function overlap is much smaller when the sample is magnetized parallel to the current, than for $M \perp J$, suppressing hopping transport through the depleted constriction region. This implies a magnetoresistance behavior that is exactly the inverse of that expected for the metallic regime and explains the increased resistance value in both the high field measurements of Fig. 6.22 (along ~50°) and the write–read experiment of Fig. 6.21.

The behavior of the device can thus be fully explained by the internal magnetic fields and the AMR coefficient as applicable to the transport regime in the constriction. We note that a further potential candidate to explain the behavior might have been the presence of a DW in the constriction as was the case in the original nanoconstrictions device discussed earlier. In such a picture, a DW occurs between the differently magnetized regions of the device in the head-to-head and tail-to-tail configuration, and would be absent in the head-to-tail configurations. Given the dimensions of the present constriction, in the absence of an external magnetic field, such a DW would not be strongly geometrically confined and one anticipates only a very low DW resistance (Bruno, 1999;

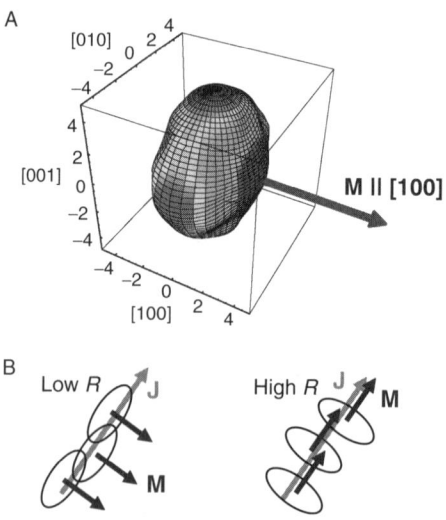

FIGURE 6.23 (A) Wavefunction of the semi-localized holes at the Mn impurities. (B) Schematic depicting the mechanism invoked to explain the large amplitude effect. (See Color Plate.)

Rüster et al., 2003). This is confirmed by a comparison of Fig. 6.21 with Fig. 6.22. The resistance values of both remanent states are in between the extreme resistance values of the homogeneously magnetized sample, which clearly does not contain any DW. The DW contribution to the constriction resistance can in the present sample thus only be a minor effect on the resistance of the remanent state and does not explain the different resistance levels in Fig. 6.22. Note also that the resistance of the head-to-head configuration, including a possible DW contribution, is lower than the resistance of the head-to-tail configuration. We can thus exclude the DW as the origin of the two remanent resistance states.

To further verify that the experiments described so far do not exhibit a DW effect, we have designed an experiment (Pappert, 2007) that would actually bring us in this limit. Using well-defined field sequences, it is possible to pin DW at the smallest part of our depleted constriction. In this case, we observe a large positive DW contribution to the device resistance exactly as that reported in Fig. 6.10. We emphasize that this is a secondary effect which occurs in addition to the main device operation described hear, and that it only occurs for special field sequences, and not as a result of simple writing sequences.

This work thus demonstrates that locally imposed magnetic anisotropies in different regions of one FS device allows for novel device designs. The orthogonally magnetized nanobars discussed in this paper are a first demonstration of the type of devices that can be fabricated using this approach. We have shown that this device exhibits two stable magnetization states, defined by the relative orientation of the two bars, which can be used for nonvolatile information storage. Measuring the constriction resistance allows for the electrical read-out of the magnetic state. We emphasized that the corresponding on/off resistance ratio can be amplified by several orders of magnitude using a partly depleted constriction and explained the physical origin of this effect. In this respect we highlighted the difference in AMR behavior between metallic and hopping transport in (Ga,Mn)As, which again should prove useful in future device design. The described mechanism has the potential to be integrated with current-induced magnetization manipulation, from both device design and dimensions perspective. This would yield an all electrical semiconductor memory cell.

18. CONCLUSION AND OUTLOOK

In the present chapter, we have shown laboratory demonstrations of various implementations of percussor devices or device elements based on FSs. These devices have shown that the combination of magnetic and semiconductor properties into a single material offers not only great versatility in novel device designs but also the potential for fundamentally new physical phenomena.

In particular, we have discussed in detail devices based on TAMR, a new magnetoresistance effect originating from the properties of tunneling into materials with a DOS having magnetic anisotropy. We have seen how a detailed study of this effect has led to several novel effects including the first observation of a metal to insulator transition induced through a change in magnetic field *direction*. It also shows promise for new device functionality based on partial (less then 180°) reversal of a magnetic layer in a scheme which may even be extendable to beyond binary storage by using the multiple interface version of the effect.

We then discussed multiple devices demonstrating many of the elements needed as building blocks for eventual spintronics-based information storage and logic paradigms based on encoding the information into the local magnetic state of specific FS device elements. All functional needs of such a device, including information storage and read out, tunable on/off ratios and electrical control have been demonstrated in device schemes with inter-compatibility.

These devices, while for the present confined to laboratory demonstrator status due to material issues with their Curie temperatures, thus form a solid basis for beginning the elaboration of more complex FS-nanostructure based circuitry, where the focus begins to shift from the functionality of the individual elements, and to the interaction of many such elements. This, along with the ongoing material work to increase T_c and thus improve the technological relevance of these schemes, will be important elements of leading edge spintronics research in the coming years.

ACKNOWLEDGMENTS

Much of the work presented here are results of the PhD activities of C. Rüster and K. Pappert and the many other students which have participated in our research team on FSs over the past six years. We thank them all for both their experimental efforts and the many interesting discussion they have stimulated during this research.

REFERENCES

Abolfath, M., Jungwirth, T., Brum, J., and MacDonald, A. (2001). *Phys. Rev. B* **63**, 054418.
Aharoni, A. (1998). *J. Appl. Phys.* **83**, 3432.
Åkerman, J. (2005). *Science* **308**, 508.
Allwood, D., Xiong, G., Faulkner, C., Atkinson, D., Petit, D., and Cowburn, R. (2005). *Science* **309**, 1688.
Altshuler, B., and Aronov, A. (1979). *Sov. Phys. JETP* **50**, 968.
Awschalom, D. D., and Flatté, M. E. (2007). *Nature Physics* **3**, 153.
Baibich, M. N., Broto, J. M., Fert, A., Dau, F. N. V., Petroff, F., Eitenne, P., Creuzet, G., Friederich, A., and Chazelas, J. (1988). *Phys. Rev. Lett.* **61**, 2472.

Binasch, G., Grünberg, P., Saurenbach, F., and Zinn, W. (1989). *Phys. Rev. B* **39**, 4828.
Bolotin, K., Kuemmeth, F., and Ralph, D. (2006). *Phys. Rev. Lett.* **97**, 127202.
Brey, L., Fernndez-Rossier, J., and Tejedor, C. (2004). *Phys. Rev. B* **70**, 235334.
Bruno, P. (1999). *Phys. Rev. Lett.* **83**, 2425.
Chantis, A. N., Belashchenko, K. D., Tsymbal, E. Y., and van Schilfgaarde, M. (2007). *Phys. Rev. Lett.* **98**, 046601.
Cowburn, R., Gray, S., Ferré, J., Bland, J., and Miltat, J. (1995). *J. Appl. Phys.* **78**, 7210.
Dietl, T., Ohno, H., Matsukura, F., Cibert, J., and Ferrand, D. (2000). *Science* **287**, 1019.
Dietl, T., Ohno, H., and Matsukura, F. (2001). *Phys. Rev. B* **63**, 195205.
DiVincenzo, D. P., Burkard, G., Loss, D., and Sukhorukov, E. (2000). "Quantum computation and spin electronics," *Quantum Mesoscopic Phenomena and Mesoscopic Devices in Microelectronics (see also condmat:9911245)*. p. 399.
Efros, A., and Shklovskii, B. (1975). *J. Phys. C* **8**, L49.
Flatté, M. E., and Vignale, G. (2001). *Appl. Phys. Lett.* **78**, 1273.
García, N., Munoz, M., and Zhao, Y. -W. (1999). *Phys. Rev. Lett.* **82**, 2923.
Giddings, A. D., Khalid, M. N., Jungwirth, T., Wunderlich, J., Yasin, S., Campion, R. P., Edmonds, K. W., Sinova, J., Ito, K., Wang, K. -Y., Williams, D., Gallagher, B. L., and Foxon, C. T. (2005). *Phys. Rev. Lett.* **94**, 127202.
Gould, C., Rüster, C., Jungwirth, T., Girgis, E., Schott, G., Giraud, R., Brunner, K., Schmidt, G., and Molenkamp, L. (2004). *Phys. Rev. Lett.* **93**, 117203.
Gould, C., Pappert, K., Rüster, C., Giraud, R., Borzenko, T., Schott, G. M., Brunner, K., Schmidt, G., and Molenkamp, L. W. (2006). *Jpn. J. Appl. Phys.* **45**, 3860.
Gould, C., Pappert, K., Schmidt, G., and Molenkamp, L. (2007). *Adv. Mater.* **19**, 323.
Hamaya, K., Moriya, R., Oiwa, A., Taniyama, T., Kitamoto, Y., and Munekata, H. (2003). *IEEE Trans. Magn.* **39**, 2785.
Hamaya, K., Taniyama, T., Koike, T., and Yamazaki, Y. (2006). *J. Appl. Phys.* **99**, 123901.
Hümpfner, S., Pappert, K., Wenisch, J., Brunner, K., Gould, C., Schmidt, G., Molenkamp, L. W., Sawicki, M., and Dietl, T. (2007). *Appl. Phys. Lett.* **90**, 102102.
Jan, J. P. (1957). "Solid State Physics" (F. Seitz and D. Turnbull, eds.), Academic Press Inc, New York.
Julliere, M. (1975). *Phys. Rev. A* **54**, 225.
Kent, A., Yu, J., Rüudiger, U., and Parkin, S. (2001). *J. Phys: Condens. Matter* **13**, R461.
Larkin, A., and Shklovskii, B. (2002). *Phys. stat. sol. (b)* **230**, 189.
Lee, M., Massey, J., Nguyen, V., and Shklovskii, B. (1999). *Phys. Rev. B* **60**, 1582.
Loss, D., and DiVincenzo, D. P. (1998). *Phys. Rev. A* **57**, 120.
Luttinger, J., and Kohn, W. (1955). *Phys. Rev B* **97**, 869.
Matos-Abiague, A., and Fabian, J. (2007). *Condmat:0702387*.
Matsukura, F., Ohno, H., Shen, A., and Sugawara, Y. (1998). *Phys. Rev. B* **57**, R2037.
Matsukura, F., Sawicki, M., Dietl, T., Chiba, D., and Ohno, H. (2004). *Physica E* **21**, 1032.
McGuire, T., and Potter, R. (1975). *IEEE Trans. Magn.* **MAG-11**, 1018.
Moser, J., Matos-Abiague, A., Schuh, D., Wegscheider, W., Fabian, J., and Weiss, D. (2006). *Condmat:0611406*.
Ohno, H., Shen, A., Matsukura, F., Oiwa, A., Endo, A., Katsumoto, S., and Iye, Y. (1996). *Appl. Phys. Lett.* **69**, 363.
Pappert, K. (2007). *PhD thesis, Würzburg University*.
Pappert, K., Schmidt, M. J., Hümpfner, S., Rüster, C., Schott, G. M., Brunner, K., Gould, C., Schmidt, G., and Molenkamp, L. W. (2006). *Phys. Rev. Lett.* **97**, 186402.
Pappert, K., Hümpfner, S., Gould, C., Wenisch, J., Brunner, K., Schmidt, G., and Molenkamp, L. (2007). *Condmat:0701478*.
Parkin, S. (2003). *U.S. Patent No. 6,834,005*.
Pollak, F. (1965). *Phys. Rev.* **138**, 618.
Raikh, M., and Ruzin, I. (1987). *Sov. Phys. JETP* **65**, 1273.

Rushforth, I., et al. (2006). *Phys. Stat. Sol. (c)* **3**, 4078.
Rüster, C., Borzenko, T., Gould, C., Schmidt, G., Molenkamp, L., Liu, X., Wojtowicz, T., Furdyna, J., Yu, Z., and Flatté, M. (2003). *Phys. Rev. Lett.* **91**, 216602.
Rüster, C., Gould, C., Jungwirth, T., Sinova, J., Schott, G. M., Giraud, R., Brunner, K., Schmidt, G., and Molenkamp, L. W. (2005a). *Phys. Rev. Lett.* **94**, 027203.
Rüster, C., Gould, C., Jungwirth, T., Girgis, E., Schott, G., Giraud, R., Brunner, K., Schmidt, G., and Molenkamp, L. (2005b). *J. Appl. Phys.* **97**, 10C506.
Sandow, B., Gloos, K., Rentzsch, R., Ionov, A., and Schirmacher, W. (2001). *Phys. Rev. Lett.* **86**, 1845.
Schmidt, G., Ferrand, D., Molenkamp, L. W., Filip, A. T., and van Wees, B. J. (2000). *Phys. Rev. B* **62**, R4790.
Schmidt, M., Pappert, K., Gould, C., Schmidt, G., Oppermann, R., and Molenkamp, L. (2007). *Condmat:0704.2028*.
Shick, A., and Mryasov, O. (2003). *Phys. Rev. B* **67**, 172407.
Tatara, G., and Kohno, H. (2004). *Phys. Rev. Lett.* **92**, 086601.
Theis, T., and Horn, P. (2003). *Phys. Today* **56**, 44.
Valet, T., and Fert, A. (1993). *Phys. Rev. B* **48**, 7099.
Vignale, G., and Flatté, M. E. (2002). *Phys. Rev. Lett.* **89**, 098302.
Viret, M., Gabureac, M., Ott, F., Fermon, C., Barreteau, C., and Guirado-Lopez, R. (2006). *Eur. Phys. Journ. B* **51**, 1.
Wang, K., Sawicki, M., Edmonds, K., Campion, R., Maat, S., Foxon, C., Gallagher, B., and Dietl, T. (2005). *Phys. Rev Lett.* **95**, 217204.
Wegrowe, J. -E., Wade, T., Hoffer, X., Gravier, L., Bonard, J. -M., and Ansermet, J. -P. (2003). *Phys. Rev. B* **83**, 104418.
Wenisch, J., Ebel, L., Gould, C., Schmidt, G., Molenkamp, L., and Brunner, K. (2006a). *MBE2006 Abstract Workbook* p. 63.
Wenisch, J., Gould, C., Ebel, L., Storz, J., Pappert, K., Schmidt, M., Kumpf, C., Schmidt, G., Brunner, K., and Molenkamp, L. (2006b). *condmat:0701479*.
West, F. (1960). *Nature* **188**, 129.
Wolf, S., Awschalom, D., Buhrman, R., Daughton, J., von Molnar, S., Roukes, M. L., Chtchelkanova, A., and Treger, D. (2001). *Science* **294**, 1488.
Wunderlich, J., Jungwirth, T., Kaestner, B., Irvine, A., Shick, A., Stone, N., Wang, K. -Y., Rana, U., Giddings, A., Foxon, C., Campion, R., Williams, D., and Gallagher, B. (2006). *Phys. Rev. Lett.* **97**, 077201.
Yamanouchi, M., Chiba, D., Matsukura, F., and Ohno, H. (2004). *Nature* **428**, 539.
Zutic, I., Fabian, J., and Sarma, S. D. (2004). *Rev. Mod. Phys.* **76**, 323.

CHAPTER 7

Quantum Structures of II–VI Diluted Magnetic Semiconductors

J. Cibert, L. Besombes, D. Ferrand, and H. Mariette

Contents
1. Magnetic and Electric Impurities in II–VI Nanostructures — 287
2. Carrier-Induced Ferromagnetism in 2D DMSs — 291
3. 0D Systems — 298
   3.1. Introduction — 298
   3.2. A single magnetic atom in a QD — 301
   3.3. Optical reading of the spin state of a single magnetic atom — 306
   3.4. Electrical control of a single magnetic atom — 310
   3.5. Conclusion — 315
4. Transport in Quantum II–VI DMS Structures — 316
5. Summary — 320
   References — 320

1. MAGNETIC AND ELECTRIC IMPURITIES IN II–VI NANOSTRUCTURES

Most of the interest in semiconductor physics is related first to our ability to dope the materials with electrically active impurities, but also, and perhaps even more, to the development of epitaxial growth, which allows

Institut Néel, CNRS-Université Joseph Fourier, BP166X, 38042 Grenoble cedex 9, France us to combine different semiconductors into heterostructures and nanostructures. In these, interfaces confine the electrons and holes, which allows us to manipulate these charge carriers using electric fields or light pulses. In diluted magnetic semiconductors (DMSs), doping with magnetically active impurities brings yet another degree of freedom.

Bulk DMSs were developed in the 1970s by introducing Mn impurities into II–VI semiconductors, and good review papers appeared in the 1980s (Furdyna, 1988; Furdyna and Kossut, 1988). These semiconductors are formed with a cation from column II (Zn, Cd, or Hg) and an anion from column VI (Te, Se, S, and more recently even O). These compounds assume the zinc-blende structure or, for the most ionic ones, the closely related wurtzite structure. Manganese impurities take the d^5 electronic configuration and substitute the cations up to 100%. The ground state is ^6S (or $^6A_1$ in cubic or hexagonal symmetry), introducing localized, isotropic spins with $S = 5/2$. If not interacting, these localized spins follow Maxwell-Boltzmann statistics, resulting in a magnetization M induced by an applied field H at temperature T given by a Brillouin function of H/T:

$$M = -xN_0 g_{Mn}\mu_B \frac{5}{2} B_{5/2}\left(\frac{5}{2}\frac{g_{Mn}\mu_B\mu_0 H}{k_B T}\right),$$

where x is the proportion of cations substituted by Mn, N_0 the density of cations in the zinc-blende or wurtzite structure, g_{Mn} (=2 to a good approximation) the Mn Landé factor, μ_B the Bohr magneton, and k_B the Boltzmann constant.

This would apply for noninteracting spins. In the actual material, antiferromagnetic "superexchange" interactions appear as soon as the Mn density is not vanishingly small. These interactions result in a reduction of the magnetization: this is quantitatively described by using a "modified Brillouin function" where x is replaced by a number of free spins x_{eff}, and the argument is $H/(T + T_{AF})$. The values of the two phenomenological parameters x_{eff} and T_{AF} are experimentally well documented in the most usual DMSs, such as $Cd_{1-x}Mn_xTe$ (Gaj et al., 1994). The number of free spins has a direct physical meaning: nearest-neighbor Mn pairs are blocked antiparallel by the strong superexchange interaction. It can be calculated from a statistics over Mn pairs and clusters either in the homogeneous material (Fatah et al., 1994; Shapira et al., 1984) or even at an interface with the nonmagnetic semiconductor (Grieshaber et al., 1996).

The description by a modified Brillouin function has proven to be very efficient at moderately low temperatures (1 to 30 K), moderate field (up to 5 T), and composition up to $x \approx 0.2$ (Gaj et al., 1994). A spin glass behavior appears at higher Mn contents in this temperature range, and at lower temperature even for a low Mn content. And nearest-neighbor pairs are no more blocked antiparallel at higher temperatures or higher fields.

It was recognized early that the key property of a DMS is the strong coupling between the bands of the semiconductor [conduction band (CB) and, even more strongly, valence band] and the localized spins. Optical spectroscopy around the bandgap reveals the so-called "giant Zeeman effect," with a spin splitting proportional to the Mn magnetization (Gaj et al., 1979). Several studies have demonstrated this proportionality and measured the strength of the coupling (Twardowski et al., 1984; Fig. 7.1).

As a result, magneto-optical spectroscopy is now a very sensitive method for measuring locally the magnetization of the Mn system. For example, it was applied (Gaj et al., 1994) to measure the diffusion of Mn across a (Cd,Mn)Te–CdTe interface during the growth by molecular beam epitaxy (MBE). Actually, the method is quite direct in semiconductors with intermediate values of the bandgap width (tellurides, selenides). In wide bandgap semiconductors, excitonic effects are important so that the splitting directly measured on the spectra is not proportional to the spin splitting of the bands (Pacuski et al., 2006).

Altogether, this excellent knowledge of the magnetic properties in (electrically) undoped II–VI DMSs, and of the coupling between the localized spins and the conduction/valence band, constitutes a very firm basis for further studies to be described below.

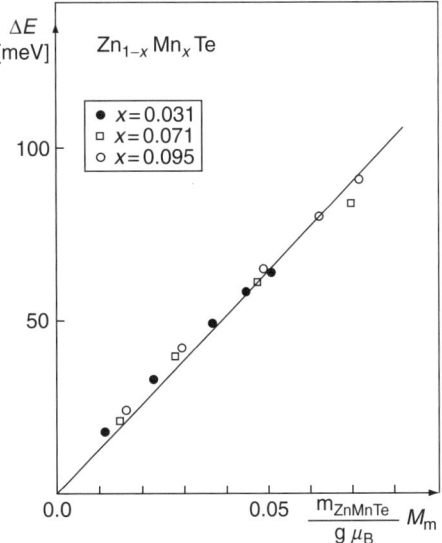

FIGURE 7.1 Splitting of the free exciton line. The straight line is the theoretical dependence calculated for $N_0(\alpha-\beta) = 1.29$ eV (Twardowski et al., 1984).

An important point is that Mn substitutes the column-II cation as an isoelectronic impurity—as opposed to the acceptor character observed in GaAs and similar III–Vs (Linnarsson et al., 1997; Schneider et al., 1987), where it gives rise to carrier-induced ferromagnetism without additional doping. Actually, doping with electrically active impurities was for a long time a major problem in II–VIs. Growth by MBE brought good solutions in some cases, at least for tellurides and selenides. Various impurities are used for n-type doping, and particularly indium (Bassani et al., 1992), iodine (Brun-Le Cunff et al., 1995; Fisher et al., 1994), and aluminum (Arnoult et al., 1999). Doping p-type is very efficient in ZnTe, using nitrogen atoms from a plasma cell (Grün et al., 1996); it is reasonably efficient in materials based on CdTe (Baron et al., 1998) and ZnSe (Fan et al., 1992; Park et al., 1990). Modulation doping is well controlled in MBE growth and it allows to build-in a two-dimensional electron gas (2DEG) or hole gas (2DHG) in a quantum well (QW). One has to pay attention to the possibility that doping causes a severe interdiffusion of the QW (Arnoult et al., 2000; Baron et al., 1997; Straßburg et al., 1998): this problem was solved by growing at lower temperature or by using electron traps that form at the surface upon oxidization (Maslana et al., 2003). The spectroscopic properties of a QW containing a 2DEG or a 2DHG were studied in details in both ZnSe- and CdTe-based structures, and the density and spin polarization of the carriers can be deduced from the spectra (Kossacki et al., 2004a).

Finally, one should note that the vapor pressure above the column-II or column-VI elements is much higher than above their compounds. As a result, stable surfaces are obtained under a flux of either the cation or the anion species—another difference with the III–Vs. This implies that atomic layer epitaxy (ALE) is feasible (Hartmann et al., 1996), as well as MBE under both cation-rich and anion-rich conditions. Cation-rich MBE growth on a (001) surface proceeds in a 2D mode (layer-by-layer) up to a sharply defined critical thickness (Cibert et al., 1990), which depends on the lattice mismatch and corresponds to a plastic relaxation. As far as the II–VI quantum dots (QDs) are concerned, the first works published for CdSe QDs in ZnSe (Xin et al., 1996) and CdTe in ZnTe (Terai et al., 1998) evidence zero-dimensional excitonic properties but did not reported *in situ* direct evidence of a spontaneous 2D–3D growth transition (the so-called Stranski-Krastanov mode). A modified procedure, which involves the reevaporation of an amorphous layer of the anion, has been developed later (Tinjod et al., 2004), in order to observe a clear 2D–3D transition and to get well-formed CdTe QDs (Tinjod et al., 2003) and CdSe QDs (Robin et al., 2006).

In what follows, we have chosen to focus onto three topics that we think to be particularly relevant in the context of spintronics: (1) carrier-induced ferromagnetism in (Cd,Mn)Te QWs, which constitute a model system for understanding the basic mechanisms, (2) the interaction of electrons and holes with a single magnetic impurity in a QD—an extreme limit for spintronics, and (3) a few examples of transport measurements in quantum DMS structures. Of course, that means that we deliberately omit a considerable body of experimental studies of II–VI DMS quantum structures, such as spectroscopic studies using the giant Zeeman effect as a tool, or studies of the dynamics of the various spins that are present and interact in such structures. We also restrict our short review to epitaxial structures, leaving apart nanocrystallites.

2. CARRIER-INDUCED FERROMAGNETISM IN 2D DMSs

Carrier-induced ferromagnetism is intensively studied on (Ga,Mn)As. Because of the peculiar position of the d levels of Mn with respect to the valence band in GaAs, Mn traps an electron from the valence band and assumes the d^5 configuration; that is, it behaves as an acceptor and carries a spin 5/2. The same configuration is obtained in (Zn,Mn)Te (Ferrand et al., 2001) or (Be,Mn)Te (Hansen et al., 2001) upon p-type doping with nitrogen or arsenic. The critical temperature is low but this allows to study carrier-induced ferromagnetism in another DMS and thus it improved our current understanding of the mechanisms involved.

Most interestingly, the fact that magnetic and electric doping are achieved by two different impurities allows remote doping of a (Cd,Mn)Te QW by doping the (Cd,Mg)Te barrier. This leads to ferromagnetism induced by a 2DHG.

A typical sample is an 8-nm-thick $Cd_{1-x}Mn_xTe$ QW, with $x = 0.02-0.05$, containing a 2DHG with a density of a few 10^{11} cm^{-2} due to nitrogen doping of the $Cd_{0.75}Mg_{0.25}Te$ barriers (Haury et al., 1997). As usual in II–VI DMSs, excitons in these QWs show a giant Zeeman effect, so that photoluminescence (PL) constitutes a very sensitive measure of the Mn magnetization in the QW (and of the susceptibility if we measure the slope with respect to the applied field). Indeed, the giant Zeeman effect is even enhanced in the presence of the 2DHG. Plotting the inverse of the susceptibility as a function of the temperature reveals a Curie-Weiss behavior, with a divergence at a finite temperature T_{CW}. Even more, when the temperature is decreased without applying a magnetic field, a single line is observed as expected above T_{CW}, but a zero field splitting appears below, confirming the presence of a spontaneous magnetization at low temperature.

The role of the 2DHG is essential. This general behavior is observed in (Cd,Mn)Te QWs containing a 2DHG transferred from near-by acceptors—either nitrogen in the barriers or from electron traps at the surface (Maslana et al., 2003). In the absence of the 2DHG, the susceptibility is smaller, the Curie-Weiss temperature is negative ($=-T_{AF}$ from the linear approximation of the modified Brillouin function), and the PL line does not split at zero field. This was checked not only in undoped samples but also by inserting the QW in a *p-i-n* structure that allows to modulate the hole density in the QW by applying an electric field across the structure (Boukari et al., 2002). That means that the magnetic properties are controlled by a bias voltage in a field effect structure (the basic structure of microelectronics). The bias voltage is as low as 1 V, lower than in previous demonstrations in III–V DMSs (Ohno et al., 2000), due to the low density of carriers in a QW and the small thickness of the whole structure. The carrier density in the QW can be changed also by proper illumination of the structure. In a QW-doped *p*-type on both sides, illumination decreases the hole density that induces a ferromagnetic to paramagnetic transition at fixed temperature (Kossacki et al., 2002). In a QW inserted in a *p-i-n* structure, illumination increases the hole density so that the spontaneous magnetization in the low-temperature phase increases (Boukari et al., 2002; Fig. 7.2).

A mean field model, used in the proposal of the experiment (Dietl et al., 1997), well describes these findings. This "Zener model" (Dietl et al., 2000; Zener, 1951) is the uniform limit of the RKKY interaction: this is justified by the low carrier density (hence long Fermi wavelength) involved in a semiconductor. It assumes a uniform magnetization of the system of localized spins and a uniform polarization of the 2DHG, and takes into account the symmetrical effect of the spin-hole coupling. This is a phenomenological model, or better, it is based on phenomenological descriptions of the two systems and of their coupling, which have been extensively validated by previous independent observations. It involves no adjustable parameters since all relevant quantities can be determined independently.

Basically, the giant Zeeman effect induces a spin splitting of the 2DHG proportional to the Mn magnetization, so that the 2DHG acquires (through the Pauli susceptibility χ_P) a spin polarization, which is also proportional to the Mn magnetization. Symmetrically, the expression of the Mn magnetization contains—in addition to the normal Zeeman effect—a term proportional to the spin polarization of the 2DHG. These two effects—the enhancement of the Pauli susceptibility through the giant Zeeman effect and the reversed effect similar to the Knight shift in NMR—are observed independently in the 2DEG contained in a (Cd,Mn) Te or (Zn,Mn)Se QW (Harris et al., 2001; König and MacDonald, 2003; Teran et al., 2002, 2003; Fig. 7.3).

When combined, these mutual effects result in a susceptibility that is further enhanced in the paramagnetic phase and diverges at a temperature T_C. Below T_C, the 2DHG is fully polarized (Dietl et al., 1997), and the Mn magnetization follows the Brillouin function, with in its argument the

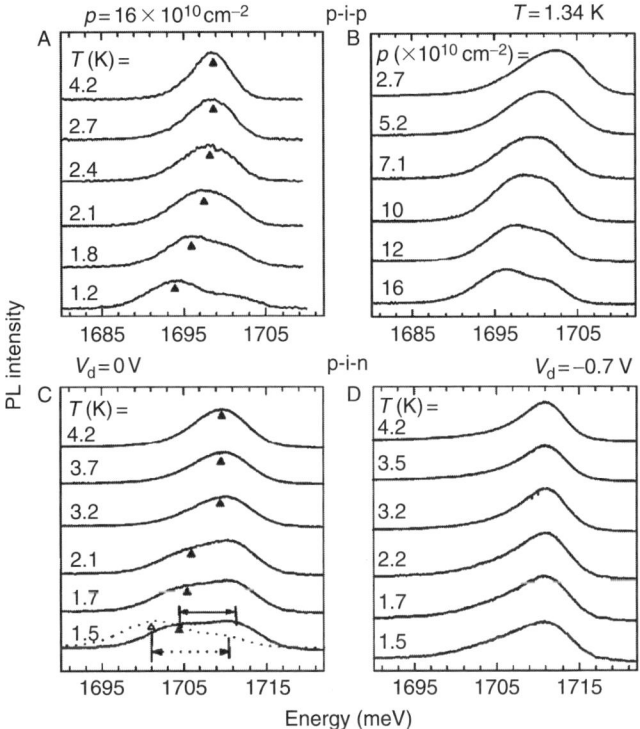

FIGURE 7.2 PL spectra for a modulation-doped p-type $Cd_{0.96}Mn_{0.04}Te$ QW located in a p-i-p structure and a modulation-doped p-type $Cd_{0.95}Mn_{0.05}Te$ QW in a p-i-n diode; (A) p-i-p structure without additional illumination (i.e., constant hole density) at various temperatures; (B) same p-i-p structure with above-barrier illumination (to reduce the hole density) at fixed temperature; (C) p-i-n structure without bias at various temperatures: the hole density is constant (solid lines) or increased (dotted line) by additional Ar-ion laser illumination; splitting and shift of the lines mark the transition to the ferromagnetic phase (D) p-i-n structure with a 0.7 V bias (depleted QW) at various temperatures. (Boukari et al., 2002).

exchange field with the fully polarized hole gas, hence a constant exchange field. As a result, it is expected that the spontaneous magnetization keeps increasing when the temperature is further decreased below T_C, but with an upward curvature. This peculiar shape, which is often attributed to disorder in 3D ferromagnetic DMSs, can thus be the signature of a full polarization of the carriers. This is particularly clear in a 2D system such as the (Cd,Mn)Te QWs, where the carrier density is low, and the saturation of the carrier gives rise to a distinct kink in the field dependence of the giant Zeeman splitting: below this kink, the susceptibility is enhanced by the collective spin-carrier behavior; at higher field,

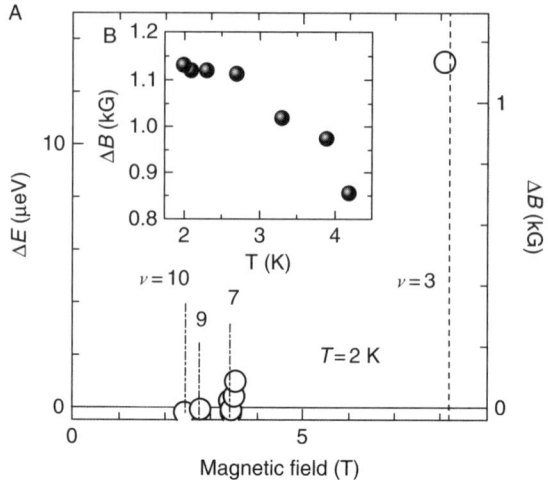

FIGURE 7.3 (A) Measured shift of the Mn^{2+} EPR line from the expected behavior. (B) Temperature dependence of the large shift observed close to filling factor $\nu = 3$ (Teran et al., 2003).

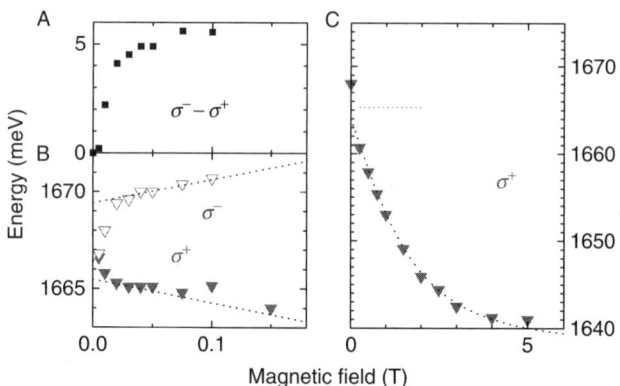

FIGURE 7.4 PL transition energies in a $Cd_{0.97}Mn_{0.03}Te$ QW with 1.6×10^{11} holes cm^{-2} versus applied field, at $T = 2K$, in the paramagnetic phase above the transition. (A) shows the energy difference between both circular polarizations, which is proportional to Mn magnetization [adapted from Cibert et al. (1998)]; (B) and (C) show different field ranges.

the localized spins follow a Brillouin function with an argument containing the sum of the applied field and the constant field due to the fully polarized carriers (Fig. 7.4).

Although evidently oversimplified, the mean field model appears particularly robust and explains qualitatively and often quantitatively

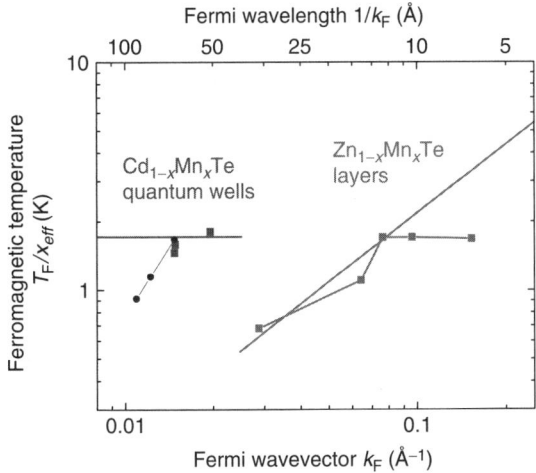

FIGURE 7.5 Normalized characteristic temperature T_F/x_{eff} measured in $Cd_{1-x}Mn_xTe$ quantum wells (circles), in (Zn,Mn)Te layers (squares) and calculated in the standard model (solid lines) [adapted from Cibert et al. (2002)].

the observations in (Cd,Mn)Te QWs. For example, the observed Curie-Weiss temperature well follows the prediction that $(T_{CW} - T_{AF})/x_{eff}$ is proportional to the Pauli susceptibility χ_P (Fig. 7.5).

In a noninteracting 2DHG, χ_P is a constant, independent of the temperature, with the effective mass as the only material-dependent parameter. In the real system, the Pauli susceptibility is enhanced (at moderate density) by carrier–carrier interactions and is reduced (at low density) by disorder-induced localization. Both effects were observed experimentally in CdTe or (Cd,Mn)Te QWs with a low Mn content, containing either an electron gas (Jusserand et al., 2003) or a hole gas (Boukari et al., 2006; Fig. 7.6). They explain at least qualitatively the experimental behavior of $(T_{CW} - T_{AF})/x_{eff}$ in QWs with an Mn content high enough to place the critical temperature in the experimentally accessible range. Their effect on the critical temperature by carrier–carrier interactions were also predicted theoretically (Kechrakos et al., 2005; Fig. 7.7).

Another observation that is well understood in the mean field model is the magnetic anisotropy, which appears to be directly related to the anisotropy of the Pauli susceptibility of the 2DHG. The mechanism has been evidenced in (Ga,Mn)As (Dietl et al., 2001) and in (Cd,Mn)Te QWs (Kossacki et al., 2004b). This case is simpler since the holes involved are close to the center of the Brillouin zone, where a textbook description of the valence band is valid. In usual (Cd,Mn)Te QWs, which are coherently grown on a (Cd,Zn)Te substrate, both confinement and strain conspire to push the heavy-hole valence band at higher energy than the light-hole

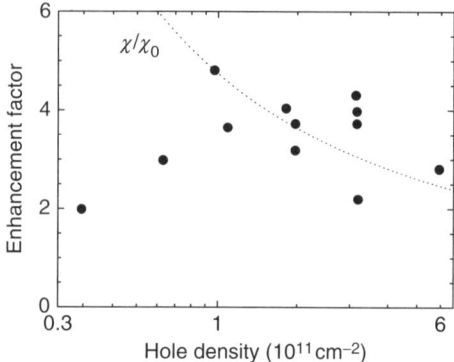

FIGURE 7.6 Enhancement factor of the spin susceptibility as the function of the carrier density, using $m^* = 0.22\,m_0$; dot line, calculation; symbols, experimental data [adapted from Boukari et al. (2006)].

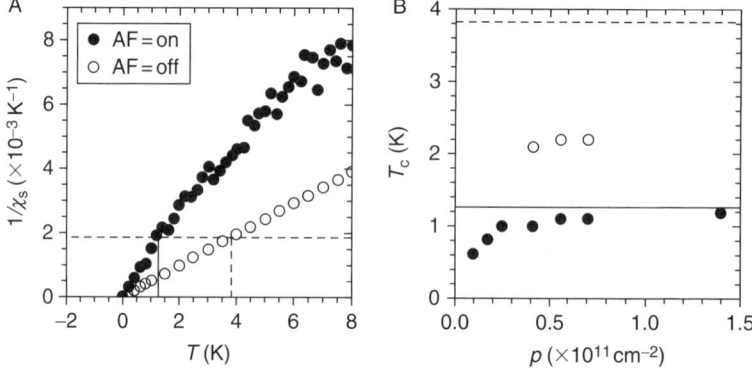

FIGURE 7.7 (A) Inverse Mn susceptibility of an undoped $Cd_{0.96}Mn_{0.04}Te$ quantum well. Open symbols, $J_{ij} = 0$; closed symbols, $J_{ij} \neq 0$. The intercept of the Monte Carlo data and the horizontal (dotted) line at $1/\chi = 1.83$ give the mean-field value of the critical temperature (T_c) for the doped system. (B) Variation of T_c with the doping level as obtained by Monte Carlo simulations. Horizontal lines are the mean-field results, as obtained from panel (A). Dashed line, $J_{ij} = 0$; solid line, $J_{ij} \neq 0$ (Kechrakos et al., 2005).

one, that is, the 2DHG contains only heavy holes. But heavy holes with small k-vector are strongly anisotropic: the spin susceptibility is nonzero only along the normal to the QW, since in-plane spin components are zero due to the orthogonal character of the orbital parts in $|\pm 3/2\rangle$ heavy holes. As a result, one observes (Fig. 7.8) a positive Curie-Weiss temperature only when the magnetic field is observed "out-of-plane," and the spontaneous magnetization is along the normal to the QW.

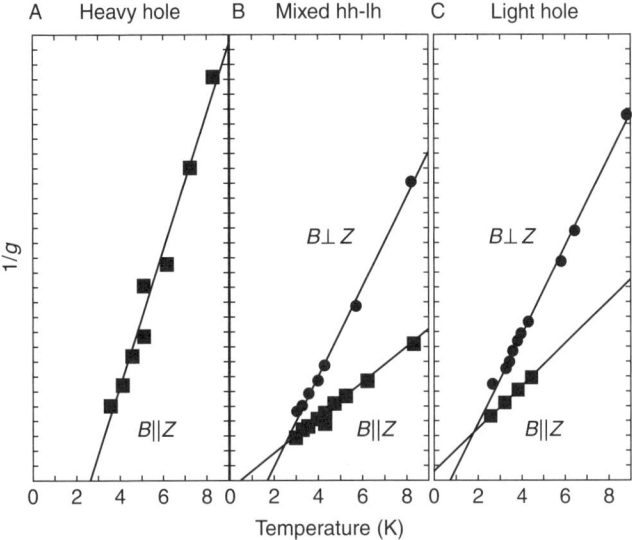

FIGURE 7.8 Curie–Weiss behavior obtained from PL measurements in magnetic field parallel ($B/\!/z$) and perpendicular ($B \perp z$) to the growth axis. The Landé factor g reported in the vertical axis is proportional to the magnetization (from Kossacki et al., 2004b).

Note that this is a true realization of an Ising interaction ($S_i^z S_j^z$) between otherwise isotropic spins. Then, by a proper design of the sample (wide QW grown on a CdTe substrate that imposes a tensile stress on the QW material), it is possible to reverse the light-hole/heavy-hole ordering (and this is easily checked on the reflectivity spectra with white light illumination, which depletes the 2DHG). In this case, one observes an easy-plane character, and the positive Curie-Weiss temperature is observed only if the magnetic field is applied in-plane (Kossacki et al., 2004b). The idea of "band engineering" is thus extended to the design of magnetic properties.

An aspect, which is evidently ignored in the mean field model, is the presence of disorder. Even in a good quality sample, disorder is present due to the random distribution of the magnetic impurity and to carrier fluctuations. The case of thick layers was addressed through multiple methods, but the 2D system appears to be more difficult to tackle with. Monte Carlo calculations (Kechrakos et al., 2005) have confirmed many aspects of the mean field model; in addition, they consider the formation of magnetic domains (the magnetization, normal to the QW, can be up or down), and show that the presence of antiferromagnetic interactions between nearest neighbors leads to a small coercive field and a smooth magnetization loop: all this is confirmed by the analysis of the PL line in the low-temperature phase (Kossacki et al., 2002). Note that the reduction of the spontaneous magnetization that is measured in 3D samples with a low carrier density, when compared to the value calculated in the mean field model (Cibert et al., 2002), is not observed in the 2D system even down to the submicrometer scale within local fluctuations of the carrier density (Maslana et al., 2004).

The nature of the domains remains an open question. The system is very original: the magnetization is weak and not saturated, antiferromagnetic interactions play a role, and in addition, the 2DHG spin polarization cannot be reversed over a distance smaller than the Fermi wavelength. The presence of domains governs the PL characteristics, and a persistent polarization is observed (Kossacki et al., 2002) under circularly polarized excitation of the PL: excitons created with one helicity are created in one type of domain, and they are trapped in these domains, which indicates that they are wider than the exciton radius (a few nanometer) and stable over at least 1 ns (a few times the exciton lifetime). However, they could not be imaged, which shows that they are either smaller than the resolution of the optical setup (a few 100 nm) or too mobile to be observed during the integration time of optical spectra (a few minutes). This motivated studies of the dynamics of the PL after a short magnetic from a small coil. While characteristic times pertinent for the relaxation of single-Mn spins are observed in the paramagnetic phase, a dramatic slowing down is observed when entering the ordered phase, up to typically 2 μs (Kossacki et al., 2006). Although an attribution to the presence of domains is quite natural, there is at present no calculation available to compare with. Another important aspect of dynamics is the possibility of collective excitations even in the paramagnetic phase and of a predicted "soft mode" behavior when approaching the critical temperature (Kavokin, 1999). A strong decrease of the Mn precession time was indeed observed through Faraday rotation (Fig. 7.9), but the agreement with the simple model was not complete (Scalbert et al., 2004).

3. 0D SYSTEMS

3.1. Introduction

As the size of magnetoelectronic devices scales down, it becomes increasingly important to understand the properties of a single magnetic atom in a solid state environment (Besombes et al., 2004; Gambardella et al., 2003; Heinrich et al., 2004; Manoharan et al., 2000; Yakunin et al., 2004). Atomic scale surface probes have been successfully used in this regard (Gambardella et al., 2003; Manoharan et al., 2000; Yakunin et al., 2004). More recently, optical probing of both magnetic (Besombes et al., 2004) and nonmagnetic (Jelezko et al., 2004) atoms in semiconductors has been demonstrated. As presented in section 2, magnetically doped semiconductors have

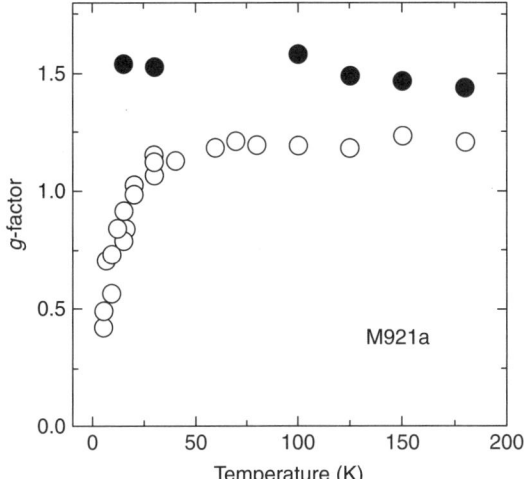

FIGURE 7.9 Effective g factor versus temperature measured in a $Cd_{0.983}Mn_{0.017}Te$ with a hole density of 3.5×10^{11} cm^{-2}. Open symbols correspond to the "soft" mode, close symbols correspond to electron spin resonance coming from the (Cd,Zn)Te substrate (Scalbert et al., 2004).

been used in the fabrication of electrically active devices that control the magnetic properties like transition temperature and coercive field (Boukari et al., 2002; Chiba et al., 2003; Ohno et al., 2000). In these devices, a macroscopic number of magnetic atoms were manipulated. We now present electrically active devices that control the charge state of an individual II–VI QD doped with a single Mn atom. Single-dot micro-PL measurements reveal that the magnetic anisotropy and spin configuration of the single Mn atom are very different depending on the charge state of the dot, which can be 0 or ±1e. Thereby, these devices are able to tune the magnetic properties of a single Mn atom embedded in a QD and represent a first step in the implementation of several proposals of electrical control of the magnetism in Mn-doped QDs (Climente et al., 2005; Efros et al., 2001; Fernandez-Rossier and Brey, 2004; Govorov and Kalameitsev, 2005a,b).

QDs doped with magnetic atoms and filled with a tunable number of carriers can behave like tunable nanomagnets. Indeed, as it has been theoretically shown, the parity of the number of confined carriers can drastically change the magnetic properties of the localized magnetic moments (Climente et al., 2005; Fernandez-Rossier and Brey, 2004; Govorov and Kalameitsev, 2005a,b; Qu and Hawrylak, 2005, 2006). As it will be presented in Section 3.2, magnetic ion-doped QDs could also be used as spin filters or spin aligner in transport experiments.

In a magnetic QD, the sp–d interaction takes place with a single carrier or a single electron–hole (e–h) pair. However, besides effects related with the carriers–Mn-ions exchange interaction such as giant Faraday rotation and a strong Zeeman shift, it was found that even a small content of Mn introduced in a II–VI semiconductor material can strongly suppress PL if the energy gap E_G exceeds the energy of the Mn internal transition. This strongly limits the study of individual DMS QDs (Chernenko *et al.*, 2005). The first studies of individual QDs doped with Mn atoms were reported by Maksimov *et al.* (2000). They studied (Cd,Mn)Te QDs inserted in CdMgTe barriers in which the optical transition energies are lower than the energy of the internal transition of the Mn atom. This suppresses the nonradiative losses due to the transfer of confined carriers to the Mn energy levels. This system allowed observing the formation of quasi-zero-dimensional magnetic polaron. Moreover, the broadening of the emission lines caused by fluctuations in the magnetic environment of the recombining e–h pair was controlled by an external magnetic field.

Another way to reduce the nonradiative losses was to introduce the magnetic atoms in the QDs barriers. This has been realized for CdSe dots in ZnMnSe barriers by Seufert *et al.* (2002). In this system, the interaction between the confined exciton and the magnetic ions is due to the spread of the wave function in the barriers and to a small diffusion of the magnetic atoms in the QDs. In these DMS structures, the response time of the paramagnetic Mn spin was extracted from the transient spectral shift of the PL caused by the dynamical spin alignment of magnetic ions incorporated in the crystal matrix. The formation of a ferromagnetically aligned spin complex was demonstrated to be surprisingly stable as compared to bulk magnetic polaron (Mackh *et al.*, 1994; Yakovlev *et al.*, 1997), even at high temperatures and high magnetic fields. The PL of a single e–h pair confined in one magnetic QD, which sensitively depends on the alignment of the magnetic ions spins, allowed to measure the statistical fluctuations of the magnetization on the nanometer scale. Quantitative access to statistical magnetic fluctuations was obtained by analyzing the linewidth broadening of the single dot emission. This optical technique allowed to address a magnetic moment of about 100 μ_B and to resolve changes in the order of a few μ_B (Bacher *et al.*, 2002; Dorozhkin *et al.*, 2003; Hundt *et al.*, 2004).

A huge effort has also been done to incorporate magnetic ions in chemically synthesized II–VI nanocrystals (Norris *et al.*, 2001). The incorporation of the magnetic impurities is strongly dependent on the growth condition and controlled by the adsorption of impurities on the nanocrystal surface during growth (Erwin *et al.*, 2005). The doping of nanocrystals with magnetic impurities also leads to interesting magneto-optical properties (Bhattacharjee and Perez-Conde, 2003) but once again, in these highly confined systems, the transfer of confined carriers to the Mn electronic levels strongly reduces their quantum efficiency and prevents the optical study of individual Mn-doped nanocrystals.

CdTe/ZnTe self-assembled QDs usually present an emission energy below the internal transition of the Mn atom and permit to introduce the magnetic atoms directly into the QD and conserve good optical properties. Up to now, however, all the experimental studies on these diluted magnetic QDs were focused on the interaction of a single carrier spin with its paramagnetic environment (large number of magnetic atoms) (Mackowski et al., 2005). We will see that CdTe/ZnTe QD structures doped with a low density of Mn atoms allow optically controlling the spin states of a single magnetic ion interacting with a single e–h pair or a single carrier.

3.2. A single magnetic atom in a QD

For the CdTe/ZnTe magnetic QDs (Fig. 7.10), a low concentration of Mn is introduced into the QDs by adjusting the density of Mn atoms in the QD layer to be close to the density of QDs ($\approx 5 \times 10^9 \text{cm}^{-2}$) (Maingault et al., 2006). The QDs are grown by MBE. A $Zn_{0.94}Mn_{0.06}Te$ barrier followed by a 10-monolayer ZnTe spacer is deposited on a ZnTe substrate. The CdTe QD layer is then deposited by ALE and capped with a ZnTe barrier (Tinjod et al., 2003). The Mn intermixing during the growth of the ZnTe spacer introduces a sparse distribution of Mn^{2+} ions in the QD layer (Grieshaber et al., 1996).

Individual Mn-doped QDs can be probed using a microspectroscopy experiment. The PL of individual QDs is excited with the 514-nm line of an argon laser or a tunable dye laser and collected through a large numerical-aperture microscope objective and aluminum shadow masks with 0.2–1.0 µm apertures. The PL is then dispersed by a 2 m additive

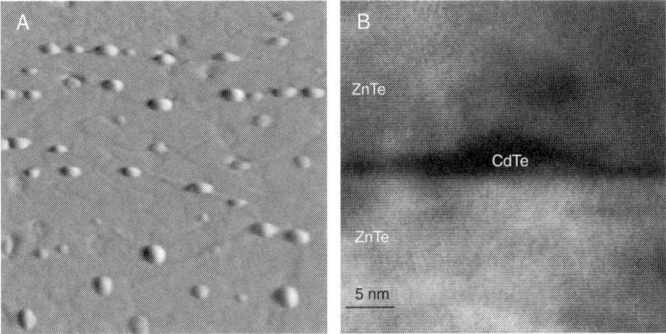

FIGURE 7.10 (A) AFM image of a CdTe surface deposited on a ZnTe substrate before deposition of a ZnTe capping layer. (B) High-resolution TEM image showing the structure of a CdTe/ZnTe quantum dot. (Maingault et al., 2006).

double monochromator and detected by a nitrogen-cooled Si CCD camera or an Si Avalanche Photodiode.

In Fig. 7.11, PL spectra of an individual Mn-doped QD are compared to those of a nonmagnetic CdTe/ZnTe reference sample. In nonmagnetic samples, narrow PL peaks (limited by the spectrometer resolution of about 50 μeV) can be resolved, each attributed to the recombination of a single e–h pair in a single QD. The emission of neutral QDs is split by the e–h exchange interaction and usually a linearly polarized doublet is observed (Besombes et al., 2000a,b). By contrast, most of the individual

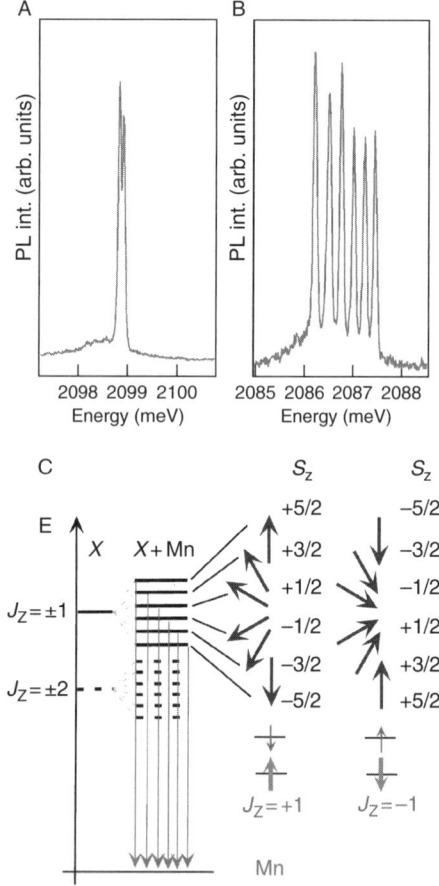

FIGURE 7.11 Low-temperature ($T=5$ K) PL spectra obtained at $B=0$ T for (A) an individual CdTe/ZnTe QD and (B) an Mn-doped QD. (C) Scheme of the energy levels of the Mn–exciton coupled system at zero magnetic field. The exciton–Mn exchange interaction shifts the energy of the exciton depending on the S_z component of the Mn spin projection.

emission peaks of magnetic single QDs are characterized by a rather large linewidth of about 0.5 meV. For some of these QDs, a fine structure can be resolved and six emission lines are clearly observed at zero magnetic field (Fig. 7.11B). The measured splitting changes from dot to dot. This fine structure splitting, as well as the broadening, is obviously related to the influence of the magnetic ions located within the spatial extent of the exciton wave function. The broadening observed in magnetic QDs has been attributed by Bacher et al. (2002) to the magnetic fluctuations of the spin projection of a *large number* of Mn^{2+} spins interacting with the confined exciton (Bacher et al., 2002). In the low-concentration Mn-doped samples, the observation of a fine structure shows that the QD exciton interacts with a single Mn^{2+} spin. In time-averaged experiments, the statistical fluctuations of a single Mn^{2+} ion ($S=5/2$) can be described in terms of populations of its six spin states quantized along the direction normal to the QD plane. The exchange interaction of the confined exciton with the Mn^{2+} ion shifts its energy depending on the Mn^{2+} spin projection, resulting in the observation of six emission lines.

QDs doped with single Mn^{2+} ions were considered theoretically in the case of spherical nanocrystals with a strong confinement (Bhattacharjee and Perez-Conde, 2003). The eigenstates resulting from the exchange coupling of the exciton and the magnetic ion were obtained by a combination of the electron, hole, and Mn^{2+} magnetic moments. In flat self-assembled QDs with a relatively weak confinement, the biaxial strains in the plane of the QD lift the degeneracy of the hole spin projections (heavy-hole/light-hole splitting). In a first approximation, this system can be described by a heavy-hole exciton confined in a symmetric QD in interaction with the six spin projections of the manganese ion (Govorov, 2004). The spin interaction part of the Hamiltonian is given by:

$$H_{int} = I_e \, \vec{\sigma} \cdot \vec{S} + I_h \, \vec{j} \cdot \vec{S} + I_{e-h} \, \vec{\sigma} \cdot \vec{j}, \qquad (7.1)$$

where $I_e(I_h)$ is the Mn–e (–h) exchange integral, I_{e-h} the e–h exchange interaction, and $\sigma(j)$ the magnetic moment of the electron (hole). The initial states of the optical transitions are obtained from the diagonalization of the spin Hamiltonian and Zeeman Hamiltonian in the subspace of the heavy-hole exciton and Mn^{2+} spin components $|\pm1/2\rangle_e|\pm3/2\rangle_h|S_z\rangle_{Mn}$, with $S_z = \pm 5/2, \pm 3/2,$ and $\pm 1/2$. Since the dipolar interaction operator does not affect the Mn d electrons, the final states involve only the Mn^{2+} states $|S_z\rangle_{Mn}$ with the same spin component (Besombes et al., 2004).

In this framework, at zero magnetic field, the QD emission presents a fine structure composed of six doubly degenerate transitions roughly equally spaced in energy. The lower energy bright states $|+1/2\rangle_e|-3/2\rangle_h|+5/2\rangle_{Mn}$ and $|-1/2\rangle_e|+3/2\rangle_h|-5/2\rangle_{Mn}$ are characterized by an antiferromagnetic coupling between the hole and the Mn^{2+} ion. The following states are associated with the Mn^{2+} spin projections $S_z = \pm 3/2$ and $\pm 1/2$ up to the higher energy states $|-1/2\rangle_e|+3/2\rangle_h| +5/2\rangle_{Mn}$ and $|+1/2\rangle_e|-3/2\rangle_h|-5/2\rangle_{Mn}$ corresponding to ferromagnetically coupled hole and manganese. In this simple model, the zero field splitting $\delta_{Mn} = 1/2(I_e - 3I_h)$ depends only on the exchange integrals I_e and I_h and is thus related to the position of the Mn^{2+} ion within the exciton wave function.

When an external magnetic field is applied in the Faraday geometry (Fig. 7.12), each PL peak is further split and 12 lines are observed, 6 in each circular polarization. As presented in Fig. 7.13, the Zeeman effect of the Mn states is identical in the initial and final state of the optical transitions and the six lines in a given polarization follow the Zeeman and diamagnetic shift of the exciton, as in a nonmagnetic QD. The parallel evolution of six lines is perturbed around 7 T in σ^- polarization by anticrossings observed for five of the lines. In addition, as the magnetic field increases, one line in each circular polarization increases in intensity and progressively dominates the spectrum.

The e–Mn^{2+} part of the interaction Hamiltonian $I_e(\vec{\sigma} \cdot \vec{S})$ couples the dark ($J_z = \pm 2$) and bright ($J_z = \pm 1$) heavy-hole exciton states. This coupling corresponds to a simultaneous electron and Mn^{2+} spin flip changing a bright exciton into a dark exciton. Because of the strain-induced splitting of light- and heavy-hole levels, a similar $Mn^{2+}-$ hole spin flip scattering is not allowed. The e–Mn^{2+} spin flip is enhanced as the corresponding levels of bright and dark excitons are brought into coincidence by the Zeeman effect. An anticrossing is observed around 7 T for five of the bright states in σ^- polarization (experiment: Fig. 7.12 and theory: Fig. 7.13). It induces a transfer of oscillator strength to the dark states. In agreement with the experimental results, in the calculations, the lower energy state in σ^- polarization ($|+1/2\rangle_e|-3/2\rangle_h|+5/2\rangle_{Mn}$) does not present any anticrossing. In this spin configuration, both the electron and the Mn^{2+} ion have maximum spin projection and a spin flip is not possible.

The minimum energy splitting at the anticrossing is directly related to the e–Mn^{2+} exchange integral I_e. For instance, the splitting measured for the higher energy line in σ^- polarization (Fig. 7.12) ΔE=150 µeV gives $I_e \approx 70$ µeV. From the overall splitting measured at zero field (1.3 meV) and with this value of I_e, we obtain $I_h \approx -150$ µeV. These values are in good agreement with values estimated from a modeling of the QD confinement by a square QW in the growth direction and a truncated parabolic potential in the QD plane. With a QW thickness $L_z = 3$ nm and a Gaussian wave function characterized by an in-plane localization parameter $\xi = 5$ nm, we obtain $I_e \approx 65$ µeV for an Mn^{2+} ion placed at the center of the QD.

However, the ratio of the exchange integral, $(3I_h)/I_e \approx -6$ for the QD presented in Fig. 7.12, does not directly reflect the ratio of the sp–d exchange constants $\beta/\alpha \approx -4$ measured in bulk (Cd,Mn)Te alloys (Furdyna, 1988). This deviation likely comes from the difference in the

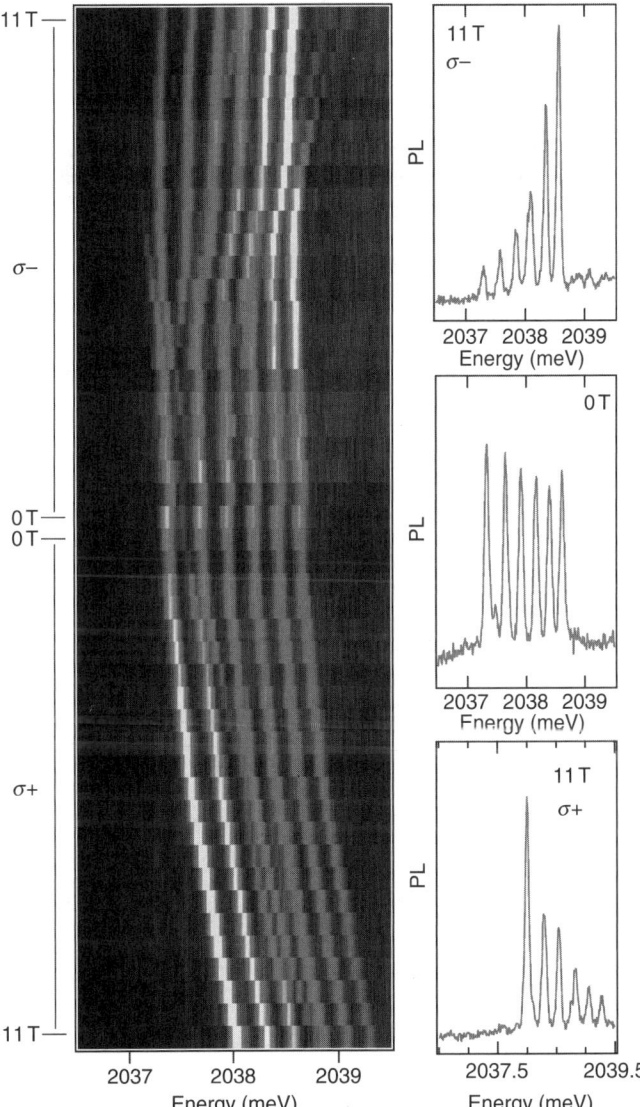

FIGURE 7.12 Magnetic field dependence of the emission of an Mn-doped QD recorded in σ^+ and σ^- polarization. Anticrossing of the bright and dark states appears around 7 T in σ^- polarization (Besombes et al., 2004).

e–Mn and h–Mn overlap expected from the difference in the electron and hole confinement length but could also be due to a change of the exchange parameters induced by the confinement (Merkulov et al., 1999). A dispersion of the zero field energy splitting observed from dot to dot is

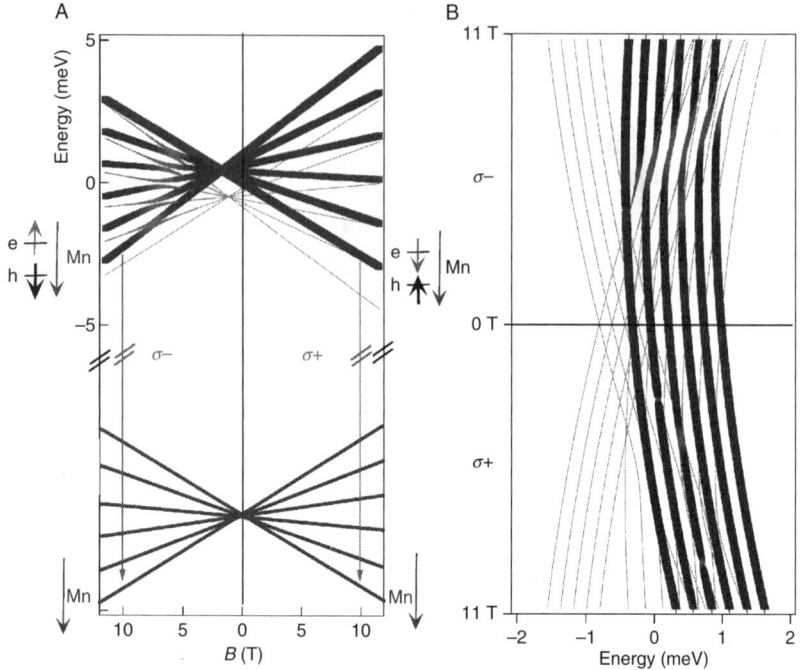

FIGURE 7.13 (A) Scheme of the energy levels of the initial and final states involved in the optical transitions of a quantum dot containing an Mn atom. (B) Modelization of the optical transitions obtained from the diagonalization of an effective spin Hamiltonian including the e–h exchange interaction, the exciton–Mn exchange interaction, the Zeeman and the diamagnetic energies. The contribution of the dark states appears in gray.

then due to a variation of the Mn–exciton overlap for different QDs. However, the spin Hamiltonian [Eq. (7.1)] does not reproduce the observed nonuniform zero field splitting (Fig. 7.11). A more accurate model has to take into account the full valence band structure and the heavy- and light-hole mixing.

3.3. Optical reading of the spin state of a single magnetic atom

As illustrated in Fig. 7.12, the relative intensities of the six emission lines observed in each circular polarization depend strongly on the applied magnetic field. The emission intensity, which is almost equally distributed over the six emission lines at zero field, is concentrated on the high-energy line of the σ^- emission and on the low-energy line of the σ^+ emission at high magnetic field. As the magnetic field increases, the Mn^{2+} ions are progressively polarized. In time-averaged experiments, the probability to observe the recombination of the bright excitons coupled with the $S_z = -5/2$ spin projection is then enhanced. Two states dominate the spectra: $|-1/2\rangle_e |+3/2\rangle_h |-5/2\rangle_{Mn}$ in the low-energy side of the σ^+ emission and $|+1/2\rangle_e |-3/2\rangle_h |-5/2\rangle_{Mn}$ in the high-energy side of the σ^- polarization. Changing the temperature of the Mn^{2+} ion will affect the distribution of the exciton emission intensities. The PL of the exciton is then a direct probe of the magnetic state of the Mn^{2+} ion.

The effective temperature of the manganese ion in the presence of the exciton, T_{Mn}, is found to depend, of course, not only on the lattice temperature but also on the laser excitation density (Fig. 7.14). For a fixed temperature and a fixed magnetic field, the asymmetry observed in the emission intensity distribution progressively disappears as the excitation intensity is increased (Fig. 7.14A and B). The variation of T_{Mn} deduced from the emission rates is presented in the inset of Fig. 7.14C (7 T) and Fig. 7.14D (0 T) as a function of the excitation density. A similar excitation intensity dependence of T_{Mn} was previously observed in semimagnetic QWs and was attributed to the heating of the Mn^{2+} ions through their spin–spin coupling with the photocreated carriers (Keller *et al.*, 2001). The photo-carriers have excess energy. Via spin-flip exchange scattering, they pass their energy to the Mn^{2+} ions and rise their spin temperature. The energy flux from the Mn to the lattice, determined by the spin lattice relaxation, will tend to dissipate this excess energy. Under steady-state photoexcitation, the resulting temperature of the magnetic ions T_{Mn} exceeds the lattice temperature. The effect of this spin–spin coupling is strongly enhanced in our system since the isolated Mn^{2+} ion is only weakly coupled to the lattice and hardly thermalize with the phonon bath (Scalbert *et al.*, 1988). Under nonresonant excitation, the injection of an exciton changes the spin distribution of the magnetic ion. As illustrated in Fig. 7.14B, at 0 T and at low excitation intensity, an asymmetry is observed in the emission intensity distribution. This polarization shows that a spin flip of the exciton–Mn system can occur during the lifetime of the exciton. The exchange interaction with the exciton acts as an effective magnetic field that splits the Mn^{2+} levels in zero-applied field, allowing a progressive polarization of its spin distribution.

Resonant excitation of e–h pairs directly in the QD limits the exciton–Mn spin relaxation. This is illustrated in Fig. 7.15B and C where the emission intensity of the ground state of an Mn-doped QD is presented as a function of the detection energy when the laser excitation energy is scanned through the resonant absorption of an excited state identified in a PLE spectra (Fig. 7.15A). This excitation energy scan reveals that the intensity distribution of the emission strongly depends on both the wavelength and the polarization of the excitation laser. This dependence shows that under resonant excitation, there is not a complete spin relaxation of

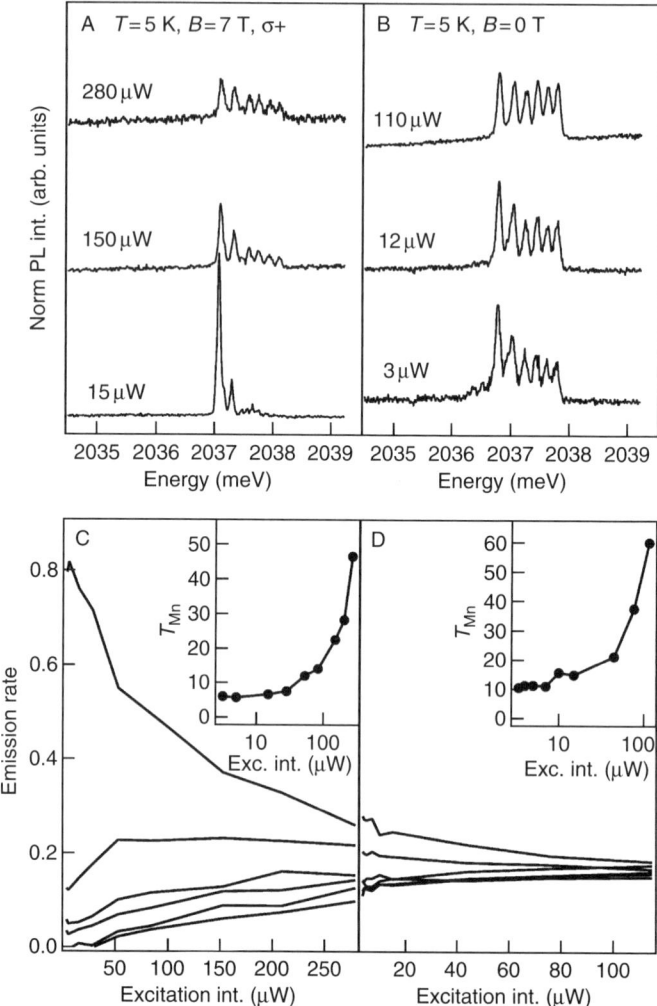

FIGURE 7.14 (A) Normalized PL (PL spectra divided by the total integrated intensity) in σ^+ polarization versus excitation intensity for a fixed temperature (5 K) and magnetic field (7 T). (B) Excitation intensity dependence of the zero magnetic field emission of a single Mn-doped QD for a fixed lattice temperature $T = 5$ K. (C) and (D) Extracted emission rates of each PL lines as a function of the excitation intensity at 7 T (C) and 0 T (D). The inset plots the extracted Mn effective temperature.

the exciton–Mn complex during the lifetime of the exciton. As shown in Fig. 7.15D, this long spin relaxation time combined with the fine structure of the excited states permits to create selectively a given spin configuration of the exciton–Mn complex by tuning the polarization and wavelength of the excitation laser.

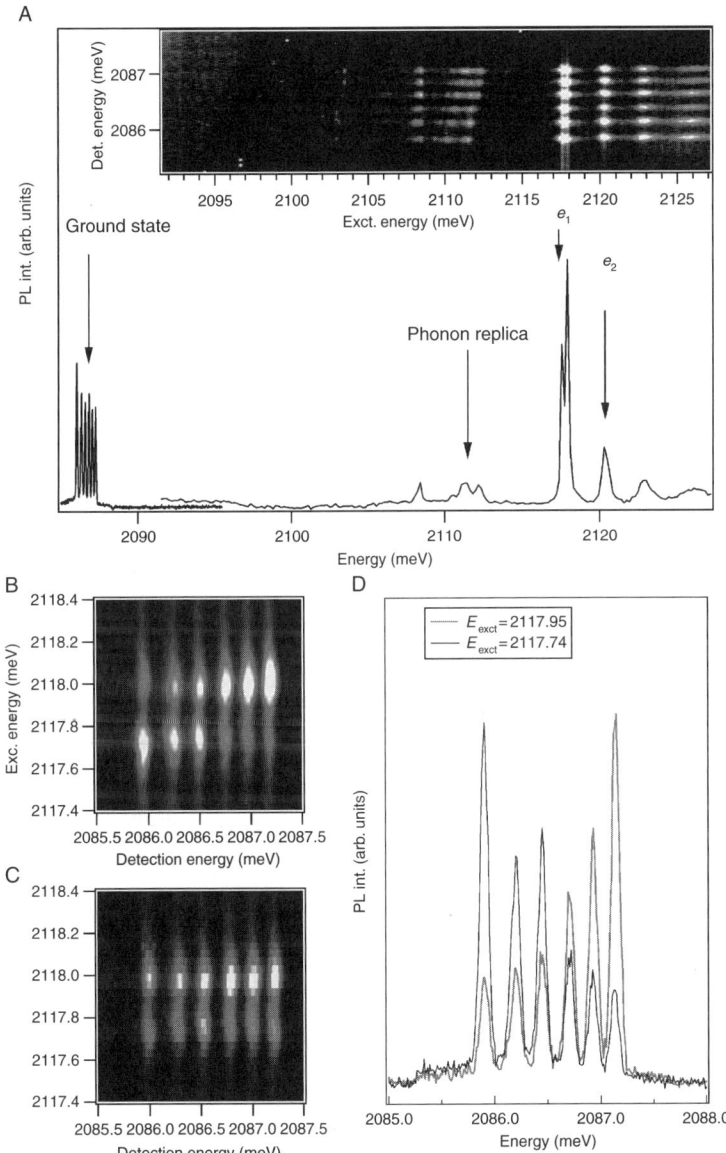

FIGURE 7.15 (A) Experimental PL and PLE spectra of an Mn-doped quantum dot exciton. The inset shows the contour plot of the multichannel PLE. (B) and (C) PLE contour plots for excited state e_1 obtained for (B) copolarized and (C) cross-polarized circular excitation and detection. (D) Resonant PL spectra obtained in copolarized circular excitation and detection for two different excitation wavelength on e_1.

3.4. Electrical control of a single magnetic atom

CdTe/ZnTe QDs can be modulation-doped p-type by the transfer of holes from the p-doped ZnTe substrate and from the surface states that act as acceptors (Besombes et al., 2002; Bhunia and Bose, 2000; Maslana et al., 2003). The occupation of the QDs by holes can be controlled by an external bias voltage V on an aluminum Schottky gate with respect to a back contact on the p-type substrate (Seufert et al., 2003). The bias-dependent emission of a nonmagnetic QD and an Mn-doped QD are presented in Fig. 7.16A and B. For increasing V, the surface level states are shifted below the ground hole level in the QDs, which results in the single-hole charging of the dots. The optically generated excitons then form charged excitons with the bias-induced extra hole in the QD. At zero bias or negative bias, the Fermi level is above the ground state and the QDs are likely to be neutral. However, the separate capture of photocreated electron or holes can sometimes charge the dots so that weak contributions of X^+ or X^- are observed in the zero bias spectra.

At zero bias, excess electrons can also be injected in the QD using resonant optical excitation into the QD levels. Under resonant excitation (energy below the bandgap of the barriers), optical transitions from delocalized valence band states to the confined electron levels will preferentially create electrons in the QD (Vasanelly et al., 2002): the probability to find an excess electron in the QD is increased. As presented in Fig. 7.16C, the negatively charged exciton emission is then seen for some discrete excitation energies. After the recombination of the charged exciton X^-, a single hole is likely to be captured to neutralize the QD and create a neutral exciton. This neutralization process is responsible of the simultaneous observation of charged and neutral species under resonant excitation (Fig. 7.16C).

The charge state of the dot can also be optically tuned (Fig. 7.16C). By combining a weak nonresonant excitation with the resonant one, a few carriers are created in the ZnTe barrier. They do not significantly contribute to the luminescence (lower PL spectra in Fig. 7.16C) but reduce the PL contribution of X^- in favor of the neutral species. This evolution is characteristic of a photodepletion mechanism in modulation-doped QDs (Hartmann et al., 2000). High-energy photoexcited e–h pairs are dissociated in the space charge region surrounding the negatively charged QDs and neutralize the QDs.

These two charge control mechanisms (bias voltage and resonant excitation combined with photodepletion) allow to independently probe the interaction between individual carriers (electron or hole) and an individual magnetic atom. Let us first consider the negatively charged exciton. Figure 7.17A presents a detail of the recombination spectrum for X^- coupled with a single Mn atom obtained at zero bias under resonant excitation. Eleven emission lines are clearly observed with intensity decreasing from the outer to the inner part of the emission structure.

A simple effective spin Hamiltonian quantitatively accounts for the emission spectrum shown in Fig. 7.17A. The emitting state in the X⁻ transition has two CB electrons and one hole coupled to the Mn. The effect of the two spin-paired electrons on the Mn is strictly zero. Thereby, the spin structure of the X⁻ state is governed by the interaction of the hole with the Mn. On the basis of previous work (Fernandez-Rossier, 2006; Govorov and Kalameitsev, 2005a,b, Koudinov et al., 2004; Kyrychenko and Kossut, 2004; Léger et al., 2005), we propose the following Hamiltonian:

$$H_{h-Mn} = I_h \left(S_z j_z + \frac{\varepsilon}{2}(j_- S_+ + j_+ S_-) \right) \quad (7.2)$$

where $\vec{S}$ is the Mn spin operator and $\vec{j}$ acts on the hole lowest energy doublet. The first term is the spin conserving or Ising exchange whereas the second is only possible if there is some heavy- and light-hole mixing (Fernandez-Rossier, 2006; Koudinov et al., 2004). From our measurements, we find that ε is small and, in a first stage, we neglect it. Later we will show its influence. The 12 eigenstates of H_{h-Mn} with $\varepsilon = 0$ are organized as 6 doublets (Fig. 7.17B) with well-defined S_z and j_z (Mn and hole spin along the z-axis). We label these states as $|S_z, j_z\rangle$. Recombination of one of the CB electrons with the hole of the X⁻ state leaves a final state with a single CB electron coupled to the Mn. The spin Hamiltonian of this system is the ferromagnetic Heisenberg model, $H_{e-Mn} = -I_e \vec{S} \cdot \vec{\sigma}_e$. The 12 eigenstates of the Mn–e complex are split into a ground state septuplet (total spin $J = 3$) and a fivefold degenerate manifold with $J = 2$. We label them all as $|J, J_z\rangle$.

Thereby, for each of the six doublets of X⁻, there are two possible final states after annihilation of an e–h pair, with either $J = 2$ or 3. From this consideration alone, we would expect 12 spectrally resolved lines. Their weight is given by both optical and spin conservation rules. Since electrons and holes reside in s and p bands, respectively, the $\Delta L = 1$ optical selection rule is immediately satisfied. The polarization of the photon imposes an additional selection rule on ΔM which is accounted for by the spin of the electron and hole. The Mn spin is not affected by the transition.

The weight of optical transitions between the initial state $|i\rangle = (\uparrow, \downarrow)_e \times |S_z, j_z\rangle$ and the final state $|f\rangle = |J, J_z\rangle$ is proportional to $|\langle f|\sum_\sigma P(\sigma, j_z) c_\sigma d_{j_z}|i\rangle|^2$ where c_σ annihilates a CB electron with spin σ and d_{j_z} annihilates a VB hole with angular momentum j_z. Here $P(\sigma, j_z)$ is given by the polarization selection rule.

Let us consider, for instance, σ^+ recombination transitions where the $(\downarrow_e, \Uparrow_h)$ e–h pair is annihilated. Each of the six doublets, characterized by their Mn spin projection S_z, can be an initial state. After the e–h annihilation, the resulting state is $|S_z, \uparrow_e\rangle$, which, in general, is not an eigenstate of $H_{e,Mn}$.

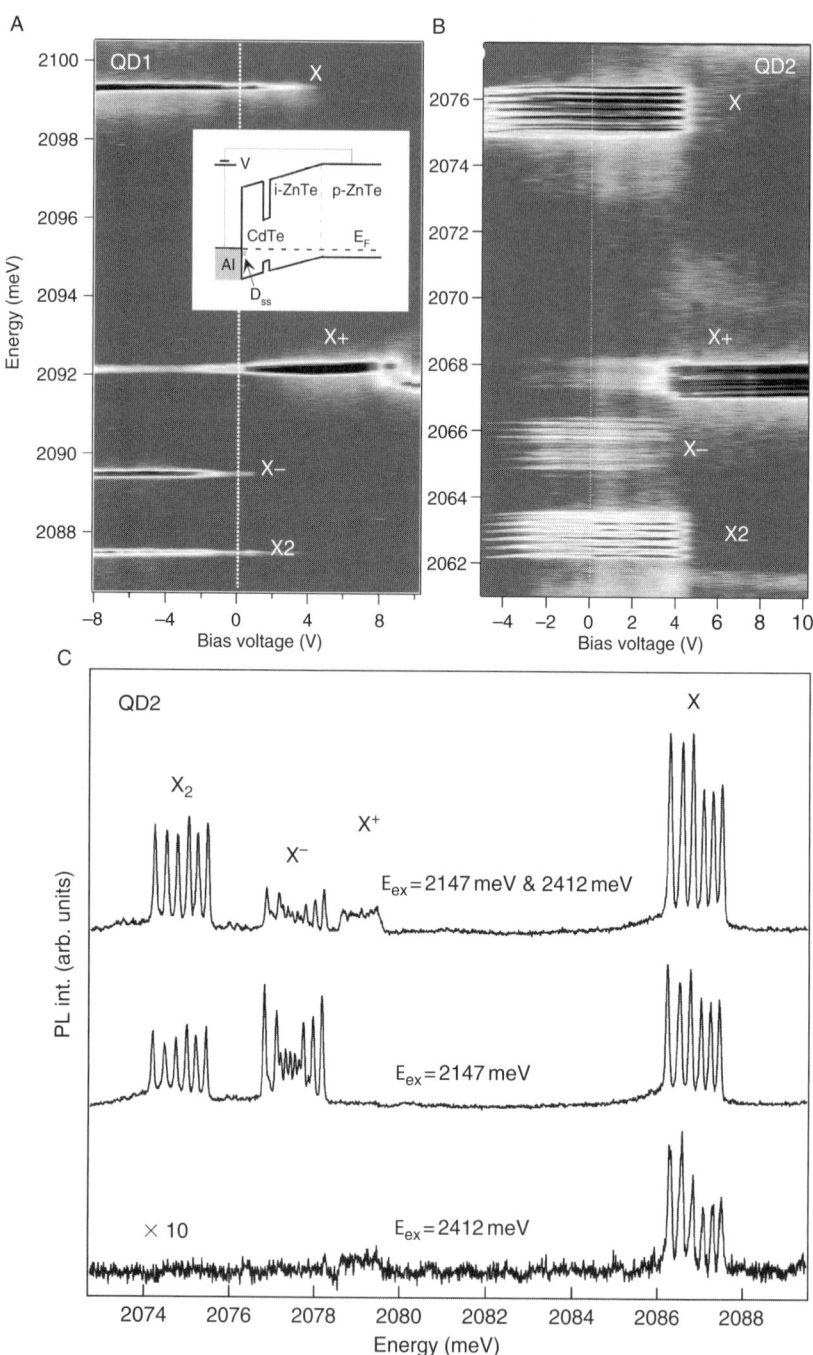

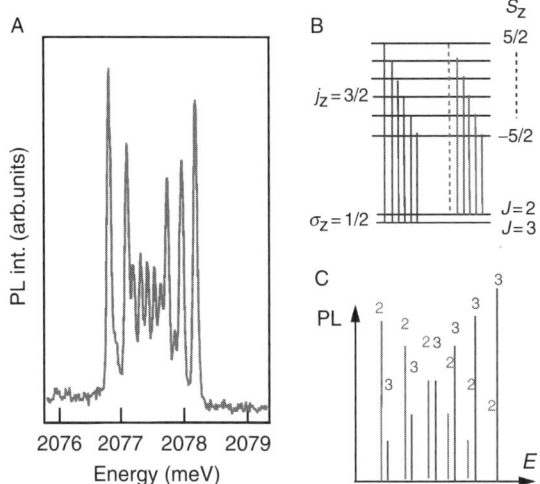

FIGURE 7.17 (A) Detail of the unpolarized emission spectrum of a negatively charged exciton (X$^-$) coupled with an Mn atom. (B) Scheme of the σ^+ optical transitions of (X$^-$,Mn) and their respective PL intensity distribution (C).

The intensity of the optical transition to a given final state $|J, J_z\rangle$ is proportional to the overlap $\langle J, J_z | S_z, \uparrow_e \rangle$, which is nothing but a Clebsh Gordan coefficient. The highest energy transition, with σ^+ polarization, would correspond to the initial state $(\uparrow, \downarrow)_e \times |+5/2, \uparrow\uparrow_h\rangle$ and a low energy final state $|J = 3, J_z\rangle$. After the photon emission, the state of the system is $|S_z = +5/2, \uparrow_e\rangle$, which is identical to $|J = 3, J_z = +3\rangle$ and thereby gives the highest optical weight (Fig. 7.17B). In contrast, emission from that initial state to $|J = 2, J_z\rangle$ is forbidden. The other five doublets have optical weights lying between 1/6 and 5/6 with both $|J = 2, J_z\rangle$ and $|J = 3, J_z\rangle$ final states. Thereby, the number of spectrally resolved lines in this model is 11.

The relative weight of the emission lines is accounted for by the model. According to the final state, the transitions belong to either the $J = 2$ or the $J = 3$ series. As the initial S_z decreases, the overlap of $|\uparrow_e S_z\rangle$ to the $J = 3 (J = 2)$ states decreases (increases). As presented in Fig. 7.17C, the PL

FIGURE 7.16 Color-scale plot of the photoluminescence intensity of (A) a nonmagnetic QD and (B) a single Mn-doped QD in a Schottky structure as a function of emission energy and bias voltage. The series of emission lines can be assigned to QD s-shell transitions, namely the recombination of the neutral exciton (X), biexciton (X$_2$), positively charged exciton (X$^+$), and negatively charged exciton (X$^-$). (C) Detail of the PL of a single Mn-doped QD under resonant excitation ($E_{ex} = 2147$ meV), nonresonant excitation ($E_{ex} = 2142$ meV), and both resonant and nonresonant excitation (Léger et al., 2006). (See Color Plate.)

of X^- can be seen as a superposition of two substructures: six lines with intensities increasing with their energy position (transitions to $J = 3$ states) and five lines with intensities decreasing with increasing their energy position (transitions to $J = 2$ states).

Reversing the role of the initial and final states, and neglecting the small coupling of two holes to the Mn spin (Besombes et al., 2005), this model should account for the emission from X^+ states. Actually, different-energy splittings are observed for the different excitonic species in the same QD. For instance, in QD3 (Fig. 7.16C), one measured $\Delta E_X = 1.23$ meV, $\Delta E_{X^-} = 0.36$ meV, and $\Delta E_{X^+} = 0.95$ meV. The energy splitting is mostly due to the Mn–h exchange coupling, which in turn is inversely proportional to the volume of the hole wave function. The difference between ΔE_{X^+} and ΔE_{X^-} indicates that a significant fraction of the confinement of the hole comes from the Coulomb attraction of the two electrons in the initial state of the X^- emission. In contrast, in the final state of the X^+ emission, there is no electron to attract the hole, resulting in a spread of the hole wave function and a smaller exchange energy. This difference appears directly in the emission structure shown in Fig. 7.16, where the peak structure of X^+ is not well resolved.

In Fig. 7.18A, we show the intensity of X^- emission as a function of the direction of a linear analyzer. It is apparent that the central lines are linearly polarized. This polarization can only be understood if we allow for some spin-flip interaction between the Mn and the hole [second term in Eq. (7.1)]. Provided that $\varepsilon \ll 1$, the effect of this interaction is small both on the wave function and on the degeneracy of all the doublets except the third, which is split, as illustrated in the inset of Fig. 7.18. The split states are the bonding and antibonding combinations of $|S_z = -1/2, \Uparrow_h\rangle$ and $|S_z = +1/2, \Downarrow_h\rangle$. These states are coupled, via linearly polarized photons, to the $|2,0\rangle$ and $|3,0\rangle$ e–Mn complex and four linearly polarized lines should be observed on the emission spectra as shown in the insets of Fig. 7.18C. Polarization directions are controlled by the distribution of strains through the Bir-Pikus Hamiltonian (Koudinov et al., 2004; Tadic et al., 2002).

We can obtain numerical values of I_h, I_e, and ε for the charged excitons comparing the transition probabilities calculated with model (2) (Fig. 7.18C) to the experimental data. The e–Mn exchange integral I_e is deduced from magneto-optics measurements on the neutral exciton (Besombes et al., 2004). I_h, the h–Mn exchange integral, is then chosen to reproduce the overall splitting of the two charged species emission structure. There is a good agreement between the experimental data (QD3 in Fig. 7.16C) and our model with $I_e = 40$ μeV, $I_h(X) = 150$ μeV, $I_h(X^+) = 95$ μeV, and $I_h(X^-) = 170$ μeV. The different values of I_h directly reflect the expected variation of the confinement of the hole. The main characteristics of the emission spectra are well reproduced, namely the

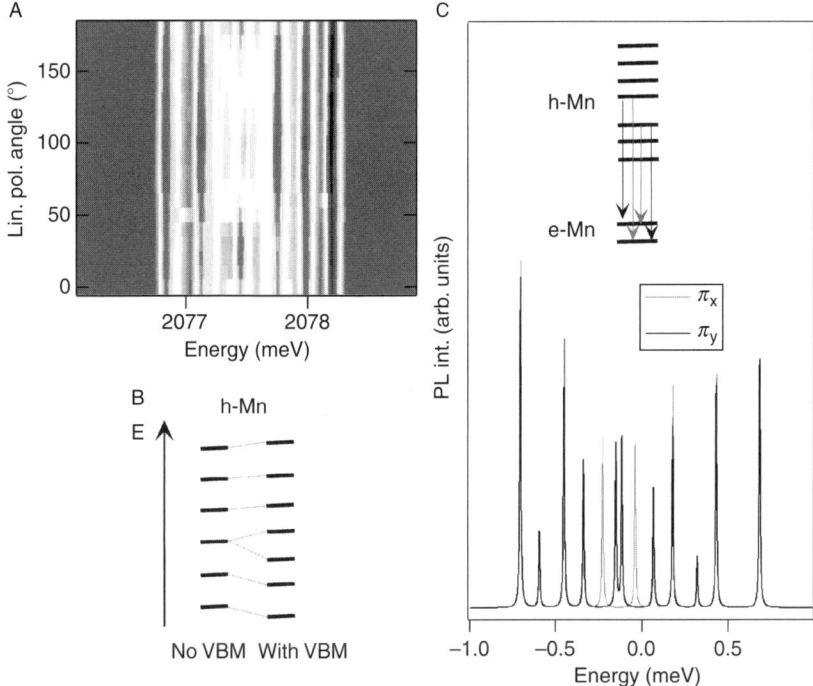

FIGURE 7.18 Color-scale plot of the dependence of the PL of (X⁻,Mn) (A) on the direction of a linear analyzer. Three lines in the center of the structure are linearly polarized. The scheme (B) presents the energy-level scheme of the h–Mn system without and with valence band mixing (VBM). (C) Calculated linearly polarized PL spectra of (X⁻,Mn) with exchange integrals I_e and I_h chosen to reproduce the overall splitting for X⁻ presented in (A). Transitions are arbitrarily broadened by 10 μeV. (See Color Plate.)

number of emission lines, their intensity distribution, and the linear polarization structure, with a slight valence band mixing coefficient $\varepsilon = 0.05$. This small value of the valence band mixing coefficient shows that h–Mn exchange interaction remains highly anisotropic.

3.5. Conclusion

We have shown that the injection of a controlled number of carriers in an individual QD permits to control the spin splitting of a single magnetic ion. The exchange interaction with a single carrier acts as an effective local magnetic field that splits the Mn levels in zero-applied magnetic field. We have fabricated a device with electrically tunable magnetic properties, in analogy with two-dimensional DMS-based electrically active heterostructures, but scaling the number of controlled magnetic atoms down to one and the size of the active region down to a few nanometers.

Micro-PL experiments allow to identify three magnetically different ground states corresponding to three charge states ($\pm 1e$ and 0) and to measure the exchange interaction of both a single electron and a single hole with a single magnetic atom. The final state of (X^+, Mn) transition (1h,1Mn) is nondegenerate in the absence of external magnetic field. This splitting of the different spin configurations should efficiently increase the spin relaxation time of both Mn and holes.

The observation of an individual spin in a QD open new possibilities in information storage. The spin of an isolated Mn atom should present a relaxation time in the millisecond range. This property could be exploited to store digital information on a single atom. The device presented here is a milestone toward the development of new memories in which an information digit would be stored on the spin state of an individual atom.

4. TRANSPORT IN QUANTUM II–VI DMS STRUCTURES

MBE allows us to grow quantum nanostructures that exhibit optical properties, including those incorporating magnetic impurities. More recently, the technology of electrical doping of II–VI compounds became reasonably mature, and many specific features of the spectroscopy of doped nanostructure were discovered in II–VI semiconductors—the most prominent example being the charged exciton (Kheng et al., 1993). The problem of contacts, particularly of ohmic contacts, remains challenging, and that explains that the transport studies in II–VI quantum structures remain scarce.

Note that quantum effects can be seen in the transport studies of thin DMS layers, such as the positive magnetoresistance due to weak localization that has been observed at weak applied field in n-type (Cd,Mn)Te (Jaroszynski et al., 1995) or p-type (Zn,Mn)Te (Ferrand et al., 2001), or Universal Conductance Fluctuations observed to be governed by the fluctuations of the system of localized spins in n-type (Cd,Mn)Te submicrometer wires (Jaroszynski et al., 1995; Fig. 7.19).

Turning to 2D systems, the problem of contacting a buried carrier gas was not satisfactorily solved and, for example, Hall effect could be measured down to only about 10 K on the 2DHG contained in a CdTe or (Cd,Mn)Te QW. Contacting a 2DEG was more successful, both in (Zn,Mn)Te- (Smorchkova et al., 1997, 1998) and (Cd,Mn)Te-based (Jaroszynski et al., 2002; Teran et al., 2002) modulation-doped structures. In both systems, the integer quantum Hall effect could be measured and explained by the interplay between the giant Zeeman effect and the formation of Landau levels. The giant Zeeman effect being strongly nonlinear in field, it can be used to induce a level crossing between two different Landau levels

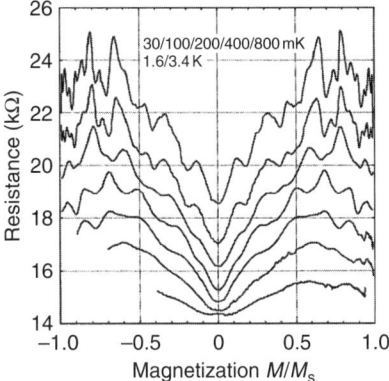

FIGURE 7.19 Magnetoresistance data observed in n-type $Cd_{0.99}Mn_{0.01}Te$ submicrometer wires plotted as a function magnetization in the units of $Ms = g\mu_B S N_0 x$ (Jaroszynski et al., 1995).

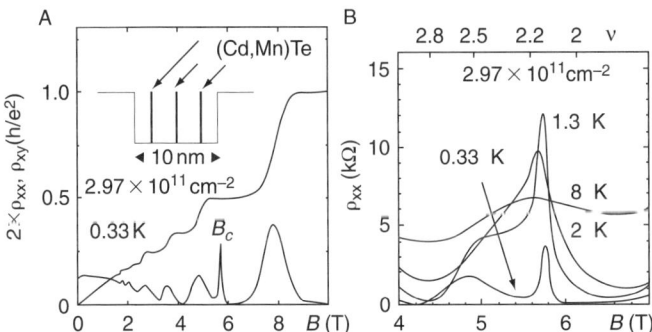

FIGURE 7.20 (A) Resistances ρ_{xx} and ρ_{xy} at $T = 0.33$ K for $n_s = 2.97 \times 10^{11}$ cm^{-2} ($V_g = 0$). Note the presence of a spike in ρ_{xx} at $B_c = 5.8$ T, shown at selected temperatures in (B). Inset: The nominal distribution of magnetic monolayers in the QW (Jaroszynski et al., 2002).

with different spin states (Harris et al., 2001; Jaroszynski et al., 2002; Knobel et al., 2002). When the crossing occurs in the vicinity of an integer value of the filling factor in n-type (Cd,Mn)Te QWs, a sharp spike was observed in the longitudinal resistance in a small range of temperatures (Fig. 7.20). This spike was shown to sign the formation of an Ising quantum Hall ferromagnet: at the crossing, only one of the two states is populated. The critical temperature was surprisingly high (about 2 K) and the critical behavior could be studied in detail (Jaroszynski et al., 2002).

Another original use of such structures that gives an interesting insight into the mechanism of Anomalous Hall Effect is described in Cumings et al. (2006). In a gated (Zn,Cd,Mn)Se QW, both the carrier

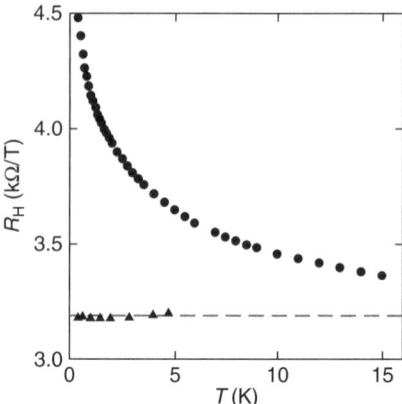

FIGURE 7.21 Hall resistance R_H as a function of temperature T. The dashed line corresponds to a Hall resistance R_H of 3190 ΩT calculated with a density of 1.96×10^{11} cm^{-2} (Cumings et al., 2006). The triangles show the results of similar Shubnikov–de Haas measurements performed over a range of T.

density and the spin polarization of the 2DEG can be controlled, and their effect on the Hall resistance analyzed and discussed with reference to various mechanisms (Fig. 7.21).

II–VI DMS have been used also to study the fundamental aspects of spin injection and spin transport in semiconductors. (Zn,Mn,Be)Se and (Zn,Mn)Te thick layers have been first used as efficient spin aligners at low temperature (Fiederling et al., 1999; Jonker et al., 2000; Schmidt et al., 2001). More recently, spin-dependent resonant tunneling has also been demonstrated with (Zn,Mn)Te/(Zn,Be)Se double-barrier tunnel junctions (Fig. 7.22), which could be good candidates to achieve in future voltage-controlled spin filters at low temperature (Gruber et al., 2001; Slobodskyy et al., 2003).

Measuring transport through a single QD was also achieved recently (Gould et al., 2006). The structure was nominally a sheet of CdSe QDs in a (Zn,Mn)Te barrier, and the current–voltage characteristics through an etched pillar displays sharp structures attributed to tunneling through a single QD. What is original in these data is that the structure splits under an applied magnetic field, with a behavior that can be convincingly ascribed to the giant Zeeman effect induced by Mn spins localized in the vicinity of the QD (it is argued that an electron confined in the QD has a probability of presence in the nearby barrier which is as high as 50%). Actually, a zero field splitting is observed also (Fig. 7.23). It is ascribed to the formation of a magnetic polaron around the electrons tunneling through the QD. A quantitative model is still lacking at the present time: the formation of the polaron is observed at a temperature as

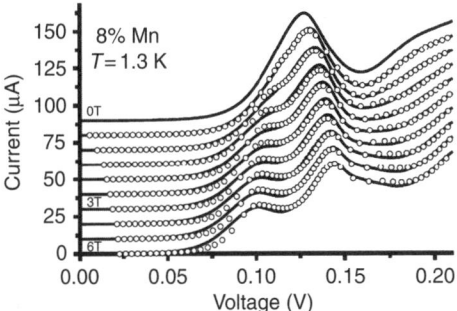

FIGURE 7.22 Experimental (lines) and modeled (circles) I–V curves for a resonant tunnel diode with a $Zn_{0.92}Mn_{0.08}Se$ quantum well. Curves taken in 0.5 T intervals from 0 to 3 T and in 1 T intervals from 3 to 6 T (Slobodskyy et al., 2003).

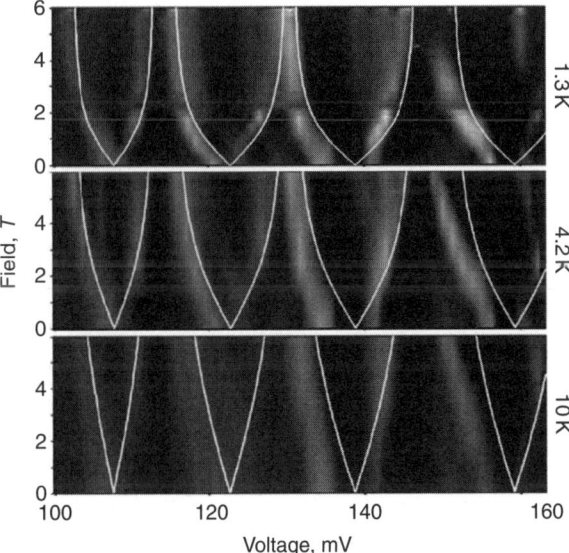

FIGURE 7.23 Color-scale image as a function of magnetic field and voltage for various bias resonances. The color scale is proportional to the voltage derivative of the current in order to better resolve the position of the resonances. The data at higher magnetic fields clearly is Brillouin-like, as evidenced by the lines plotting Brillouin functions. However, at fields below 500 mT, the behavior departs from a Brillouin function, with the splitting becoming constant and remaining finite even at zero field (Gould et al., 2006).

high as 5 K, much higher than the temperature expected from an estimation of the interaction between the Mn spins around the dot and the electron in the dot. A more collective behavior invoking the polarization of the current in the barrier is suggested.

These transport experiments in quantum structures of II–VI DMSs are few; however, their example shows that solving the technological problems of making ohmic contacts to II–VI DMS structures allows one to fabricate samples with extremely original properties.

5. SUMMARY

In this chapter, we demonstrate some advantages of an important fact concerning the II–VI DMSs that Mn substitutes the column-II cations as an isoelectronic impurity, as opposed to the acceptor character observed in III–V ones. As a consequence, manganese impurities take the d^5 electronic configuration introducing localized, isotropic spins with $S = 5/2$. Moreover, these magnetic atoms can substitute the cations in the II–VI matrix up to 100% at the optimal growth temperature. It allows then to tune independently the reservoir of spins and the reservoir of carriers.

Actually, doping with electrically active impurities was for a long time a major problem in II–VIs. Thanks to the MBE technology, significant progresses have been made in the fabrication of heterostructures and nanostructures based on these materials as well as in their electrical doping. We focused here on three topics considered as most relevant in the frame of semiconductor spintronics:

1. Carrier-induced ferromagnetism in (Cd,Mn)Te QWs, which constitutes a model system for understanding the basic mechanisms of magnetic transition in DMS.
2. The interaction of electrons and holes with a single magnetic impurity in a QD—an extreme limit for spintronics: by controlling the charge state of the QD (neutral and ±1e), it is possible to tune the magnetic behavior of the Mn atom. This opens the route to a single-spin memory in which the information digit would be stored on the spin state of an individual atom.
3. A few examples of transport measurements in quantum DMS structures. After the first results in these magnetotransport measurements, the next challenge would be to improve the electrical contacts in order to control and develop new transport properties in the II–VI quantum structures.

REFERENCES

Arnoult, A., Ferrand, D., Huard, V., Cibert, J., Grattepain, C., Saminadayar, K., Bourgognon, C., Wasiela, A., and Tatarenko, S. (1999). *J. Cryst. Growth* **201/202**, 715.
Arnoult, A., Cibert, J., Tatarenko, S., and Wasiela, A. (2000). *J. Appl. Phys.* **87**, 3777.

Bacher, G., Maksimov, A. A., Schömig, H., Kulakovskii, V. D., Welsch, M. K., Forchel, A., Dorozhkin, P. S., Chernenko, A. V., Lee, S., Dobrowolska, M., and Furdyna, J. K. (2002). *Phys. Rev. Lett.* **89**, 127201.
Baron, T., Kany, F., Saminadayar, K., Magnea, N., and Cox, R. T. (1997). *Appl. Phys. Lett.* **70**, 2963.
Baron, T., Saminadayar, K., and Magnea, N. (1998). *J. Appl. Phys.* **83**, 1354.
Bassani, F., Tatarenko, S., Saminadayar, K., Magnea, N., Cox, R. T., Tardot, A., and Grattepain, C. (1992). *J. Appl. Phys.* **72**, 2927.
Besombes, L., Kheng, K., and Martrou, D. (2000a). *Phys. Rev. Lett.* **85**, 425.
Besombes, L., Kheng, K., and Martrou, D. (2000b). *J. Crystal Growth* **214**, 742.
Besombes, L., Kheng, K., Marsal, L., and Mariette, H. (2002). *Phys. Rev. B* **65**, 121314.
Besombes, L., Léger, Y., Maingault, L., Ferrand, D., Mariette, H., and Cibert, J. (2004). *Phys. Rev. Lett.* **93**, 207403.
Besombes, L., Léger, Y., Maingault, L., Ferrand, D., Mariette, H., and Cibert, J. (2005). *Phys. Rev. B* **71**, 161307.
Bhattacharjee, A. K., and Perez-Conde, J. (2003). *Phys. Rev. B* **68**, 045303.
Bhunia, S., and Bose, D. N. (2000). *J. Appl. Phys.* **87**, 2931.
Boukari, H., Kossacki, P., Bertolini, M., Ferrand, D., Cibert, J., Tatarenko, S., Wasiela, A., Gaj, J. A., and Dietl, T. (2002). *Phys. Rev. Lett.* **88**, 207204.
Boukari, H., Perez, F., Ferrand, D., Kossacki, P., Jusserand, B., and Cibert, J. (2006). *Phys. Rev. B* **73**, 115320.
Brun-Le Cunff, D., Baron, T., Daudin, B., Tatarenko, S., and Blanchard, B. (1995). *Appl. Phys. Lett.* **67**, 965.
Chernenko, A. V., Dorozhkin, P. S., Kulakovskii, V. D., Brichkin, A. S., Ivanov, S. V., and Toropov, A. A. (2005). *Phys. Rev. B* **72**, 045302.
Chiba, D., Yamanouchi, M., Matsukura, F., and Ohno, H. (2003). *Science* **301**, 943.
Cibert, J., Gobil, Y., Dang, L. S., Tatarenko, S., Feuillet, G., Jouneau, P. H., and Saminadayar, K. (1990). *Appl. Phys. Lett.* **56**, 292.
Cibert, J., Kossacki, P., Haury, A., Ferrand, D., Wasiela, A., Merle d'Aubigné, Y., Arnoult, A., Tatarenko, S., and Dietl, T. (1998). *Proceedings of the 24th International Conference on the Physics of Semiconductors*, Jerusalem, Israel, (Gershoni, D., ed.), p. 51. World Scientific, Singapore.
Cibert, J., Ferrand, D., Boukari, H., Tatarenko, S., Wasiela, A., Kossacki, P., and Dietl, T. (2002). *Physica E* **13**, 489.
Climente, J. I., Korkusiski, M., Hawrylak, P., and Planelles, J. (2005). *Phys. Rev. B* **71**, 125321.
Cumings, J., Moore, L. S., Chou, H. T., Ku, K. C., Xiang, G., Crooker, S. A., Samarth, N., and Goldhaber-Gordon, D. (2006). *Phys. Rev. Lett.* **96**, 196404.
Dietl, T., Haury, A., and Merle d'Aubigné, Y. (1997). *Phys. Rev. B* **55**, R3347.
Dietl, T., Ohno, H., Matsukura, F., Cibert, J., and Ferrand, D. (2000). *Science* **287**, 1019.
Dietl, T., Ohno, H., and Matsukura, F. (2001). *Phys. Rev. B* **63**, 195205.
Dorozhkin, P. S., Chernenko, A. V., Kulakovskii, V. D., Brichkin, A. S., Maksimov, A. A., Schoemig, H., Bacher, G., Forchel, A., Lee, S., Dobrowolska, M., and Furdyna, J. K. (2003). *Phys. Rev. B* **68**, 195313.
Efros, A. L., Rashba, E. I., and Rosen, M. (2001). *Phys. Rev. Lett.* **87**, 206601.
Erwin, S. C., Zu, L., Haftel, M. I., Efros, A. L., Kennedy, T. A., and Norris, D. J. (2005). *Nature* **436**, 91.
Fan, Y., Han, J., He, L., Saraie, J., Gunshor, R. L., Hagerott, M., Jeon, H., Nurmikko, A. V., Hua, G. C., and Otsuka, N. (1992). *Appl. Phys. Lett.* **61**, 3160.
Fatah, J. M., Piorek, T., Harrison, P., Stirner, T., and Hagston, W. E. (1994). *Phys. Rev. B* **49**, 10341.
Fernandez-Rossier, J. (2006). *Phys. Rev. B* **73**, 045301.
Fernandez-Rossier, J., and Brey, L. (2004). *Phys. Rev. Lett.* **93**, 117201.

Ferrand, D., Cibert, J., Wasiela, A., Bourgognon, C., Tatarenko, S., Fishman, G., Andrearczyk, T., Jaroszynski, J., Kolesnik, S., Dietl, T., Barbara, B., and Dufeu, D. (2001). *Phys. Rev. B* **63**, 85201.
Fiederling, R., Keim, M., Reuscher, G., Ossau, W., Schmidt, G., Waag, A., and Molenkamp, L. W. (1999). *Nature (London)* **402**, 787.
Fisher, F., Waag, A., Bilger, G., Litz, T. h., Scholl, S., Schmitt, M., and Landwehr, G. (1994). *J. Crystal Growth* **141**, 93.
Furdyna, J. K. (1988). *J. Appl. Phys.* **64**, R29.
Furdyna, J. K., and Kossut, J. (1988). *In* "Semiconductors and Semimetals," vol. 25, edied by, Academic Press, Boston.
Gaj, J. A., Planel, R., and Fishman, G. (1979). *Solid State Commun.* **29**, 435.
Gaj, J. A., Grieshaber, W., Bodin, C., Cibert, J., Feuillet, G., Merle d'Aubigné, Y., and Wasiela, A. (1994). *Phys. Rev. B* **50**, 5512.
Gambardella, P., Rusponi, S., Veronese, M., Dhesi, S. S., Grazioli, C., Dallmeyer, A., Cabria, I., Zeller, R., Dederichs, P. H., Kern, K., Carbone, C., and Brune, H. (2003). *Science* **300**, 1130.
Gould, C., Slobodskyy, A., Supp, D., Slobodskyy, T., Grabs, P., Hawrylak, P., Qu, F., Schmidt, G., and Molenkamp, L. W. (2006). *Phys. Rev. Lett.* **97**, 017202.
Govorov, A. O. (2004). *Phys. Rev. B* **70**, 035321.
Govorov, A. O., and Kalameitsev, A. V. (2005a). *Phys. Rev. B* **71**, 035338.
Govorov, A. O., and Kalameitsev, A. V. (2005b). *Phys. Rev. B* **72**, 075359.
Grieshaber, W., Haury, A., Cibert, J., Merle d'Aubigné, Y., Wasiela, A., and Gaj, J. A. (1996). *Phys. Rev. B* **53**, 4891.
Gruber, T. H., Keim, M., Fiederling, R., Reuscher, G., Ossau, W., Schmidt, G., Molenkamp, L. W., and Waag, A. (2001). *Appl. Phys. Lett.* **78**, 1101.
Grün, M., Haury, A., Cibert, J., and Wasiela, A. (1996). *J. Appl. Phys.* **79**, 7386.
Hansen, L., Ferrand, D., Richter, G., Thierley, M., Hork, V., Schwarz, N., Reuscher, G., Schmidt, G., and Molenkamp, L. (2001). *Appl. Phys. Lett.* **79**, 3125.
Harris, J. G. E., Knobel, R., Maranowski, K. D., Gossard, A. C., Samarth, N., and Awschalom, D. D. (2001). *Phys. Rev. Lett.* **86**, 4644.
Hartmann, J. M., Cibert, J., Kany, F., Mariette, H., Charleux, M., Alleyson, P., Langer, R., and Feuillet, G. (1996). *J. Appl. Phys.* **80**, 6257.
Hartmann, A., Ducommun, Y., Kapon, E., Hohenester, U., and Molinari, E. (2000). *Phys. Rev. Lett.* **84**, 5648.
Haury, A., Wasiela, A., Arnoult, A., Cibert, J., Dietl, T., Merle d'Aubigné, Y., and Tatarenko, S. (1997). *Phys. Rev. Lett.* **79**, 511.
Heinrich, A. J., Gupta, A., Lutz, C. P., and Eigler, D. M. (2004). *Science* **306**, 466.
Hundt, A., Puls, J., and Henneberger, H. (2004). *Phys. Rev. B* **69**, 121309.
Jaroszynski, J., Wróbel, J., Sawicki, M., Kaminska, E., Skoskiewicz, T., Karczewski, G., Wojtowicz, T., Piotrowska, A., Kossut, J., and Dietl, T. (1995). *Phys. Rev. Lett.* **75**, 3170; and erratum *Phys. Rev. Lett.* **76**, 1556.
Jaroszynski, J., Andrearczyk, T., Karczewski, G., Wróbel, J., Wojtowicz, T., Papis, E., Kaminska, E., Piotrowska, A., Popovic, D., and Dietl, T. (2002). *Phys. Rev. Lett.* **89**, 266802.
Jelezko, F., Gaebel, T., Popa, I., Gruber, A., and Wrachtrup, J. (2004). *Phys. Rev. Lett.* **92**, 076401.
Jonker, B. T., Park, Y. D., Bennett, B. R., Cheong, H. D., Kioseoglou, G., and Petrou, A. (2000). *Phys. Rev B* **62**, 8180.
Jusserand, B., Perez, F., Richards, D. R., Karczewski, G., Wojtowicz, T., Testelin, C., Wolverson, D., and Davies, J. J. (2003). *Phys. Rev. Lett.* **91**, 086802.
Kavokin, K. V. (1999). *Phys. Rev. B* **59**, 9822.
Kechrakos, D., Papanikolaou, N., Trohidou, K. N., and Dietl, T. (2005). *Phys. Rev. Lett.* **94**, 127201.

Keller, D., Yakovlev, D. R., König, B., Ossau, W., Gruber, T. H.., Waag, A., Molenkamp, L. W., and Scherbakov, A. V. (2001). *Phys. Rev. B* **65**, 035313.
Kheng, K., Cox, R. T., Merle d'Aubigné, Y., Bassani, F., Saminadayar, K., and Tatarenko, S. (1993). *Phys. Rev. Lett.* **71**, 1752.
Knobel, R., Samarth, N., Harris, J. G. E., and Awschalom, D. D. (2002). *Phys. Rev. B* **65**, 235327.
Kossacki, P., Kudelski, A., Gaj, J. A., Cibert, J., Tatarenko, S., Ferrand, D., Wasiela, A., Deveaud, B., and Dietl, T. (2002). *Physica E* **12**, 344.
Kossacki, P., Boukari, H., Bertolini, M., Ferrand, D., Cibert, J., Tatarenko, S., Gaj, J. A., Deveaud, B., Ciulin, V., and Potemski, M. (2004a). *Phys. Rev. B* **70**, 195337.
Kossacki, P., Pacuski, W., Maślana, W., Gaj, J. A., Bertolini, M., Ferrand, D., Bleuse, J., Tatarenko, S., and Cibert, J. (2004b). *Physica E* **21**, 943.
Kossacki, P., Ferrand, D., Goryca, M., Nawrocki, M., Pacuski, W., Maślana, W., Tatarenko, S., and Cibert, J. (2006). *Physica E* **32**, 454.
Koudinov, A. V., Akimov, I. A., Kusrayev, Y. G., and Henneberger, F. (2004). **70**, 241305.
Kyrychenko, F. V., and Kossut, J. (2004). *Phys. Rev. B* **70**, 205317.
König, J., and MacDonald, A. H. (2003). *Phys. Rev. Lett.* **91**, 077202.
Léger, Y., Besombes, L., Maingault, L., Ferrand, D., and Mariette, H. (2005). *Phys. Rev. B* **72**, 241309.
Léger, Y., Besombes, L., Fernandez-Rossier, J., Maingault, L., and Mariette, H. (2006). *Phys. Rev. Lett.* **97**, 107401.
Linnarsson, M., Janzén, E., Monemar, B., Kleverman, M., and Thilderkvist, A. (1997). *Phys. Rev. B* **55**, 6938.
Mackh, G., Ossau, W., Yakovlev, D. R., Waag, A., Landwehr, G., Hellmann, R., and Gobel, E. O. (1994). *Phys. Rev. B* **49**, 10248.
Mackowski, S., Gurung, T., Jackson, H. E., Smith, L. M., Karczewski, G., and Kossut, J. (2005). *Appl. Phys. Lett.* **87**, 072502.
Maingault, L., Besombes, L., Léger, Y., Bougerol, C., and Mariette, H. (2006). *Appl. Phys. Lett.* **89**, 193109.
Maksimov, A. A., Bacher, G., McDonald, A., Kulakovskii, V. D., Forchel, A., Becker, C. R., Landwehr, G., and Molenkamp, L. W. (2000). *Phys. Rev. B* **62**, R7767.
Manoharan, H. C., Lutz, C. P., and Eigler, D. M. (2000). *Nature* **403**, 512.
Maslana, W., Bertolini, M., Boukari, H., Kossacki, P., Ferrand, D., Gaj, J. A., Tatarenko, S., and Cibert, J. (2003). *Appl. Phys. Lett.* **82**, 1875.
Maslana, W., Kossacki, P., Plochocka, P., Golnik, A., Gaj, J. A., Ferrand, D., Bertolini, M., Tatarenko, S., and Cibert, J. (2004). In International Conference on the Physics of Semiconductors, Flagstaff (USA), AIP Conference Proceedings, Volume 772 Melville, New York, 2005, (J. Menendez and C. Van de Walle, eds.), p. 1299.
Merkulov, I. A., Yakovlev, D. R., Keller, A., Ossau, W., Geurts, J., Waag, A., Landwehr, G., Karczewski, G., Wojtowicz, T., and Kossut, J. (1999). *Phys. Rev. Lett.* **83**, 1431.
Norris, D. J., Yao, N., Charnock, F. T., and Kennedy, T. A. (2001). *Nano Lett.* **1**, 3.
Ohno, H., Chiba, D., Matsukura, F., Omiya, T., Abe, E., Dietl, T., Ohno, Y., and Ohtani, K. (2000). *Nature* **408**, 944.
Pacuski, W., Ferrand, D., Cibert, J., Deparis, C., Gaj, J. A., Kossacki, P., and Morhain, C. (2006). *Phys. Rev. B* **73**, 035214.
Park, R. M., Trofer, M. B., Rouleau, C. M., DePuydt, J. M., and Haase, M. A. (1990). *Appl. Phys. Lett.* **57**, 2127.
Qu, F., and Hawrylak, P. (2005). *Phys. Rev. Lett.* **95**, 217206.
Qu, F., and Hawrylak, P. (2006). *Phys. Rev. Lett.* **96**, 157201.
Robin, I. C., André, R., Bougerol, C., Aichele, T., and Tatarenko, S. (2006). *Appl. Phys. Lett.* **88**, 233103.
Scalbert, D., Cernogora, J., and Benoit à la Guillaume, C. (1988). *Solid State Commun.* **66**, 571.

Scalbert, D., Teppe, F., Vladimirova, M., Tatarenko, S., Cibert, J., and Nawrocki, M. (2004). *Phys. Rev. B* **70**, 245304.
Schmidt, G., Richter, G., Grabs, P., Gould, C., Ferrand, D., and Molenkamp, L. W. (2001). *Phys. Rev. Lett.* **87**, 227203.
Schneider, J., Kaufmann, U., Wilkening, W., Baeumler, M., and Köhl, F. (1987). *Phys. Rev. Lett.* **59**, 240.
Seufert, J., Bacher, G., Scheibner, M., Forchel, A., Lee, S., Dobrowolska, M., and Furdyna, J. K. (2002). *Phys. Rev. Lett.* **88**, 027402.
Seufert, J., Rambach, M., Bacher, G., Forchel, A., Passow, T., and Hommel, D. (2003). *Appl. Phys. Lett.* **82**, 3946.
Shapira, Y., Foner, S., Ridgley, D. H., Dwight, K., and Wold, A. (1984). *Phys. Rev. B* **30**, 4021.
Slobodskyy, A., Gould, C., Slobodskyy, T., Becker, C. R., Schmidt, G., and Molenkamp, L. W. (2003). *Phys. Rev. Lett.* **90**, 246601.
Smorchkova, I. P., Samarth, N., Kikkawa, J. M., and Awschalom, D. D. (1997). *Phys. Rev. Lett.* **78**, 3571.
Smorchkova, I. P., Samarth, N., Kikkawa, J. M., and Awschalom, D. D. (1998). *Phys. Rev. B* **58**, R4238.
Straßburg, M., Kuttler, M., Stier, O., Pohl, U. W., Bimberg, D., Behringer, M., and Hommel, D. (1998). *J. Cryst. Growth* **184–185**, 465.
Tadic, M., Peeters, F. M., and Janssens, K. L. (2002). *Phys. Rev. B* **65**, 165333.
Terai, Y., Kuroda, S., Takita, K., Okuno, T., and Masumoto, Y. (1998). *Appl. Phys. Lett.* **73**, 3757.
Teran, F. J., Potemski, M., Maude, D. K., Andrearczyk, T., Jaroszynski, J., and Karczewski, G. (2002). *Phys. Rev. Lett.* **88**, 186803.
Teran, F. J., Potemski, M., Maude, D. K., Plantier, D., Hassan, A. K., Sachrajda, A., Wilamowski, Z., Jaroszynski, J., Wojtowicz, T., and Karczewski, G. (2003). *Phys. Rev. Lett.* **91**, 077201.
Tinjod, F., Gilles, B., Moehl, S., Kheng, K., and Mariette, H. (2003). *Appl. Phys. Lett.* **82**, 4340.
Tinjod, F., Robin, I. C., André, R., Kheng, K., and Mariette, H. (2004). *J. Alloys Comp.* **371**, 63.
Twardowski, A., Swiderski, P., von Ortenberg, M., and Pauthenet, R. (1984). *Solid State Commun.* **50**, 509.
Vasanelly, A., Ferreira, R., and Bastard, G. (2002). *Phys. Rev. Lett.* **89**, 216804.
Xin, S. H., Wang, P. D., Yin, A., Kim, C., Dobrowolska, M., Merz, J. L., and Furdyna, J. K. (1996). *Appl. Phys. Lett.* **69**, 3884.
Yakovlev, D. R., Kavokin, K. V., Merkulov, I. A., Mackh, G., Ossau, W., Hellmann, R., Gobel, E. O., Waag, A., and Landwehr, G. (1997). *Phys. Rev. B* **56**, 9782.
Yakunin, A. M., Silov, A. Y., Koenraad, P. M., Tang, J. M., Flatté, M. E., Roy, W. V., Boeck, J. D., and Wolter, J. H. (2004). *Phys. Rev. Lett.* **92**, 216806.
Zener, C. (1951). *Phys. Rev.* **81**, 440.

CHAPTER 8

Magnetic Impurities in Wide Band-gap III–V Semiconductors

Agnieszka Wolos and **Maria Kaminska**

Contents			
	1.	Introduction	326
	2.	Diluted Magnetic Semiconductors	330
	3.	Nature of Mn Impurity in III–V Semiconductors	334
		3.1. Ionized acceptor, $Mn^{2+}(d^5)$	335
		3.2. Neutral acceptor, $Mn^{3+}(d^4)$	340
		3.3. Neutral acceptor, $Mn^{2+}(d^5)$+hole	344
		3.4. Localization radius of the hole bound to $Mn^{2+}(d^5)$ center in III–V semiconductors	345
		3.5. Location of Mn acceptor level in III–V semiconductors band structure—lattice relaxation	348
		3.6. Lower charge states of Mn in wide band gap III–V semiconductors	350
	4.	Magnetic Interactions in III–V DMSs with Mn	352
		4.1. Magnetic properties of wurtzite GaN with Mn	352
		4.2. Exchange integrals in GaN doped with Mn	355
	5.	GaN-Based DMSs	357
		5.1. GaN diluted with Fe	357
		5.2. GaN diluted with Cr	359
	6.	Internal Reference Rule for Transition Metal Ions—Case of GaN	361
	7.	Summary and Conclusions	362
	Acknowledgments		364
	References		364

Institute of Experimental Physics, University of Warsaw, Hoza 69, 00–681 Warsaw, Poland

1. INTRODUCTION

The name of magnetic impurities refers to impurities of transition metals or rare earths, since these species introduce magnetic moments to semiconductor host lattice. The characteristic feature of transition metal atoms is partially filled d sub-shell. The most common transition metals are placed in the fourth row of the periodic table of elements and their electron configuration is $[Ar]3d^n4s^2$ (where n changes from 1 for scandium to 8 for nickel; for the last in row copper the configuration is $3d^{10}4s^1$). Other transition metals are in the fifth, sixth, or seventh period; their electron configuration is analogical—all of them have an incomplete d sub-shell—and they differ in the number of filled shells. Being introduced into semiconductors they keep d electrons localized at the center, providing valence s electrons to bonds. Electrons from the incomplete d sub-shell are sources of non-vanishing spin moment and, in some cases, also of non-vanishing orbital moment. Both spin and orbital moments constitute a source of localized magnetic moment of the center. Unpaired electrons originating from the d orbit and configured according to Hund's rules provide especially high spin-related magnetic moment in the case of half-filled d^5 sub-shell of total spin $S = 5/2$. Rare earth's name is often used to describe lanthanoids and less common actinoids which have partially filled f sub-shell. Electron configuration of lanthanoids is $[Xe]4f^n6s^2$. Localized electrons of rare earth impurities, originating from f orbit, introduce magnetic moments to the host lattice. In this case, the highest magnetic moment related to spin corresponds to f^7 configuration of total spin $S = 7/2$. Transition metal d or rare earth f localized electrons often introduce deep energy states within host semiconductor energy gap, and therefore they strongly influence optical and electrical properties of the host semiconductor. High magnetic moment of these impurities, and interactions of their d or f electrons with conduction and valence band electrons, leads also to high sensitivity of electron transport and optical properties on external magnetic field.

In the 1970s and 1980s, there were many studies that were performed on bulk semiconductor crystals doped with transition metals. Transition metals were always common unintentional dopants, and at that time they were also introduced intentionally in order to change optical or electrical properties of host semiconductor. Information about configurations of transition metal impurities in semiconductor lattice and about energy levels they created in the range of energy gap of a particular semiconductor was of strong demand as these properties decided about application possibilities—transition metal impurities could lead to improvement or degradation of working semiconductor-based devices. Studies of transition metal impurities allowed for understanding of radiative and non-radiative recombination properties of impure semiconductor, mechanism of compensation of shallow impurities, Fermi level position, change of non-equilibrium carrier life time, scattering of drifting free carriers, etc. Especially important was such knowledge about transition metal impurities in silicon, number one material of electronics, and in III–V semiconductors, mainly in GaAs, since they were basic optoelectronic materials. Classical example of commonly studied transition metal impurity was chromium in GaAs, which having deep acceptor level at GaAs midgap could compensate shallow (often unintentional) donors, stabilize Fermi level close to the center of energy gap, and provide material of semi-insulating properties, necessary for substrates used for further epitaxial growth of device structures. Indeed, for some time GaAs:Cr was used as a semi-insulating substrate material in commercial applications, until it was substituted by GaAs less contaminated with shallow donors, in which compensation of residual shallow acceptors (mostly carbon) was achieved by native defect EL2.

We fruitfully profit nowadays from the heritage of these studies, when searching for semiconductors useful for spintronics applications. We have a broad knowledge about the characteristic features and specific trends caused by the presence of transition metals in semiconductors, mentioning just the following one. The energy positions of acceptor $A^{-/0}$ and donor $D^{0/+}$ levels of transition metals in respect to semiconductor valence and conduction band edges are universal if the relative position of band edges are shifted according to band offsets determined from adequate heterostructure studies (Langer et al., 1988; Zunger 1986). This rule allows for predicting the position of a particular transition metal's energy levels within a particular semiconductor band structure. Figure 8.1 shows an example of experimentally determined location of Mn, Fe, and Cr acceptor levels in a group of III–V semiconductors with a clear trend of aligning the charge transfer levels in GaAs, InP, and GaP. Implication of the reference rule for transition metals located in GaN will be discussed later.

Studies of transition metal impurities performed in 1970s and 1980s also allowed working out methods useful for determination of transition metal impurity configurations in semiconductors. The very important characterization technique occurred to be measurements of electron spin resonance (ESR). Observation of resonance transition between effective spin levels split in external magnetic field and studies of dependence of the resonances on the magnetic field orientation in respect to crystal axis helped in identification of transition metal-related centers, determination of their symmetry, and the nature of their closest neighborhood. The other valuable methods were optical, absorption or luminescence, often accompanied with use of external fields—either magnetic field or pressure. Transition metals give rise to two kinds of optical transitions: photoionization and intracenter. Photoionization processes are connected with a change of local charge state of the transition metal center (either by

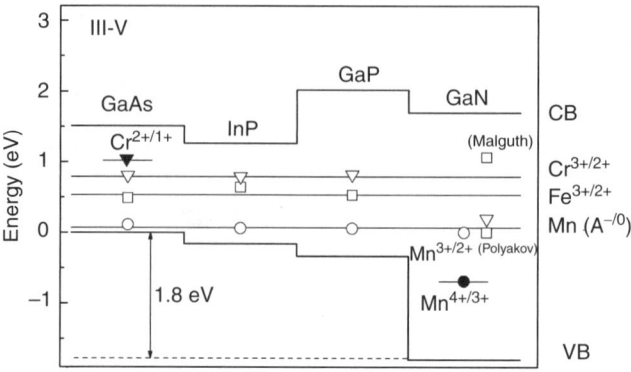

FIGURE 8.1 Location of Mn (dots), Fe (squares), and Cr (triangles) acceptor or donor levels within the band structure of a series of III–V semiconductors. Band offsets between GaAs, InP, and GaP after (Langer et al., 1988), between GaAs and GaN after (Vurgaftman et al., 2001). Energy gaps are taken from Madelung's handbook (2004). Location of Mn levels in GaAs, InP, and GaP after (Tarhan et al., 2003), in GaN after (Wolos et al., 2004a) and (Han et al., 2005). Location of Fe acceptor level in GaAs, InP, GaP (Shanabrook et al., 1983), in GaN (Malguth et al., 2006a) and alternatively (Polyakov et al., 2003a). Location of Cr acceptor level in GaAs (Szawelska and Allen, 1979), InP (Fung and Nicholas, 1981), GaP (Halliday et al., 1986), and GaN (Polyakov et al., 2003a). Solid lines marking position of the impurity levels are guides for an eye.

adding one electron from the valence band or by losing one on account of the conductive band). Studies of such transitions allowed determining energy level position within the semiconductor band structure, corresponding to transition metal impurity. Localized character of impurity led typically to strong electron-lattice coupling, which manifested itself by lattice relaxation (mostly symmetric), accompanying the photoionization process. These resulted in a significant difference between electron and optical (equal electron plus lattice relaxation) energies of the impurity level. The intracenter transitions are connected with change of electron distribution in the vicinity of impurity, without change of a total local charge. Localized character of wave function also resulted, in this case, in electron-lattice coupling of typically smaller relaxation energy than in the case of photoionization process. Moreover, lattice relaxation was often non-symmetric (so called Jahn–Teller effect) and obviously depended on the electron wave function symmetry. Intracenter transition created typically broad absorption or luminescence bands accompanied by fine structure with zero phonon line corresponding to transition between zero oscillator ground and excited level. Intracenter transitions measured at external field helped with identification of the center and disclosed main interactions related to it. Especially valuable in this aspect were patterns of splitting of the fine structures under external fields.

In parallel to the studies of semiconductors containing transition metal ions in concentration on impurity level, intense research activity had been developed in the area of so called diluted magnetic semiconductors (DMSs), which at that time were II–VI, II–V, or IV–VI semiconductors with high (typically of the order of a few atomic percent) amount of magnetic ions substituted for nonmagnetic cations (for review see *Semiconductors and Semimetals*, 1988). The very first extensively studied DMS was $Hg_{1-x}Mn_xTe$ which showed characteristic for this class of materials anomalous magneto-optical (Bastard et al., 1978; Pastor et al., 1979) and magneto-transport (Jaczynski et al., 1979) properties. They were explained as a result of exchange interaction of $Mn^{2+}(d^5)$ ions with conduction band and valence band electrons, which will be discussed in more details in the following section.

Studies of semiconductors doped with magnetic ions, as well as knowledge accomplished about DMSs, happened to be very needful in the last decade in context of search for materials interesting for a newborn branch of electronics, named spintronics. Spintronics' intention is to exploit the spin of carriers in electronic devices and the expected advantage of spintronic devices over conventional electronic ones should be nonvolatility, increased data processing speed, increased transistor density, and decreased power consumption. The most ambitious goal of spintronics is development of quantum computing. First, the already working and genially simple spintronic devices are based on classical ferromagnetic materials—ferromagnetic metals. Sandwich type structures made of such metals with nonmagnetic metals in-between and similar structures with thin insulating layers in-between—called spin valves, working on principle of gigantic magneto-resistance phenomenon (Baibich et al., 1988; Binasch et al., 1989), are nowadays commonly commercially used in read heads of magnetic memories and in magnetic random memories. However, the real challenge is to build all-semiconductor spintronic devices, for which the first step is to fabricate semiconductors that can display electrically tunable ferromagnetism at temperatures above room temperature and can be easily integrated with conventional electronic devices. The first choice in the search for such materials are different DMSs. This class of materials has been extended in the last years also to III–V semiconductors diluted with magnetic ions. Manganese, iron, chromium, and gadolinium creating centers with d or f sub-shells filled close to half, are sources of high spin and are most commonly used as magnetic ions in DMSs.

In the following review, we shall describe general properties of DMSs caused by the presence of magnetic ions. Then we will discuss the nature of Mn impurity in III–V semiconductors. Special attention will be devoted to GaN:Mn since GaMnN has been, in the recent years, the most extensively studied DMS material for possible spintronic application.

The results of the research performed on GaN diluted with iron or chromium will also be presented. Finally, we shall present the summary and conclusions.

2. DILUTED MAGNETIC SEMICONDUCTORS

Chemical compounds containing transition metals or rare earths are often obviously magnetic materials. Some of them are also semiconductors. Implementation of such magnetic semiconductors in devices could provide new type of control, not only over charge transport of n- or p-type but also over quantum property of electron—its spin state. Ferromagnetism and semiconductor properties coexist in manganese or europium chalcogenides and Cr spinels (Medvedkin et al., 2000; Ohno and Matsukura, 2001), which have been studied for about 40 years. However, like in the case of ferromagnetic metals, there is a problem with their incorporation into integrated circuits. Their crystal structure is very different from traditional semiconductors used in electronics today. Moreover, their critical temperature is quite low, they are technologically difficult, and expensive. Therefore, DMSs, which are based on traditional semiconductors used in electronic devices, are the most promising materials for spintronic applications, providing that there is hope for their ferromagnetism at temperatures above room temperature. More than 20 years ago ferromagnetism of DMSs was demonstrated for the first time. It was found that PbSnMnTe DMS could behave ferromagnetically, although only at liquid helium temperature (Story et al., 1986). In spite of commercial uselessness of PbSnMnTe, the very important and interesting outcome of its studies was discovery of free carrier contribution to ferromagnetic ordering and dependence of Curie temperature, T_C, on free carrier concentration. It occurred that the two systems present in DMSs, namely band carriers and localized magnetic moments, interact and lead to several interesting properties. One of them is the creation of ferromagnetic phase. This finding gave hope that playing with semiconductor parameters may eventually lead to a DMS performing as a ferromagnetic at room temperature.

Up to the late 1980s, the best known and the most extensively studied DMSs were diluted magnetic II–VI compounds (zinc, mercury and cadmium sulfides, selenides, and tellurides) with manganese, iron, cobalt, chromium, and vanadium, which substitute cations in the semiconductor lattice—for review see Furdyna, 1988. In such semiconductors, the valence of cations matches that of magnetic ions, so extra impurities are needed to make the DMS material of n- or p-type. Solubility of transition metals in II–VI compounds occurred to be quite high, and $A_{1-x}Mn_xB$ (where A and B stands for group II and VI elements, respectively) ternary alloys of zinc blende structure, with contents x of Mn up to above 0.8 in some cases, were obtained. It is remarkable in that sense that the stable crystal structures of MnS, MnSe, and MnTe are not zinc blende. Solubility of other than Mn transition metals was a bit lower, and the value of solubility decreased in the following order: Mn, Fe, and then Cr, but still diluted II–VI compounds with x of the order of a few percent of transition metals other than Mn were possible to be grown.

Wide studies of magnetic interactions taking place in $A_{1-x}Mn_xB$ DMSs showed that they were dominated by d–d antiferromagnetic superexchange between manganese $Mn^{2+}(d^5)$ ions, with almost a total contribution from indirect exchange interaction mediated by anions. Therefore, these interactions involved valence band, as properties of this band are dominated by anions. Depending on the amount of manganese and temperature, paramagnetic, antiferromagnetic, or spin glass behavior was observed. Ferromagnetic ordering due to non-direct interactions between manganese $Mn^{2+}(d^5)$ ions, mediated by free carriers, was found at helium temperatures only (Ferrand et al., 2001; Jaroszynski et al., 2002).

Although the above discussed diluted magnetic II–VI are not considered as semiconductors potentially interesting for spintronics—it would be difficult to have them ferromagnetic at room temperature—their studies showed several spectacular consequences of interactions between localized magnetic moments of d electrons with band carriers (s conduction band electrons and p valence band holes), so called sp–d exchange interactions. In the virtual-crystal and molecular-field approximations (Gaj et al., 1978, 1979), these interactions are characterized by α and β exchange integrals for the s-like Γ_6 electrons and p-like Γ_8 holes, respectively:

$$\alpha \equiv \langle S| J^{sp-d} |S\rangle / \Omega_0,$$

$$\beta \equiv \langle X| J^{sp-d} |X\rangle / \Omega_0,$$

where J^{sp-d} is the free carrier–magnetic ion sp–d exchange coupling constant and Ω_0 is the volume of an elementary cell. The band structure of DMSs is modified strongly by the sp–d exchange interactions. In wide-gap $A_{1-x}Mn_xB$ alloys, exchange term in total Hamiltonian, describing band structure, leads to magnetic splitting much greater than the Landau and spin splittings predicted by sp band theory for nonmagnetic semiconductor. Therefore, it is justified to simplify the formula describing behavior of the bottom of the conduction band (of Γ_6 symmetry) and the top of the valence band (of Γ_8 symmetry) (at $k=0$) neglecting Landau and spin splittings. The respective energies can be approximated by the formulas:

—for the valence band

$$E\left(\left|+\frac{3}{2}\right\rangle\right) = \frac{1}{2}N_0\beta x \langle S_z \rangle$$

$$E\left(\left|+\frac{1}{2}\right\rangle\right) = \frac{1}{6}N_0\beta x \langle S_z \rangle$$

$$E\left(\left|-\frac{1}{2}\right\rangle\right) = -\frac{1}{6}N_0\beta x \langle S_z \rangle$$

$$E\left(\left|-\frac{3}{2}\right\rangle\right) = -\frac{1}{2}N_0\beta x \langle S_z \rangle$$

—for the conduction band

$$E\left(\left|-\frac{1}{2}\right\rangle\right) = E_g - \frac{1}{2}N_0\alpha x \langle S_z \rangle$$

$$E\left(\left|+\frac{1}{2}\right\rangle\right) = E_g + \frac{1}{2}N_0\alpha x \langle S_z \rangle$$

where E_g is the energy gap.

The above described splittings of the valence and conduction bands are qualitatively identical to "normal" spin splittings, which occur at magnetic field in nonmagnetic semiconductors. However, in the case of DMSs the magnitude of splitting is extremely large, and this magnitude is determined primarily by magnetization because of its dependence on $\langle S_z \rangle$. This means that the splitting observed for DMS at low magnetic field could be observed for its non-magnetic counterpart at many times higher magnetic field. The obtained values of the exchange constant $N_0\alpha$ and $N_0\beta$ for II–VI DMSs with Mn were of opposite sign ($\alpha > 0$, $\beta < 0$), the magnitude of $N_0\alpha$ was typically of the order of 0.2 eV, the absolute value of $N_0\beta$ was larger (of the order of 1 eV), and both constants were insensitive to Mn concentration.

The scheme of conduction and valence band levels at $k = 0$ in a magnetic field, assuming positive sign of $N_0\alpha$ and negative sign for $N_0\beta$, is shown in Fig. 8.2.

The consequence of such extremely large Zeeman splitting of conduction and valence band levels due to sp–d interaction was that magneto-optical and magneto-transport properties of diluted magnetic II–VI compounds were found to be qualitatively different from those observed in their nonmagnetic counterparts. Especially well known is the effect of giant Zeeman splitting measured on excitons, and the giant Faraday

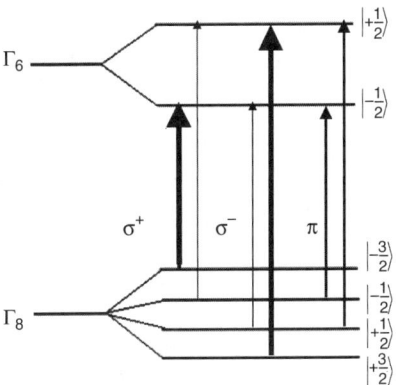

FIGURE 8.2 Scheme of splitting of the conduction (Γ_6) and the valence band (Γ_8) at $k = 0$ for wide gap diluted magnetic II–VI compounds in magnetic field. Arrows show electric-dipole allowed transitions of different polarizations, circular σ (in the plane perpendicular to the applied magnetic field) and linear π (electric field in the plane parallel to the applied magnetic field). In the case of circular polarization thicker arrows indicate stronger transitions.

rotation, resulting from large difference in refraction index for two opposite circular polarizations of light in the Faraday geometry in vicinity of the band edge (Gaj, 1988).

The breakthrough in the studies of semiconductor ferromagnetism was discovery of ferromagnetic properties at relatively high temperatures in technologically important III–V compounds InAs (Munekata et al., 1989; Ohno et al., 1992) and GaAs (Ohno et al., 1996) diluted with manganese (record high Curie temperature T_C is about 173 K for epitaxial GaMnAs (Wang et al., 2005)). Explanation of the magnetic ordering, leading to ferromagnetism in these compounds, was made in frame of Zener's model, originally proposed for transition metals in 1950 (Zener, 1951). It assumed ferromagnetic correlation, mediated by holes originating from shallow acceptors in the ensemble of localized spins of magnetic impurities (Dietl et al., 2000). The model explained quantitatively very well the ferromagnetism observed in GaMnAs. It implied that high spin Mn ions and high concentration of free carriers were both necessary to achieve ferromagnetism of a DMS. The model showed that Curie temperature increases as a second power of exchange integral, and also that smaller value of spin-orbit splitting between the Γ_8, Γ_7 valence bands leads to higher T_C. It predicted that Curie temperature for a specific semiconductor depends on free hole concentration p and $Mn^{2+}(d^5)$ mole fraction x, according to the formula:

$$T_C = Cxp^{1/3},$$

where C is a constant specific to host material, dependent on the p–d exchange integral.

Since the free carriers seemed to be crucial for ferromagnetic ordering, advantage of the above mentioned diluted magnetic III–V compounds over II–VI compounds was that in the former case Mn substituting for trivalent Ga or In stayed in d^5 high spin configuration, and at the same time provided free hole to the valence band. In II–VI compounds Mn with its $4s^2$ electrons matched for divalent cations, and although also had favorable high spin of 5/2, it did not introduce carriers to bands. Therefore, extrinsic co-doping was necessary to generate carrier-mediated manganese spin–spin interactions. However, problems with efficient co-doping with shallow impurities and commonly stronger spin–orbit interactions in comparison with III–V do not give too much hope that ferromagnetism at room temperature could be attainable in II–VI compounds. Exceptions are oxides, among them ZnMnO, which has started to be intensively investigated recently, since in ZnO oxygen-related valence band carriers undergo very small spin–orbit interaction. However, generally diluted magnetic III–V semiconductors seemed to be more promising for achieving high temperature ferromagnetism. Especially distinguishable should be GaN:Mn because of small spin–orbit coupling and small lattice parameter leading to strong p–d hybridization; therefore, one could expect high exchange constant $N_0\beta$. The model of Dietl *et al.*, 2000 predicted high Curie temperature, about 400 K, for GaMnN containing 5% of Mn in high spin 5/2 configuration and $3.5 \times 10^{20} \mathrm{cm}^{-3}$ hole concentration.

In order to better understand the magnetic interactions in GaMnN and other III–V based semiconductors diluted with Mn, it was necessary to undertake study of Mn impurity in different III–V semiconductors and try to find general trends of its behavior. Results of such studies are the subject of the next section.

3. NATURE OF Mn IMPURITY IN III–V SEMICONDUCTORS

The electron configuration of a free Mn atom is $[\mathrm{Ar}]3d^54s^2$. In III–V semiconductors, Mn substitutes the cation site thus giving three electrons to crystal bonds. Depending on the compensation ratio Mn can exist in different charge states and electron configurations. Most common are the ionized $\mathrm{Mn}^{2+}(d^5)$ and neutral Mn^{3+} acceptors, often denoted as A^- and A^0, respectively. The latter center corresponds to $\mathrm{Mn}^{2+}(d^5)$+hole configuration, and the localization of the hole varies depending on a compound. We will show in the following sections that, in case of GaAs:Mn, the hole is very much like a hydrogenic hole bound on negatively charged $\mathrm{Mn}^{2+}(d^5)$, while in case of GaN localized configuration of neutral acceptor $\mathrm{Mn}^{3+}(d^4)$ is realized. Presence of lower charge state $\mathrm{Mn}^{4+}(d^3)$, an ionized donor, besides A^- and A^0 states of Mn impurity, has also been suggested in wide band gap group-III nitrides.

3.1. Ionized acceptor, $Mn^{2+}(d^5)$

Ionized acceptor $Mn^{2+}(d^5)$ with five tightly bound electrons on Mn d-shell is the most common configuration of Mn in III–V semiconductors. It has been observed in GaAs (Almeleh and Goldstein, 1962; Fedorych et al., 2002; Szczytko et al., 1999), GaP (Kreissl et al., 1996), InP (Macieja et al., 2003; Sun et al., 1992), GaN and AlN (Baranov et al., 1997; Graf et al., 2003). Following Hund's rule, the spin of the $Mn^{2+}(d^5)$ ion equals to $S = 5/2$, and the orbital momentum $L = 0$. $Mn^{2+}(d^5)$ ion in III–V semiconductors gives rise to the ESR signal with g-factor close to $g = 2$.

The ESR spectrum of the $Mn^{2+}(d^5)$ ion in a wurtzite GaN single-crystals grown by high-pressure technique (Krukowski et al., 2001) is shown in Fig. 8.3(A). The signal consists of 30 resonance lines, which are well separated for magnetic field B parallel to GaN c-axis. In this configuration, the lines are clearly divided into five fine structure groups, each of them consisting of six hyperfine components. At arbitrary orientations of the magnetic field, the separation between lines becomes smaller and the lines merge. Fine structure of the spectrum originates from $\Delta M_S = 1$

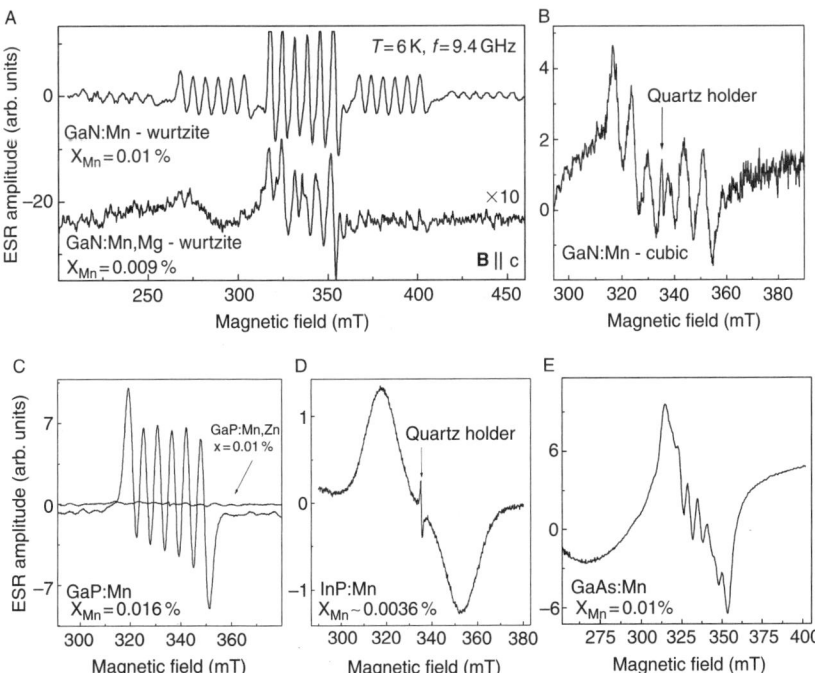

FIGURE 8.3 ESR spectra of manganese in a $Mn^{2+}(d^5)$ charge state in (A) hexagonal GaN, (B) cubic GaN, (C) GaP, (D) InP, and (E) GaAs, measured at the microwave frequency of 9.4 GHz. Co-doping with acceptors leads to decrease of the $Mn^{2+}(d^5)$ amplitude-panels (A) and (C).

transitions of the electron spin system with $S = 5/2$, while hyperfine structure lines are due to interaction of the electron spin system with the natural ^{55}Mn isotope having spin $I = 5/2$. In paramagnetic resonance, the transitions with $\Delta M_I = 0$ are observed. Transition with nuclear spin flip are also allowed, but they stand for nuclear magnetic resonance which occurs at much lower frequencies.

The spectrum shown in Fig. 8.3(A) can be described by a standard spin Hamiltonian with: g-factor $g = 1.998 \pm 0.001$, axial crystal-field parameter $D = (252 \pm 1)$ G·$g\mu_B$, and a hyperfine constant $A = (70 \pm 1)$ G·$g\mu_B$. The parameter D describes a zero-field splitting of a Mn^{2+}(d^5) ground state. Due to its high value, it gives the main contribution to the angular anisotropy of the resonance spectrum. A comprehensive analysis of the Mn^{2+}(d^5) resonance in GaN as well as in AlN epilayers has been performed by Graf et al. (2003).

Table 8.1 compares Hamiltonian parameters of bulk GaN:Mn with literature data obtained for epitaxial films (Graf et al., 2003) and bulks (Baranov et al., 1997). It may be interesting to point out a large difference in value of parameter D between bulk samples and epitaxial layers. GaN layers, grown usually on a sapphire substrate, suffer biaxial strain due to the lattice mismatch to the substrate. Parameter D decreases its value with increasing lattice distortion. The value obtained in this work for high-pressure grown unstrained crystals is close to the value obtained earlier for bulk samples by Baranov et al. (1997). It is still higher than that for nominally relaxed epitaxial films. It may suggest that the epilayers which are relaxed at the room temperature, still suffer strain after cooling down to the helium temperature, at which ESR experiment is performed.

ESR spectra of Mn^{2+}(d^5) in cubic crystals: GaN, GaP, InP, and GaAs are shown in Figs. 8.3(B)–(E), respectively. Cubic GaN is an epilayer grown by molecular beam epitaxy (Fay et al., 2005; Foxon et al., 2005; Novikov et al., 2005), while GaP, InP, and GaAs are bulks obtained by Czochralski method. Mn^{2+}(d^5) resonance in cubic crystals is built of six hyperfine structure lines. An angular anisotropy of the spectra originates in this case from the cubic crystal-field described by parameter a; in case of epilayers also from the axial anisotropy induced by lattice mismatch to the substrate (Fedorych et al., 2002). The angular anisotropy is much smaller than that for wurtzite crystals, and in many cases smaller than the resonance line width. The spectra shown in Figs. 8.3(B)–(E) are characterized by the hyperfine constant $A = (58 \pm 1)$ G·$g\mu_B$ and a g-factor $g = 2.002 \pm 0.001$ for GaP, $g = 2.003 \pm 0.001$ for InP, and $A = (54 \pm 1)$ G·$g\mu_B$ and $g = 2.005 \pm 0.001$ for GaAs, respectively. Lack of a hyperfine structure in the spectrum of InP:Mn is due to the interaction of Mn electron spin system with indium isotope ^{115}In having high nuclear spin $I = 9/2$ (Sun et al., 1992), which leads to the broadening of the component lines.

TABLE 8.1 g-factor, cubic and axial crystal-field parameters a and D, respectively, and a hyperfine constant A determined for $Mn^{2+}(d^5)$ ion in GaN, GaP, InP, and GaAs

Mn^{2+} in	g	$|D|(G \cdot g\mu_B)$	$|A|(G \cdot g\mu_B)$	$|a|(G \cdot g\mu_B)$	Reference
hex-GaN bulk	1.999	257	75		Baranov et al., 1997
hex-GaN bulk	1.998 ± 0.001	252 ± 1	70 ± 1		this work
hex-GaN relaxed film	1.9994 ± 0.0008	236 ± 2	69 ± 1		Graf et al., 2003
hex-GaN strained film	1.9994 ± 0.0008	218 ± 2	69 ± 1		Graf et al., 2003
cubic-GaN film	1.999 ± 0.001		68 ± 1	<10	this work
GaP bulk	2.002 ± 0.001		58 ± 1		this work
InP bulk	2.003 ± 0.001				this work
GaAs bulk	2.005 ± 0.001		54 ± 1		this work

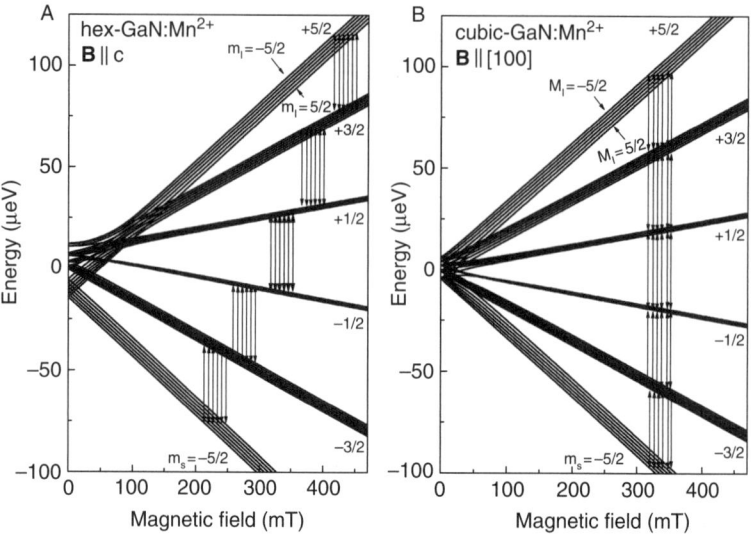

FIGURE 8.4 Energy level diagrams of the $Mn^{2+}(d^5)$ ground state in hexagonal (A) and cubic (B) GaN versus the external magnetic field. Arrows mark the allowed ESR transitions at the microwave frequency of 9.4 GHz. Spin Hamiltonian parameters used in calculations are: (A) $g = 1.998$, $D = -252$ $G \cdot g\mu_B$, $A = -70$ $G \cdot g\mu_B$, $a = 10$ $G \cdot g\mu_B$, (B) $g = 1.999$, $A = -68$ $G \cdot g\mu_B$, $a = 10$ $G \cdot g\mu_B$, respectively.

Figure 8.4 shows an energy structure diagram of a $Mn^{2+}(d^5)$ ground state in a magnetic field, calculated using spin Hamiltonian parameters for wurtzite and cubic GaN, respectively. The ground state of $Mn^{2+}(d^5)$ in a wurtzite crystal is built of three doublet states, split at zero magnetic field in ratio $2D:4D$. This axial zero-field splitting is, however, not clearly visible in Fig. 8.4, due to the additional hyperfine interaction. At high magnetic field the fine structure levels are characterized by m_S quantum numbers: $-5/2, \ldots, +5/2$, respectively. Each level is additionally split into six hyperfine components characterized by m_I quantum number: $-5/2, \ldots, +5/2$, respectively. ESR lines are observed when the applied magnetic field meets the resonance condition. Five groups of fine structure lines appear in the spectra (see Fig. 8.3(A)).

Lack of an axial symmetry and small value of cubic parameter a leads to an unresolved zero-field splitting of $Mn^{2+}(d^5)$ ground state in cubic crystals. Consequently, all fine structure resonances are observed simultaneously at the same magnetic field, and only hyperfine components can be resolved. This is clearly visible in ESR spectra of cubic GaN:Mn, GaP:Mn, and GaAs:Mn shown in Fig. 8.3.

Parameter a, due to its small value, has been usually only estimated from the ESR spectra. The upper limit for a given for GaAs:Mn is about $(3.5 \div 15)$ $G \cdot g\mu_B$ (Almeleh and Goldstein, 1962; Fedorych et al., 2002), for

ZnO:Mn about 10 G·$g\mu_B$ (Schneider and Sircar, 1962). In case of GaN:Mn, we can give the upper limit for a, above which the broadening of the component lines at arbitrary orientations of the magnetic field should be observed, $a < 10G \cdot g\mu_B$.

The amplitude of $Mn^{2+}(d^5)$ resonance in III–V semiconductors decreases after co-doping with acceptors, thus lowering the Fermi level. This effect has been observed while co-doping GaN:Mn with Mg acceptor (Wolos et al., 2004a) (see also Fig. 8.3(A)), or for co-doping GaP:Mn with Zn acceptor, see Fig. 8.3(C). Unlike in II–VI semiconductors, in materials from a III–V group the Mn acceptor level is located in host crystal band gap, allowing thus to obtain different charge states of Mn depending on the compensation. Mutual position of the Fermi level with respect to the Mn acceptor level decides about the fraction of manganese in a $Mn^{2+}(d^5)$ charge state, thus about the amplitude of the $Mn^{2+}(d^5)$ resonance.

It has been shown that some of $Mn^{2+}(d^5)$ resonances in III–V semiconductors are photo-sensitive. Under the illumination, exceeding the ionization energy of $Mn^{3+}(d^4)$ to GaN valence band, it was possible to increase the amplitude of the $Mn^{2+}(d^5)$ signal; thus the number of Mn in 2+ charge state, by a factor of two (Wolos et al., 2004a), Fig. 8.5. Such a photo-ESR experiment performed on highly resistive GaN:Mn,Mg served as one more argument for identification of Mn photoionization band.

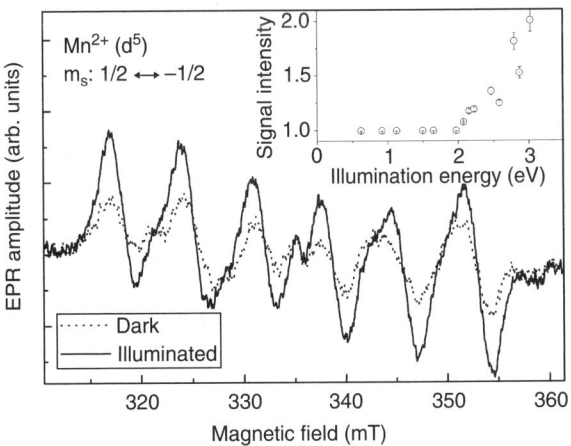

FIGURE 8.5 Photo-ESR experiment. Central transition m_s: $1/2 \leftrightarrow -1/2$ of the $Mn^{2+}(d^5)$ resonance in bulk GaN:Mn,Mg, measured in dark (dotted line), and after UV illumination (solid line). Increase of the signal amplitude by factor of 2 is visible. The inset shows spectral dependence of the signal intensity on the illumination. The measured photoionization onset coincides well with the band observed by optical absorption, identified as a transition: $Mn^{3+}(d^4) + h\nu \rightarrow Mn^{2+}(d^5) + hole_{VB}$ (Wolos et al., 2004a).

3.2. Neutral acceptor, $Mn^{3+}(d^4)$

$Mn^{3+}(d^4)$ with four electrons on Mn d-shell has been identified as a neutral configuration of Mn in nitrides, GaN and AlN (Graf et al., 2002; Korotkov et al., 2002; Wolos et al., 2004a,b). Spin of the $Mn^{3+}(d^4)$ ion in a ground state equals to $S = 2$, an orbital momentum $L = 2$. ESR resonance for this configuration has not been yet detected, although in principle the effective spin Hamiltonian parameters (as estimated by Marcet et al., 2006) should allow observation of some transitions due to $Mn^{3+}(d^4)$ in the range of standard ESR spectrometers. However, the resonance line width is expected here to be very sensitive to any strain fluctuations, as usual for non-Kramers ions, what limits the possibility of studying these resonances to high-quality samples only.

Optical absorption experiment provides an easier way to observe $Mn^{3+}(d^4)$, as this configuration is responsible for the appearing of the strong $^5T_2 \rightarrow {}^5E$ intracenter band with zero-phonon line at the energy of about 1.4 eV in both GaN and AlN, see Fig. 8.6 (Graf et al., 2002; Korotkov et al., 2002; Marcet et al., 2006; Wolos et al., 2004a,b). This transition is actually equivalent to the well known intracenter transition of $Cr^{2+}(d^4)$ in II–VI semiconductors (Kaminska et al., 1979).

The $Mn^{3+}(d^4)$ absorption band in GaN is built of a pronounced zero-phonon line followed by a high-energy sideband, resembling the

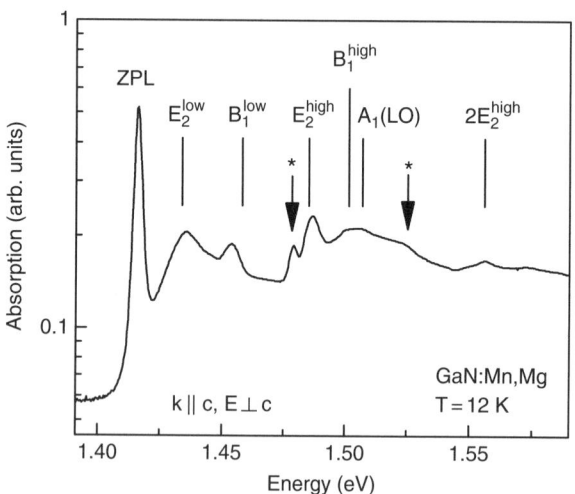

FIGURE 8.6 Absorption band due to the intracenter transition $^5T_2 \rightarrow {}^5E$ within $Mn^{3+}(d^4)$ configuration in GaN single crystal grown by high-pressure technique. The zero-phonon line at 1.41 eV is followed by a high-energy band resembling the structure of GaN phonon density of states, peaked at the Γ - point GaN phonons: E_2^{low}, B_1^{low}, E_2^{high}, B_1^{high}, $A_1(LO)$. Asterisks mark unidentified peaks.

spectrum of GaN phonon density of states. The sideband is peaked at the energy of Γ-point GaN phonons, which are E_2^{low}, B_1^{low}, E_2^{high}, E_1^{high}, and $A_1(LO)$, for k-vector of the incident light parallel and electric field vector E perpendicular to GaN c-axis. For $k \perp c$ and $E \parallel c$ the coupling with $A_1(TO)$ phonon dominates. Double or even triple phonon replicas can be also recognized at higher energies.

High intensity of the zero-phonon line with respect to the phonon replicas spectrum indicates weak electron–lattice coupling for this transition. The Huang–Rhys factor has been estimated here as equal between 0.6 for coupling with a phonon of $A_1(TO)$ symmetry, up to 1.1 for coupling with E_2^{low} phonon (or a Mn local vibration mode) (Korotkov *et al.*, 2002; Wolos *et al.*, 2004b).

The 1.4-eV band has also been recently observed in luminescence with an above band gap excitation (Zenneck *et al.*, 2007). This observation was rather unusual because typically presence of Mn in III–V compounds leads to killing of any luminescence. The mechanism of such behavior is not fully understood yet.

The 1.4-eV intracenter transition band in GaN is weakly sensitive to the magnetic field. Shift of the zero-phonon line, in magnetic field as high as 22 T, is of the order of 1 meV, and the maximum splitting in Faraday configuration for σ^+ and σ^- polarizations is less than 0.13 meV (Wolos *et al.*, 2002, 2004b). The most characteristic feature of the absorption versus the magnetic field is a step-like shift of the zero-phonon line, which occurs for Faraday configuration, at the magnetic field of about 7 T. This behavior can be explained based on the crystal-field theory, including crystal-field interaction, static Jahn–Teller distortion, spin–orbit interaction, and trigonal distortion due to the wurtzite structure of GaN. Detailed analysis was performed by Wolos *et al.* (2004b). Marcet *et al.* (2006) repeated the measurements and argued dynamic Jahn–Teller effect for this transition.

Figure 8.7 shows the calculated energy schemes of the ground state $^5T_2$ and first excited state ^5E of $Mn^{3+}(d^4)$ ion in magnetic field, for B parallel and perpendicular to c, respectively, according to the model proposed by Wolos *et al.* (2004b). $^5T_2$ state is split in magnetic field into five levels, which at high magnetic field are characterized by m_s quantum numbers $-2, \ldots, +2$, respectively. At low magnetic field, due to the spin–orbit interaction, the energy levels have unidentified spin and orbital quantum numbers. This refers to the ^5E excited state as well. At high magnetic fields the absorption transition occurs between the $^5T_2$ level with $m_s = -2$ and the two lowest levels of ^5E state having both $m_s = -2$. The two transitions can be distinguished in σ^+ and σ^- polarizations. Slope in the magnetic field of the respective energy levels corresponding to states of $^5T_2$ and ^5E symmetry is almost equal, which explains small shift of the zero-phonon line observed while sweeping the magnetic field. At low magnetic field the situation is more complicated, as both the ground and excited state

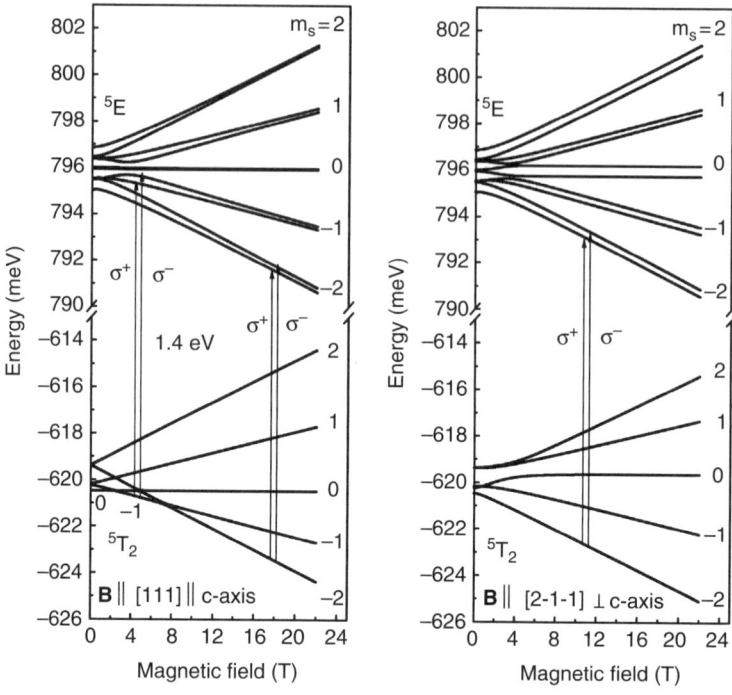

FIGURE 8.7 Energy level diagram for $^5T_2$ and ^5E states of $Mn^{3+}(d^4)$ in GaN versus the magnetic field. Arrows mark dominating transitions in σ^+ and σ^- polarizations, respectively.

energy levels have mixed orbital symmetry and spin quantum numbers. Thus the selection rules allow transition from all occupied $^5T_2$ to all empty ^5E levels. In particular, crossing of the lowest $^5T_2$ levels at about 7 T is responsible for step-like behavior of the zero-phonon line position in Faraday configuration. Unfortunately, line width of the component lines building the zero-phonon line reported by Wolos et al. (2004b) was comparable to the zero-field splitting of the ground state, which did not allow to track the position of component lines in the magnetic field with a satisfactory precision.

The crystal-field splitting of $Mn^{3+}(d^4)$ ion, $10Dq = 1.4$ eV, observed in nitrides, is much higher than that of $Cr^{2+}(d^4)$ in II–VI or III–V materials, which is 0.5–0.6 eV or 0.8–0.9 eV, respectively. The high value determined for GaN and AlN can be accounted for small lattice constant and high bond ionicity in group III nitrides.

The crystal-field model presented above can successfully describe anisotropy of the magnetization observed in GaN:Mn,Mg (Gosk et al., 2005), which will be discussed later.

Intensity of the 1.4-eV absorption can be conveniently used for determination of the fraction of manganese being in the $Mn^{3+}(d^4)$ charge state. Figure 8.8 shows the absorption coefficient integrated between the energy of 1.3 and 2.0 eV for two samples with different $Mn^{3+}(d^4)$ concentrations. $Mn^{3+}(d^4)$ fraction was determined from comparison of the secondary ion mass spectroscopy data, giving the concentration of Mn in any electron configuration or even lattice location, and the ESR and magnetization measurements (Gosk et al. 2005; Wolos et al., 2004a), which allowed to distinguish contributions from substitutional $Mn^{2+}(d^5)$ and $Mn^{3+}(d^4)$ configurations, respectively. The integrated absorption coefficient scales linearly with the $Mn^{3+}(d^4)$ concentration with a scaling factor, which in this case is an integrated optical cross section, equal to the $2010 \pm 50 cm^{-1} eV/at.\%$. The optical cross section determined at the maximum of the phonon sideband, at 1.5 eV, equals to $7030 \pm 70 cm^{-1}/at.\%$. Note, that the integrated optical cross section is more than order of magnitude higher than that determined by Marcet et al. (2006).

$Mn^{3+}(d^4)$ configuration has been suggested by EPR also for GaP (Kreissl et al., 1996), however further spectroscopic investigations indicate other configuration of neutral Mn acceptor in this material, the delocalized $Mn^{2+}(d^5)$+hole configuration (Tarhan et al., 2003). This controversy will be discussed in the next section.

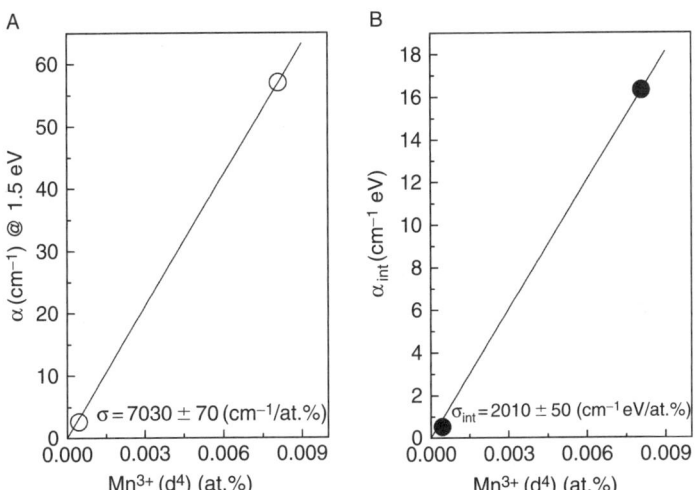

FIGURE 8.8 (A) Absorption coefficient at the maximum of the GaN:$Mn^{3+}(d^4)$ phonon sideband (at 1.5 eV), and (B) absorption coefficient integrated over the whole absorption band area (between 1.3 eV and 2.0 eV) versus $Mn^{3+}(d^4)$ concentration.

3.3. Neutral acceptor, $Mn^{2+}(d^5)$+hole

$Mn^{2+}(d^5)$+hole configuration is built of a negatively charged $Mn^{2+}(d^5)$ ion and a band hole, weakly bound on a delocalized orbit. $Mn^{2+}(d^5)$+hole is neither a localized nor effective mass center, it represents a group of impurities having an intermediate character. As it will be discussed here, localization radius of the hole varies depending on a host crystal lattice.

$Mn^{2+}(d^5)$+hole configuration gives rise to the absorption spectrum appearing near the ionization energy of Mn acceptor in GaAs, InP, and GaP (Chapman and Hutchinson, 1967; Lambert et al., 1985; Tarhan et al., 2003). The spectrum is built of a series of sharp lines followed by the excitation continuum. Tarhan et al. have noticed a similarity of this spectrum to the Lyman series, originating from transitions from $1s$ to various p-type states. Appearance of the spectrum allowed to determine, with spectroscopic precision, the ionization energy of Mn acceptor to the host crystal valence band, which equals to 0.11232 eV, 0.22004 eV, and 0.38775 eV for GaAs, InP, and GaP, respectively. Although the separation between subsequent excitation lines can be well described by the effective mass theory (Baldereschi and Lipari, 1973), the ionization energy of the acceptor is far from the effective mass approximation. For comparison, typical shallow acceptor energies are 0.03 eV, 0.045 eV, and 0.055 eV for GaAs, InP, and GaP, respectively (Madelung, 2004).

The potential produced by charged impurity, at distances from the impurity much larger than lattice constant, is well described by Coulomb potential screened by dielectric constant of the host crystal. At the distances comparable to the lattice constant, the potential strongly deviates from Coulombic behavior, due to the presence of the impurity core. Since the wave functions of the excited states of the impurity is extended over distances of many lattice constants, the deviations from Coulombic behavior affect mostly the ground state of the impurity, giving origin to the so called "chemical shift", i.e. difference between ionization energy of theoretical hydrogenic-like center and the real one. The more localized is the center, the more sensitive it is to the chemical shift. This is also a case of Mn acceptor in group III–V semiconductors. Here, binding of a hole to negatively charged $Mn^{2+}(d^5)$ center results from the short-range core potential, rather than a gradual Coulomb potential, arising from the impurity charge (Bebb, 1969; Brown et al., 1973; Woodbury and Blakemore, 1973). The ground state of such a moderately deep acceptor can be described within classical delta-function Lucovsky model or a quantum defect theory (Bebb, 1969). The excited states of Mn acceptor are, however, well characterized by Coulombic potential, resulting in appearance of the Lyman excitation lines in absorption spectra, with energy separation matching exactly theoretical hydrogenic-like ones,

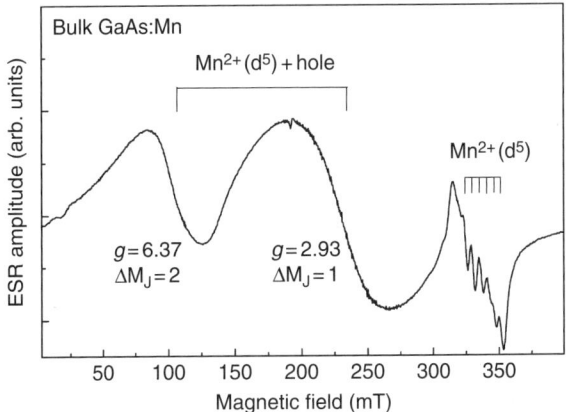

FIGURE 8.9 Electron spin resonance of $Mn^{2+}(d^5)$+hole center in GaAs, together with the $Mn^{2+}(d^5)$ resonance. Microwave frequency 9.4 GHz.

and also these of shallow impurities (Chapman and Hutchinson, 1967; Lambert et al., 1985; Tarhan et al., 2003).

$Mn^{2+}(d^5)$+hole configuration has been also observed by ESR. In GaAs two resonance lines with g-factors $g = 2.93 \pm 0.02$ and $g = 6.37 \pm 0.02$ (Fig. 8.9) have been attributed to $\Delta M_J = 1$ and $\Delta M_J = 2$ transitions of spin system with total angular momentum equal to $J = 1$. According to the model proposed by Schneider et al. (1987) the $Mn^{2+}(d^5)$+hole center is built of antiferromagnetically coupled $Mn^{2+}(d^5)$ ion having spin $S = 5/2$ and a valence band hole having total angular momentum $J_h = 3/2$. The ESR signal appears exclusively in p-type GaAs samples, with Mn concentration not exceeding 10^{19} cm^{-3} (Szczytko et al., 1999). Lack of the resonance in samples with higher Mn concentration may suggest that $Mn^{2+}(d^5)$+hole description of a neutral configuration of Mn is valid only for impurity doping limit, until wave functions of Mn acceptors begin to overlap. It is worth to notice, that the same concerns observation of the Lyman absorption spectra, as the bands were recorded only for very diluted samples.

p-type GaP:Mn and InP:Mn samples with low Mn concentration were also investigated for the $Mn^{2+}(d^5)$+hole resonance, however the signal has not yet been detected in the standard range of the ESR experiment.

3.4. Localization radius of the hole bound to $Mn^{2+}(d^5)$ center in III–V semiconductors

One of the methods to study localization radii of impurities is to measure the hopping conductivity, as the magnitude of this transport mechanism depends strongly on the overlap of hole (or electron) wave functions, thus

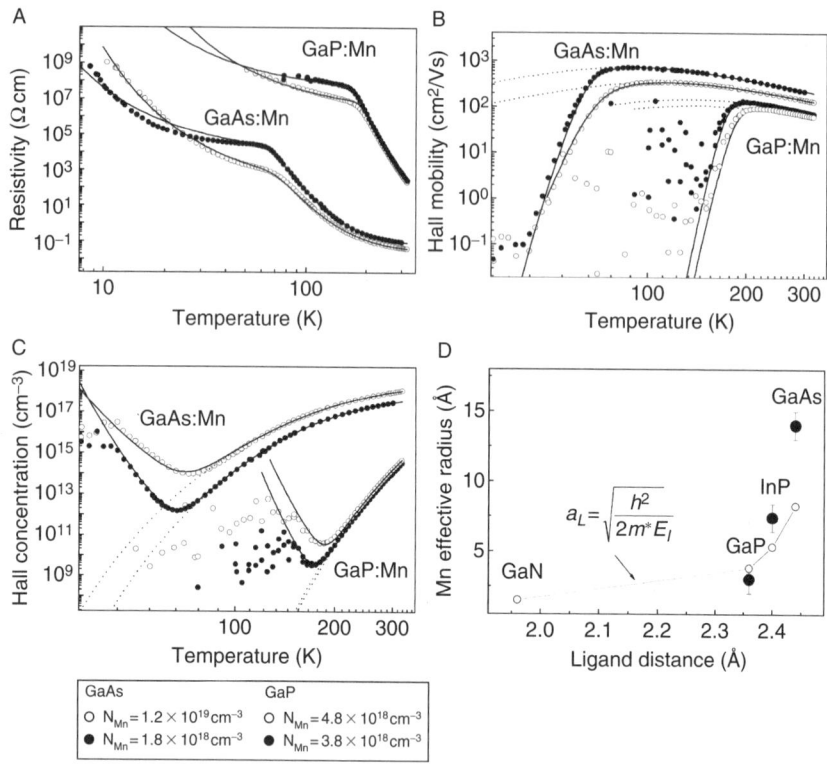

FIGURE 8.10 (A) Resistivity, (B) Hall mobility, and (C) Hall concentration measured versus the temperature for GaAs and GaP doped with Mn. Dots are experimental, solid lines are fits of the resistivity, effective mobility and effective concentration, respectively, assuming two transport mechanisms, nearest neighbor hopping and p-type conductivity in the host crystal valence band. Dotted lines show contribution of the latter. (D) An effective radius of Mn acceptor determined from the magnitude of hopping conductance for GaAs (this work), InP (Korona et al., 2006), and GaP (this work)—full dots. Open dots show effective radius calculated within the Lucovsky, or quantum defect formalism. The prediction of the Mn acceptor radius in GaN falls out of the range of hydrogen-like approximation.

on both the localization radius and an average distance between the impurity centers. Hopping conductivity was investigated in Mn-doped GaAs, InP, and GaP (Blakemore et al., 1973; Korona et al., 2006; Pawlowski et al., 2005; Wolos et al., 2007; Wolos et al., to be published; Woodbury and Blakemore, 1973). Figure 8.10 shows Hall data obtained for GaAs and GaP versus the temperature. All curves display common characteristic features. At high temperatures electric transport is dominated by p-type conductivity in the host crystal valence band, with activation energy close to the ionization energy of Mn acceptor. While lowering the temperature a rapid drop of the mobility and apparent increase of the Hall concentration appear, which is a signature of two co-existing mechanisms, band and hopping transport. Hopping conductivity gains domination at low temperatures, at which Hall concentration and mobility could not be measured any more. On the other hand, the resistivity could be measured in the whole range of temperatures, having a characteristic kink at the temperature at which hopping conductance begins to compete with band transport. This temperature equals to about 60 K for measured GaAs samples and about 110 K for GaP samples. It moves towards higher temperatures with increasing localization of the Mn impurity.

Solid lines shown in Fig. 8.10 are theoretical fits: (A) of the resistivity assuming band conductivity at high temperatures and nearest neighbor hopping at low temperatures, (B) of the effective mobility assuming scattering on phonons and charged impurities for band transport, and (C) of the effective concentration. For details of hopping transport theory see a review of Shklovskii and Efros (1984).

Performing a fitting procedure to the Hall data for a set of samples with varied Mn concentration one can obtain an effective radius of the Mn acceptor, which equals to 14 Å in GaAs, 7 Å in InP (Korona et al., 2006), and 3 Å in GaP, respectively. A clear trend of increasing a localization of the neutral Mn acceptor in III–V semiconductors with decreasing host crystal lattice constant can be seen in Fig. 8.10(D). The trend is consistent, actually, with Lucovsky or quantum defect models (Bebb, 1969), which both require the radius of a defect wave function in a form of $a_L = \sqrt{\hbar^2/2m^*E_I}$, dependent on the defect ionization energy E_I, which is gradually increasing when proceeding from GaAs to GaP. As a consequence of the observed trend one can see that the localization of Mn acceptor in wide band gap group-III nitrides with high Mn ionization energy (equal to 1.8 eV for GaN and 2.6 eV for AlN) falls beyond the hydrogen-like approximation. Mn in group-III nitrides is a well localized deep defect, which can be described by the crystal-field model, as it has been shown in previous chapters.

The performed studies showed, therefore, that neutral Mn acceptor changes its configuration from $Mn^{2+}(d^5)$+hole of localization radius decreasing in series from GaAs through InP to GaP, up to fully localized $Mn^{3+}(d^4)$ center in case of GaN. Excited states of the hole bound to $Mn^{2+}(d^5)$ have hydrogenic-like character in GaAs, InP, and GaP, with energy levels described well by effective mass theory for shallow acceptors. Only in case of GaAs:Mn ESR revealed lines which could be described as originating from ground state of $Mn^{2+}(d^5)$+valence band ($S_{3/2}$) hole. It seems probable that lack of such lines in case of InP:Mn and GaP:Mn results from much more localized Mn neutral acceptor. Its ground state cannot be described as $Mn^{2+}(d^5)$+valence band hole. It is worth recalling the results of Kreissl et al. (1996) mentioned before, where authors observed ESR resonance ascribed to $Mn^{3+}(d^4)$ configuration of neutral Mn acceptor in GaP. This result may indicate that the ground state of neutral Mn acceptor in GaP is more like localized $Mn^{3+}(d^4)$ than effective mass hole bound to $Mn^{2+}(d^5)$.

Mahadevan and Zunger (2004a,b) and Zhao et al., 2004 have already predicted theoretically the change of localization of the hole in the host series GaN→GaP→GaAs→GaSb. Using first principle calculations they found that while in GaN the hole resided in the crystal-field level localized primarily on Mn, in GaAs and GaSb the hole resided primarily on the Mn neighboring anions. Direct spatial mapping of the wave-function of a hole bound to a Mn acceptor in GaAs was done by Yakunin et al., 2003 and Celebi et al., 2008, using cross-sectional scanning tunneling microscopy. They found that in the neutral configuration manganese acceptor is seen as a cross-like feature which they attributed to a hole weakly bound to the Mn ion forming the $(Mn^{2+}(d^5)$+hole) complex. Recently, Celebi and Koenraad (2008, private communication) compared features of Mn neutral acceptor in GaAs and InP seen in scanning tunneling microscopy experiments and found definitively higher localization of the bound hole in the case of InP. These results support the picture of Mn neutral acceptor in III–V compounds as a $Mn^{2+}(d^5)$ center with a bound hole, localization of which depends on a compound.

The strong localization of a hole on $Mn^{2+}(d^5)$ ion in the case of GaN, resulting in localized $Mn^{3+}(d^4)$ center, should have consequences on magnetic interactions leading to potential ferromagnetic ordering. In the recent paper of Dietl (2008) it is acknowledged that the virtual-crystal and molecular-field approximations break down and cannot be applied to GaMnN system. Therefore, more adequate generalized alloy theory, introduced first by Tworzydlo (1994), is proposed by Dietl (2008) to describe diluted magnetic nitrides and oxides, and especially to explain abnormally small exciton splittings observed for these compounds (Pacuski et al., 2006, 2007, 2008; Przezdziecka et al., 2006).

3.5. Location of Mn acceptor level in III–V semiconductors band structure—lattice relaxation

Appearance of the Lyman absorption spectrum due to excitation of Mn acceptor in GaAs, InP, and GaP allowed to determine with high precision the location of Mn acceptor level in band structure of these materials. As already mentioned, the values of 0.11232 eV, 0.22004 eV, and 0.38775 eV with respect to the host crystal valence band maximum were obtained, respectively. In a case of nitrides, for which deep character of Mn impurity was defined, the ionization energy was determined based on the analysis of unstructured photoionization bands (Graf et al., 2002; Wolos et al., 2004a), which resulted in values $E_{VB} + 1.8$ eV and $E_{VB} + 2.6$ eV for GaN and AlN, respectively. In a case of GaN significant role of lattice relaxation on the appearing of photoionization onsets was pointed out.

In n-type GaN electron transitions from occupied Mn acceptor level to GaN conduction band, $Mn^{2+}(d^5) + h\nu \rightarrow Mn^{3+}(d^4) + e_{CB}$, whereas in highly resistive GaN transitions from valence band to empty Mn acceptor level, $Mn^{3+}(d^4) + h\nu \rightarrow Mn^{2+}(d^5) + hole_{VB}$, were observed in both optical absorption and photocurrent (Graf et al., 2002; Wolos et al., 2004a). The optical cross section at about the band maximum for the transition involving conduction band equals to $\sigma_{CB} = (6 \pm 2) \times 10^{-18}$ cm^2, while for the transition involving valence band equals to about 2×10^{-17} cm^2. These values are typical for the photoionization transitions between localized d states and s-type conduction band (forbidden) or p-type valence bands (allowed) (Stoneham, 1985). The onsets of both electron transitions from and to the $Mn^{2+/3+}$ level do not add up to the energy of GaN band gap, indicating strong coupling to the lattice. The relaxation energy determined from temperature dependence of the Mn photoionization band in n-type GaN occurred to be as much as 0.38 eV (Wolos et al., 2004a), which is consistent with the theory. The calculations of the electron structure of Mn in GaN performed within the local spin density approximation by Boguslawski and Bernholc (2005) show strong influence of lattice relaxation on the energy of Mn-related electron levels. The calculated equilibrium Mn-N bond length changes by as much as 4% for the sequence of $Mn^{2+}(d^5)$–$Mn^{5+}(d^2)$, being close to the ideal Ga-N value only for neutral $Mn^{3+}(d^4)$ configuration. The energy gained by displacements from the unrelaxed to the equilibrium positions ranges from 0.5 to 0.05 eV, depending on the charge state involved (0.5 eV for $Mn^{3+}(d^4)$ to $Mn^{2+}(d^5)$ transition). The structural investigations of GaMnN also show the increased Mn-N bond length with respect to the Ga-N lattice constant (Bacewicz et al., 2003; Giraud et al., 2004; Sato et al., 2002; Soo et al., 2001), which is consistent with $Mn^{2+}(d^5)$ charge state.

Taking into account lattice relaxation, the location of $Mn^{3+/2+}$ acceptor level in GaN band structure was determined from temperature dependence of the photoionization absorption band in n-type GaN by Wolos et al. (2004a). The derived position of the level, with respect to the bottom of GaN conduction band, was equal to 1.8 ± 0.1eV in this paper. However, to be more precise one needs to take into account the Burstein–Moss effect present in these highly n-type samples, which shifts up both the fundamental absorption edge and the onset of Mn photoionization by about 0.1 eV (Teisseyre et al., 1994). Thus we get location of $Mn^{3+/2+}$ acceptor level at 1.7 eV below the bottom of GaN conduction band, or 1.8 eV above the GaN valence band maximum (assuming GaN

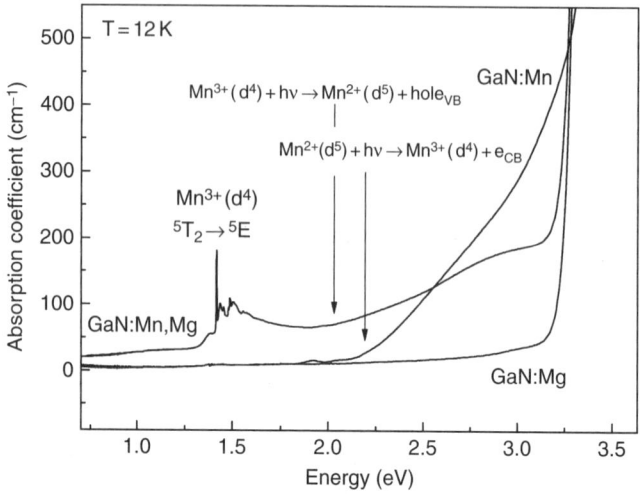

FIGURE 8.11 Optical absorption spectra of bulk GaN samples: n-type GaN:Mn, insulating GaN:Mn,Mg, and a reference GaN:Mg. Bulk samples obtained by high-pressure technique are unintentionally contaminated with high concentration of oxygen donors, making them highly conductive. Arrows mark photoionization onsets.

energy gap derived from the excitonic luminescence equal to 3.5 eV (Madelung, 2004)). The small correction for the Burstein–Moss effect is, however, in a range of the experimental error.

Figure 8.11 shows some examples of optical absorption spectra obtained for n-type and highly resistive GaN:Mn, with identified photoionization and interacenter transitions of Mn. Figure 8.12 shows the configuration–coordinate diagram for the photoionization transitions of Mn in GaN determined from the optical absorption experiments. Table 8.2 summarizes the obtained values of $Mn^{2+/3+}$ level positions in band structure of different III–V semiconductors.

3.6. Lower charge states of Mn in wide band gap III–V semiconductors

As it has been argued earlier, deep character of Mn impurity in nitrides, GaN and AlN, was established in contrast to the delocalized configuration of Mn in GaAs, InP, or even in GaP.

In GaN of n-type Mn exists in $Mn^{2+}(d^5)$ configuration, as shown by ESR investigations. While lowering the Fermi level, $Mn^{3+}(d^4)$ configuration appears, with $^5T_2$–^5E intracenter transition at 1.4 eV. The deep $Mn^{2+/3+}$ acceptor level was allocated in the midgap of GaN, 1.7 eV below the bottom of the conduction band.

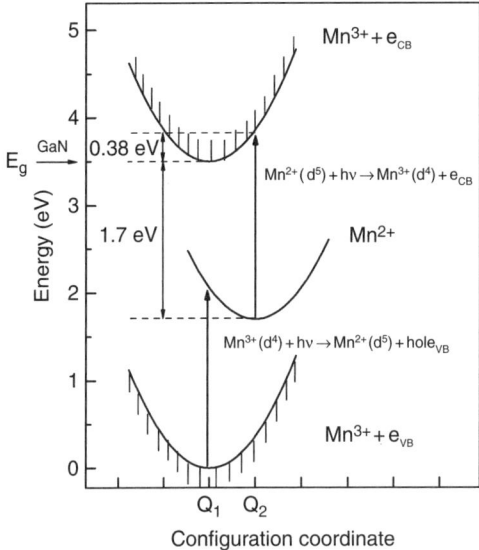

FIGURE 8.12 Configuration coordinate diagram of Mn photoionization transitions in GaN.

TABLE 8.2 Ionisation energies of Mn acceptor in a series of III–V semiconductors

Compound	Ionization energy of Mn acceptor to the valence band (eV)	Reference
GaAs	0.11232	Tarhan et al., 2003
InP	0.22004	Tarhan et al., 2003
GaP	0.38775	Tarhan et al., 2003
GaN	1.8	Wolos et al., 2004a
AlN	2.6	Graf et al., 2003

Due to the large value of the band gap energy in GaN (3.5 eV; Madelung, 2004) lower charge states of Mn were also expected to be located in the gap. The calculations performed by Boguslawski and Bernholc (2005) show that even the d^2 configuration, $Mn^{5+}(d^2)$, can be present in GaN with Fermi level lowered to the valence band. In Mg co-doped epilayers, Han et al. (2004, 2005) observed a photoluminescence band build of a series of sharp lines at the energy of about 1 eV, which they attributed to the $^4T_2$–$^4T_1$ intracenter transition within $Mn^{4+}(d^3)$ configuration. The photoluminescence excitation spectroscopy data allowed the authors to allocate the $Mn^{3+/4+}$ charge transfer level at 1.11 eV above the GaN valence band maximum.

4. MAGNETIC INTERACTIONS IN III–V DMSs WITH Mn

Investigations of Mn doped GaN have been motivated by the search for a DMS keeping the ferromagnetic ordering up to the room temperature and above. Theoretical considerations made within the Zener model of carrier-mediated ferromagnetism gave very optimistic predictions for the Curie temperature of GaMnN, of the order of 400 K, as far as 5% of Mn in a high-spin $Mn^{2+}(d^5)$ configuration can be incorporated, together with 10^{20} cm^{-3} of holes in GaN valence band (Dietl et al., 2000). The carrier-mediated ferromagnetism gives a unique possibility to control and utilize magnetic properties of a semiconductor by applying voltage or driving electric currents, as it has been shown on the example of GaMnAs or InMnAs (Ohno et al., 1999, 2000). No wonder then, that many further experimental efforts have been devoted to the investigations of GaMnN in this context. The works resulted in the determination of deep character of Mn impurity in wurtzite GaN with its numerous possible charge states which depend on the compensation with donors or acceptors. In point of view of these results it is evident, that the desired co-existence of $Mn^{2+}(d^5)$ and holes cannot be achieved in bulk GaN under thermal equilibrium conditions. Below a brief review of current state of knowledge of magnetic properties of Mn-doped GaN is given.

4.1. Magnetic properties of wurtzite GaN with Mn

For low Mn concentrations, up to 0.1–0.2 at. % of Mn, when the spins of magnetic ions are well separated from each other, paramagnetic behavior in GaN:Mn is observed by magnetization measurements using superconducting quantum interference device (SQUID). Depending on the Fermi level position, paramagnetic behavior is due to the $Mn^{2+}(d^5)$ or $Mn^{3+}(d^4)$ configuration of Mn (Gosk et al., 2005; Wolos et al., 2004a; Zajac et al., 2001a,b).

Paramagnetism due to $Mn^{2+}(d^5)$ is observed in n-type samples. The magnetization versus magnetic field can be described by isotropic Brillouin function with $S = 5/2$. Both the hyperfine splitting and the zero-field splitting of $Mn^{2+}(d^5)$ ground state due to the wurtzite structure of GaN, as shown in a section devoted to ESR (see Fig. 8.4), are too small to be detected by SQUID. Thus the ground state of d^5 configuration in magnetic field is usually approximated by the sequence of six equidistant levels, for which the Brillouin description of magnetization is suitable.

Paramagnetism due to $Mn^{3+}(d^4)$ ion in GaN, which appears when the Fermi level is lowered to the GaN midgap, can be described assuming the structure of the $Mn^{3+}(d^4)$ ground state shown in Fig. 8.7. One can see that the scheme of energy levels for magnetic field perpendicular to GaN

c-axis differs from that for parallel configuration, which consequently leads to the anisotropy of magnetization. This was described in details by Gosk et al. (2005). Similar origin of magnetization anisotropy due to the d^4 configuration of Cr^{2+} in II–VI semiconductors was observed previously by Mac et al. (1994) and Herbich et al. (1998).

Antiferromagnetic d–d coupling can be observed in n-type GaN samples having higher Mn content, starting from about 0.2 at. % of Mn (Zajac et al., 2001b). The magnetization of such GaMnN samples saturates more slowly in magnetic field than in a case of non-interacting ions. The magnetization versus magnetic field can be described by a modified Brillouin function, with an effective temperature and an effective Mn concentration (Gaj et al., 1979). For antiferromagnetic coupling, the effective temperature is higher than the temperature of measurement, and the effective concentration is lower than the actual Mn content. The antiferromagnetic ordering was also confirmed by magnetization versus temperature measurements, yielding unambiguously the Curie–Weiss temperature of a negative value −12.5 K for sample containing as much as 9% of Mn (Zajac et al., 2001b).

There are many reports in literature concerning observation of ferromagnetic response of GaMnN samples with Curie temperatures ranging from about 10 K up to above room temperature. First reports (Baik et al., 2004; Overberg et al., 2001; Reed et al., 2001; Sasaki et al., 2002; Sonoda et al., 2002; Theodoropoulou et al., 2001, and many more) appeared shortly after the theoretical paper of Dietl et al. (2000). Most of the reports rely however on SQUID measurements, where the separation of contributions from different magnetic phases present in one sample is often a difficult task. As soon as the permanent problem of the homogeneity of GaMnN samples became apparent, the observed high-temperature ferromagnetism has been often related to Mn-rich regions in the GaN matrix (Dhar et al., 2003; Martinez-Criado et al., 2005; Soo et al., 2001; Zajac et al., 2003). Careful studies of magnetization showed that the investigated material was often multiphase, performing as paramagnetic and ferromagnetic simultaneously (Zajac et al., 2003). It also occurred that only a small part of Mn present in the studied samples was responsible for their ferromagnetic behavior. Identification of ferromagnetic phase was a problem, since X-ray method was in many cases not sensitive enough to detect phases other than GaN, although magnetization indicated multi-phase material. There were some expectations that precipitates of MnN_4 may occur in highly doped GaN and cause ferromagnetic behavior; however, no straightforward proof of this was found, as well as that MnN_4 may be the only phase responsible for the detected ferromagnetism.

Recently, new explanation of ferromagnetic contribution to magnetization has been proposed—so called spinodal decomposition mechanism, leading to existence of regions with low and high concentrations of Mn constituent (Dietl, 2007). According to this explanation GaMnN with high

Mn content, higher than average, is ferromagnetic, whereas GaMnN of lower Mn content is responsible for paramagnetic contribution to total magnetization. The existence of spinodal decomposition is supported by both theory and experiment. *Ab initio* calculations have revealed tendency of DMS to form nonrandom alloys (Sato *et al.*, 2005; van Schilfgaarde and Mryasov, 2001). On the other hand, the presence of Mn-rich nanocrystals in hexagonal lattice of GaMnN has been detected by means of spatially resolved X-ray diffraction (Martinez-Criado *et al.*, 2005). The theoretical predictions, as well as experimental observations of Mn-rich nanocrystals coherent with GaN, may lead to conclusion that such formations are responsible for the observed ferromagnetism in GaMnN system.

However, there is still an open question about the possible mechanism leading to ferromagnetism in these coherent, highly Mn-rich regions embedded in GaN lattice. The greatest problem in fulfillment of assumptions of carrier-mediated models of ferromagnetism is technological difficulty with efficient p-type doping of GaN. GaN shallow acceptors have relatively high energy, greater than 200 meV (Amano *et al.*, 1989), and it would be practically impossible to achieve co-doping resulting in 10^{20} cm^{-3} free hole concentration. Therefore, we still need theoretical explanation of high temperature ferromagnetic phase/phases present in GaMnN system. The other problem has been raised by Franceschetti *et al.* (2006, 2007). They showed theoretically that when Mn-rich clusters are formed due to for example spinodal decomposition of the GaMnAs alloy, the clusters could be only weakly coupled (super-paramagnetism), leading to a significant decrease of Curie temperature compared to the homogeneous random alloy. However, for higher Mn concentration strong magnetic coupling of clusters could be achieved, resulting in increase of T_C. Therefore, one should be aware that the behavior of T_C in non-homogenous DMS depends crucially not only on ferromagnetic ordering within individual clusters, but also on the shape, size and spatial distribution of Mn-rich clusters. This understanding seems to be a very important step in controlling magnetic properties of GaMnN by means of controlling composition, size, and density of ferromagnetic inclusions. Such control may let the use of GaMnN as spintronic material.

After a number of experiences with multiphase (Ga,Mn,N) material, described above, an intrinsic ferromagnetism was argued in GaMnN epilayers containing 6.3% of Mn, predominantly in Mn^{3+}(d^4) charge state (Sarigiannidou *et al.*, 2006). Based on the SQUID measurements supplemented by X-ray magnetic circular dichroism analysis, the authors determined the spontaneous magnetic moment 2.4 μ_B per Mn at 2 K and a Curie temperature of 8 K for that material. Strong precautions have been taken in that case to avoid secondary phases in the samples. The mechanism of the magnetic ordering has not yet been clarified. An alternative model for the carrier-mediated ferromagnetism can be provided by a double exchange mechanism, which assumes spin-conserved hopping in the Mn impurity band located in the GaN mid-gap, leading to the ordering of the Mn spins (Sato and Katayama-Yoshida, 2001; Sato et al., 2003, 2004, 2005). The predicted Curie temperatures are lower than in the case of Zener's model, which actually can explain the low-temperature ferromagnetism argued for GaMnN. Forthcoming experimental papers devoted to GaMnN seem to follow this idea, showing, e.g. by X-ray magnetic circular dichroism that a hybridized d^4-d^5 configuration of Mn is weakly ferromagnetically coupled (Freeman et al., 2007).

4.2. Exchange integrals in GaN doped with Mn

Based on the presented studies it is evident that under thermal equilibrium conditions, high-spin $Mn^{2+}(d^5)$ configuration may coexist with electrons in GaN conduction band, not with holes in GaN valence band.

In bulk GaN single crystals, unintentional contamination with oxygen donors leads to n-type material with electron concentration reaching 10^{20} cm^{-3}, which only slightly decreases when co-doping with small concentration of Mn, 0.2 at.% is performed. The resulting material is paramagnetic and the ESR signal due to $Mn^{2+}(d^5)$ can be observed (Wolos et al., 2003). The spin relaxation times of $Mn^{2+}(d^5)$ studied by ESR showed linear dependence of the longitudinal relaxation rate $1/T_1$ on temperature, which is characteristic for Korringa spin relaxation. This mechanism depends on the magnitude of exchange interaction, allowing determination of the s–d integral. It is important to note here, that the bottle-neck effect could be excluded in this case, as the whole fine and hyperfine structure of the resonance was well resolved. The effective exchange of spin excitations between localized Mn and carrier spins in the bottle-neck phenomenon would lead to averaging and vanishing of both fine and hyperfine structures. Analysis of the $1/T_1$ relaxation rate allowed to determine the absolute value of the s–d exchange integral $|N_0\alpha| = 14$ meV for GaN, which was more than ten times smaller than typical values found for II–VI materials. This low value practically excludes possibility of obtaining high Curie temperature in n-type GaN:Mn.

Exchange interaction between conduction electrons and paramagnetic ions, s–d, is of Coulombic (potential) type. Electron orbitals s and d, taking part in this interaction, are localized on the same lattice site. Therefore, this kind of interaction should be of ferromagnetic type and practically independent on host lattice. In the whole family of II–VI semiconductors diluted with Mn the s–d exchange integral $N_0\alpha$ is always positive (ferromagnetic type) and close to 200 meV (Furdyna, 1988; Kossut, 1988; Twardowski et al., 1993), so that this value is often regarded as universal. Thus the result obtained for GaN:Mn was surprising.

To explain this discrepancy we have pointed out the difference in the nature of $Mn^{2+}(d^5)$ ion in II–VI and III–V semiconductors (Wolos et al., 2003). While in II–VI family the $Mn^{2+}(d^5)$ is neutral, in III–V materials it is negatively charged, what results in Coulomb repulsion of conduction electrons, which may lead to the attenuation of the s–d exchange interaction. A similar problem was encountered in a case of other semiconductor of III–V family, GaAs with Mn (Heimbrodt et al., 2001; Stern et al., 2007). Here as well, the available experimental data point out to small value of the $N_0\alpha$, equal approximately to 20 meV. There is no clarity, however, about the sign of the constant as experiments give so far contradictory results. This is the next interesting problem requiring clarification in both theory and experiment.

Exchange interaction p–d is different from the s–d discussed above. Valence band of semiconductors originates mostly from p orbitals of anion sublattice. Since the d electron is anchored at the cation sublattice the p–d exchange interaction is of kinetic type, and comes from virtual presence of p electron on d magnetic ion level or d electron in p band. Therefore, the p–d exchange interaction depends very much on the Mn ion surrounding, which means on the host lattice. Important factor influencing the value of $N_0\beta$ is the mutual energy position of valence band and d orbital level, which influences the degree of p–d hybridization. Indeed, for II–VI compounds diluted with Mn, the p–d exchange interaction $N_0\beta$ changed from about -0.5 eV to -1.8 eV depending on the compound. It was always negative (antiferromagnetic type).

Excitonic giant Zeeman splitting was observed in GaMnN, containing predominantly $Mn^{3+}(d^4)$ (Pacuski et al., 2007) and thus allowing the authors to determine the difference of exchange integrals between $Mn^{3+}(d^4)$ and free carriers in GaN equal to $N_0(\alpha-\beta) = -1.2$ eV. This value corresponds to the positive sign of $N_0\beta$, thus to the ferromagnetic interaction between $Mn^{3+}(d^4)$ ions and GaN holes, which is a situation different from II–VI DMSs. However, there is a difference in charge state of Mn in the considered cases. Antiferromagnetic p–d exchange in II–VI DMSs refers to $Mn^{2+}(d^5)$ configuration, whereas $Mn^{3+}(d^4)$ configuration dominates in the measured GaN. Theoretical model predicts possibility of positive exchange integral p–d for DMS with d shell filled in less than half with electrons (Blinowski and Kacman, 1991). Experimental support was obtained for II–VI DMSs diluted with Cr (Mac et al., 1996) and V (Mac et al., 2000). Assuming $|N_0\alpha|$ for GaMnN equal approximately to 15 meV, as determined in experiments mentioned above, one gets $N_0\beta$ of about $+1.2$ eV.

Recently, a new explanation for such value of $N_0\beta$ was given by Dietl (2008), who linked small and positive value of apparent exchange splitting $N_0\beta^{app}$ to strong p–d coupling and, therefore, breaking of the virtual crystal approximation.

5. GaN-BASED DMSs

After initial burst of papers on GaMnN, and further disappointment with the results showing unambiguously that such one phase material ferromagnetic at above room temperature is unlikely, there have been attempts to dilute GaN with other than Mn magnetic ions. Theoretical predictions showed that GaN diluted with Fe (Blinowski and Kacman, 2001) Cr or V (Sato and Katayama-Yoshida, 2001) were again promising candidates for room temperature ferromagnetic semiconductor. The first problem yet was still of technological kind—solubility of transition metals in III–V compounds is much lower than in II–VI. The content of manganese, solubility of which is typically the highest among transition metal ions, was occasionally about 10% in GaN, whereas in reproducible technological processes this amount did not exceed 3% (Zajac et al., 2003). Experience with DMSs based on II–VI compounds showed lower solubility for Fe and Cr than Mn, so one could expect problems with high content of Fe or Cr in GaN.

Starting from 2002 attempts to obtain ferromagnetic GaN diluted with either Fe or Cr were undertaken (Bonanni et al., 2006, 2007; Gosk et al., 2003; Hashimoto et al., 2002; Liu et al., 2004; Rudzinski et al., 2006; Singh et al., 2005; Vaudo et al., 2003). In order to understand possible ferromagnetic ordering in such materials it seems important to understand properties of these transition metals present at low concentration, similarly to the widely described above case of GaN diluted with Mn.

5.1. GaN diluted with Fe

Electron configuration of free Fe atom is [Ar] $3d^6 4s^2$. The neutral configuration in III–V semiconductors is thus $Fe^{3+}(d^5)$. Iron was found in $Fe^{3+}(d^5)$ or $Fe^{2+}(d^6)$ substitutional configuration, depending on Fermi level position. The ESR signal of $Fe^{3+}(d^5)$ is characterized by crystal-field parameters $g_\parallel = 1.990$, $g_\perp = 1.997$, $D = 763 G \cdot g\mu_B$, $|a - F| = 56 G \cdot g\mu_B$, and $a = 51 G \cdot g\mu_B$ (Maier et al., 1994). The effective spin Hamiltonian parameters may slightly differ, depending on the growth technology and, in a case of epilayers, also the on used substrate (Baranov et al., 1997; Gehlhoff et al., 2006). The intracenter transition within $Fe^{3+}(d^5)$ configuration, $^4T_1(G) \rightarrow {}^6A_1(S)$, was found in luminescence at the energy of 1.3 eV, with a phonon sideband resembling the density of GaN phonon states (Maier et al., 1994). This identification was further supported by studies of photoluminescence at magnetic field (Niedzwiadek et al., 2007).

The location of $Fe^{3+/2+}$ acceptor level was initially assigned to about 0.28 eV below the bottom of GaN conduction band (Heitz et al., 1997). Later electron transport investigations of GaN doped with Fe showed, however, the activation energy of 0.5–0.6 eV for such material (Kordos et al., 2004; Polyakov et al., 2003b; Vaudo et al., 2003), which required revision of the location of the iron acceptor level within GaN band gap. The subsequently proposed position based on the optical absorption and photoluminescence excitation spectroscopy data assigned the $Fe^{3+/2+}$ level to about 0.68 eV below the bottom of GaN conduction band (precisely 2.863 eV above the GaN valence band maximum) (Malguth et al., 2006a). An alternative position of the level was located even more deeply, 1.8 eV above the GaN valence band maximum (Polyakov et al., 2003a). This assignment was based, however, on the optical absorption measurements performed on thin epilayers, what relates to weak absorption and thus small precision in determination of the band onset. In both cases of Malguth et al. (2006a) and Polyakov et al. (2003a) the possible lattice relaxation effects accompanying photoionization processes were not investigated.

The negatively charged acceptor $Fe^{2+}(d^6)$ configuration was initially suggested in GaN samples compensated with donors, based on the appearance of the Van Vleck component in the magnetization (Gosk et al., 2003; Przybylinska et al., 2006). Recently, the internal transition, $^5E \rightarrow {}^5T_2$, within $Fe^{2+}(d^6)$ substitutional configuration was also discovered at the energy of 0.39 eV (Malguth et al., 2006a, b; Niedzwiadek et al., 2007), which actually is the same spectral region as in the case of analogical transitions observed for GaAs, InP, and GaP (see for example review by Bishop, 1986). Lower charge states of Fe have not yet been detected; however, due to the wide band gap of GaN the $Fe^{4+/3+}$ level is expected still to be located within the gap (Boguslawski and Bernholc, 2008, private communication).

In the recent paper of Pacuski et al., 2008 the ferromagnetic sign and a reduced magnitude in comparison to the value expected from Zener's model (Dietl et al., 2000) of the p–d exchange energy $N_0\beta = +0.5 \pm 0.2$ eV was determined. The authors explained this value as a result of a significant renormalization of the valence band exchange splitting, due to localized character of Fe center, as described theoretically by Dietl (2008).

Maximum Fe content in GaN grown by different methods was typically less than 1%. Paramagnetic and ferromagnetic components of magnetization were observed for such samples (Gosk et al., 2003). Paramagnetic component was of either Van Vleck-type or Brillouin-type, depending on Fermi level position and domination of iron in $Fe^{2+}(d^6)$ or $Fe^{3+}(d^5)$ substitutional configuration, respectively. The origin of the ferromagnetic phase remains unknown. However, the situation reminds that of GaMnN, in such a sense that X-ray technique could not often detect alien phases other than GaN. In the explanation of such behavior the same arguments as in the case of GaMnN can be used—ferromagnetism originating from spinodal decomposition of GaFeN material, and existence of Fe-rich ferromagnetic regions, coherent with GaN lattice. What should, however, be admitted is that the nature of ferromagnetism in such Fe-rich regions is still unknown.

In some cases of GaN epilayers highly doped with Fe it was possible to observe, by ferromagnetic resonance, precipitation of different Fe compounds (Przybylinska et al., to be published). A few magnetically ordered phases could be detected in samples cut out from the same GaFeN wafer, showing no shape anisotropy, and a pronounced magneto-crystalline anisotropy of cubic or trigonal symmetry. With other words, the magnetic precipitates were spherical inclusions of cubic or trigonal structure embedded in the wurtzite host crystal lattice. Chemical composition of precipitates could be determined in a few cases by comparing the obtained anisotropy constants with values known from literature. In this way it was possible to determine presence of metallic Fe, magnetite Fe_3O_4, and hematite Fe_2O_3 precipitates.

Doping of GaN with Fe results in pinning of the Fermi level in the vicinity of the Fe acceptor level, located, as it was mentioned above, in the upper part of GaN band gap. Such samples are characterized by high resistivity, reaching almost 1 GΩ cm at room temperature. This property of Fe-doping is now being utilized to compensate unintentional donors, particularly in hydride vapor phase epitaxy layers meant as native substrates for GaN-based electronic and optoelectronic devices (Heikman et al., 2002; Vaudo et al., 2003). It seems that we observe the come-back of transition metals in the old role as compensating centers of residual dopants, well known from much earlier studies.

It has been shown recently that deep Fe centers in GaN act also as predominant nonradiative efficient recombination channel, leading to photo-induced carrier lifetime as short as a few picoseconds. This observation opens a new possibility of application of GaN:Fe in fast photoswitches (Marcinkevicius et al., 2007).

5.2. GaN diluted with Cr

The electron configuration of a free Cr atom is $[Ar]3d^44s^2$ with its neutral configuration in III–V semiconductors being $Cr^{3+}(d^3)$. Investigations of GaN doped with Cr were up to now focused on magnetic properties, thus it is very little known about electron levels and internal transitions related to Cr in GaN. Comparing trends observed for GaAs, InP, and GaP one can expect Cr to form a deep acceptor level, $Cr^{2+/3+}$, in the upper part of GaN band gap. Cr-implanted GaN epilayers, as reported by Polyakov et al. (2003a), show the presence of absorption band which may be related to the photoionization of substitutional Cr, from which the authors allocate the $Cr^{2+/3+}$ acceptor level around 2 eV above GaN valence band maximum. However the precision of this assignment needs to be improved.

The internal transition of $Cr^{2+}(d^4)$ in GaN has not yet been detected. This transition is well known in II–VI semiconductors, where it appears at the energy of about 0.6 eV. The position of the band is only slightly dependent on the lattice constant of the host crystal, shifting to higher energies when decreasing nearest neighbor distance. It can be observed in both absorption and luminescence (Kaminska et al., 1979). The band is built of a zero phonon line and a phonon sideband, intensity of which in many cases exceeds the intensity of the zero phonon line, indicating strong coupling to lattice vibrations. In III–V semiconductors, the $Cr^{2+}(d^4)$ intracenter transition is shifted to higher energies, appearing at above 0.8 eV in GaAs and GaP (Clerjaud et al., 1980; Williams et al., 1982).

The electron structure of the $Cr^{2+}(d^4)$ ion in a host semiconductor is equivalent to that of $Mn^{3+}(d^4)$, described previously. It is worth noting that $Mn^{3+}(d^4)$ intracenter transition appears in group III-nitrides at much higher energy of 1.4 eV (Graf et al., 2002; Wolos et al. 2004b), which can be explained by a small lattice constant, thus stronger interactions, in these materials. Observing trends visible for intracenter transitions of d^4 configurations in II–VI and III–V materials one can expect the $Cr^{2+}(d^4)$ crystal-field transition to appear in GaN between 0.8 and 1.4 eV (see Fig. 8.13).

In the studies of Cr doped GaN, most of the effort was dedicated to obtain GaCrN with the highest possible Cr content. Up to 3% Cr concentration was achieved (Liu et al., 2004). Ferromagnetism at over 900K in such material was reported, with magnetic properties changing as a

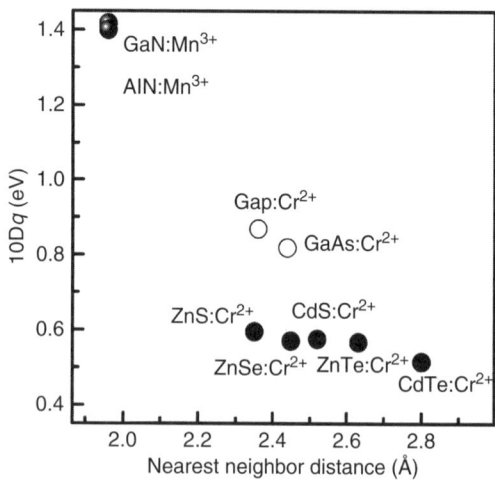

FIGURE 8.13 Crystal field transitions of $Cr^{2+}(d^4)$ in II–VI (Kaminska et al., 1979; Vallin et al., 1970), III–V semiconductors (Clerjaud et al., 1980; Williams et al., 1982), and $Mn^{3+}(d^4)$ in AlN (Graf et al., 2002) and GaN (Graf et al., 2002; Wolos et al., 2004b).

function of Cr concentration. Large variations in the magnetic properties of GaCrN were observed with growth temperature and post-growth annealing (Newman et al., 2006; Singh et al., 2005). The authors showed that the location of Cr sites in GaCrN lattice plays a crucial role in magnetic behavior, and that substitutional Cr is involved in creation of ferromagnetism. They proposed double exchange mechanism, resulting from hopping between near-midgap substitutional Cr impurity bands, as possible origin of ferromagnetic ordering. Optical properties of GaCrN have been also studied (Hashimoto et al., 2004) and showed appearance of photoemission peak at energy below GaN band gap, at 3.29 eV for 1.5% Cr content. This peak has been assigned to band-to-band transitions in GaCrN. However other studies performed on bulk and epitaxial GaN diluted with Cr indicated huge nonhomogeneity of the material. The energy position of the luminescence peak depended strongly on the place on the samples. These results could be explained by formation of micro islands of $Ga_{1-x}Cr_xN$ alloy with fluctuating Cr composition x (Jaworek et al., 2005). Magnetization measurements of such samples showed ferromagnetism above room temperature, however concentration of magnetically active Cr was always much lower than total Cr concentration (Liu et al., 2004). What needs to be stressed, some of the samples showed pure ferromagnetic phase by magnetization measurement. This could indicate that GaCrN alloy was indeed ferromagnetic. Evidently, GaCrN needs further studies in order to improve homogeneity of the obtained material as well as to prove double exchange or other mechanism as responsible for ferromagnetic ordering. This would be the first case of ferromagnetism in DMS with no free carrier mediation.

6. INTERNAL REFERENCE RULE FOR TRANSITION METAL IONS—CASE OF GaN

Investigations of electron levels introduced by transition metal ions into GaN are at the moment in an initial stage. Most of the experimental works were devoted to Mn, as it was the most promising dopant in a context of high temperature ferromagnetism. The knowledge about other impurities, Fe and particularly Cr is still incomplete.

The experimentally determined positions of charge transfer levels of Mn, Fe, and Cr in GaN, as well as in GaAs, GaP, and InP have been plotted in Fig. 8.1. It is well established, that the internal reference rule holds for transition metal donor or acceptor levels within a group of III–V (or II–VI) semiconductors (Langer et al., 1988; Zunger 1986). Questionable is, however, whether the rule is still valid for GaN in alignment with other III–V semiconductors. The experimentally determined location of Mn acceptor level seems to be consistent with the rule, on the other hand the Fe and Cr do not fit particularly well to the scheme. It has been argued by Malguth et al. (2006a), that the internal reference rule cannot be used in this case due to large electronegativity of N in comparison with other group V-elements. For the same reason (large electronegativity), however, the change of a charge state of transition metal impurity can be accompanied in GaN by large lattice relaxation, e.g., reaching almost 0.4 eV for $Mn(d^5)$–$Mn(d^4)$ transition (Wolos et al., 2004a). Thus the photoionization onsets, from which the position of acceptor levels are derived, can be significantly shifted with respect to the charge transfer level position. The problem of lattice relaxation was investigated experimentally so far only for Mn in GaN, and subtracting lattice relaxation energy of this center results in nice alignment of its charge transfer level in GaN with other III–V semiconductors. In a case of Fe, the calculated Franck–Condon energies, which can be related directly to the measured energies of lattice relaxation are less than for Mn impurity (Boguslawski and Bernholc, 2008, private communication). In any case, this problem needs to be clarified in order to draw the conclusion about validity of the internal reference rule for GaN with other III–V semiconductors.

7. SUMMARY AND CONCLUSIONS

The conclusions coming out from broad studies performed on group III–V DMSs strongly indicate presence of ferromagnetic phase in all compounds diluted with manganese, iron, or chromium. Results obtained for GaMnAs could be explained by one phase ferromagnetic material. Originally, ferromagnetic ordering of Mn spins was connected with mediation by holes from GaAs valence band (Dietl et al., 2000). Curie temperature of GaMnAs depended on Mn content as well as hole concentration. However, recently there are some suggestions that the holes in GaMnAs reside in impurity band (Alvarez and Dagotto, 2003; Berciu and Bhatt 2001; Mahadevan and Zunger, 2004a; Rokhinson et al., 2007) and it is probable that holes from Mn band, rather than from GaAs valence band, are responsible for ferromagnetism of GaMnAs. The main problem with commercialization of this ferromagnetic semiconductor is its transition to paramagnetic phase still well below 300K, and not much hope for increase of this value. In the case of GaN diluted with manganese or iron ions, ferromagnetic phase is responsible only for part of its magnetic properties. Precipitation of alien phase with its own crystalline structure within GaN, or regions rich in transition metal, resulting from spinodal decomposition to mutually coherent phases, seems to explain creation of multi- (at least two: ferromagnetic and paramagnetic) component material. However, the accounting for ferromagnetism of one of these components with well understood origin of ferromagnetism is still an open issue. The interesting system seems to be GaN diluted with chromium. For this material one phase, ferromagnetic above room temperature, was often observed. The explanation of ferromagnetic ordering cannot follow in this case the well established for DMS model of carrier mediated exchange interactions within transition metal ensemble because of lack of holes in high concentration. The proposed double-exchange mechanism (Liu et al., 2004) needs further experimental support. From the point of view of possible application - still technological effort is necessary in order to improve homogeneity of this material.

Studies of group III–V DMSs showed that, for better understanding of possible magnetic ordering in DMSs, it is important to undertake research of semiconductors with transition metals of concentrations on impurity level. In this chapter it is shown clearly, mostly on the example of GaN:Mn material. The broad studies of Mn impurity in GaN: magnetic, optical and electron transport allowed to understand why the conditions of Zener's model of ferromagnetism are difficult to be fulfilled in bulk GaMnN. While studying manganese impurity in GaN, several new results connected with behavior of transition metal dopant in ionic semiconductors were obtained. One of them is a pretty high energy of lattice relaxation accompanying impurity charge transfer transitions. It was shown for the case of GaN:Mn that only when taking into account lattice relaxation, the internal reference rule for transition metal ions may be kept valid. Electron transport measurements performed on different III–V materials, doped with Mn of different concentrations, allowed determining neutral acceptor configuration of manganese in these compounds (Wolos et al., 2007). It occurred that atomic structure of such center changes in series from GaAs through InP and GaP, to GaN. In the case of GaAs, manganese stays in d^5 state with hole bound to it, having localization radius of about 14 Å. This hole can be well described by wave function of GaAs valence band. Such picture is strongly supported by optical absorption with hydrogenic-like series of lines corresponding to optical transition of Mn-bound hole to its excited states (Tarhan et al., 2003), as well as by ESR signal ascribed to GaAs valence band hole bound to Mn center (Schneider et al., 1987). In the case of InP and GaP, localization radius of hole bound to Mn in d^5 configuration is smaller, and it varies from 7 to 4 Å, respectively. For both InP and GaP no ESR signal due to transitions within ground state of valence band hole bound to Mn in d^5 configuration was detected, what may reflect loss of valence band character of this hole in its ground state. On the other hand, hydrogenic-like series of lines in optical absorption were observed in both cases, proving hydrogenic-like character of the excited states. The extreme case is neutral acceptor configuration of manganese in GaN, for which well-localized d^4 configuration was found. Absorption related to this configuration is entirely different than in the case of three other compounds, and can be explained in frame of crystal-field theory, with localized wave function restricted to the area in-between manganese ligands. With such a picture in mind one should reconsider Zener's model of ferromagnetism in semiconductors, since, not always, valence band-like holes may mediate exchange interactions in III–V DMSs. It seems justified in the case of GaMnAs material, but not necessarily in the case of Mn in InP or GaP, for which more localized holes should be considered. In the case of GaMnN the input for Zener's model is Mn in d^4 or d^3, and holes originating from additional shallow acceptors. Since, as it was mentioned before, it would be technologically very difficult to provide enough holes by co-doping of GaN for efficient mediation, Zener's model is not adequate for explanation of ferromagnetism observed in diluted magnetic GaN. Ferromagnetic phases detected by means of magnetization measurements need another theoretical description, although spinodal decomposition may account for existence of magnetic ion-rich phase. The problem of changes in localization of neutral Mn acceptor was acknowledged in theoretical papers of Mahadevan and Zunger (2004a, 2004b) as well as in the recent paper of Dietl (2008), who proposed a revision of the Zener's model describing ferromagnetic ordering in semiconductors (Dietl et al., 2000).

It was found that when growing of GaMnN or GaFeN with high content of transition metals, secondary phases of ferromagnetic character are easily created. The exact origin of these phases, as well as mechanism of ferromagnetism of them, needs further work. Gaining knowledge about their nature, as well as gaining control over their size and density, could help with engineering of a new and interesting class of materials—semiconductor ferromagnetic composites.

ACKNOWLEDGMENTS

We would like to acknowledge many years of collaboration in the subject of this chapter with M. Bockowski, J. Gosk, I. Grzegory, A. Hruban, K. Korona, M. Palczewska, M. Piersa, G. Strzelecka, A. Twardowski, D. Wasik, Z. Wilamowski, A. Wysmolek, and M. Zajac. Without the efforts of these people in the growth and characterization of the special materials, and the many discussions held, the presented results would not be possible to obtain. This work was partially supported by the Polish Ministry of Science and High Education by Grant No. N202052 32/1189 and the project MTKD-CT-2005-029671.

REFERENCES

Almeleh, N., and Goldstein, B. (1962). *Phys. Rev.* **128**, 1568.
Alvarez, G., and Dagotto, E. (2003). *Phys. Rev. B* **68**, 045202.
Amano, H., Kito, M., Hiramatsu, K., and Akasaki, I. (1989). *Jpn. J. Appl. Phys.* **28**, L2112.
Bacewicz, R., Filipowicz, J., Podsiadlo, S., Szyszko, T., and Kaminski, M. (2003). *J. Phys. Chem. Solids* **64**, 1469.
Baibich, M. N., Broto, J. M., Fert, A., Nguyen Van Dau, F., Petroff, F., Etienne, P., Creuzet, G., Friederich, A., and Chazelas, J. (1988). *Phys. Rev. Lett.* **61**, 2472.

Baik, J. M., Shon, Y., Kang, T. W., and Lee, J.-L. (2004). *Appl. Phys. Lett.* **84**, 1120.
Baldereschi, A., and Lipari, N. O. (1973). *Phys. Rev. B* **8**, 2697.
Baranov, P. G., Ilyin, I. V., and Mokhov, E. N. (1997). *Solid. State Comm.* **101**, 611.
Bastard, G., Rigaux, C., Guldner, Y., and Mycielski, J. (1978). *J. Phys.* **39**, 87.
Bebb, H. B. (1969). *Phys. Rev.* **185**, 1116.
Berciu, M., and Bhatt, R. N. (2001). *Phys. Rev. Lett.* **87**, 107203.
Binasch, G., Grunberg, P., Saurenbach, F., and Zinn, W. (1989). *Phys. Rev. B* **39**, 4828.
Bishop, S. G. (1986). In "Deep Centers in Semiconductors" (S. T. Pantelides, ed.), p. 541. Gordon and Breach, New York.
Brown, W. J., Jr., Woodbury, D. A., and Blakemore, J. S. (1973). *Phys. Rev. B* **8**, 5664.
Blakemore, J. S., Brown, W. J., Jr., Stass, M. L., and Woodbury, D. A. (1973). *J. Appl. Phys.* **44**, 3352.
Blinowski, J., and Kacman, P. (1991). *Acta Phys. Pol. A* **80**, 295.
Blinowski, J., and Kacman, P. (2001). *Acta Phys. Pol. A* **100**, 343.
Boguslawski, P., and Bernholc, J. (2005). *Phys. Rev. B* **72**, 115208.
Bonanni, A., Simbrunner, C., Wegscheider, M., Przybylinska, H., Wolos, A., Sitter, H., and Jantsch, W. (2006). *Phys. Stat. Sol. (b)* **243**, 1701.
Bonanni, A., Kiecana, M., Simbrunner, C., Li, T., Sawicki, M., Wegscheider, M., Quast, M., Przybylinska, H., Navarro-Quezada, A., Jakiela, R., Wolos, A., Jantsch, W., *et al.* (2007). *Phys. Rev. B* **75**, 125210.
Chapman, R. A., and Hutchinson, W. G. (1967). *Phys. Rev. Lett.* **18**, 443.
Celebi, C., Koenraad, P. M., Silov, A. Y., Van Roy, W., Monakhov, A. M., Tang, J. M., and Flatte, M. E. (2008). *Phys. Rev. B* **77**, 075328.
Clerjaud, B., Hennel, A. M., and Martinez, G. (1980). *Solid State Comm.* **33**, 983.
Dhar, S., Brandt, O., Trampert, A., Daweritz, L., Friedland, K. J., Ploog, K. H., Keller, J., Beschoten, B., and Guntherodt, G. (2003). *Appl. Phys. Lett.* **82**, 2077.
Dietl, T. (2007). *J. Phys. Condens. Matter* **19**, 165204.
Dietl, T. (2008). *Phys. Rev. B* **77**, 085208.
Dietl, T., Ohno, H., Matsukura, F., Cibert, J., and Ferrand, D. (2000). *Science* **287**, 1019.
Fedorych, O. M., Hankiewicz, E., Wilamowski, Z., and Sadowski, J. (2002). *Phys. Rev. B* **66**, 045201.
Ferrand, D., Cibert, J., Wasiela, A., Bourgognon, C., Tatarenko, S., Fishman, G., Andrearczyk, T., Jaroszynski, J., Kolesnik, S., Dietl, T., Barbara, B., and Dufeu, D. (2001). *Phys. Rev. B* **63**, 085201.
Fay, M. W., Han, Y., Brown, P. D., Novikov, S. V., Edmonds, K. W., Campion, R. P., Gallagher, B. L., and Foxon, C. T. (2005). *Appl. Phys. Lett.* **87**, 31902.
Foxon, C. T., Novikov, S. V., Zhao, L. X., Edmonds, K. W., Giddings, A. D., Wang, K. Y., Campion, R. P., Staddon, C. R., Fay, M. W., Han, Y., Brown, P. D., Sawicki, M., *et al.* (2005). *J. Cryst. Growth* **278**, 685.
Franceschetti, A., Barabash, S. V., Osorio-Guillen, J., Zunger, A., and van Schilfgaarde, M. (2006). *Phys. Rev. B* **74**, 241303.
Franceschetti, A., Zunger, A., and van Schilfgaarde, M. (2007). *J. Phys. Condens. Matter* **19**, 242203.
Freeman, A. A., Edmonds, K. W., Farley, N. R. S., Novikov, S. V., Campion, R. P., Foxon, C. T., Gallagher, B. L., Sarigiannidou, E., and van der Laan, G. (2007). *Phys. Rev. B* **76**, 081201(R).
Fung, S., and Nicholas, R. J. (1981). *J. Phys. C* **14**, 2135.
Furdyna, J. K. (1988). *J. Appl. Phys.* **64**, R29.
Gaj, J. A., Ginter, J., and Galazka, R. R. (1978). *Phys. Stat. Sol. (b)* **89**, 655.
Gaj, J. A., Planel, R., and Fishman, G. (1979). *Solid State Comm.* **29**, 435.
Gaj, J. (1988). In "Semiconductors and Semimetals" (R. K. Willardson, A. C. Beer, Treatise eds; J. K. Furdyna, and J. Kossut, Volume eds.) Vol. 25, p. 275. Academic Press, Boston.

Gehlhoff, W., Azamat, D., and Hoffmann, A. (2006). *Phys. Stat. Sol. B* **243**, 1687.
Giraud, R., Kuroda, S., Marcet, S., Bellet-Amalric, E., Biquard, X., Barbara, B., Fruchard, D., Ferrand, D., Cibert, J., and Mariette, H. (2004). *Europhys. Lett.* **65**, 553.
Gosk, J., Zajac, M., Byszewski, M., Kaminska, M., Szczytko, J., Twardowski, A., Strojek, B., and Podsiadlo, S. (2003). *J. Supercon. Novel Magn.* **16**, 79.
Gosk, J., Zajac, M., Wolos, A., Kaminska, M., Twardowski, A., Grzegory, I., Bockowski, M., and Porowski, S. (2005). *Phys. Rev. B* **71**, 094432.
Graf, T., Gjukic, M., Brandt, M. S., Stutzmann, M., and Ambacher, O. (2002). *Appl. Phys. Lett.* **81**, 5159.
Graf, T., Gjukic, M., Hermann, M., Brandt, M. S., and Stutzmann, M. (2003). *Phys. Rev. B* **67**, 165215.
Hashimoto, M., Zhou, Y.-K., Kanamura, M., and Asahi, H. (2002). *Solid State Comm.* **122**, 37.
Hashimoto, M., Tanaka, H., Asano, R., Hasegawa, S., and Asahi, H. (2004). *Appl. Phys. Lett.* **84**, 4191.
Halliday, D. P., Ulrici, W., and Eaves, L. (1986). *J. Phys. C* **19**, L683.
Han, B., Korotkov, R. Y., Wessels, B. W., and Ulmer, M. P. (2004). *Appl. Phys. Lett.* **84**, 5320.
Han, B., Wessels, B. W., and Ulmer, M. P. (2005). *Appl. Phys. Lett.* **86**, 042505.
Heikman, S., Keller, S., DenBaars, S. P., and Mishra, U. K. (2002). *Appl. Phys. Lett.* **81**, 439.
Heimbrodt, W., Hartmann, T., Klar, P. J., Lampalzer, M., Stolz, W., Volz, K., Schaper, A., Treutmann, W., Krug von Nidda, H.-A., Loidl, A., Ruf, T., and Sapega, V. F. (2001). *Phys. E* **10**, 175.
Heitz, R., Maxim, P., Eckey, L., Thurian, P., Hoffmann, A., Broser, I., Pressel, K., and Meyer, B. K. (1997). *Phys. Rev. B* **55**, 4382.
Herbich, M., Mac, W., Twardowski, A., Ando, K., Shapira, Y., and Demianiuk, M. (1998). *Phys. Rev. B* **58**, 1912.
Jaczynski, M., Kossut, J., and Galazka, R. R. (1979). *Phys. Stat. Sol. (b)* **88**, 73.
Jaroszynski, J., Andrearczyk, T., Karczewski, G., Wrobel, J., Wojtowicz, T., Papis, E., Kaminska, E., Piotrowska, A., Popovic, D., and Dietl, T. (2002). *Phys. Rev. Lett.* **89**, 266802.
Jaworek, M., Wysmolek, A., Zajac, M., Gosk, J., Kaminska, M., and Twardowski, A. (2005). XXXIV International School on the Physics of Semiconducting Compounds, Jaszowiec 2005, Abstract Book, p. 188.
Kaminska, M., Baranowski, J. M., Uba, S. M., and Vallin, J. T. (1979). *J. Phys. C* **12**, 2197.
Kordos, P., Morvic, M., Betko, J., Novak, J., Flynn, J., and Brandes, G. R. (2004). *Appl. Phys. Lett.* **85**, 5616.
Korona, K. P., Wysmolek, A., Kaminska, M., Twardowski, A., Piersa, M., Palczewska, M., Strzelecka, G., Hruban, A., Kuhl, J., Adamovicius, R., and Krotkus, A. (2006). *Phys. B* **382**, 220.
Korotkov, R. Y., Gregie, J. M., and Wessels, B. W. (2002). *Appl. Phys. Lett.* **80**, 1731.
Kossut, J. (1988). *Semiconduct. Semimet.* **25**, 183.
Kreissl, J., Ulrici, W., El-Metoui, M., Vasson, A.-M., Vasson, A., and Gavaix, A. (1996). *Phys. Rev. B* **54**, 10508.
Krukowski, S., Bockowski, M., Lucznik, B., Grzegory, I., Porowski, S., Suski, T., and Romanowski, Z. (2001). *J. Phys. Condens. Matter* **13**, 8881.
Lambert, B., Clerjaud, B., Naud, C., Deveaud, B., Picoli, G., and Toudic, Y. (1985). Thirteenth International Conference on Defects in Semiconductors. *In* "Metall. Soc." p. 1141. AIME, Warrendale, PA, USA.
Langer, J. M., Delerue, C., Lannoo, M., and Heinrich, H. (1988). *Phys. Rev. B* **38**, 7723.
Liu, H. X., Wu, S. Y., Singh, R. K., Lin, G., Smith, D. J., Newman, N., Dilley, N. R., Montes, L., and Simmonds, M. B. (2004). *Appl. Phys. Lett.* **85**, 4076.
Mac, W., Twardowski, A., Eggenkamp, P. J. T., Swagten, H. J. M., Shapira, Y., and Demianiuk, M. (1994). *Phys. Rev. B* **50**, 14144.
Mac, W., Twardowski, A., and Demianiuk, M. (1996). *Phys. Rev. B* **54**, 5528.

Mac, W., Herbich, M., Twardowski, A., and Demianiuk, M. (2000). *Semiconduct. Sci. Technol.* **15**, 748.
Macieja, B., Korona, K. P., Piersa, M., Witowski, A. M., Wasik, D., Wysmolek, A., Strzelecka, G., Hruban, A., Surma, B., Palczewska, M., Kaminska, M., and Twardowski, A. (2003). *Acta Phys. Pol. A* **103**, 637.
Madelung, O. (2004). "Semiconductors: Data handbook," Springer-Verlag, Berlin.
Mahadevan, P., and Zunger, A. (2004a). *Phys. Rev. B* **69**, 115211.
Mahadevan, P., and Zunger, A. (2004b). *Appl. Phys. Lett.* **85**, 2860.
Maier, K., Kunzer, M., Kaufmann, U., Schneider, J., Monemar, B., Akasaki, I., and Amano, H. (1994). *Mater. Sci. Forum* **143–147**, 93.
Malguth, E., Hoffmann, A., Gelhoff, W., Gelhausen, O., Phillips, M. R., and Xiu, X. (2006a). *Phys. Rev. B* **74**, 165202.
Malguth, E., Hoffmann, A., and Xu, X. (2006b). *Phys. Rev. B* **74**, 165201.
Marcet, S., Ferrand, D., Halley, D., Kuroda, S., Mariette, H., Gheeraert, E., Teran, F., Sadowski, M., Galera, R. M., and Cibert, J. (2006). *Phys. Rev. B* **74**, 125201.
Marcinkevicius, S., Aggerstam, T., Pinos, A., Linnarsson, M., and Laurducloss, S. (2007). Presented at Semiconducting and Insulating Materials Conference SIMC XIV, Fayetteville, AR, USA.
Martinez-Criado, G., Somogyi, A., Ramos, S., Campo, J., Tucoulou, R., Salome, M., Susini, J., Hermann, M., Eickhoff, M., and Stutzmann, M. (2005). *Appl. Phys. Lett.* **86**, 131927.
Medvedkin, G. A., Ishibashi, T., Hayata, T. N., Hasegawa, Y., and Sato, K. (2000). *Jpn. J. Appl. Phys.* **39**, L949.
Munekata, H., Ohno, H., von Molnar, S., Segmüller, A., Chang, L. L., and Esaki, L. (1989). *Phys. Rev. Lett.* **63**, 1849.
Newman, N., Wu, S. Y., Liu, H. X., Medvedeva, J., Gu, L., Singh, R. K., Yu, Z. G., Krainsky, I. L., Krishnamurthy, S., Smith, D. J., Freeman, A. J., and van Schilfgaarde, M. (2006). *Phys. Stat. Sol. (a)* **203**, 2729.
Niedzwiadek, A., Wysmolek, A., Wasik, D., Szczytko, J., Kaminska, M., Twardowski, A., Sadowski, M. L., Potemski, M., Clerjaud, B., Pastuszka, B., Lucznik, B., and Grzegory, I. (2007). *Phys. B–Condens. Matter* **401–402**, 458.
Novikov, S. V., Edmonds, K. W., Zhao, L. X., Giddings, A. D., Wang, K. Y., Campion, R. P., Staddon, C. R., Fay, M. W., Han, Y., Brown, P. D., Sawicki, M., Gallagher, B. L., *et al.* (2005). *J. Vac. Sci. Technol. B* **23**, 1294.
Ohno, H., and Matsukura, F. (2001). *Solid State Comm.* **117**, 179.
Ohno, H., Munekata, H., Penney, T., von Molnar, S., and Chang, L. L. (1992). *Phys. Rev. Lett.* **68**, 2664.
Ohno, H., Shen, A., Matsukura, F., Oiwa, A., Endo, A., Katsumoto, S., and Iye, Y. (1996). *Appl. Phys. Lett.* **69**, 363.
Ohno, H., Chiba, D., Matsukura, F., Omiya, T., Abe, E., Dietl, T., Ohno, Y., and Ohtani, K. (2000). *Nature* **408**, 994.
Ohno, Y., Young, D. K., Beschoten, B., Matsukura, F., Ohno, H., and Awschalom, D. D. (1999). *Nature* **402**, 790.
Overberg, M. E., Abernathy, C. R., Pearton, S. J., Theodoropoulou, N. A., McCarthy, K. T., and Hebard, A. F. (2001). *Appl. Phys. Lett.* **79**, 1312.
Pacuski, W., Ferrand, D., Cibert, J., Deparis, C., Gaj, J. A., Kossacki, P., and Morhain, C. (2006). *Phys. Rev. B* **73**, 035214.
Pacuski, W., Ferrand, D., Cibert, J., Gaj, J. A., Golnik, A., Kossacki, P., Marcet, S., Sarigiannidou, E., and Mariette, H. (2007). *Phys. Rev. B* **76**, 165304.
Pacuski, W., Kossacki, P., Ferrand, D., Golnik, A., Cibert, J., Wegscheider, M., Navarro-Quezada, A., Bonanni, A., Kiecana, M., Sawicki, M., and Dietl, T. (2008). *Phys. Rev. Lett.* **100**, 037204.
Pastor, K., Grynberg, M., and Galazka, R. R. (1979). *Phys. Stat. Sol. (b)* **29**, 739.

Pawlowski, M., Piersa, M., Wolos, A., Palczewska, M., Strzelecka, G., Hruban, A., Gosk, J., Kaminska, M., and Twardowski, A. (2005). *Acta Phys. Pol. A* **108**, 825.
Polyakov, A. Y., Smimov, N. B., Govorkov, A. V., Pashkova, N. V., Shlensky, A. A., Pearton, S. J., Overberg, M. E., Abernathy, C. R., and Zavada, J. M. (2003a). *J. Appl. Phys.* **93**, 5388.
Polyakov, A. Y., Smimov, N. B., Govorkov, A. V., and Pearton, S. J. (2003b). *Appl. Phys. Lett.* **83**, 3314.
Przezdziecka, E., Kaminska, E., Kiecana, M., Sawicki, M., Klopotowski, L., Pacuski, W., and Kossut, J. (2006). *Solid State Comm.* **139**, 541.
Przybylinska, H., Bonanni, A., Wolos, A., Kiecana, M., Sawicki, M., Dietl, T., Malissa, H., Simbrunner, C., Wegscheider, M., Sitter, H., Rumpf, K., Granitzer, P., *et al.* (2006). *Mater. Sci. Eng.* **126**, 222.
Reed, M. L., El-Masry, N. A., Stadelmaier, H. H., Ritums, M. K., Reed, M. J., Parker, C. A., Roberts, J. C., and Bedair, S. M. (2001). *Appl. Phys. Lett.* **79**, 3473.
Rokhinson, L. P., Lyanda-Geller, Y., Ge, Z., Shen, S., Liu, X., Dobrowolska, M., and Furdyna, J. K. (2007). *Aixiv: Condens.-Mater.* 0707.2416.
Rudzinski, M., Desmaris, V., van Hal, P. A., Weyher, J. L., Hageman, P. R., Dynefors, K., Rodle, T. C., Jos, H. F. F., Zirath, H., and Larsen, P. K. (2006). *Phys. Stat. Sol. (c)* **3**, 2231.
Sarigiannidou, E., Wilhelm, F., Monroy, E., Galera, R. M., Bellet-Amalric, E., Rogalev, A., Goulon, J., Cibert, J., and Mariette, H. (2006). *Phys. Rev. B* **74**, 041306(R).
Sasaki, T., Sonoda, S., Yamamoto, Y., Suga, K., Shimizu, S., Kindo, K., and Hori, H. (2002). *J. Appl. Phys.* **91**, 7911.
Sato, K., and Katayama-Yoshida, H. (2001). *Jpn. J. Appl. Phys.* **40**, L485.
Sato, K., Dederichs, P. H., Katayama-Yoshida, H., and Kudrnovsky, J. (2003). *J. Phys. B* **340–342**, 863.
Sato, K., Schweika, W., Dederichs, P. H., and Katayama-Yoshida, H. (2004). *Phys. Rev. B* **70**, 201202(R).
Sato, K., Katayama-Yoshida, H., and Dederichs, P. H. (2005). *Jpn. J. Appl. Phys.* **44**, L948.
Sato, M., Tanida, H., Kato, K., Sasaki, T., Yamamoto, Y., Sonoda, S., Shimizu, S., and Hori, H. (2002). *Jpn. J. Appl. Phys. Part 1* **41**, 4513.
Schneider, J., and Sircar, S. R. (1962). *Z. Naturfosch. A* **17A**, 570.
Schneider, J., Kaufmann, U., Wilkening, W., Baeumler, M., and Kohl, F. (1987). *Phys. Rev. Lett.* **59**, 240.
Semiconductors and Semimetals (1988). Willardson, R. K. and Beer, A. C., Treatise editors; Furdyna, J. K. and Kossut, J., Volume editros, Vol. 25. Academic Press, Boston.
Shanabrook, B. V., Klein, P. B., and Bishop, S. G. (1983). *Physica* **116B**, 444.
Shklovskii, B. I., and Efros, A. L. (1984). Electronic Properties of Doped Semiconductors. *In* "Springer Series in Solid-State Sciences" (M. Cardona, ed.), Vol 45, p. 25. Springer-Verlag, Berlin, Heidelberg, New York, Tokyo.
Singh, R. K., Wu, S. Y., Liu, H. X., Gu, L., Smith, D. J., and Newman, N. (2005). *Appl. Phys. Lett.* **86**, 012504.
Sonoda, S., Shimizu, S., Sasaki, T., Yamamoto, Y., and Hori, H. (2002). *J. Cryst. Growth* **237**, 1358.
Soo, Y. L., Kioseoglou, G., Kim, S., Huang, S., Kao, Y. H., Kuwabara, S., Owa, S., Kondo, T., and Munekata, H. (2001). *Appl. Phys. Lett.* **79**, 3926.
Stern, N. P., Myers, R. C., Poggio, M., Gossard, A. C., and Awschalom, D. D. (2007). *Phys. Rev. B* **75**, 045329.
Stoneham, A. M. (1985). "Theory of Defects in Solids." Clarendon Press, Oxford.
Story, T., Galazka, R. R., Frankel, R. B., and Wolff, P. A. (1986). *Phys. Rev. Lett.* **56**, 777.
Sun, H. J., Peale, R. E., and Watkins, G. D. (1992). *Phys. Rev. B* **45**, 8310.
Szawelska, H. R., and Allen, J. W. (1979). *J. Phys. C* **12**, 3359.

Szczytko, J., Twardowski, A., Swiatek, K., Palczewska, M., Tanaka, M., Hayashi, T., and Ando, K. (1999). *Phys. Rev. B* **60**, 8304.
Tarhan, E., Miotkowski, I., Rodriguez, S., and Ramdas, A. K. (2003). *Phys. Rev. B* **67**, 195202.
Teisseyre, H., Perlin, P., Suski, T., Grzegory, I., Porowski, S., Jun, J., Pietraszko, A., and Moustakas, T. D. (1994). *J. Appl. Phys.* **76**, 2429.
Theodoropoulou, N., Hebard, A. F., Overberg, M. E., Abernathy, C. R., Pearton, S. J., Chu, S. N. G., and Wilson, R. G. (2001). *Appl. Phys. Lett.* **78**, 3475.
Twardowski, A., Swagten, H. J. M., and de Jonge, W. J. M. (eds) (1993). "Jain M., Co-based II–VI semimagnetic semiconductors, in II–VI Semiconductor Compounds." p. 227. World Scientific Publishing, Singapore.
Tworzydlo, J. (1994). *Phys. Rev. B* **50**, 14591.
van Schilfgaarde, M., and Mryasov, O. N. (2001). *Phys. Rev. B* **63**, 233205.
Vallin, Y. T., Slack, G. A., Roberts, S., and Hughes, A. E. (1970). *Phys. Rev. B* **2**, 4313.
Vaudo, R. P., Xu, X., Salant, A., Malearne, J., and Brandes, G. R. (2003). *Phys. Stat. Sol. (a)* **200**, 18.
Vurgaftman, I., Meyer, J. R., and Ram-Mohan, L. R. (2001). *J. Appl. Phys.* **89**, 5815.
Wang, K. Y., Campion, R. P., Edmonds, K. W., Sawicki, M., Dietl, T., Foxon, C. T., and Gallagher, B. L. (2005). AIP Conf. Proc. **772**, ICPS-27, (J. Menendez, ed.), p. 333. Melville, New York.
Williams, P. J., Eaves, L., Simmonds, P. E., Henry, M. O., Lightowlers, E. C., and Uihlein, C. (1982). *J. Phys. C* **15**, 1337.
Wolos, A., Kossacki, P., Golnik, A., Kaminska, M., Gaj, J. A., Twardowski, A., Grzegory, I., Bockowski, M., and Porowski, S. (2002). *Acta Phys. Pol. A* **102**, 695.
Wolos, A., Palczewska, M., Wilamowski, Z., Kaminska, M., Twardowski, A., Bockowski, M., Grzegory, I., and Porowski, S. (2003). *Appl. Phys. Lett.* **83**, 5428.
Wolos, A., Palczewska, M., Zajac, M., Gosk, J., Kaminska, M., Twardowski, A., Bockowski, M., Grzegory, I., and Porowski, S. (2004a). *Phys. Rev. B* **69**, 115210.
Wolos, A., Wysmolek, A., Kaminska, M., Twardowski, A., Bockowski, M., Grzegory, I., Porowski, S., and Potemski, M. (2004b). *Phys. Rev. B* **70**, 245202.
Wolos, A., Zajac, M., Gosk, J., Korona, K., Wasik, D., Wysmolek, A., Palczewska, M., Grzegory, I., Bockowski, M., Piersa, M., Strzelecka, G., Hruban, A., *et al.* (2007). AIP Conf. Proc. **893**, ICPS-28, (W. Jantsch and F. Schaffler, eds), p. 1201. Melville, New York.
Woodbury, D. A., and Blakemore, J. S. (1973). *Phys. Rev.* **8**, 3803.
Yakunin, A. M., Silov, A. Y., Koenraad, P. M., Van Roy, W., De Boeck, J., and Wolter, J. H. (2003). *Superlat. Microst.* **34**, 539.
Zajac, M., Doradzinski, R., Gosk, J., Szczytko, J., Lefeld-Sosnowska, M., Kaminska, M., Twardowski, A., Palczewska, M., Grzanka, E., and Gebicki, W. (2001a). *Appl. Phys. Lett.* **78**, 1276.
Zajac, M., Gosk, J., Kaminska, M., Twardowski, A., Szyszko, T., and Podsiadlo, S. (2001b). *Appl. Phys. Lett.* **79**, 2432.
Zajac, M., Gosk, J., Grzanka, E., Kaminska, M., Twardowski, A., Stojek, B., Szyszko, T., and Podsiadlo, S. (2003). *J. Appl. Phys.* **93**, 4715.
Zener, C. (1951). *Phys. Rev.* **81**, 440.
Zenneck, J., Niermann, T., Mai, D., Roever, M., Kocan, M., Malindretos, J., Seibt, M., Rizzi, A., Kaluza, N., and Hardtdegen, H. (2007). *J. Appl. Phys.* **101**, 063504.
Zhao, Y.-J., Mahadevan, P., and Zunger, A. (2004). *Appl. Phys. Lett.* **84**, 3753.
Zunger, A. (1986). In "Solid State Physics" (H. Ehrenreich and D. Turnbull, eds), Vol. 39, p. 275. Academic Press, New York.

Exchange Interactions and Nanoscale Phase Separations in Magnetically Doped Semiconductors

Tomasz Dietl

Contents		
	1. Introduction	372
	2. Substitutional Transition Metal Impurities in Semiconductors	375
	2.1. TM impurity-related levels	375
	2.2. TMs as isoelectronic impurities	377
	2.3. TMs as carrier traps	377
	2.4. TMs as carrier dopant	378
	2.5. TMs as resonant impurities	379
	2.6. Charge transfer states	379
	2.7. Magnetic moments	380
	3. Origin of Exchange Interactions between Carriers and Localized Spins	381
	4. Effects of sp–d(f) Exchange Interactions	382
	4.1. Splitting of extended states: Weak coupling	382
	4.2. Splitting of extended states: Strong coupling	383
	4.3. Spin-disorder scattering	390
	4.4. Kondo effect	391
	4.5. Magnetic polarons	391
	5. Exchange Interactions between Effective Mass Carriers	392
	5.1. Intra-band exchange interaction	392

Institute of Physics, Polish Academy of Sciences and ERATO Semiconductor Spintronics Project, al. Lotników 32/46, PL 02-668 Warszawa, Poland
Institute of Theoretical Physics, University of Warsaw, ul. Hoża 69, PL 00-681 Warszawa, Poland

5.2. Electron-hole exchange interaction 392
5.3. Exchange interactions between electrons
and bound holes 393
5.4. Exchange interactions between band and
bound holes 397
6. Models of Ferromagnetic Spin–Spin Interactions in
Semiconductors 397
  6.1. Superexchange 397
  6.2. Double exchange 398
  6.3. Zener/RKKY models 398
7. p–d Zener Model of Carrier-Mediated
Ferromagnetism 399
  7.1. The model 399
  7.2. Application to DMS films 402
  7.3. Application to modulated structures of DMSs 405
  7.4. Beyond colinear magnetic ground state 406
8. Effects of Disorder And Localization on
Carrier-Mediated Ferromagnetism 407
  8.1. Quantum localization 407
  8.2. Weak disorder 410
  8.3. Critical region 411
  8.4. Critical scattering near MIT 413
  8.5. Strongly localized regime 415
9. Effects of Nonrandom Distribution of Magnetic Ions 416
  9.1. Observations of spinodal decomposition 416
  9.2. Controlling spinodal decomposition by
  inter-ion Coulomb interactions 419
  9.3. Possible functionalities 420
10. Is Ferromagnetism Possible in Semiconductors with
No Magnetic Elements? 422
11. Summary 423
Acknowledgments 425
References 425

1. INTRODUCTION

Previous chapters, describing low-temperature studies carried out on transition-metal doped semiconductors, demonstrate the unprecedented progress that has been accomplished over the last decade in exploring physical phenomena and device concepts in previously unavailable combinations of quantum structures and ferromagnetism in semiconductors (Cibert *et al.*, 2008; Gould *et al.*, 2008; Jungwirth *et al.*, 2008; Matsukura *et al.*, 2008; Yu *et al.*, 2008). Remarkably, as clear from these surveys, experimental research on dilute ferromagnetic semiconductors has been prompted and/or assisted by successful quantitative theoretical modeling which has also encompassed properties that are usually regarded as the domain of experimental studies, such as magnetic anisotropy or the Gilbert dumping constant.

However, despite the massive achievements, the control and understanding of dilute magnetic semiconductors (DMSs) have emerged as the most controversial and challenging field of today's material science and condensed matter physics. We become increasingly aware that these systems, similarly to colossal magnetoresistance and superconducting oxides, may exhibit nanoscale electronic phase separations, driven by competing interactions, electronic correlation, and disorder inherent to carrier-doped materials. At the same time, it is more and more obvious that the electronic phase separations are entangled in these dilute systems with chemical and/or crystalographic phase separations. Accordingly, a meaningful description of DMSs requires detailed information on the magnetic ion incorporation and distribution on both atomic and mesoscopic scales. Here, the questions that have been addressed in other chapters concern the substitutional versus interstitial location (Yu et al., 2008) as well as preferential magnetic ion pairing which may lead to self-organized aggregation of magnetic nanostructures (Katayama-Yoshida et al., 2008; Tanaka et al., 2008).

In general, because of the limited solubility of magnetic elements in semiconductors, nanoparticles containing a large concentration of magnetic ions nucleate and precipitate during the growth or processing, assuming rather a chemical composition and crystallographic structure imposed by the host and unlisted in materials compendia, than the form of inclusions of a known magnetic system. In either case, the assembled nanocrystals can be uniformly distributed over the film volume, accumulate at the surface or interface, or extends in the form of nanocolumns along the growth direction. As discussed in this and next chapters, a number of functionalities have already been demonstrated (Tanaka et al., 2008) or predicted (Katayama-Yoshida et al., 2008) for such multi-component systems. Importantly, we start to realize the existence of a relationship between electronic properties, such as the Fermi-level position during the growth and processing, and the resulting incorporation (Yu et al., 2008) and distribution of magnetic ions.

The list of experimental challenges would be certainly incomplete if we were not recalled that a highly sensitive SQUID magnetometer is necessary to detect small signals produced by thin films only weakly doped by magnetic elements. Actually, the magnetic moment of DMS films is often inferior to that coming from typical sample holders, substrates, magnet remanence, or that produced by nonideal signal-processing software. Furthermore, the magnetic response of the sample can be contaminated by the presence of foreign magnetic nanoparticles. Such nanoparticles may originate from residual impurities in the growth chamber or source materials as well as can be introduced by post-growth processing. It is now appreciated that a further progress in the field requires the application and the development of element-specific and spin-sensitive three dimensional (3D) nanoscopic tools as well as the critical assessment whether the employed characterization method leaves the probed samples intact.

On the theoretical front, conceptual difficulties of charge transfer magnetic insulators and strongly correlated disordered metals are combined in these systems with intricate properties of heavily doped semiconductors and semiconductor alloys, such as Anderson-Mott localization, defect creation by self-compensation mechanisms, spinodal decomposition, and break down of the virtual-crystal approximation. Accordingly, as indicated in the next chapter (Katayama-Yoshida et al., 2008), these materials have posed an enormous challenge to first principles methods, particularly those involving standard local spin-density approximation (LSDA) and coherent-potential approximation (CPA). It becomes increasingly obvious that inaccuracies of these approximations, such as improper treatment of strong correlation at transition metal atoms, errors in bandgaps and d-level positions, as well as the inadequate description of the Anderson-Mott localization, have resulted in incorrect predictions of the electronic and magnetic ground states, generally overestimating the tendency towards the metallic and ferromagnetic phases. In appears that reliable *ab initio* computational studies of ferromagnetism in semiconductors and oxides should explain simultaneously magnetic *and* electric characteristics of these systems.

This contribution is designed to rather provide some essential theoretical background for detailed discussions of ferromagnetic DMSs presented in previous chapters (Cibert et al., 2008; Gould et al., 2008; Jungwirth et al., 2008; Matsukura et al., 2008; Yu et al., 2008) than as a comprehensive survey of the theoretical DMS literature, already completed by others (Jungwirth et al., 2006a). In particular, we refrain purposely from discussing results of *ab initio* computations which are reviewed in the next chapter (Katayama-Yoshida et al., 2008) as well as from presenting those numerous theoretical models of DMS ferromagnetism, which have been put forward but not yet quantitatively verified against experimental findings.

In the next section we take over the question of levels introduced by substitutional magnetic impurities discussed in the previous chapter (Wolos and Kaminska, 2008). It is probably safe to state that all DMSs studied so far can be classified as magnetic isolators, in a sense that the states derived from the open 3d or 4f shells of substitutional transition metal (TM) and rare earth (RE) atoms do not contribute to the Fermi volume but give rise to localized magnetic moments of the magnitude and anisotropy determined by the Hund's rule, the intra-center spin–orbit interaction, and the coupling to neighboring atoms. In this perspective, the so-called impurity limit is valid over the entire range of relevant magnetic ion concentrations. We then discuss the coupling of effective mass carriers to the localized spins, emphasizing non-perturbative effects which appear when the coupling gets strong. As an example of a recent development we call attention to the role of exchange interactions among the carriers, often equally important as the coupling to the localized spins. At the same time, we delegate the readers to one of the previous chapters (Yu et al., 2008) for the survey of effects associated with the presence of *interstitial* magnetic impurities in DMSs.

A large part of this chapter is devoted to the origin and consequences of carrier-mediated ferromagnetic spin–spin interactions. It is explained why the disorder-free p–d Zener model captures the relevant physics as long as the holes stay delocalized or weakly localized. This brings us to the important question of the interplay between carrier localization and carrier-mediated ferromagnetism. We argue that the dominant effect that shows up on crossing the insulator-to-metal transition is the appearance of large spatial fluctuations in the carrier density, which lead to the corresponding highly nonuniform texture of TM or RE spin polarization. In the subsequent section, we describe why the assumption about a random distribution of magnetic ions has to be abandoned in many cases. We discuss methods allowing to control the magnetic ion aggregation and indicate a number of applications proposed for these multicomponent and multifunctional systems. Finally, we address the intriguing question about the possibility of high-temperature ferromagnetic ordering in materials containing no magnetic elements.

2. SUBSTITUTIONAL TRANSITION METAL IMPURITIES IN SEMICONDUCTORS

2.1. TM impurity-related levels

It has been long appreciated that the physics of particular DMSs is determined by the position of the states derived from the open d or f shells with respect to the band-edges as well as by the degree of the mixing between the local and band states (Dietl, 1981; Wolos and Kaminska, 2008). In particular, the fascinating properties of Mn-based DMSs described in the other chapters, such as carrier-controlled ferromagnetism as well as giant magnetooptical and magnetotransport phenomena, are known to originate from spin-dependent hybridization between the open Mn 3d shells and the valence band states (Cibert et al., 2008; Gould et al., 2008; Jungwirth et al., 2008; Matsukura et al., 2008). In particular, the present understanding of states associated with Mn in GaAs comes to a large extent from informative magnetic resonance

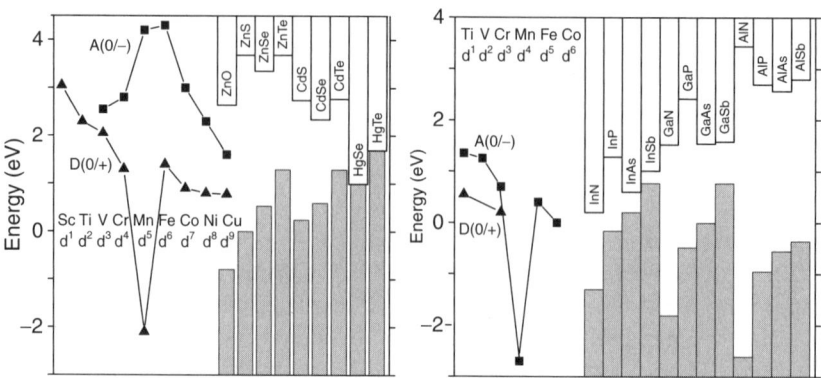

FIGURE 9.1 Approximate positions of levels derived from d shells of transition metals relative to the conduction and valence band-edges of II–VI (left panel) and III–V (right panel) compounds. By triangles the d^n/d^{n-1} donor and by squares the d^n/d^{n+1} acceptor states are denoted. If a donor (acceptor) state overlaps with the conduction band (valence band) shallow donor-like (acceptor-like) states are formed in the gap (not shown). Owing to a large magnitude of p–d hybridization in oxides, sulphides, and nitrides, a transition metal impurity can bind a hole in a charge transfer state (not shown) [adapted from Langer et al. (1988); Zunger (1986)].

(Fedorych et al., 2002; Szczytko et al., 1999b) and photoemission (Hwang et al., 2005; Mizokawa et al., 2002; Rader et al., 2004) studies.

Figure 9.1 shows schematically the positions of ground-state levels derived from the 3d shells of substitutional transition metal (TM) impurities occupied by different numbers of electrons in respect to the band energies of the host II–VI and III–V compounds. In Fig. 9.1, the levels labeled "donors" denote the minimum ionization energy of the magnetic electrons ($TM^{2+} \rightarrow TM^{3+}$ or $d^n \rightarrow d^{n-1}$) in respects to the band states, whereas the "acceptors" correspond to the maximum affinity energy ($TM^{2+} \rightarrow TM^{1+}$ or $d^n \rightarrow d^{n+1}$). Owing to a relatively large on-site correlation energy U, the acceptor state (which can trap an electron) resides at much higher energies than the donor state (which can donate an electron or in other words trap a hole). The difference between the two is the on-d-shell Coulomb (Hubbard) repulsion energy U in the semiconductor matrix.

It is important to realize that the electric field produced by neighboring host atoms as well hybridization with the band states shift and split the TM atomic d levels into twofold degenerate e_g and threefold degenerate t_2 states in the tetrahedral environment. Further symmetry lowering may be caused by the Jahn-Teller effect. This has to be taken into account on the equal footing with the intra-impurity Coulomb and exchange interactions in order to determine the impurity spin state at given charge state as well as the symmetry (e_g vs t_2) of the states corresponding to the levels shown in Fig. 9.1 (Kacman, 2001; Zunger, 1986). At the same time, the presence of the magnetic ion can affect so strongly neighboring atoms that additional levels, built up from host wave functions, may appear in the bandgap, as explained below.

However, despite hybridization between local and band states, it appears that all DMSs studied so far can be classified as magnetic isolators, in which the states derived from the open magnetic shells (3d or 4f) contribute neither to the Fermi volume nor to the conductivity but give rise to localized magnetic moments. In terms of the Anderson magnetic impurity model (Anderson, 1961) this means that the hybridization effect is weaker than the on-site Coulomb interactions among the localized electrons, though—according to the Anderson–Haldane insight (Haldane and Anderson, 1976)—the hybridization reduces the magnitude of U in solids. In addition to the Mott-Hubbard mechanism, the localization of magnetic electrons is enhanced further on by an interplay of disorder and correlation (Anderson-Mott localization), which destroys coherent effects specific to, for instance, heavy fermion systems.

2.2. TMs as isoelectronic impurities

In the case of Mn, in which the 3d shell is half filled, the d-like donor state lies deep in the valence band, whereas the acceptor level resides high in the conduction band, so that $U \approx 7$ eV according to photoemission and inverse photoemission studies. Thus, Mn-based DMSs can be classified as charge transfer insulators, $E_g < U$. Since the d orbitals contribute barely to the bonding in this case, Mn is divalent and there is no energy gain by forming the nearest neighbor Mn pairs (Kuroda et al., 2007). Accordingly, high-quality random alloys (II, Mn) VI up to large concentrations of Mn can be grown. Since the $Mn^{2+/3+}$ state resides in the valence band, the Mn ion remains in the 2+ charge state. This means that the Mn does not supply and does not trap any carriers in II–VI materials, a rather unique situation, according to Fig. 9.1. Nevertheless, there is a spin-dependent coupling between effective mass carriers and localized spins, which leads to strong magnetooptical and magnetotransport phenomena that can be studied quantitatively owing to the possibility of the independent control of spin and carrier densities. However, as discuss below, in materials with a short bond length, such as oxides, the coupling between TM 3d shells and neighboring anion p ligands is so strong that a hole trap can appear in the bandgap.

2.3. TMs as carrier traps

According to Fig. 9.1, the acceptor and donor states derived from the 3d levels appear in the bandgap in a number of cases. It is still unknown how large should be the concentration of the magnetic constituent to observe

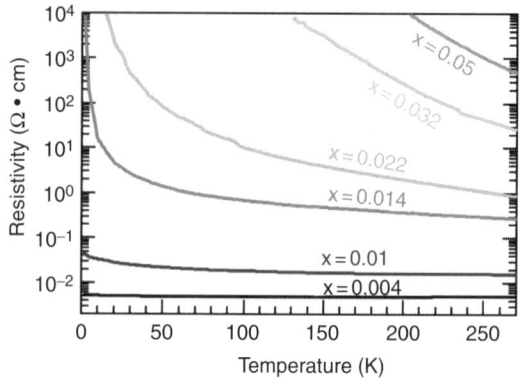

FIGURE 9.2 Temperature dependence of resistivity in p-$Zn_{1-x}Cr_xTe$ with various Cr content x and Nitrogen concentration $p \approx 7 \times 10^{20}$ cm^{-3} showing trapping of holes by Cr ions [after Ozaki et al. (2005)].

an isolator-to-metal transition involving the open d levels in such systems. In fact, the presence of mid-gap electron and hole traps is exploited to fabricate semi-insulating materials, such as GaAs:Cr, InP: Fe, or CdTe:V. This trapping ability is illustrated in Fig. 9.2 which shows how the resistivity increases with the Cr content x in p-$Zn_{1-x}Cr_xTe$ doped by 10^{20} N acceptors per cm^3.

2.4. TMs as carrier dopant

If the donor-like levels derived from the 3d shells are degenerate with the conduction band, the TM impurities donate the electrons to the conduction band. Actually, these electrons will reside on shallow donor states created by the Coulomb potentials of the parent TM impurities. To this class belongs Se in CdSe (Głód et al., 1994). Similarly, if the acceptor-like 3d states are degenerated with the valence band, shallow acceptor states appear in the bandgap, the case of Mn in GaAs (Schneider et al., 1987). Such DMSs exhibit electronic properties specific to n-type and p-type semiconductors, say, CdSe:In and GaAs:Zn, respectively, including the presence of the metal-insulator transition occurring if the distance between the carriers is 2.5 times longer than the Bohr radius of the isolated single impurity, inversely proportional of square root of the corresponding binding energy (Edwards and Sienko, 1978). As already discussed (Wolos and Kaminska, 2008), the Mn acceptor is relatively shallow (effective-mass-like) and, therefore, built up mostly from the host wave functions, in the case of III–V antimonides and arsenides, while the corresponding state becomes deeper in the case of phosphides and nitrides. This indicates that the so-called central-cell corrections (or the chemical shift) increase on going from the antimonides, through arsenides and phosphides to the nitrides, presumably because the p–d hybridization increases when the anion–cation distance diminishes (Dietl et al., 2002).

2.5. TMs as resonant impurities

An interesting question arises what happens if the concentration of the impurities is so large that not only the insulator-to-metal transition has occurred but also the Fermi energy has reached the resonant level. Guided by the physics of resonant states in metals one may expect a dramatic drop of carrier's mobility associated with the appearance of resonant scattering. This question was addressed in the case of HgSe doped with Fe acting as a resonant donor in this zero-gap semiconductor (Mycielski, 1988). It was found that the inter-site Coulomb interactions between carriers residing on the impurity atoms play a crucial role (Wilamowski et al., 1990). In particular, the inter-site interactions lead to the formation of the Efros-Shklovskii Coulomb gap in the density of the impurity states, which decouples the impurity and extended states. This makes resonant scattering of band carriers to be entirely inefficient. At the same time, scattering by ionized impurities is reduced, as the inter-site interactions impose spatial ordering of the impurity charges. Accordingly, record high electron mobilities are observed when the Fermi level is pinned by resonant states (Wilamowski et al., 1990). As shown in Fig. 9.3, the values up to 2×10^7 cm^2/Vs were observed in the zero-gap p-Hg$_{0.93}$Mn$_{0.07}$Te in which acceptor levels, presumably associated with residual Cu impurities, are degenerate with the conduction band, so that the Fermi level resides close to the neutrality point under appropriately adjusted hydrostatic pressure (Sawicki et al., 1983).

2.6. Charge transfer states

It has been suggested (Benoit à la Guillaume et al., 1992; Dietl, 2008a; Dietl et al., 2002) that if the local potential associated with the magnetic atom is particularly strong, an additional level built up of the host wave functions appears in the gap, even if the TM acts as an isoelectronic impurity, such as Mn in ZnO. Such states are reminiscent of the Zhang-Rice singlet consisting of a hole bound by a Cu ion on oxygen ligands in high T_C superconductors. Actually, the relevant physics is also similar to that encountered in the case of the so-called mismatched semiconductor alloys, for instance Ga(As,N), where isoelectronic N impurities can have bound states in GaAs, particularly under hydrostatic pressure (Kent and Zunger, 2001; Wu et al., 2002). Similarly, in some DMSs, the

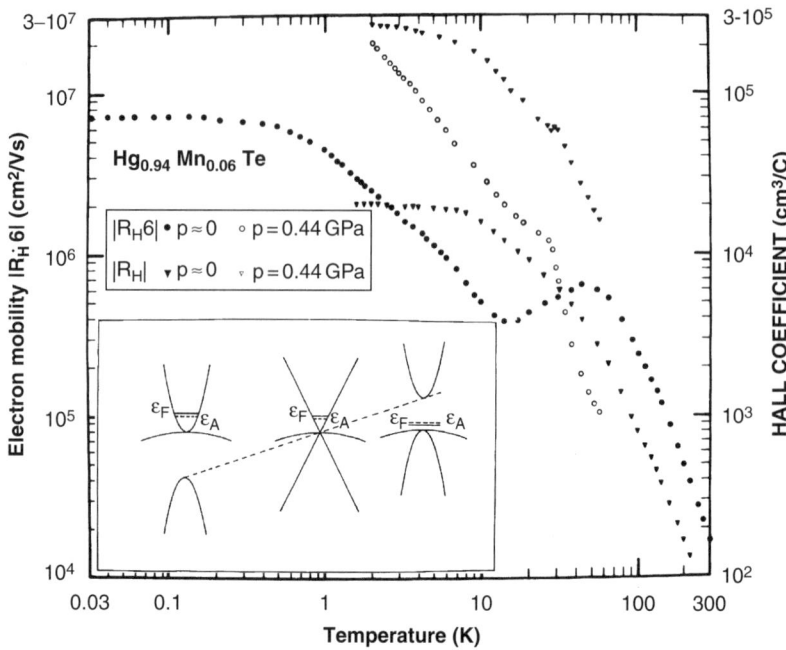

FIGURE 9.3 Temperature dependence of the Hall coefficient $|R_H|$ (triangles) and mobility $|R_H\sigma|$ (dots) without and under hydrostatic pressure p of 0.44 GPa in p-Hg$_{0.94}$Mn$_{0.06}$Te. Inset shows the evolution of the band structure with the pressure that leads (around 0.5 GPa) to the degeneracy of the hole and conduction bands (middle panel). Together with the pinning of the Fermi level by acceptors levels, this leads to a mobility value exceeding 2×10^7 cm^2/Vs below 2 K (full dots). A mobility minimum at 20 K is caused by interband phonon scattering [after Sawicki et al. (1983)].

TM impurity can trap a hole on the ligands, examples being Mn in ZnO and Fe in GaN (Pacuski et al., 2008). So, a charge transfer state appears in the bandgap, despite that the corresponding d-like donor states, Mn^{2+}/Mn^{3+} and Fe^{3+}/Fe^{4+}, respectively, reside in the valence band according to Fig. 9.1.

2.7. Magnetic moments

According to Hund's rule the total spin $S = 5/2$ and the total orbital momentum $L = 0$ for the d^5 shell in the ground state. The lowest excited state d^{*5} corresponds to $S = 3/2$ and its optical excitation energy is about 2 eV. Thus, if there are no interactions between the Mn spins, their magnetization is described by the paramagnetic Brillouin function for $S = 5/2$. Furthermore, spin–orbit interaction and Jahn-Teller effect control positions and splittings of the levels in the case of ions with $L \neq 0$. If the resulting ground state is a magnetically inactive singlet there is no permanent magnetic moment associated with the ion, the case of Fe^{2+}, whose magnetization is of the Van Vleck type at low temperatures.

3. ORIGIN OF EXCHANGE INTERACTIONS BETWEEN CARRIERS AND LOCALIZED SPINS

According to the above discussion, a good starting point for the description of DMSs is the Vonsovskii model, according to which the electron states can be divided into two categories (Dietl, 1981): (i) localized magnetic d or f shells and (ii) extended band states built up of s, p, and sometimes d atomic orbitals. The key aspect of DMSs is the existence of a strong spin-dependent coupling of the effective mass carriers to the localized d electrons, first discovered and examined in (Cd,Mn)Te (Gaj et al., 1978; Komarov et al., 1977) and (Hg,Mn)Te (Bastard et al., 1978; Jaczyński et al., 1978) and reviewed later on (Furdyna and Kossut, 1988). Neglecting non-scalar corrections that can appear for ions with nonzero orbital momentum, $L \neq 0$, this interaction assumes the Kondo form,

$$H_K = \sum_i \left[V\left(\vec{r} - \vec{R}^{(i)}\right) - I\left(\vec{r} - \vec{R}^{(i)}\right) \vec{s}\,\vec{S}^{(i)} \right], \quad (9.1)$$

where $V(\vec{r} - \vec{R}^{(i)})$ is a spin-independent potential change introduced by the TM impurity and $I(\vec{r} - \vec{R}^{(i)})$ is a short-range exchange energy operator between the carrier spin $\vec{s}$ and the TM spin $\vec{S}$ localized at $\vec{R}^{(i)}$. In general, $V(\vec{r} - \vec{R}^{(i)})$ contains a long-range screened Coulomb part, present if the magnetic impurity is not isoelectronic, and a short-range alloy potential $V_s(\vec{r} - \vec{R}^{(i)})$, known as a central-cell correction or a band offset in the impurity and alloy nomenclature, respectively.

We generalize for the spin-dependent case the effective mass approximation that has been successfully employed for a number of semiconductors alloys. When incorporated to the kp scheme, the effect of H_K is described by matrix elements $\langle u_i|V_s|u_i\rangle$ and $\langle u_i|I|u_i\rangle$, where u_i are the Kohn-Luttinger amplitudes of the corresponding band extremum. In the case of carriers at the Γ point of the Brillouin zone in zinc-blende DMSs, there are four relevant matrix elements $V = \langle u_c|V_s|u_c\rangle$ and $\alpha = \langle u_c|I|u_c\rangle$; $W = \langle u_v|V_s|u_v\rangle$ and $\beta = \langle u_v|I|u_v\rangle$ involve s-type and p-type periodic parts of the Bloch wave functions, respectively.

There are two mechanisms contributing to the spin-dependent part of the Kondo coupling I (Bhattacharjee et al., 1983; Dietl, 1981, 1994; Kacman, 2001): (i) the exchange part of the Coulomb interaction between the effective mass and localized electrons, called the potential exchange; (ii) the spin-dependent hybridization between the band and local states, leading via the Schrieffer-Wolf transformation to the so-called kinetic exchange. Importantly, this mechanism contributes also significantly to the magnitude of V_s (Benoit à la Guillaume et al., 1992).

Since in the case of zinc-blende DMSs there is no hybridization between the Γ_6 states and the levels derived from the d shell (e_g and t_2 states), the s–d coupling is determined by the direct potential exchange. The experimentally determined magnitudes are of the order of $\alpha N_0 \approx 0.25$ eV, somewhat reduced comparing to the value deduced from the energy difference between $S1/2$ states of the free singly ionized Mn atom $3d^5 4s^1$, $\alpha N_0 = 0.39$ eV, mostly because of the spread of the electron wave function between both cation and anion sites. In contrast, there is a strong hybridization between Γ_8 and t_2 states, which affects their relative position, and leads to a large magnitude of $|\beta N_0| \approx 1$ eV. If the relevant effective mass state is above the t_{2g} level (the case of, for example, Mn-based DMSs, $\beta < 0$ but otherwise β can be positive [the case of, for example, $Zn_{1-x}Cr_xSe$ (Mac et al., 1993)].

It should be emphasized that the role of hybridization depends crucially on the symmetry of band states and the parity of the local state. Nevertheless, the above scenario holds to a good approximation in the case of the wurtzite DMSs (Arciszewska and Nawrocki, 1986), where the local symmetry only weakly deviates from the zinc-blende case. In contrast, the description of the carrier–spin interaction has had to be reconsidered entirely in, for example, rock-salt DMSs, such as (Pb,Mn)Te (Dietl, et al., 1994) as well as when RE atoms are involved, as in (Pb,Eu)Se (Dietl, 1994) and (Ga, Gd) N (Dalpian and Wei, 2005; Liu et al., 2008). Finally, we note that appropriately far away from the band extremum, the kp method will cease to work and, additionally, the term $\exp(i\vec{k}\cdot\vec{r})$ in the carrier wave function—responsible for hybridization in magnetically doped metals—will generate additional hybridization with the open magnetic shells (Dalpian and Wei, 2006).

4. EFFECTS OF sp–d(f) EXCHANGE INTERACTIONS

4.1. Splitting of extended states: Weak coupling

The next chapters of this volume describe a number of effects brought about by the presence of the exchange coupling between band carries and localized spins, as given in Eq. (9.1). For instant, giant splitting of bands and excitonic levels, proportional to magnetization of localized spins, results in spectacular optical and magnetooptical properties of II–VI DMSs (Cibert et al., 2008). According to the p–d Zener model (Dietl et al., 2000), to be discussed in detail later on, the exchange splitting of bands accounts also for carrier-mediated ferromagnetic spin–spin interactions leading to many outstanding properties and functionalities of these systems (Gould et al., 2008; Jungwirth et al., 2008; Matsukura et al., 2008).

The effect of magnetic doping on electronic states can often be treated within the time-honored virtual-crystal (VCA) and molecular-field (MFA) approximations, employed so successfully to a range of semiconductor and DMS alloys, such as (Al,Ga)As and (Cd,Mn)Te, respectively (Dietl, 1994). It is obvious theoretically and confirmed experimentally (Gaj et al., 1979) that VCA and MFA can be applied to, for instance (Cd,Mn)Te, as the change in potential introduce by an individual Mn impurity in CdTe is weak and does not produce any bound states. Interestingly, however, VCA and MFA can be applied also to, for example, (Ga,Mn)As despite that a single Mn impurity acts as an acceptor whose levels cannot be obtained from a perturbation theory. The key observation here is that according to the recent insight such a weak coupling approach is valid as long as the total screened potential associated with magnetic impurities and carriers does not produce any bound states (Dietl, 2008a). This explains why the use of VCA and MFA (Dietl et al., 2000; Jungwirth et al., 2008) leads to quantitative correct results in the case of ferromagnetic (Ga,Mn)As films.

In terms of VCA and MFA, that is within the first order perturbation theory for classical spins, the bandgap difference between an $A_{1-x}TM_xB$ alloy and AB compound is $x(V-W)N_0$ where N_0 is the cation concentration. Similarly, $-xN_0\alpha\langle S_z\rangle$ and $-xN_0\beta\langle S_z\rangle$ are the values of spin splitting in the conduction and valence band, respectively, for the temperature and magnetic field-dependent TM spin polarization $\langle S_z(T,\vec{H})\rangle$. In general, however, the spin subband splittings should be obtained from a multi-band kp method or a multi-orbital tight-binding approximation containing terms describing the sp–d(f) exchange as well as the effects of the spin–orbit interaction and the direct influence of the magnetic field $\vec{H}$ on the band states (Furdyna and Kossut, 1988). The magnitude of spin splittings obtained in this way depends in a complex way on the spin polarization, magnetic field, and carrier wave vector even within the VCA and MFA.

4.2. Splitting of extended states: Strong coupling

Figure 9.4 shows values of the p–d exchange integral to βN_0 as determined by photoemission and x-ray absorption (XAS) studies in various DMSs employing a variant of the Anderson impurity model for the

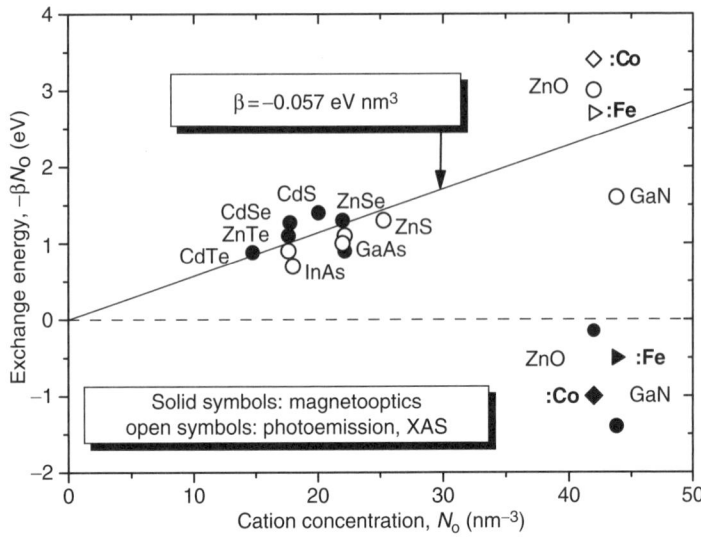

FIGURE 9.4 Energy $N_0\beta$ of the p–d exchange interaction determined experimentally for various DMSs as a function of the unit cell volume $1/N_0$. The values shown by solid symbols were determined from excitonic splittings in the magnetic field in (Cd,Mn)Te (Gaj et al., 1979), (Zn,Mn)Te (Twardowski et al., 1984a), (Zn,Mn)Se (Twardowski et al., 1984b), (Cd,Mn)S (Benoit à la Guillaume et al., 1992), (Zn,Co)O (Pacuski et al., 2006), (Zn,Mn)O (Przeździecka et al., 2006), (Ga,Mn)N (Pacuski et al., 2007), (Ga,Fe)N (Pacuski et al., 2008), and from band splittings in the magnetic field in (Ga,Mn)As (Szczytko et al., 1999a). The empty symbols denote the values evaluated from photoemission and x-ray absorption data (XAS) for (Cd,Mn)VI (Mizokawa and Fujimori, 1993), (Zn,TM)O (Okabayashi et al., 2004), (In, Mn)As (Okabayashi et al., 2002), (Ga,Mn)As (Okabayashi et al., 1999), (Ga,Mn)N (Hwang et al., 2005). Circles denote data for Mn-, triangles for Fe-, and diamonds for Co-doped ZnO and GaN. Solid line corresponds to a constant value of β.

description of the data (Hwang et al., 2005; Mizokawa et al., 2002; Okabayashi et al., 2004). The magnitudes of βN_0 show the phenomenological chemical trend, $\beta N_0 \sim N_0$ (Blinowski et al., 2001; Dietl et al., 2001b, 2002). Moreover, they are in agreement with the values determined from magnetooptical studies for tellurides, selenides, and arsenides within VCA and MFA. Surprisingly, however, the determined values of βN_0 for oxides and nitrides show either opposite sign and/or much reduced amplitude comparing to those stemming from photoemission and XAS as well as expected from the chemical trends. For instance, a detailed examination of the giant spin splitting of free excitons in $Ga_{1-x}Mn_xN$ leads to $\beta N_0 = +1.4 \pm 0.3$ eV (Pacuski et al., 2007). Similarly, the study of bound excitons in $Zn_{1-x}Mn_xO$ implies $|\beta N_0| \approx 0.1$ eV (Przeździecka et al., 2006). The contradiction in question is not limited to Mn-based nitrides and oxides—it is also evident for $Zn_{1-x}Co_xO$, where $\beta N_0 = -3.4$ eV according to XAS and in agreement with the chemical trends (Blinowski et al., 2001), while the investigation of the free exciton splitting results in $\beta N_0 \approx -0.6$ eV or 1 eV, depending on the assumed ordering of the valence band subbands (Pacuski et al., 2006). Particularly relevant in the present context is exciton splittings in (Ga,Fe)N, as in GaN, in contrast to ZnO, the actual ordering of valence subbands is settled, so that the sign of the apparent exchange energy $N_0\beta^{(app)}$ can be unambiguously determined from polarization-resolved magnetooptical spectra. Furthermore, unlike Mn, Fe in GaN is an isoelectronic impurity with the simple d^5 configuration (Bonanni et al., 2007; Malguth et al., 2006) allowing a straightforward interpretation of the data. As shown in Fig. 9.5, all three excitons A, B, and C are resolved in high-quality MOVPE (Ga,Fe)N layers (Bonanni et al., 2007), whose splittings in the magnetic fields point to $N_0\beta = +0.5 \pm 0.2$ eV (Pacuski et al., 2008).

The above findings have been explained by evaluating the splitting of extended states abandoning VCA and MFA (Dietl, 2008a). While, as noted in the previous section, these approximations describe very well the giant exciton splittings in tellurides (Gaj et al., 1979), they have already been called into question in the case of $Cd_{1-x}Mn_xS$ (Benoit à la Guillaume et al., 1992). In this system, unexpected dependencies of the bandgap and of the apparent exchange integral $\beta^{(app)}$ on x, obtained from magnetooptical studies, have been explained by circumventing VCA and MFA either employing a non-perturbative Wigner-Seitz-type model (Benoit à la Guillaume et al., 1992) or generalizing the alloy theory to DMSs (Tworzydlo, 1994). It has been found that the physics of the problem is governed by the ratio of a characteristic magnitude of the total magnetic impurity potential U to its critical value $U_c < 0$ at which a bound state starts to form. In particular, the weak coupling regime, where VCA and MFA apply, corresponds to $U/U_c \ll 1$. By modeling the total potential of the isoelectronic magnetic impurity with a square well of radius b one obtains (Benoit à la Guillaume et al., 1992; Tworzydlo, 1994),

$$U/U_c = 6m^*[W - (S+1)\beta/2]/(\pi^3\hbar^2 b), \qquad (9.2)$$

where the bare valence band offset $WN_0 = dE_v(x)/dx$, S is the impurity spin, and m^* is the effective mass of a particle with the spin $s = 1/2$ assumed to reside in a simple parabolic band.

By inserting to Eq. (9.2) parameters expected for $Cd_{1-x}Mn_x$Te, one obtains Dietl (2008a) $0.20 \lesssim U/U_c \lesssim 0.33$, in agreement with the notion that VCA and MFA can be applied to this system. In the case of $Cd_{1-x}Mn_xS$, the coupling strength increases to $0.77 \lesssim U/U_c \lesssim 1.25$. Hence, in agreement with experimental findings, $Cd_{1-x}Mn_xS$ represents a marginal case, in which the local spin-dependent potential introduced by the

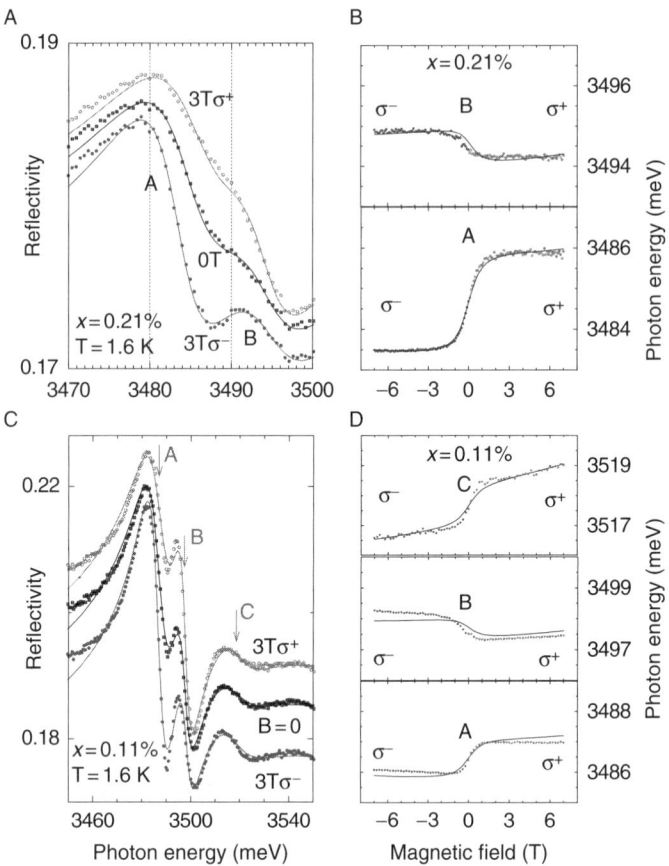

FIGURE 9.5 Reflectivity of $Ga_{1-x}Fe_xN$ in Faraday configuration at 1.6 K for $x = 0.21\%$ (A) and 0.11% (C) where particularly well-resolved excitons A, B, and C are visible in σ^- polarization. (B, D) Field-induced exciton shifts of the excitons determined from the polariton model (points) compared to the expectations of the exciton model (solid lines). The determined values of the exchange parameters are $N_0\beta^{(app)} = +0.5 \pm 0.2$ eV and $N_0\alpha^{(app)} = +0.1 \pm 0.2$ eV [after Pacuski et al. (2008)].

magnetic ion is too weak to bind the hole, but too strong to be described by VCA and MFA (Benoit à la Guillaume et al., 1992; Tworzydlo, 1994). Even greater magnitudes of U/U_c can be expected for nitrides and oxides, as a_0 is smaller, while both m^* and W tend to be larger comparing to compounds of heavier anions. In the case of $Ga_{1-x}Mn_xN$, one obtains $0.96 \lesssim U/U_c \lesssim 1.6$ while for $Zn_{1-x}Mn_xO$, one finds $2.0 \lesssim U/U_c \lesssim 3.3$ Dietl (2008a). Interestingly, the above evaluation of U/U_c remains valid for other transition metal impurities since as long as all d states of t_2 symmetry are singly occupied

(Mn, Fe, Co), the p–d hybridization energy and, hence, βS does not depend on S to a first approximation (Blinowski et al., 2001; Mizokawa and Fujimori, 1997). Although the quoted values of U/U_c are subject of uncertainty stemming from the limited accuracy of the input parameters as well as from the approximate treatment of the hole band structure, it can be concluded that magnetically doped nitrides and oxides are in the strong coupling regime, $U/U_c > 1$.

The effect of magnetic ions on a single band-edge particle ($k = 0$) with the spin $s = 1/2$ has been evaluated (Dietl, 2008a) within a generalized alloy theory developed by Tworzydło (Tworzydlo, 1994). The theory is built for non-interacting, randomly distributed, and fluctuating quantum spins $\vec{S}$ characterized by the field-induced averaged polarization $\langle S_z(T,H) \rangle$ and neglecting the direct effect of the magnetic field H on the band states. The potential of the individual impurities is modeled (Benoit à la Guillaume et al., 1992) by the square-well potential containing both spin-dependent (exchange) and spin-independent (chemical shift) contributions, as specified in Eq. (9.2). Accordingly, the theory can be applied to both magnetic and nonmagnetic alloys. However, its accuracy is questionable if the impurity potential has a complex shape leading to the appearance of resonant states or if the impurity is not isoelectronic, so that the potential contains a long-range Coulomb term. The band-edge particle self-energy $\Sigma_{s_z}(\omega)$ is derived from the Matsubara formalism by summing up an infinite series of diagrams for the irreducible self-energy in the average t-matrix approximation (ATA) (Tworzydło, 1995). As a result of thermal and quantum fluctuations, $\Sigma_{s_z}(\omega)$ contains weighted contributions corresponding to two spin orientations, $\Sigma_0(S)$ and $\Sigma_0(-S-1)$.

The quantity of interest is the spectral density of states,

$$A_{s_z}(\omega) = -\frac{1}{\pi} \text{Im} \frac{1}{\omega + i\gamma - \Sigma_{s_z}(\omega)}, \quad (9.3)$$

whose maxima provide the position of the band-edge $\tilde{E}_0$ in the presence of spin and chemical disorder as a function of $\langle S_z(T, H) \rangle$ and where γ is an extrinsic broadening of band states.

Most of the relevant experiments for magnetically doped nitrides and oxides have been carried out at low concentrations $x \lesssim 3\%$, where the effects of the interactions among localized spins, neglected in the present approach, are not yet important. In this case, $\langle S_z(T, H) \rangle$ can be determined from magnetization measurements or from the partition function of the spin hamiltonian for the relevant magnetic ion. For higher x, it is tempting to take into account the effects of the short-range antiferromagnetic superexchange according to the standard recipe (Dietl et al., 2001b; Gaj et al., 1979), that is, via replacement of x by x_{eff} and T by $T + T_{\text{AF}}$, where $T_{\text{AF}}(x,T) > 0$ and $x_{\text{eff}}(x,T) < x$. However, this procedure is rather inaccurate, as the effect of spin up and spin down impurities does not cancel in the strong coupling case, so that antiferromagnetically aligned pairs give a nonzero contribution to the shift and broadening of carrier states.

Figure 9.6 a presents a grey scale plot of the spectral density $A_{s_z}(\omega)$ as a function of the coupling strength in the absence of spin polarization, $\langle S_z \rangle = 0$, so that the spin degeneracy is conserved, $A_{1/2}(\omega) = A_{-1/2}(\omega)$. In the weak coupling limit, $U/U_c \ll 1$, the band-edge position $\tilde{E}_0$ is obviously given by $-xN_0W$. For higher U/U_c a downward shift of $\tilde{E}_0$ is visible, particularly rapid for $U/U_c > 1$, when the magnetic ion potential is strong enough to bind a hole, even if the magnetic ion is an isoelectronic impurity. This approach shows, therefore, how the charge transfer states, discussed in Sec. 2.6, emerge. The appearance of such a bandgap state in the strong coupling limit has also been revealed within the dynamic mean-field theory (Chattopadhyay et al., 2001), by a numerical diagonalization of the alloy problem (Bouzerar et al., 2007) and tight-binding model (Popescu et al., 2006). The small magnetic polaron formed in this way, reminiscent of the Zhang-Rice singlet, will be built also of k states away from the Brillouin zone center (Dietl et al., 2002). The interpretation of the gap level in terms of the charge transfer state implies, in line with the photoemission findings, that d levels of TM impurities reside at about

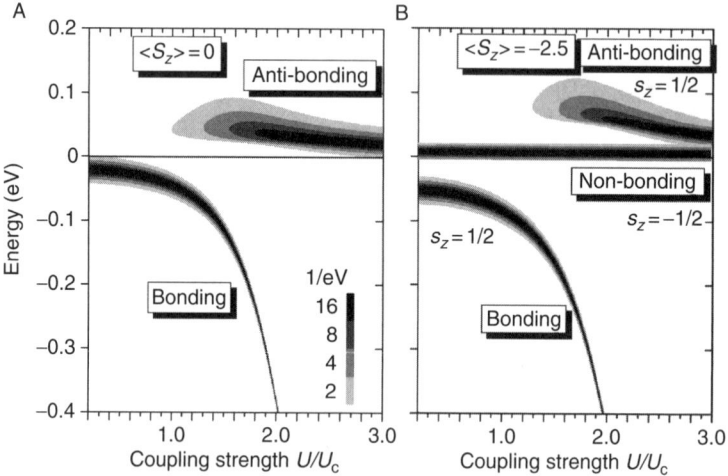

FIGURE 9.6 Energy distribution (grey scale) of the spectral density of states $A_{s_z}(\omega)$ at the band-edge ($k = 0$) as a function of the coupling strength U/U_c between the spin 1/2 particle and the system of unpolarized (A) and spin-polarized (B) randomly distributed magnetic impurities. The magnitude of U/U_c is changed by the range of the impurities' potential b. The remaining material parameters are taken as $S = 5/2$, $x = 1\%$, $m^* = 1.3m_0$, $W = 0.05$ eV nm^3, $\beta = -0.057$ eV nm^3, and $\gamma = 3$ meV [after Dietl (2008a)].

2 eV lower than implied by DFT computations within LSDA. This issue is now being considered by many groups, and various post-DFT computation schemes are being proposed to improve the reliability of first principles methods (Rinke et al., 2007).

The presence of the bandgap hole traps introduced by the magnetic ions will lead to strong hole localization in p-type samples. In the n-type case, in turn, these charge transfer states should be visible in photoionization experiments. Indeed, if there is no hole on a magnetic impurity, a photon can transfer an electron from surrounding bonds to the conduction band leaving a trapped hole behind, $d^n \to d^n + h + e$. The corresponding subgap absorption has indeed been observed in both n-$Ga_{1-x}Mn_xN$ (Graf et al., 2003; Wolos et al., 2004a) and $Zn_{1-x}Mn_xO$ (Kittilstved et al., 2006) but assigned in those works to $d^5 \to d^4 + e$ transitions. In $Ga_{1-x}Mn_xN$ samples, in which donors were partly compensated, intra-center excitations corresponding to this level were also detected at 1.4 eV (Korotkov et al., 2002) and analyzed in considerable details (Marcet et al., 2006; Wolos et al., 2004a,b), again assuming that the d^4 state is probed. While symmetry considerations cannot discriminate between the $d^5 + h$ and d^4 many-electron configurations, the larger crystal-field splitting and the smaller Huang-Rhys factor in $Ga_{1-x}Mn_xN$, compared with their values in (II, Cr) VI compounds (Marcet et al., 2006; Wolos et al., 2004b) point to a relatively large localization radius expected for the $d^5 + h$ case. The reader is referred to the chapter devoted to impurity levels for further details on this disputed issue (Wolos and Kaminska, 2008).

According to Fig. 9.6A, when U/U_c increases beyond one, the spectral density is gradually transferred from the bonding state discussed above to an anti-bonding state appearing above the band-edge energy E_0 expected within VCA and MFA. This state is identified with the actual valence band-edge for $U/U_c > 1$, whose position $\tilde{E}_0$ determines, for example, the onset of interband optical transitions and the free exciton energy.

As shown in Fig. 9.6B, the saturating magnetic field, such that $\langle S_z(T,H) \rangle = -2.5$, produces a downward and upward shift of the bonding and anti-bonding states (which correspond to $s_z = 1/2$), respectively, and leads to the appearance of a non-bonding state ($s_z = -1/2$), whose energy is virtually independent of U/U_c. Remarkably, the latter resides below the anti-bonding level, which means that the character of the apparent p–d exchange changes from antiferromagnetic for $U/U_c < 1$ to ferromagnetic for $U/U_c > 1$. Interestingly, this sign reversal of spin splitting of extended states, appearing when the coupling strength is large enough to produce bound states, can also be inferred from Fig. 9.1 of Bouzerar et al. (2007), where the alloy hamiltonian was diagonalized numerically. Importantly, according to both experimental and theoretical studies, the giant splitting of extended states exhibits a linear dependence on magnetization of localized spins also in the strong coupling limit (Dietl, 2008a), which allows to describe the optical data in terms of an apparent exchange integral $\beta^{(app)}$.

Thus, the generalized alloy theory explains qualitatively the surprising sign and magnitude of $\beta^{(app)}$ stemming from magnetooptical studies on nitrides and oxides, as quoted above. Furthermore, in view of this development, it appears understandable why the value of $N_0(\beta - \alpha) = 2.3 \pm 0.6$ was derived from excitonic magnetoreflectivity spectra for $Ga_{1-x}Mn_xAs$ in the strongly localized limit ($x < 0.001$) (Szczytko et al., 1996): in this regime, Mn bound states produced partly by the Mn Coulomb potential are not washed out by many body effects. Accordingly, the exchange splitting of the valence band states assumes an anti-crossing character resulting in the ferromagnetic sign of $\beta^{(app)}$.

The generalized alloy theory may also serve for describing nonmagnetic dilute alloys such as $GaAs_{1-x}N_x$, where isoelectronic N impurities form bound states, particularly under hydrostatic pressure. The relevant experimental results for such mismatched alloys are often analyzed in terms of the so-called band anti-crossing model (Wu et al., 2002). It is assumed within this approach that the impurities introduce a resonant level and that the coupling strength of this level to the relevant band is proportional to x. The resulting band diagram is then similar to that of Fig. 9.6A. The non-perturbative theory of the carrier coupling to a system of randomly distributed potential wells exposed above provides a microscopic explanation how the anti-crossing behavior and related properties can emerge, even if the single impurity potential does not introduce a resonant level.

4.3. Spin-disorder scattering

A difference between the full potential introduced by magnetic ions and that taken into account within the VCA and MFA leads to alloy and spin disorder scattering by spatial and thermodynamic fluctuations of magnetization. The generalized alloy theory presented above describes the corresponding shift and broadening of the band-edge. The generalization of the theory for nonzero k-vectors would allow to find out the effect of the magnetic impurities on the carrier effective mass and scattering rate for an arbitrary large magnitude of the coupling.

In the weak coupling limit, according to the fluctuation–dissipation theorem, the energy shift and scattering rate brought about by thermodynamic magnetization fluctuations are proportional to $T\chi(T)$, where $\chi(T)$ is the magnetic susceptibility of the localized spins (Dietl, 1994; Dietl et al., 1991; Rys et al., 1967). Except to the vicinity of ferromagnetic phase transitions, a direct contribution of spin-disorder scattering to momentum relaxation turns out to be small. In contrast, this scattering mechanism controls the spin lifetime of effective mass carriers in DMSs, as evidenced by studies of universal conductance fluctuations (Jaroszyński et al., 1998), line-width of spin–flip Raman scattering (Dietl et al., 1991), and optical pumping efficiency (Krenn et al., 1989). Particularly important is spin-disorder scattering in the vicinity of the metal-to-insulator transition (MIT), where its role is enhanced by fluctuations in the local density of states. This issue is discussed in Section. 8.4.

4.4. Kondo effect

So far we have considered spin-disorder scattering in one-electron approximation disregarding a possibility that the s,p–d interaction can result in the formation of a spin cloud of carriers screening the impurity spin at low temperatures. The Kondo effect, extensively studied in the case of dilute magnetic metals, exists only for an antiferromagnetic coupling between carriers and localized spins. Thus, the effect can, in principle, be expected for the Fermi sea of band holes coupled to localized impurity spins. In general, however, the Kondo temperature,

$$T_K \approx |\varepsilon_F| \exp[-3/(|\beta|\rho_F)]/k_B, \qquad (9.4)$$

is much lower in semiconductors than in metals because of smaller magnitudes of both Fermi energy $|\varepsilon_F|$ and one-particle density-of-states at the Fermi level for spin excitations, ρ_F. Accordingly, the observation of the Kondo effect would require DMS samples in which from the one hand the hole density is large enough to preclude localization phenomena to show up and, on the other, the density of magnetic impurities is sufficiently small to insure that the Curie-Weiss temperature describing the competing effect of the superexchange or the carrier-mediated interactions, is much lower than T_K.

4.5. Magnetic polarons

Bound magnetic polaron (BMP), that is a bubble of spins ordered ferromagnetically by the exchange interaction with an effective mass carrier in a localized state, modifies optical, transport, and thermodynamic properties of otherwise paramagnetic DMSs. The spin bubble appears not only inside the localization radius of an occupied impurity or quantum dot state but also around a trapped exciton, as the polaron formation time is typically shorter than the exciton lifetime (Dietl et al., 1995). The BMP formation enhances the binding energy and reduces the localization radius. The resulting spontaneous carrier spin splitting is proportional to the magnitude of local magnetization, which is built up by two effects: the molecular field of the localized carrier and thermodynamic fluctuations of magnetization (Dietl, 1983, 1994; Dietl and

Spałek, 1982, 1983). The fluctuating magnetization leads to dephasing and enlarges width of optical lines.

Typically, in paramagnetic 2D and 3D systems, the spins alone cannot localize itinerant carriers but in the 1D case the polaron is stable even without any pre-localizing potential (Benoit à la Guillaume, 1993). In contrast, a free magnetic polaron—a delocalized carrier accompanied by a traveling cloud of polarized spins—is expected to exist only in magnetically ordered phases. This is because coherent tunneling of quasi-particles dressed by spin polarization is hampered, in disordered magnetic systems, by a smallness of quantum overlap between magnetizations in neighboring space regions. Interestingly, theory of BMP can readily be applied for examining effects of the hyperfine coupling between nuclear spins and carriers in localized states, imposed by the impurity potential or the confinement in a quantum dot.

5. EXCHANGE INTERACTIONS BETWEEN EFFECTIVE MASS CARRIERS

5.1. Intra-band exchange interaction

Spin phenomena associated with the presence of magnetic ions are often entangled in DMSs with sizable exchange interactions between effective mass carriers. In particular, the exchange interactions within the carrier liquid, which—according to the Stoner model—account for the ferromagnetism of transition metals, assist the p–d Zener mechanism in DMSs (Dietl et al., 1997, 2001b; Jungwirth et al., 1999). Quantitatively, the Stoner effect, described by the Fermi liquid Landau parameter A_F, increases the Curie temperature by about 20% in (Ga,Mn)As (Jungwirth et al., 1999) and by a factor of about two in the case of p-(Cd,Mn)Te quantum wells (Boukari et al., 2002; Haury et al., 1997), indicating that the Zener and Stoner mechanisms are equally important in this low-dimensional system. Actually, the quantum Hall ferromagnetism observed in quantum wells of n-(Cd,Mn)Te (Jaroszyński et al., 2002) is driven almost entirely by the exchange interactions of the electrons residing in Landau levels. Interestingly, the same parameter A_F describes the magnitude of spin-sensitive quantum corrections to the conductivity of disordered systems (Altshuler and Aronov, 1985; Dietl et al., 1997; Lee and Ramakrishnan, 1985).

5.2. Electron-hole exchange interaction

A number of crucial information on DMSs come from inter-band optical and magnetooptical experiments. Under these conditions, in addition to the coupling of photo-carriers to TM spins, one may expect spin-dependent effects associated with the electron-hole exchange interaction. The presence of the electron-hole exchange is particularly noticeable in oxides (Pacuski et al., 2006) and nitrides (Pacuski et al., 2007, 2008), where the exciton Bohr radius is small and, thus, the electron-hole interaction affects substantially the spin splitting of exciton lines.

One also may expect the existence of exchange interactions between minority photo-carriers and spin-polarized majority carriers in doped materials. As described in the next section, such interactions have been found to affect the spin splitting of photoelectrons in (Ga,Mn)As with so low Mn concentrations that holes stayed localized on Mn acceptors. *A priori*, however, the exchange interaction between photoelectrons and delocalized holes should take place, too.

5.3. Exchange interactions between electrons and bound holes

The presence of sizable electron-hole exchange, as witnessed by exciton spectroscopy, suggests that one should expect the existence of a spin-dependent coupling between photoelectrons and holes in p-type DMSs. It has been shown that this effect can explain an anomalous sign and magnitude of the apparent s–d coupling reported in series of papers (Myers et al., 2005; Poggio et al., 2005; Stern et al., 2007). In these works comprehensive studies of spin precession of photoelectrons in quantum wells of highly dilute paramagnetic $Ga_{1-x}Mn_xAs$ ($x \leq 0.13\%$) were reported. The description of the electron spin splitting $\hbar\omega_s$ as a function of the magnetic field and temperature was carried out taking into account the contribution from the s–d exchange interaction and the band Zeeman term according to,

$$\hbar\omega_s(T, H) = -xN_0\alpha\langle S_z(T - \Theta_p, H)\rangle + g^*\mu_B H. \quad (9.5)$$

The spin polarization $\langle S_z \rangle$ (T, H) was described by the Brillouin function for $S = 5/2$, where Θ_p together with $N_0\alpha$ and g^* were treated as adjustable parameters. The values of $N_0\alpha$ obtained in this way are shown in Fig. 9.7 as a function of the ground-state subband energy E_e in $Ga_{1-x}Mn_xAs$ quantum well (Stern et al., 2007). There are two interesting findings reveled by these studies. First, the exchange coupling is found to be antiferromagnetic, $N_0\alpha < 0$. Second, the magnitude of the coupling $|N_0\alpha|$ increases with E_e. Actually, the latter can be semi-quantitatively explained (Stern et al., 2007) within the *kp* theory of DMSs. According to this theory, when the energy distance to the bottom of the Γ_6 band increases, the *periodic* part of the electron function acquires a p-type component, for which the coupling to the Mn spins is antiferromagnetic, an effect observed and examined in (Hg,Mn)Te (Bastard et al., 1978) and (Cd,Mn)Te (Merkulov et al., 1999). An additional E_e-dependent contribution may come from hybridization of

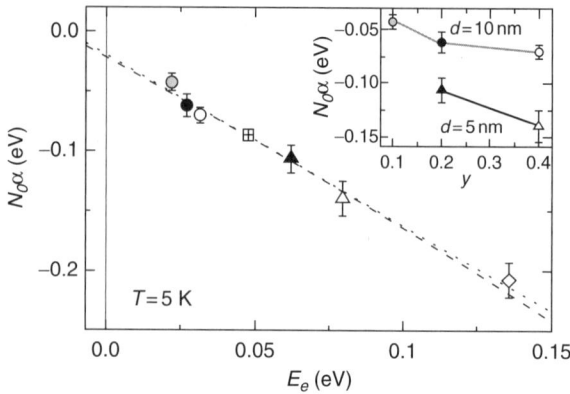

FIGURE 9.7 Exchange energy $N_0\alpha$ plotted as a function of electron kinetic energy for quantum wells $Ga_{1-x}Mn_xAs/Ga_{1-y}Al_yAs$ with width $d = 10$ nm (circles), $d = 7.5$ nm (squares), $d = 5$ nm (triangles), and $d = 3$ nm (diamonds); data for $y = 0.4$ (open symbols) are from Myers et al. (2005), while $y = 0.2$ (black) and $y = 0.1$ (grey) from Stern et al. (2007). The dotted line is the linear approximation and the dashed line is an envelope function calculation after Merkulov et al. (1999). The inset compares $N_0\alpha$ for quantum wells of different barrier height (different y but the same width d [after Stern et al. (2007)].

the carrier *envelop* function with the TM d states (Dalpian and Wei, 2006). In either case, however, the ferromagnetic s–d coupling is expected in the limit $E_e \to 0$, as established unambiguously for II–VI DMSs. Surprisingly, the data (Myers et al., 2005; Poggio et al., 2005; Stern et al., 2007) imply the *antiferromagnetic* $N_0\alpha_{be} = -20 \pm 6$ meV for photoelectrons at the band-edge in (Ga,Mn)As, as seen in Fig. 9.7. This finding has not been explained by the recent theory (Dalpian and Wei, 2006), which leads to $N_0\alpha = 0.3$ eV, as expected for the s–d exchange in tetrahedrally coordinated DMSs, as discussed above.

The starting point of the approach (Śliwa and Dietl, 2007) that elucidates the sign and magnitude of $N_0\alpha_{be}$ is the realization that the density of Mn acceptors in the studied quantum wells of GaAs (Poggio et al., 2005; Stern et al., 2007) was more than one order of magnitude lower than the critical value corresponding to the insulator-to-metal transition and the onset of the hole-mediated ferromagnetism in this system. Furthermore, a relatively high growth temperature resulted in a small concentration of compensating defects. Accordingly, the conduction band photoelectrons interacted with complexes consisting of both Mn and hole spin, $d^5 + h$, which are bound by the electrostatic potential and mutually coupled by a strong antiferromagnetic p–d exchange interaction.

The Mn acceptor complex can be described within the tight-binding approximation (Yakunin et al., 2003) or in terms of the Baldareschi-Lipari spherical model as proposed (Bhattacharjee and Benoit à la Guillaume, 2000) for GaAs:Mn and more recently employed to study impurity band effects (Fiete *et al.*, 2005). Within the latter, the Mn acceptor Hamiltonian for the magnetic field along z direction reads,

$$H = \varepsilon \vec{J} \cdot \vec{S} + \mu_B H(g_{Mn} S_z + g_h J_z), \quad (9.6)$$

where $\varepsilon = 5$ meV is the experimentally determined p–d exchange energy in the Mn acceptor (Bhattacharjee and Benoit à la Guillaume, 2000), $g_{Mn} = 2.0$, and $g_h = 0.75$ is the evaluated Landé factor of the hole bound to a Mn acceptor (Śliwa and Dietl, 2007). From the corresponding density matrix $\rho = \exp[-H/(k_B T)]$, the magnitudes of $\langle S_z \rangle_{T,H}$ and $\langle J_z \rangle_{T,H}$ are determined. Within the MFA, the exchange splitting of the conduction band-edge becomes

$$\hbar \omega_s(T, H) = x N_0 \left[-\alpha \langle S_z \rangle_{T,H} + J_{eh} \langle J_z \rangle_{T,H} \right] + g^* \mu_B H, \quad (9.7)$$

where the second term arises from the coupling between the electron spin and the hole angular momentum characterized by the s–p exchange energy $N_0 J_{eh}$.

In general, a short-range J_{eh}^{SR} term and a long-range J_{eh}^{LR} term contribute to the electron-hole exchange (Pikus and Bir, 1971). The form of corresponding exchange integrals for the case of a band-edge electron and a hole bound to a Mn ion were derived (Śliwa and Dietl, 2007), and found to be described by parameters available from previous studies of the free excitons. In particular, employing the known values of the exchange (Blackwood *et al.*, 1994) and the longitudinal–transverse splitting (Ekardt, 1977) of the free excitons in GaAs, one obtains (Śliwa and Dietl, 2007) $J_{eh}^{SR} = -0.28$ eV and $J_{eh}^{LR} = -0.23$ eV, implying $J_{eh}^{SR} = -0.51$ eV. For these values, the s–p-exchange dominates over the s–d interaction, which explains apparently antiferromagnetic coupling in the weakly compensated GaAs:Mn.

Adopting the above value of $N_0 J_{eh}$ eV and $N_0 \alpha = 0.219$ eV one obtains the field dependence of electron spin splitting shown in Fig. 9.8. We recall that the data on the photoelectron precession frequency (Myers *et al.*, 2005; Poggio *et al.*, 2005; Stern *et al.*, 2007) were interpreted neglecting the presence of the bound holes ($J_{eh} = 0$) as well as by treating both $N_0 \alpha$ and T in the Brillouin function for $S = 5/2$ as adjustable parameters (Myers *et al.*, 2005; Poggio *et al.*, 2005; Stern *et al.*, 2007). Proceeding in the same way one can describe our theoretical results very well with $N_0 \alpha^{(app)} = -20$ meV, $T_{eff} = 22$ K, as shown by dashed line in Fig. 9.8. Thus, the theory (Śliwa and Dietl, 2007) explains why the small antiferromagnetic apparent exchange energy $N_0 \alpha^{(app)} = -20 \pm 6$ meV and enhanced temperature $T_{eff} = 20 \pm 10$ K were found experimentally (Poggio *et al.*, 2005; Stern *et al.*, 2007).

In view of the theory in question, it would be remarkable to perform similar experiments on nonmagnetic p-type semiconductors on the

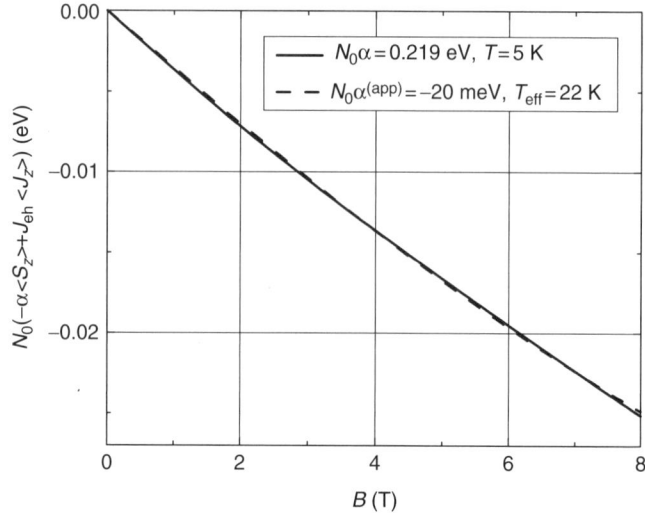

FIGURE 9.8 Theoretical values of the electron spin-splitting energies $\hbar\omega_s(B)/x$ (solid line) computed as a function of the magnetic field at 5 K. Dashed line represents fitting to the solid line obtained by treating the apparent s–d exchange energy $N_0\alpha^{(app)}$ and temperature T_{eff} as adjustable parameters within the model that neglects the presence of the electron-hole exchange interaction ($J_{eh} = 0$) [after Śliwa and Dietl (2007)].

insulating side of the insulator-to-metal transition, where a large exchange splitting of the conduction band by the bound holes is predicted by the present theory. It would also be interesting to put forward an *ab initio* approach capturing such an effect.

When the bound hole concentration is diminished by donor compensation, the relative importance of the s–p exchange decreases. This can be the case of a $Ga_{1-x}Mn_xAs$ sample with $x = 0.1\%$, where $N_0\alpha^{(app)} = +23$ meV, according to spin–flip Raman scattering (Heimbrodt *et al.*, 2001). Even a lower value $|N_0\alpha| = 14 \pm 4$ meV was found by analyzing the effect of the electrons on the Mn longitudinal relaxation time T_1 in n-$Ga_{1-x}Mn_xN$ with $x \leq 0.2\%$ (Wołoś *et al.*, 2003). The interpretation of the data was carried out (Wołoś *et al.*, 2003) neglecting possible effects of the relaxation-time bottleneck (Barnes, 1981), which increases the apparent T_1. It can be shown, however, that for the expected magnitudes of electron spin–flip scattering times in wurtzite GaN:Mn (Majewski, 2005; Sawicki *et al.*, 1986), this effect leads to an underestimation of the $|N_0\alpha|$ by less than 15%. On the contrary, it was found earlier (Śliwa and Dietl, 2005) that the presence of positively charged donors shifts the electron wave function away from negatively charged Mn acceptors, which results in a rather strong reduction in the magnitude of the apparent s–d exchange integral in the relevant range of Mn concentrations in n-(Ga,Mn)N and compensated (Ga,Mn)As (Śliwa and Dietl, 2005).

5.4. Exchange interactions between band and bound holes

As described above, there is a sizable effect of the exchange interaction between photoelectrons and holes residing on Mn acceptors in (Ga,Mn) As on the insulator side of the MIT. Similarly, a spin-dependent coupling may exist between photoholes and bound holes in these systems, contributing to the observed magnitude of magnetization-dependent free exciton splitting in (Ga,Mn)As (Szczytko *et al.*, 1996) and (Ga,Mn)N (Pacuski *et al.*, 2007). It has been suggested that this p–p coupling might be ferromagnetic (Szczytko *et al.*, 1996) but certainly further studies are necessary to quantify the effect.

6. MODELS OF FERROMAGNETIC SPIN–SPIN INTERACTIONS IN SEMICONDUCTORS

6.1. Superexchange

As a result of the aforementioned sp–d exchange interaction, the electrons residing in the sp bands are either attracted to or repulsed by a given magnetic ion depending on their spin orientation. This results in a spatial separation of spin-down and spin-up electrons if the bands are entirely occupied, as in insulators or intrinsic semiconductors. Such a separation leads clearly to an antiferromagnetic interaction between neighboring localized spins, a mechanism known as superexchange. Indeed, in the absence of holes, localized spins are antiferromagnetically coupled in DMSs.

However, the case of europium chalcogenides (e.g., EuS) and chromium spinels (e.g., $ZnCr_2Se_4$) implies that superexchange is not always antiferromagnetic and that ferromagnetism is not always related to the presence of free carriers, albeit despite a large magnetic ion concentration the Curie temperature T_C does not exceed 100 K in these compounds. In the case of rock-salt Eu compounds, there appears to be a competition between antiferromagnetic cation–anion–cation and ferromagnetic cation–cation superexchange (Wachter, 1979). The latter can be traced back to the ferromagnetic s–f coupling, and the presence of s–f hybridization, which is actually stronger than the p–f hybridization due to symmetry reasons (Dietl, 1994; Wachter, 1979). In such a situation, the lowering of the conduction band associated with the ferromagnetic order enhances the energy gain due to hybridization. The Cr-spinels represents the case, in which the d orbitals of the two cations are not coupled to the same p orbital, which results in a ferromagnetic superexchange. In fact, a theoretical suggestion has been put forward that superexchange in Cr-based and V-based II–VI compounds can lead to a weak ferromagnetic spin–spin interactions (Blinowski *et al.*, 1996).

6.2. Double exchange

This mechanism operates if the width of the carrier band is smaller than the exchange energy I, a situation expected for bands formed from d states. Here, spin ordering facilitates carrier hopping over the d states, so that the ferromagnetic transition is driven by the lowering of the carrier energy due to an increase in V. Accordingly, in such systems spin ordering is always accompanied by a strong increase in the conductivity, an effect leading to the so-called colossal magnetoresistance. This is the case of manganites, such as (La, Sr) MnO_3, where Sr doping introduces holes in the Mn d band. Their ferromagnetic order, surviving up to 350 K, is brought by the double-exchange interaction involving on-site Hund's ferromagnetic spin coupling and hopping of d electrons between neighbor Mn^{3+} and Mn^{4+} ions.

6.3. Zener/RKKY models

When the carrier band is partly filled, the sp–d(f) interaction implies the appearance of spin-polarized carrier clouds around each localized spin in extrinsic DMSs. Since the spins of all carriers can assume the same direction if the band is unfilled, a ferromagnetic ordering can emerge, as noted by Zener in the 1950s in the context of magnetic metals. This ordering can be viewed as driven by lowering of the carriers' energy associated with their redistribution between spin subbands split in energy apart by the exchange interaction. A more detailed quantum treatment indicates, however, that the sign of the interaction between localized spin oscillates with their distance according to the celebrated Ruderman-Kittel-Kasuya-Yosida (RKKY) model. Nevertheless, the RKKY and Zener models lead to the same values of T_C in the mean-field approximation (Dietl *et al.*, 1997), which should be valid as long as the carrier concentration is smaller than that of the localized spins.

In europium and chromium chalcogenides discussed above, the presence of carriers can affect T_C but is not necessary for the appearance of the ferromagnetic order. In contrast, in the case of Mn-based IV–VI (Story *et al.*, 1986), III–V (Ohno *et al.*, 1992, 1996), and II–VI (Ferrand *et al.*, 2001; Haury *et al.*, 1997) DMSs the ferromagnetism can be observed provided that the hole concentration is sufficiently high. Particularly fascinating is the case of $Ga_{1-x}Mn_xAs$, where T_C reaches 180 K for x as low as 8% (Olejník *et al.*, 2008).

7. p–d ZENER MODEL OF CARRIER-MEDIATED FERROMAGNETISM

7.1. The model

According to the p–d Zener model (Dietl et al., 2000), the exchange splitting of the valence band accounts for carrier-mediated ferromagnetic interactions between diluted localized spins. Within this model spin polarization of localized magnetic moments produces a spin splitting and, thus, a redistribution of holes between spin subbands. At sufficiently low temperatures, the resulting lowering of the holes' energy overcompensates an increase in the spin-free energy associated with short-range superexchange and spin entropy, leading to the ferromagnetic order, as shown schematically in Fig. 9.9. At the same time, owing to small values of the s–d exchange integral and density of states, no ferromagnetic order induced by electrons is expected in DMSs above 1 K (Dietl et al., 1997). By the same token, small values of sp–f exchange integrals (Dietl, 1994) make the carrier-mediated ferromagnetism of DMSs containing randomly distributed rare earth impurities rather improbable.

Interestingly, a variant of the p–d Zener model explains the origin of ferromagnetism in double perovskite compounds, such as Sr_2CrReO_6, where the magnitudes of T_C attain 625 K, despite that the distance between localized spins is as large as 0.6–0.7 nm (Serrate et al., 2007). This shows a potential of this mechanism to support high-temperature ferromagnetism, possibly also in DMSs, where the distance between *second* nearest neighbor cations is smaller than 0.5 nm in GaN and ZnO.

There are seven main paradigms behind the applicability of the p–d Zener model for the description of ferromagnetism in tetrahedrally coordinated DMSs containing the coupled systems of carriers and randomly distributed substitutional TM impurities (Dietl et al., 2000):

- Electrons that occupy levels derived from the open d-shells give rise to localized spins, and do not contribute to the Fermi volume and charge transport.
- Since exchange integrals involve overlaps of the wave functions, the presence of delocalized or weakly localized carriers is necessary for the existence of sufficiently long-range couplings between diluted spins. In other words, the ferromagnetism of DMSs vanishes abruptly when the carriers become strongly localized.
- The p–d coupling (in contrast to the s–d coupling) is strong enough to result in hole-mediated ferromagnetic interactions that can overcompensate antiferromagnetic superexchange. However, the strong p–d coupling shifts the MIT and, therefore, the onset of ferromagnetism to high hole densities.

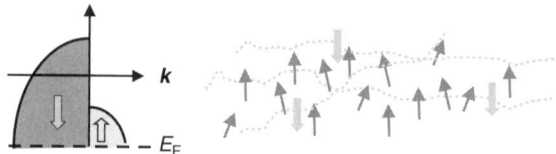

FIGURE 9.9 Pictorial presentation of carrier-mediated ferromagnetism in p-type DMSs, a model proposed originally by Zener for metals. Owing to the p–d exchange interaction, ferromagnetic ordering of localized spins (smaller arrows) leads to the spin splitting of the valence band. The corresponding redistribution of the carriers between spin subbands lowers energy of the holes, which at sufficiently low temperatures overcompensates an increase of the free energy associated with a decrease of Mn entropy. (See Color Plate.)

- Similarly to other thermodynamic properties, ferromagnetic characteristics do not show any critical behavior across the MIT and can be, to a certain approximation, described by disorder-free models even at criticality. This is in contrast to transport phenomena whose behavior is dominated by quantum localization and correlation effects in the carrier density range relevant to ferromagnetism.
- To quantify relevant ferromagnetic properties, such as T_C, magnetic anisotropy and stiffness, the peculiarities of the valence band, including relativistic effects, should be carefully taken into account, for example, via a multi-band kp method or multi-orbital tight-binding approximation.
- Owing to a long-range character of spin–spin interactions as well as large magnitudes of magnetic anisotropy and spin stiffness specific to the coupling mediated by holes, the mean-field approximation describes adequately materials properties.
- Because of self-compensation effects and solubility limits specific to doped semiconductor alloys, the actual incorporation and distribution of magnetic ions have to be carefully assessed.

It is convenient to apply the p–d Zener model by introducing the functional free energy density, $\mathcal{F}[\vec{M}(\vec{r})]$. The choice of the local magnetization $\vec{M}(\vec{r})$ as an order parameter means that the spins are treated as classical vectors, and that spatial disorder inherent to magnetic alloys is neglected. In the case of magnetic semiconductors $\mathcal{F}[\vec{M}(\vec{r})]$ consists of two terms, $\mathcal{F}[\vec{M}(\vec{r})] = \mathcal{F}_s[\vec{M}(\vec{r})] + \mathcal{F}_c[\vec{M}(\vec{r})]$, which describe, for a given magnetization profile $\vec{M}(\vec{r})$, the free energy densities of the Mn spins in the absence of any carriers and of the carriers in the presence of the Mn spins, respectively. A visible asymmetry in the treatment of the carries and of the spins corresponds to an adiabatic approximation: the dynamics of the spins in the absence of the carriers is assumed to be much slower than that of the carriers. Furthermore, in the spirit of the virtual-crystal and molecular-field approximations, the classical continuous field $\vec{M}(\vec{r})$ controls the effect of the spins upon the carriers. Now, the thermodynamics of the system is described by the partition function Z, which can be obtained by a functional integration of the Boltzmann factor $\exp(-\int d\vec{r}\, \mathcal{F}[\vec{M}(\vec{r})]/k_B T)$ over all magnetization profiles $\vec{M}(\vec{r})$, an approach developed for bound magnetic polarons (Dietl, 1983; Dietl and Spałek, 1983), and directly applicable for spin physics in quantum dots as well. In the mean-field approximation, which should be valid for spatially extended systems and long-range spin–spin interactions, a term corresponding to the minimum of $\mathcal{F}[\vec{M}(\vec{r})]$ is assumed to determine Z with a sufficient accuracy.

If energetics is dominated by spatially uniform magnetization $\vec{M}$, the spin part of the free energy density in the magnetic field $\vec{H}$ can be written in the form (Świerkowski and Dietl, 1988)

$$\mathcal{F}_S[\vec{M}] = \int_0^{\vec{M}} d\vec{M}_0 \vec{h}(\vec{M}_0) - \vec{M}\vec{H}. \quad (9.8)$$

Here, $\vec{h}(\vec{M}_0)$ denotes the inverse function to $\vec{M}_0(\vec{h})$, where $\vec{M}_0$ is the available experimentally macroscopic magnetization of the spins in the absence of carriers in the field h and temperature T. In DMSs, it is usually possible to parameterize $M_0(h)$ by the Brillouin function $B_S(T, H)$ that takes the presence of intrinsic short-range antiferromagnetic interactions into account. Near T_C and for $H = 0$, M is sufficiently small to take $M_0(T,h) = \chi(T)h$, where $\chi(T)$ is the magnetic susceptibility of localized spins in the absence of carriers.

Under these conditions,

$$\mathcal{F}_S[M] = M^2/2\chi(T), \quad (9.9)$$

which shows that the increase of $\mathcal{F}_S$ with M slows down with lowering temperature, where $\chi(T)$ grows. Turning to $\mathcal{F}_c[M]$ we note that owing to the giant Zeeman splitting of the bands proportional to M, the energy of the carriers, and thus $\mathcal{F}_c[M]$, decreases with $|M|$, $\mathcal{F}_c[M] - \mathcal{F}_c[0] \sim -M^2$. Accordingly, a minimum of $\mathcal{F}[M]$ at nonzero M may develop in $H = 0$ at sufficiently low temperatures signalizing the appearance of a ferromagnetic order.

7.2. Application to DMS films

The present authors and coworkers (Dietl et al., 2000) found that the minimal hamiltonian necessary to describe properly effects of the complex structure of the valence band in tetrahedrally coordinated semiconductors upon $\mathcal{F}_c[M]$ is the Luttinger 6×6 kp model supplemented by the p–d exchange contribution taken in the virtual crystal and molecular field approximations,

$$H_{pd} = \beta \vec{s}\vec{M}/g\mu_B. \qquad (9.10)$$

This term leads to spin splittings of the valence subbands, whose magnitudes—because of the spin–orbit coupling—depend on the hole wave vectors $\vec{k}$ in a complex way even for spatially uniform magnetization $\vec{M}$. It would be technically difficult to incorporate such effects to the RKKY model, as the spin–orbit coupling leads to non-scalar terms in the spin–spin Hamiltonian. At the same time, the indirect exchange associated with the virtual spin excitations between the valence subbands, the Bloembergen-Rowland mechanism, is automatically included. The model allows for strain, confinement, and was developed for both zinc blende and wurtzite materials (Dietl et al., 2001b). Furthermore, the direct influence of the magnetic field on the hole spectrum can be taken into account (Dietl et al., 2001b; Śliwa and Dietl, 2006). Carrier–carrier spin correlation is described by introducing a Fermi-liquid-like parameter A_F (Dietl et al., 1997; Haury et al., 1997; Jungwirth et al., 1999), which enlarges the Pauli susceptibility of the hole liquid. No disorder effects are taken into account on the ground that their influence on thermodynamic properties is relatively weak except for strongly localized regime. Having the hole energies, the free energy density $\mathcal{F}_c[\vec{M}]$ is evaluated according to the procedure suitable for Fermi liquids of arbitrary degeneracy.

By minimizing $\mathcal{F}[\vec{M}] = \mathcal{F}_S[\vec{M}] + \mathcal{F}_c[\vec{M}]$ with respect to $\vec{M}$ at given T, H, and hole concentration p, Mn spin magnetization $M(T, H)$ is obtained as a solution of the mean-field equation,

$$\vec{M}(T,H) = x_{eff}N_0 g\mu_B SB_S\left[g\mu_B\left(-\partial \mathcal{F}_c[\vec{M}]/\partial \vec{M} + \vec{H}\right)/k_B(T+T_{AF})\right], \qquad (9.11)$$

where the carrier energy and entropy as well as peculiarities of the valence band structure, such as the presence of various hole subbands, anisotropy, and spin–orbit coupling, are hidden in $\mathcal{F}_c[\vec{M}]$. In this formula, x_{eff} and T_{AF} are the effective concentration of magnetic ions and the Curie-Weiss temperature, respectively, which serve to parameterize the magnetic susceptibility $\tilde{\chi}(T)$ and magnetization $M_0(T, H)$ in the absence of carriers. In the case of (Ga,Mn)As and related compounds, where magnetic ions occupy both substitutional and interstitial positions as well as are partly passivated at the surface, x_{eff} denote the concentration of the substitutional Mn—only such ions are strongly coupled to the valence band holes (Blinowski and Kacman, 2003).

Near the Curie temperature T_C and at $H = 0$, where M is small, we expect $\mathcal{F}_c[M] - \mathcal{F}_c[0] \sim -M^2$. It is convenient to parameterize this dependence by a generalized carrier spin susceptibility $\tilde{\chi}_c$, which is related to the magnetic susceptibility of the carrier liquid according to $\tilde{\chi}_c = A_F(g^*\mu_B)^2 \chi_c$. In terms of $\tilde{\chi}_c$,

$$\mathcal{F}_c[M] = \mathcal{F}_c[0] - A_F\tilde{\chi}_c\beta^2 M^2/2(g\mu_B)^2. \qquad (9.12)$$

By expanding $B_S(M)$ for small M one arrives to the mean-field formula for $T_C = T_F - T_{AF}$, where T_F is given by

$$T_F = x_{\text{eff}} N_0 S(S+1) A_F \tilde{\chi}_c(T_C)\beta^2/3k_B. \qquad (9.13)$$

For an arbitral form of the spin susceptibility $\tilde{\chi}(T)$, the Curie temperature T_C is a solution of the equation,

$$A_F \tilde{\chi}_c(T_C)\tilde{\chi}(T)\beta^2 = 1. \qquad (9.14)$$

The model can readily be generalized to various dimensions d (Aristov, 1997; Dietl et al., 1997) as well as to the case, when $\vec{M}$ is not spatially uniform in the ground state, the case of spin density waves expected in 1D and, perhaps, 2D systems. For a strongly degenerated carrier liquid $|\varepsilon_F|/k_B T \gg 1$, as well as neglecting the spin–orbit interaction

$$\tilde{\chi}_c = A_F \rho_d/4, \qquad (9.15)$$

where ρ_d is the total density-of-states for intra-band charge excitations,

$$\rho_d = \frac{4m^* \sum_n \left(k_F^{(n)}\right)^{d-2}}{(4\pi)^{d/2}\Gamma(d/2)\hbar^2}, \qquad (9.16)$$

which in 1D and 2D depends on the number of the occupied subbands n, whereas in the 3D case, $d = 3$, is given by $\rho = m^*_{\text{DOS}} k_F/\pi^2\hbar^2$. In this case and for $A_F = 1$, T_F assumes the well-known form, derived already in the context of carrier-mediated nuclear ferromagnetism in the 1940s (Fröhlich and Nabarro, 1940). It is worth emphasizing that $\tilde{\chi}_c$ is *not* controlled by the one-particle density of states (involved in, for instance, tunneling), which shows a Coulomb anomaly at the Fermi level evolving into the Coulomb gap in the strongly localized regime.

In general $\tilde{\chi}_c$ has to be determined numerically by computing $\mathcal{F}_c[M]$ for a given band structure and degeneracy of the carrier liquid. The approximation of strong degeneracy (i.e., the neglect of the entropy of the carrier liquid) has been found to be valid in (Ga,Mn)As (Dietl et al., 2001b) but not in the case of p-(Cd,Mn)Te quantum wells, where the carrier entropy lowers T_C at small hole densities (Boukari et al., 2002), the effect visible in Fig. 9.14, to be discussed in the next section.

The same formalism, in addition to T_C and Mn magnetization $M(T, H)$, as discussed above, also provides quantitative information on spin polarization and magnetization of the hole liquid (Dietl et al., 2001b; Śliwa and Dietl, 2006). Furthermore, it can be exploited to describe chemical trends (Dietl et al., 2000, 2001b) as well as micromagnetic (Dietl, 2004), optical (Dietl et al., 2001b; Jungwirth et al., 2006a), and transport properties (Jungwirth et al., 2008) of ferromagnetic DMSs. In particular, a detailed theoretical analysis of anisotropy energies and anisotropy fields in films of (Ga,Mn)As was carried out for a number of experimentally important cases within the p–d Zener model (Abolfath et al., 2001; Dietl et al., 2001b). The cubic anisotropy as well as uniaxial anisotropy under biaxial epitaxial strain were examined as a function of the hole concentration p. Both shape and magneto-crystalline anisotropies were taken into account. The perpendicular and in-plane orientation of the easy axis is expected for the compressive and tensile strain, respectively, provided that the hole concentration is sufficiently small. However, according to theory, a reorientation of the easy axis direction is expected at higher hole concentrations. Furthermore, in a certain concentration range the character of magnetic anisotropy is found to depend on the magnitude of spontaneous magnetization, that is, on the temperature. The computed phase diagram for the reorientation transition compared to the experimental results for a (Ga,Mn)As film is shown in Fig. 9.10. In view that theory is developed with no adjustable parameters the agreement between experimental and computed concentrations and temperature corresponding to the reorientation transition is very good. Furthermore, the computed magnitudes of the anisotropy field H_u (Dietl et al., 2001b) are consistent with the available findings for both compressive and tensile strain. Furthermore, the theory describes correctly the width of stripe domains appearing in films with perpendicular magnetic anisotropy Dietl et al. (2001a).

The issue how various corrections to the mean-field p–d Zener model (Dietl et al., 2000) affect theoretical values of T_C was examined in some details for (Ga,Mn)As (Brey and Gómez-Santos, 2003; Jungwirth et al., 2005; Popescu et al., 2006; Timm and MacDonald, 2005). According to the present insight, the overall picture remains quantitatively valid as long as holes are delocalized or weakly localized. As shown in Fig. 9.11, in this regime the theory describes satisfactorily experimental data for (Ga,Mn)As provided that the effective Mn concentration x_{eff} is carefully assessed.

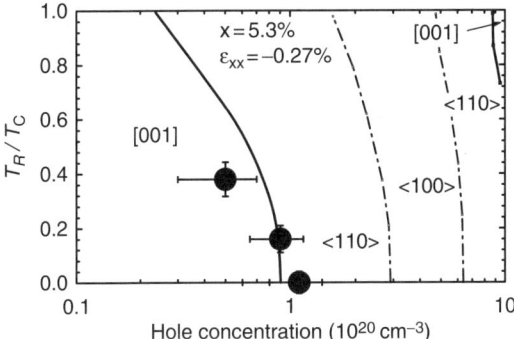

FIGURE 9.10 Experimental (full points) and computed values (thick lines) of the ratio of the reorientation to Curie temperature for the transition from perpendicular to in-plane magnetic anisotropy. Dashed lines mark expected temperatures for the reorientation of the easy axis between ⟨100⟩ and ⟨110⟩ in-plane directions [after Sawicki et al. (2005)].

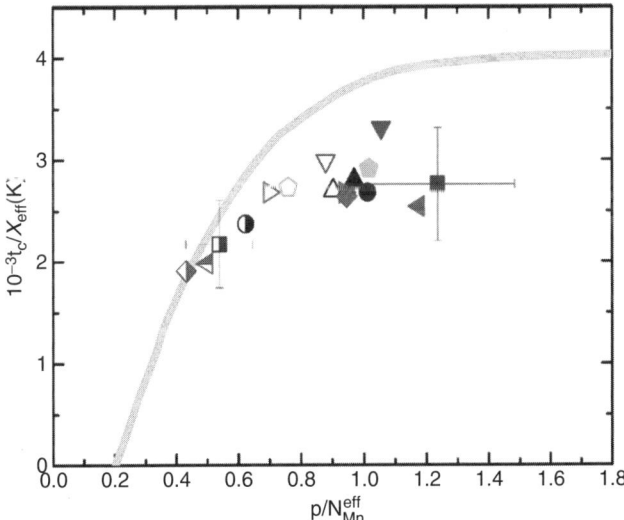

FIGURE 9.11 Experimental Curie temperatures versus hole density p relative to effective concentration of Mn moments, $N_{Mn} = 4x_{eff}/a_0^3$, where x_{eff} is the content of Mn in the substitutional positions and a_0 is the lattice constant. Grey line is theoretical computed within the tight-binding and coherent potential approximations [after Jungwirth et al. (2005)].

7.3. Application to modulated structures of DMSs

The discovery of carrier-induced ferromagnetism in zinc-blende III–V and II–VI compounds has made it possible to consider physical phenomena and device concepts for previously unavailable combinations of quantum structures and magnetism in semiconductors. Indeed, it has already been demonstrated that various modulated structures of (Ga,Mn)As show functionalities relevant for spintronic devices (Gould et al., 2008; Jungwirth et al., 2008; Matsukura et al., 2008).

Because of paramount importance of interfaces as well of Rashba and Dresselhaus terms, spin properties of modulated semiconductor structures cannot be meaningfully modeled employing a standard kp theory. Accordingly, an empirical multi-orbital tight-binding theory of multilayer structures has been developed (Oszwałdowski et al., 2006; Sankowski and Kacman, 2005; Sankowski et al., 2006, 2007). The employed procedure describes properly the carrier dispersion in the entire Brillouin zone and takes into account the presence of magnetic ions in the virtual-crystal and molecular-field approximations. Furthermore, since the phase coherence and spin diffusion lengths are comparable in these devices and, moreover, they are typically longer than the length of the active region, the formulation of spin transport model in terms of the Boltzmann distribution function f for particular spin orientations is not appropriate. Recently, theory that combines a Landauer-Büttiker formalism with tight-binding approximation has been developed (Sankowski et al., 2006, 2007; Van Dorpe et al., 2005). In contrast to the standard kp method (Brey et al., 2004; Petukhov et al., 2002), this theory, in which $sp^3d^5s^*$ orbitals are taken into account, describes properly the interfaces and inversion symmetry breaking as well as the band dispersion in the entire Brillouin zone, so that the essential for the spin-dependent tunneling Rashba and Dresselhaus terms as well as the tunneling via $\vec{k}$ points away from the zone center are taken into account.

The approach in question, developed with no adjustable parameters, provided information on sign of the interlayer coupling (Sankowski and Kacman, 2005), explained experimentally observed large magnitudes of both electron current spin polarization up to 70% in the (Ga,Mn)As/ n-GaAs Zener diode (Van Dorpe et al., 2005) and TMR of the order of 300% in a (Ga,Mn)As/GaAs/(Ga,Mn)As trilayer structure (Sankowski et al., 2006), as shown in Fig. 9.12. Furthermore, theory reproduced a fast decrease of these figures with the device bias as well as it indicated that the magnitude of TAMR should not exceed 10% under usual strain conditions and for hole densities corresponding to the metal side of the MIT (Sankowski et al., 2007). A similar model was employed to examine an intrinsic domain-wall resistance in (Ga,Mn)As (Oszwałdowski et al., 2006).

7.4. Beyond colinear magnetic ground state

A number of effects have been identified, which may lead to deviations from a simple ferromagnetic spin order in carrier-controlled diluted ferromagnetic semiconductors even if the spatial distribution of magnetic ions is uniform. In particular, spin-density waves may appear as the ground state

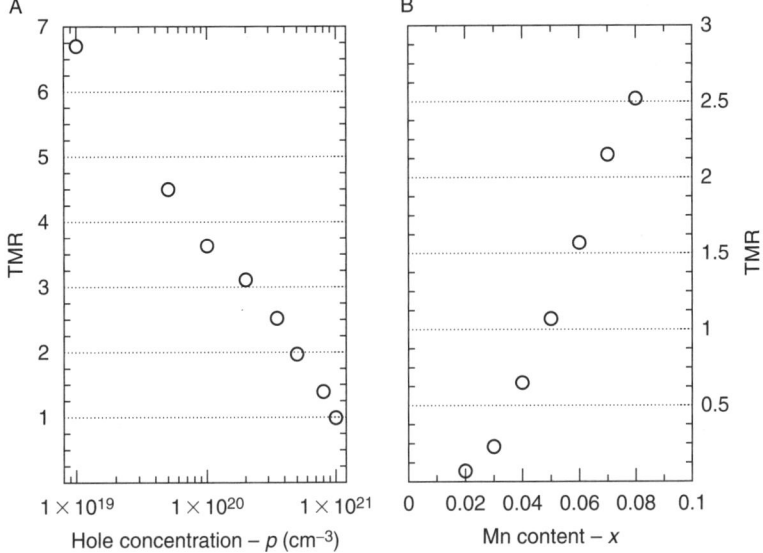

FIGURE 9.12 Difference in resistance for antiparallel and parallel magnetization orientations normalized by resistance for parallel orientation for tunneling structures p-Ga$_{1-x}$Mn$_x$As/GaAs/p-Ga$_{1-x}$Mn$_x$As as a function of (A) hole concentration p (for $x = 0.08$); (B) Mn content x (for $p = 3.5 \times 10^{20}$ cm^{-3}) in the limit of small bias voltage [after Sankowski et al. (2006)].

in the case of 1D and, possibly, 2D systems (Dietl et al., 1999). Another proposal involves canted ferromagnetism stemming from a non-scalar form of spin–spin interactions, brought about by spin–orbit coupling (Zaránd and Jankó, 2002), though a large value of saturation magnetization in (Ga,Mn)As indicates that the effect is not large (Jungwirth et al., 2006b). Finally, a competition between long-range ferromagnetic interactions and intrinsic short-range antiferromagnetic interactions (Kępa et al., 2003) may affect the character of magnetic order (Kechrakos et al., 2005). It appears that the effect is more relevant in II–VI DMSs than in III–V DMSs where Mn centers are ionized, so that the enhanced hole density at closely lying Mn pairs may compensate antiferromagnetic interactions (Dietl et al., 2001b).

8. EFFECTS OF DISORDER AND LOCALIZATION ON CARRIER-MEDIATED FERROMAGNETISM

8.1. Quantum localization

All doped semiconductors undergo a MIT when disorder increases. Empirically, the MIT occurs when an average distance between carriers becomes two and half times greater than the Bohr radius $a_B = \hbar/(2E_I m^*)^{1/2}$, where

E_I is the relevant impurity binding energy (Edwards and Sienko, 1978). This rule indicates that the enhancement of E_I by the magnetic polaron effect (Sawicki et al., 1986) and p–d hybridization (Dietl, 2008a; Dietl et al., 2002) may shift the MIT toward higher hole concentrations in p-type DMSs. More generally, the Anderson-Mott MIT sets-in when $k_F \ell \lesssim 1$, where ℓ is the mean free path calculated within the lowest order perturbation theory, usually smaller than that resulting from the value of conductivity (Altshuler and Aronov, 1985; Fukuyama, 1985; Lee and Ramakrishnan, 1985). Thus, the insulator regime is approached not only when the acceptor density is reduced but also when alloy scattering and/or spin disorder scattering increases or the film is depleted of holes by compensating donors, electrostatic gates, or hole transfer to either surface states or neighbor undoped layers.

Comparing to GaAs doped with shallow acceptors such as Carbon, the Mn impurity introduces a stronger local potential stemming from sizable p–d hybridization. This enlarges the critical hole concentration n_c corresponding to the MIT, perhaps by one order of magnitude, comparing to, for example, GaAs:C. This shift of n_c is presumably smaller in InSb:Mn and InAs:Mn but even larger in GaP:Mn and GaN:Mn where the hole binding energy reaches $E_I \approx 1$ eV due to the short bond length and the resulting sizable p–d hybridization (Dietl, 2008a; Dietl et al., 2002; Wolos and Kaminska, 2008). As emphasized in Sec. 4.2, the strong coupling gives rise to a Zhang-Rice-like bound state even if there is no Coulomb potential associated with the magnetic ion. The corresponding large magnitudes of scattering potentials will shift the MIT to higher hole concentrations, as shown schematically in Fig. 9.13.

It is still a major challenge to describe quantitatively effects of both disorder and carrier–carrier correlation near the Anderson-Mott transition, even in nonmagnetic semiconductors. In this regime the physics of charge transport is dominated by quantum localization phenomena originating from interference of scattered waves and interference of carrier–carrier interaction amplitudes (Altshuler and Aronov, 1985; Belitz and Kirkpatrick, 1994; Dietl, 1994; Fukuyama, 1985; Lee and Ramakrishnan, 1985). These two interference manifestations have to be considered on equal footing. As long as temperature broadening is smaller than the Fermi energy, $k_B T < |E_F|$, the corresponding quantum corrections dominate over the Drude-Boltzmann conductivity in the entirely hole density range relevant to carrier-controlled ferromagnetism and account for the observed disappearance of conductance at the critical point at low temperatures. While a theoretical description of the corresponding critical exponents is feasible, the evaluation of the conductance value is a subject of uncertainty because the magnitude of quantum terms depends on the form of four-point correlation functions ("coperons" and "diffusons") at high momenta and energies, which is unknown and, presumably, not

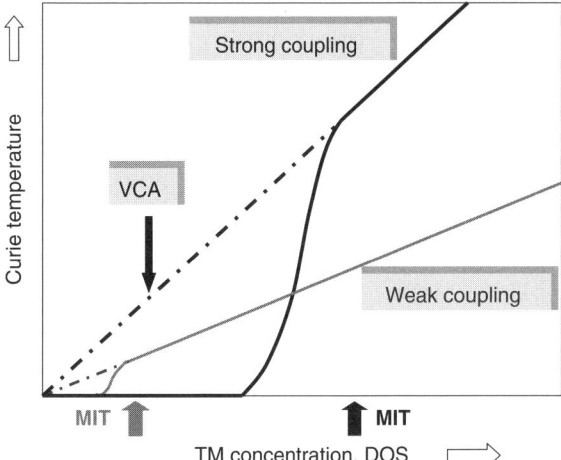

FIGURE 9.13 Schematic dependence of T_C on the magnetic ion concentration and density of hole states at the Fermi level for a weak and a strong coupling. Higher values of T_C are predicted within VCA and MFA for the strong coupling. However, the region where the holes are localized and do not mediate the spin–spin interaction is wider in the strong coupling case [after Dietl (2008a)].

universal. Nevertheless, the presence of these terms can be revealed by their specific dependencies on the dimensionality, magnetic field, spin splitting, spin-dependent scattering, temperature, and frequency, which can be examined quantitatively as they are determined by "cooperon" and "diffuson" poles at low momenta and energies.

As discussed in details elsewhere (Dietl, 2008b), a characteristic positive magnetoconductance of (Ga,Mn)As $\Delta \sigma \sim B^{1/2}$ has been explained quantitatively in terms of the orbital single-particle interference phenomenon (Matsukura et al., 2004). Furthermore, it has been found theoretically that this phenomenon affects the magnitude of anomalous Hall conductance (Dugaev et al., 2001). At the same time, the correlation effects account for a low-temperature downturn of conductance in (Ga,Mn)As (Dietl, 2008b; Honolka et al., 2007; Matsukura et al., 2004; Neumaier et al., 2008), a minimum of conductance at T_c (Dietl, 2008b), and a Coulomb gap in the one-particle density-of-states at the Fermi level (Pappert et al., 2006). We may add that the question how quantum localization affects anomalous Hall effect has already been addressed (Dugaev et al., 2001). In view of the documented importance of the quantum localization effects in carrier-control ferromagnetic semiconductors, a fitting of the a. c. conductivity values by the Drude formula may lead to highly misleading results. Also, it would be interesting to find out how the quantum terms depend on the orientation of magnetization against the direction of the current and crystal axes, in other words, to what extend anisotropic magnetoresistance (Jungwirth et al., 2008) is determined by quantum localization phenomena.

The presence of the quantum effects in charge transport indicates that the localization occurs due to subtle interference of scattering amplitudes originating from a collective effect of many scattering centers on the band carriers. According to this insight, the Curie temperature along with other thermodynamic properties should retain the magnitude evaluated for the band carriers across the MIT (Dietl et al., 2000), a behavior sketched in Fig. 9.13. Guided by this expectation, we now discuss how the carrier-mediated ferromagnetism looks like on the metallic side of the MIT, where disorder can be regarded as weak, at the criticality, and in the strongly localized regime.

8.2. Weak disorder

In materials in question, scattering of band holes is brought about by spin-dependent and spin-independent potentials provided by magnetic and nonmagnetic impurities and defects. Since particular scattering centers can simultaneously be source of Coulomb, spin disorder, and alloy scattering and, moreover, spatial positions of impurities can be correlated, interference between scattering amplitudes should be carefully taken into account when evaluating the mean free path ℓ (Timm, 2003).

Within the standard RKKY model, the carrier-mediated ferromagnetic interactions are limited to spin–spin distances smaller than ℓ (Matsukura et al., 1998). However, such an approach is valid when both ℓ and average distance between the spins are much larger than the inverse of the Fermi wave vector k_F, neither fulfilled in p-type DMSs.

In the framework of the Zener model, effects of disorder can be taken into account by allowing for scattering broadening of the density-of-states entering to $\tilde{\chi}_c$ (Dietl et al., 1997). This broadening, proportional to the scattering rate, leads to band tailing, clearly seen in tunneling and optical experiments, notably in (Ga,Mn)As (Szczytko et al., 2001). The disorder-induced decrease in the magnitude of the density-of-states and the enhanced localization work together to reduce T_C when disorder increases at a given carrier concentration. Phenomenologically, there is a clear relationship between conductance σ at say, 50 K, and T_C of p-type DMSs at constant density of magnetic ions.

Figure 9.14 shows that T_C in modulation-doped p-type (Cd,Mn)Te quantum wells diminishes when decreasing the hole density, the dependence described quite satisfactorily theoretically by allowing for a gaussian broadening of the density-of-states (Boukari et al., 2002). This behavior is also reproduced by Monte-Carlo simulations, in which hole eigen-energies are evaluated at each Monte-Carlo sweep (Kechrakos et al., 2005),

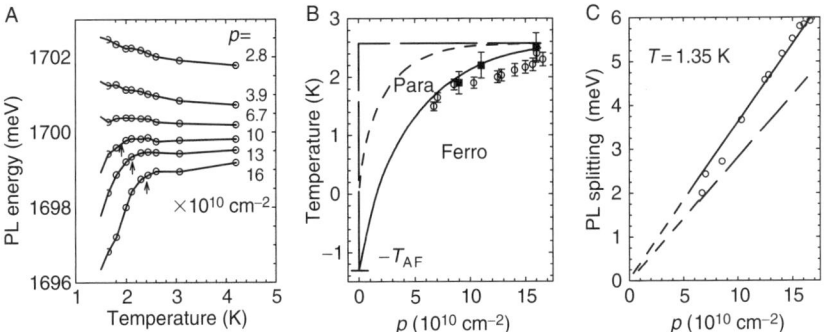

FIGURE 9.14 (A) Example of Moss-Burstein shift between PLE and PL spectra serving to determine the hole density p. (B) Temperature dependence of the low-energy peak of the PL spectra for selected values of hole densities (changed by illumination) for a p-type $Cd_{0.96}Mn_{0.04}Te$ QW. Arrows indicate the points taken as indicating the critical temperature of the ferromagnetic transition (T_C). (C) T_C versus hole density for two samples [circles, same as (B), and squares]. The line shows T_C values calculated within the mean field model and for various assumptions about the hole spin susceptibility: the dashed line is for a 2D degenerate Fermi liquid, the dotted line takes into account the effect of nonzero temperature on a clean 2D liquid, and the solid line is obtained assuming additionally the gaussian broadening of the density-of-states. (D) Zero-field splitting of the PL line at 1.35 K, as a function of the hole density. The solid straight line is drawn through the data points; the dashed line is calculated within the mean-field model neglecting the effects of hole–hole correlation on the optical spectra. Both lines cross zero as shown by the dotted lines [after Boukari et al. (2002)].

as depicted in Fig. 9.15. These simulations demonstrated also a somewhat fortunate quantitative accuracy of the mean-field approximation when the carrier-mediated ferromagnetism and intrinsic superexchange compete. As already mentioned, the Stoner enhancement factor A_F of about 2 is implied by a comparison of theoretical and experimental data, a magnitude expected for the 2D case and hole densities in question.

8.3. Critical region

In has been found that the Curie temperature, like other thermodynamic properties, does not show up any critical behavior on crossing the MIT (Ferrand et al., 2001; Matsukura et al., 1998). The noncritical persistence of ferromagnetism into the insulator side of the MIT has been explained by the present author and coworkers (Dietl et al., 2000, 1997; Ferrand et al., 2001) in terms of the scaling theory of the MIT in doped semiconductors (Belitz and Kirkpatrick, 1994; Lee and Ramakrishnan, 1985). Within this scenario, the hole localization length, which diverges at the MIT, remains much greater than an average distance between the acceptors for the

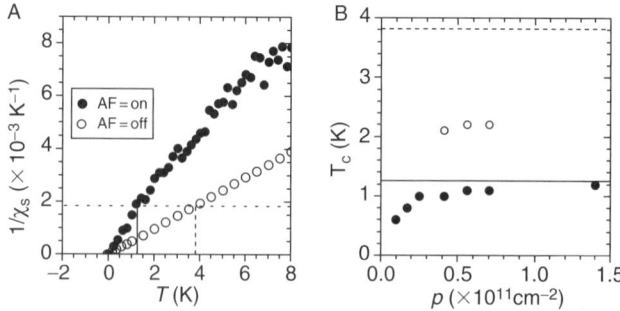

FIGURE 9.15 (A) Inverse Mn susceptibility of an *undoped* $Cd_{0.96}Mn_{0.04}Te$ quantum well as obtained by Monte-Carlo simulations without, $J_{ij} = 0$ (open symbols), and with antiferromagnetic interactions, $J_{ij} \neq 0$ (closed symbols). The intercept of the Monte-Carlo data and the horizontal (dotted) line gives the mean-field value of the Curie temperature (T_C) for the hole doped system according to Eqs. (9.14, 9.15) for $d = 2, n = 1$. (B) Variation of T_C with hole doping level as obtained by Monte-Carlo simulations. Horizontal lines are the mean-field results, as obtained from (A). Dashed line (upper part of the panel): $J_{ij} = 0$, solid line: $J_{ij} \neq 0$ [after Kechrakos et al. (2005)].

range of the hole densities, where signatures of ferromagnetism persist. Accordingly, the holes can be regarded as delocalized at the length scale relevant for the coupling between magnetic ions. Hence, the spin–spin exchange interactions are effectively mediated by the itinerant carriers, so that the p–d Zener or RKKY model can be applied also on the insulator side of the MIT.

At the same time, however, large mesoscopic fluctuations in the local value of the density-of-states are expected near the MIT. As a result, nanoscale phase separation into paramagnetic and ferromagnetic regions takes place below and in the vicinity of the apparent Curie temperature. The paramagnetic phase persists down to the lowest temperatures in the locations that are *not* visited by the holes or characterized by a low value of the blocking temperature T_B. The ferromagnetic order develops in the regions, where the carrier liquid sets long-range ferromagnetic correlation between the randomly distributed TM spins. According to this model, the portion of the material encompassing the ferromagnetic bubbles, and thus the magnitude of the saturated ferromagnetic moment, grows with the net acceptor concentration, extending over the whole sample on the metallic side of the MIT.

There is a growing amount of experimental results indicating that the model outlined in the previous paragraph is qualitatively correct. In particular, for samples on the insulator side of MIT, the field dependence of magnetization shows the presence of superimposed ferromagnetic and paramagnetic contributions in both (Ga,Mn)As (Oiwa et al., 1997) and p-(Zn,Mn)Te (Ferrand et al., 2001). Interestingly, the paramagnetic component is less visible in the anomalous Hall effect data, presumably because it probes merely the regions visited by the carriers (Ferrand et al., 2001). At the same time, colossal negative magnetoresistance is observed, leading to the field-induced insulator-to-metal transition in samples with the appropriate acceptor densities (Ferrand et al., 2001). The enhanced conductance in the magnetic field can be linked to the ordering of ferromagnetic bubbles and to the alignment of the spins in the paramagnetic regions. Remarkably, the corresponding effects have recently been found in modulation-doped quantum well of (Cd,Mn)Te, where no localization of carriers by individual ionized impurities and, thus, no formation of bound magnetic polarons is expected (Jaroszyński et al., 2007). The question whether the holes bound by individual acceptors or rather the holes residing in weakly localized states mediate ferromagnetism in DMSs on the insulating side of the MIT was also addressed by inelastic neutron scattering in (Zn,Mn)Te:P (Kępa et al., 2003). In that work, the difference in the nearest neighbor Mn pairs exchange energy J_1 in the presence and in the absence of the holes was determined. The hole-induced contribution to J_1 was found to be by a factor of four smaller than that calculated under the assumption that the holes reside on individual acceptors. By contrast, if the hole states are assumed to be metallic-like at length scale of the nearest neighbor distance, the calculated value is smaller than the experimental one by a factor of 1.5, a discrepancy well within combine uncertainties in the input parameters to theory and experimental determination.

Comparing to p-(Zn,Mn)Te, where T_C exceeds barely 5 K, T_C in (Ga,Mn)As reaches rather fast a 15–20 K level (Myers et al., 2006). This difference is associated with a higher value of the density-of-states at n_c and the absence of competing antiferromagnetic interactions in (Ga,Mn)As (Dietl et al., 2001b). We should, however, mention that the determination of T_C is by no means univocal in this case of cluster ferromagnetism. Actually, ferromagnetic order starts to develop at $T^* > T_C$ in the regions where the local carrier density is large enough to support long-range ferromagnetic correlations between randomly distributed Mn spins (Mayr et al., 2002). As could be expected for samples on the insulator side of the MIT, in which the disorder-driven fluctuations of the local density-of-states are particularly large, the field (Oiwa et al., 1997) and temperature (Myers et al., 2006) dependence of magnetization shows that even at $T \ll T_C$ only a fraction of spins is aligned ferromagnetically.

8.4. Critical scattering near MIT

As argued above, a characteristic feature of carrier-controlled ferromagnetic semiconductors at the localization boundary is the presence of randomly oriented ferromagnetic bubbles, which start to develop at

$T^* > T_C$. Together with critical thermodynamic fluctuations that develop at $T \to T_C^+$, they account for a resistance maximum near T_C and the associated negative magnetoresistance, as shown in Fig. 9.16 (Matsukura et al., 1998). Both effects disappear deeply in the metallic and isolating phases. The underlying physics is analogous to that accounting for similar anomalies, albeit at much lower temperatures, in n-type DMSs at the localization boundary, as discussed in detail elsewhere (Dietl, 2008c; Jaroszyński et al., 2007). According to this insight, the presence of randomly oriented ferromagnetic bubbles has two consequences. First, they constitute potential barriers of the order of the Fermi energy, which diminish conductance. Second, they give rise to efficient spin-disorder scattering of the carriers. Such scattering reduce "anti-localization" corrections to conductivity near the Anderson-Mott transition (Altshuler and Aronov, 1985; Belitz and Kirkpatrick, 1994; Fukuyama, 1985; Lee and Ramakrishnan, 1985). In particular, spin-disorder scattering, once more efficient than spin–orbit scattering, destroys "anti-localization" quantum corrections to the conductivity associated with the one-particle interference effect ("cooperon" channel) (Timm et al., 2005) and with the particle-hole triplet ("diffuson") channel (Jaroszyński et al., 2007). While both phenomena increase the resistance value upon approaching T_C, the

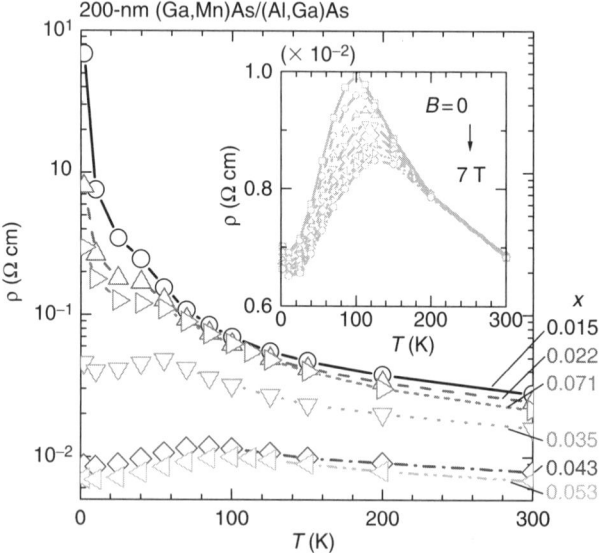

FIGURE 9.16 Temperature dependence of resisitivity for $Ga_{1-x}Mn_xAs$ films on the both sides of the metal-insulator transition. Resistance maximum in the vicinity of the Curie temperature and the associated negative magnetoresistance (inset) are seen [after Matsukura et al. (1998)].

latter—resulting from disorder-modified carrier correlation—is usually quantitatively more significant (Altshuler and Aronov, 1985; Belitz and Kirkpatrick, 1994; Fukuyama, 1985; Lee and Ramakrishnan, 1985). Once on cooling T_C is crossed, which means that spontaneous magnetization begins to develop, the ferromagnetic bubbles start to order. This leads to a reduction in spin disorder scattering resulting in a resistance maximum at T_C, as the "anti-localization" terms are reduced by spin splitting (that appears below T_C) about 30% less effectively than by spin disorder scattering.

As expected within the above model, the resistance maximum tends to disappear deeply in the insulator phase (Myers et al., 2006) (where ferromagnetic bubbles occupy only a small portion of the sample) as well as deeply in the metallic phase (Potashnik et al., 2001) (where ferromagnetic alignment is uniform and quantum localization unimportant). In the metallic region, the resistance is approximately constant down to T_C and gradually decreases at lower temperatures. The Drude-Boltzmann approach taking into account the spin-splitting-induced carrier redistribution between spin subbands describes satisfactorily the data in this range (López-Sancho and Brey, 2003). Furthermore, the disappearance of the resistance maximum away from MIT appears to suggest that contributions expected from critical scattering (Omiya et al., 2000) and magnetic-impurity (Goennenwein et al., 2005; Nagaev, 2001; Yuldashev et al., 2003) models are quantitatively unimportant, a conclusion indirectly supported by the fact that rather large values of the exchange integral β had to be assumed to fit the temperature dependence of resistance near T_C in (Ga,Mn)As samples close to MIT within those models (Omiya et al., 2000; Yuldashev et al., 2003).

8.5. Strongly localized regime

Experimentally, the ferromagnetism vanishes in the strongly localized regime, where the carrier localization length becomes of the order of the Bohr radius of the relevant single impurity. This is seen when varying the hole concentration in (Ga,Mn)As (Myers et al., 2006), where T_C vanishes rather abruptly away from the MIT. The strong coupling specific to nitrides and oxides, shifts the MIT to higher hole concentrations than those attained in samples studied till now. According to our insight (Dietl et al., 2000, 2001b), in the absence of delocalized or weakly localized holes, no ferromagnetism is expected for randomly distributed diluted spins. Indeed, recent studies of (Ga,Mn)N indicate that in samples containing up to 6% of Mn, holes stay strongly localized and, accordingly, T_C below 10 K is experimentally revealed (Edmonds et al., 2005; Sarigiannidou et al., 2006). An interesting intermediate situation is realized in (Ga,Mn)P, where T_C reaches 60 K (Scarpulla et al., 2005). Because of smaller value of the Bohr radius a_B in (Ga,Mn)P comparing to (Ga,Mn)As (Wolos and Kaminska, 2008) and, therefore, a greater degree of localization, the magnitude of T_C is expected to be substantially smaller in phosphides than in arsenides at given Mn concentration.

As depicted in Fig. 9.13, at sufficiently high hole densities, an insulator-to-metal transition is expected, even in the strong coupling case. It has been predicted (Dietl, 2008a)—but not yet verified experimentally—that once the MIT is reached, which means that bound states are washed out by many body screening, high values of T_C expected within the VCA and MFA (Dietl et al., 2000) should emerge.

9. EFFECTS OF NONRANDOM DISTRIBUTION OF MAGNETIC IONS

We have so far assumed that the localized spins are distributed randomly. Nevertheless, as we have argued, nonuniformities of magnetization occur due to large fluctuations in the local density of carrier states near MIT. However, it turns out that in a number of DMSs, the distribution of magnetic ions is highly nonrandom.

The question about the relation between high T_C and the magnetic ion distribution has been around for many years (Dietl, 2003). In particular, as pointed out elsewhere (Dietl, 2005), both hexagonal and zinc-blende Mn-rich (Ga,Mn)As nanocrystals are clearly seen in annealed (Ga,Mn)As (Moreno et al., 2002; Yokoyama et al., 2005). These two cases correspond to precipitation of other phases and spinodal decomposition, respectively. As emphasized in recent reviews (Bonanni, 2007; Dietl, 2005, 2007; Katayama-Yoshida et al., 2007b), the alloy decomposition into nanoregions with low and high concentrations of magnetic ions constitutes a generic property of many DMSs, occurring if the mean density of the magnetic constituent exceeds the solubility limit at given growth or thermal processing conditions. Importantly, the host can stabilize magnetic nanostructures in a new crystallographic and/or chemical form, not as yet listed in materials compendia.

9.1. Observations of spinodal decomposition

It is well known that phase diagrams of a number of alloys exhibit a solubility gap in a certain concentration range. This may lead to a spinodal decomposition referred to above. If the concentration of one of the alloy constituents is small, it may appear in a form of coherent nanocrystals embedded by the majority component. For instance, such a spinodal decomposition occurs in (Ga, In) N (Farhat and Bechstedt, 2002) where In-rich quantum-dot like regions are embedded by the In poor matrix.

However, according to the pioneering *ab initio* work (van Schilfgaarde and Mryasov, 2001) and subsequent developments (Kuroda *et al.*, 2007; Sato *et al.*, 2005), particularly strong tendency to form nonrandom alloy occurs in the case of many DMSs. For instance, the evaluated gain in energy by bringing two Ga-substitutional Mn atoms together is $E_d =$ 120 meV in GaAs and 300 meV in GaN, and reaches 350 meV in the case of Cr pair in GaN (van Schilfgaarde and Mryasov, 2001).

Since spinodal decomposition does not usually involve any precipitation of other crystallographic phases, it is not easily detectable experimentally. Nevertheless, its presence was found by electron transmission microscopy (TEM) in (Ga,Mn)As (Moreno *et al.*, 2002; Yokoyama *et al.*, 2005), where coherent zinc-blende Mn-rich (Mn,Ga)As metallic nanocrystals led to the apparent Curie temperature up to 360 K (Yokoyama *et al.*, 2005). Furthermore, according to results summarized in Fig. 9.17, coherent hexagonal nanocrystals were detected by spatially resolved x-ray diffraction in (Ga,Mn)N (Martinez-Criado *et al.*, 2005) and by TEM with energy dispersive spectroscopy (EDS) in (Al,Cr)N, (Ga,Cr)N (Gu *et al.*, 2005), and (Ga,Fe)N (Bonanni *et al.*, 2007, 2008). The same technique revealed spinodal decomposition in (Zn,Cr)Te (Kuroda *et al.*, 2007; Sreenivasan *et al.*, 2007) confirming indirectly the early suggestion that the presence of ferromagnetism in (Zn,Cr)Se points to a nonrandom Cr distribution (Karczewski *et al.*, 2003). Finally, we mention the case of (Ge,Mn). For this DMS, one group (Bougeard *et al.*, 2006) observed Mn-rich regions in the form of nanodots of diameter of 5 nm, whereas another team (Devillers *et al.*, 2007; Jamet *et al.*, 2006) found periodically arranged nanocolumns of diameter of 3 nm, which extend from the substrate to the surface, as shown in Fig. 9.18. Actually, a tendency for the nanocolumn formation was also reported for (Al,Cr)N (Gu *et al.*, 2005).

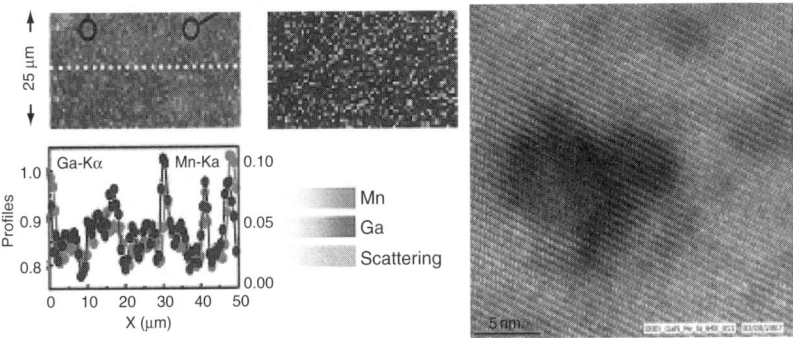

FIGURE 9.17 Evidences for spinodal decomposition in (Ga,Mn)N [synchrotron radiation micro-probe—left panel (Martinez-Criado *et al.*, 2005)] and in (Ga,Fe)N [transmission electron microscope (Bonanni *et al.*, 2007, 2008)]. (See Color Plate.)

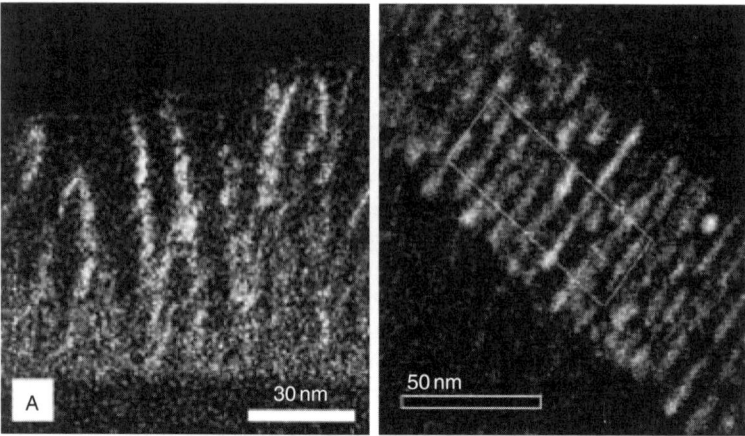

FIGURE 9.18 Evidences for self-organized assembling of nano-columns rich in the magnetic constituent obtained from energy dispersive spectroscopy (EDS) in (Al,Cr)N (Gu et al., 2005) and (Ge,Mn) (Jamet et al., 2006).

This demonstrates that growth conditions can serve to control nanocrystal shapes. Interestingly, these two kinds of nanocrystal forms were reproduced by Monte-Carlo simulations (Katayama-Yoshida et al., 2007b, 2008).

Since magnetic nanocrystals assembled by spinodal decomposition assume the form imposed by the matrix, it is *a priori* unknown of whether they will be metallic or insulating as well as of whether they will exhibit ferromagnetic, ferrimagnetic, or antiferromagnetic spin order. However, owing to the large concentration of the magnetic constituent within the nanocrystals, their spin ordering temperature is relatively high, typically above the room temperature. It is obvious that ferromagnetic or ferrimagnetic nanocrystals will lead to spontaneous magnetization and magnetic hysteresis up to the blocking temperature (Goswami et al., 2005; Shinde et al., 2004) Interestingly, uncompensated spins at the surface of antiferromagnetic nanocrystals can also result in a sizable value of spontaneous magnetization up to the usually high Néel temperature (Eftaxias and Trohidou, 2005). It has recently been suggested that the ferromagnetic-like behavior of (Zn,Co)O (Chambers et al., 2006) in which no precipitates of other crystallographic phases are detected, originates from coherently in-built nanocrystals of wurtzite antiferromagnetic CoO (Dietl et al., 2007) or ferromagnetic CoZn (Kaspar et al., 2008). Such nanocrystals can aggregate under appropriate growth conditions or thermal treatment.

It is, therefore, legitimate to suppose that coherent nanocrystals with a large concentration of the magnetic constituent and a sufficiently large magnitude of magnetic anisotropy account for high apparent Curie temperatures detected in a number of DMSs. This model explains a long-staying puzzle about the origin of ferromagnetic-like response in DMSs, in which the average concentration of magnetic ions is below the percolation limit for the nearest neighbor coupling and, at the same time, the itinerant hole density is too low to mediate an efficient long-range exchange interaction. Remarkably, the presence of magnetically active metallic nanocrystals leads to enhanced magnetotransport (Jamet et al., 2006; Kuroda et al., 2007; Shinde et al., 2004; Ye et al., 2003) and magnetooptical (Kuroda et al., 2007; Yokoyama et al., 2005) properties over a wide spectral range. This opens doors for various applications of such hybrid systems provided that methods for controlling nanocrystal characteristics and distribution would be elaborated.

9.2. Controlling spinodal decomposition by inter-ion Coulomb interactions

So far, the most rewarding method of controlling the self-organized growth of coherent nanocrystals or quantum dots has exploited strain fields generated by lattice mismatch at interfaces of heterostructures (Stangl et al., 2004). Remarkably, it becomes possible to fabricate highly ordered 3D dot crystals under suitable spatial strain anisotropy (Stangl et al., 2004).

It has been suggested that outstanding properties of magnetic ions in semiconductors can be utilized for the control of self-organized assembly of magnetic nanocrystals (Dietl and Ohno, 2006). As discussed in Sec. 2.1, such a trapping alters the charge state of the magnetic ions and, hence, affect their mutual Coulomb interactions (Dietl and Ohno, 2006; Kuroda et al., 2007; Ye and Freeman, 2006). Accordingly, co-doping of DMSs with shallow acceptors or donors, during either growth or post-growth processing, modifies E_d and thus provides a mean for the control of ion aggregation. Indeed, the energy of the Coulomb interaction between two elementary charges residing on the nearest neighbor cation sites in the GaAs lattice is 280 meV. This value indicates that the Coulomb interaction can preclude the aggregation, as the change of energy associated with the bringing two Mn atoms in (Ga,Mn)As is $E_d = -120$ meV.

It is evident that the model in question should apply to a broad class of DMSs as well to semiconductors and insulators, in which a constituent, dopant, or defect can exist in different charge states under various growth conditions. As important examples we consider (Ga,Mn)N (Kane et al., 2006; Reed et al., 2005) and (Zn,Cr)Te (Kuroda et al., 2007; Ozaki et al., 2005, 2006) in which remarkable changes in ferromagnetic characteristics on co-doping with shallow impurities have recently been reported. In particular, a strong dependence of saturation magnetization M_s at 300 K on co-doping with Si donors and Mg acceptors has been found for

(Ga,Mn)N with an average Mn concentration $x_{Mn} \approx 0.2\%$ (Reed et al., 2005) Both double exchange and superexchange are inefficient at this low Mn concentration and for the mid-gap Mn level in question. At the same time, the model of nanocrystal self-organized growth explains readily why M_s goes through a maximum when Mn impurities are in the neutral Mn^{3+} state, and vanishes if co-doping by the shallow impurities makes all Mn atoms to be electrically charged.

Particularly relevant in this context are data for (Zn,Cr)Te, where strict correlation between co-doping, magnetic properties, and magnetic ion distribution has been put into evidence (Kuroda et al., 2007). It has been found that the apparent Curie temperature $T_C^{(app)}$ and the aggregation of Cr-rich nanocrystals depend dramatically on the concentration of shallow Nitrogen acceptors in (Zn,Cr)Te (Kuroda et al., 2007; Ozaki et al., 2005). Actually, $T_C^{(app)}$ decreases monotonically when the concentration x_N of nitrogen increases, and vanishes when x_{Cr} and x_N become comparable. This supports the model as in ZnTe the Cr donor level resides about 1 eV above the nitrogen acceptor state. Accordingly, for $x_N \approx x_{Cr}$ all Cr atoms become ionized and the Coulomb repulsion precludes the nanocrystal formation, as evidenced by Cr spatial mapping (Kuroda et al., 2007). At the same time, the findings are not consistent with the originally proposed double-exchange mechanism (Ozaki et al., 2005) as undoped ZnTe is only weakly p-type, so that T_C should be small for either $x_N \approx 0$ or $x_N \approx x_{Cr}$, and pick at $x_N \approx x_{Cr}/2$, not at $x_N \approx 0$. Importantly, when native acceptors are compensated by Iodine donor doping, both $T_C^{(app)}$ and inhomogeneity in the Cr distribution attains a maximum (Kuroda et al., 2007), a behavior shown in Fig. 9.19.

Finally, we mention the case of Mn-doped GaAs, InAs, GaSb, and InSb. As argued in the previous section, owing to a relatively shallow character of Mn acceptors and a large Bohr radius, the holes reside in the valence band in these systems. Thus, the Mn atoms are negatively charged, which—according to the model in question—reduces their clustering, and makes it possible to deposit, by low-temperature epitaxy, a uniform alloy with a composition beyond the solubility limit. Co-doping with shallow donors, by reducing the free-carrier screening, will enhance repulsions among Mn, and allow one to fabricate homogenous layers with even greater x_{Mn}. On the contrary, co-doping by shallow acceptors, along with the donor formation by a self-compensation mechanism (Yu et al., 2002), will enforce the screening and, hence, lead to nanocrystal aggregation.

9.3. Possible functionalities

Over the recent years we have learnt a lot about the effects of solubility limits and self-compensation as well as on the role of the strong coupling occurring in materials with a small bond length, the issues emphasized

Exchange Interactions and Nanoscale Phase Separations 421

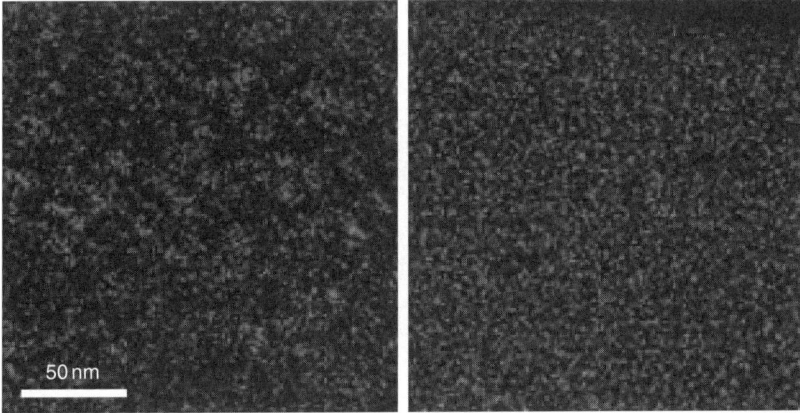

FIGURE 9.19 Cr distribution, as obtained from EDS, in (Zn,Cr)Te showing the dependence of spinodal decomposition on co-doping with I (left panel) and N (right panel); the maker corresponds to 50 nm [after (Kuroda et al., 2007)].

originally as experimental challenges for the high-temperature hole-controlled ferromagnetism in DMSs (Dietl et al., 2000). A progress in these fields should lead to an increase of T_C over 180 K in (Ga,Mn)As (Olejník et al., 2008; Wang et al., 2005), so far the highest confirmed value found for uniform tetrahedrally coordinated DMSs.

However, extensive experimental and computational search for multifunctional materials has already resulted in the development of semiconductor and oxide systems which exhibit surprisingly stable ferromagnetic signatures, often up to well above the room temperature, despite the absence of itinerant holes and a small content of magnetic elements. As discussed here, the ferromagnetism of these compounds, and the associated magnetooptical and magnetotransport functionalities, stem from the presence of magnetic nanocrystals embedded coherently in the nonmagnetic or paramagnetic matrix. Depending on the host and transition metal combination, the assembled nanocrystals are either metallic or insulating. Importantly, their shape (nanodots vs nanocolumns) and size can be altered by growth parameters and co-doping during the epitaxy, allowing one to fabricate *in situ*, for example, TMR and Coulomb blockade devices.

It is clear that the findings in question open the door for a variety of novel room-temperature applications. In particular, the media in question can be employed for low-power high-density magnetic memories, including spin-injection magnetic random access memories and race-track 3D domain-wall-based memories (Parkin et al., 2008). If sufficiently high TMR will be found, one can envisage the field programmable logic

(TMR-based connecting/disconnecting switches) (Reiss and Meyners, 2006) and even all-magnetic logic, characterized by low power consumption and radiation hardness. Furthermore, a combination of a strong magnetic circular dichroism and weak losses suggest possible applications as optical isolators (Amemiya et al., 2006) as well as 3D tunable photonic crystals and spatial light modulators for advanced photonic applications (Park et al., 2002). Furthermore, coherently grown metallic nanostructures may serve as building blocks for all-metallic nanoelectronics and for high-quality nanocontacts in nanoelectronics, optoelectronics, and plasmonics as well as constitute media for thermoelectric applications (Katayama-Yoshida et al., 2007a). Worth mentioning is also the importance of hybrid semiconductor/ferromagnetic systems in various proposals of scalable quantum processors (Loss and DiVincenzo, 1998) and magnetic logic (Dery et al., 2007).

10. IS FERROMAGNETISM POSSIBLE IN SEMICONDUCTORS WITH NO MAGNETIC ELEMENTS?

Organic ferromagnets (Fujiwara et al., 2005) and quantum Hall ferromagnets (Jungwirth et al., 2000) demonstrate that ferromagnetism is possible in materials without magnetic ions, albeit the corresponding Curie temperatures are rather low, below 20 K. It has also been suggested that a robust ferromagnetism can appear in certain zinc-blende metals, such as CaAs, driven by a Stoner instability in the narrow heavy hole band (Geshi et al., 2005). Some other exciting possibilities are discussed elsewhere in this volume (Katayama-Yoshida et al., 2008).

Other researchers relate the presence of unexpected high-temperature ferromagnetism to magnetic moments associated rather with defects than with magnetic impurities. There is, however, a number of qualitative indications against the persistence of ferromagnetism up to above room temperature in semiconductors, oxides, or carbon derivatives incorporating no magnetic elements and, at the same time, containing only a low carrier density, as witnessed by a relatively high resistivity.

It has been known for a long time that a number of defects or nonmagnetic impurities form localized paramagnetic centers in nonmetals. Actually, the oxygen molecule O_2 carries the spin $S = 1$. However, it is not easy to find a mechanism which would generate a ferromagnetic interaction between spins in nonmetallic solids, which—in view of high T_C in question—has to be rather strong. Furthermore, it should be long range, as the concentration of the relevant centers is low, according to a small magnitude of the saturation magnetization. Moreover, in the absence of carriers, exchange interactions between spins are typically not only short-range but also merely antiferromagnetic (Bhatt and Lee, 1982). In line with these arguments, the solid oxygen is an antiferromagnetic insulator. Only in rather few cases, a compensation of various exchange contributions leads to net ferromagnetic interactions between localized spins. Celebrated examples are actually ferromagnetic semiconductors, such as EuO and $CdCr_2Se_4$, in which despite the absence of carriers the coupling between highly localized f and d spins is ferromagnetic. However, the Curie temperature barely exceeds 100 K in these materials though the spin concentration and magnetic moments are rather high.

As discussed earlier in this chapter, a sizable exchange interaction is expected between localized and delocalized sp electrons, say, between valence-band holes trapped by acceptors and electrons in the conduction band. In principle, such an interaction could mediate a coupling between the spins associated with the localized states. However, under thermal equilibrium conditions we do not expect the coexistence of localized and delocalized sp carriers—an insulator-to-metal transition, which takes place when an average distance between the localized centers is 2.5 times larger than their localization radius, annihilates usually magnetic moments residing on defects or impurities.

In view of the facts above it is not surprising that it is hard to invent a sound model explaining high Curie temperatures in materials in which, say, 5% of sites contains a spin, and only short-range exchange interactions can operate. At this stage it appears more natural to assume that a small number of magnetic nanoparticles which escaped from the detection procedure account for the ferromagnetic-like behavior of nominally nonmagnetic insulators and semiconductors.

Finally, we note that a spontaneous magnetic moment reported for the case of some nonmagnetic metallic nanostructures has been assigned to orbital magnetism, enhanced by a spin–orbit interaction (Hernando *et al.*, 2006). We also note that it has been recently suggested that the spin–orbit coupling within the Fermi liquid may lead to long-range spin-dependent interactions (Gangadharaiah *et al.* 2008).

11. SUMMARY

We have discussed the origin of the spin-dependent sp–d(f) coupling in DMSs as well as emphasized that it should be considered on equal footing with sp–sp exchange interactions in a number of experimentally important cases. Furthermore, it becomes increasingly clear that the effect of magnetic impurities on hole extended states changes dramatically if the ions give rise to hole bound states. At the same time, no hole-mediated ferromagnetism operates in the regime where holes are strongly localized.

According to results summarized in this chapter, semiconductors which exhibit ferromagnetic features, such as spontaneous magnetization, can be grouped into five classes:

1. Diluted ferromagnetic semiconductors such as (Ga,Mn)As, heavily doped p-(Zn,Mn)Te, and related systems containing randomly distributed substitutional magnetic ions and delocalized or weakly localized holes. In these solid solutions the theory built on p–d Zener's model of hole-mediated ferromagnetism and on either the Kohn-Luttinger kp theory or the multiorbital tight-binding approach describes qualitatively, and often quantitatively, thermodynamic, micromagnetic, optical, and transport properties. Moreover, the understanding of these materials has provided a basis for the development of novel methods enabling magnetization manipulation and switching.
2. Carrier-doped DMSs in which a competition between long-range ferromagnetic and short-range antiferromagnetic interactions and/or the proximity of the localization boundary lead to an electronic nanoscale phase separation driven by competing interactions and/or mesoscopic fluctuations in the density of carrier states. These materials exhibit characteristics similar to colossal magnetoresistance oxides.
3. Alloys showing chemical nanoscale chemical phase separation into the regions with small and large concentrations of the magnetic ions. Here, high-temperature ferromagnetic-like properties are determined by the regions with high ion concentrations, whose crystal and chemical structure is imposed by the host.
4. Composite materials in which precipitation or contamination by nanoparticles of magnetic metal or of a ferromagnetic, ferrimagnetic, or antiferromagnetic compound account for ferromagnetic-like characteristics.
5. Finally, in magnetic semiconductors, such as EuO or $CdCr_2Se_4$, where most of cation sites is occupied by magnetic ions, short-range super- or double-exchange can, in specific cases, result in a ferromagnetic ordering.

While semiconductors belonging to the first and fifth class show magnetic and micromagnetic properties specific to standard ferromagnets, materials encompassed by remaining classes exhibit rather characteristics of superparamagnetic systems. Importantly, the incorporation and spatial distribution of magnetic ions and, therefore, the relative importance of behaviors specific to particular classes, depend on the growth conditions and co-doping by shallow impurities. This explains a variety of magnetic properties observed in a given material system as well as elucidates why the magnetic response is often a superposition of ferromagnetic, superparamagnetic, and paramagnetic contributions. At the same time, the sensitivity to growth conditions and co-doping may serve to control the self-organized assembly of the magnetic nanocrystals. It has already been demonstrated that the use of embedded metallic and semiconducting nanocrystals can enhance the performance of various commercial devices, such as flash memories and low-current quantum-dot lasers as well as, by allowing to manipulate single charges and single spins in a solid state environment, has opened new research directions. As we have underlined in this chapter, a number of novel functionalities have been proposed for magnetic and metallic nanocrystals obtained during DMSs growth or processing.

With no doubts, however, the question about the origin of high temperature ferromagnetism in semiconductors and oxides as well as the interplay between antiferromagnetic couplings, ferromagnetism, carrier localization, and superconductivity in transition metal compounds remain among the most intriguing topics in today's condensed matter physics and materials science. We may expect a fast development of novel tools for 3D element-specific and spin-sensitive analyses at the nanoscale, which will brought a number of unexpected developments in the years ahead.

ACKNOWLEDGMENTS

The works of the author described in this chapter were supported in part by ERATO project of Japan Science and Technology Agency, NANOSPIN E. C. project (FP6-2002-IST-015728), Humboldt Foundation, and carried out in collaboration with W. Pacuski, M. Sawicki, and C. Śliwa in Warsaw, as well as with groups of H. Ohno in Sendai, S. Kuroda in Tsukuba, K. Trohidou in Athens, J. Cibert and D. Ferrand in Grenoble, J. Jaroszyński and D. Popović in Tallahassee, L. Molenkamp in Würzburg, B. Gallagher in Nottingham, and A. Bonanni in Linz.

REFERENCES

Abolfath, M., Jungwirth, T., Brum, J., and MacDonald, A. (2001). *Phys. Rev. B* **63**, 054418.
Altshuler, B. L., and Aronov, A. G. (1985). In "Electron-Electron Interactions in Disordered Systems" (A. L. Efros and M. Pollak, eds.), p. 1. North Holland, Amsterdam.
Amemiya, T., Shimizu, H., Nakano, Y., Hai, P. N., Yokoyama, M., and Tanaka, M. (2006). *Appl. Phys. Lett.* **89**, 021104.
Anderson, P. (1961). *Phys. Rev.* **124**, 41.
Arciszewska, M., and Nawrocki, M. (1986). *J. Phys. Chem. Solids* **47**, 309.
Aristov, D. N. (1997). *Phys. Rev. B* **55**, 8064.
Barnes, S. (1981). *Adv. Phys.* **30**, 801.
Bastard, G., Rigaux, C., Guldner, Y., Mycielski, J., and Mycielski, A. (1978). *J. de Physique (Paris)* **39**, 87.
Belitz, D., and Kirkpatrick, T. R. (1994). *Rev. Mod. Phys.* **66**, 261.
Benoit à la Guillaume, C. (1993). *Phys. Stat. Solidi (b)* **175**, 369.
Benoit à la Guillaume, C., Scalbert, D., and Dietl, T. (1992). *Phys. Rev. B* **46**, 9853(R).
Bhatt, R. N., and Lee, P. A. (1982). *Phys. Rev. Lett.* **48**, 344.
Bhattacharjee, A. K., and Benoit à la Guillaume, C. (2000). *Solid State Commun.* **113**, 17.
Bhattacharjee, A. K., Fishman, G., and Coqblin, B. (1983). *Physica B + C* **117**, 449.
Blackwood, E., Snelling, M. J., Harley, R. T., Andrews, S. R., and Foxon, C. T. B. (1994). *Phys. Rev. B* **50**, 14246.

Blinowski, J., and Kacman, P. (2003). *Phys. Rev. B* **67,** 121204.
Blinowski, J., Kacman, P., and Majewski, J. A. (1996). *Phys. Rev. B* **53,** 9524.
Blinowski, J., Kacman, P., and Dietl, T. (2001). MRS Spintronics Symposium Proc. **69,** F6.9.
Bonanni, A. (2007). *Semicond. Sci. Technol.* **22,** R41.
Bonanni, A., Kiecana, M., Simbrunner, C., Li, T., Sawicki, M., Wegscheider, M., Quast, M., Przybylinska, H., Navarro-Quezada, A., Jakiela, R., and Wolos, A.Jantsch, W., *et al.* (2007). *Phys. Rev. B* **75,** 125210.
Bonanni, A., Navarro-Quezada, A., Li, T., Wegscheider, M., Matej, Z., Holy, V., Lechner, R., Bauer, G., Kiecana, M., Sawicki, M., and Dietl, T. (2008). arXiv:0804.3324v1.
Bougeard, D., Ahlers, S., Trampert, A., Sircar, N., and Abstreiter, G. (2006). *Phys. Rev. Lett.* **97,** 237202.
Boukari, H., Kossacki, P., Bertolini, M., Ferrand, D., Cibert, J., Tatarenko, S., Wasiela, A., Gaj, J. A., and Dietl, T. (2002). *Phys. Rev. Lett.* **88,** 207204.
Bouzerar, R., Bouzerar, G., and Zimai, T. (2007). *Europhys. Lett.* **78,** 67003.
Brey, L., and Gómez-Santos, G. (2003). *Phys. Rev. B* **68,** 115206.
Brey, L., Fernández-Rossier, J., and Tejedor, C. (2004). *Phys. Rev. B* **70,** 235334.
Chambers, S. A., Droubay, T. C., Wanga, C. M., Rosso, K. M., Heald, S. M., Schwartz, D., Kittilstved, K. R., and Gamelin, D. (2006). *Mater. Today* **9**(11), 28.
Chattopadhyay, A., Das Sarma, S., and Millis, A. J. (2001). *Phys. Rev. Lett.* **87,** 227202.
Cibert, J., *et al.* (2008). Quantum structures of II-VI diluted magnetic semiconductors. *In* "Spintronics" (T. Dietl, D. D. Awschalom, M. K., and H. Ohno, eds.), pp., this volume. Elsevier, Amsterdam.
Dalpian, G. M., and Wei, S. H. (2005). *Phys. Rev. B* **72,** 115201.
Dalpian, G. M., and Wei, S. H. (2006). *Phys. Rev. B* **73,** 245204.
Dery, H., Dalal, P., Cywiński, L., and Sham, L. J. (2007). *Nature* **447,** 573.
Devillers, T., Jamet, M., Barski, A., Poydenot, V., Bayle-Guillemaud, P., Bellet-Amalric, E., Cherifi, S., and Cibert, J. (2007). *Phys. Rev. B* **76,** 205306.
Dietl, T. (1981). Semimagnetic semiconductors in high magnetic fields. *In* "Physics in High Magnetic Fields" (S. Chikazumi and N. Miura, eds.), p. 344. Springer, Berlin.
Dietl, T. (1983). *J. Mag. Mag. Mat.* **38,** 34.
Dietl, T. (1994). Diluted magnetic semiconductors. *In* "Handbook of Semiconductors" (S. Mahajan, ed.), **3B,** p. 1251. North Holland, Amsterdam.
Dietl, T. (2003). *Nat. Mater.* **2,** 646.
Dietl, T. (2004). *J. Phys. Condens. Matter* **16,** 5471.
Dietl, T. (2005). Spintronics and ferromagnetism in wide-band-gap semiconductors. *In*, "Proceedings 27th International Conference on the Physics of Semiconductors" (J. Menèndez and C. G. Van de Walle, eds.), p. 56. Flagstaff, USA 2004 AIP, Melville.
Dietl, T. (2007). *J. Phys. Condens. Matter* **19,** 165204.
Dietl, T. (2008a). *Phys. Rev. B* **77,** 085208.
Dietl, T. (2008b). *J. Phys. Soc. Jpn.* **77,** 031005.
Dietl, T. (2008c). *J. Appl. Phys.* **103,** 07D111.
Dietl, T., and Ohno, H. (2006). *Mater. Today* **9,** 18.
Dietl, T., and Spałek, J. (1982). *Phys. Rev. Lett.* **48,** 355.
Dietl, T., and Spałek, J. (1983). *Phys. Rev. B* **28,** 1548.
Dietl, T., Sawicki, M., Isaacs, E., Dahl, M., Heiman, D., Graf, M., Gubarev, S., and Alov, D. L. (1991). *Phys. Rev. B* **43,** 3154.
Dietl, T., Śliwa, C., Bauer, G., and Pascher, H. (1994). *Phys. Rev. B* **49,** 2230.
Dietl, T., Peyla, P., Grieshaber, W., and Merle d'Aubigné, Y. (1995). *Phys. Rev. Lett.* **74,** 474.
Dietl, T., Haury, A., and Merle d'Aubigné, Y. (1997). *Phys. Rev. B* **55,** R3347.
Dietl, T., Cibert, J., Ferrand, D., and Merle d'Aubigné, Y. (1999). *Mater. Sci. Engin.* **63,** 103.
Dietl, T., Ohno, H., Matsukura, F., Cibert, J., and Ferrand, D. (2000). *Science* **287,** 1019.
Dietl, T., König, J., and MacDonald, A. H. (2001a). *Phys. Rev. B* **64,** 241201.

Dietl, T., Ohno, H., and Matsukura, F. (2001b). *Phys. Rev. B* **63**, 195205.
Dietl, T., Matsukura, F., and Ohno, H. (2002). *Phys. Rev. B* **66**, 033203.
Dietl, T., Andrearczyk, T., Lipińska, A., Kiecana, M., Tay, M., and Wu, Y. (2007). *Phys. Rev. B* **76**, 155312.
Dugaev, V. K., Crépieux, A., and Bruno, P. (2001). *Phys. Rev. B* **64**, 104411.
Edmonds, K. W., Novikov, S. V., Sawicki, M., Campion, R. P., Staddon, C. R., Giddings, A. D., Zhao, L. X., Wang, K. Y., Dietl, T., Foxon, C. T., and Gallagher, B. L. (2005). *Appl. Phys. Lett.* **86**, 152114.
Edwards, P. P., and Sienko, M. J. (1978). *Phys. Rev. B* **17**, 2575.
Eftaxias, E., and Trohidou, K. N. (2005). *Phys. Rev. B* **71**, 134406.
Ekardt, W. (1977). *Solid State Commun.* **22**, 531.
Farhat, M., and Bechstedt, F. (2002). *Phys. Rev. B* **65**, 075213.
Fedorych, O. M., Hankiewicz, E. M., Wilamowski, Z., and Sadowski, J. (2002). *Phys. Rev. B* **66**, 045201.
Ferrand, D., Cibert, J., Wasiela, A., Bourgognon, C., Tatarenko, S., Fishman, G., Andrearczyk, T., Jaroszyński, J., Koleśnik, S., Dietl, T., Barbara, B., and Dufeu, D. (2001). *Phys. Rev. B* **63**, 085201.
Fiete, G. A., Zaránd, G., Damle, K., and Moca, C. P. (2005). *Phys. Rev. B* **72**, 045212.
Fröhlich, F., and Nabarro, F. (1940). *Proc. Roy. Soc. London A* **175**, 382.
Fujiwara, M., Kambe, T., and Oshima, K. (2005). *Phys. Rev. B* **71**, 174424.
Fukuyama, H. (1985). *In* "Electron-Electron Interactions in Disordered Systems" (A. L. Efros and M. Pollak, eds.), p. 155. North Holland, Amsterdam.
Furdyna, J., and Kossut, J. (1988). Diluted magnetic semiconductors. *In* "Semiconductor and Semimetals" (S. Chikazumi and N. Miura, eds.), Vol. 25. Academic Press, New York.
Gaj, J. A., Gałązka, R. R., and Nawrocki, M. (1978). *Solid State Commun.* **25**, 193.
Gaj, J. A., Planel, R., and Fishman, G. (1979). **29**, 435.
Gangadharaiah, S., Sun, J., and Starykh, O. A. (2008). *Phys. Rev. Lett.* **100**, 156402.
Geshi, M., Kusakabe, K., Tsukamoto, H., and Suzuki, N. (2005). *AIP Conf. Proc.* **772**, 327.
Głód, P., Dietl, T., Fromherz, T., Bauer, G., and Miotkowski, I. (1994). *Phys. Rev. B* **49**, 7797.
Goennenwein, S. T. B., Russo, J., Morpurgo, A. F., Klapwijk, T. M., van Roy, W., and de Boeck, J. (2005). *Phys. Rev. B* **B**, 193306.
Goswami, R., Kioseoglou, G., Hanbicki, A. T., van't Erve, O. M. J., Jonker, B. T., and Spanos, G. (2005). *Appl. Phys. Lett.* **86**, 032509.
Gould, C., et al. (2008). Spintronic nanodevices. *In* "Spintronics" (T. Dietl, D. D. Awschalom, M. K., and H. Ohno, eds.), pp. this volume. Elsevier, Amsterdam.
Graf, T., Goennenwein, S. T. B., and Brandt, M. S. (2003). *Phys. Status Solidi B* **239**, 277.
Gu, L., Wu, S., Liu, H., Singh, R., Newman, N., and Smith, D. (2005). *J. Magn. Magn. Mater.* **290–291**, 1395.
Haldane, F. D. M., and Anderson, P. W. (1976). *Phys. Rev. B* **13**(6), 2553.
Haury, A., Wasiela, A., Arnoult, A., Cibert, J., Tatarenko, S., Dietl, T., and Merle d'Aubigné, Y. (1997). *Phys. Rev. Lett.* **79**, 511.
Heimbrodt, W., Hartmann, T., Klar, P. J., Lampalzer, M., Stolz, W., Volz, K., Schaper, A., Treutmann, W., Krug von Nidda, H. A., Loidl, A., Ruf, T., and Sapega, V. F. (2001). *Physica E* **10**, 175.
Hernando, A., Crespo, P., and García, M. A. (2006). *Phys. Rev. Lett.* **96**, 057206.
Honolka, J., Masmanidis, S., Tang, H. X., Awschalom, D. D., and Roukes, M. L. (2007). *Phys. Rev. B* **75**(24), 245310.
Hwang, J. I., Ishida, Y., Kobayashi, M., Hirata, H., Takubo, K., Mizokawa, T., Fujimori, A., Okamoto, J., Mamiya, K., Saito, Y., Muramatsu, Y., Ott, H., et al. (2005). *Phys. Rev. B* **72**, 085216.
Jaczyński, M., Kossut, J., and Gałązka, R. R. (1978). *Phys. Stat. Sol. (b)* **88**, 73.

Jamet, M., Barski, A., Devillers, T., Poydenot1, V., Dujardin, R., Bayle-Guillmaud, P., Rotheman, J., Bellet-Amalric, E., Marty, A., Cibert, J., Mattana, R., and Tatarenko, S. (2006). *Nat. Mater.* **5,** 653.
Jaroszyński, J., Wróbel, J., Karczewski, G., Wojtowicz, T., and Dietl, T. (1998). *Phys. Rev. Lett.* **80,** 5635.
Jaroszyński, J., Andrearczyk, T., Karczewski, G., Wróbel, J., Wojtowicz, T., Papis, E., Kamińska, E., Piotrowska, A., Popovic, D., and Dietl, T. (2002). *Phys. Rev. Lett.* **89,** 266802.
Jaroszyński, J., Andrearczyk, T., Karczewski, G., Wróbel, J., Wojtowicz, T., Popović, D., and Dietl, T. (2007). *Phys. Rev. B* **76,** 045322.
Jungwirth, T., Atkinson, W., Lee, B., and MacDonald, A. H. (1999). *Phys. Rev. B* **59,** 9818.
Jungwirth, T., Atkinson, W., Lee, B., and MacDonald, A. (2000). *Physica E* **6,** 794.
Jungwirth, T., Wang, K., Mašek, J., Edmonds, K., König, J., Sinova, J., Polini, M., Goncharuk, N., MacDonald, A., Sawicki, M., Campion, R., Zhao, L., et al. (2005). *Phys. Rev. B* **72,** 165204.
Jungwirth, T., Sinova, J., Mašek, J., Kučera, J., and MacDonald, A. H. (2006a). *Rev. Mod. Phys.* **78,** 809.
Jungwirth, T., Wang, K. Y., Mašek, J., Edmonds, K. W., König, J., Sinova, J., Polini, M., Goncharuk, N., MacDonald, A. H., Sawicki, M., Campion, R. P., Zhao, L. X., et al. (2006b). *Phys. Rev. B* **73,** 165205.
Jungwirth, T., et al. (2008). Transport properties of ferromagnetic semiconductors. In "Spintronics" (T. Dietl, D. D. Awschalom, M. K., and H. Ohno, eds.), pp. this volume. Elsevier, Amsterdam.
Kacman, P. (2001). *Semicond. Sci. Technol.* **16,** R25.
Kane, M. H., Strassburg, M., Fenwick, E. W., Asghar, A., Payne, A. M., Gupta, S., Song, Q., Zhang, Z. J., Dietz, N., Summers, C. J., and Ferguson, I. T. (2006). *J. Cryst. Growth* **287,** 591.
Karczewski, G., Sawicki, M., Ivanov, V., Rüster, C., Grabecki, G., Matsukura, F., Molenkamp, L. W., and Dietl, T. (2003). *J. Supercond./Novel Magn.* **16,** 55.
Kaspar, T. C., Droubay, T., Heald, S. M., Engelhard, M. H., Nachimuthu, P., and Chambers, S. A. (2008). *Phys. Rev. B* **77,** 201303(R).
Katayama-Yoshida, H., Fukushima, T., Dinh, V. A., and Sato, K. (2007a). *Jpn. J. Appl. Phys.* **46,** L777.
Katayama-Yoshida, H., Sato, K., Fukushima, T., Toyoda, M., Kizaki, H., Dinh, A. V., and Dederichs, P. H. (2007b). *Phys. Stat. Sol. (a)* **204,** 15.
Katayama-Yoshida, H., et al. (2008). Computational nano-materials design for the wide bandgap and high-Tc semiconductor spintronics. In "Spintronics" (T. Dietl, D. D. Awschalom, M. K., and H. Ohno, eds.), pp. this volume. Elsevier, Amsterdam.
Kępa, H., Khoi, L. V., Brown, C. M., Sawicki, M., Furdyna, J. K., Giebułtowicz, T. M., and Dietl, T. (2003). *Phys. Rev. Lett.* **91,** 087205.
Kechrakos, D., Papanikolaou, N., Trohidou, K. N., and Dietl, T. (2005). *Phys. Rev. Lett.* **94,** 127201.
Kent, P. R. C., and Zunger, A. (2001). *Phys. Rev. B* **64,** 115208.
Kittilstved, K. R., Liu, W. K., and Gamelin, D. R. (2006). *Nat. Mater.* **5,** 291.
Komarov, A., Ryabchenko, S., Terletskii, O., Zheru, I., and Ivanchuk, R. (1977). *Sov. Phys. JETP* **46,** 318.
Korotkov, R. Y., Gregie, J. M., and Wessels, B. W. (2002). *Appl. Phys. Lett.* **80,** 1731.
Krenn, H., Kaltenegger, K., Dietl, T., Spałek, J., and Bauer, G. (1989). *Phys. Rev. B* **39,** 10918.
Kuroda, S., Nishizawa, N., Takita, K., Mitome, M., Bando, Y., Osuch, K., and Dietl, T. (2007). *Nat. Mater.* **6,** 440.
Langer, J., Delerue, C., Lannoo, M., and Heinrich, H. (1988). *Phys. Rev. B* **38,** 7723.
Lee, P. A., and Ramakrishnan, T. V. (1985). *Rev. Mod. Phys.* **57,** 287.
Liu, L., Yu, P. Y., Ma, Z., and Mao, S. S. (2008). *Phys. Rev. Lett.* **100,** 127203.
López-Sancho, M., and Brey, L. (2003). *Phys. Rev. B* **68,** 113201.

Loss, D., and DiVincenzo, D. P. (1998). *Phys. Rev. A* **57**, 120.
Mac, W., Nguyen The Khoi, A., Twardowski, A., Gaj, J. A., and Demianiuk, M. (1993). *Phys. Rev. Lett.* **71**, 2327.
Majewski, J. A. (2005). *Acta Phys. Polon. A* **108**, 777.
Malguth, E., Hoffmann, A., Gehlhoff, W., Gelhausen, O., Phillips, M., and Xu, X. (2006). *Phys. Rev. B* **74**, 165202.
Marcet, S., Ferrand, D., Halley, D., Kuroda, S., Mariette, H., Gheeraert, E., Teran, F. J., Sadowski, M. L., Galera, R. M., and Cibert, J. (2006). *Phys. Rev. B* **74**.
Martinez-Criado, G., Somogyi, A., Ramos, S., Campo, J., Tucoulou, R., Salome, M., Susini, J., Hermann, M., Eickhoff, M., and Stutzmann, M. (2005). *Appl. Phys. Lett.* **86**, 131927.
Matsukura, F., et al. (2008). Spintronic properties of ferromagnetic semiconductors. In "Spintronics" (T. Dietl, D. D. Awschalom, M. K., and H. Ohno, eds.), pp. this volume. Elsevier, Amsterdam.
Matsukura, F., Ohno, H., Shen, A., and Sugawara, Y. (1998). *Phys. Rev. B* **57**, R2037.
Matsukura, F., Sawicki, M., Dietl, T., Chiba, D., and Ohno, H. (2004). *Physica E* **21**, 1032.
Mayr, M., Alvarez, G., and Dagotto, E. (2002). *Phys. Rev. B* **65**, 241202.
Merkulov, I. A., Yakovlev, D. R., Keller, A. W., Ossau, J. G., Waag, A., Landwehr, G., Karczewski, G., Wojtowicz, T., and Kossut, J. (1999). *Phys. Rev. Lett.* **83**, 1431.
Mizokawa, T., and Fujimori, A. (1993). *Phys. Rev. B* **48**, 14150.
Mizokawa, T., and Fujimori, A. (1997). *Phys. Rev. B* **56**, 6669.
Mizokawa, T., Nambu, T., Fujimori, A., Fukumura, T., and Kawasaki, M. (2002). *Phys. Rev. B* **66**, 085209.
Moreno, M., Trampert, A., Jenichen, B., Däweritz, L., and Ploog, K. H. (2002). *J. Appl. Phys.* **92**, 4672.
Mycielski, A. (1988). *J. Appl. Phys.* **63**, 3279.
Myers, R. C., Poggio, M., Stern, N. P., Gossard, A. C., and Awschalom, D. D. (2005). *Phys. Rev. Lett.* **95**, 017204.
Myers, R. C., Sheu, B. L., Jackson, A. W., Gossard, A. C., Schiffer, P., Samarth, N., and Awschalom, D. D. (2006). *Phys. Rev. B* **74**, 155203.
Nagaev, E. L. (2001). *Phys. Rep.* **346**, 387.
Neumaier, D., Schlapps, M., Wurstbauer, U., Sadowski, J., Reinwald, M., Wegscheider, W., and Weiss, D. (2008). *Phys. Rev. B* **77**, 041306.
Ohno, H., Munekata, H., Penney, T., von Molnár, S., and Chang, L. L. (1992). *Phys. Rev. Lett.* **68**, 2664.
Ohno, H., Shen, A., Matsukura, F., Oiwa, A., Endo, A., Katsumoto, S., and Iye, Y. (1996). *Appl. Phys. Lett.* **69**, 363.
Oiwa, A., Katsumoto, S., Endo, A., Hirasawa, M., Iye, Y., Ohno, H., Matsukura, F., Shen, A., and Sugawara, Y. (1997). *Solid State Commun.* **103**, 209.
Okabayashi, J., Kimura, A., Mizokawa, T., Fujimori, A., Hayashi, T., and Tanaka, M. (1999). *Phys. Rev. B* **59**, R2486.
Okabayashi, J., Mizokawa, T., Sarma, D. D., Fujimori, A., Slupinski, T., Oiwa, A., and Munekata, H. (2002). *Phys. Rev. B* **65**, 161203.
Okabayashi, J., Ono, K., Mizuguchi, M., Oshima, M., Gupta, S. S., Sarma, D. D., Mizokawa, T., Fujimori, A., Yuri, M., Chen, C. T., Fukumura, T., Kawasaki, M., et al. (2004). *J. Appl. Phys.* **95**, 3573.
Olejník, K., Owen, M. H. S., Novák, V. J., Mašek, A. C. I., Wunderlich, J., and Jungwirth, T. (2008). *Phys Rev. B* **78**, 054403.
Omiya, T., Matsukura, F., Dietl, T., Ohno, Y., Sakon, T., Motokawa, M., and Ohno, H. (2000). *Physica E* **7**, 976.
Oszwałdowski, R., Majewski, J. A., and Dietl, T. (2006). *Phys. Rev. B* **74**, 153310.
Ozaki, N., Okabayashi, I., Kumekawa, T., Nishizawa, N., Marcet, S., Kuroda, S., and Takita, K. (2005). *Appl. Phys. Lett.* **87**, 192116.

Ozaki, N., Nishizawa, N., Marcet, S., Kuroda, S., Eryu, O., and Takita, K. (2006). *Phys. Rev. Lett.* **97**, 037201.
Pacuski, W., Ferrand, D., Cibert, J., Deparis, C., Gaj, J. A., Kossacki, P., and Morhain, C. (2006). *Phys. Rev. B* **73**, 035214.
Pacuski, W., Ferrand, D., Cibert, J., Gaj, J. A., Golnik, A., Kossacki, P., Marcet, S., Sarigiannidou, E., and Mariette, H. (2007). *Phys. Rev. B* **76**, 165304.
Pacuski, W., Kossacki, P., Ferrand, D., Golnik, A., Cibert, J., Wegscheider, M., Navarro-Quezada, A., Bonanni, A., Kiecana, M., Sawicki, M., and Dietl, T. (2008). *Phys. Rev. Lett.* **100**, 037204.
Pappert, K., Schmidt, M. J., Humpfner, S., Ruster, C., Schott, G. M., Brunner, K., Gould, C., Schmidt, G., and Molenkamp, L. W. (2006). *Phys. Rev. Lett.* **97**, 186402.
Park, Y. D., Hanbicki, A. T., Erwin, S. C., Hellberg, C. S., Sullivan, J. M., Mattson, J. E., Ambrose, T. F., Wilson, A., Spanos, G., and Jonker, B. T. (2002). *Science* **295**, 651.
Parkin, S. S. P., Hayashi, M., and Thomas, L. (2008). *Science* **320**, 190.
Petukhov, A. G., Chantis, A. N., and Demchenko, D. O. (2002). *Phys. Rev. Lett.* **89**, 107205.
Pikus, G. E., and Bir, G. L. (1971). *Zh. Eksper. Teor. Fiz.* **60**, 195.
Poggio, M., Myers, R. C., Stern, N. P., Gossard, A. C., and Awschalom, D. D. (2005). *Phys. Rev. B* **72**, 235313.
Popescu, F., Yildirim, Y., Alvarez, G., Moreo, A., and Dagotto, E. (2006). *Phys. Rev. B* **73**, 075206.
Potashnik, S. J., Ku, K. C., Chun, S. H., Berry, J. J., Samarth, N., and Schiffer, P. (2001). *Appl. Phys. Lett.* **79**, 1495.
Przeździecka, E., Kamińska, E., kiecana, M., Sawicki, M., Kłopotowski, Ł., Pacuski, W., and Kossut, J. (2006). *Solid State Commun.* **139**, 541.
Rader, O., Pampuch, C., Shikin, A. M., Okabayashi, J., Mizokawa, T., Fujimori, A., Hayashi, T., Tanaka, M., Tanaka, A., and Kimura, A. (2004). *Phys. Rev. B* **69**, 075202.
Reed, M. J., Arkun, F. E., Berkman, E. A., Elmasry, N. A., Zavada, J., Luen, M. O., Reed, M. L., and Bedair, S. M. (2005). *Appl. Phys. Lett.* **86**, 102504.
Reiss, G., and Meyners, D. (2006). *Appl. Phys. Lett.* **88**, 043505.
Rinke, P., Qteish, A., Neugebauer, J., and Scheffler, M. (2007). *Ψk Newsletters* **79**, 163.
Rys, F., Helman, J. S., and Baltensperger, W. (1967). *Phys. Kondens. Materie* **6**, 107.
Sankowski, P., and Kacman, P. (2005). *Phys. Rev. B* **71**, 201303(R).
Sankowski, P., Kacman, P., Majewski, J., and Dietl, T. (2006). *Physica E* **32**, 375.
Sankowski, P., Kacman, P., Majewski, J., and Dietl, T. (2007). *Phys. Rev. B* **75**, 045306.
Sarigiannidou, E., Wilhelm, F., Monroy, E., Galera, R. M., Bellet-Amalric, E., Rogalev, A., Goulon, J., Cibert, J., and Mariette, H. (2006). *Phys. Rev. B* **74**, 041306(R).
Sato, K., Katayama-Yoshida, H., and Dederichs, P. H. (2005). *Japanan. J. Appl. Phys.* **44**, L948.
Sawicki, M., Dietl, T., Plesiewicz, W., Sękowski, P., Śniadower, L., Baj, M., and Dmowski, L. (1983). Influence of an acceptor state on transport in zero-gap $Hg_{1-x}Mn_xTe$. In "Application of High Magnetic Fields in Physics of Semiconductors" (G. Landwehr, ed.), p. 382. Springer, Berlin.
Sawicki, M., Dietl, T., Kossut, J., Igalson, J., Wojtowicz, T., and Plesiewicz, W. (1986). *Phys. Rev. Lett.* **56**, 508.
Sawicki, M., Wang, K. Y., Edmonds, K. W., Campion, R., Staddon, C., Farley, N., Foxon, C., Papis, E., Kamińska, E., Piotrowska, A., Dietl, T., and Gallagher, B. (2005). *Phys. Rev. B* **71**, 121302.
Scarpulla, M. A., Cardozo, B. L., Farshchi, R., Hlaing Oo, W. M., McCluskey, M. D., Yu, K. M., and Dubon, O. D. (2005). *Phys. Rev. Lett.* **90**, 207204.
Schneider, J., Kaufmann, U., Wilkening, W., Baeumler, M., and Köhl, F. (1987). *Phys. Rev. Lett.* **59**, 240.
Serrate, D., Teresa, J. M. D., and Ibarra, M. R. (2007). *J. Phys. Cond. Matter* **19**, 023201.
Shinde, S. R., Ogale, S. B., Higgins, J. S., Zheng, H., Millis, A. J., Kulkarni, V. N., Ramesh, R., Greene, R. L., and Vankatesan, T. (2004). *Phys. Rev. Lett.* **92**, 166601.

Śliwa, C., and Dietl, T. (2005). arXiv:cond-mat/0505126.
Śliwa, C., and Dietl, T. (2006). *Phys. Rev. B* **74**, 245215.
Śliwa, C., and Dietl, T. (2007). arXiv:0707.3542.
Sreenivasan, M. G., Teo, K. L., Cheng, X. Z., Jalil, M. B. A., Liew, T., Chong, T. C., Du, A. Y., Chan, T. K., and Osipowicz, T. (2007). *J. Appl. Phys.* **102**, 053702.
Stangl, J., Holý, V., and Bauer, G. (2004). *Rev. Mod. Phys.* **76**, 725.
Stern, N. P., Myers, R. C., Poggio, M., Gossard, A. C., and Awschalom, D. D. (2007). *Phys. Rev. B* **75**, 045329.
Story, T., Gałązka, R. R., Frankel, R. B., and Wolff, P. A. (1986). *Phys. Rev. Lett.* **56**, 777.
Świerkowski, L., and Dietl, T. (1988). *Acta Phys. Polon. A* **73**, 431.
Szczytko, J., Stachow, A., Mac, W., Twardowski, A., Belca, P., and Tworzydlo, J. (1996). *Acta Physica Polonica* **90**, 951.
Szczytko, J., Mac, W., Twardowski, A., Matsukura, F., and Ohno, H. (1999a). *Phys. Rev. B* **59**, 12935.
Szczytko, J., Twardowski, A., Swiatek, K., Palczewska, M., Tanaka, M., Hayashi, T., and Ando, K. (1999b). *Phys. Rev. B* **60**, 8304.
Szczytko, J., Bardyszewski, W., and Twardowski, A. (2001). *Phys. Rev. B* **64**, 075306.
Tanaka, M., et al. (2008). Properties and functionalities of mnas/iii-v hybrid and composite structures. In "Spintronics" (T. Dietl, D. D. Awschalom, M. K., and H. Ohno, eds.), pp. this volume. Elsevier, Amsterdam.
Timm, C. (2003). *J. Phys. Condens. Matter* **15**, R1865.
Timm, C., and MacDonald, A. (2005). *Phys. Rev. B* **71**, 155206.
Timm, C., Raikh, M. E., and von Oppen, F. (2005). *Phys. Rev. Lett.* **94**, 036602.
Twardowski, A., Swiderski, P., von Ortenberg, M., and Pauthenet, R. (1984a). *Solid State Commun.* **50**, 509.
Twardowski, A., von Ortenberg, M., Demianiuk, M., and Pauthenet, R. (1984b). *Solid State Commun.* **51**, 849.
Tworzydło, J. (1994) *Phys. Rev. B* **50**, 14591
Tworzydło, J. (1995). *Acta Phys. Polon. A* **94**, 821.
Van Dorpe, P., Van Roy, W., De Boeck, J., Borghs, G., Sankowski, P., Kacman, P., Majewski, J. A., and Dietl, T. (2005). *Phys. Rev. B* **72**, 205322.
van Schilfgaarde, M., and Mryasov, O. N. (2001). *Phys. Rev. B* **63**, 233205.
Wachter, P. (1979). In "Handbook on the Physics and Chemistry of Rare Earth" (K. A. Gschneidner,, Jr and L. Eyring, eds.), p. 507, Vol. 2. North-Holland, Amsterdam.
Wang, K., Edmonds, K., Campion, R., Gallagher, B., Foxon, T., Sawicki, M., Dietl, T., Bogusławski, P., and Jungwirth, T. (2005). *In*: "Proceedings 27th International Conference on the Physics of Semiconductors" (J. Menendez and C. G. Van de Walle, eds.), p. 333, Flagstaff, USA 2004 AIP, Melville.
Wilamowski, Z., Świątek, K., Dietl, T., and Kossut, J. (1990). *Solid State Commun.* **74**, 833.
Wolos, A., et al. (2008). Magnetic impurities in semiconductors. In "Spintronics" (T. Dietl, D. D. Awschalom, M. K., and H. Ohno, eds.), pp. this volume. Elsevier, Amsterdam.
Wołoś, A., Palczewska, M., Wilamowski, Z., Kamińska, M., Twardowski, A., Boćkowski, M., Grzegory, I., and Porowski, S. (2003). *Appl. Phys. Lett.* **83**, 5428.
Wolos, A., Palczewska, M., Zajac, M., Gosk, J., Kaminska, M., Twardowski, A., Bockowski, M., Grzegory, I., and Porowski, S. (2004a). *Phys. Rev. B* **69**, 115210.
Wolos, A., Wysmolek, A., Kaminska, M., Twardowski, A., Bockowski, M., Grzegory, I., Porowski, S., and Potemski, M. (2004b). *Phys. Rev. B* **70**, 245202.
Wu, J., Shan, W., and Walukiewicz, W. (2002). *Semicond. Sci. Technol.* **17**, 860.
Yakunin, A., Silov, A., Koenraad, P., Wolter, J., Roy, W. V., and Boeck, J. D. (2003). *Superlatt. and Microstruct.* **34**, 539.
Ye, L. H., and Freeman, A. J. (2006). *Phys. Rev. B* **73**, 081304(R).

Ye, S., Klar, P. J., Hartmann, T., Heimbrodt, W., Lampalzer, M., Nau, S., Torunski, T., Stolz, W., Kurz, T., Krug von Nidda, H. A., and Loidl, A. (2003). *Appl. Phys. Lett.* **83**, 3927–3929.

Yokoyama, M., Yamaguchi, H., Ogawa, T., and Tanaka, M. (2005). *J. Appl. Phys.* **97**, 10D317.

Yu, K. M., Walukiewicz, W., Wojtowicz, T., Kuryliszyn, I., Liu, X., Sasaki, Y., and Furdyna, J. K. (2002). *Phys. Rev. B* **65**, 201303.

Yu, K. M., et al. (2008). Fermi level effects on Mn incorporation in III-Mn-V ferromagnetic semiconductors. *In* "Spintronics" (T. Dietl, D. D. Awschalom, M. K., and H. Ohno, eds.), pp. this volume. Elsevier, Amsterdam.

Yuldashev, S. U., Im, H., Yalishev, V. S., Park, C. S., Kang, T. W., Lee, S., Sasaki, Y., Liu, X., and Furdyna, J. K. (2003). *Appl. Phys. Lett.* **82**, 1206.

Zaránd, G., and Jankó, B. (2002). *Phys. Rev. Lett.* **89**, 047201.

Zunger, A. (1986). Electronic structure of 3d transition-atom impurities in semiconductors. *In* "Solid State Physics" (F. Seitz and D. Turnbull, eds.), Vol. 39, pp. 275–464. Academic Press, New York.

CHAPTER 10

Computational Nano-Materials Design for the Wide Band-Gap and High-T_C Semiconductor Spintronics

Hiroshi Katayama-Yoshida,*,† Kazunori Sato,*
Tetsuya Fukushima,* Masayuki Toyoda,*
Hidetoshi Kizaki,*,† and An van Dinh*

Contents		
	1. Introduction	433
	2. Magnetic Mechanism, T_C, and Unified Physical Picture	436
	3. Spinodal Nano-Decomposition and Nano-Spintronics Applications	444
	4. New Class of Oxide Spintronics Without a 3d TM	450
	5. Conclusion	452
	References	452

1. INTRODUCTION

For the realization of ultrafast switching (THz), high-integration (Tera-bits/in.^{2}), and energy-saving (nonvolatile) semiconductor nano-spintronics to go beyond Si-CMOS technology (Wolf et al., 2001), we need (1) to realize a high Curie temperature (T_C) in diluted magnetic

---

* The Institute of Scientific and Industrial Research, Osaka University, Ibaraki, Mihogaoka 8-1, Osaka 567-0047, Japan
† Department of Materials Engineering Science, Graduate School of Engineering Science, Osaka University, Toyonaka, Machikaneyama 1-3, Osaka 560-8531, Japan

semiconductors (DMS) ($T_C > 1000$ K), (2) to develop a new fabrication method for 100 Tbits/in.^{2} semiconductor nano-magnets by the self-organization, and (3) to develop a colossal magnetic response controlled by the electric field or photonic excitations. To do so, we have developed a computational nano-materials design system for semiconductor spintronics, which is called the CMD® System at Osaka University (see Fig. 10.1). On the basis of a quantum simulation, we can analyze the physical mechanisms of ferromagnetism in the DMS. Then, we can design new functional nano-materials based on an integration of the physical mechanisms proposed by quantum simulations. We can verify the functionality by the quantum simulations. We analyze the reason why the new designed materials do not satisfy our demands, and then we propose another designed candidate as an alternative. Using this circulation of design processes, finally, we design the desired new functional materials, and then we offer an experimental group to fabricate and realize the designed new functional materials. If the experimentally fabricated materials do not satisfy our demands, we can determine the reason why the newly fabricated materials do not satisfy our final demands. Then, on the basis of the analysis, we can design more appropriate new functional materials by using the CMD® System.

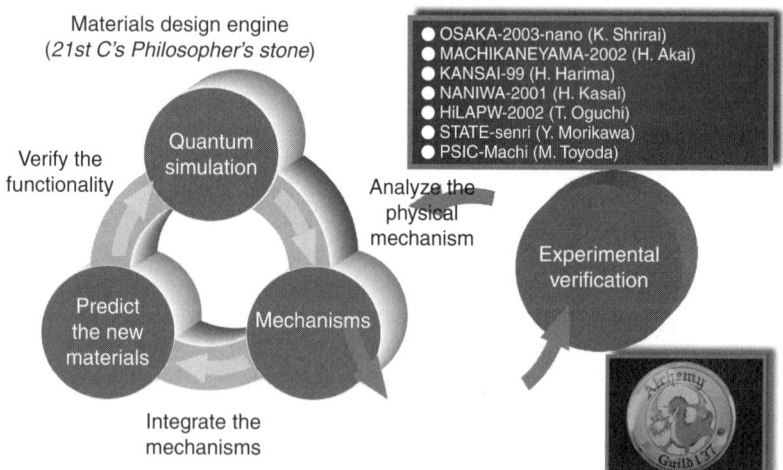

FIGURE 10.1 Computational nano-materials design system (CMD® System) developed in Osaka University by the CMD® consortium. The CMD® system includes FLAPW (HiLAPW), KKR-CPA-LDA (Machkaneyama-2002), KKR-CPA-SIC-LDA (PSIC-Machi), Ultrasoft pseudopotential molecular dynamics (STATE-senri), Car-Parrinello-type molecular dynamics (OSAKA-2003-nano), and quantum-dynamics simulation for the chemical reaction (NANIWA-2001).

The wide band-gap semiconductors such as ZnTe, ZnS, ZnSe, ZnO, and GaN-based DMS system is a disordered system, where Zn or Ga atoms are randomly replaced by transition-metal (TM) impurities. We treat the substitutional disorder by using the Korringa–Kohn–Rostoker coherent-potential approximation (KKR-CPA) method (Akai, 1998; Akai and Dederichs, 1993) with the local density approximation (LDA) and the self-interaction corrected LDA (SIC-LDA) to go beyond the LDA (Dinh *et al.*, 2006; Toyoda *et al.*, 2006a,b). We use the KKR-CPA program package MACHIKANEYAMA-2002 developed by Akai (http://sham.phys.sci.osaka-u.ac.jp/kkr/). In the KKR-CPA, an effective medium, which describes the configuration average of the disordered system, is calculated self-consistently within the single-site approximation (Shiba, 1971). The LDA is well known to be sometimes insufficient for correlated electron systems such as ZnO, where the LDA band gap is $\sim$1 eV contrary to the experimentally observed band gap of 3.3 eV (Toyoda *et al.*, 2006a,b). In order to go beyond the LDA, we have developed a new method to take into account the self-interaction correction to the LDA (SIC-LDA), where the band gap of ZnO is 2.8 eV (Dinh *et al.*, 2006; Toyoda *et al.*, 2006a,b).

On the basis of calculations by using KKR-CPA-LDA (Akai,1998; Akai and Dederichs, 1993; http://sham.phys.sci.osaka-u.ac.jp/kkr/) and KKR-CPA-SIC-LDA (Dinh *et al.*, 2006; Toyoda *et al.*, 2006a,b), we propose a unified physical picture of the magnetism in the wade band-gap II–VI- and III–V-based DMS, where *Zener's double-exchange mechanism* (Akai, 1998; Sato and Katayama-Yoshida, 2000, 2001, 2002; Sato *et al.*, 2003, 2004) and the *super-exchange interaction mechanisms* (Goodenough, 1955; Kanamori, 1959) are dominant (Sato and Katayama-Yoshida, 2000, 2001a,b,c, 2002; Sato *et al.*, 2003, 2004a,b). In a homogeneous system, we calculate almost exactly the Curie temperature (T_C) by using a Monte Carlo simulation (Bergqvist *et al.*, 2004; Sato *et al.*, 2004), combined with the magnetic force theorem, and obtain good agreement with the experiments [T_C *and photoemission spectroscopy (PES)*]. In an inhomogeneous system, we propose a three-dimensional (*3D*) *Dairiseki phase* (Katayama-Yoshida *et al.*, 2007a,b,c; Sato *et al.*, 2005, 2006) and a *1D Konbu phase* (Fukushima *et al.*, 2006; Katayama-Yoshida *et al.*, 2007a,b,c) caused by spinodal nano-decomposition. Spinodal nano-decomposition is responsible for the high-T_C [or high blocking (B) temperature (T_B), (Sato *et al.*, 2007)] phases in real DMS. We design a position control method that uses seeding (*the top-down nanotechnology*) and a shape control method that uses the vapor pressure in self-organization (*the bottom-up nanotechnology*) (Fukushima *et al.*, 2006; Katayama-Yoshida *et al.*, 2007a,b,c). We show the self-organized fabrication method for nano-magnets with 100 Tbits/in.^{2} densities by using thermal nonequilibrium crystal growth methods such as molecular-beam epitaxcy (MBE), metal-organic chemical-vapor deposition (MOCVD), or metal-organic vapor-phase epitaxcy (MOVPE) (Fukushima *et al.*, 2006; Katayama-Yoshida *et al.*, 2007a,b,c).

2. MAGNETIC MECHANISM, T_C, AND UNIFIED PHYSICAL PICTURE

In the framework of the KKR-CPA method, the concept of a disordered local moment (DLM) state is conveniently used to describe the paramagnetic state at finite temperature (Akai, 1998). The DLM state is well known to describe the paramagnetic state of a ferromagnet above T_C, and by combining the mean-field approximation (MFA) (Dietl, 2002; Jungwirth et al., 2006; Koenig et al., 2002), the random phase approximation (RPA) (Bouzerar et al., 2003; Pajda et al., 2001), and Monte Carlo simulation (MCS, almost exact) (Bergqvist et al., 2004; Sato et al., 2004a,b) with the magnetic force theorem, we can estimate T_C from first principles (Fukushima et al., 2004; Sato et al., 2003, 2004, 2007). This approach was used for a systematic study on the origin of the ferromagnetism in DMS. In Fig. 10.2, the calculated total energy difference per one formula unit between the ferromagnetic (FM) state and the paramagnetic state (spin-glass state) is shown for V-, Cr-, Mn-, Fe-, Co-, and Ni-doped (see

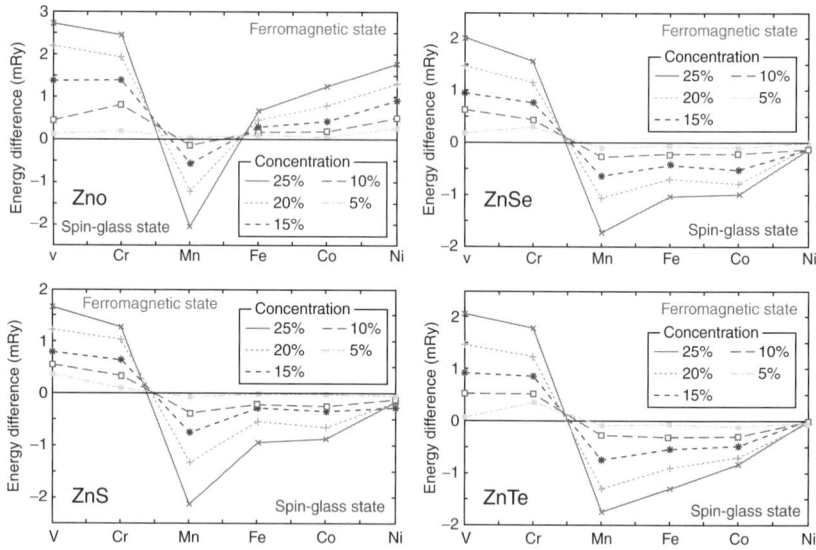

FIGURE 10.2 Calculated total energy difference between the spin-glass states and the ferromagnetic states for diluted magnetic semiconductors (DMS) in ZnO, ZnSe, ZnS, and ZnTe, calculated by using the local density approximation, as a function of 3d transition-metal impurity concentration. A positive value means that the ferromagnetic state is more stable than the spin-glass state. V and Cr show ferromagnetism due to Zener's double-exchange mechanism; on the other hand, Mn shows an antiferromagnetic spin-glass state due to the antiferromagnetic superexchange interaction for the d^5 configuration based on the Kanamori–Goodenough rule (Goodenough, 1955; Kanamori, 1959).

Fig. 10.2) II–VI compound semiconductors, such as ZnO, ZnS, ZnSe, and ZnTe (Sato and Katayama-Yoshida, 2000, 2001, 2002). The positive energy difference indicates that the FM state is more stable than the paramagnetic state. The FM state is stable for the first-half of the 3d TM series, whereas the paramagnetic state is stable for the latter-half. The paramagnetic state is most stable in Mn-doped II–VI DMS. In order to elucidate the mechanism that stabilizes the FM states in DMS, we calculate the density of states (DOS) in ZnSe-based DMS. In Figs. 10.3–10.5, the total DOS per unit cell and the partial DOS of 3d states per TM atom at the TM site obtained by using the LDA (Sato and Katayama-Yoshida, 2000, 2001, 2002) and the SIC-LDA (Dinh et al., 2006; Toyoda et al., 2006a,b) are shown for the FM state, compared with the results from PES (Ishida et al., 2004; Kobayashi et al., cond-mat; Mizokawa et al., 2001). The calculated total and partial V 3d DOS of FM (Zn,V)O with 5% V-doping obtained by the LDA and the SIC-LDA (PSIC-LDA) are compared with the results from PES obtained by Ishida et al. (2004). In the SIC-LDA, the energy positions of the deep-impurity band and the Zn-3d state are in good agreement with the PES,

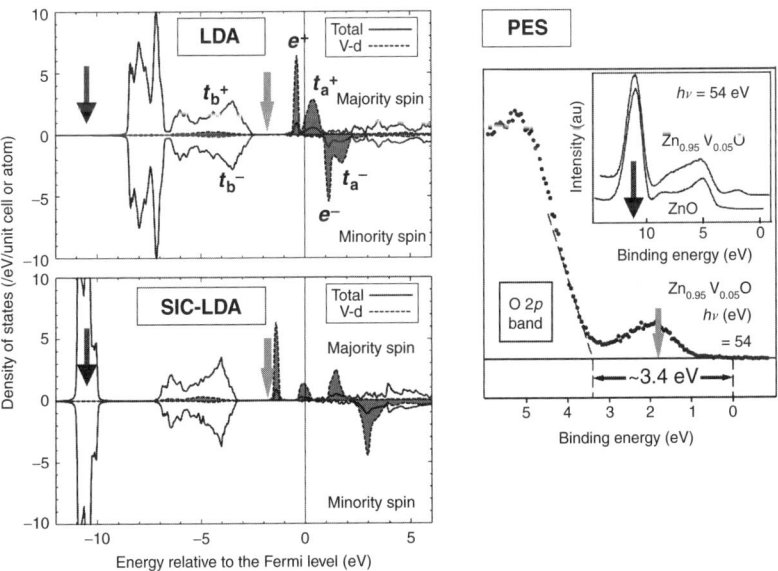

FIGURE 10.3 Calculated total and partial V 3d density of states of ferromagnetic (Zn,V)O with 5% V-doping calculated by using the local density approximation (LDA) and self-interaction corrected-LDA (SIC-LDA) (PSIC-LDA), compared with the results of photoemission spectroscopy (PES) obtained by Ishida et al. (2004). The grey arrow corresponds to the deep-impurity band and the black arrow corresponds to the Zn-3d state observed by PES. A partially occupied deep-impurity band in the band gap stabilizes the ferromagnetism due to Zener's double-exchange mechanism.

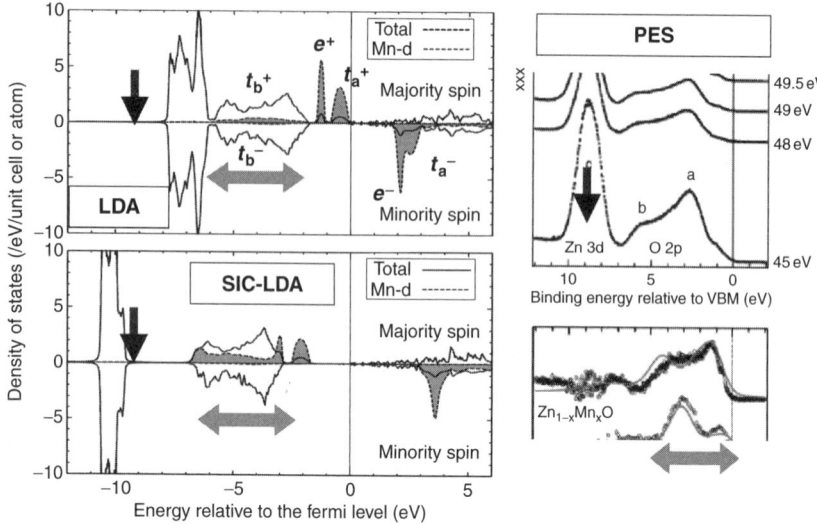

FIGURE 10.4 Calculated total and partial Mn-3d density of states of (Zn,Mn)O with 5% Mn doping calculated by using the local density approximation (LDA) and self-interaction corrected-LDA (SIC-LDA) (PSIC-LDA), compared with the results of photoemission spectroscopy (PES) by Mizokawa et al. (2001). The grey arrow corresponds to the deep-impurity band in the valence band observed by PES and the black arrow corresponds to Zn-3d state observed by PES. Since the Fermi level is located in the band gap, we have no partially occupied deep-impurity band to stabilize the ferromagnetism caused by Zener's double-exchange mechanism. The antiferromagnetic superexchange interaction for d^5 configuration dominates the antiferromagnetic spin-glass (paramagnetic) state. The SIC-LDA is quantitatively consistent with the PES data, contrary to the LDA.

contrary to the LDA. Both the LDA and the SIC-LDA indicate that the partially occupied deep-impurity band in the band gap stabilizes the ferromagnetism due to the Zener's double-exchange mechanism (see Fig. 10.3).

The total and partial Mn-3d DOS of (Zn,Mn)O with 5% Mn-doping calculated by using the LDA and the SIC-LDA (PSIC-LDA) are compared with the results of PES obtained by Mizokawa et al. (2001) and shown in Fig. 10.4. The agreement of the deep-impurity bands in the valence band and the Zn-3d state observed using the PES and the SIC-LDA is reasonable. Since the Fermi level is located in the band gap, we have no partially occupied deep-impurity band to stabilize the ferromagnetism caused by Zener's double-exchange mechanism. The antiferromagnetic (AF) super-exchange interaction for the d^5 configuration dominates the AF spin-glass (*paramagnetic*) state. The SIC-LDA is quantitatively consistent with the PES, contrary to the LDA.

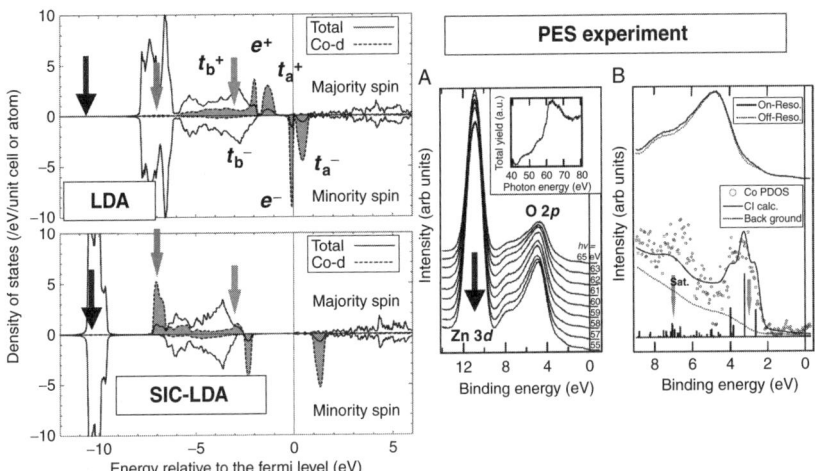

FIGURE 10.5 Calculated total and partial Co-3d density of states of ferromagnetic (Zn, Co)O with 5% Co doping calculated by using the local density approximation (LDA) and self-interaction corrected-LDA (SIC-LDA) (PSIC-LDA), compared with the results of photoemission spectroscopy (PES) obtained by Kobayashi *et al.* (cond-mat). The grey arrow corresponds to the resonant deep-impurity band in the valence band observed by PES, the black arrow corresponds to Zn-3d state observed by PES, and the white arrow is the satellite two-hole bound states with a resonance in the valence band. Since the Fermi level is located in the band gap, we have no partially occupied deep-impurity band to stabilize the ferromagnetism caused by Zener's double-exchange mechanism. However, the ferromagnetic superexchange interaction for the d^7 configuration dominates the ferromagnetic interaction. The SIC-LDA is quantitatively consistent with PES, contrary to the LDA.

The total and partial Co-3d DOS of FM (Zn,Co)O with 5% Co-doping calculated by using the LDA and the SIC-LDA (PSIC-LDA) are compared with the results of PES obtained by Kobayashi *et al.* (cond-mat) and shown in Fig. 10.5. Agreement between the SIC-LDA and the PES is also reasonable for the Zn-3d state and the satellite of two-hole bound states with resonance in the valence band. Since the Fermi level is located in the band gap, we have no partially occupied deep-impurity band to stabilize the ferromagnetism caused by Zener's double-exchange mechanism. However, the FM superexchange interaction for the d^7 configuration dominates the FM interaction. The SIC-LDA is quantitatively consistent with the PES, contrary to the LDA.

In order to discuss the T_C from the first-principle calculations, we have calculated the effective exchange interaction projected on the classical Heisenberg model by using the magnetic force theorem in (Zn,Cr)Te as a function of distance. As Liechtenstein *et al.* (1987) proposed, we calculate the total energy change due to infinitesimal rotations of the two magnetic moments embedded at sites i and j in the effective CPA medium. The total energy change is mapped on the classical Heisenberg model to deduce the effective exchange interactions. We can calculate the T_C by using MFA (Dietl, 2002; Jungwirth et al., 2006; Koenig et al., 2002), RPA (Bouzerar et al., 2003; Pajda et al., 2001), and MCS (Bergqvist et al., 2004; Sato et al., 2004) for (Zn,Cr)Te (Fukushima et al., 2004). Available experimental values (Read et al., 2001; Koenig et al., 2002) are also shown in Fig. 10.6. In a homogeneous system, the T_C calculated by using the MCS is almost exactly in agreement with the experimental values (Read et al., 2001; Koenig et al., 2002). In the wide band-gap II–VI compound semiconductors, the exchange interaction is strong, but short-ranged: therefore, it is still very difficult to go beyond $T_C > 1000$ K due to the percolation problem (Stauffer and Aharony, 1994) in DMS.

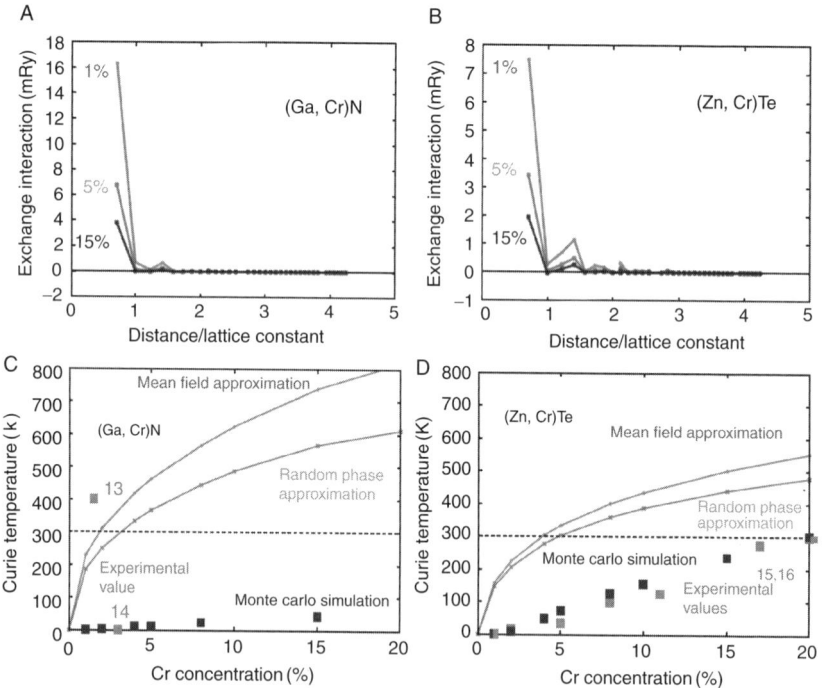

FIGURE 10.6 Calculated effective exchange interaction projected on the Heisenberg model of (A) (Ga,Cr)N and (B) (Zn,Cr)Te as a function of distance. (C–D) T_C calculated by using the mean-field approximation, the random phase approximation, and a Monte Carlo simulation for (Ga,Cr)N and (Zn,Cr)Te. Available experimental values (Reed et al., 2001) are also shown (*squares*). In the homogeneous system, T_C calculated by using a Monte Carlo simulation (*squares*) is almost exact and is in good agreement with the experimental values.

In each case of II–VI-based DMS in the LDA and the SIC-LDA (see Figs. 10.3–10.5), between the valence band and the conduction band, we find clear impurity bands originating from the 3d states of TM impurities. These impurity bands show large exchange splitting, and they are gradually occupied as the atomic number of the impurity is increased. Because the exchange splitting (Δ_X) is larger than the crystal field splitting (Δ_{CF}), as shown in Figs. 10.3–10.5, and 10.7, all TMs are in the high-spin states due to the highly localized nonbonding e-state (Sato and Katayama-Yoshida, 2000, 2001, 2002; Sato et al., 2003, 2004). In particular, the half-metallic DOS are realized in V-, Cr-, Fe-, and Ni-doped ZnTe (Sato and Katayama-Yoshida, 2000, 2001, 2002). Figure 10.4 shows that the Mn impurity has a d^5 electron configuration due to the substitution of Zn^{2+} for Mn^{2+}. In this case, we suggest that the AF superexchange interaction (Kanamori, 1959; Goodenough, 1955) between Mn ions stabilizes the paramagnetic state.

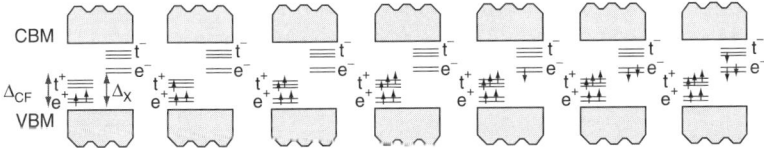

II-VI	$Ti^{2+}(3d^2)$	$V^{2+}(3d^3)$	$Cr^{2+}(3d^4)$	$Mn^{2+}(3d^5)$	$Fe^{2+}(3d^6)$	$Co^{2+}(3d^7)$	$Ni^{2+}(3d^8)$
III-V	$V^{3+}(3d^2)$	$Cr^{3+}(3d^3)$	$Mn^{3+}(3d^4)$	$Fe^{3+}(3d^5)$	$Co^{3+}(3d^6)$	$Ni^{3+}(3d^7)$	$Cu^{3+}(3d^8)$
Double-exchange interaction	▬	FM	FM	▬	FM	▬	FM
Super-exchange interaction	FM	AF	AF	AF	AF	FM/AF	AF

FIGURE 10.7 Electron configuration of magnetic impurities in diluted magnetic semiconductors (DMS) with a high-spin ground state, where exchange splitting (Δ_X) between up (+) and down (−) spin states is larger than the crystal field splitting (Δ_{CF}) between t_2 (t) and e states (e), and chemical trend of magnetism in II–VI and III–V compound semiconductor-based DMS. Ferromagnetic (FM) and antiferromagnetic (AF) indicate the magnetic interactions for ferromagnetic, and antiferromagnetic, states, respectively. The partially occupied deep-impurity band stabilizes the ferromagnetism due to Zener's double-exchange interaction caused by the kinetic energy gain of electron or hole hopping. The superexchange interaction caused by virtual charge transfer is dominated by the Kanamori–Goodenough rule (Goodenough, 1955; Kanamori, 1959), where d^2 and d^7 configurations contribute to the ferromagnetic interaction and others are all antiferromagnetic interactions. Since Zener's double-exchange interaction is always stronger than the superexchange interaction, Zener's double-exchange interaction dominates the magnetisms in II–VI-based DMS.

On the other hand, V^{2+}, Cr^{2+}, Fe^{2+}, and Ni^{2+} have d^3, d^4, d^6, and d^8 electronic configurations, respectively; therefore, the 3d impurity bands of up-spin states or down-spin states are not fully occupied. In these cases, the 3d electron in the partially occupied 3d orbitals of TM is allowed to hop to 3d orbitals of neighboring TM when the neighboring TM ions have parallel magnetic moments. As a result, the d electron lowers its kinetic energy by hopping in the FM state. This is the so-called Zener's FM double-exchange mechanism (Sato and Katayama-Yoshida, 2002; Sato et al., 2003, 2004). This kind of energy gain is not expected when neighboring TMs have antiparallel magnetic moments; therefore, Zener's double-exchange interaction is an FM interaction. Akai (1998) has already pointed out this mechanism in his paper on the magnetism in (In,Mn)As. The FM double-exchange interaction does not work for d^2, d^5, and d^7 cases because the Fermi level is located in the gap of the impurity band for these cases (Sato and Katayama-Yoshida, 2000, 2001, 2002; Sato et al., 2003, 2004).

According to the Kanamori–Goodenough rule (Kanamori, 1959; Goodenough, 1955), the superexchange becomes FM for particular combinations of 3d orbitals, for example, for the d^2 and d^7 cases in T_d symmetry. The electron configuration of TM impurities in II–VI DMS and the chemical trend of the magnetism are summarized in Fig. 10.7. This general rule agrees very well with the available experimental data and with recent experimental verifications. As shown in Fig. 10.7, the competition between Zener's double-exchange mechanism and the superexchange interaction mechanism determines the magnetic state of the II–VI DMS. As shown in Fig. 10.3, most of the DMS belong to the deep-impurity band in II–VI DMS, where Zener's double-exchange mechanism dominates; therefore, Dietl's theory based on Zener's p–d exchange mechanism (Dietl, 2002, 2002; Jungwirth et al., 2006, 2006; Koenig et al., 2002, 2002) cannot apply only for II–VI DMS, but is adequate only for a limited case such as (Ga,Mn)Sb with very shallow impurity levels in the band gap and with highly localized 3d states in the deep valence band (Sato and Katayama-Yoshida, 2000, 2001, 2002; Sato et al., 2003, 2004).

In order to discuss the doping dependence of the magnetism, we depict the calculated total energy difference between the spin-glass states (*local magnetic moment disordered state*) and the FM states in (Zn,Mn)O, (Zn,Fe)O, (Zn,Co)O, and (Zn,Ni)O as a function of donor (Ga-donor doping) and acceptor (N-acceptor doping) codoping with 3d TM impurity in the LDA (see Fig. 10.8.). The positive value means that FM state is more stable than the spin-glass state. Hole-doping by the N-acceptor stabilizes the ferromagnetism in (Zn,Mn)O, whereas electron-doping by the Ga-donor stabilizes the ferromagnetism in (Zn,Fe)O, (Zn,Co)O, and (Zn,Ni)O. For spintronics device application, we can use the spin control method by applying a gate voltage to the II–VI DMS.

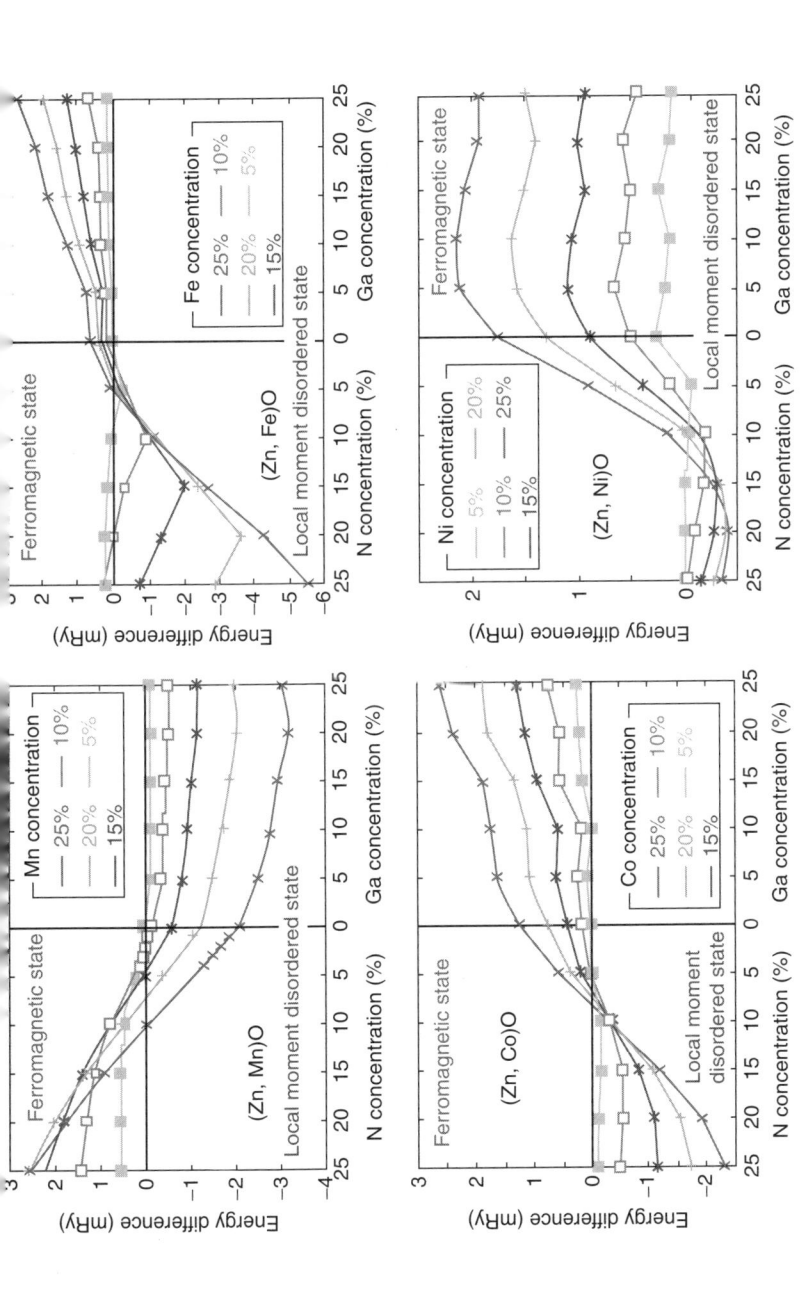

FIGURE 10.8 Calculated total energy differences between the spin-glass states (local magnetic moment disordered state) and the ferromagnetic states in (Zn,Mn)O, (Zn,Fe)O, (Zn,Co)O, and (Zn,Ni)O as functions of the donor (Ga-acceptor doping) and the acceptor (N-acceptor doping) codoping in the LDA. A positive value means that the ferromagnetic state is more stable than the spin-glass state. Hole-doping by an N-acceptor stabilizes the ferromagnetism in (Zn,Mn)O; on the other hand, electron-doping by a Ga-donor stabilizes the ferromagnetism in (Zn,Fe)O, (Zn,Co)O, and (Zn,Ni)O.

3. SPINODAL NANO-DECOMPOSITION AND NANO-SPINTRONICS APPLICATIONS

As shown in Section 10.2, we have succeeded in reproducing the experimental T_C in (Zn,Cr)Te accurately from first principles. These systems are supposed to be homogeneous because these materials are grown by using low-temperature MBE to control the atomic diffusion in order to avoid phase separation. In contrast to this success, agreement between the theory and the experiments is not satisfactory in wide band-gap DMS, such as ZnO-based and nitride DMS, so a more realistic description of wide band-gap DMS is needed. For example, in general, DMS systems have a solubility gap, and they show phase separation (or *spinodal decomposition*) in thermal equilibrium; however, in previous theoretical approaches, a homogeneous impurity distribution is supposed.

In Fig. 10.9, we depict the calculated mixing energy of (Zn,Cr)Te obtained by using the KKR-CPA-LDA method, and most of the calculated mixing energy in the II–VI and III–V DMS shows a convexity in the mixing energy as a function of the concentration (Katayama-Yoshida *et al.*, 2007a,b,c; Sato *et al.*, 2007a,b,c). These results are consistent with the experimental evidence that these systems show phase separation in thermal equilibrium and that wide band-gap II–VI DMS show stronger phase separation compared with the III–V arsenide. The chemical trends of convexity of the mixing energy can be understood by using the band-gap energy difference where the larger band gap gives larger convexity in the mixing energy, and the mixing energies of (Al,Mn)Sb, (In,Mn)Sb, and (Ga,Mn)Sb have very small convexity or concavity (Katayama-Yoshida *et al.*, 2007; Sato and Katayama-Yoshida, 2007). We can expect a nanoscale-size spinodal decomposition by using a thermal nonequilibrium crystal growth technique such as low-temperature MBE, MOCVD, and MOVPE, in which we can control the atomic diffusion on the surface of the crystal growth (Fukushima *et al.*, 2006; Katayama-Yoshida *et al.*, 2007a,b,c; Sato *et al.*, 2005, 2006).

Figure 10.10 depicts the calculated effective chemical pair interactions between i and j sites, projected on Ising model, of $V_{ij} = (V_{ij}^{Cr,Cr} + V_{ij}^{Zn,Zn} - 2V_{ij}^{Zn,Cr})$ as a function of distance measured by using the lattice constant in (Zn,Cr)Te. Negative interactions in V_{ij} mean that the attractive chemical interaction between the Cr impurities in ZnTe and the short-range interactions are due to the deep impurity band in the wide band gap. These are consistent with the steep convexity in the mixing energy in (Zn,Cr)Te. The attractive interaction accelerates the spinodal nano-decomposition under thermal nonequilibrium crystal growth conditions.

Here, we focus on the spinodal nano-decomposition in II–VI DMS, such as (Zn,Cr)Te and (Zn,Co)O, and study how the inhomogeneous impurity distribution affects the ferromagnetism. The spinodal nano-decomposition

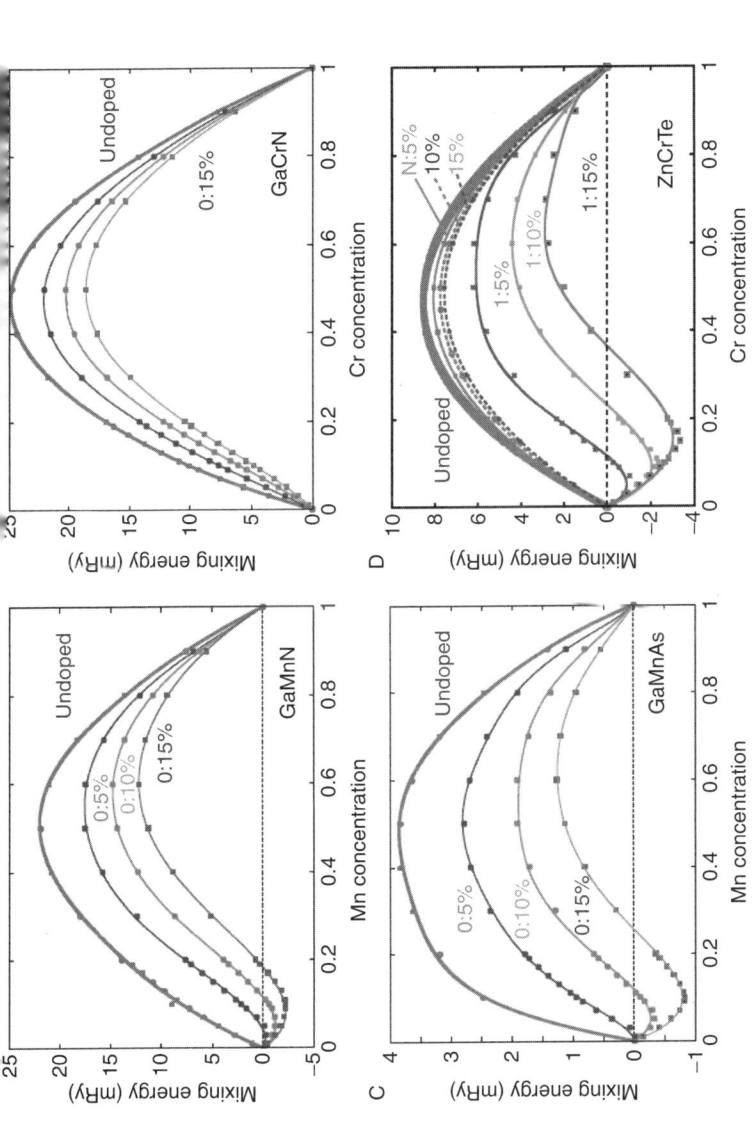

FIGURE 10.9 Calculated mixing energy of (Ga,Mn)N, (Ga,Cr)N, (Ga,Mn)As, and (Zn,Cr)Te upon Mn, Cr doping with the codoping of donors (O, I) and acceptors (N). The mixing energy is calculated as a function of magnetic impurity concentration. We can reduce the mixing energy by codoping in (A)–(D) for all the cases on compensation. For (Zn,Cr)Te (see panel D), the iodine (I) donor or nitrogen (N) acceptor is considered as codopants to control the convexity (*with spinodal nano-decomposition or phase separation*) and concavity (*without spinodal nano-decomposition*) of the mixing energy. The concentrations of codopants are 5%, 10%, and 15%. A negative mixing energy means that we can have alloying in thermal equilibrium. The lines are polynomial fitting curves to calculated points.

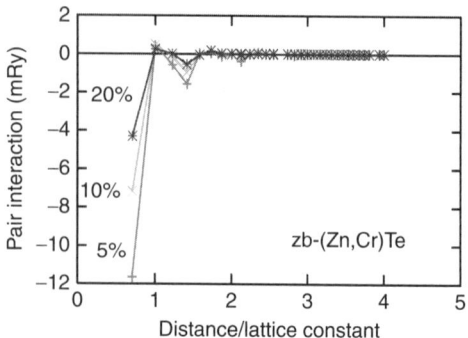

FIGURE 10.10 Calculated effective chemical pair interactions between i and j sites, projected on the Ising model, of $V_{ij} = (V_{ij}^{Cr,Cr} + V_{ij}^{Zn,Zn} - 2V_{ij}^{Zn,Cr})$ as a function of distance measured by using the lattice constant in (Zn,Cr)Te. Negative interactions in V_{ij} mean the attractive interaction between the Cr impurities in ZnTe and the calculation indicates short-range interactions.

in II–VI DMS is simulated by applying the Monte Carlo method for the Ising model by using the calculated chemical pair interactions in Fig. 10.10. In the MCS of the spinodal nano-decomposition, we have fixed the structure of host semiconductors; therefore, we did not take into account the local lattice relaxation. The Curie temperatures of the simulated spinodal nano-decomposition phases are calculated within the RPA taking disorder into account (Katayama-Yoshida et al., 2007a,b,c; Sato et al., 2005, 2006). This method gives us the almost same value of the T_C as was obtained by using the MCS.

In Fig. 10.11, the simulation results for spinodal nono-decomposition in 3D diffusion of TM impurities in (Zn,Cr)Te are summarized. Figure 10.11 shows the initial random configuration of (Zn,Cr)Te, where $T_C = 200$ K due to the percolation path is achieved by 20% Cr [the percolation limit (Stauffer and Aharony, 1994) due to the nearest-neighbor interaction is 20% for (Zn,Cr)Te]. Because of the attractive interactions, (Zn,Cr)Te shows spinodal nano-decomposition in the MCS (Sato et al., 2005, 2006) and the complete percolation network is connected, so that the T_C is enhanced dramatically up to 450 K.

The $T_C = 60$ K for the case of the low concentration ($\sim$5%) is below the percolation limit in Fig. 10.11. We can expect spinodal nano-decomposition to form small clusters [called *3D Dairiseki phase* (Sato et al., 2005, 2006), the meaning is marble in Japanese] with strong FM coupling inside the cluster, but weak FM coupling between the small clusters due to the short-range FM interactions. Therefore, the system shows a superparamagnetic state or very low T_C ($T_C = 20$ K) due to missing the percolation path in the whole crystal (Fig. 10.11). For the investigated concentration of 5%, the system is superparamagnetic or

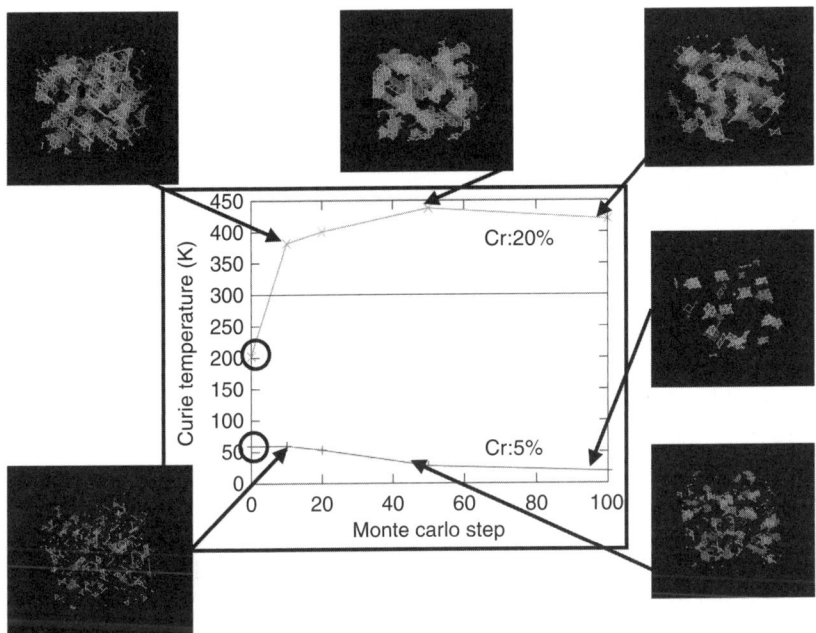

FIGURE 10.11 Monte Carlo simulation of T_C and three-dimensional (3D) spinodal nano-decomposition phase (called the *3D Dairiseki Phase*) of (Zn,Cr)Te (Cr 5% and 20%). For a low concentration of 5% Cr, T_C decreases with increasing spinodal nano-decomposition due to the small cluster formation and loss of the ferromagnetic percolation path between the clusters due to the short-range ferromagnetic exchange interaction by Zener's double-exchange interaction; then, T_C decrease and reaches the paramagnetic state ($T_C = 0$). For a high concentration of 20% Cr, just above the percolation limit, T_C increases dramatically up to 450 K due to an increased number of the percolation paths. We plot the Cr atoms in this figure, they are bonded by lines with nearest neighbor Cr atoms. The circles indicate T_C without spinodal nano-decomposition (*completely random cases*).

low T_C ($T_C = 20$ K) because small isolated clusters are formed in DMS due to the spinodal nano-decomposition (*3D Dairiseki phase*) (Sato et al., 2005, 2006). This discussion and the simulation are limited only to 3D diffusion of 3d TM impurities in DMS.

Using the layer-by-layer crystal growth method in atomic layer epitaxy (ALE) by using a thermal nonequilibrium crystal growth technique, such as MBE, MOCVD, or MOVPE, we can control the 2D diffusion of 3d TM impurities on the crystal growth surface (Katayama-Yoshida, 2007). We can simulate the spinodal nano-decomposition in the case of layer-by-layer crystal growth by using an MCS (Fukushima et al., 2006; Katayama-Yoshida et al., 2007a,b,c). By introducing a layer-by-layer crystal growth condition in our simulations, we find that a quasi-1D nano-superstructures (called a pseudo-*1D Konbu phase*) (Fukushima et al., 2006; Katayama-Yoshida et al., 2007a,b,c) of impurities are formed in (Zn,Cr)Te and (Ga,Mn)N, even for a low concentration (<5%), as shown in Fig. 10.12(B) and (D). MCS for the magnetization process of the decomposition phases indicates that the superparamagnetic blocking temperature (T_B) could be higher because the activation energy required to flip the magnetization becomes larger for the decomposition phases (Sato et al., 2007). As shown in Fig. 10.13, by simulation, we show that we can specify the position of nano-structures in the 1D *Konbu phase* by putting seeds at desired positions before the MBE process (Fukushima et al., 2006; Katayama-Yoshida et al., 2007a,b,c). Moreover, by controlling the impurity concentration during MBE growth, we can control the shape of the nano-superstructures, as shown in Fig. 10.13. These simulations suggest a new way to fabricate 100 Tbits/in.^{2} densities of a nano-magnet array with a controlled shape (Fukushima et al., 2006; Katayama-Yoshida et al., 2007a,b,c).

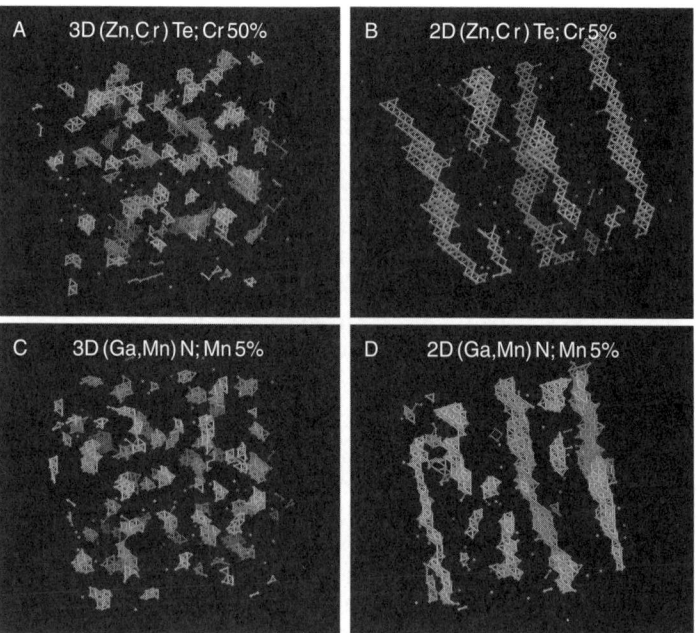

FIGURE 10.12 (A) Three-dimensional (3D) self-organized *Dairiseki phase* of (Zn,Cr)Te (Cr 5%; $T_C = 0$ K) and (B) the 2D spinodal nano-decomposition of (Zn,Cr)Te (Cr 5%), called the *1D Konbu phase*, with a high blocking temperature even for the superparamagnetic state ($T_C = 0$ K). (C) Self-organized *3D Dairiseki phase* of (Ga,Mn)N (Mn 5%; $T_C = 0$ K) and (D) the 2D spinodal nano-decomposition of (Ga,Mn)N (Mn 5%), called the *1D Konbu phase*. We plot Cr or Mn atoms in gray color and they are bonded by lines with nearest neighbor Cr or Mn atoms.

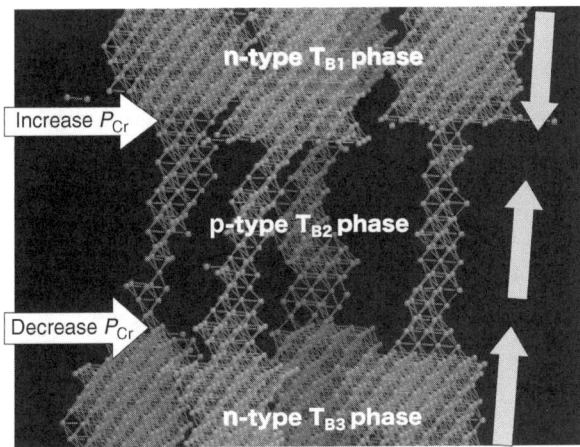

FIGURE 10.13 Monte Carlo simulation of spinodal nano-decomposition in the layer-by-layer crystal growth condition for the fabrication of a nano-spintronics device and formation of one-dimensional (1D) nano-superstructure magnets (*1D Konbu phase*) with some designed structures in the ZnTe matrixes by spinodal nano-decomposition in (Zn,Cr)Te. Cr impurities are shown by gray dots and a pair of nearest-neighbor Cr is bonded by a line. We prepare the nano-seeding by the 25 periodic Cr atoms by using nano-sphere lithography on the substrate; then, we simulate the layer-by-layer crystal growth by using a Monte Carlo simulation within the condition of 2D diffusion of Cr impurities on the seeded ZnTe surface. The shape control was done by reducing the vapor pressure of Cr atoms during the layer-by-layer crystal growth (*lower white arrow*) and by increasing (*upper white arrow*) the vapor pressure of Cr atoms during the 2D layer-by-layer crystal growth. We can fabricate the three different connected nano-magnets with different high blocking temperatures (T_B) and different directions of magnetization (*indicated by gray arrows*). Using the spin current, we can flip the magnetization in the central nano-magnets. The density of these nano-magnets is 100 Tbits/in.^{2}.

We can control the mixing energy of (Zn,Cr)Te by codoping (Yamamoto and Katayama-Yoshida, 1997), as was shown in Fig. 10.9, as a function of the magnetic impurity concentration. The iodine (I) donor or nitrogen (N) acceptor is considered for (Zn,Cr)Te as a codopant to control the convexity (*with spinodal nano-decomposition or phase separation*) and concavity (*without spinodal nano-decomposition*) of the mixing energy. The concentrations of codopants are 5%, 10%, and 15%. Negative mixing energy means that we can make continuous alloying in thermal equilibrium. If compensating impurities are introduced into these DMS, the mixing energy shows a gradual transition from a convex to a concave concentration dependence, resulting in a negative mixing energy of magnetic impurities for low concentrations. This observation suggests that the codoping increases the solubility limit of magnetic impurities in DMS; thus, high concentration doping of magnetic impurities into DMS becomes possible. Kuroda *et al.* (2007) and Ozaki *et al.* (2006) reported that they had observed enhanced spinodal decomposition upon I and Cr codoping in ZnTe; however, their observation is opposite compared with our theoretical prediction, where I and Cr codoping reduces the spinodal decomposition, increases the solubility of Cr with I codoping due to the reduction of the mixing energy, and finally reaches a negative value (see Fig. 10.9). If compensating donors are introduced, the mixing energy is lowered, and as a result, the spinodal nano-decomposition is suppressed, leading to a higher solubility limit. Roughly speaking, the mixing energy curve shows a minimum around the same concentration of codopants. For the low concentration range where the mixing energy is negative, we can expect uniform mixing even under thermal equilibrium. Thus, a very high solubility of magnetic impurities is expected in (Zn,Cr)Te by using the codoping method with I. By the codoping, the hole states in the t_2-impurity band induced by magnetic Cr impurities are occupied by electrons from codoped I donors. The energy gain associated with codoping comes from this electron transfer from I donors. This intuitive picture is partly supported by the calculations of nitrogen (N) codoping into (Zn,Cr)Te. N behaves as an acceptor in ZnTe and does not compensate for hole states in the impurity band induced by Cr. Therefore, there is no electron transfer from N acceptors and no energy gain, resulting in negligible changes in the mixing energy, as shown in Fig. 10.9. N codoping holds the convexity in the mixing energy; therefore, spinodal decomposition still occurs with Cr and N codoping. This is contrary to I and Cr codoping where the convexity in the mixing energy disappears.

For the realization of semiconductor nano-spintroics, we should develop the colossal magnetic response controlled by the electric field or photonic excitations. The colossal magnetic response caused by (1) 3D self-organized *Dairiseki phase* of (Zn,Co)O or (2) the *1D Konbu phase* caused by 2D spinodal nano-decomposition of (Zn,Co)O may be possible using the spinodal nano-decomposition. By applying a positive gate voltage, as shown in Fig. 10.14, we can make n-type doping through the gate (see Fig. 10.9). Because of the spatial (R) distribution of the local band gap caused by the spinodal nano-decomposition, the doped electrons can accumulate near the Co high-concentration region, so we can realize a colossal magnetic response and a colossal transition by using electron-doped ferromagnetism.

4. NEW CLASS OF OXIDE SPINTRONICS WITHOUT A 3d TM

Our recent proposal includes transparent FM DMS with 3d TM-doped delafossite $CuAlO_2$ (Kizaki et al., 2005), FM DMS without TM impurities such as C- or N-doped CaO, BaO, MgO, SrO (Dinh et al., 2005), and SiO_2

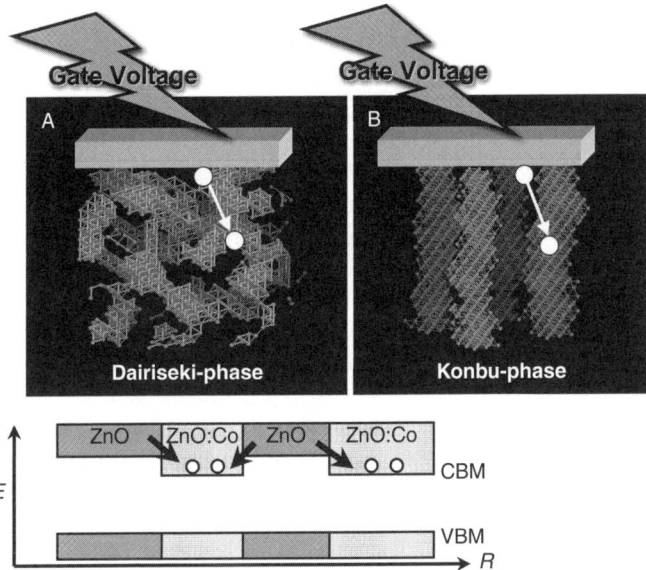

FIGURE 10.14 (A) Colossal magnetic response caused by the three-dimensional (3D) self-organized *Dairiseki phase* in (Zn,Co)O and (B) *1D Konbu phase* caused by 2D spinodal nano-decomposition in (Zn,Co)O. Applying a positive gate voltage, we can make n-type doping through the gate. Because of the spatial (R) distribution of the local band gap caused by the spinodal nano-decomposition, the doped electrons can be accumulated near the Co high-concentration region; therefore, we can expect a colossal magnetic response and a phase transitions by electron doping through the gate.

(Kenmochi et al., 2004) and an electronic structure obtained by using the SIC-LDA (Dinh et al., 2006). For the realization of ferromagnetism, Friedel's virtual bound state in the 3d TM impurity is not necessary to stabilize the ferromagnetism; however, we need a partially occupied and highly correlated deep-impurity band in the wide band gap of the oxides. These systems have a local magnetic moment by satisfying the condition $U [= E(N + 1) + E(N - 1) - 2E(N)] > W$, where the width ($W$) of the deep impurity band is approximately proportional to c for low concentrations (c) of the impurity, and $E(N)$ is the total energy of the impurity state with N electron occupation. By choosing the impurity concentration, we can satisfy the condition of $U > W$ and partial occupation of the deep-impurity band. This FM mechanism is the same as Zener's double-exchange mechanism, which we have already discussed. We await experimental verification of the present computational nano-materials design.

5. CONCLUSION

On the basis of the state-of-the-art *ab initio* electronic structure calculation done by using the KKR-CPA-LDA and the *SIC-LDA* to go beyond the LDA, we propose a unified physical picture of wide band-gap II–VI and III–V DMS such as ZnO, ZnS, ZnSe, ZnTe, and GaN. *Zener's double-exchange interaction* and *superexchange interaction mechanisms* are dominant in the magnetic magnetism. In the homogeneous system, the calculated Curie temperature (T_C) obtained by using an MCS and electronic structure is in good agreement with the experiment (T_C and PES). In the inhomogeneous system, we propose a *3D Dairiseki phase* and a *1D Konbu phase* caused by spinodal nano-decomposition. These are responsible for the high-T_C (or -T_B) phases in DMS. We design a growth position and shape-control method for nano-magnets in the superstructure by using self-organization. Finally, on the basis of the FM mechanism of Zener's double-exchange mechanism, we design a new class of half-metallic ferromagnets without 3d TM impurities, such as C- or N-doped CaO, BaO, MgO, SrO, and SiO_2.

ACKNOWLEDGMENTS

This research was partially supported by a Grant-in-Aid for Scientific Research in Priority Areas "Quantum Simulators and Quantum Design" (No. 17064014), a Grand-in-Aid for Scientific Research for young researchers, Japan Science and Technology Agency (CREST), New Energy and Industrial Technology Development Organization Nanotech, the 21st Century Center of Excellence, and the Japan Society for Promotion of Science core-to-core program "Computational Nano-materials Design." This research was also supported in part by the National Science Foundation under grant No. PHY99-07949 (Kavli Institute for Theoretical Physics SPINTRONICS06 program, University of California, Santa Barbara), the Kansai Research Foundation for technology promotion, the Foundation of C&C Promotion, the Murata Science Foundation, and the Inoue Foundation for Science.

REFERENCES

Akai, H. (1998). *Phys. Rev. Lett.* **81**, 3002.
Akai, H. http://sham.phys.sci.osaka-u.ac.jp/kkr/.
Akai, H., and Dederichs, P. H. (1993). *Phys. Rev. B* **47**, 8739.
Bergqvist, L., Eriksson, O., Kudrnovsk'y, J., Drchal, V., Korzhavyi, P., and Turek, I. (2004). *Phys. Rev. Lett.* **93**, 137202.
Bouzerar, G., Kudrnovsky, J., Bergqvist, L., and Bruno, P. (2003). *Phys. Rev. B.* **68**, 81203(R).
Dietl, T. (2002). *Semicond. Sci. Technol.* **17**, 377.
Dinh, V. A., Sato, K., and Katayama-Yoshida, H. (2005). *Solid State Commun.* **136**, 1.
Dinh, V. A., Toyoda, M., Sato, K., and Katayama-Yoshida, H. (2006). *J. Phys. Soc. Jpn.* **75**, 093705.
Fukushima, T., Sato, K., and Katayama-Yoshida, H. (2004). *Jpn. J. Appl. Phys.* **43**, L1416.

Fukushima, T., Sato, K., Katayama-Yoshida, H., and Dederichs, P. H. (2006). *Jpn. J. Appl. Phys.* **45**, L416.
Goodenough, J. B. (1955). *Phys. Rev.* **100**, 564.
Ishida, Y., Hwang, J. I., Kobayashi, M., Fujimori, A., Saeki, H., Tabata, H., and Kawai, T. (2004). *Physica B* **351**, 304.
Jungwirth, T., Sinova, J., Ma'sek, J., Ku'cera, J., and MacDonald, A. H. (2006). *Rev. Mod. Phys.* **78**, 809.
Kanamori, J. (1959). *J. Phys. Chem. Solids* **10**, 87.
Katayama-Yoshida, H., Sato, K., Fukushima, T., Toyoda, M., Kizaki, H., Dinh, V. A., and Dederichs, P. H. (2007a). *J. Magn. Magn. Mater.* **310**, 2070.
Katayama-Yoshida, H., Sato, K., Fukushima, T., Toyoda, M., Kizaki, H., Dinh, V. A., and Dederichs, P. H. (2007b). *Phys. Stat. Sol. A* **204**, 15.
Katayama-Yoshida, H., Fukushima, T., and Sato, K. (2007c). *Jpn. J. Appl. Phys.* **46**, L777.
Kenmochi, K., Dinh, V. A., Sato, K., Yanase, A., and Katayama-Yoshida, H. (2004). *J. Phys. Soc. Jpn.* **73**, 2952.
Kizaki, H., Sato, K., Yanase, A., and Katayama-Yoshida, H. (2005). *Jpn. J. Appl. Phys.* **44**, L1187.
Kobayashi, M., et al. cond-mat/0505387.
Koenig, J., Schliemann, J., Jungwirth, T., and MacDonald, A. H. (2002). In "Electronic Structure and Magnetism of Complex Materials" (D. J. Singh and D. A. Papa, eds.), p. 163. constantopoulos, Springer, New York.
Kuroda, S., Nishizawa, N., Takita, K., Mitome, M., Bando, Y., Osuch, K., and Dietl, T. (2007). *Nature Mater.* **6**, 440.
Liechtenstein, A. I., Katsnelson, M. I., Antropov, V. P., and Gubanov, V. A. (1987). *J. Magn. Magn. Mater.* **67**, 65.
Mizokawa, T., Nambu, T., Fujimori, A., Fukumura, T., and Kawasaki, M. (2001). *Phys. Rev. B* **65**, 85209.
Ozaki, N., Nishizawa, N., Marcet, S., Kuroda, S., Eryu, O., and Takita, K. (2006). *Phys. Rev. Lett.* **97**, 037201.
Pajda, M., Kudrnovsky, J., Turek, I., Drchal, V., and Bruno, P. (2001). *Phys. Rev. B* **64**, 174402.
Reed, M. L., et al. (2001). *Appl. Phys. Lett.* **79**, 3473; Thaler, G. T., et al. (2002). *Appl. Phys. Lett.* **80**, 3964; Theodoropoulou, N. A., et al. (2002). *Appl. Phys. Lett.* **80**, 3475; Overberg, M. E., et al. (2001). *Appl. Phys. Lett.* **79**, 1312; Ploog, K. H., et al. (2003). *J. Vac. Sci. Technol. B* **21**, 1756; Matsukura, F., et al. (1998). *Phys. Rev. B* **57**, R2037; Edmonds, K. W., et al. (2002). *Appl. Phys. Lett.* **81**, 4991; Ku, C., et al. (2003). *Appl. Phys. Lett.* **82**, 2302; Edmonds, K. W., et al. (2004). *Phys. Rev. Lett.* **92**, 37201; Hashimoto, M., et al. (2003). *J. Cryst. Growth* **251**, 327; Yamaguchi, K., et al. (2004). *Jpn. J. Appl. Phys.* **43**, L1312; Saito, H. et al. (2003). *Phys. Rev. Lett.* **90**, 207202; Ozaki, N. et al. (2004). *Phys. Stat. Sol. (c)* **1**, 957.
Sato, K., and Katayama-Yoshida, H. (2000). *Jpn. J. Appl. Phys.* **39**, L555.
Sato, K., and Katayama-Yoshida, H. (2001a). *Jpn. J. Appl. Phys.* **40**, L485.
Sato, K., and Katayama-Yoshida, H. (2001b). *Jpn. J. Appl. Phys.* **40**, L334.
Sato, K., and Katayama-Yoshida, H. (2001c). *Jpn. J. Appl. Phys.* **40**, L651.
Sato, K., and Katayama-Yoshida, H. (2002). *Semicond. Sci. Technol.* **17**, 367.
Sato, K., and Katayama-Yoshida, H. (2007). *Jpn. J. Appl. Phys.* **46**, L1120.
Sato, K., Dederichs, P. H., and Katayama-Yoshida, H. (2003). *Europhys. Lett.* **61**, 403.
Sato, K., Dederichs, P. H., Katayama-Yoshida, H., and Kudrnovsky, J. (2004a). *J. Phys.: Condens. Matter* **16**, S5491.
Sato, K., Schweika, W., Dederichs, P. H., and Katayama-Yoshida, H. (2004b). *Phys. Rev. B* **70**, 201202(R).
Sato, K., Katayama-Yoshida, H., and Dederichs, P. H. (2005). *Jpn. J. Appl. Phys.* **44**, L948.
Sato, K., Dederichs, P. H., and Katayama-Yoshida, H. (2006). *Physica B* **376/377**, 639.
Sato, K., Fukushima, T., and Katayama-Yoshida, H. (2007a). *Jpn. J. Appl. Phys.* **46**, L682.

Sato, K., Dederichs, P. H., and Katayama-Yoshida, H. (2007b). *J. Phys. Soc. Jpn.* **76**, 024717–1.
Shiba, H. (1971). *Prog. Theor. Phys.* **46**, 77.
Stauffer, D., and Aharony, A. (1994). Introduction to Percolation Theory, revised 2nd edn. Taylor and Francis, London.
Toyoda, M., Akai, H., Sato, K., and Katayama-Yoshida, H. (2006a). *Physica B* **376**, 647.
Toyoda, M., Akai, H., Sato, K., and Katayama-Yoshida, H. (2006b). *Phys. Stat. Soli. (c)* **3**, 4155.
Wolf, S. A., Awschalom, D. D., Buhrman, R. A., Daughton, J. M., von Molnar, S., Roukes, M. L., Chtchelkanova, A. Y., and Treger, D. M. (2001). *Science* **294**, 1488.
Yamamoto, T., and Katayama-Yoshida, H. (1997). *Jpn. J. Appl. Phys.* **36**, L180.

CHAPTER 11

Properties and Functionalities of MnAs/III–V Hybrid and Composite Structures

Masaaki Tanaka,[*,†] **Masafumi Yokoyama,**[*] **Pham Nam Hai,**[*] **and Shinobu Ohya**[*,‡]

Contents		
	1. Introduction	455
	2. Fabrication and Structure of GaAs:MnAs Nano-Particles	456
	3. Large Magnetoresistance at Room Temperature	458
	4. Spin Dependent Tunneling Transport Properties in III–V Based Heterostructures Containing GaAs:MnAs	463
	5. Properties of Zinc-Blende Type and NiAs-Type MnAs Nano-Particles	471
	6. Magneto-Optical Device Applications	478
	Acknowledgments	483
	References	484

1. INTRODUCTION

Adding magnetic or spin-related functions to semiconductor-based materials and devices is a new challenge for researchers in the field of solid-state physics and electronics (Jungwirth et al., 2006; Zutic et al., 2004). Advanced fabrication technologies, especially molecular beam epitaxy (MBE),

---

[*] Department of Electronic Engineering, The University of Tokyo, 7-3-1 Hongo, Bunkyo-ku, Tokyo 113-8656, Japan
[†] Institute for Nano Quantum Information Electronics, The University of Tokyo
[‡] PRESTO, Japan Science and Technology Agency, 4-1-8 Honcho, Kawaguchi-shi, Saitama 332-0012, Japan

are offering a new opportunity to explore in this direction, because they have made it possible to fabricate epitaxial hybrid structures having both functions of semiconductors and of magnetic materials. Among them are (i) ferromagnetic alloy semiconductors ($In_{1-x}Mn_x$)As (Munekata et al., 1989; Ohno et al., 1992) and ($Ga_{1-x}Mn_x$)As (Hayashi et al., 1997; Ohno, 1999; Ohno et al., 1996; Tanaka, 1998, 2000; Van Esch et al., 1997), (ii) ferromagnet/ semiconductor heterostructures such as MnAs/GaAs (Akeura et al., 1995, 1996; Tanaka, 1995; Tanaka et al., 1994, 1999), and (iii) GaAs:MnAs composite nanoparticle structures (Akinaga et al., 2000a,b; De Boeck et al., 1996). Here, MnAs is a ferromagnetic metal with a NiAs-type hexagonal crystal structure and its Curie temperature is 313 K. These materials have very good compatibility with III–V semiconductor systems, together with good crystalline quality and atomic-scale controllability of the layer thickness. For instance, one can observe large spin-dependent transport phenomena in (GaMn)As based ferromagnetic semiconductor heterostructures (Ruster et al., 2005; Tanaka and Higo, 2001). Also, these magnetic semiconductor materials were found to have large magneto-optical effects (Ando et al., 1998; Kuroiwa et al., 1998). The main obstacle in spintronics applications of these ferromagnetic alloy semiconductors is their relatively low Curie temperature (<300 K). For application to practical devices, it is necessary to understand and control their properties, and enlarge these effects at room temperature. On the other hand, GaAs(III–V):MnAs composite nanoparticle structures have similar spin dependent properties even at room temperature, together with good compatibility with III–V systems.

In this article, we review our recent studies on the properties and functionalities of hybrid and composite structures consisting of ferromagnetic MnAs nano-scale particles and III–V (GaAs, AlAs, InAlAs) semiconductors.

2. FABRICATION AND STRUCTURE OF GaAs:MnAs NANO-PARTICLES

The fabrication process of MnAs nano-particles embedded in GaAs consists of two steps, MBE growth of ($Ga_{1-x}Mn_x$)As at low temperature and subsequent annealing (Akinaga et al., 2000; De Boeck et al., 1996; Tanaka et al., 2001). First, ($Ga_{1-x}Mn_x$)As alloy semiconductor films with Mn content x ranging from 0.01 to 0.10 were grown by MBE at 250 °C. The film thickness can be varied from 1 nm to a few micrometers. Then, with an As_4 flux kept supplying, the ($Ga_{1-x}Mn_x$)As films were annealed in the MBE growth chamber at 500–700 °C, during which MnAs nanoscale particles are formed in GaAs by phase separation.

In order to check the crystal structure and quality of this GaAs:MnAs nanoparticle system, cross-sectional transmission electron microscopy (TEM)

Properties and Functionalities of MnAs/III–V Hybrid and Composite Structures

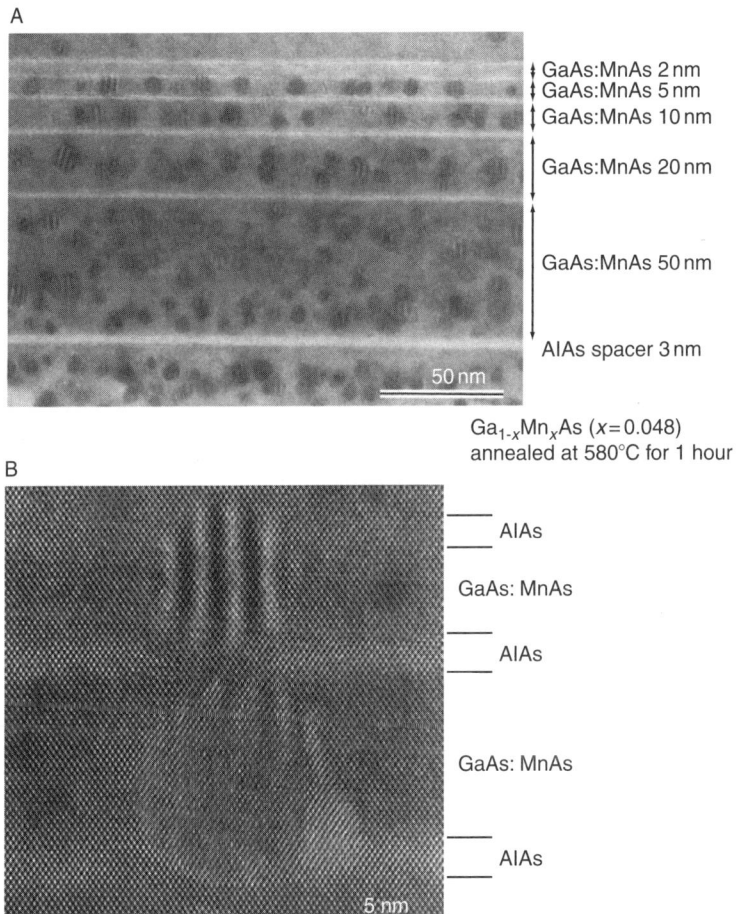

FIGURE 11.1 (A) Bright field and (B) High-resolution TEM lattice images with [$\bar{1}10$] projection. The as-grown sample structure prior to the annealing was a heterostructure consisting of 2–50-nm thick (GaMn)As layers with $x = 0.048$ separated by 3-nm thick AlAs markers grown at 250 °C. The sample was annealed at 580 °C after the MBE growth. MnAs particles of NiAs-type crystal structure with diameters of about 5–10 nm were formed in the GaAs (or GaMnAs with very low Mn content) layers.

images were taken from some of our samples. Figure 11.1 shows (A) a bright field and (B) a high-resolution lattice image with [$\bar{1}10$] projection. The as-grown sample structure prior to the annealing was a heterostructure consisting of 2–50-nm thick (Ga$_{1-x}$Mn$_x$)As layers with $x = 0.048$ separated by 3-nm thick AlAs markers grown at 250 °C. The sample in Fig. 11.1 was annealed at 580 °C for 1 h after the growth. MnAs particles of NiAs-type crystal structure with diameters of about 5–10 nm were formed in the GaAs (or GaMnAs with very low Mn content) layers. In the lattice image of Fig. 11.1B, one can see that two MnAs particles with diameters of about 5 and 10 nm of hexagonal crystal structure was embedded in the zinc-blende (ZB) lattice of GaAs/AlAs, without any dislocations. The lattice relationship between MnAs and GaAs was found to be MnAs(0001)//GaAs(111)B. Although the crystal structure and lattice parameters of MnAs and GaAs are very different, the GaAs:MnAs material is dislocation free and very compatible with III–V heterostructures, as in the case of MnAs/GaAs heterostructures (Tanaka, 1995; Tanaka et al., 1994, 1999).

We have characterized a series of GaAs:MnAs layers whose structure is similar to Fig. 11.1A, and found that the size, density, and uniformity of MnAs particles can be controlled to some extent:

(1) The average size of MnAs particles becomes larger when the annealing temperature is higher.
(2) The density of MnAs particles becomes larger when the Mn content x of the original $(Ga_{1-x}Mn_x)As$ layer is higher.
(3) The uniformity of the size becomes better when the original $(Ga_{1-x}Mn_x)As$ layer is thinner.

An example of (3) can be observed in the 5-nm thick GaAs:MnAs layer sandwiched by 3-nm thick AlAs layers in Fig. 11.1A, where the diameter of the MnAs particles are 5 nm almost uniformly. This means that growth of MnAs particles during annealing is limited by the existence of nonmagnetic semiconductor (AlAs) layers on both sides.

3. LARGE MAGNETORESISTANCE AT ROOM TEMPERATURE

In the following, we show large positive magnetoresistance at room temperature in a GaAs:MnAs thin film, in which MnAs nano-particles were embedded in a GaAs matrix. The MR ratio of the GaAs:MnAs granular thin film reached more than 600%, when a bias voltage of 110 V was applied to the film (Yokoyama et al., 2006).

Huge MR effects at room temperature were reported so far; the MR ratio was beyond 320,000% with a magnetic field of 2,000 Oe in a MnSb/GaAs granular hybrid system (Akinaga et al., 2000) and more than 1,000,000% with a magnetic field of 15,000 Oe in a Au/GaAs Schottky diode system (Sun et al., 2004). These MR effects are magnetic-field-sensitive phenomenon, thus they are expected to be used as magnetic-field-sensitive sensor devices. However, since both MnSb/GaAs hybrid granular systems and Au/GaAs Schottky diode systems are fabricated on GaAs surfaces, and their MR effects rely on the surface structure, they may be unstable or vulnerable. If stable metal-semiconductor hybrid systems can be integrated into semiconductor-based devices, such large MR effects would be used more effectively.

The GaAs:MnAs granular material, in which ferromagnetic MnAs nano-particles with 5–10 nm in diameter are embedded in a GaAs matrix, is promising metal-semiconductor hybrid structures because of large magneto-optical effects and good compatibility with existing III–V technology. Previously, negative MR with a MR ratio of 1.5% at 30 K with a magnetic field of 1 T was also reported in GaAs:MnAs granular films (Akinaga et al., 1998). In p^+-GaAs/GaAs:MnAs/p^+-GaAs structures, both negative MR (MR ratio = −80%) at low temperature (<20 K) and positive MR (MR ratio = +115%) at 20 K with a magnetic field of 7 T were observed (Wellmann et al., 1998). In this section, we show that a GaAs:MnAs granular thin film exhibits a large positive MR at room temperature and the MR ratio is more than 600% at a lower magnetic field ($B < 0.5$ T).

The film preparation process is the following. First, a 200-nm thick $Ga_{1-x}Mn_xAs$ (Mn content x was 7 at.%) thin film, in which Be of 2.5×10^{19} cm^{-3} was doped, was grown on a semi-insulating GaAs(001) substrate at 280 °C by low-temperature (LT) MBE. Then, the $Ga_{1-x}Mn_xAs$ thin film was annealed at 600 °C in the MBE growth chamber in an As$_4$ flux for 10 min to form a GaAs:MnAs granular thin film, in which MnAs nano-particles are precipitated in a GaAs matrix by phase separation. The surface of the GaAs:MnAs thin film was observed by atomic force microscopy (AFM), as shown in Fig. 11.2, which is a 500 nm×500 nm scan size image and the vertical-axis scale is 10 nm/div. A root mean square (RMS) of the surface roughness along the vertical axis was estimated to be 0.097 nm. The film surface was very smooth and no segregation of MnAs particles was found on the surface. This means that, unlike the case of

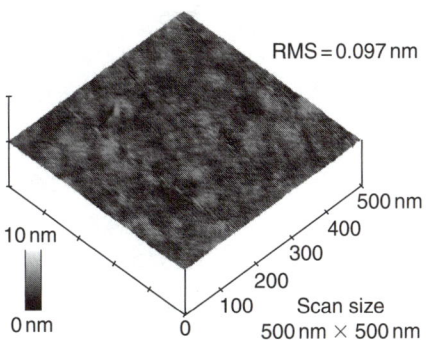

FIGURE 11.2 Atomic force microscopy (AFM) image of the GaAs:MnAs granular thin film surface. The scan size was 500 nm×500 nm. The vertical-axis scale was 10 nm/div. A root mean square of the surface roughness in the vertical axis was estimated to be 0.097 nm.

MnSb/GaAs and Au/GaAs hybrid systems, the current flows not on the surface but inside the granular layer in the GaAs:MnAs granular thin film in magneto-transport measurements.

We carried out magnetic circular dichroism (MCD) measurements of the GaAs:MnAs granular thin film to confirm that MnAs nano-particles were formed in the GaAs matrix. Figure 11.3A shows MCD spectra measured at 50 K (a dotted curve) and 300 K (a solid curve) with a

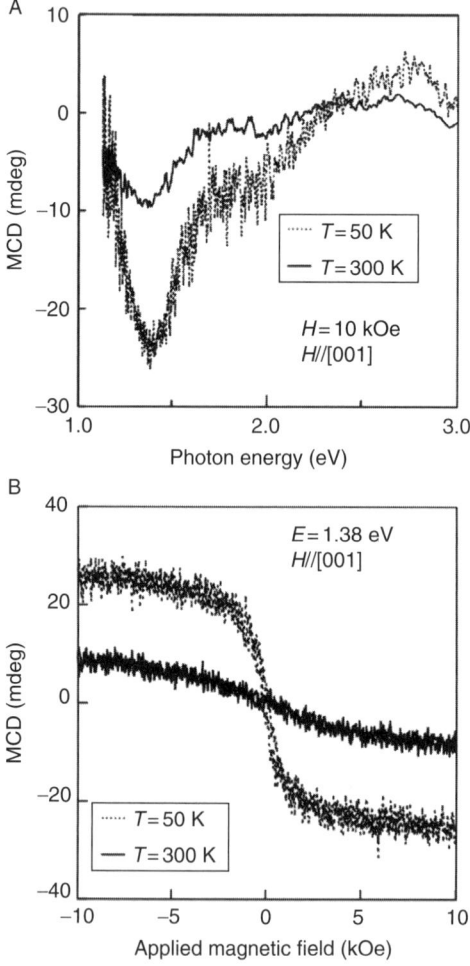

FIGURE 11.3 (A) Magnetic circular dichroism (MCD) spectra of the GaAs:MnAs granular thin film at 50 K (a dotted curve) and 300 K (a solid curve) measured with a magnetic field of 1 T (=10 kOe) applied perpendicularly to the film surface. (B) Magnetic field dependence of MCD intensity at 50 K (a dotted curve) and 300 K (a solid curve) measured with a magnetic field of 1 T applied perpendicularly to the film surface.

magnetic field of 1 T (=10 kOe) applied perpendicular to the film surface. The spectral features with a broad peak around 1.42 eV are those of the typical MCD spectra of GaAs:MnAs granular thin films (Akinaga et al., 2000; Tanaka et al., 2001). Figure 11.3B shows the magnetic field dependence of the MCD intensity measured at 50 K (a dotted curve) and 300 K (a solid curve) with a magnetic field of 1 T applied perpendicular to the film surface. Superparamagnetic behavior was observed at 300 K. These MCD results indicate that MnAs nano-particles were formed in the GaAs matrix.

In order to carry out magneto-transport measurements, the GaAs:MnAs granular thin film was patterned into a bar of 500 μm in length and 100 μm in width by wet-etching. Current–voltage (I–V) characteristics and MR effects were measured by a two-point-probe method at room temperature, as shown in Fig. 11.4. Ohmic contacts were made by In solders. A d.c. voltage was supplied to the film from 0 to 110 V, and then from 110 to 0 V. Without magnetic field, the current increased linearly for the voltage up to 8 V, as shown by a solid curve in Fig. 11.4. Then, the current gradually saturated up to 96 V. When the voltage was above 96 V, the current jumped from 0.16 to 0.70 mA. On the other hand, when a magnetic field of 1 T was applied

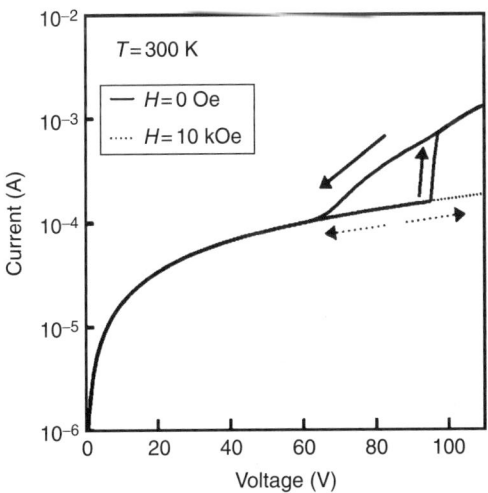

FIGURE 11.4 Current-voltage (I–V) characteristics of the GaAs:MnAs granular thin film measured by two-point-probe method at room temperature in a d.c. bias voltage range between 0 and 110 V. When a magnetic field was 0 T (a solid curve), the I–V characteristics showed a current jump at 96 V and hysteretic behavior. When a magnetic field of 1 T (=10 kOe) was applied perpendicular to the film (a dotted curve), no current jump was observed in this voltage range.

perpendicular to the film, no jump was observed in the *I–V* characteristics of the GaAs:MnAs granular thin film, as shown by a dotted curve in Fig. 11.4. Similar phenomena were reported in the MnSb/GaAs and Au/GaAs hybrid system (Akinaga *et al.*, 2000; Sun *et al.*, 2004). In the Au/GaAs hybrid system, the *I-V* characteristics could be interpreted by the velocity controlled high field transport mechanism and by the impact ionization model (Neumann, 2001; Paracchini and Dallacasa, 1989; Sun *et al.*, 2004). In the present GaAs:MnAs granular thin film, if a voltage could be biased much larger than 96 V, the second jump of the current could also be observed.

Next, the magnetic field dependence of the current was measured by a two-point-probe method at room temperature, as shown in Fig. 11.5. A magnetic field was applied perpendicular to the current direction, first sweeping from 0 to 1 T, second 1 to –1 T, and finally –1 to 1 T, with a fixed d.c. voltage. In Fig. 11.5A, when the voltage was 110 V, which was above the jump (96 V) of the *I–V* curve in Fig. 11.4, a reversible large current change was observed. On the other hand, in Fig. 11.5B, when the voltage

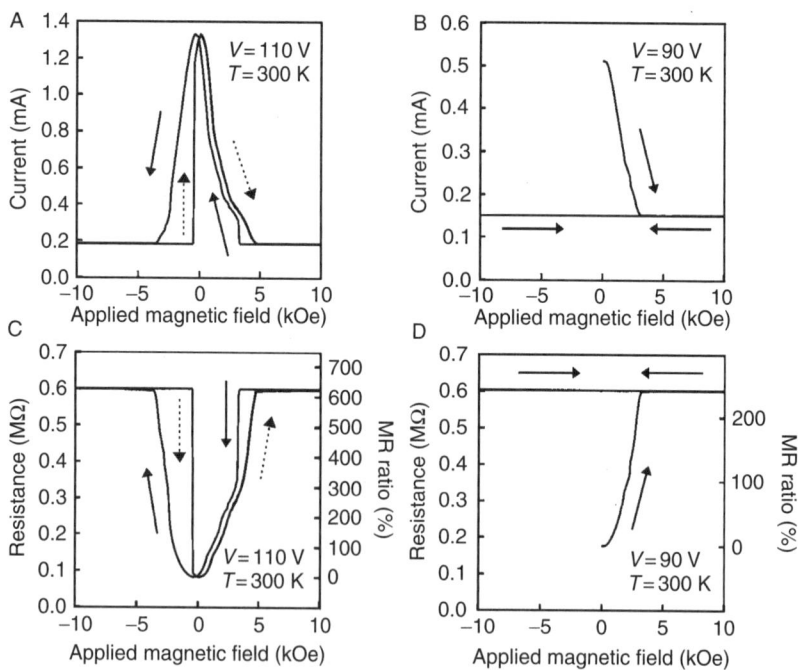

FIGURE 11.5 Magneto-transport characteristics of the GaAs:MnAs granular thin film at room temperature. Magnetic field dependence of the current at a constant bias voltage of (A) 110 V and (B) 90 V. Magnetic field dependence of the resistance at a constant bias voltage of (C) 110 V and (D) 90 V.

was 90 V, which was in the hysteresis region of the I–V curves in Fig. 11.4, an irreversible large current change was observed. Figure 11.5C and D show the magnetic field dependence of the resistance at the voltages of 110 and 90 V, respectively. A large positive MR change was observed, when the voltage was 110 V, and the MR ratio was estimated to be 626%, when the MR ratio was defined by $[R(H)-R(0)]/R(0)\times 100\%$, where $R(H)$ and $R(0)$ are the resistances at a magnetic field of H and 0, respectively. This positive MR effect is similar to the giant magnetoresistance effect of the other metal-semiconductor hybrid systems (Akinaga et al., 2000; Sun et al., 2004). The large positive MR effect in the GaAs:MnAs granular thin film would be caused by magnetic-field-dependent avalanche breakdown phenomena which were observed in Au/GaAs Schottky diodes. In this case, the depletion layers spread three dimensionally near the interfaces between the MnAs nano-particles and a GaAs matrix might also play important roles for a large MR effect in a GaAs:MnAs granular thin film.

To summarize this section, a GaAs:MnAs granular thin film exhibited a large positive MR effect at room temperature measured by a two-point-probe method, the MR ratio reached more than 600% at a d.c. bias voltage of 110 V. The origin of this large positive MR change would be the same as the MR effect in Au/GaAs and MnSb/GaAs. Since this large MR effect is obtained in a GaAs:MnAs granular thin film which can be easily integrated in a variety of III–V heterostructures, it can be more versatile and expected to be applied to magnetic-field-sensitive switching and sensor devices.

4. SPIN DEPENDENT TUNNELING TRANSPORT PROPERTIES IN III–V BASED HETEROSTRUCTURES CONTAINING GaAs:MnAs

In the following section, we show spin dependent tunneling transport properties in epitaxial single-crystal magnetic tunnel junctions (MTJs) consisting of magnetic electrodes (ferromagnetic MnAs and GaAs:MnAs granular material) and a AlAs/GaAs III–V tunnel barrier, grown by MBE on GaAs(001) substrates. Clear tunneling magnetoresistance (TMR) was observed from 7 K up to room temperature on both bias voltage polarities. This indicates that GaAs:MnAs nano-particles can be used as a spin injecting source and as a spin detector in semiconductor based spintronics devices (Hai et al., 2006a,b). Unique bias dependence and AlAs tunnel barrier thickness dependence of the TMR ratio will be described.

Our MTJ structures were grown by MBE on p^+ GaAs(001) substrates. The schematic structure and the growth process are shown in Fig. 11.6A. In this heterostructure, the GaAs:MnAs layer and the MnAs thin film act

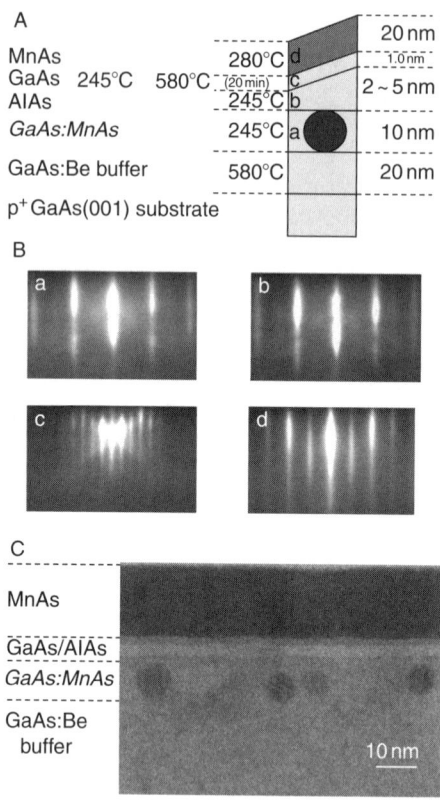

FIGURE 11.6 (A) Schematic structure and growth temperature of magnetic tunnel junctions (MTJs) consisting of MnAs thin film (20 nm)/GaAs (1 nm)/AlAs ($d = 2$–5 nm)/ GaAs:MnAs (10 nm) grown on a p$^+$ GaAs (001) substrate. (B) Reflection high energy electron diffraction (RHEED) patterns during the growth process. The incident electron beam is parallel to the GaAs [$\bar{1}$10] azimuth. (C) Transmission electron microscopy (TEM) image of a sample with $d \sim 3$ nm. MnAs nano-particles with 5–10 nm in diameters were formed in the 10-nm thick GaAs matrix.

as ferromagnetic electrodes and the AlAs/GaAs double layer acts as a tunnel barrier. Firstly, we grew a 20-nm thick Be-doped GaAs buffer layer on a p$^+$GaAs(001) substrate at 580 °C. After cooling the substrate temperature to 245 °C, we grew Ga$_{1-x}$Mn$_x$As ($x = 0.048$–0.089, 5–10 nm), AlAs ($d = 2$–5 nm), GaAs (1 nm), from the bottom to the top. We changed the thickness of the AlAs tunnel barrier layer by linearly moving a shutter in front of the substrate during the growth of the AlAs layer. Then, the structure was annealed at 580 °C for 20 min in the MBE growth chamber, during which phase separation occurred in the GaMnAs layer and MnAs nano-particles were formed in the GaAs matrix. Finally, a 20-nm thick type-A MnAs thin film was grown at 280 °C as a top electrode. After completing the growth, post growth annealing was carried out at 370–400 °C for 1–10 min to improve the structural and magnetic properties of the top MnAs film. Figure 11.6B shows the reflection high energy electron diffraction (RHEED) patterns with the GaAs [$\bar{1}$10] azimuth during the growth of each layer. The RHEED showed streaky patterns during the whole growth, revealing the smooth surface of every layer. Note that the RHEED pattern showed clear (2 × 4) reconstruction of the 1-nm thick GaAs layer during the annealing process at 580 °C, even when phase separation occurred in the GaMnAs layer, indicating that diffusion of Mn to the surface is minimal. Figure 11.6C shows a cross-sectional transmission electron microscope (TEM) image of a sample with $d \approx 3$ nm. MnAs nano-particles with about 10 nm in diameter were formed in the 10-nm thick GaAs matrix. A high-resolution TEM image showed that although the crystal structure of the MnAs nano-particles (NiAs type hexagonal) is different from that of the GaAs matrix, the GaAs matrix and the AlAs/GaAs barrier keep high quality ZB type crystal structure without any dislocation. Note that the diameter of the MnAs nano-particles are determined by the thickness of GaMnAs sandwiched between nonmagnetic layers as shown in Fig. 11.1 in Section 2, thus it is uniformly 10 nm as shown in Fig. 11.6C.

The sample then was processed by standard photolithography to fabricate mesa shaped MTJs with 20–200 μm in diameter for tunneling transport measurements. Figure 11.7 shows a TMR loop of a MTJ device with 20 μm in diameter measured at 7 K with a bias voltage of 300 mV. The structure of this MTJ is schematically shown in the inset of the figure.

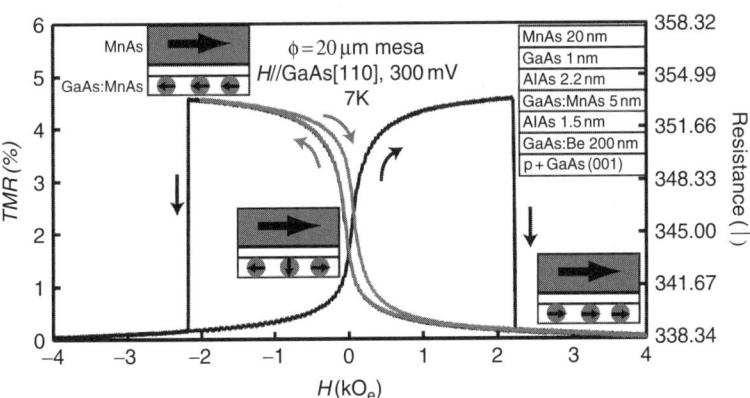

FIGURE 11.7 Tunneling magnetoresistance (TMR) at 7 K with a bias voltage of 300 mV of a MTJ, whose structure is (from the bottom to the top) AlAs (1.5 nm)/GaAs:MnAs (5 nm, $x = 0.048$)/AlAs (2.2 nm)/GaAs (1 nm)/MnAs (20nm). The diameter of the MTJ was 20 μm. The solid and gray curves are major and minor loops, respectively. The magnetic field was applied in the film plane and parallel to the GaAs [110] and the MnAs [$\bar{1}\bar{1}$20] axis.

The magnetic field was applied along the easy magnetization axis [1̄1̄20] of type-A MnAs, which is parallel to the in-plane GaAs[110] azimuth. Clear TMR characteristics are observed both in major and minor loops. The resistance jump in the major loop at ±2.2 kOe corresponds to the abrupt magnetization reversal of the top MnAs electrode. The hysteresis of the minor loop indicates that the thermal magnetization fluctuation of the MnAs nano-particles is quenched. The appearance of TMR shows that even a 5 nm nanoparticle can work as a spin injector. The TMR ratio at 7 K is 4.5%, which is higher than that of MnAs based MTJs reported earlier (Sugahara and Tanaka, 2002).

Figure 11.8 shows the bias voltage dependence of the TMR ratio. Here at the positive bias, electrons are transported form the substrate to the surface. Up to +300 mV, the TMR increases monotonously and reaches a maximum value of 4.5% at +300 mV. Over +300 mV, the TMR decreases slowly but the bias voltage V_{half} at which the TMR is reduced by half is surprisingly as high as 1,200 mV. This V_{half} value is much higher than that (~40 mV) of ferromagnetic semiconductor based MTJs (Chiba et al., 2004; Ohya et al., 2006, 2007a). The peak of TMR at +300 mV may reflect a local maximum of spin-polarization at an energy level above the Fermi level of MnAs, which is predicted by theoretical calculations (Pangulur et al., 2003; Ravindran et al., 1999). The similar behavior is also observed at the negative biases. The appearance of TMR at negative biases indicates that GaAs:MnAs nano-particles can also work as a spin detector. Figure 11.9 shows the temperature dependence of TMR. The TMR is observed up to room temperature and remains 0.15% at 300 K.

Next, we studied the AlAs tunnel barrier thickness dependence of the TMR properties. Figure 11.10A show the dependence of RA (tunnel resistance × junction area) on the AlAs barrier thickness d measured at 7 K. In this series of experiments, the thickness of the original $(Ga_{1-x}Mn_x)$As layer was 10 nm and x was 0.089, thus the diameter of the

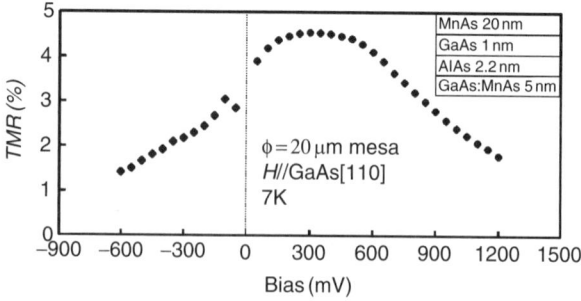

FIGURE 11.8 Bias voltage dependence of TMR of the same MTJ (Fig. 11.7) at 7 K. The magnetic field was applied in the film plane and parallel to the GaAs[110] azimuth. The V_{half} is as high as 1,200 mV.

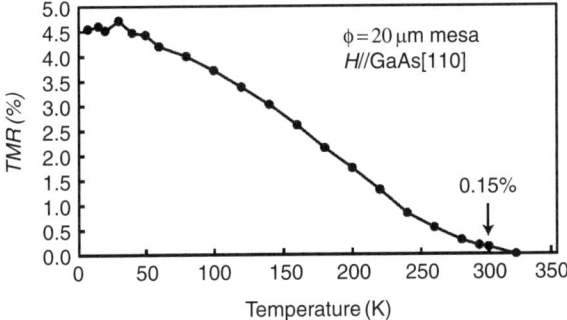

FIGURE 11.9 The temperature dependence of the TMR ratio of the same MTJ (Fig. 11.7). TMR was observed up to room temperature.

MnAs particles is nearly uniformly 10 nm. Here, RA was measured at zero magnetic field in nearly parallel magnetization, after the magnetic field of 4 kOe was applied and then swept to zero. The RA increases exponentially with increasing d, indicating that the AlAs layer works as a tunnel barrier and there is no extrinsic leak current. We emphasize that a high quality tunnel barrier is very important to achieve reliable spin injection in semiconductor spin-electronic devices. In MnAs thin film/ III–V/MnAs thin film MTJs, tunneling through the midgap defect band in the GaAs or AlAs barrier resulted in small TMR ratios (<2%) or a negative TMR (Garcia et al., 2005; Sugahara and Tanaka, 2002). In contrast to those cases, the AlAs barrier in our structure has very good crystallinity, because monocrystalline epitaxial III–V layers can be easily grown on GaAs:MnAs layers and the diffusion of Mn atoms into the barrier during the annealing process seems to be suppressed. The dashed line in Fig. 11.10A shows the best-fitted $\log(RA_{WKB})$-d calculated using the WKB approximation $RA_{WKB} = R_0 A \exp[(2/\hbar)\sqrt{2m^*U}d]$, where R_0 is a constant, A is the junction area, m^* is the effective electron mass in the tunnel barrier, and U is the effective barrier height. From the gradient of the log (RA_{WKB})-d line, we deduce $m^*U = 0.057m_o$ [kg·eV], where m_o is the free electron mass. If we take the tunneling effective electron mass $m^* = 0.09m_o$ of AlAs (Brozak et al., 1990), then $U = 0.63$ eV. This value is close to but smaller than the reported values (~0.8 eV) of the barrier height between MnAs and III–V semiconductors (Garcia et al., 2006; Sugahara and Tanaka, 2002).

Figure 11.10B shows TMR major and minor loops of the MTJ with $d = 2.9$ nm measured at 7 K with a bias voltage of $+20$ mV. Here at the positive bias, electrons are transported from the substrate to the surface. The magnetic field was applied along the easy magnetization axis of the MnAs thin film, that is [$\bar{1}\bar{1}20$] which is parallel to GaAs [110].

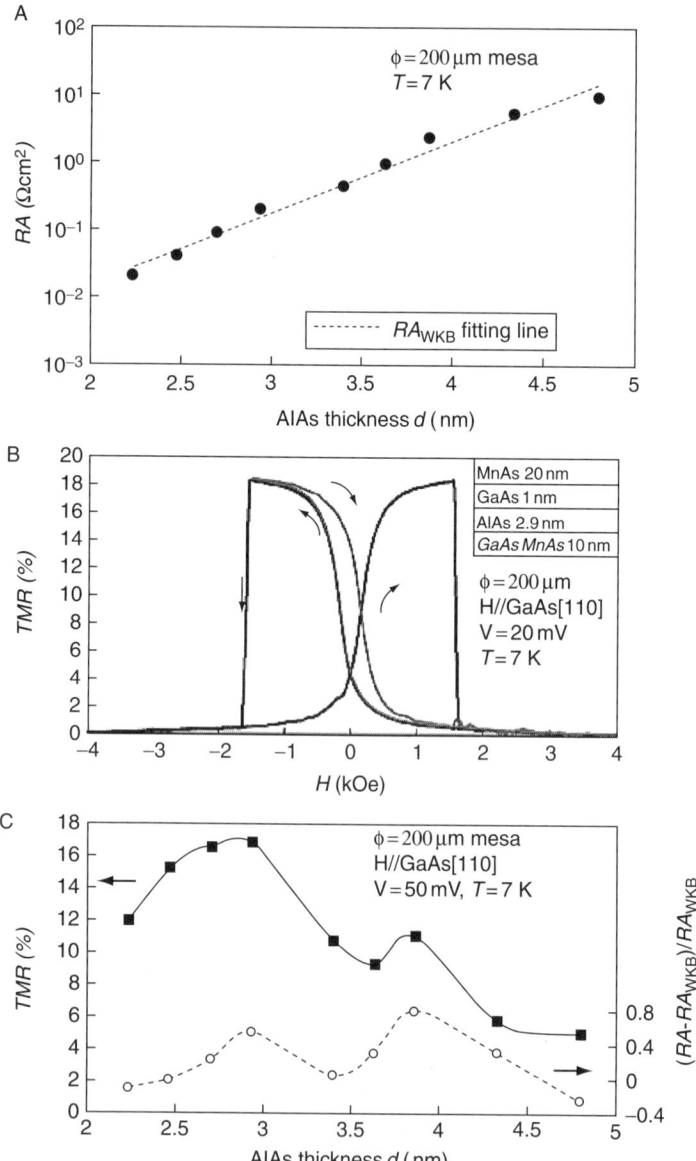

FIGURE 11.10 (A) Dependence of *RA* (tunnel resistance × junction area) on AlAs thickness *d* measured at 7 K. Dashed line shows the best-fitted log(RA_{WKB}) using the WKB approximation. (B) TMR major loop (black) and minor loop (gray) of the MTJ with $d = 2.9$ nm measured at 7 K with a bias voltage of 20 mV. The magnetic field was applied in plane along the easy magnetization axis [1120] of the MnAs thin film. (C) Dependence of the TMR ratio and ($RA-RA_{WKB}$)/RA_{WKB} on the AlAs barrier thickness *d*, which show the same oscillatory behavior with two peaks at $d = 2.9$ nm and $d = 3.9$ nm.

The resistance jumps in the major loop at ±1.55 kOe correspond to the abrupt magnetization reversal of the top MnAs electrode. The appearance of TMR indicates that MnAs nano-particles can work as a spin injector. The TMR ratio reached 18.3%, which is much larger than the previously reported values of MnAs/III–V/MnAs based MTJs (Garcia et al., 2005; Sugahara and Tanaka, 2002). At a negative bias voltage of –20 mV, a TMR ratio of 15.9% was observed. The appearance of TMR at both positive and negative biases indicates that MnAs nano-particles can work as a spin injector and a spin detector. No systematic variation of the coercive force of MnAs nano-particles and MnAs thin film was observed in a number of MTJs with different AlAs barrier thickness.

Solid squares in Fig. 11.10C shows the dependence of the TMR ratio on the AlAs barrier thickness d, measured at 7 K with a positive bias of 50 mV. The TMR ratios show a complicated oscillatory behavior with two peaks at $d = 2.9$ nm and $d = 3.9$ nm. The oscillation and the positions of peaks were reproduced in other MTJs. Open circles in Fig. 11.10C show the d-dependence of the relative difference of RA from the WKB approximation value, $(RA - RA_{WKB})/RA_{WKB}$. The oscillation behavior of $(RA - RA_{WKB})/RA_{WKB}$ is the same as that of the TMR ratio, suggesting that the electron tunneling is associated with the quantum interference in the AlAs barrier.

In order to understand the origin of the oscillation of TMR and RA, we analyze the wave functions of electrons in the AlAs tunnel barrier. Figure 11.11 shows the complex band structure of the AlAs barrier calculated by the sp^{3}s* tight binding method, where k_z is the wave number along the tunneling direction. The origin of energy ($E = 0$) is set at the valence-band maximum. In the calculation, we assume $k_{//} = (k_x, k_y) = 0$ at which the tunneling probability is largest, where $k_{//}$ is the wave vector parallel to the growth plane. In Fig. 11.11, k_z is shown in π/a unit, where a is the lattice constant of AlAs. The black solid (dashed) curves correspond to the purely real (purely imaginary) states, while the gray solid (dashed) curves correspond to the real (imaginary) part of the complex states, respectively. The conduction-band wave functions at the Γ valley are s-like, while the wave functions at X_1 and X_3 valley are sps*-like. In the following argument, we assume that the wave functions of the incident electrons from MnAs are mostly composed of the s-like component but still have a very small contribution of other components including the sps*-like one. If the Fermi level of MnAs $E_{F\text{-}MnAs}$ is somewhere in the middle of the indirect band gap of AlAs, the wave function of a tunneling electron inside the AlAs tunnel barrier at a small bias must be the purely imaginary Γ_1 state, which has the same symmetry with the incident wave function and decays most slowly. In this case, the barrier height becomes $E_{\Gamma 1} - E_{F\text{-}MnAs}$, where $E_{\Gamma 1}$ is the energy of the Γ_1 minimum of the AlAs conduction band. Thus, the barrier height would be larger than 1 eV,

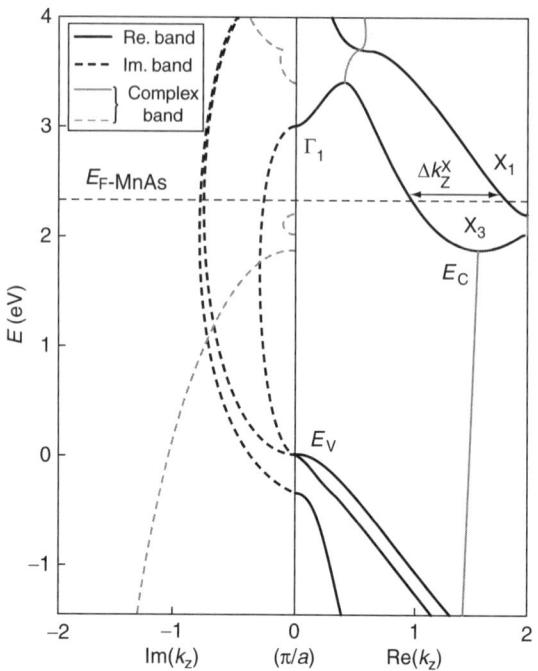

FIGURE 11.11 Complex band structure of the AlAs barrier calculated by the sp^{3}s* tight binding method. The origin of energy ($E = 0$) is set at the valence-band maximum. In the calculation, we assume $k_{//} = 0$. The wave number k_z along the tunnel direction is shown in π/a unit, where a is the lattice constant of AlAs. The black solid (dashed) curves correspond to the purely real (purely imaginary) states, while the gray solid (dashed) curves correspond to the real (imaginary) part of the complex states, respectively. The Fermi level of MnAs $E_{F\text{-MnAs}}$ is assumed to be at 2.37 eV so that the barrier height $E_{\Gamma1} - E_{F\text{-MnAs}}$ is 0.63 eV.

which is not consistent with our experimental result. Then, we assume that $E_{F\text{-MnAs}}$ is at 2.37 eV so that the barrier height $E_{\Gamma1} - E_{F\text{-MnAs}}$ is 0.63 eV, which was estimated from the data of Fig. 11.10A. In this case, there are two real k_z^{X1} and k_z^{X3} states corresponding to the X_1 and X_3 band, respectively, to which the very small sps*-like component of the incident wave function can connect. The transport path through the X states results in a leaky current through the tunneling barrier and may explain the deviation of the tunnel resistances from the WKB approximation values. It is known that the interference of these two real X valley-related states results in an oscillatory wave function with a period of $2\pi/((k_z^{X1} - k_z^{X3}))$ in GaAs/AlAs/GaAs tunnel junctions (Boykin and Harris, 1992; Ogawa et al., 1998; Rousseau et al., 1989). When the X states cancel out each other at the opposite site of the tunnel barrier, the leaky path is blocked thus the tunnel resistance and the TMR ratio take maxima.

However, if the X states strengthen each other at the opposite site of the tunnel barrier, the leaky path is largest thus the tunnel resistance and the TMR ratio take minima. In this way, the $(RA-RA_{WKB})/RA_{WKB}$ and the TMR ratio is likely to oscillate with increasing the AlAs thickness d. The TMR peak-to-peak value of $\Delta d = 1$ nm in Fig. 11.10C corresponds to $\Delta k_z^X = k_z^{X1} - k_z^{X3} = 1.1\pi/a$, which is close to the calculated result of $0.8\pi/a$. While our model can explain the experimental results semi-quantitatively, full understanding of electron tunneling in the single crystal MnAs/III–V/MnAs requires further theoretical investigations taking into account the symmetry of electron wave functions at the Fermi level of MnAs and their connections with the corresponding states in the barrier.

To summarize this section, in single crystal MTJs of MnAs/GaAs/AlAs/GaAs:MnAs nano-particles, spin dependent tunneling was clearly observed with very large V_{half} of 1,200 mV. We have observed an oscillatory behavior of TMR ratio with increasing the AlAs barrier thickness. The TMR oscillation can be explained by tunneling through the Γ–X mixing state of the AlAs tunnel barrier. The value of TMR ratios in our MTJs is largest among MnAs based MTJs, indicating that the ferromagnetic GaAs:MnAs nanoparticle system is promising for spin injection and detection in semiconductor based spintronic devices.

5. PROPERTIES OF ZINC-BLENDE TYPE AND NiAs-TYPE MnAs NANO-PARTICLES

In the following section, we describe comparative studies on Zinc-Blende (ZB) type MnAs nano-particles and conventional NiAs-type MnAs nano-particles. ZB-type MnAs particles were found to be formed by annealing (GaMn)As at relatively low temperature of 500 °C. Magnetic and magneto-optical properties of this GaAs:MnAs granular material with ZB-type MnAs particles were different from those of the GaAs:MnAs with NiAs-type hexagonal MnAs particles. The Curie temperature T_C of the ZB-type MnAs was estimated to be 360 K, which is higher than T_C (313 K) of the NiAs-type hexagonal MnAs (Yokoyama et al., 2005).

It is theoretically predicted that ZB-type MnAs is a half-metallic ferromagnet, by calculation of the band structure and density of states (Ogawa et al., 1999; Sanvito and Hill, 2000; Shirai et al., 1998). However, it is difficult to fabricate the ZB-type MnAs experimentally, because the ZB-type MnAs does not exist in the phase diagram (Massalski et al., 1990) thus it is thermodynamically unstable. Recently, it has been reported that extremely small ZB-type Mn(Ga)As particles with 2–3 nm in diameter embedded in a GaAs matrix can be fabricated by annealing GaMnAs in a MBE chamber under an As_4 flux (Moreno et al., 2002), and that ZB-type MnAs nanoscale dots were grown on a GaAs substrate (Ono *et al.,* 2002). However, there have been no reports concerning the magnetic and magneto-optical properties and other details of ZB-type MnAs particles. In this section, we describe the fabrication, magnetic, and magneto-optical properties of ZB-type MnAs particles embedded in a GaAs matrix.

The fabrication process of GaAs:MnAs granular thin films is as follows: First, a 1-μm thick $Ga_{0.924}Mn_{0.076}As$ thin film with an Mn concentration of 0.076 was grown at 250 °C on a semi-insulating GaAs(001) substrate by MBE. Then, the as-grown film was annealed *ex situ* at the temperature ranging from 500 to 650 °C under a nitrogen atmosphere to form MnAs nano-particles embedded in a GaAs matrix. Annealing conditions of the samples were listed in Table 11.1. The as-grown $Ga_{0.924}Mn_{0.076}As$ samples were annealed at T_a (°C) for 10 min. T_a and R (°C/min) are the maximum temperature and a ramp rate during the heat cycle, respectively. For comparison, an as-grown $Ga_{0.924}Mn_{0.076}As$ sample was also prepared under the same growth conditions.

The structural properties of the samples were characterized by X-ray diffraction (XRD) $\theta-2\theta$ scans and cross-sectional TEM. Figure 11.12 shows XRD spectra of the as-grown $Ga_{0.924}Mn_{0.076}As$ and GaAs:MnAs granular films. Although there is no sharp peak which is clearly assigned as ZB-type MnAs particles, some peaks at 33.05–33.20 ° in the spectra of GaAs:MnAs (sample A–D) include the information of the GaAs:MnAs granular layers containing MnAs particles. The XRD spectra suggest that GaAs:MnAs granular layers have been fabricated under all the annealing conditions examined here, and that the epilayer peaks from the GaAs:MnAs granular samples got closer to the substrate GaAs(004) peak with increasing the annealing temperature T_a. Furthermore, whereas the epilayer peak of the as-grown $Ga_{0.924}Mn_{0.076}As$ is on the lower angle side of the GaAs substrate peak, the epilayer peaks of GaAs:MnAs are on the higher angle side when the annealing temperature was higher than 500 °C.

Figure 11.13 shows a cross-sectional high resolution TEM lattice image of sample A and diffraction patterns of a MnAs particle (at the position

TABLE 11.1 GaAs:MnAs granular films (samples A–D) were fabricated by annealing 1 μm thick $Ga_{0.924}Mn_{0.076}As$ thin films under the conditions shown in this table

Sample no.	T_a (°C)	R (°C/min)
A	500	5
B	500	50
C	600	5
D	600	50

Here, T_a is the annealing temperature (maximum temperature during the heat cycle), and the R is the temperature ramp rate during heating and cooling. All the samples were annealed at T_a (°C) for 10 min.

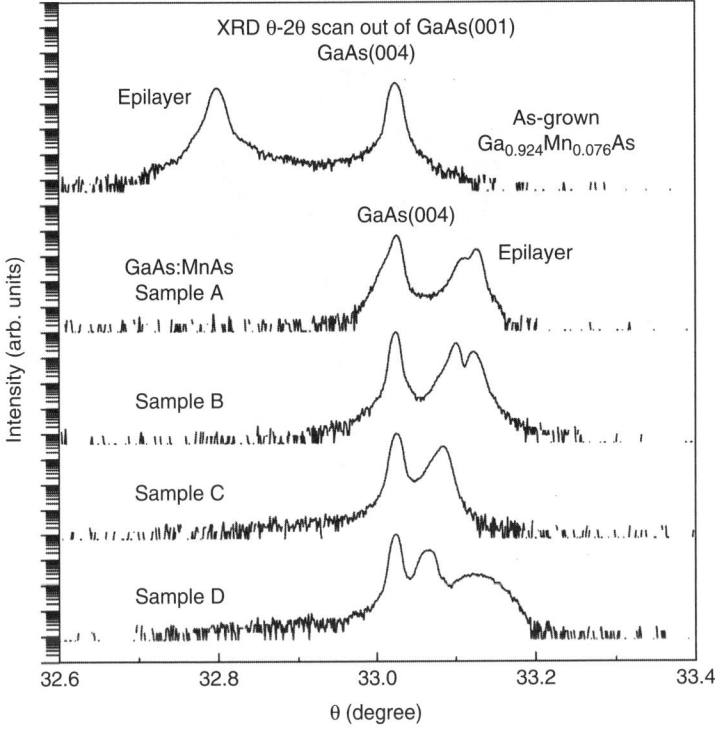

FIGURE 11.12 XRD θ−2θ scans for the (004) reflection from an as-grown 1 μm thick $Ga_{0.924}Mn_{0.076}As$ film and 1 μm thick GaAs:MnAs granular films (sample A–D) annealed under different annealing conditions of Table 11.1.

of *1) and a GaAs matrix (*2). As can be seen in Fig. 11.13, the whole lattice image is coherent and the crystal structure of the MnAs particle has ZB-type symmetry, and no dislocations are seen at the interfaces between the MnAs particles and the GaAs matrix. In the case of NiAs-type hexagonal MnAs particles embedded in GaAs, the TEM lattice image of MnAs is different from that of GaAs, and there are dislocations or distorted lattice at the interfaces between the hexagonal MnAs particles and the ZB-type GaAs matrix. The diffraction patterns of the MnAs particle (*1) and the GaAs matrix (*2) also indicate that their crystal structures are the same ZB-type.

The magnetization of the samples was investigated by alternating gradient force magnetometry (AGFM), and superconducting quantum interference device (SQUID) measurements. In Fig. 11.14A, solid and dotted curves show the magnetization of sample A (GaAs:MnAs of ZB-type) and sample C (GaAs:MnAs of NiAs-type) measured at 300 K by AGFM with a magnetic field applied perpendicular to the film plane.

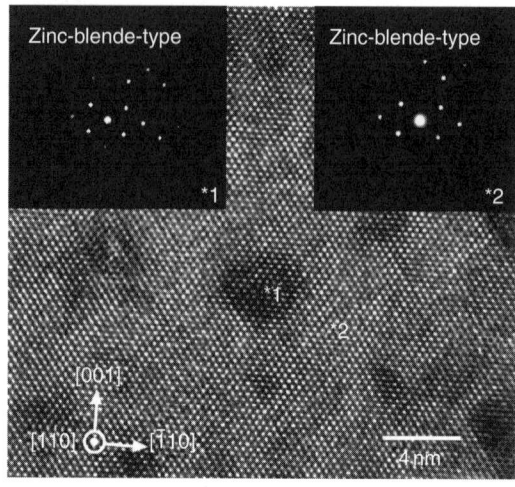

FIGURE 11.13 A TEM lattice image and two diffraction patterns of sample A: Diffraction patterns are taken from a MnAs particle (*1) and a GaAs matrix (*2).

The magnetization ($M–H$) measurements revealed that samples A, B, C, and D are superparamagnetic or ferromagnetic with very small coercivity, and that the saturation magnetizations per the whole film volume of samples A and B were 4 and 2 emu/cm^3, respectively, which were much smaller than those of other two samples (17 emu/cm^3 for sample C, 14 emu/cm^3 for sample D). Assuming that all Mn atoms contribute to form MnAs and the magnetization comes only from the MnAs particles and that the lattice constant of the ZB-type MnAs is 0.598 nm (Ogawa et al., 1999; Ohno, 1999; Ohno et al., 1996; Shirai et al., 1998), the magnetization per Mn atom was estimated to be 0.3 μ_B, 0.1 μ_B, 1.2 μ_B, and 1.0 μ_B for sample A, B, C, and D, respectively, where μ_B is the Bohr magnetron. Here, considering the fact that the magnetization per Mn atom of NiAs-type hexagonal MnAs thin films is 2.2–2.6 μ_B at room temperature (Das et al., 2003; Nakane et al., 2004), this reduction of saturation magnetization in sample A and B is attributed to the ZB-type crystal structure of the MnAs particles with 2–3 nm in diameter.

Figure 11.14B shows the temperature dependence of magnetization ($M–T$) of the as-grown Ga$_{0.924}$Mn$_{0.076}$As sample (broken line), sample A (solid line, GaAs:MnAs of ZB-type), and sample C (dotted line, GaAs: MnAs of NiAs-type), measured by SQUID with a magnetic field applied perpendicular to the film plane. The temperature range and the applied field were 10–70 K and 1,000 Oe for the as-grown Ga$_{0.924}$Mn$_{0.076}$As sample, and 10–350 K and 2,000 Oe for sample A and sample C, respectively. T_C of the ZB-type MnAs particles in sample A was above 350 K and estimated to be 360 K by extrapolating from the M-T curve, whereas T_C

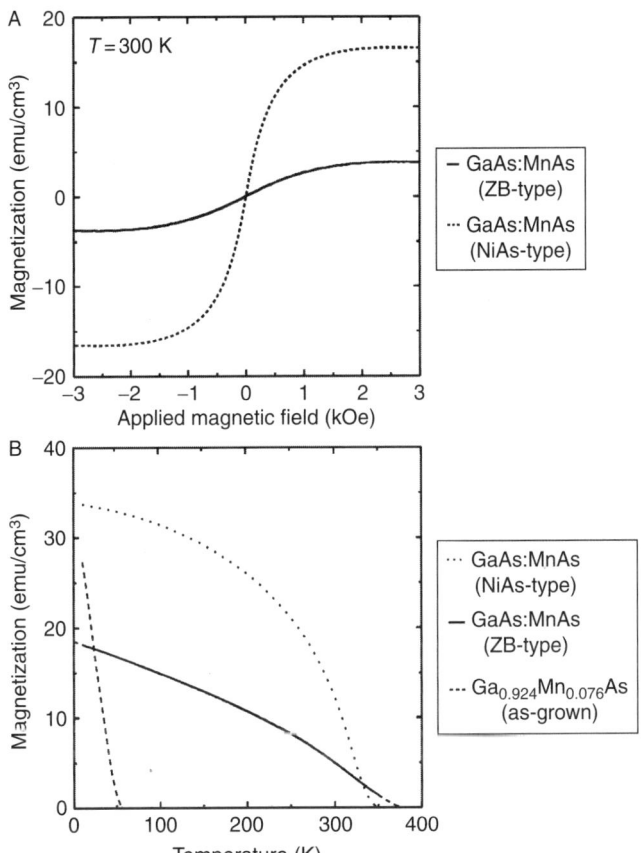

FIGURE 11.14 (A) Magnetization of sample A (solid line, GaAs:MnAs of ZB-type), and sample C (dotted line, GaAs:MnAs of NiAs-type), measured at 300 K with a magnetic field applied perpendicular to the film plane, respectively. (B) Temperature dependence of magnetization of the as-grown $Ga_{0.924}Mn_{0.076}$ As sample (broken line) and sample A (solid line), and sample C as a reference (dotted line), with magnetic fields of 1,000, 2,000, and 2,000 Oe perpendicular to the film plane, respectively.

of the as-grown $Ga_{0.924}Mn_{0.076}As$ sample was 50 K. It was reported that the T_C of GaMnAs has recently reached 173 K (Jungwirth *et al.*, 2005; Ohya *et al.*, 2007), and it is known that T_C of the NiAs-type hexagonal MnAs is 310–320 K (Mira *et al.*, 2002; Sanvito and Hill, 2000). T_C (= 360 K) of the ZB-type MnAs is remarkably higher than that of the NiAs-type hexagonal MnAs.

Finally, we investigated the magneto-optical properties of the GaAs: MnAs granular films. Figure 11.15 shows Kerr ellipticity spectra of

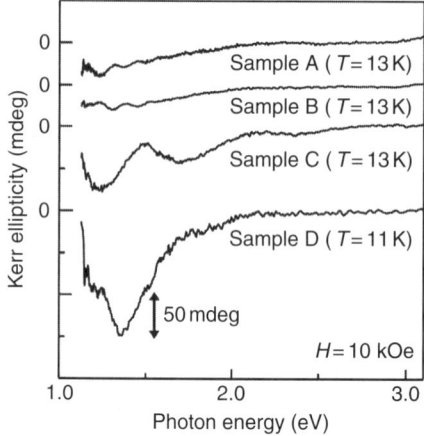

FIGURE 11.15 Kerr ellipticity spectra of sample A at 13 K, B at 13 K, C at 13 K, and D at 11 K, with a magnetic field of 1 T (=10 kOe) applied perpendicular to the film plane, respectively.

samples A, B, C, and D measured in a spectral range of 400–1,100 nm (1.13–3.1 eV) at 11–13 K. Samples C and D have a strong negative peak at 1.2 and 1.4 eV, respectively. However, samples A and B did not have such peak at the same energy point. The Kerr ellipticity spectra of samples A and B with ZB MnAs particles are broader and their intensities are lower than those of samples C and D. Figure 11.16 shows (A) optical reflectivity, (B) Kerr rotation, and (C) Kerr ellipticity spectra of GaAs:MnAs of ZB-type (sample A), and (D) optical reflectivity, (E) Kerr rotation, and (F) Kerr ellipticity spectra of a reference sample of GaAs:MnAs of NiAs-type (fabricated by annealing a 300-nm thick $Ga_{0.944}Mn_{0.056}As$ film *in situ* under an As_4 atmosphere in an MBE chamber at 630–650 °C) measured in a spectral range of 200–1,100 nm (1.13–6.2 eV) at room temperature. The Kerr ellipticity and the Kerr rotation spectra of sample A and the reference GaAs:MnAs of NiAs-type were measured with a magnetic field of 1 T (=10 kOe) applied perpendicular to the film plane. The optical reflectivity of both samples is similar to that of the host GaAs matrix. However, the magneto-optical intensity and magneto-optical spectra of the GaAs:MnAs of ZB-type are quite different form those of GaAs:MnAs of NiAs-type. The reason for this difference is not fully understood, but probably because the ZB-type MnAs particle size is smaller (2–3 nm) and the electronic structure of the ZB-type MnAs is different from that of NiAs-type hexagonal MnAs.

To summarize this section, we have shown the structural, magnetic, and magneto-optical properties of GaAs:MnAs granular thin films, which were fabricated by annealing $Ga_{0.924}Mn_{0.076}$ As thin films under various conditions. ZB-type MnAs particles were formed in the matrix of

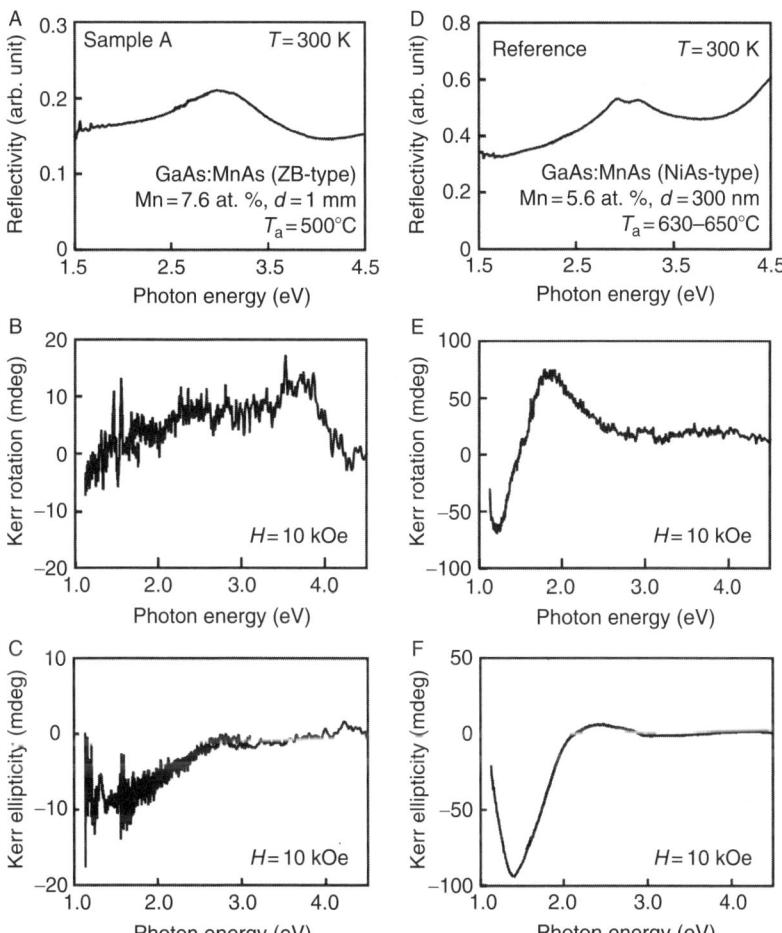

FIGURE 11.16 (A) Optical reflectivity, (B) Kerr rotation, and (C) Kerr ellipticity spectra of the GaAs:MnAs of ZB-type (sample A), and (D) optical reflectivity, (E) Kerr rotation, and (F) Kerr ellipticity spectra of a reference sample of GaAs:MnAs of NiAs-type, which was fabricated by annealing a 300-nm thick $Ga_{0.944}Mn_{0.056}As$ film *in situ* at 630–650 °C. All the spectra were measured at 300 K. (B), (C), (E), and (F) were measured with a magnetic field of 1 T (=10 kOe) applied perpendicular to the film plane.

GaAs by annealing $Ga_{0.924}Mn_{0.076}As$ at relatively low temperature of 500 °C. The magnetic and magneto-optical properties of the GaAs:MnAs granular films with ZB-type MnAs particles were different form those of GaAs:MnAs granular films with NiAs-type hexagonal MnAs particles. In particular, we found that T_C of the ZB-type MnAs particles is ~360 K, which is significantly higher than that of the NiAs-type hexagonal MnAs.

6. MAGNETO-OPTICAL DEVICE APPLICATIONS

As one of the device applications, we proposed and theoretically analyzed semiconductor-waveguide-type optical isolators, which are based on the nonreciprocal loss/gain in the magneto-optical waveguide having MnAs nano-particles embedded in an InAlAs matrix (Shimizu and Tanaka, 2002a,b). The whole device structure is grown on an InP substrate and the operation wavelength is 1.55 μm. In the TM mode, more than 119 dB/cm of isolation is predicted. Furthermore, we proposed a semiconductor-waveguide-type optical isolator for the TE mode, which can realize 36 dB/cm of isolation. Since the proposed waveguide-type optical isolators are composed of all semiconductor-based materials, they can be easily integrated with III–V based optoelectronic devices such as edge-emitting laser diodes.

Optical isolators are the devices which block unwanted reflected light, and are indispensable for stable operation of semiconductor lasers. A commercially available optical isolator is composed of a Faraday rotator and two linear polarizers. At present, ferrimagnetic garnet bulk crystals are used for discrete optical isolators. They are not compatible with semiconductor laser diodes and optical waveguides, because of the difference in size, materials, and device shape, between semiconductor lasers and optical isolators. Semiconductor-waveguide-type optical isolators, which are compatible with edge-emitting semiconductor lasers, are desired for smaller device modules and integration with III–V optoeletronic devices. Waveguide-type optical isolators using garnet crystals were reported in the past (Ando et al., 1988; Fujita et al., 2000; Levy et al., 1996; Shintaku, 1998; Sugimoto et al., 1999; Yokoi et al., 1999). However, since it is impossible to grow garnet crystals on semiconductor substrates, monolithic integration of optical isolators with III–V optoelectronic devices has been considered very difficult.

Here, unlike the conventional optical isolators using Faraday rotation, we focus on semiconductor-waveguide-type optical isolators based on nonreciprocal refractive index change. The nonreciprocal refractive index change occurs in the light of TM mode which propagates in a magneto-optical planar waveguide where the magnetization of the film is aligned transverse to the light propagation direction in the film plane (Yokoi et al., 2000; Zaets and Ando, 1999). This phenomenon is based on the transverse magneto-optical Kerr effect inside the magneto-optical planar waveguide.

The GaAs:MnAs granular material, in which MnAs nano-particles are embedded in a III–V semiconductor (GaAs) matrix, is fully compatible with GaAs/AlGaAs heterostructures and exhibits large magneto-optical effects at room temperature (Akinaga et al., 2000; Shimizu et al., 2001; Tanaka et al., 2001). Therefore, the III–V:MnAs nanoparticle system is suitable for realizing semiconductor-waveguide-type optical isolators.

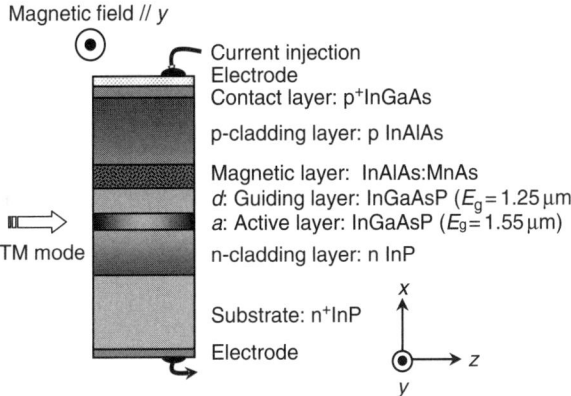

FIGURE 11.17 Waveguide-type optical isolator structure for TM mode operation proposed in this study. The light propagates in the InGaAsP core (active) layer along the z-direction. The InAlAs:MnAs layer is magnetized along the y-direction.

Here, we propose and analyze a waveguide-type optical isolator using MnAs nano-particles integrated on an InP substrate, as shown in Fig. 11.17. The device consists of a bottom electrode, a n^+-InP substrate, a n-InP cladding layer (the refractive index $n = 3.16$), an InGaAsP active layer whose bandgap wavelength E_g is 1.55 µm ($n = 3.53$), an InGaAsP guiding layer ($E_g = 1.25$ µm, $n = 3.36$), a magnetic layer of InAlAs:MnAs (MnAs nano-particles are embedded in InAlAs), a p-InAlAs cladding layer ($n = 3.22$), a p^+-InGaAs contact layer, and a top electrode. As a magneto-optical layer, we adopted InAlAs:MnAs which is lattice-matched to the InP substrate.[1] The operation wavelength λ is set at 1.55 µm, and the operation temperature is set at 300 K, throughout this section. Since it is expected that the optical loss of InAlAs:MnAs is smaller than that of ferromagnetic metals and it is possible to overgrow high quality semiconductor heterostructures on top of the InAlAs:MnAs layer, we can set the InAlAs:MnAs magneto-optical layer near the InGaAsP core (active) layer of the waveguide; thus a strong magneto-optical effect is expected. The magneto-optical effect (Faraday rotation and Faraday ellipticity) of the MnAs nanoparticle system is independent of the host semiconductor matrix.[1] We estimated the off-diagonal element of the dielectric-permeability tensor ε_{xz} to be 0.028–0.00576 i, using the Faraday rotation θ_F and ellipticity η_F ($\theta_F = -0.034°$ and $\eta_F = 0.186°$) and the extinction coefficient κ ($\kappa = 0.08$) of a 200-nm thick GaAs:MnAs

---

[1] We confirmed that by annealing (InAlMn)As grown by low-temperature molecular-beam epitaxy on an InP substrate, the InAlAs:MnAs nanocluster material was formed and it shows superparamagnetism and a large magneto-optical effect (Yokoyama, J. Crystal Growth 2007).

film (Shimizu and Tanaka, 2002), and the refractive index of InAlAs at the wavelength of 1.55 µm.

The operation principle is as follows. In the TM mode, a nonreciprocal loss shift is brought about by the reflection of light at the interface between the InAlAs:MnAs layer and the InGaAsP layer (transverse Kerr effect). The propagation loss for the forward propagating light is compensated by the optical gain of the InGaAsP active layer owing to the current injection, whereas the propagation loss for the backward propagating light still remains. Therefore the optical isolator operation is realized.

The nonreciprocal loss/gain shift $\Delta\kappa_{\text{eff,TM}}$ for the TM mode was calculated by solving Maxwell's equations of the planar waveguide. One can obtain an eigen value Eq. (11.1) by satisfying the boundary conditions of the electromagnetic field,

$$\frac{d^2}{dx^2}H_y + k_0^2\left(\frac{n^4 + \varepsilon_{xz}^2}{n^2} - n_{\text{eff,TM}}^2\right)H_y = 0, \tag{11.1}$$

where, H_y is the y component of the magnetic field vector of the TM mode, k_0 is the wave number in the vacuum, and n is the refractive index of the each layer.

By solving Eq. (11.1), we can obtain the effective refractive index $n_{\text{eff,TM}}$ of the waveguide. Then, $\Delta\kappa_{\text{eff,TM}}$ was given by calculating the imaginary part,

$$\Delta\kappa_{\text{eff,TM}} = \text{Im}[n_{\text{eff,TM forward}}] - \text{Im}[n_{\text{eff,TM backward}}], \tag{11.2}$$

where $n_{\text{eff,TM backward}}$ can be obtained by reversing the sign of the off-diagonal term ε_{xz} of the dielectric-permeability tensor.

Figure 11.18 shows InGaAsP guiding layer thickness (d) dependence of the isolation and the required internal gain of the InGaAsP active layer when the InGaAsP active (core) layer thickness $a = 250$ nm. It was found that with increasing d, the isolation and the required internal gain of the InGaAsP active layer decrease. This is because the penetration of the electromagnetic field into the magnetic layer is decreased with increasing d. Furthermore, to investigate the optimum structure for our waveguide-type optical isolator, we calculated the figure of merit, which is a parameter defined as the isolation divided by the required internal gain based on Fig. 11.18. The figure of merit showed a local maximum of 0.095 dB at $d = 150$ nm. Therefore, $d = 150$ nm gives the optimum structure of our device when a is 250 nm. In this case, the isolation and the internal gain required for compensating the loss of the forward propagating light were calculated to be 222 dB/cm and 2,330 cm^{-1}, respectively. The isolation of 222 dB/cm corresponds to the device length of 1.35 mm to obtain 30 dB isolation. This device size is small enough for practical application.

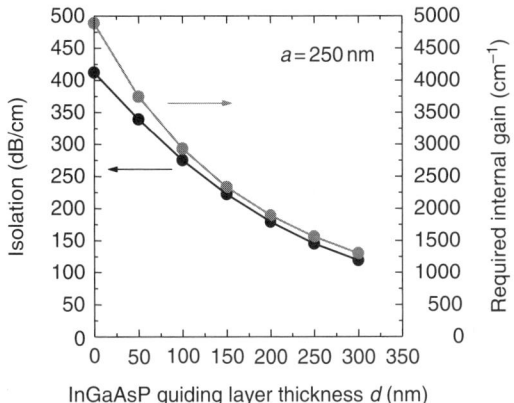

FIGURE 11.18 InGaAsP guiding layer thickness (*d*) dependence of the isolation and the required internal gain of the InGaAsP active layer. The wavelength λ was set at 1.55 μm, and the InGaAsP active layer thickness (*a*) was 250 nm.

The required internal gain of 2,330 cm^{-1} is rather large, but within the achievable range using III–V heterostructures. Here, for optical isolator operation, we have to adopt tensile-strained quantum wells as the active layer, which amplifys the TM mode selectively as discussed in ref. (Zaets and Ando, 1999), because the propagation loss for the TM mode is larger than that for the TE mode in the planar waveguide of Fig. 11.17.

The semiconductor-waveguide-type optical isolator discussed above operates in the TM mode. However, many semiconductor laser diodes usually operate in the TE mode. Therefore, we also propose and analyze a waveguide-type optical isolator for the TE mode, as presented below. For realizing the TE mode operation in a waveguide-type optical isolator, the magnetic field vector (H_x) of the TE mode light (E_y, H_x, H_z) has to be aligned parallel to the magnetization vector of the magneto-optical material (Fujita et al., 2000). To realize this alignment, the magento-optical layer has to be put at one side of the waveguide and the magnetic field has to be applied perpendicular to the waveguide along the x direction. We propose a device structure of the TE mode waveguide-type optical isolator, as shown in Fig. 11.19.

The nonreciprocal loss/gain shift for the TE mode was calculated by the procedure explained as follows. It is difficult to solve the Maxwell's equations in the three dimensional waveguide of Fig. 11.19, thus we adopted the perturbation theory (Morse and Feshbach, 1953) and the effective refractive index method (Streifer and Kapon, 1979). Note that "TE mode" described here means "TE-like mode" whose dominant components of the propagating light are E_y and H_x in the three-dimensional waveguide.

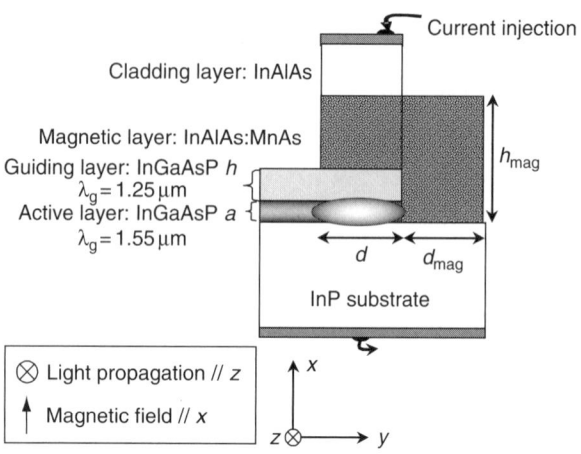

FIGURE 11.19 Waveguide-type optical isolator structure for TE mode operation proposed in this study. The light propagates along the z-direction in the InGaAsP core (active) layer nearby the InAlAs:MnAs magnetic layer. The magnetization and magnetic field is along the x-direction.

In the perturbation theory, the nonreciprocal refractive index change $\Delta n_{\mathrm{eff,TE}}$ for the TE mode is derived as follows:

$$\Delta n_{\mathrm{eff,TE}} = -\frac{i}{2k_0} \frac{\iint (\varepsilon_{yz}/\varepsilon) H_x^* (\partial H/\partial y) dx dy}{\iint |H_x|^2 dx dy} \qquad (11.3)$$

When the InGaAsP core (active) layer thickness $a = 0.2$ μm, the guiding layer thickness $h = 0.3$ μm, the ridge width $d = 1.1$ μm, the width and height of the InAlAs:MnAs layer $d_{\mathrm{mag}} = h_{\mathrm{mag}} = 1$ μm, the isolation and the required internal gain for the TE mode light at the wavelength $\lambda = 1.55$ μm were calculated to be 36 dB/cm and 730 cm^{-1}, respectively. The isolation of 36 dB/cm corresponds to the device length of 8.3 mm to obtain 30 dB isolation. From the calculation described above, it was found that with decreasing d, the nonreciprocal refractive index change $\Delta n_{\mathrm{eff,TE}}$ for the TE mode increased. This is because the overlap of the electromagnetic field of the TE mode with the magnetic layer of the sidewall is increased with decreasing d. Figure 11.20 shows ridge width (d) dependence of the isolation for the TE mode at $\lambda = 1.55$ μm when $a = 0.2$ μm, $h = 0.3$ μm, $d_{\mathrm{mag}} = h_{\mathrm{mag}} = 1$ μm. The optimum device structure for the TE mode waveguide-type optical isolator can be obtained by the decrease of d down to 1 μm. Since the single mode in the y direction of Fig. 11.19 is realized for $d > \sim 1$ μm, the optimum value of d is around 1.1 μm. This cut off condition gives the limitation of isolation (the maximum isolation is 39 dB/cm at $d = 1.0$ μm) for the TE mode in the waveguide geometry of Fig. 11.19, which is smaller than that (>119 dB/cm) for the TM mode.

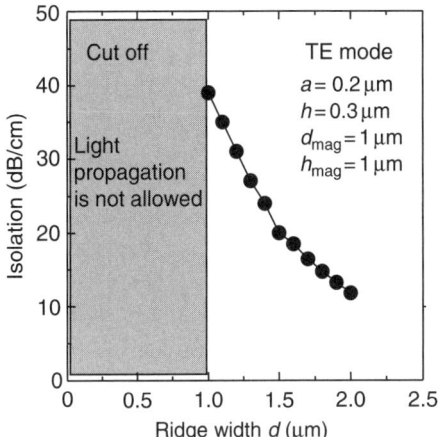

FIGURE 11.20 Ridge width d dependence of the isolation for the TE mode, when $a = 0.2$ μm, $h = 0.3$ μm, $d_{Mag} = h_{Mag} = 1$ μm.

To summarize this section, we proposed and analyzed the semiconductor-waveguide-type optical isolators with MnAs nano-particles for both the TM and TE modes. Advantages of our waveguide-type optical isolators compared with conventional waveguide-type optical isolators are follows. (1) Since the optical absorption loss of the MnAs particle system is smaller than those of the ferromagnetic metals, it is possible to put the MnAs nanoparticle magnetic layer near the core (active) layer of the waveguide, thus a strong magneto-optical effect can be obtained. (2) Due to the excellent compatibility of the MnAs particle system with nonmagnetic semiconductor heterostructures, the device structure can be flexibly designed. Experimentally, semiconductor optical isolators for TE and TM mode operations, based on nonreciprocal loss induced by ferromagnetic Fe and MnAs thin films, respectively, have been fabricated recently (Shimizu et al., 2004; Amemiya et al., 2006, 2007). These devices are expected to be monolithically integrated with waveguide based optical devices on an InP substrate.

ACKNOWLEDGMENTS

The authors thank Dr. H. Shimizu, Mr. T. Amemiya, Prof. Y. Nakano, and Prof. S. Sugahara for fruitful collaboration and discussion. This work was partly supported by the SORST and PRESTO Programs of JST, IT Program of RR2002, R&D for Next-Generation IT, Grant-in-Aids for Scientific Research of MEXT, Kurata Memorial Hitachi Science & Technology Foundation, and by the Special Coordination Funds for Promoting Science and Technology. M.Y. and P-N.H. thank the financial support from the JSPS Fellowships for Young Scientists.

REFERENCES

Akeura, K., Tanaka, M., Ueki, M., and Nishinaga, T. (1995). *Appl. Phys. Lett.* **67,** 3349.
Akeura, K., Tanaka, M., Nishinaga, T., and De Boeck, J. (1996). *J. Appl. Phys.* **79,** 4957.
Akinaga, H., De Boeck, J., Borghs, G., Miyanishi, S., Asamitsu, A., Van Roy, W., Tomioka, Y., and Kuo, L. H. (1998). *Appl. Phys. Lett.* **72,** 3368.
Akinaga, H., Miyanishi, S., Tanaka, K., Van Roy, W., and Onodera, K. (2000a). *Appl. Phys. Lett.* **76,** 97.
Akinaga, H., Mizuguchi, M., Ono, K., and Oshima, M. (2000b). *Appl. Phys. Lett.* **76,** 357.
Amemiya, T., Shimizu, H., Nakano, Y., Hai, P. N., Yokoyama, M., and Tanaka, M. (2006). *Appl. Phys. Lett.* **89,** 021104.
Amemiya, T., Shimizu, H., Hai, P. N., Yokoyama, M., Tanaka, M., and Nakano, Y. (2007). *Jpn. J. Appl. Phys.* **46,** 205.
Ando, K., Okoshi, T., and Koshizuka, N. (1988). *Appl. Phys. Lett.* **53,** 4.
Ando, K., Hayashi, T., Tanaka, M., and Twardowski, A. (1998). *J. Appl. Phys.* **83,** 6548.
Boykin, T. B., and Harris, J. S. (1992). *J. Appl. Phys.* **72,** 988.
Brozak, G., de Andrada e Silva, E. A., Sham, L. J., DeRosa, F., Miceli, P., Schwarz, S. A., Harbison, J. P., Florez, L. T., and Allen, S. J. (1990). *Phys. Rev. Lett.* **64,** 471.
Chiba, D., Matsukura, F., and Ohno, H. (2004). *Physica E* **21,** 966.
Das, A. K., Pampuch, C., Ney, A., Hesjedal, T., Däweritz, L., Koch, R., and Ploog, K. H. (2003). *Phys. Rev. Lett.* **91,** 087203.
De Boeck, J., Oesterholt, R., Van Esch, A., Bender, H., Bruynseraede, C., Van Hoof, C., and Borghs, G. (1996). *Appl. Phys. Lett.* **68,** 2744.
Fujita, J., Levy, M., Osgood, R. M., Jr., Wilkens, L., and Dotsch, H. (2000). *Appl. Phys. Lett.* **76,** 2158.
Garcia, V., Jaffres, H., Eddrief, M., Marangolo, M., Etgens, V. H., and George, J.-M. (2005). *Phys. Rev. B* **72,** 081303.
Garcia, V., Marangolo, M., Eddrief, M., Jaffrès, H., George, J.-M., and Etgens, V. H. (2006). *Phys. Rev. B* **73,** 035308.
Hai, P. N., Yokoyama, M., Ohya, S., and Tanaka, M. (2006a). *Appl. Phys. Lett.* **89,** 242106.
Hai, P. N., Yokoyama, M., Ohya, S., and Tanaka, M. (2006b). *Physica E* **32,** 416.
Hayashi, T., Tanaka, M., Nishinaga, T., Shimada, H., Tsuchiya, H., and Ootuka, Y. (1997). *J. Cryst. Growth* **175/176,** 1063.
Jungwirth, T., Wang, K. Y., Maek, J., Edmonds, K. W., Jürgen, K., Jairo, S.Polini, M., Goncharuk, N. A., MacDonald, A. H., Sawicki, M., Rushforth, A. W., Campion, R. P., et al. (2005). *Phys. Rev. B* **72,** 165204.
Jungwirth, T., Sinova, J., Macek, J., Kucera, J., and MacDonald, A. H. (2006). *Rev. Mod. Phys.* **78,** 809; Also, see the Special Issue on Spintronics, (2007). *IEEE Trans. Electron Dev.* **54** (5).
Kuroiwa, T., Yasuda, T., Matsukura, F., Shen, A., Ohno, Y., Segawa, Y., and Ohno, H. (1998). *Electron. Lett.* **34,** 190.
Levy, M., Osgood, R. M., Jr., Hegde, H., Cadieu, F. J., and Fratello, V. J. (1996). *IEEE Photon. Tech. Lett.* **8,** 903.
Massalski, T. B., Okamoto, H., Subramanian, P. R., and Kacprzak, L., (eds.). (1990). "Binary Alloy Phase Diagram" Vol. 1, 2nd edition, p. 295 and references therein. Akademic Society for Metals, Metals Park, OH.
Mira, J., Rivadulla, F., Rivas, J., Fondado, A., Caciuffo, R., Carsughi, F., Guidi, T., and Goodenough, J. B. (2002). *Condens. Matter* **1,** 0201478.
Moreno, M., Trampert, A., Jenichen, B., Däweriz, L., and Ploog, K. H. (2002). *J. Appl. Phys.* **92,** 4672.
Morse, P. M., and Feshbach, H. (1953). *"Method of Theoretical Physics."* New York: McGraw-Hill, New York.
Munekata, H., Ohno, H., von Molner, S., Segmuller, A., Chang, L. L., and Esaki, L. (1989). *Phys. Rev. Lett.* **63,** 1849.

Nakane, R., Sugahara, S., and Tanaka, M. (2004). *J. Appl. Phys.* **95,** 6558.
Neumann, A. (2001). *J. Appl. Phys.* **90,** 1.
Ogawa, M., Sugano, T., and Miyoshi, T. (1998). *Solid State Electron.* **42,** 1527.
Ogawa, T., Shirai, M., Suzuki, N., and Kitagawa, I. (1999). *J. Magn. Magn. Mater.* **196–197,** 428.
Ohno, H. (1999). *J. Magn. Magn. Mater.* **200,** 110.
Ohno, H., Munekata, H., Penny, T., von Molnar, S., and Chang, L. L. (1992). *Phys. Rev. Lett.* **68,** 2664.
Ohno, H., Shen, A., Matsukura, F., Oiwa, A., Endo, A., Katsumoto, S., and Iye, Y. (1996). *Appl. Phys. Lett.* **69.**
Ohya, S., Hai, P. N., Mizuno, Y., and Tanaka, M. (2006). *Phys. State Sol. (c)* **3,** 4184.
Ohya, S., Hai, P. N., Mizuno, Y., and Tanaka, M. (2007a). *Phys. Rev. B* **75,** 155328.
Ohya, S., Ohno, K., and Tanaka, M. (2007b). *Appl. Phys. Lett.* **90,** 112503.
Ono, K., Okabayashi, J., Mizuguchi, M., Oshima, M., Fujimori, A., and Akinaga, H. (2002). *J. Appl. Phys.* **91,** 8088.
Pangulur, R. P., Tsoi, G., Nadgorny, B., Chun, S. H., Samarth, N., and Mazin, I. I. (2003). *Phys. Rev. B* **68,** 201307.
Paracchini, C., and Dallacasa, V. (1989). *Solid State Commun.* **69,** 49.
Ravindran, P., Delin, A., James, P., Johasson, B., Wills, J. M., Ahuja, R., and Eriksson, O. (1999). *Phys. Rev. B* **59,** 15680.
Rousseau, K. V., Wang, K. L., and Schulman, J. N. (1989). *Appl. Phys. Lett.* **54,** 1341.
Ruster, C., Gould, C., Jungwirth, T., Sinova, J., Schott, G. M., Giraud, R., Brunner, K., Schmidt, G., and Molenkamp, L. W. (2005). *Phys. Rev. Lett.* **94,** 027203.
Sanvito, S., and Hill, N. A. (2000). *Phys. Rev. B* **62,** 15553.
Shimizu, H., Miyamura, M., and Tanaka, M. (2001). *Appl. Phys. Lett.* **78,** 1523.
Shimizu, H., and Tanaka, M. (2002a). *Appl. Phys. Lett.* **81,** 5246.
Shimizu, H., and Tanaka, M. (2002b). *Physica E* **13,** 597.
Shimizu, H., and Nakano, Y. (2004). *Jpn. J. Appl. Phys.* **43,** L1561.
Shintaku, T. (1998). *Appl. Phys. Lett.* **73,** 1946.
Shirai, M., Ogawa, T., Kitagawa, I., and Suzuki, N. (1998). *J. Magn. Magn. Mater.* **177–181,** 1383–1384.
Streifer, W., and Kapon, E. (1979). *Appl. Opt.* **18,** 3724.
Sugahara, S., and Tanaka, M. (2002). *Appl. Phys. Lett.* **80,** 1969.
Sugimoto, N., Shintaku, T., Tate, A., Terui, H., Shimokozono, M., Kubota, E., Ishii, M., and Inoue, Y. (1999). *IEEE Photon. Tech. Lett.* **11,** 355.
Sun, Z. G., Mizuguchi, M., and Akinaga, H. (2004). *Jpn. J. Appl. Phys.* **43,** 2101.
Tanaka, M. (1995). *J. Mater. Sci. Eng.* **B31,** 117.
Tanaka, M. (1998). *J. Vac. Sci. Technol.* **B16,** 2267.
Tanaka, M. (2000). *J. Vac. Sci. Technol.* **A18,** 1247.
Tanaka, M., and Higo, Y. (2001). *Phys. Rev. Lett.* **87,** 026602/1–4.
Tanaka, M., Harbison, J. P., Sands, T., Cheeks, T. L., and Rothberg, G. M. (1994). *J. Vac. Sci. Technol.* **B12,** 1091.
Tanaka, M., Saito, K., and Nishinaga, T. (1999). *Appl. Phys. Lett.* **74,** 64.
Tanaka, M., Shimizu, H., and Miyamura, M. (2001). *J. Cryst. Growth* **227–228,** 839.
Van Esch, A., Van Bockstal, L., De Boeck, J., Verbanck, G., van Steenbergen, A. S., Wellmann, P. J., Grietens, B., Bogaerts, R., Herlach, F., and Borghs, G. (1997). *Phys. Rev.* **B56,** 13103.
Wellmann, P. J., Garcia, J. M., Feng, J. L., and Petroff, P. M. (1998). *Appl. Phys. Lett.* **73.**
Yokoi, H., Mizumoto, T., Takano, T., and Shinjo, N. (1999). *Appl. Opt.* **38,** 7409.
Yokoi, H., Mizumoto, T., Shinjo, N., Futakuchi, N., and Nakano, Y. (2000). *Appl. Opt.* **39,** 6158.
Yokoyama, M., Yamaguchi, H., Ogawa, T., and Tanaka, M. (2005). *J. Appl. Phys.* **97,** 10D317.
Yokoyama, M., Ogawa, T., Nazmul, A. M., and Tanaka, M. (2006). *J. Appl. Phys.* **99,** 08D502.
Yokoyama, M., Ohya, S., and Tanaka, M. (2007). *J. Cryst. Growth* **301/302,** 627–630.
Zaets, W., and Ando, K. (1999). *IEEE Photon. Tech. Lett.* **11,** 1012.
Zutic, I., Fabian, J., and Das Sarma, S. (2004). *Rev. Mod. Phys.* **76,** 323.

INDEX

A

Aharonov–Casher effect (ACE), 46
Alternating gradient force magnetometry (AGFM), 473
Anderson magnetic impurity model, 377, 383
Anisotropic magnetoresistance (AMR), (Ga,Mn)As
  Coulomb blockade
    charging energy contribution, 185–186
    conductance oscillations, 188–189
    field-sweep and rotation measurements, 188, 190
    high field effects, 187–188
    magnetization reorientation, 189, 192
    opposite transistor characteristics, 191, 193
    resistance variation magnitude, 188, 191
    schematic and CB conductance, 184–185
    Zeeman coupling, 187
  ohmic regime AMR
    Corbino disk and Hall bar devices, 174–175
    Corbino measurements, 173–174
    disordered valence-band description, 170
    field-induced changes, 172
    hole scattering rates simulations, 173
    lithographic patterning, 175–176
    longitudinal and transverse AMR, 174
    magnetization and external magnetic field vectors, 168–169
    microscopic mechanism, 170–171
    Mott's model, 167–168
    noncrystalline AMR spherical bands, 173
    phenomenological decomposition, 168–170
  tunneling effects
    amplification effect, 180–181
    anisotropic density states, 183–184
    kinetic exchange models, 178–179
    Landauer–Büttiker transport theory calculation, 183
    low-field ac MR measurement, 182–183
    MR measurements, 180
    signalling and devices, 177–178
Anisotropic strain relaxation
  finite element simulation, 274–275
  grazing incidence X-ray diffraction (GIXRD), 276
  transport and magnetization, 274, 277
Anomalous Hall effect (AHE), 46, 61–64
Au/GaAs Schottky diode systems, 458
Average t-matrix approximation (ATA), 387

B

Band anti-crossing model, 390
Bandgap semiconductor. *See* Diamond
Bloch wave functions, 381
Bloembergen-Rowland mechanism, 402
Boltzmann factor, 401
Bound magnetic polaron (BMP), 391–392
Bulk inversion symmetry (BIA), 56

C

Carrier-induced ferromagnetism, DMSs
  Curie–Weiss behavior, 296–297
  domains and soft mode behavior, 298
  mean field model and Weiss temperature, 294–295
  Monte Carlo calculations, 297–298
  PL transition energies, 293–294
  spin susceptibility enhancement factor, 295–296
  (Cd,Mn)Te quantum wells, 291–292
  Zener model and giant Zeeman effect, 292–293
Carrier-mediated ferromagnetism, p–d Zener model
  applicability paradigms, 399–400
  dilute magnetic semiconductor (DMS) films

Carrier-mediated ferromagnetism, p–d Zener model (*cont.*)
  anisotropy analysis, 404
  generalized carrier spin susceptibility, 403
  mean-field equation, 402–403
  phase diagram for (Ga,Mn)As film, 404–405
  spin–orbit coupling, 402
  disorder and localization effects
    critical region, 411–413
    critical scattering, MIT, 413–415
    quantum localization, 407–410
    strongly localized regime, 415–416
    weak disorder, 410–411
  DMS modulated structures, 405–406
  functional free energy density, 400–401
  mean-field approximation, 401
  origin of ferromagnetism, 399
(Cd,Mn)Te quantum wells
  Curie–Weiss behavior, 296–297
  domains and soft mode behavior, 298
  mean field model and Weiss temperature, 294–295
  Monte Carlo calculations, 297–298
  PL transition energies, 293–294
  spin susceptibility enhancement factor, 295–296
  Zener model and giant Zeeman effect, 292–293
CdTe/ZnTe magnetic quantum dots
  exchange integral ratio, 304–305
  exchange interaction, 303
  spin configuration, 304
  TEM image, 301
  vs. Mn-doped QDs, 301–303
Computational nano-materials design system
  KKR-CPA-LDA, 435
  magnetic mechanism, T_C, and unified physical picture
    effective exchange interaction calculation, 440
    electron configuration of magnetic impurities, 441
    KKR-CPA method, 436
    (Zn,Co)O, total and partial Co-3d DOS calculation, 439
    (Zn,Mn)O, total and partial Mn-3d DOS calculation, 438
    (Zn,V)O, total and partial V 3d DOS calculation, 437
  paramagnetic *vs.* ferromagnetic states, 436–437
  spin-glass *vs.* ferromagnetic states, 442–443
  Zener's double-exchange mechanism, 441–442
  oxide spintronics without 3d TM, 450–451
  semiconductor nanospintronics realization, 433–434
  spinodal nano-decomposition and nano-spintronics applications
    colossal magnetic response, 450–451
    convexity and concavity control, 449
    2D and 3D diffusion, 447–448
    1D Konbu phase, 448–449
    3d TM impurities, 447
    effective chemical pair interaction calculation, 444, 446
    mixing energy calculation, 444–445
    mixing energy control, codoping, 449–450
    molecular-beam epitaxy (MBE) process, 448
    Monte Carlo simulation of T_C, 446–447
    wide band-gap DMS, 444
Constriction resistance, 281
Current-induced domain wall switching, (Ga,Mn)As alloys
  depinning/pinning events, 197–198
  external field and negative current ramps, 196–197
  L-shaped geometry, 194–195
  L-shaped microbars phenomenology, 195–196
  magneto-optical Kerr-effect imaging, 192, 194
  scanning electron micrographs, 193–194
  switching mechanism, 198–199

D

Density of states (DOS) calculation
  Co-doping, 439
  Mn-doping, 438
  V-doping, 437–438
Diamond, single spins
  anisotropic interactions
    Bloch equations, 31
    electronic ground state, 30
    energy diagram *vs.* magnetic field, 34
    level avoided crossing (LAC), 30–31
    steady-state solution, 33
  coupled spins, 38–40

NV center
  confocal fluorescence microscopy techniques, 29–30
  electronic ground state, 28–29
  optical transitions, 29
  spin manipulation and coherence
    Rabi nutations and oscillations, 35–36
    timing diagram, 36–37
Diluted magnetic semiconductors (DMSs)
  band structure, 331–332
  Curie temperature, 333
  in electronic devices, 330
  GaMnAs
    amphoteric defect model, 101
    Be codoping, 115–118
    Curie temperature limit, 101–103
    extrinsic doping, 114–115
    film thickness reducing effects, 108–113
    heterostructures Modulation doping, 118–122
    low-temperature annealing, 103–108
    magnetic characterization, 97–98
    Mn lattice site locations, 92–93
    Mn location determination, 95
    T-MBE growth, 93–95
  GaN diluted with Cr
    crystal field transitions, 360–361
    electron configuration, 359
    electron structure, 360
  GaN diluted with Fe
    electron configuration, 357
    Fermi level, 359
    negatively charged acceptor, 358
    paramagnetic and ferromagnetic components, 358–359
  III–V semiconductors, 334
  II–VI compounds, 330–331
  $II_{1-x}Mn_xV$ alloys
    curie temperature and hole concentration saturation, 92
    hole concentration determination, 6–97
    hole-mediated ferromagnetism, 91–92
    self-compensation, 122–125
  magnetic interactions
    exchange integrals, 355–357
    magnetic properties, 352–355
Dirac equation, 48–49
DMSs. See Diluted magnetic semiconductors
Drude-Boltzmann conductivity, 408
Dynamic mean-field theory, 388

E

Efros-Shklovskii Coulomb gap, 379
Electro-optic modulator (EOM), 9
Exchange interactions, magnetically doped semiconductors
  carriers and localized spins, 381–382
  effective mass carriers
    band and bound holes, 397
    electron-hole exchange interaction, 392–393
    electrons and bound holes, 393–397
    intra-band exchange interaction, 392
  sp–d(f) effects
    bound magnetic polaron (BMP), 391–392
    Kondo effect, 391
    spin-disorder scattering, 390–391
    weak coupling extended states, 382–383

F

Faraday ellipticity, 479
Faraday rotation, 478–479
Ferromagnetic semiconductors (Ga,Mn)As
  atomic structure, 138–139
  extraordinary magnetotransport, 167–199
  impurity-band to valence-band crossover, 139–154
  Mn impurity concentration, 139
  quantum interference and interaction effects, 154–162
  resistivity, critical contribution, 162–166
  memory devices
    lithographic anisotropy control, 277
    local anisotropy control, 277–278
Ferromagnetic spin–spin interaction models, 397–398

G

(Ga,Mn)As alloys
  amphoteric defect model, 101
  Coulomb blockade AMR
    charging energy contribution, 185–186
    conductance oscillations, 188–189
    field-sweep and rotation measurements, 188, 190
    high field effects, 187–188
    magnetization reorientation, 189, 192

(Ga,Mn)As alloys (*cont.*)
  opposite transistor characteristics, 191, 193
  resistance variation magnitude, 188, 191
  schematics and CB conductance, 184–185
  Zeeman coupling, 187
current-induced domain wall switching
  depinning/pinning events, 197–198
  external field and negative current ramps, 196–197
  L-shaped geometry, 194–195
  L-shaped microbars phenomenology, 195–196
  magneto-optical Kerr-effect imaging, 192, 194
  scanning electron micrographs, 193–194
  switching mechanism, 198–199
Curie temperature limit, 101–103
double donar defects, 95
elemental sources and fluxes, 93–94
Fermi level manipulation
  Be codoping, 115–118
  extrinsic doping, 114–115
  heterostructures modulation doping, 118–121
film thickness reducing effects, $Ga_{1-x}Mn_xAs$
films thickness, 109
  Mn (PIXE) and GaAs (RBS) angular scans, 110
  Mn distribution, 112–113
  Mn_I outdiffusion, 110–111
  Mn 2p x-ray absorption spectroscopic measurements, 111–112
interstitial and substitutional Mn concentrations, 99–100
lattice site, 98
low-temperature annealing
  acceptor compensating center, 108
  conductivity and the Curie temperature, 105
  double-peak feature, 106–108
  magnetization, 105–106
  RBS and PIXE angular scans, 107
  theoretical magnetoimpurity model, 105
  zero-field resistivity, 104–105
Mn impurity-band to valence-band
  absolute conductance values, 154
  acceptor function and doping, 139–140
  ac conductivity, 144–145
  crossover regime, 141
  dc conductivity, 145–147
  diagonalization calculations, 152–153
  high-doped the impurity band, 148
  high *vs.* low doped insulating (Ga,Mn)As materials, 149, 151
  hole binding potential, 140
  infrared absorption measurements, 149–151
  inter-valence-band transitions, 151–152
  mean impurity separation, 142–143
  short-range potential model, 140–141
  sum rule, 153–154
  temperature-dependent resistivity, 147–148
ohmic regime AMR
  Corbino disk and Hall bar devices, 174–175
  Corbino measurements, 173–174
  disordered valence-band description, 170
  field-induced changes, 172
  hole scattering rates simulations, 173
  lithographic patterning, 175–176
  longitudinal and transverse AMR, 174
  magnetization and external magnetic field vectors, 168–169
  microscopic mechanisms, 170–171
  Mott's model, 167–168
  noncrystalline AMR spherical bands, 173
  phenomenological decomposition, 168–170
quantum interference and interaction effects
  Aharonov–Bohm (AB) effect, 154
  electronûelectron interaction quantum correction, 155–156
  magneto-crystalline anisotropies, 156–157
  semiclassical theory, 154–155
  UCFs and WL corrections, 155
  weak anti-localization (WAL) effect, 155
reflection high-energy electron diffraction (RHEED) pattern, 94
resistivity, critical contribution
  dr/dT singularity and Curie temperatures, 164–166
  magnetization M(T), 162–163

Index

temperature r(T) data, 163
spinodal decomposition, 94–95
tetrahedral interstitial sites, 98–99
tunneling AMR
 amplification effect, 180–181
 anisotropic density states, 183–184
 kinetic exchange models, 178–179
 Landauer–Büttiker transport theory calculation, 183
 low-field ac MR measurement, 182–183
 MR measurements, 180
 TAMR signalling and devices, 177–178
 unstructured bar and constricted nanodevice schematics, 181–182
GaN based diluted magnetic semiconductors
 diluted with Cr
  crystal field transitions, 360–361
  electron configuration, 359
  electron structure, 360
 diluted with Fe
  electron configuration, 357
  Fermi level, 359
  negatively charged acceptor, 358
  paramagnetic and ferromagnetic components, 358–359
 internal reference rule, 361–362
Generalized alloy theory, 387, 390
Grazing incidence X-ray diffraction (GIXRD), 276

H

Hanle effect, 25–26
Heisenberg model, 439–440
Hopping conductivity, 345–346

I

II–VI diluted magnetic semiconductors (DMSs)
 Mn system magnetization
  carrier-induced ferromagnetism and MBE growth, 290
  free exciton line splitting, 289
  giant Zeeman effect, 289
  II–VI quantum dots (QDs), 290
  Maxwell-Boltzmann statistics, 288
  modified Brillouin function, 288–289
 quantum transport
  current–voltage characteristics, 318–319
  integer quantum Hall effect, 316–317

spin injection and spin transport, 318
universal conductance fluctuations, 316
Ising model, 444, 446

J

Jahn-Teller effect, 328, 376, 380

K

Kanamori–Goodenough rule, 436, 441–442
Kerr ellipticity spectra, 475–477
Kerr rotation, 476–477
 conduction-band energy level, 6
 distributed Bragg reflectors (DBRs), 7
 photoluminescence, 7–8
KKR-CPA method. See Korringa–Kohn–Rostoker coherent-potential approximation method
Kondo coupling, 382
Kondo effect, 391
Korringa–Kohn–Rostoker coherent-potential approximation method
 disordered local moment (DLM) state, 436
 mixing energy calculation, 444
 wide band-gap semiconductors, 435
Kubo diagrammatic formalism, 47

L

Layer-by-layer crystal growth method, 447, 449

M

Magnetically doped semiconductors
 carrier-mediated ferromagnetism, p–d Zener model
  applicability paradigms, 399–400
  beyond colinear magnetic ground state, 406–407
  dilute magnetic semiconductor (DMS) films, 402–405
  disorder and localization effects, 407–416
  DMS modulated structures, 405–406
  functional free energy density, 400–401
  mean-field approximation, 401
  origin of ferromagnetism, 399
 exchange interactions
  carriers and localized spins, 381–382
  effective mass carriers, 392–397

Magnetically doped semiconductors (cont.)
  sp–d(f) effects, 382–392
  ferromagnetic spin–spin interaction
    models, 397–398
  ferromagnetism without magnetic
    elements, 422–423
  magnetic ion nonrandom distribution
    magnetooptical and magnetotransport
      functionalities, 420–422
    spinodal decomposition controlling,
      419–421
    spinodal decomposition observations,
      416–419
  spontaneous magnetization, 424
  substitutional transition metal impurities
    carrier dopant, 378–379
    carrier traps, 377–378
    charge transfer states, 379–380
    magnetic moments, 380–381
    resonant impurities, 379
    TM impurity-related levels, 375–377
  substitutional transition metal impurities
    isoelectronic impurities, 377
Magnetic circular dichroism (MCD)
  measurements, 460–461
Magnetic force theorem
  Heisenberg model, 439
  Monte Carlo simulation, 435–436
Magnetic ion nonrandom distribution
  magnetooptical and magnetotransport
    functionalities, 420–422
  spinodal decomposition controlling
    co-doping, 419–420
    Coulomb interaction, 419
    Cr distribution, 420–421
    nanocrystal self-organized growth
      model, 420
  spinodal decomposition observations
    coherent hexagonal
      nanocrystals, 417
    ferromagnetic-like response,
      DMS, 419
    magnetic nanocrystals, 418
    nanocolumn formation, 417–418
    phase diagrams of alloys, 416
Magnetic tunnel junctions (MTJs)
  annealing process, 465
  bias voltage dependence, 466
  complex band structure, AlAs barrier
    calculation, 469–471
  dependence of TMR ratio, 468–469
  high-resolution TEM image, 464–465
  RA dependence, 466–468
  schematic structure and growth process,
    463–464
  temperature dependence, 466–467
  TMR loops, 467–469
  tunneling magnetoresistance (TMR)
    characteristics, 465–466
Magnetooptical planar waveguide, 478
Maxwell's equations, 480
Mean-field approximation (MFA), 436, 440
Mean-field equation, 402
Memory devices, ferromagnetic
  semiconductor
  lithographic anisotropy control, 277
  local anisotropy control
    nonvolatile mode, 277–278
    SEM photograph, 278
Metal-insulator transition, 254–255
MnAs/III–V hybrid and composite
  structures
  GaAs:MnAs nano-particles
    crystal structure and quality, 456–458
    fabrication process, 456
    layer characteristics, 458
  GaAs:MnAs thin film
    current-voltage (I–V) characteristics,
      461–462
    film preparation process, 459–460
    granular material, 459
    magnetic circular dichroism (MCD)
      measurements, 460–461
    magnetic field dependence
      measurements, 462–463
    magnetic-field-sensitive phenomenon,
      458
    magneto-transport characteristics, 462
    metal-semiconductor hybrid system,
      458–459
  magneto-optical device applications
    advantages of waveguide-type optical
      isolators, 483
    Faraday rotation and Faraday
      ellipticity, 479
    InGaAsP guiding layer, isolation
      dependence, 480–481
    nonreciprocal loss/gain shift, 481
    nonreciprocal refractive index
      change, 478
    operation principle, 480
    ridge width dependence of isolation,
      482–483
    TE mode operation, 481–482

TM mode operation, 478–479
waveguide geometry, 482
spin dependent tunneling transport
properties
annealing process, 465
bias voltage dependence, 466
complex band structure, AlAs barrier
calculation, 469–471
dependence of TMR ratio, 468–469
high-resolution TEM image, 464–465
magnetic tunnel junctions (MTJs)
structures, 463
RA dependence, 466–468
schematic structure and growth
process, 463–464
temperature dependence, 466–467
TMR loops, 467–459
tunneling magnetoresistance (TMR)
characteristics, 465–466
zinc-blende type and NiAs-type MnAs
nanoparticles properties
annealing conditions, 472
GaAs:MnAs granular thin films
fabrication process, 471–472
half-metallic ferromagnet, 471
high resolution TEM lattice image,
472–474
magnetization measurements,
473–475
magneto-optical properties, 475–476
optical reflectivity, Kerr rotation, and
Kerr ellipticity spectra, 476–477
structural properties, 472
X-ray diffraction (XRD) spectra,
472–473
Mn doped II–VI quantum dots
CdTe/ZnTe self-assembled, 301
magnetic atoms introduction, 300
single-dot micro-PL measurements, 299
tunable nanomagnets and sp–d
interaction, 299–300
Mn impurities, III–V semiconductors
ionisation energies of, 351
ionized acceptor $Mn^{2+}(d^5)$
energy level diagrams of, 338
ESR spectra of, 335–336
exchange interaction, 356
Hamiltonian parameters, 336, 338
$vs.$ II–VI semiconductors, 339
lattice relaxation, 349
localization radius, $Mn^{2+}(d^5)$
exchange interaction, 356

Hall mobility and Hall concentration,
347
hopping conductivity, 346
quantum defect models, 347
resistivity, 346–347
lower charge states, 350–351
magnetic interactions
exchange integrals, 355–357
magnetic properties, 352–355
neutral acceptor $Mn^{3+}(d^4)$
absorption band, 340–341
absorption coefficient, 343
energy level diagram, 341–342
Huang–Rhys factor, 341
optical absorption experiment, 340
neutral acceptor $Mn^{2+}(d^5)$+hole
configuration
chemical shift, 344
electron spin resonance, 344–345
n-type GaN
electron transitions, 349
optical absorption spectra, 350
MnSb/GaAs hybrid granular systems, 458
$III_{1-x}Mn_xV$ alloys
synthesis routes
ion implantation and pulsed laser
melting (II-PLM) process, 126
LT-MBE growth phase, 125–126
Mn ion implantation, 126–127
TC increasing strategies, 129–130
Molecular-field (MFA) approximation
spin-disorder scattering, 390
splitting of extended states
strong coupling, 384–386
weak coupling, 383
Monte Carlo simulation (MCS), 410, 418
magnetic force theorem, 435–436
spinodal nano-decomposition, 449
Mott-Hubbard mechanism, 377

N

Nanobars
anisotropic strain relaxation
finite element simulation, 274–275
grazing incidence X-ray diffraction
(GIXRD), 276
transport and magnetization, 274, 277
device fabrication, 278–279
magnetic characterization
magnetic anisotropy, 271
SEM structures, 269
SQUID magnetization data, 270–271

Nanobars (cont.)
 magnetic states
  magnetic field line patterns, 280
  writing process, 279
  resistance signal origin
   locally imposed magnetic anisotropies, 283
   magnetoresistance behavior, 281
   wave-function geometry, 281–282
  transport characterization
   lithographically imposed anisotropy, 274
   magnetoresistance scans, 271–273
Nanoconstriction devices
 current-induced switching
  magnetoresistance spin-valve signal, 267
  resistance vs. voltage curves, 266–267
  SEM and schematic diagram, 265–266
 etching process, 264
 lithographic anisotropy control
  in-plane magnetic anisotropy, 268
  strain relaxation, 269
 magnetoresistance, 262–263
 nanofabrication, 260
 spin accumulation-induced resistance, 262
Non-perturbative Wigner-Seitz-type model, 385
n-type GaN semiconductors
 electron transitions, 349
 optical absorption spectra, 350

O

On-d-shell Coulomb (Hubbard) repulsion energy (U), 376
One dimension systems
 atomic scale surface probes, 298
 CdTe/ZnTe magnetic QDs
  exchange integral ratio, 304–305
  exchange interaction, 303
  spin configuration, 304
  TEM image, 301
  vs. Mn-doped QDs, 301–303
 Mn doped II–VI QD
  CdTe/ZnTe self-assembled QDs, 301
  magnetic atoms introduction, 300
  single-dot micro-PL measurements, 299
  tunable nanomagnets and sp–d interaction, 299–300
 single magnetic atom, electrical control
  different-energy splittings, 314
  h–Mn exchange interaction, 314–315
  negatively charged exciton emission spectrum, 310–313
  nonmagnetic QD and Schottky structure, 310–311
  optical transition scheme, 312–313
  resonant and nonresonant excitation, 310–311
 single magnetic atom spin state
  e–h pairs resonant excitation, 307–308
  laser excitation density, 307
  Mn-doped quantum dot exciton, PL spectra, 309
Optical reflectivity, 476–477
Optical Stark effect, semiconductor quantum dots
 exciton transition energy, 14
 intensity and detuning dependence, 17
 perturbation theory, 13–14
 spin manipulation, 15
 ultrafast optical pulses, 13
 ultrafast π rotations, 16

P

Paramagnetic Brillouin function, 380
p–d Zener model, carrier-mediated ferromagnetism
 applicability paradigms, 399–400
 beyond colinear magnetic ground state, 406–407
 dilute magnetic semiconductor (DMS) films
  anisotropy analysis, 404
  generalized carrier spin susceptibility, 403
  mean-field equation, 402–403
  phase diagram for (Ga,Mn)As film, 404–405
  spin–orbit coupling, 402
 disorder and localization effects
  critical region, 411–413
  critical scattering, MIT, 413–415
  quantum localization, 407–410
  strongly localized regime, 415–416
  weak disorder, 410–411
 DMS modulated structures, 405–406
 functional free energy density, 400–401

mean-field approximation, 401
origin of ferromagnetism, 399

R

Random phase approximation (RPA), 436, 440

S

Schrieffer-Wolf transformation, 382
Self-interaction corrected local density approximation
  vs. photoemission spectroscopy (PES), 437–439
  wide band-gap semiconductors, 435
semiclassical Boltzmann equation (SBE)
  electron functions, 65
  golden rules, 65–66
  kinetic equation, 67–69
  mechanisms, 69–71
  side-jump mechanism, 67
  transport theory, 64
  velocity renormalizations, 69
Semiconductor quantum dots, single electron spins
  Faraday effect, 5
  Kerr rotation (KR)
  conduction-band energy level, 6
  distributed Bragg reflectors (DBRs), 7
  photoluminescence, 7–8
  optical selection rules
  band structure, 4
  optical spin injection, 5
  optical Stark effect
    exciton transition energy, 14
    intensity and detuning dependence, 17
    perturbation theory, 13–14
    spin manipulation, 15
    ultrafast optical pulses, 13
    ultrafast π rotations, 16
  photoluminescence (PL), 3–4
  spin coherence time, 18
  spin dynamics
    electro-optic modulator (EOM), 9
    KR spectra, 10
    magnetic field dependence, 11
    nuclear spin polarizations, 12
Semiconductor quantum wells, magnetic spins
  dynamic magnetic polarization, 23–24
  electron-hole spin states, 18–19
  exchange splitting, 25
  optical spin centers, 19
  optical transitions, 20–21
  spin dynamics, 25–27
  zero-field optical control
    circular polarization, 23
    magnetic ion spin orientation, 22
    spectral splitting, 21
Semiconductor-waveguide-type optical isolators
  advantages, 483
  Faraday rotation and Faraday ellipticity, 479
  InGaAsP guiding layer, 480–481
  nonreciprocal loss/gain shift, 481
  nonreciprocal refractive index change, 478
  operation principle, 480
  ridge width dependence of isolation, 482–483
  TE mode operation, 481–482
  TM mode operation, 478–479
  waveguide geometry, 482
SHE and AHE coupled systems
  heuristic picture
    2D electronic eigenstates, 73–74
    electron spin and coupling term, 75
    Fermi momenta and energy, 76
    skew-scattering, 76–77
  numerical studies
    edge spin-polarization, 78–79
    light emitting diodes (LEDs), 77–78
    optical-based experiments, 77
  semiclassical Boltzmann equation (SBE)
    electron functions, 65
    golden rules, 65–66
    kinetic equation, 67–69
    mechanisms, 69–71
    side-jump mechanism, 67
    transport theory, 64
    velocity renormalizations, 69
  skew scattering, 73
  two dimensions, 71–72
SIC-LDA. See Self-interaction corrected local density approximation
Single-electron transistor (SET), (Ga,Mn)As
  Coulomb blockade AMR
    charging energy contribution, 185–186
    conductance oscillations, 188–189
    field-sweep and rotation measurements, 188, 190
    high field effects, 187–188
    magnetization reorientation, 189, 192

Single-electron transistor (SET), (Ga,Mn)As (*cont.*)
  opposite transistor characteristics, 191, 193
  resistance variation magnitude, 188, 191
  schematics and CB conductance, 184–185
  Zeeman coupling, 187
Spin-disorder scattering, 390–391
Spin dynamics
  quantum dots
    electro-optic modulator, 9
    KR spectra, 10
    magnetic field dependence, 11
    nuclear spin polarizations, 12
  quantum wells, 25–27
Spin–flip Raman scattering, 391
Spin Hall effect (SHE), 46, 61–64
Spin-orbit coupling systems
  band structure
    8-band Kane model, 52–53
    8-band Kohn–Luttinger, 53–56
    effective mass, 56–61
    k.p hamiltonians, 51–52
  relativistic origins
    bispinorand spinor, 49
    compton wavelength, 51
    Dirac equation, 48–49
    physically transparent, 50–51
  SHE and AHE, 61–64
    heuristic picture, 73–77
    numerical studies, 77–79
    semiclassical Boltzmann equation (SBE), 64–71
    skew scattering, 73
    two dimensions, 71–72
  topological berry's system
    ACE calculations, 82–85
    AC phases, 80–81
    coherent nanoscale device, 80
    HgTe ring structure, 82
    rashba coupling, 81
Spin–spin Hamiltonian, 402
Spintronics nanodevices
  anisotropic strain relaxation, 274–277
  device fabrication, 278–279
  magnetic characterization, 269–271
  magnetic states, 279–280
  nanoconstrictions devices
    agnetoresistance, 262–263
    anofabrication, 260

    current-induced switching, 265–267
    etching process, 264
    lithographic anisotropy control, 268–269
    pin accumulation-induced resistance, 262
    resistance signal origin, 281–283
    transport characterization, 271–274
  tunneling anisotropic magnetoresistance (TAMR)
    amplification, 252–253
    characteristics, 245–246
    construction of, 244–245
    differential conductance, 253–254
    electronic structure, 248
    ferromagnetic metal break junctions, 259
    magnetic and DOS anisotropies, 247–248
    metallic transport properties, 256
    multiple interfaces, 249–250
    quantum interference effects, 255–256
    switching mechanism, 247
    tunneling geometry, 259–260
    uniaxial anisotropy, 256–257
    volatile and nonvolatile operations modes, 251–252
Spin valves, 329
Splitting of extended state, strong coupling
  anti-bonding state, 389
  free exciton splitting, 384–385
  generalized alloy theory, 390
  grey scale spectral density, 388–389
  p–d exchange interaction, 383–384
  spectral density of states, 387–388
  VCA and MFA, 385–386
Stoner model, 392
Substitutional transition metal impurities
  carrier dopant, 378–379
  carrier traps, 377–378
  charge transfer states, 379–380
  isoelectronic impurities, 377
  magnetic moments, 380–381
  resonant impurities, 379
  TM impurity-related levels
    donors and acceptors, 376
    ground-state levels, 375–376
    hybridization, 376–377
    magnetic isolators, 377
Superconducting quantum interference device (SQUID) measurement, 473–74

Index

T

Topological berry's, spin-orbit coupling systems
 ACE calculations
  2-d electron system, 82
  experimental and theoretical data, 84–85
  multichannel rings, 82–83
  AC phases, 80–81
  coherent nanoscale device, 80
  rashba coupling, 81
Transverse magneto-optical Kerr effect, 478
Tunneling anisotropic magnetoresistance (TAMR)
 (Ga,Mn)As
  amplification effect, 180–181
  anisotropic density states, 183–184
  kinetic exchange models, 178–179
  Landauer–Büttiker transport theory calculation, 183
  low-field a.c. MR measurement, 182–183
  MR measurements, 180
  signalling and devices, 177–178
 characteristics, 245–246
 construction of, 244–245
 correlated effects
  amplification, 252–253
  conductance vs. voltage curves, 258
  differential conductance, 253–254
  metallic transport properties, 256
  quantum interference effects, 255–256
  uniaxial anisotropy, 256–257
 electronic structure, 248
 ferromagnetic metal break junctions, 259
 magnetic and DOS anisotropies, 247–248
 multiple interfaces
  DW-mediated reorientations, 250
  structure construction, 249
  sweeping function, 249–250
 nanodevices, 260
 switching mechanism, 247
 tunneling geometry, 259–260
 volatile and nonvolatile operations modes, 251–252
Two-dimensional diluted magnetic semiconductors (2D DMSs)
 carrier-induced ferromagnetism
  Curie–Weiss behavior, 296–297
  domains and soft mode behavior, 298
  mean field model and Weiss temperature, 294–295
  Monte Carlo calculations, 297–298
  PL transition energies, 293–294
  spin susceptibility enhancement factor, 295–296
  (Cd,Mn)Te QWs, 291–292
  Zener model and giant Zeeman effect, 292–293

V

Van Vleck type magnetization, 381
Virtual-crystal (VCA) approximation
 spin-disorder scattering, 390
 splitting of extended states
  strong coupling, 384–386
  weak coupling, 383
Volatile TAMR, 251–252
Vonsovskii model, 381

W

Wide band-gap III–V semiconductors, Mn impurities
 GaN diluted with Cr
  crystal field transitions, 360–361
  electron configuration, 359
  electron structure, 360
 GaN diluted with Fe
  electron configuration, 357
  Fermi level, 359
  negatively charged acceptor, 358
  paramagnetic and ferromagnetic components, 358–359
  internal reference rule, 361–362
  ionisation energies of, 351
  ionized acceptor $Mn^{2+}(d^5)$
   energy level diagrams of, 338
   ESR spectra of, 335–336
   exchange interaction, 356
   Hamiltonian parameters, 336, 338
   vs. II–VI semiconductors, 339
  lattice relaxation, 349
  localization radius, $Mn^{2+}(d^5)$
   exchange interaction, 356
   Hall mobility and Hall concentration, 347
   hopping conductivity, 346
   quantum defect models, 347
   resistivity, 346–347
  lower charge states, 350–351
  magnetic interactions
   exchange integrals, 355–357
   magnetic properties, 352–355

Wide band-gap III–V semiconductors, Mn
  impurities (cont.)
 neutral acceptor $Mn^{3+}(d^4)$
  absorption band, 340–341
  absorption coefficient, 343
  energy level diagram, 341–342
  Huang–Rhys factor, 341
  optical absorption experiment, 340
 neutral acceptor $Mn^{2+}(d^5)$+hole
   configuration
  chemical shift, 344
  electron spin resonance, 344–345
 n-type GaN
  electron transitions, 349
  optical absorption spectra, 350
 WKB approximation, 13–15

Z

Zeeman splitting, 356
Zener/Ruderman-Kittel-Kasuya-Yosida
  (RKKY) models, 398
Zener's double-exchange mechanism
  vs. superexchange interaction
  mechanism, 441–442
 wide band-gap based DMS, 435
Zinc blende systems, 56

Contents of Volumes in This Series

Volume 1 Physics of III–V Compounds

C. Hilsum, Some Key Features of III–V Compounds
F. Bassani, Methods of Band Calculations Applicable to III–V Compounds
E. O. Kane, The k-p Method
V. L. Bonch-Bruevich, Effect of Heavy Doping on the Semiconductor Band Structure
D. Long, Energy Band Structures of Mixed Crystals of III–V Compounds
L. M. Roth and P. N. Argyres, Magnetic Quantum Effects
S. M. Puri and T. H. Geballe, Thermomagnetic Effects in the Quantum Region
W. M. Becker, Band Characteristics near Principal Minima from Magnetoresistance
E. H. Putley, Freeze-Out Effects, Hot Electron Effects, and Submillimeter Photoconductivity in InSb
H. Weiss, Magnetoresistance
B. Ancker-Johnson, Plasma in Semiconductors and Semimetals

Volume 2 Physics of III–V Compounds

M. G. Holland, Thermal Conductivity
S. I. Novkova, Thermal Expansion
U. Piesbergen, Heat Capacity and Debye Temperatures
G. Giesecke, Lattice Constants
J. R. Drabble, Elastic Properties
A. U. Mac Rae and G. W. Gobeli, Low Energy Electron Diffraction Studies
R. Lee Mieher, Nuclear Magnetic Resonance
B. Goldstein, Electron Paramagnetic Resonance
T. S. Moss, Photoconduction in III–V Compounds
E. Antoncik and J. Tauc, Quantum Efficiency of the Internal Photoelectric Effect in InSb
G. W. Gobeli and I. G. Allen, Photoelectric Threshold and Work Function
P. S. Pershan, Nonlinear Optics in III–V Compounds
M. Gershenzon, Radiative Recombination in the III–V Compounds
F. Stern, Stimulated Emission in Semiconductors

Volume 3   Optical Properties of III–V Compounds

M. Hass, Lattice Reflection
W. G. Spitzer, Multiphonon Lattice Absorption
D. L. Stierwalt and R. F. Potter, Emittance Studies
H. R. Philipp and H. Ehrenveich, Ultraviolet Optical Properties
M. Cardona, Optical Absorption Above the Fundamental Edge
E. J. Johnson, Absorption Near the Fundamental Edge
J. O. Dimmock, Introduction to the Theory of Exciton States in Semiconductors
B. Lax and J. G. Mavroides, Interband Magnetooptical Effects
H. Y. Fan, Effects of Free Carries on Optical Properties
E. D. Palik and G. B. Wright, Free-Carrier Magnetooptical Effects
R. H. Bube, Photoelectronic Analysis
B. O. Seraphin and H. E. Benett, Optical Constants

Volume 4   Physics of III–V Compounds

N. A. Goryunova, A. S. Borchevskii and D. N. Tretiakov, Hardness
N. N. Sirota, Heats of Formation and Temperatures and Heats of Fusion of Compounds of $A^{III}B^{V}$
D. L. Kendall, Diffusion
A. G. Chynoweth, Charge Multiplication Phenomena
R. W. Keyes, The Effects of Hydrostatic Pressure on the Properties of III–V Semiconductors
L. W. Aukerman, Radiation Effects
N. A. Goryunova, F. P. Kesamanly, and D. N. Nasledov, Phenomena in Solid Solutions
R. T. Bate, Electrical Properties of Nonuniform Crystals

Volume 5   Infrared Detectors

H. Levinstein, Characterization of Infrared Detectors
P. W. Kruse, Indium Antimonide Photoconductive and Photoelectromagnetic Detectors
M. B. Prince, Narrowband Self-Filtering Detectors
I. Melngalis and T. C. Hannan, Single-Crystal Lead-Tin Chalcogenides
D. Long and J. L. Schmidt, Mercury-Cadmium Telluride and Closely Related Alloys
E. H. Putley, The Pyroelectric Detector
N. B. Stevens, Radiation Thermopiles
R. J. Keyes and T. M. Quist, Low Level Coherent and Incoherent Detection in the Infrared
M. C. Teich, Coherent Detection in the Infrared
F. R. Arams, E. W. Sard, B. J. Peyton and F. P. Pace, Infrared Heterodyne Detection with Gigahertz IF Response
H. S. Sommers, Jr., Macrowave-Based Photoconductive Detector
R. Sehr and R. Zuleeg, Imaging and Display

Volume 6 Injection Phenomena

M. A. Lampert and R. B. Schilling, Current Injection in Solids: The Regional Approximation Method
R. Williams, Injection by Internal Photoemission
A. M. Barnett, Current Filament Formation
R. Baron and J. W. Mayer, Double Injection in Semiconductors
W. Ruppel, The Photoconductor-Metal Contact

Volume 7 Application and Devices

Part A

J. A. Copeland and S. Knight, Applications Utilizing Bulk Negative Resistance
F. A. Padovani, The Voltage-Current Characteristics of Metal-Semiconductor Contacts
P. L. Hower, W. W. Hooper, B. R. Cairns, R. D. Fairman, and D. A. Tremere, The GaAs Field-Effect Transistor
M. H. White, MOS Transistors
G. R. Antell, Gallium Arsenide Transistors
T. L. Tansley, Heterojunction Properties

Part B

T. Misawa, IMPATT Diodes
H. C. Okean, Tunnel Diodes
R. B. Campbell and Hung-Chi Chang, Silicon Junction Carbide Devices
R. E. Enstrom, H. Kressel, and L. Krassner, High-Temperature Power Rectifiers of $GaAs_{1-x}P_x$

Volume 8 Transport and Optical Phenomena

R. J. Stirn, Band Structure and Galvanomagnetic Effects in III–V Compounds with Indirect Band Gaps
R. W. Ure, Jr., Thermoelectric Effects in III–V Compounds
H. Piller, Faraday Rotation
H. Barry Bebb and E. W. Williams, Photoluminescence I: Theory
E. W. Williams and H. Barry Bebb, Photoluminescence II: Gallium Arsenide

Volume 9 Modulation Techniques

B. O. Seraphin, Electroreflectance
R. L. Aggarwal, Modulated Interband Magnetooptics
D. F. Blossey and Paul Handler, Electroabsorption
B. Batz, Thermal and Wavelength Modulation Spectroscopy
I. Balslev, Piezooptical Effects
D. E. Aspnes and N. Bottka, Electric-Field Effects on the Dielectric Function of Semiconductors and Insulators

Volume 10 Transport Phenomena

R. L. Rhode, Low-Field Electron Transport
J. D. Wiley, Mobility of Holes in III–V Compounds
C. M. Wolfe and G. E. Stillman, Apparent Mobility Enhancement in Inhomogeneous Crystals
R. L. Petersen, The Magnetophonon Effect

Volume 11 Solar Cells

H. J. Hovel, Introduction; Carrier Collection, Spectral Response, and Photocurrent; Solar Cell Electrical Characteristics; Efficiency; Thickness; Other Solar Cell Devices; Radiation Effects; Temperature and Intensity; Solar Cell Technology

Volume 12 Infrared Detectors (II)

W. L. Eiseman, J. D. Merriam, and R. F. Potter, Operational Characteristics of Infrared Photodetectors
P. R. Bratt, Impurity Germanium and Silicon Infrared Detectors
E. H. Putley, InSb Submillimeter Photoconductive Detectors
G. E. Stillman, C. M. Wolfe, and J. O. Dimmock, Far-Infrared Photoconductivity in High Purity GaAs
G. E. Stillman and C. M. Wolfe, Avalanche Photodiodes
P. L. Richards, The Josephson Junction as a Detector of Microwave and Far-Infrared Radiation
E. H. Putley, The Pyroelectric Detector – An Update

Volume 13 Cadmium Telluride

K. Zanio, Materials Preparations; Physics; Defects; Applications

Volume 14 Lasers, Junctions, Transport

N. Holonyak, Jr., and M. H. Lee, Photopumped III–V Semiconductor Lasers
H. Kressel and J. K. Butler, Heterojunction Laser Diodes
A. Van der Ziel, Space-Charge-Limited Solid-State Diodes
P. J. Price, Monte Carlo Calculation of Electron Transport in Solids

Volume 15 Contacts, Junctions, Emitters

B. L. Sharma, Ohmic Contacts to III–V Compounds Semiconductors
A. Nussbaum, The Theory of Semiconducting Junctions
J. S. Escher, NEA Semiconductor Photoemitters

Volume 16 Defects, (HgCd)Se, (HgCd)Te

H. Kressel, The Effect of Crystal Defects on Optoelectronic Devices
C. R. Whitsett, J. G. Broerman, and C. J. Summers, Crystal Growth and Properties of $Hg_{1-x}Cd_x$ Se Alloys

M. H. Weiler, Magnetooptical Properties of $Hg_{1-x}Cd_x$ Te Alloys

P. W. Kruse and J. G. Ready, Nonlinear Optical Effects in $Hg_{1-x}Cd_x$ Te

Volume 17 CW Processing of Silicon and Other Semiconductors

J. F. Gibbons, Beam Processing of Silicon

A. Lietoila, R. B. Gold, J. F. Gibbons, and L. A. Christel, Temperature Distributions and Solid Phase Reaction Rates Produced by Scanning CW Beams

A. Leitoila and J. F. Gibbons, Applications of CW Beam Processing to Ion Implanted Crystalline Silicon

N. M. Johnson, Electronic Defects in CW Transient Thermal Processed Silicon

K. F. Lee, T. J. Stultz, and J. F. Gibbons, Beam Recrystallized Polycrystalline Silicon: Properties, Applications, and Techniques

T. Shibata, A. Wakita, T. W. Sigmon and J. F. Gibbons, Metal-Silicon Reactions and Silicide

Y. I. Nissim and J. F. Gibbons, CW Beam Processing of Gallium Arsenide

Volume 18 Mercury Cadmium Telluride

P. W. Kruse, The Emergence of ($Hg_{1-x}Cd_x$) Te as a Modern Infrared Sensitive Material

H. E. Hirsch, S. C. Liang, and A. G. White, Preparation of High-Purity Cadmium, Mercury, and Tellurium

W. F. H. Micklethwaite, The Crystal Growth of Cadmium Mercury Telluride

P. E. Petersen, Auger Recombination in Mercury Cadmium Telluride

R. M. Broudy and V. J. Mazurczyck, (HgCd) Te Photoconductive Detectors

M. B. Reine, A. K. Soad, and T. J. Tredwell, Photovoltaic Infrared Detectors

M. A. Kinch, Metal-Insulator-Semiconductor Infrared Detectors

Volume 19 Deep Levels, GaAs, Alloys, Photochemistry

G. F. Neumark and K. Kosai, Deep Levels in Wide Band-Gap III–V Semiconductors

D. C. Look, The Electrical and Photoelectronic Properties of Semi-Insulating GaAs

R. F. Brebrick, Ching-Hua Su, and Pok-Kai Liao, Associated Solution Model for Ga-In-Sb and Hg-Cd-Te

Y. Ya. Gurevich and Y. V. Pleskon, Photoelectrochemistry of Semiconductors

Volume 20 Semi-Insulating GaAs

R. N. Thomas, H. M. Hobgood, G. W. Eldridge, D. L. Barrett, T. T. Braggins, L. B. Ta, and S. K. Wang, High-Purity LEC Growth and Direct Implantation of GaAs for Monolithic Microwave Circuits

C. A. Stolte, Ion Implantation and Materials for GaAs Integrated Circuits

C. G. Kirkpatrick, R. T. Chen, D. E. Holmes, P. M. Asbeck, K. R. Elliott, R. D. Fairman, and J. R. Oliver, LEC GaAs for Integrated Circuit Applications

J. S. Blakemore and S. Rahimi, Models for Mid-Gap Centers in Gallium Arsenide

Volume 21 Hydrogenated Amorphous Silicon

Part A

J. I. Pankove, Introduction

M. Hirose, Glow Discharge; Chemical Vapor Deposition

Y. Uchida, di Glow Discharge

T. D. Moustakas, Sputtering

I. Yamada, Ionized-Cluster Beam Deposition

B. A. Scott, Homogeneous Chemical Vapor Deposition

F. J. Kampas, Chemical Reactions in Plasma Deposition

P. A. Longeway, Plasma Kinetics

H. A. Weakliem, Diagnostics of Silane Glow Discharges Using Probes and Mass Spectroscopy

L. Gluttman, Relation between the Atomic and the Electronic Structures

A. Chenevas-Paule, Experiment Determination of Structure

S. Minomura, Pressure Effects on the Local Atomic Structure

D. Adler, Defects and Density of Localized States

Part B

J. I. Pankove, Introduction

G. D. Cody, The Optical Absorption Edge of a-Si: H

N. M. Amer and W. B. Jackson, Optical Properties of Defect States in a-Si: H

P. J. Zanzucchi, The Vibrational Spectra of a-Si: H

Y. Hamakawa, Electroreflectance and Electroabsorption

J. S. Lannin, Raman Scattering of Amorphous Si, Ge, and Their Alloys

R. A. Street, Luminescence in a-Si: H

R. S. Crandall, Photoconductivity

J. Tauc, Time-Resolved Spectroscopy of Electronic Relaxation Processes

P. E. Vanier, IR-Induced Quenching and Enhancement of Photoconductivity and Photoluminescence

H. Schade, Irradiation-Induced Metastable Effects

L. Ley, Photoelectron Emission Studies

Part C

J. I. Pankove, Introduction

J. D. Cohen, Density of States from Junction Measurements in Hydrogenated Amorphous Silicon

P. C. Taylor, Magnetic Resonance Measurements in a-Si: H

K. Morigaki, Optically Detected Magnetic Resonance

J. Dresner, Carrier Mobility in a-Si: H

T. Tiedje, Information About Band-Tail States from Time-of-Flight Experiments

A. R. Moore, Diffusion Length in Undoped a-S: H

W. Beyer and J. Overhof, Doping Effects in a-Si: H

H. Fritzche, Electronic Properties of Surfaces in a-Si: H

C. R. Wronski, The Staebler-Wronski Effect

R. J. Nemanich, Schottky Barriers on a-Si: H

B. Abeles and T. Tiedje, Amorphous Semiconductor Superlattices

Part D

J. I. Pankove, Introduction

D. E. Carlson, Solar Cells

G. A. Swartz, Closed-Form Solution of I–V Characteristic for a s-Si: H Solar Cells

I. Shimizu, Electrophotography

S. Ishioka, Image Pickup Tubes

P. G. Lecomber and W. E. Spear, The Development of the a-Si: H Field-Effect Transistor and its Possible Applications

D. G. Ast, a-Si: H FET-Addressed LCD Panel

S. Kaneko, Solid-State Image Sensor

M. Matsumura, Charge-Coupled Devices

M. A. Bosch, Optical Recording

A. D'Amico and G. Fortunato, Ambient Sensors

H. Kulkimoto, Amorphous Light-Emitting Devices

R. J. Phelan, Jr., Fast Decorators and Modulators

J. I. Pankove, Hybrid Structures

P. G. LeComber, A. E. Owen, W. E. Spear, J. Hajto, and W. K. Choi, Electronic Switching in Amorphous Silicon Junction Devices

Volume 22 Lightwave Communications Technology

Part A

K. Nakajima, The Liquid-Phase Epitaxial Growth of InGaAsP

W. T. Tsang, Molecular Beam Epitaxy for III–V Compound Semiconductors

G. B. Stringfellow, Organometallic Vapor-Phase Epitaxial Growth of III–V Semiconductors

G. Beuchet, Halide and Chloride Transport Vapor-Phase Deposition of InGaAsP and GaAs

M. Razeghi, Low-Pressure, Metallo-Organic Chemical Vapor Deposition of $Ga_xIn1-xAsP1-y$ Alloys

P. M. Petroff, Defects in III–V Compound Semiconductors

Part B

J. P. van der Ziel, Mode Locking of Semiconductor Lasers

K. Y. Lau and A. Yariv, High-Frequency Current Modulation of Semiconductor Injection Lasers

C. H. Henry, Special Properties of Semi Conductor Lasers

Y. Suematsu, K. Kishino, S. Arai, and F. Koyama, Dynamic Single-Mode Semiconductor Lasers with a Distributed Reflector

W. T. Tsang, The Cleaved-Coupled-Cavity (C^3) Laser

Part C

R. J. Nelson and N. K. Dutta, Review of InGaAsP InP Laser Structures and Comparison of Their Performance

N. Chinone and M. Nakamura, Mode-Stabilized Semiconductor Lasers for 0.7–0.8- and 1.1–1.6-μm Regions

Y. Horikoshi, Semiconductor Lasers with Wavelengths Exceeding 2 μm

B. A. Dean and M. Dixon, The Functional Reliability of Semiconductor Lasers as Optical Transmitters

R. H. Saul, T. P. Lee, and C. A. Burus, Light-Emitting Device Design
C. L. Zipfel, Light-Emitting Diode-Reliability
T. P. Lee and T. Li, LED-Based Multimode Lightwave Systems
K. Ogawa, Semiconductor Noise-Mode Partition Noise

Part D

F. Capasso, The Physics of Avalanche Photodiodes
T. P. Pearsall and M. A. Pollack, Compound Semiconductor Photodiodes
T. Kaneda, Silicon and Germanium Avalanche Photodiodes
S. R. Forrest, Sensitivity of Avalanche Photodetector Receivers for High-Bit-Rate Long-Wavelength Optical Communication Systems
J. C. Campbell, Phototransistors for Lightwave Communications

Part E

S. Wang, Principles and Characteristics of Integrable Active and Passive Optical Devices
S. Margalit and A. Yariv, Integrated Electronic and Photonic Devices
T. Mukai, A. Yamamoto, and T. Kimura, Optical Amplification by Semiconductor Lasers

Volume 23 Pulsed Laser Processing of Semiconductors

R. F. Wood, C. W. White and R. T. Young, Laser Processing of Semiconductors: An Overview
C. W. White, Segregation, Solute Trapping and Supersaturated Alloys
G. E. Jellison, Jr., Optical and Electrical Properties of Pulsed Laser-Annealed Silicon
R. F. Wood and G. E. Jellison, Jr., Melting Model of Pulsed Laser Processing
R. F. Wood and F. W. Young, Jr., Nonequilibrium Solidification Following Pulsed Laser Melting
D. H. Lawndes and G. E. Jellison, Jr., Time-Resolved Measurement During Pulsed Laser Irradiation of Silicon
D. M. Zebner, Surface Studies of Pulsed Laser Irradiated Semiconductors
D. H. Lowndes, Pulsed Beam Processing of Gallium Arsenide
R. B. James, Pulsed CO_2 Laser Annealing of Semiconductors
R. T. Young and R. F. Wood, Applications of Pulsed Laser Processing

Volume 24 Applications of Multiquantum Wells, Selective Doping, and Superlattices

C. Weisbuch, Fundamental Properties of III–V Semiconductor Two-Dimensional Quantized Structures: The Basis for Optical and Electronic Device Applications
H. Morkoç and H. Unlu, Factors Affecting the Performance of (Al,Ga)As/GaAs and (Al,Ga)As/InGaAs Modulation-Doped Field-Effect Transistors: Microwave and Digital Applications
N. T. Linh, Two-Dimensional Electron Gas FETs: Microwave Applications
M. Abe et al., Ultra-High-Speed HEMT Integrated Circuits
D. S. Chemla, D. A. B. Miller and P. W. Smith, Nonlinear Optical Properties of Multiple Quantum Well Structures for Optical Signal Processing
F. Capasso, Graded-Gap and Superlattice Devices by Band-Gap Engineering
W. T. Tsang, Quantum Confinement Heterostructure Semiconductor Lasers
G. C. Osbourn et al., Principles and Applications of Semiconductor Strained-Layer Superlattices

Volume 25 Diluted Magnetic Semiconductors

W. Giriat and J. K. Furdyna, Crystal Structure, Composition, and Materials Preparation of Diluted Magnetic Semiconductors

W. M. Becker, Band Structure and Optical Properties of Wide-Gap $Al_{1-x}IIMn_xBIV$ Alloys at Zero Magnetic Field

S. Oseroff and P. H. Keesom, Magnetic Properties: Macroscopic Studies

T. Giebultowicz and T. M. Holden, Neutron Scattering Studies of the Magnetic Structure and Dynamics of Diluted Magnetic Semiconductors

J. Kossut, Band Structure and Quantum Transport Phenomena in Narrow-Gap Diluted Magnetic Semiconductors

C. Riquaux, Magnetooptical Properties of Large-Gap Diluted Magnetic Semiconductors

J. A. Gaj, Magnetooptical Properties of Large-Gap Diluted Magnetic Semiconductors

J. Mycielski, Shallow Acceptors in Diluted Magnetic Semiconductors: Splitting, Boil-off, Giant Negative Magnetoresistance

A. K. Ramadas and R. Rodriquez, Raman Scattering in Diluted Magnetic Semiconductors

P. A. Wolff, Theory of Bound Magnetic Polarons in Semimagnetic Semiconductors

Volume 26 III–V Compound Semiconductors and Semiconductor Properties of Superionic Materials

Z. Yuanxi, III–V Compounds

H. V. Winston, A. T. Hunter, H. Kimura, and R. E. Lee, InAs-Alloyed GaAs Substrates for Direct Implantation

P. K. Bhattacharya and S. Dhar, Deep Levels in III–V Compound Semiconductors Grown by MBE

Y. Ya. Gurevich and A. K. Ivanov-Shits, Semiconductor Properties of Supersonic Materials

Volume 27 High Conducting Quasi-One-Dimensional Organic Crystals

E. M. Conwell, Introduction to Highly Conducting Quasi-One-Dimensional Organic Crystals

I. A. Howard, A Reference Guide to the Conducting Quasi-One-Dimensional Organic Molecular Crystals

J. P. Pouqnet, Structural Instabilities

E. M. Conwell, Transport Properties

C. S. Jacobsen, Optical Properties

J. C. Scolt, Magnetic Properties

L. Zuppiroli, Irradiation Effects: Perfect Crystals and Real Crystals

Volume 28 Measurement of High-Speed Signals in Solid State Devices

J. Frey and D. Ioannou, Materials and Devices for High-Speed and Optoelectronic Applications

H. Schumacher and E. Strid, Electronic Wafer Probing Techniques

D. H. Auston, Picosecond Photoconductivity: High-Speed Measurements of Devices and Materials

J. A. Valdmanis, Electro-Optic Measurement Techniques for Picosecond Materials, Devices and Integrated Circuits

J. M. Wiesenfeld and R. K. Jain, Direct Optical Probing of Integrated Circuits and High-Speed Devices
G. Plows, Electron-Beam Probing
A. M. Weiner and R. B. Marcus, Photoemissive Probing

Volume 29 Very High Speed Integrated Circuits: Gallium Arsenide LSI

M. Kuzuhara and T. Nazaki, Active Layer Formation by Ion Implantation
H. Hasimoto, Focused Ion Beam Implantation Technology
T. Nozaki and A. Higashisaka, Device Fabrication Process Technology
M. Ino and T. Takada, GaAs LSI Circuit Design
M. Hirayama, M. Ohmori, and K. Yamasaki, GaAs LSI Fabrication and Performance

Volume 30 Very High Speed Integrated Circuits: Heterostructure

H. Watanabe, T. Mizutani, and A. Usui, Fundamentals of Epitaxial Growth and Atomic Layer Epitaxy
S. Hiyamizu, Characteristics of Two-Dimensional Electron Gas in III–V Compound Heterostructures Grown by MBE
T. Nakanisi, Metalorganic Vapor Phase Epitaxy for High-Quality Active Layers
T. Nimura, High Electron Mobility Transistor and LSI Applications
T. Sugeta and T. Ishibashi, Hetero-Bipolar Transistor and LSI Application
H. Matsuedo, T. Tanaka, and M. Nakamura, Optoelectronic Integrated Circuits

Volume 31 Indium Phosphide: Crystal Growth and Characterization

J. P. Farges, Growth of Discoloration-Free InP
M. J. McCollum and G. E. Stillman, High Purity InP Grown by Hydride Vapor Phase Epitaxy
I. Inada and T. Fukuda, Direct Synthesis and Growth of Indium Phosphide by the Liquid Phosphorous Encapsulated Czochralski Method
O. Oda, K. Katagiri, K. Shinohara, S. Katsura, Y. Takahashi, K. Kainosho, K. Kohiro, and R. Hirano, InP Crystal Growth, Substrate Preparation and Evaluation
K. Tada, M. Tatsumi, M. Morioka, T. Araki, and T. Kawase, InP Substrates: Production and Quality Control
M. Razeghi, LP-MOCVD Growth, Characterization, and Application of InP Material
T. A. Kennedy and P. J. Lin-Chung, Stoichiometric Defects in InP

Volume 32 Strained-Layer Superlattices: Physics

T. P. Pearsall, Strained-Layer Superlattices
F. H. Pollack, Effects of Homogeneous Strain on the Electronic and Vibrational Levels in Semiconductors
J. Y. Marzin, J. M. Gerárd, P. Voisin, and J. A. Brum, Optical Studies of Strained III–V Heterolayers
R. People and S. A. Jackson, Structurally Induced States from Strain and Confinement
M. Jaros, Microscopic Phenomena in Ordered Superlattices

Volume 33 Strained-Layer Superlattices: Material Science and Technology

R. Hull and J. C. Bean, Principles and Concepts of Strained-Layer Epitaxy

W. J. Shaff, P. J. Tasker, M. C. Foisy, and L. F. Eastman, Device Applications of Strained-Layer Epitaxy

S. T. Picraux, B. L. Doyle, and J. Y. Tsao, Structure and Characterization of Strained-Layer Superlattices

E. Kasper and F. Schaffer, Group IV Compounds

D. L. Martin, Molecular Beam Epitaxy of IV–VI Compounds Heterojunction

R. L. Gunshor, L. A. Kolodziejski, A. V. Nurmikko, and N. Otsuka, Molecular Beam Epitaxy of I–VI Semiconductor Microstructures

Volume 34 Hydrogen in Semiconductors

J. I. Pankove and N. M. Johnson, Introduction to Hydrogen in Semiconductors

C. H. Seager, Hydrogenation Methods

J. I. Pankove, Hydrogenation of Defects in Crystalline Silicon

J. W. Corbett, P. Déak, U. V. Desnica, and S. J. Pearton, Hydrogen Passivation of Damage Centers in Semiconductors

S. J. Pearton, Neutralization of Deep Levels in Silicon

J. I. Pankove, Neutralization of Shallow Acceptors in Silicon

N. M. Johnson, Neutralization of Donor Dopants and Formation of Hydrogen-Induced Defects in n-Type Silicon

M. Stavola and S. J. Pearton, Vibrational Spectroscopy of Hydrogen-Related Defects in Silicon

A. D. Marwick, Hydrogen in Semiconductors: Ion Beam Techniques

C. Herring and N. M. Johnson, Hydrogen Migration and Solubility in Silicon

E. E. Haller, Hydrogen-Related Phenomena in Crystalline Germanium

J. Kakalios, Hydrogen Diffusion in Amorphous Silicon

J. Chevalier, B. Clerjaud, and B. Pajot, Neutralization of Defects and Dopants in III–V Semiconductors

G. G. DeLeo and W. B. Fowler, Computational Studies of Hydrogen-Containing Complexes in Semiconductors

R. F. Kiefl and T. L. Estle, Muonium in Semiconductors

C. G. Van de Walle, Theory of Isolated Interstitial Hydrogen and Muonium in Crystalline Semiconductors

Volume 35 Nanostructured Systems

M. Reed, Introduction

H. van Houten, C. W. J. Beenakker, and B. J. Wees, Quantum Point Contacts

G. Timp, When Does a Wire Become an Electron Waveguide?

M. Búttiker, The Quantum Hall Effects in Open Conductors

W. Hansen, J. P. Kotthaus, and U. Merkt, Electrons in Laterally Periodic Nanostructures

Volume 36 The Spectroscopy of Semiconductors

D. Heiman, Spectroscopy of Semiconductors at Low Temperatures and High Magnetic Fields
A. V. Nurmikko, Transient Spectroscopy by Ultrashort Laser Pulse Techniques
A. K. Ramdas and S. Rodriguez, Piezospectroscopy of Semiconductors
O. J. Glembocki and B. V. Shanabrook, Photoreflectance Spectroscopy of Microstructures
D. G. Seiler, C. L. Littler, and M. H. Wiler, One- and Two-Photon Magneto-Optical Spectroscopy of InSb and $Hg1-xCdx$ Te

Volume 37 The Mechanical Properties of Semiconductors

A.-B. Chen, A. Sher, and W. T. Yost, Elastic Constants and Related Properties of Semiconductor Compounds and Their Alloys
D. R. Clarke, Fracture of Silicon and Other Semiconductors
H. Siethoff, The Plasticity of Elemental and Compound Semiconductors
S. Guruswamy, K. T. Faber, and J. P. Hirth, Mechanical Behavior of Compound Semiconductors
S. Mahajan, Deformation Behavior of Compound Semiconductors
J. P. Hirth, Injection of Dislocations into Strained Multilayer Structures
D. Kendall, C. B. Fleddermann, and K. J. Malloy, Critical Technologies for the Micromatching of Silicon
J. Matsuba and K. Mokuya, Processing and Semiconductor Thermoelastic Behavior

Volume 38 Imperfections in III/V Materials

U. Scherz and M. Scheffler, Density-Functional Theory of sp-Bonded Defects in III/V Semiconductors
M. Kaminska and E. R. Weber, E12 Defect in GaAs
D. C. Look, Defects Relevant for Compensation in Semi-Insulating GaAs
R. C. Newman, Local Vibrational Mode Spectroscopy of Defects in III/V Compounds
A. M. Hennel, Transition Metals in III/V Compounds
K. J. Malloy and K. Khachaturyan, DX and Related Defects in Semiconductors
V. Swaminathan and A. S. Jordan, Dislocations in III/V Compounds
K. W. Nauka, Deep Level Defects in the Epitaxial III/V Materials

Volume 39 Minority Carriers in III–V Semiconductors: Physics and Applications

N. K. Dutta, Radiative Transition in GaAs and Other III–V Compounds
R. K. Ahrenkiel, Minority-Carrier Lifetime in III–V Semiconductors
T. Furuta, High Field Minority Electron Transport in p-GaAs
M. S. Lundstrom, Minority-Carrier Transport in III–V Semiconductors
R. A. Abram, Effects of Heavy Doping and High Excitation on the Band Structure of GaAs
D. Yevick and W. Bardyszewski, An Introduction to Non-Equilibrium Many-Body Analyses of Optical Processes in III–V Semiconductors

Volume 40 Epitaxial Microstructures

E. F. Schubert, Delta-Doping of Semiconductors: Electronic, Optical and Structural Properties of Materials and Devices

A. Gossard, M. Sundaram, and P. Hopkins, Wide Graded Potential Wells

P. Petroff, Direct Growth of Nanometer-Size Quantum Wire Superlattices

E. Kapon, Lateral Patterning of Quantum Well Heterostructures by Growth of Nonplanar Substrates

H. Temkin, D. Gershoni, and M. Panish, Optical Properties of $Ga_{1-x}In_xAs/InP$ Quantum Wells

Volume 41 High Speed Heterostructure Devices

F. Capasso, F. Beltram, S. Sen, A. Pahlevi, and A. Y. Cho, Quantum Electron Devices: Physics and Applications

P. Solomon, D. J. Frank, S. L. Wright and F. Canora, GaAs-Gate Semiconductor-Insulator-Semiconductor FET

M. H. Hashemi and U. K. Mishra, Unipolar InP-Based Transistors

R. Kiehl, Complementary Heterostructure FET Integrated Circuits

T. Ishibashi, GaAs-Based and InP-Based Heterostructure Bipolar-Transistors

H. C. Liu and T. C. L. G. Sollner, High-Frequency-Tunneling Devices

H. Ohnishi, T. More, M. Takatsu, K. Imamura, and N. Yokoyama, Resonant-Tunneling Hot-Electron Transistors and Circuits

Volume 42 Oxygen in Silicon

F. Shimura, Introduction to Oxygen in Silicon

W. Lin, The Incorporation of Oxygen into Silicon Crystals

T. J. Schaffner and D. K. Schroder, Characterization Techniques for Oxygen in Silicon

W. M. Bullis, Oxygen Concentration Measurement

S. M. Hu, Intrinsic Point Defects in Silicon

B. Pajot, Some Atomic Configuration of Oxygen

J. Michel and L. C. Kimerling, Electrical Properties of Oxygen in Silicon

R. C. Newman and R. Jones, Diffusion of Oxygen in Silicon

T. Y. Tan and W. J. Taylor, Mechanisms of Oxygen Precipitation: Some Quantitative Aspects

M. Schrems, Simulation of Oxygen Precipitation

K. Simino and I. Yonenaga, Oxygen Effect on Mechanical Properties

W. Bergholz, Grown-in and Process-Induced Effects

F. Shimura, Intrinsic/Internal Gettering

H. Tsuya, Oxygen Effect on Electronic Device Performance

Volume 43 Semiconductors for Room Temperature Nuclear Detector Applications

R. B. James and T. E. Schlesinger, Introduction and Overview

L. S. Darken and C. E. Cox, High-Purity Germanium Detectors

A. Burger, D. Nason, L. Van den Berg, and M. Schieber, Growth of Mercuric Iodide

X. J. Bao, T. E. Schlesinger, and R. B. James, Electrical Properties of Mercuric Iodide
X. J. Bao, R. B. James, and T. E. Schlesinger, Optical Properties of Red Mercuric Iodide
M. Hage-Ali and P. Siffert, Growth Methods of CdTe Nuclear Detector Materials
M. Hage-Ali and P. Siffert, Characterization of CdTe Nuclear Detector Materials
M. Hage-Ali and P. Siffert, CdTe Nuclear Detectors and Applications
R. B. James, T. E. Schlesinger, J. Lund, and M. Schieber, Cd1−xZnxTe Spectrometers for Gamma and X-Ray Applications
D. S. McGregor, J. E. Kammeraad, Gallium Arsenide Radiation Detectors and Spectrometers
J. C. Lund, F. Olschner, and A. Burger, Lead Iodide
M. R. Squillante and K. S. Shah, Other Materials: Status and Prospects
V. M. Gerrish, Characterization and Quantification of Detector Performance
J. S. Iwanczyk and B. E. Patt, Electronics for X-ray and Gamma Ray Spectrometers
M. Schieber, R. B. James and T. E. Schlesinger, Summary and Remaining Issues for Room Temperature Radiation Spectrometers

Volume 44 II–IV Blue/Green Light Emitters: Device Physics and Epitaxial Growth

J. Han and R. L. Gunshor, MBE Growth and Electrical Properties of Wide Bandgap ZnSe-based II–VI Semiconductors
S. Fujita and S. Fujita, Growth and Characterization of ZnSe-based II–VI Semiconductors by MOVPE
E. Ho and L. A. Kolodziejski, Gaseous Source UHV Epitaxy Technologies for Wide Bandgap II–VI Semiconductors
C. G. Van de Walle, Doping of Wide-Band-Gap II–VI Compounds – Theory
R. Cingolani, Optical Properties of Excitons in ZnSe-Based Quantum Well Heterostructures
A. Ishibashi and A. V. Nurmikko, II–VI Diode Lasers: A Current View of Device Performance and Issues
S. Guha and J. Petruzello, Defects and Degradation in Wide-Gap II–VI-based Structure and Light Emitting Devices

Volume 45 Effect of Disorder and Defects in Ion-Implanted Semiconductors: Electrical and Physiochemical Characterization

H. Ryssel, Ion Implantation into Semiconductors: Historical Perspectives
You-Nian Wang and Teng-Cai Ma, Electronic Stopping Power for Energetic Ions in Solids
S. T. Nakagawa, Solid Effect on the Electronic Stopping of Crystalline Target and Application to Range Estimation
G. Miller, S. Kalbitzer, and G. N. Greaves, Ion Beams in Amorphous Semiconductor Research
J. Boussey-Said, Sheet and Spreading Resistance Analysis of Ion Implanted and Annealed Semiconductors
M. L. Polignano and G. Queirolo, Studies of the Stripping Hall Effect in Ion-Implanted Silicon
J. Sroemenos, Transmission Electron Microscopy Analyses
R. Nipoti and M. Servidori, Rutherford Backscattering Studies of Ion Implanted Semiconductors
P. Zaumseil, X-ray Diffraction Techniques

Volume 46 Effect of Disorder and Defects in Ion-Implanted Semiconductors: Optical and Photothermal Characterization

M. Fried, T. Lohner, and J. Gyulai, Ellipsometric Analysis
A. Seas and C. Christofides, Transmission and Reflection Spectroscopy on Ion Implanted Semiconductors
A. Othonos and C. Christofides, Photoluminescence and Raman Scattering of Ion Implanted Semiconductors. Influence of Annealing
C. Christofides, Photomodulated Thermoreflectance Investigation of Implanted Wafers. Annealing Kinetics of Defects
U. Zammit, Photothermal Deflection Spectroscopy Characterization of Ion-Implanted and Annealed Silicon Films
A. Mandelis, A. Budiman, and M. Vargas, Photothermal Deep-Level Transient Spectroscopy of Impurities and Defects in Semiconductors
R. Kalish and S. Charbonneau, Ion Implantation into Quantum-Well Structures
A. M. Myasnikov and N. N. Gerasimenko, Ion Implantation and Thermal Annealing of III–V Compound Semiconducting Systems: Some Problems of III–V Narrow Gap Semiconductors

Volume 47 Uncooled Infrared Imaging Arrays and Systems

R. G. Buser and M. P. Tompsett, Historical Overview
P. W. Kruse, Principles of Uncooled Infrared Focal Plane Arrays
R. A. Wood, Monolithic Silicon Microbolometer Arrays
C. M. Hanson, Hybrid Pyroelectric-Ferroelectric Bolometer Arrays
D. L. Polla and J. R. Choi, Monolithic Pyroelectric Bolometer Arrays
N. Teranishi, Thermoelectric Uncooled Infrared Focal Plane Arrays
M. F. Tompsett, Pyroelectric Vidicon
T. W. Kenny, Tunneling Infrared Sensors
J. R. Vig, R. L Filler, and Y. Kim, Application of Quartz Microresonators to Uncooled Infrared Imaging Arrays
P. W. Kruse, Application of Uncooled Monolithic Thermoelectric Linear Arrays to Imaging Radiometers

Volume 48 High Brightness Light Emitting Diodes

G. B. Stringfellow, Materials Issues in High-Brightness Light-Emitting Diodes
M. G. Craford, Overview of Device Issues in High-Brightness Light-Emitting Diodes
F. M. Steranka, AlGaAs Red Light Emitting Diodes
C. H. Chen, S. A. Stockman, M. J. Peanasky, and C. P. Kuo, OMVPE Growth of AlGaInP for High Efficiency Visible Light-Emitting Diodes
F. A. Kish and R. M. Fletcher, AlGaInP Light-Emitting Diodes
M. W. Hodapp, Applications for High Brightness Light-Emitting Diodes
J. Akasaki and H. Amano, Organometallic Vapor Epitaxy of GaN for High Brightness Blue Light Emitting Diodes
S. Nakamura, Group III–V Nitride Based Ultraviolet-Blue-Green-Yellow Light-Emitting Diodes and Laser Diodes

Volume 49 Light Emission in Silicon: from Physics to Devices

D. J. Lockwood, Light Emission in Silicon

G. Abstreiter, Band Gaps and Light Emission in Si/SiGe Atomic Layer Structures

T. G. Brown and D. G. Hall, Radiative Isoelectronic Impurities in Silicon and Silicon-Germanium Alloys and Superlattices

J. Michel, L. V. C. Assali, M. T. Morse, and L. C. Kimerling, Erbium in Silicon

Y. Kanemitsu, Silicon and Germanium Nanoparticles

P. M. Fauchet, Porous Silicon: Photoluminescence and Electroluminescent Devices

C. Delerue, G. Allan, and M. Lannoo, Theory of Radiative and Nonradiative Processes in Silicon Nanocrystallites

L. Brus, Silicon Polymers and Nanocrystals

Volume 50 Gallium Nitride (GaN)

J. I. Pankove and T. D. Moustakas, Introduction

S. P. DenBaars and S. Keller, Metalorganic Chemical Vapor Deposition (MOCVD) of Group III Nitrides

W. A. Bryden and T. J. Kistenmacher, Growth of Group III–A Nitrides by Reactive Sputtering

N. Newman, Thermochemistry of III–N Semiconductors

S. J. Pearton and R. J. Shul, Etching of III Nitrides

S. M. Bedair, Indium-based Nitride Compounds

A. Trampert, O. Brandt, and K. H. Ploog, Crystal Structure of Group III Nitrides

H. Morkoç, F. Hamdani, and A. Salvador, Electronic and Optical Properties of III–V Nitride based Quantum Wells and Superlattices

K. Doverspike and J. I. Pankove, Doping in the III-Nitrides

T. Suski and P. Perlin, High Pressure Studies of Defects and Impurities in Gallium Nitride

B. Monemar, Optical Properties of GaN

W. R. L. Lambrecht, Band Structure of the Group III Nitrides

N. E. Christensen and P. Perlin, Phonons and Phase Transitions in GaN

S. Nakamura, Applications of LEDs and LDs

I. Akasaki and H. Amano, Lasers

J. A. Cooper, Jr., Nonvolatile Random Access Memories in Wide Bandgap Semiconductors

Volume 51A Identification of Defects in Semiconductors

G. D. Watkins, EPR and ENDOR Studies of Defects in Semiconductors

J.-M. Spaeth, Magneto-Optical and Electrical Detection of Paramagnetic Resonance in Semiconductors

T. A. Kennedy and E. R. Claser, Magnetic Resonance of Epitaxial Layers Detected by Photoluminescence

K. H. Chow, B. Hitti, and R. F. Kiefl, µSR on Muonium in Semiconductors and Its Relation to Hydrogen

K. Saarinen, P. Hautojärvi, and C. Corbel, Positron Annihilation Spectroscopy of Defects in Semiconductors

R. Jones and P. R. Briddon, The Ab Initio Cluster Method and the Dynamics of Defects in Semiconductors

Volume 51B Identification Defects in Semiconductors

G. Davies, Optical Measurements of Point Defects

P. M. Mooney, Defect Identification Using Capacitance Spectroscopy

M. Stavola, Vibrational Spectroscopy of Light Element Impurities in Semiconductors

P. Schwander, W. D. Rau, C. Kisielowski, M. Gribelyuk, and A. Ourmazd, Defect Processes in Semiconductors Studied at the Atomic Level by Transmission Electron Microscopy

N. D. Jager and E. R. Weber, Scanning Tunneling Microscopy of Defects in Semiconductors

Volume 52 SiC Materials and Devices

K. Järrendahl and R. F. Davis, Materials Properties and Characterization of SiC

V. A. Dmitiriev and M. G. Spencer, SiC Fabrication Technology: Growth and Doping

V. Saxena and A. J. Steckl, Building Blocks for SiC Devices: Ohmic Contacts, Schottky Contacts, and p-n Junctions

M. S. Shur, SiC Transistors

C. D. Brandt, R. C. Clarke, R. R. Siergiej, J. B. Casady, A. W. Morse, S. Sriram, and A. K. Agarwal, SiC for Applications in High-Power Electronics

R. J. Trew, SiC Microwave Devices

J. Edmond, H. Kong, G. Negley, M. Leonard, K. Doverspike, W. Weeks, A. Suvorov, D. Waltz, and C. Carter, Jr., SiC-Based UV Photodiodes and Light-Emitting Diodes

H. Morkoç, Beyond Silicon Carbide! III–V Nitride-Based Heterostructures and Devices

Volume 53 Cumulative Subjects and Author Index Including Tables of Contents for Volumes 1–50

Volume 54 High Pressure in Semiconductor Physics I

W. Paul, High Pressure in Semiconductor Physics: A Historical Overview

N. E. Christensen, Electronic Structure Calculations for Semiconductors Under Pressure

R. J. Neimes and M. I. McMahon, Structural Transitions in the Group IV, III–V and II–VI Semiconductors Under Pressure

A. R. Goni and K. Syassen, Optical Properties of Semiconductors Under Pressure

P. Trautman, M. Baj, and J. M. Baranowski, Hydrostatic Pressure and Uniaxial Stress in Investigations of the EL2 Defect in GaAs

M. Li and P. Y. Yu, High-Pressure Study of DX Centers Using Capacitance Techniques

T. Suski, Spatial Correlations of Impurity Charges in Doped Semiconductors

N. Kuroda, Pressure Effects on the Electronic Properties of Diluted Magnetic Semiconductors

Volume 55 High Pressure in Semiconductor Physics II

D. K. Maude and J. C. Portal, Parallel Transport in Low-Dimensional Semiconductor Structures

P. C. Klipstein, Tunneling Under Pressure: High-Pressure Studies of Vertical Transport in Semiconductor Heterostructures

E. Anastassakis and M. Cardona, Phonons, Strains, and Pressure in Semiconductors

F. H. Pollak, Effects of External Uniaxial Stress on the Optical Properties of Semiconductors and Semiconductor Microstructures

A. R. Adams, M. Silver, and J. Allam, Semiconductor Optoelectronic Devices

S. Porowski and I. Grzegory, The Application of High Nitrogen Pressure in the Physics and Technology of III–N Compounds

M. Yousuf, Diamond Anvil Cells in High Pressure Studies of Semiconductors

Volume 56 Germanium Silicon: Physics and Materials

J. C. Bean, Growth Techniques and Procedures

D. E. Savage, F. Liu, V. Zielasek, and M. G. Lagally, Fundamental Crystal Growth Mechanisms

R. Hull, Misfit Strain Accommodation in SiGe Heterostructures

M. J. Shaw and M. Jaros, Fundamental Physics of Strained Layer GeSi: Quo Vadis?

F. Cerdeira, Optical Properties

S. A. Ringel and P. N. Grillot, Electronic Properties and Deep Levels in Germanium-Silicon

J. C. Campbell, Optoelectronics in Silicon and Germanium Silicon

K. Eberl, K. Brunner, and O. G. Schmidt, $Si_{1-y}C_y$ and $Si_{1-x-y}Ge_2C_y$ Alloy Layers

Volume 57 Gallium Nitride (GaN) II

R. J. Molnar, Hydride Vapor Phase Epitaxial Growth of III–V Nitrides

T. D. Moustakas, Growth of III–V Nitrides by Molecular Beam Epitaxy

Z. Liliental-Weber, Defects in Bulk GaN and Homoepitaxial Layers

C. G. Van de Walk and N. M. Johnson, Hydrogen in III–V Nitrides

W. Götz and N. M. Johnson, Characterization of Dopants and Deep Level Defects in Gallium Nitride

B. Gil, Stress Effects on Optical Properties

C. Kisielowski, Strain in GaN Thin Films and Heterostructures

J. A. Miragliotta and D. K. Wickenden, Nonlinear Optical Properties of Gallium Nitride

B. K. Meyer, Magnetic Resonance Investigations on Group III–Nitrides

M. S. Shur and M. Asif Khan, GaN and AlGaN Ultraviolet Detectors

C. H. Qiu, J. I. Pankove, and C. Rossington, II–V Nitride-Based X-ray Detectors

Volume 58 Nonlinear Optics in Semiconductors I

A. Kost, Resonant Optical Nonlinearities in Semiconductors

E. Garmire, Optical Nonlinearities in Semiconductors Enhanced by Carrier Transport

D. S. Chemla, Ultrafast Transient Nonlinear Optical Processes in Semiconductors

M. Sheik-Bahae and E. W. Van Stryland, Optical Nonlinearities in the Transparency Region of Bulk Semiconductors

J. E. Millerd, M. Ziari, and A. Partovi, Photorefractivity in Semiconductors

Volume 59   Nonlinear Optics in Semiconductors II

J. B. Khurgin, Second Order Nonlinearities and Optical Rectification

K. L. Hall, E. R. Thoen, and E. P. Ippen, Nonlinearities in Active Media

E. Hanamura, Optical Responses of Quantum Wires/Dots and Microcavities

U. Keller, Semiconductor Nonlinearities for Solid-State Laser Modelocking and Q-Switching

A. Miller, Transient Grating Studies of Carrier Diffusion and Mobility in Semiconductors

Volume 60   Self-Assembled InGaAs/GaAs Quantum Dots

Mitsuru Sugawara, Theoretical Bases of the Optical Properties of Semiconductor Quantum Nano-Structures

Yoshiaki Nakata, Yoshihiro Sugiyama, and Mitsuru Sugawara, Molecular Beam Epitaxial Growth of Self-Assembled InAs/GaAs Quantum Dots

Kohki Mukai, Mitsuru Sugawara, Mitsuru Egawa, and Nobuyuki Ohtsuka, Metalorganic Vapor Phase Epitaxial Growth of Self-Assembled InGaAs/GaAs Quantum Dots Emitting at 1.3 µm

Kohki Mukai and Mitsuru Sugawara, Optical Characterization of Quantum Dots

Kohki Mukai and Mitsuru Sugawara, The Photon Bottleneck Effect in Quantum Dots

Hajime Shoji, Self-Assembled Quantum Dot Lasers

Hiroshi Ishikawa, Applications of Quantum Dot to Optical Devices

Mitsuru Sugawara, Kohki Mukai, Hiroshi Ishikawa, Koji Otsubo, and Yoshiaki Nakata, The Latest News

Volume 61   Hydrogen in Semiconductors II

Norbert H. Nickel, Introduction to Hydrogen in Semiconductors II

Noble M. Johnson and Chris G. Van de Walle, Isolated Monatomic Hydrogen in Silicon

Yurij V. Gorelkinskii, Electron Paramagnetic Resonance Studies of Hydrogen and Hydrogen-Related Defects in Crystalline Silicon

Norbert H. Nickel, Hydrogen in Polycrystalline Silicon

Wolfhard Beyer, Hydrogen Phenomena in Hydrogenated Amorphous Silicon

Chris G. Van de Walle, Hydrogen Interactions with Polycrystalline and Amorphous Silicon–Theory

Karen M. McManus Rutledge, Hydrogen in Polycrystalline CVD Diamond

Roger L. Lichti, Dynamics of Muonium Diffusion, Site Changes and Charge-State Transitions

Matthew D. McCluskey and Eugene E. Haller, Hydrogen in III–V and II–VI Semiconductors

S. J. Pearton and J. W. Lee, The Properties of Hydrogen in GaN and Related Alloys

Jörg Neugebauer and Chris G. Van de Walle, Theory of Hydrogen in GaN

Volume 62   Intersubband Transitions in Quantum Wells: Physics and Device Applications I

Manfred Helm, The Basic Physics of Intersubband Transitions

Jerome Faist, Carlo Sirtori, Federico Capasso, Loren N. Pfeiffer, Ken W. West, Deborah L. Sivco, and Alfred Y. Cho, Quantum Interference Effects in Intersubband Transitions

H. C. Liu, Quantum Well Infrared Photodetector Physics and Novel Devices

S. D. Gunapala and S. V. Bandara, Quantum Well Infrared Photodetector (QWIP) Focal Plane Arrays

Volume 63 Chemical Mechanical Polishing in Si Processing

Frank B. Kaufman, Introduction
Thomas Bibby and Karey Holland, Equipment
John P. Bare, Facilitization
Duane S. Boning and Okumu Ouma, Modeling and Simulation
Shin Hwa Li, Bruce Tredinnick, and Mel Hoffman, Consumables I: Slurry
Lee M. Cook, CMP Consumables II: Pad
François Tardif, Post-CMP Clean
Shin Hwa Li, Tara Chhatpar, and Frederic Robert, CMP Metrology
Shin Hwa Li, Visun Bucha, and Kyle Wooldridge, Applications and CMP-Related Process Problems

Volume 64 Electroluminescence I

M. G. Craford, S. A. Stockman, M. J. Peansky, and F. A. Kish, Visible Light-Emitting Diodes
H. Chui, N. F. Gardner, P. N. Grillot, J. W. Huang, M. R. Krames, and S. A. Maranowski, High-Efficiency AlGaInP Light-Emitting Diodes
R. S. Kern, W. Götz, C. H. Chen, H. Liu, R. M. Fletcher, and C. P. Kuo, High-Brightness Nitride-Based Visible-Light-Emitting Diodes
Yoshiharu Sato, Organic LED System Considerations
V. Bulović, P. E. Burrows, and S. R. Forrest, Molecular Organic Light-Emitting Devices

Volume 65 Electroluminescence II

V. Bulović and S. R. Forrest, Polymeric and Molecular Organic Light Emitting Devices: A Comparison
Regina Mueller-Mach and Gerd O. Mueller, Thin Film Electroluminescence
Markku Leskelä, Wei-Min Li, and Mikko Ritala, Materials in Thin Film Electroluminescent Devices
Kristiaan Neyts, Microcavities for Electroluminescent Devices

Volume 66 Intersubband Transitions in Quantum Wells: Physics and Device Applications II

Jerome Faist, Federico Capasso, Carlo Sirtori, Deborah L. Sivco, and Alfred Y. Cho, Quantum Cascade Lasers
Federico Capasso, Carlo Sirtori, D. L. Sivco, and A. Y. Cho, Nonlinear Optics in Coupled-Quantum-Well Quasi-Molecules
Karl Unterrainer, Photon-Assisted Tunneling in Semiconductor Quantum Structures
P. Haring Bolivar, T. Dekorsy, and H. Kurz, Optically Excited Bloch Oscillations–Fundamentals and Application Perspectives

Volume 67 Ultrafast Physical Processes in Semiconductors

Alfred Leitenstorfer and Alfred Laubereau, Ultrafast Electron-Phonon Interactions in Semiconductors: Quantum Kinetic Memory Effects

Christoph Lienau and Thomas Elsaesser, Spatially and Temporally Resolved Near-Field Scanning Optical Microscopy Studies of Semiconductor Quantum Wires

K. T. Tsen, Ultrafast Dynamics in Wide Bandgap Wurtzite GaN

J. Paul Callan, Albert M.-T. Kim, Christopher A. D. Roeser, and Eriz Mazur, Ultrafast Dynamics and Phase Changes in Highly Excited GaAs

Hartmut Hang, Quantum Kinetics for Femtosecond Spectroscopy in Semiconductors

T. Meier and S. W. Koch, Coulomb Correlation Signatures in the Excitonic Optical Nonlinearities of Semiconductors

Roland E. Allen, Traian Dumitrică, and Ben Torralva, Electronic and Structural Response of Materials to Fast, Intense Laser Pulses

E. Gornik and R. Kersting, Coherent THz Emission in Semiconductors

Volume 68 Isotope Effects in Solid State Physics

Vladimir G. Plekhanov, Elastic Properties; Thermal Properties; Vibrational Properties; Raman Spectra of Isotopically Mixed Crystals; Excitons in LiH Crystals; Exciton–Phonon Interaction; Isotopic Effect in the Emission Spectrum of Polaritons; Isotopic Disordering of Crystal Lattices; Future Developments and Applications; Conclusions

Volume 69 Recent Trends in Thermoelectric Materials Research I

H. Julian Goldsmid, Introduction

Terry M. Tritt and Valerie M. Browning, Overview of Measurement and Characterization Techniques for Thermoelectric Materials

Mercouri G. Kanatzidis, The Role of Solid-State Chemistry in the Discovery of New Thermoelectric Materials

B. Lenoir, H. Scherrer, and T. Caillat, An Overview of Recent Developments for BiSb Alloys

Citrad Uher, Skutterudities: Prospective Novel Thermoelectrics

George S. Nolas, Glen A. Slack, and Sandra B. Schujman, Semiconductor Clathrates: A Phonon Glass Electron Crystal Material with Potential for Thermoelectric Applications

Volume 70 Recent Trends in Thermoelectric Materials Research II

Brian C. Sales, David G. Mandrus, and Bryan C. Chakoumakos, Use of Atomic Displacement Parameters in Thermoelectric Materials Research

S. Joseph Poon, Electronic and Thermoelectric Properties of Half-Heusler Alloys

Terry M. Tritt, A. L. Pope, and J. W. Kolis, Overview of the Thermoelectric Properties of Quasicrystalline Materials and Their Potential for Thermoelectric Applications

Alexander C. Ehrlich and Stuart A. Wolf, Military Applications of Enhanced Thermoelectrics

David J. Singh, Theoretical and Computational Approaches for Identifying and Optimizing Novel Thermoelectric Materials

Terry M. Tritt and R. T. Littleton, IV, Thermoelectric Properties of the Transition Metal Pentatellurides: Potential Low-Temperature Thermoelectric Materials

Franz Freibert, Timothy W. Darling, Albert Migliori, and Stuart A. Trugman, Thermomagnetic Effects and Measurements

M. Bartkowiak and G. D. Mahan, Heat and Electricity Transport Through Interfaces

Volume 71 Recent Trends in Thermoelectric Materials Research III

M. S. Dresselhaus, Y.-M. Lin, T. Koga, S. B. Cronin, O. Rabin, M. R. Black, and G. Dresselhaus, Quantum Wells and Quantum Wires for Potential Thermoelectric Applications

D. A. Broido and T. L. Reinecke, Thermoelectric Transport in Quantum Well and Quantum Wire Superlattices

G. D. Mahan, Thermionic Refrigeration

Rama Venkatasubramanian, Phonon Blocking Electron Transmitting Superlattice Structures as Advanced Thin Film Thermoelectric Materials

G. Chen, Phonon Transport in Low-Dimensional Structures

Volume 72 Silicon Epitaxy

S. Acerboni, ST Microelectronics, CFM-AGI Department, Agrate Brianza, Italy

V.-M. Airaksinen, Okmetic Oyj R&D Department, Vantaa, Finland

G. Beretta, ST Microelectronics, DSG Epitaxy Catania Department, Catania, Italy

C. Cavallotti, Dipartimento di Chimica Fisica Applicata, Politecnico di Milano, Milano, Italy

D. Crippa, MEMC Electronic Materials, Epitaxial and CVD Department, Operations Technology Division, Novara, Italy

D. Dutartre, ST Microelectronics, Central R&D, Crolles, France

Srikanth Kommu, MEMC Electronic Materials inc., EPI Technology Group, St. Peters, Missouri

M. Masi, Dipartimento di Chimica Fisica Applicata, Politecnico di Milano, Milano, Italy

D. J. Meyer, ASM Epitaxy, Phoenix, Arizona

J. Murota, Research Institute of Electrical Communication, Laboratory for Electronic Intelligent Systems, Tohoku University, Sendai, Japan

V. Pozzetti, LPE Epitaxial Technologies, Bollate, Italy

A. M. Rinaldi, MEMC Electronic Materials, Epitaxial and CVD Department, Operations Technology Division, Novara, Italy

Y. Shiraki, Research Center for Advanced Science and Technology (RCAST), University of Tokyo, Tokyo, Japan

Volume 73 Processing and Properties of Compound Semiconductors

S. J. Pearton, Introduction

Eric Donkor, Gallium Arsenide Heterostructures

Annamraju Kasi Viswanatli, Growth and Optical Properties of GaN

D. Y. C. Lie and K. L. Wang, SiGe/Si Processing

S. Kim and M. Razeghi, Advances in Quantum Dot Structures

Walter P. Gomes, Wet Etching of III–V Semiconductors

Volume 74 Silicon-Germanium Strained Layers and Heterostructures

S. C. Jain and M. Willander, Introduction; Strain, Stability, Reliability and Growth; Mechanism of Strain Relaxation; Strain, Growth, and TED in SiGeC Layers; Bandstructure and Related Properties; Heterostructure Bipolar Transistors; FETs and Other Devices

Volume 75 Laser Crystallization of Silicon

Norbert H. Nickel, Introduction to Laser Crystallization of Silicon

Costas P. Grigoropoidos, Seung-Jae Moon and Ming-Hong Lee, Heat Transfer and Phase Transformations in Laser Melting and Recrystallization of Amorphous Thin Si Films

Robert Černý and Petr Přikryl, Modeling Laser-Induced Phase-Change Processes: Theory and Computation

Paulo V. Santos, Laser Interference Crystallization of Amorphous Films

Philipp Lengsfeld and Norbert H. Nickel, Structural and Electronic Properties of Laser-Crystallized Poly-Si

Volume 76 Thin-Film Diamond I

X. Jiang, Textured and Heteroepitaxial CVD Diamond Films
Eberhard Blank, Structural Imperfections in CVD Diamond Films
R. Kalish, Doping Diamond by Ion-Implantation
A. Deneuville, Boron Doping of Diamond Films from the Gas Phase
S. Koizumi, n-Type Diamond Growth
C. E. Nebel, Transport and Defect Properties of Intrinsic and Boron-Doped Diamond
Miloš Nesládek, Ken Haenen and Milan Vaněček, Optical Properties of CVD Diamond
Rolf Sauer, Luminescence from Optical Defects and Impurities in CVD Diamond

Volume 77 Thin-Film Diamond II

Jacques Chevallier, Hydrogen Diffusion and Acceptor Passivation in Diamond
Jürgen Ristein, Structural and Electronic Properties of Diamond Surfaces
John C. Angus, Yuri V. Pleskov and Sally C. Eaton, Electrochemistry of Diamond
Greg M. Swain, Electroanalytical Applications of Diamond Electrodes
Werner Haenni, Philippe Rychen, Matthyas Fryda and Christos Comninellis, Industrial Applications of Diamond Electrodes
Philippe Bergonzo and Richard B. Jackman, Diamond-Based Radiation and Photon Detectors
Hiroshi Kawarada, Diamond Field Effect Transistors Using H-Terminated Surfaces
Shinichi Shikata and Hideaki Nakahata, Diamond Surface Acoustic Wave Device

Volume 78 Semiconducting Chalcogenide Glass I

V. S. Minaev and S. P. Timoshenkov, Glass-Formation in Chalcogenide Systems and Periodic System

A. Popov, Atomic Structure and Structural Modification of Glass

V. A. Funtikov, Eutectoidal Concept of Glass Structure and Its Application in Chalcogenide Semiconductor Glasses

V. S. Minaev, Concept of Polymeric Polymorphous-Crystalloid Structure of Glass and Chalcogenide Systems: Structure and Relaxation of Liquid and Glass

Volume 79 Semiconducting Chalcogenide Glass II

M. D. Bal'makov, Information Capacity of Condensed Systems

A. Česnys, G. Juška and E. Montrimas, Charge Carrier Transfer at High Electric Fields in Noncrystalline Semiconductors

Andrey S. Glebov, The Nature of the Current Instability in Chalcogenide Vitreous Semiconductors

A. M. Andriesh, M. S. Iovu and S. D. Shutov, Optical and Photoelectrical Properties of Chalcogenide Glasses

V. Val. Sobolev and V. V. Sobolev, Optical Spectra of Arsenic Chalcogenides in a Wide Energy Range of Fundamental Absorption

Yu. S. Tver'yanovich, Magnetic Properties of Chalcogenide Glasses

Volume 80 Semiconducting Chalcogenide Glass III

Andrey S. Glebov, Electronic Devices and Systems Based on Current Instability in Chalcogenide Semiconductors

Dumitru Tsiulyanu, Heterostructures on Chalcogenide Glass and Their Applications

E. Bychkov, Yu. Tveryanovich and Yu. Vlasov, Ion Conductivity and Sensors

Yu. S. Tver'yanovich and A. Tverjanovich, Rare-earth Doped Chalcogenide Glass

M. F. Churbanov and V. G. Plotnichenko, Optical Fibers from High-purity Arsenic Chalcogenide Glasses

Volume 81 Conducting Organic Materials and Devices

Suresh C. Jain, Magnus Willander and Vikram Kumar,
Introduction; Polyacetylene; Optical and Transport Properties; Light Emitting Diodes and Lasers; Solar Cells; Transistors

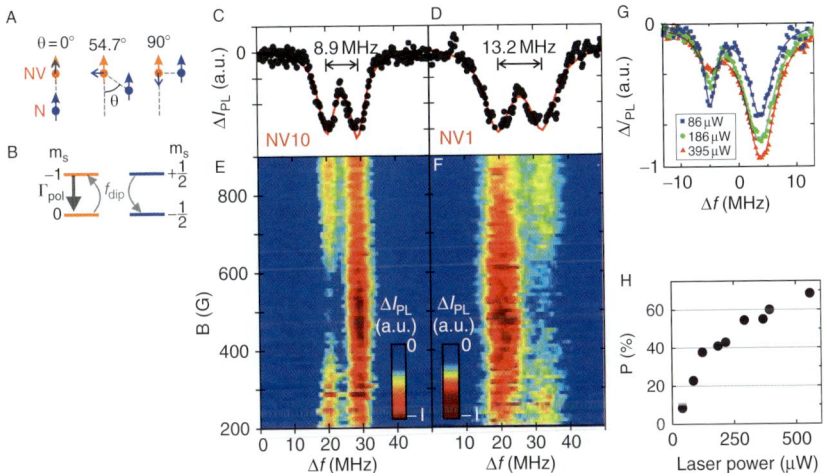

Maiken H. Mikkelsen et al., Figure 1.20 (A) Schematic diagram showing the direction of B_{dip} for an NV center coupled with a single N defect for the limiting cases of $\theta = 0°$, $54.7°$, and $90°$. (B) Energy diagram of the spin dynamics that results from dipolar coupling between an NV center and an N spin. (C) and (D) show ESR spectra taken as a function of B for NV10 and NV1, respectively. The spectra for each value of B are concatenated and plotted with a 2D color scale (E) and (F). (G) Plot of ESR spectra for three different values of laser illumination. (H) Plot of polarization versus laser power. Figure is modified from Hanson et al. (2006b).

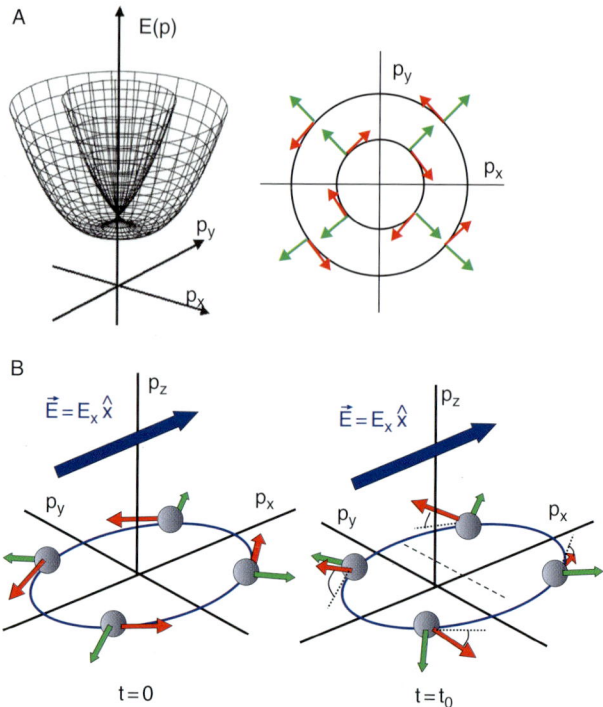

Jairo Sinova and A. H. MacDonald Figure 2.5 (A) The 2D electronic eigenstates in a Rashba spin–orbit coupled system are labeled by momentum (green arrows). For each momentum, the two eigenspinors point in the azimuthal direction. (B) In the presence of an electric field, the Fermi surface is displaced an amount $|eE_x t_0/\hbar|$ at time t_0 (shorter than typical scattering times). While moving in momentum space, electrons experience an effective torque which tilts the spins up for $p_y > 0$ and down for $p_y < 0$, creating a spin-current in the y-direction (Sinova et al., 2004a).

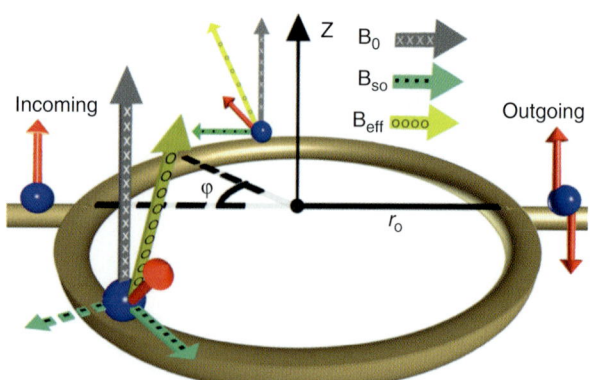

Jairo Sinova and A. H. MacDonald Figure 2.7 One channel ring of radius r_0 subject to Rashba coupling in the presence of an additional magnetic field $\mathbf{B}_o$. Electron (hole) spin traveling around the ring acquires phase due to the applied out-of-plane magnetic field (gray arrow) and the Rashba in-plane magnetic field (momentum dependent, green arrows) caused by the spin–orbit interaction.

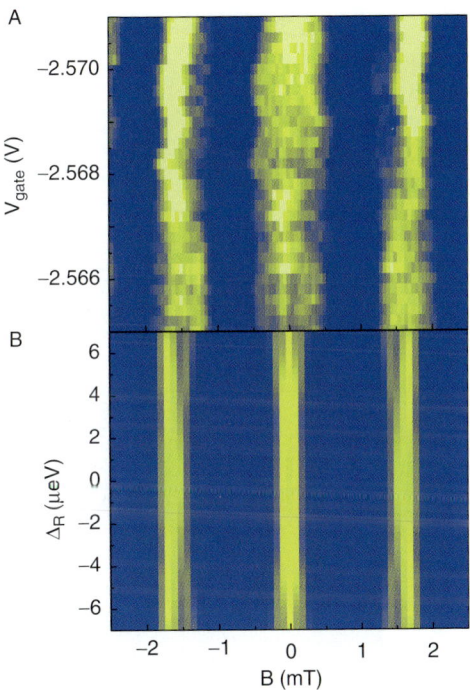

Jairo Sinova and A. H. MacDonald Figure 2.8 (A) Near the symmetry point where the Rashba coupling vanishes the interference pattern is unperturbed by the spin–orbit interaction. (B) The theoretical calculations for a 6-channel ring show consistent results for the corresponding range of Δ_{Rashba}. Yellow and blue correspond to conductance maxima and minima, respectively (Konig *et al.*, 2006).

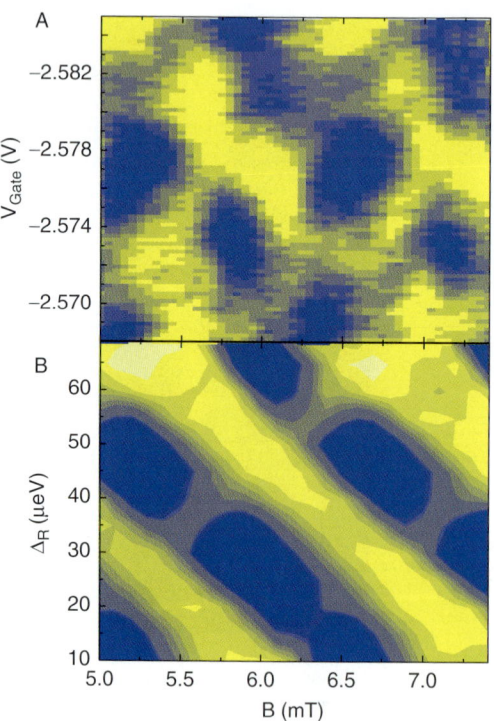

Jairo Sinova and A. H. MacDonald Figure 2.9 When the spin–orbit interaction is modified via the gate voltage, a shift of the conductance maxima (yellow) can be observed due to the Aharonov–Casher phase (A). In figure (B), the theoretical results for the conductance in a 6-channel ring as a function of the Rashba energy and B_{ext} are shown. The scaling of the y-axis allows a direct comparison of the experimental and theoretical data (Konig et al., 2006).

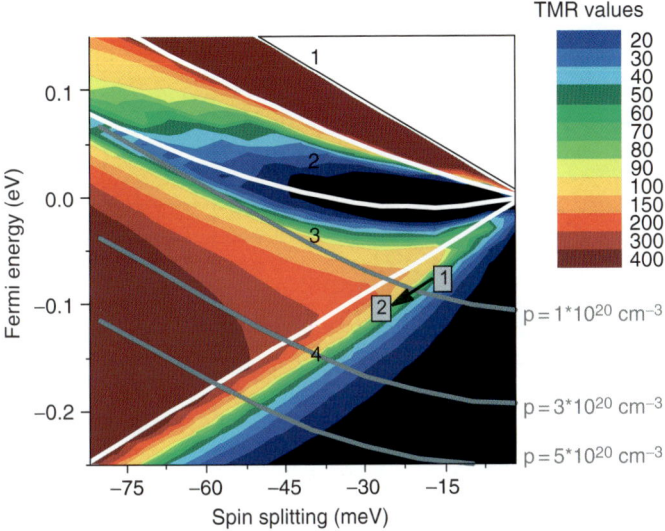
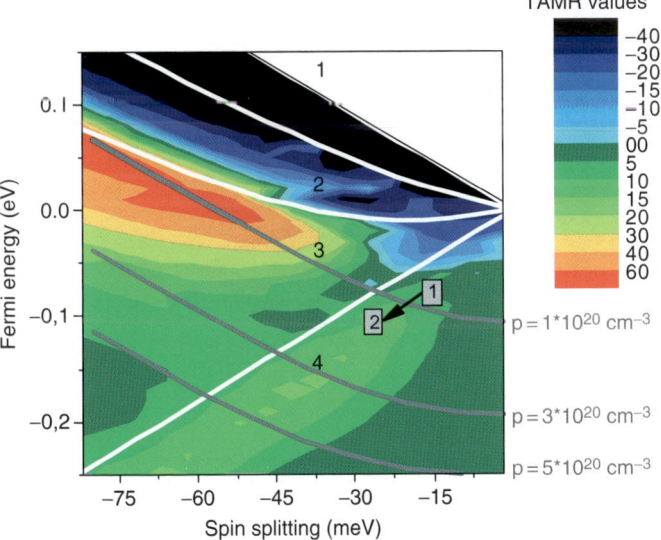

T. Jungwirth et al., Figure 4.33 Tunnel MR values (A) and tunnel AMR values (B) represented as a function of the Fermi and spin splitting energy for a 6 nm (In,Ga)As barrier with a band offset of 450 meV. White lines represent the four bands at the center of the Brillouin zone. Gray lines indicate the Fermi energy for different hole concentrations (from Elsen et al., 2007).

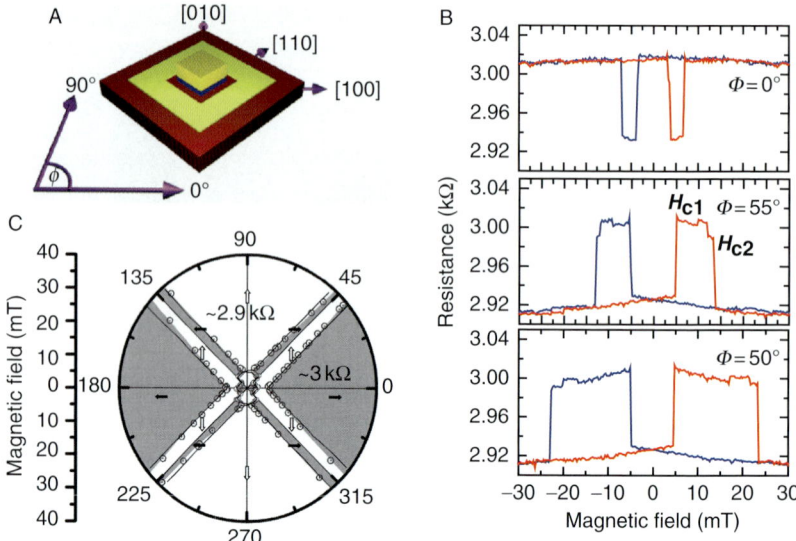

C. Gould et al., Figure 6.1 (A) Schematic of the first TAMR device. (B) Magnetoresistance curves at various in-plane field angles, showing the spin-valve–like behavior characteristic of TAMR. (C) Polar plot summarizing both the data and the model. The device exhibits high (low) resistance if the magnetization is along $\pm[100]$ ($\pm[010]$). The open triangles are experimental extracted switching fields H_{c1} and H_{c2} data and the lines fits to the model. The arrows indicate the magnetization direction, and the dark shading indicates regions of high resistance.

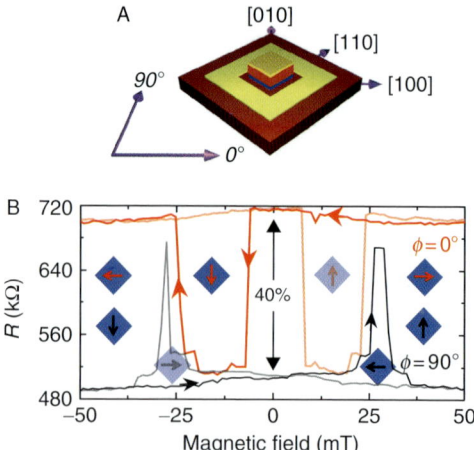

C. Gould et al., Figure 6.2 (B) Magnetoresistance curves along the 40° direction for the double TAMR structure depicted schematically in (A). The arrows in the diamonds indicate the direction of magnetization of *both* layers at each point in the spin-valve curve.

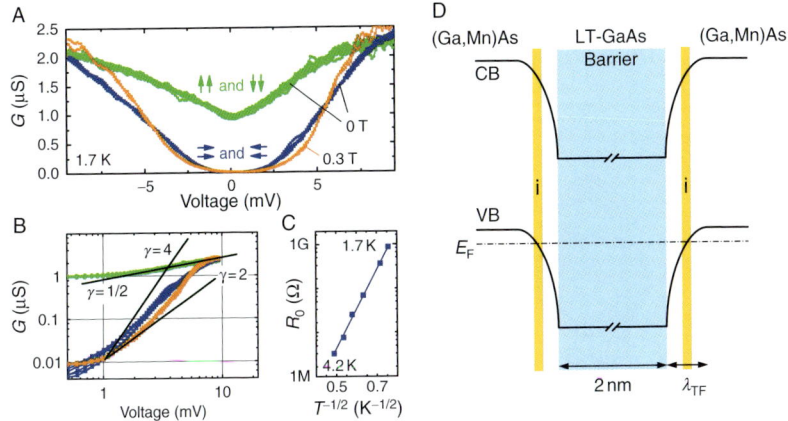

C. Gould et al., Figure 6.6 (A) Differential conductance–voltage curves at 1.7 K including sets of curves belonging to the two magnetization states at $B = 0$, and one set at $B = 300$ mT along $\phi = 3$ and $6°$. (B) Log–log plot of same data. (C) Temperature dependence of the zero bias resistance of the high resistance state. (D) Schematic of the band bending caused by the barrier.

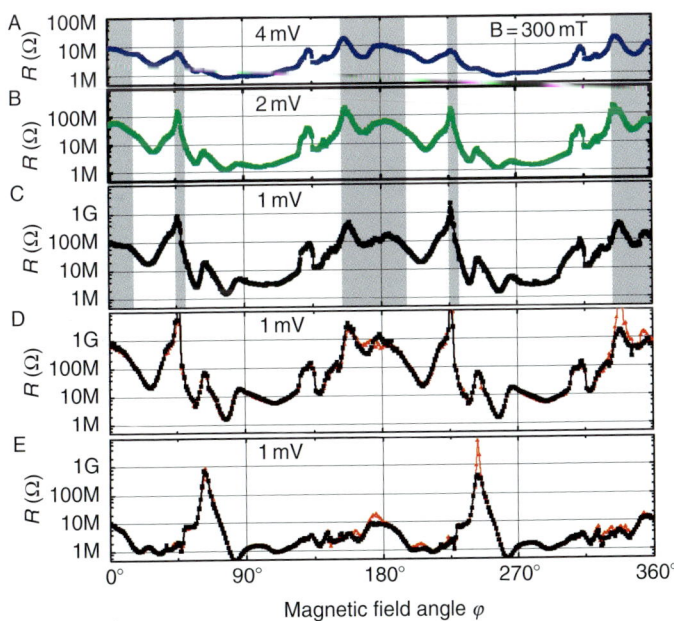

C. Gould et al., Figure 6.7 Resistance of the sample at 1.7 K and 300 mT as a function of magnetic field direction under various bias and on different cooldowns. (D) and (E) each show the results of two separate measurement, confirming reproducibility.

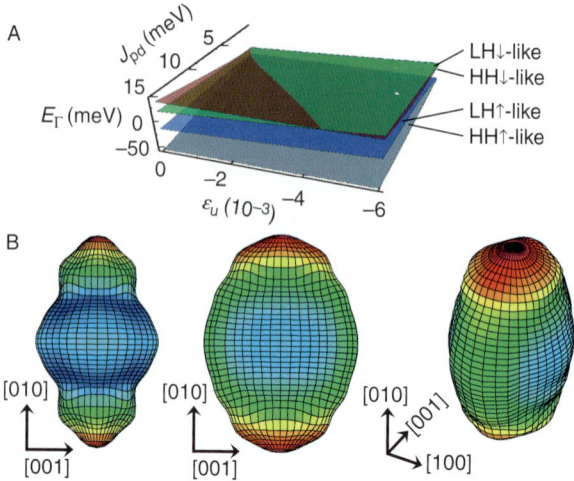

C. Gould et al., Figure 6.8 (A) Γ point energy of the four top bands as a function of strain ε_u and pd-exchange. (B) Extents of a hole bound to a Mn-impurity. [001] is the growth direction and ε_u is along [010]. Left: Magnetization (**M** ∥ [100]). Middle: (**M** ∥ [010]). Right: Perspective view for **M** ∥ [010].

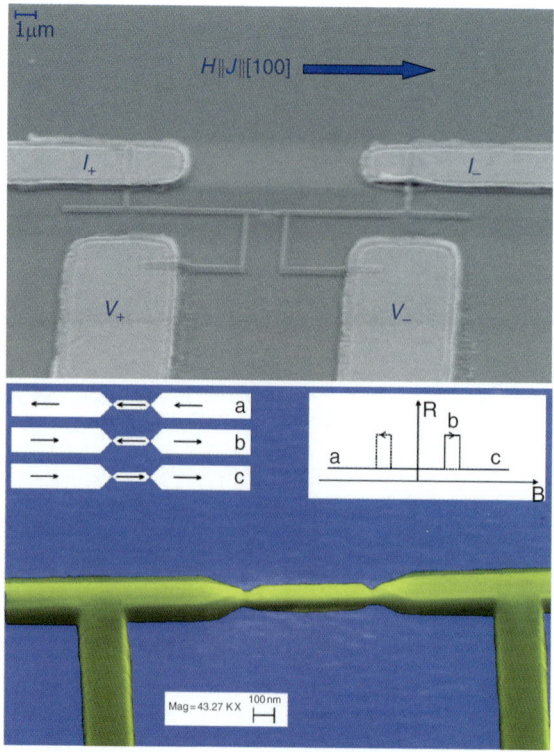

C. Gould et al., Figure 6.9 SEM micrographs of the nanoconstrictions device, with insets schematically depicting the expected behavior.

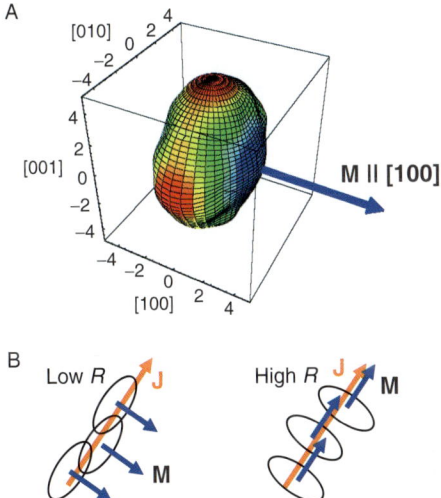

C. Gould et al., Figure 6.23 (A) Wavefunction of the semi-localized holes at the Mn impurities. (B) Schematic depicting the mechanism invoked to explain the large amplitude effect.

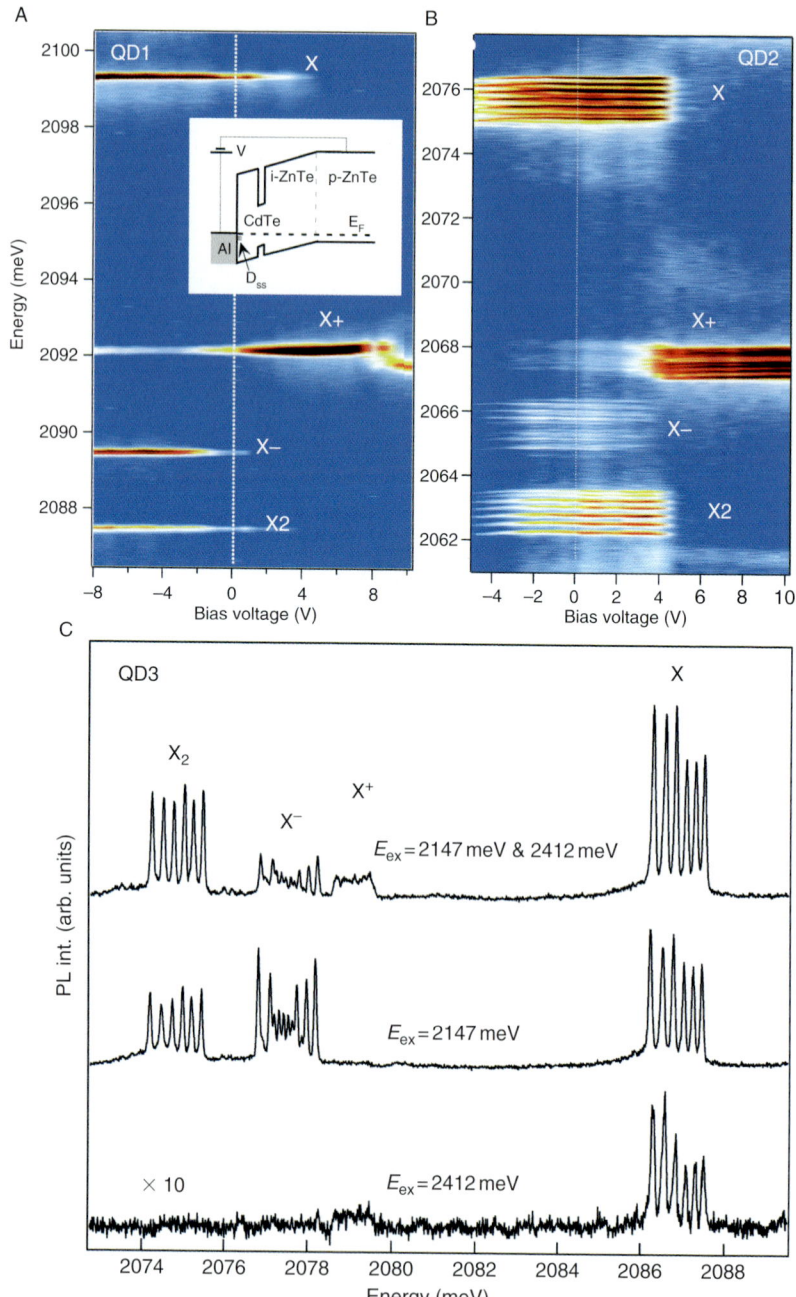

J. Cibert et al., Figure 7.16 Color-scale plot of the photoluminescence intensity of (A) a nonmagnetic QD and (B) a single Mn-doped QD in a Schottky structure as a function of emission energy and bias voltage. The series of emission lines can be assigned to QD s-shell transitions, namely the recombination of the neutral exciton (X), biexciton (X_2), positively charged exciton (X^+), and negatively charged exciton (X^-). Detail of the PL of a single Mn-doped QD under resonant excitation ($E_{ex} = 2147$ meV), nonresonant excitation ($E_{ex} = 2142$ meV), and both resonant and nonresonant excitation (Léger et al., 2006).

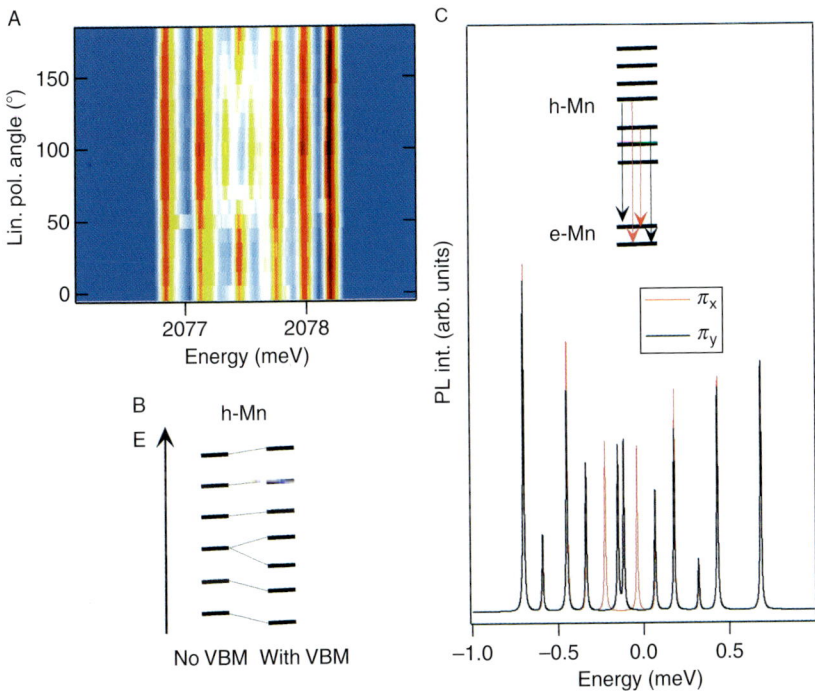

J. Cibert et al., Figure 7.18 Color-scale plot of the dependence of the PL of (X^-,Mn) (A) on the direction of a linear analyzer. Three lines in the center of the structure are linearly polarized. The scheme (B) presents the energy-level scheme of the h–Mn system without and with valence band mixing (VBM). (C) Calculated linearly polarized PL spectra of (X^-,Mn) with exchange integrals I_e and I_h chosen to reproduce the overall splitting for X^- presented in (A). Transitions are arbitrarily broadened by 10 µeV.

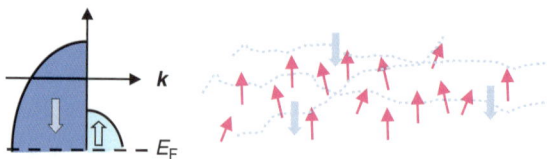

Tomasz Dietl, Figure 9.9 Pictorial presentation of carrier-mediated ferromagnetism in p-type DMSs, a model proposed originally by Zener for metals. Owing to the p–d exchange interaction, ferromagnetic ordering of localized spins (smaller arrows) leads to the spin splitting of the valence band. The corresponding redistribution of the carriers between spin subbands lowers energy of the holes, which at sufficiently low temperatures overcompensates an increase of the free energy associated with a decrease of Mn entropy.

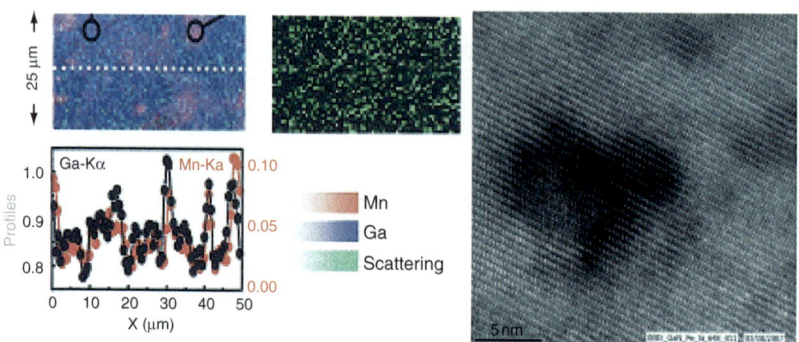

Tomasz Dietl, Figure 9.17 Evidences for spinodal decomposition in (Ga,Mn)N [synchrotron radiation micro-probe—left panel (Martinez-Criado et al., 2005)] and in (Ga,Fe)N [transmission electron microscope (Bonanni et al., 2007, 2008)].